Handbook of Bioplastics and Biocomposites Engineering Applications

Scrivener Publishing
100 Cummings Center, Suite 541J
Beverly, MA 01915-6106

Publishers at Scrivener
Martin Scrivener (martin@scrivenerpublishing.com)
Phillip Carmical (pcarmical@scrivenerpublishing.com)

Handbook of Bioplastics and Biocomposites Engineering Applications

2nd Edition

Edited by

Inamuddin

Faculty of Engineering and Technology, Aligarh Muslim University, Aligarth, India

and

Tariq Altalhi

Department of Chemistry, Taif University, Taif, Saudi Arabia

Scrivener Publishing

WILEY

This edition first published 2023 by John Wiley & Sons, Inc., 111 River Street, Hoboken, NJ 07030, USA and Scrivener Publishing LLC, 100 Cummings Center, Suite 541J, Beverly, MA 01915, USA

For more information about Scrivener publications please visit www.scrivenerpublishing.com.

1st edition (2011), 2nd edition (2023)

Wiley Global Headquarters
111 River Street, Hoboken, NJ 07030, USA

For details of our global editorial offices, customer services, and more information about Wiley products visit us at www.wiley.com.

Library of Congress Cataloging-in-Publication Data

ISBN 978-1-119-16013-7

Cover image: Pixabay.Com
Cover design by Russell Richardson

Set in size of 11pt and Minion Pro by Manila Typesetting Company, Makati, Philippines

Printed in the USA

Contents

Preface

Bioplastics are strong materials that are being researched as a possible replacement for petroleum-based traditional plastics to make them more environmentally friendly. They are made from renewable resources and sometimes naturally recycled through biological processes, conserving natural resources and reducing CO_2 emissions. Bioplastics are employed in many industries, including food packaging, composting bags, agriculture, hygiene, and so on. Biobased materials come in a variety of forms, including biomass, biodegradable and/or renewable polymers, and polymers made by natural or genetically engineered microorganisms. Polymer blends and composites are frequently investigated to improve the properties of bioplastics and to replace a large number of unsuitable versatile polymers. Inorganic reinforcement (e.g., calcium carbonate, montmorillonite, kaolinite, sepiolite, talc), natural fiber (both wood and plant fibers), and various types of fillers (graphene nanosheets, carbon nanotubes) have all been investigated for polymer composites or biocomposites.

The development of smart or intelligent biocomposite materials that are more versatile, more survivable, adaptable, and ecologically friendly has gotten a lot of attention in recent decades and has become a crucial research topic for all scientific disciplines ranging from technology to engineering. Smart items, which will be capable of being active, clever, sharp, stylish, and sophisticated, may already be found in a wide range of human life and expertise. Furthermore, numerous scholars have investigated the use of their potential in a variety of technical applications that might benefit the entire globe.

The 2nd edition of the *Handbook of Bioplastics and Biocomposites Engineering Applications* discusses a wide range of technologies, classifications concerned with bioplastics and biocomposites with their applications in various paradigms including the engineering segment. Chapters discuss the biobased materials; recycling of bioplastics; biocomposites modeling; different biomedical and engineering-based applications including optical devices, smart materials, cosmetics, drug delivery, clinical, electrochemical, industrial, flame retardant, sports, packaging, disposables, biomass, and various approaches for sustainability are summarized. This is a useful guide for engineers, faculty, researchers, students from academics and laboratories which are linked to bioplastics and biocomposites, their useful properties, classifications, and varied applications including the engineering sector. The summaries of the work reported in the following 30 chapters are as follows:

Chapter 1 comprises a wide presentation of bioplastics, detailing the principal categories. The rheological, optical, mechanical, and electrical properties of many polymer bioplastic systems are reviewed together with their engineering fields, like sensors, energy, bioengineering, and "green" electronics.

Chapter 2 distinguishes different biobased materials according to their degradability nature. Biobased materials are frequently associated with being solely made of biomass, but sometimes they are non-biodegradable. Therefore, this chapter discusses the classification of biobased materials by branching them into biodegradable and non-biodegradable.

Chapter 3 gives a brief discussion about bioplastics. Plastics are mostly of petrochemical origin and are difficult to degrade. An alternative to petrochemical-based plastics may be biodegradable and are bioplastic. Here different types of bioplastics, their production by various methods, characterizations, and applications have been discussed in detail. Bioplastic waste management strategies have also been highlighted.

Chapter 4 discusses the prediction of the mechanical behavior of biocomposite materials in some structural applications before production. This chapter first introduces the biocomposite materials and their different processing techniques. In continuation, it will present the main models for the biocomposites' mechanical properties prediction based on the relationship between the parameters affecting the material performances.

Chapter 5 details information on biocomposites, that are an amalgamation of two or more natural polymers, reinforced to make a potential composite material. Process modeling techniques mainly involve injection molding, compression molding, and extrusion methods. These methods make use of specialized techniques for reinforcing polymers to make potential biocomposites that can be used for various medical and industrial applications.

Chapter 6 summarizes the application of microbial technology in the production of bioplastics. After a brief introduction to the basic principles of microbial technology, the chapter discusses at length the use of fermentation processes employing natural and genetically modified microorganisms for the production of bioplastics or its monomeric building blocks.

Chapter 7 details different bioplastics such as starch, polyhydroxyalkanoates, polylactide, cellulose, and chitosan, which can replace traditional plastics. Some properties (gas barrier, biocompatibility, thermostability) and applications are also discussed.

Chapter 8 focuses on epoxidized plant oils and their properties were determined. Important characteristics of these epoxidized plant oils were analyzed with multi-criteria decision-making methods, taking into account the attributes that are important in the production of suitable bioplastics. Entropy and Copras were used as analysis methods.

Chapter 9 discusses various biomedical applications of bioplastics in tissue engineering, fabrication of implants, antimicrobial agents, and biological control of pests. In addition, different properties of bioplastics, including mechanical, electrical, magnetic, optical, and biological properties, are discussed.

Chapter 10 goes into detail about bioplastics as a material that will be very useful in the future. Cosmetics are highly valued but are easily damaged. In the present environment, cosmetic packaging necessitates sustainable solutions, and research has focused on modifying bio-based and biodegradable polymers to meet stringent requirements for cosmetic preservation while retaining stability and biodegradability.

Chapter 11 discusses the use of biodegradable polymers in drug delivery systems. Polymers, both natural and synthetic, are discussed. Nanocarriers and three-dimensional printing devices, as well as intravenous, transdermal, oral, and ocular applications, are presented. Finally, conclusions and projections for the future are addressed.

Chapter 12 focuses on microorganism-derived bioplastics like PHB, PLA, PLGA, PCL as "Biologically safe", due to their unique physicochemical properties *viz.* renewability, bio-compatibility, and thermo-plasticity. These have potential clinical applications such as tissue engineering, drug delivery, implants, prostheses, etc. Overall, this chapter emphasizes the microorganisms-derived bioplastics, their properties, and biomedical applications.

Chapter 13 describes different types of methods to isolate cellulose in varied dimensions from micro to nano. The fabrication procedures of biocomposites from the discussed cellulose varieties are also included in the chapter. Lastly, the diversified applications of the biocomposites are mentioned as well.

Chapter 14 presents some of the recent relevant findings related to the employment of biomaterials in the biomedical engineering (BME) domain with highlights concerning their prospective in different applications (tissue engineering, artificial organs, implants, auxiliary materials). New biomaterials can be designed using a wide selection of polymers, natural and synthetic, in various combinations.

Chapter 15 summarizes the environmental pollution caused due to plastic and presents the possible solutions to overcome it. Additionally, details on the bioplastic and their classifications are mentioned. The use of various bioplastics in sports, particularly hockey and football, is addressed. In addition, industries such as toys, camping, and so on are explored.

Chapter 16 discusses the recent development of biocomposites as active and intelligent food packaging materials. Discussions will focus on biobased materials enhanced with plant extracts, essential oils, and inorganic compounds; to develop biocomposites that can maintain the freshness of food products and prevent microbial spoilage.

Chapter 17 explores the biological manifestations of many possible materials sourced from starch, cellulose, tree gums, chitosan, carrageenan, and xanthan as biodegradable, lightweight, economical, reusable, and ecofriendly biofoams for being an alternative to petroleum-based polymers. The packaging industry inspires to use and behold natural and sustainable materials in many potential appliances.

Chapter 18 focuses on the characteristics of biobased and biodegradable packaging materials, describing the features of the main biopolymers used. The chapter explores details of some functional properties and their impact on food preservation and how they can be modified with the incorporation of additives such as fibers, nanoparticles, and antimicrobial agents.

Chapter 19 discusses the concepts, types, performance, and functions of bioplastic-based nanocomposites, i.e., bio nanocomposites; as well as their development and applications for packaging. In addition, public concerns and regulations of nanostructured materials, especially for food packaging use are also addressed.

Chapter 20 presents the importance of bioplast and its types such as starch-based, aliphatic polyester, bioderived polyethylene. The applications of bioplast in food packaging, bioremediation, 3D printing, and drug delivery are also discussed in detail. This chapter highlights the advantages and future feasibility of bioplast over conventional plastics.

Chapter 21 gives a brief description of smart materials technologies that can respond to different external stimuli in different ways such as pH-sensitive materials, thermosensitive materials, antimicrobial materials, and thermochromic materials.

Chapter 22 presents an overview of the recent advancements in the development of optical devices from different bio-derived materials. The possibilities of polysaccharides, polypeptides, and polynucleotides in the development of optical devices are discussed in detail. These devices will provide better barrier transport, catalytic ability, and high electrical conductivity.

Chapter 23 explains the divergent types of biocomposite and bioplastic materials utilized in fundamental electrochemical measurements such as sensors, biosensors, fuel cells, corrosion inhibitors, and supercapacitors. The chapter mainly focuses on the electrochemical contributions of bio-sourced materials such as chitosan, cellulose, wool, silk, pectin.

Chapter 24 represents the types of biofibers that are commonly used in biocomposites as reinforcing agents in detail. Furthermore, the physical and chemical modification methods and their effects on the mechanical and thermal properties of the biofibers are discussed. Various applications of biocomposites in the industry are also presented.

Chapter 25 is a detailed overview of the latest literature on flame retardant biocomposites and bioplastics. Emphasis is given on the preparation methods and applications of these flame retardant biocomposites and bioplastics. Amongst the preparation methods, the incorporation of environmentally friendly flame-retardant additives into biocomposites and bioplastics is mostly discussed.

Chapter 26 discusses the importance of thermosetting polymer materials, which are biobased and biodegradable. The newly synthesized biocompatible materials show better engineering applications with the production of biothermoset from renewable sources.

Chapter 27 discusses the public perception of replacing traditional plastics with bioplastics. The procurement of bio-plastics is tied to awareness campaigns of recycling, their reuse as well as refusal. This chapter investigates the emotions and elements that influence sustainable purchasing, as well as public knowledge and attitudes toward sustainable purchasing in various nations.

Chapter 28 summarizes the most recent research on biopots (biodegradable planting pots) made of bioplastics, which are commonly reinforced with plant or wood fibres. Bioplastics could be a viable substitute to traditional petro-based pots for horticulture and floriculture containers, in terms of plant development and functionality.

Chapter 29 details the preferences of bio-based polymers because of environmental sustainability. The innovative and translational approaches are being adopted for applications ranging from food packaging to drug delivery. The characteristics, biodegradation, regulations of national/international agencies, and futuristic approaches are elaborated with usage patterns in the current scenario.

Chapter 30 explains how the advent of eco-friendly materials in the polymer and plastic industry has urged the use of bioplastics. However, the single-use of various items has endangered the natural resources, emphasizing the recycling of bioplastics. It has elaborated the model of recycling of bioplastics along with its impact on the circular economy.

We would like to deeply thank all our contributors for their time and expertise and parting in this important and timely 2nd edition.

The Editors
October 2022

Part I

BIOPLASTICS, SYNTHESIS AND PROCESS TECHNOLOGY

1

An Introduction to Engineering Applications of Bioplastics

Andreea Irina Barzic

"Petru Poni" Institute of Macromolecular Chemistry, Grigore Ghica Voda Alley, Iași, Romania

Abstract
In the past decades, our society has known a considerable benefit upon utilization of plastics. Such developments have led to the draining of the fossil fuels, and inadequate disposal of the related products produce harmful effects to the environment and also to the health of the living beings. To diminish as much as possible the unwanted consequences, researchers made efforts to find novel plastic materials, which are characterized by renewable, biocompatible, and biodegradable features. This chapter is devoted to make a general presentation of bioplastics, emphasizing, their main categories and properties. The biobased plastics are ideal for minimizing the dependency on petroleum-based compounds, whereas they avoid the storage of persistent waste and lower the emission of carbon dioxide and other toxic residues in the environment. The chapter offers a general overview on some key physical properties of bioplastics and their utilization in applications in certain engineering areas, such as sensors, energy, bioengineering, and "green" electronics.

***Keywords*:** Bioplastics, classification, properties, applications

1.1 Introduction

Plastics are regarded as one of the most valuable types of materials owing to their remarkable versatility, lightweight and relatively low cost of production or processing [1]. They are encountered in a broad area of applications, ranging from electrical components (insulators, circuits supports), food packaging, biomedical equipment (gloves, syringes, gowns, tubes, masks, coverings, wound dressing), transportation (cars, watercraft, aircraft parts), and other products from our daily life (utensils, computers, cell phones, and watches) [2–5]. The majority of the synthetic plastics are achieved from petroleum and related products. Most of these materials do not display eco-compatible properties, and for this reason, they raise a huge concern from environmental point of view. The excessive utilization of plastics over the decades has led to generation of enormous accumulation of wastes, which are frequently thrown away as municipal solid garbage [6]. Therefore, there is an urgent necessity to find alternatives that use non–petroleum-derived and sustainable feed stocks [7].

Email: irina_cosutchi@yahoo.com

Inamuddin and Tariq Altalhi (eds.) Handbook of Bioplastics and Biocomposites Engineering Applications: 2nd Edition, (3–22)

The problem of the preservation of a green environment has been brought in the attention of the scientific and industrial communities that have been joining their efforts to formulate solutions, which account for biobased and biodegradable plastics [8, 9].

Bioplastics can be defined as biodegradable and biocompatible plastic materials acquired based on renewable biomass resources, like food waste, straw, plants, and wood residues [1, 10]. By using them, it is possible to diminish the needs of the human society on depleting fossil fuels, and in this way, transition to a low-carbon economy become possible [11, 12]. Even if this category of compounds is offering timely and essential demands for environmental sustainability, they did not earn the appropriate place in top of material science. This could be explained by considering that the synthetic counterparts have better properties than those of the bioplastics. Therefore, at global level, the chemists, physicists, and engineers are continuously working on the development of the procedures to enhance the performance of the bioplastics [13–15].

This chapter is aiming to address some basic aspects concerning bioplastics, such as the main categories and the physical properties in solution and solid phase, namely rheological, thermal, optical, mechanical, and electrical characteristics. Furthermore, the chapter also describes how the aforementioned aspects are interfering in the development of current engineering applications of the bioplastics. The chapter reviews recent advances in the field of bioplastics engineering, which undoubtedly might provide novel market opportunities.

1.2 Classification of Bioplastics

According to the literature references [1, 10, 16, 17], the classification of bioplastics can be accomplished by accounting on several factors, such as:

- *the degradable behavior*: based on this criterion, bioplastics can be broadly imparted in bioderived and/or biodegradable materials. However, a more recent classification imparts bioplastics into non-bioderived and biodegradable, bioderived and biodegradable and bioderived and nonbiodegradable. Materials are considered biodegradable when they can be fragmented under the action of microbial agents (fungi, bacteria, and enzymes). When a full microbial activity for assimilation of the material is taking place in around 180 days in a compost medium, it is then regarded as compostable. Materials are considered biobased when the main attention is given to the origin of the carbon building units and less focus is ascribed to the final product life. It is paramount to state that all biobased compounds are not generally displaying biodegradable features, and similarly, not all biodegradable substances can be regarded as biobased;
- *the source used in the production routes*: based on this criterion, it is possible to identify three prime sources, namely petroleum sources, renewable sources, and mixed sources. Conventional plastics are produced by employing either petroleum or natural gas sources, being known as petroleum-based biodegradable polymers. Another category of bioplastics can be obtained by utilizing some resources from nature, like polysaccharides (starch, cellulose, chitosan), proteins (gluten, gelatin). Other routes involve chemical

preparation using sugars or lipids gathered out of vegetal or animal sources, or biologically yielded via fermentation of the specific compounds. They are called renewable resource-based polymers and here belong the following sorts of polymers: protein-containing plastics, certain aliphatic polyesters (polylactic acid [PLA], poly-3-hydroxybutyrate), polyamide 11, polyhydroxy-urethanes, lipid-based polymers or bioderived polyethylene;
- *thermal behavior*: as in the case of polymers, there are thermosets (strengthens when heated) and thermoplastics (remolded or heated after the primer forming and no chemical modifications occur).

Each category of the above-described bioplastics has different combination of thermal, mechanical, optical, and morphological properties as a function of the used source for their production.

1.3 Physical Properties

Bioplastics are characterized by different analytical means to discern if they meet the demands imposed by applications in engineering domain. In the next paragraphs, a brief presentation of some primary techniques of characterization used in material science is made. Also, the parameters of interest for practical situations and related properties for the current technical applicative domains are reviewed.

1.3.1 Rheological Properties

The rheological behavior and viscoelastic character of many bioplastics derived from biopolymers are reported in the literature [18–23]. The shear flow data are relevant for discerning the regions where the viscosity is dependent or not on the applied deformation to the samples. In addition to this, viscoelasticity tests are very useful for the clarification of the microstructure changes in the presence of other components, like polymers, plasticizers, or micro/nanofillers introduced in the bioplastic matrix. Also, when processing the material under variable temperature conditions, one must elucidate how this affects the shear flow properties. The bioplastic solutions response to shear deformation can be regarded as the key point in producing products under different shapes starting from fibrous membranes, films, up to tapes or more complex features. For instance, the fluid polymer matrix has to respond in a specific way to the imposed deformation during the flow process in order to attain proper leveling of the solution. Therefore, the shear flow rheological functions must be correlated with the processing technology applied to the biopolymer fluid. So, for obtaining tapes, membranes or other products, it is paramount to understand what concentration domain must be used for processing, and this must be chosen in agreement with polymer solubility and molecular weight. For example, if higher mechanical strength is desires, the polymer solution must be placed in the regime of semidiluted concentration where chain entanglements occur. This is facilitated by using polymers of high molecular weight. In addition, solubility is a key element that dictates not only the processability but also the morphology of the material in solid state. Thermodynamics of polymer solutions offers an alternative to produce porous architectures, which are relatively facile to acquire if a mixture of solvent/nonsolvent are used. On the other hand, the achievement of films or coatings having uniform thickness is strongly related to the bioplastic solution response to shear deformation. When subjecting a polymer fluid to spin coating, an inherent shearing field is acting on the sample that gradually covers the rotating disk. Depending on the viscosity variation with the shear rate, the sample might lead or not to proper films using this technique. As a result, the shearing interfering during the spin coating of the bioplastic solution must be of such level, so that the sample would behave as a Newtonian fluid. In this way, the thickness profile of the bioplastic sample will remain uniform along the radial position. Another relevant aspect is that some cellulose-based bioplastic at high concentrations are acquiring liquid crystalline behavior. Rheology in this situation is a useful method to highlight the critical concentration where the fluid becomes anisotropic. According to literature data [22, 23], hydroxypropyl cellulose is becoming a chiral nematic liquid crystal at a specific concentration depending on the used solvent and molecular weight. The viscosity versus concentration dependence displays a maximum, which at shear rates close to zero can be considered the point where the isotropic phase is converted into a lyotropic one. This peculiar behavior can be exploited in many engineering applications, such as alignment layers for nematic liquid crystals in displays, in anisotropic foils for controlling optical retardation is optoelectronic devices or in temperature sensors.

1.3.2 Optical Properties

The interaction of the bioplastics with electromagnetic waves can be described by means of optical analysis methods. For example, some materials can undergo photo-degradation processes under exposure to sunlight or in other situations it is desirable to utilize optically transparent or highly absorbent layers for manufacturing components for optoelectronic devices [24–26]. More recent trends indicate that sensing properties of bioplastics via optical characteristics is a hot topic of great interest in development of biophotonic application for chemical recognition and disease detection [27]. Optical biosensors are mainly operating through the optical responses (modification of intensity, wavelength, or polarization state) generated by chemical or biological variations that can be attributed to appearance of a physiological problem [27]. The accuracy and selectivity of the optical responses can be affected by functionalization of biorecognition units from the optical instrument, for example optical waveguides or an interface sustaining surface plasmonic resonance [28, 29]. In other cases, introduction of inorganic particles in bioplastic polymers can lead to nonlinear optical properties, such as nonlinear optical index and nonlinear optical susceptibility [24]. Such aspects are important in production of elements for photonic devices. Optical properties of bioplastics can be also exploited for pharmaceutical purposes [30, 31]. For instance, bioplastics based on cellulose can be employed in the preparation of wound dressing. Depending on the system specific parameters, it is possible to attain prolonged release ability, which in turn can be controlled by UV radiations [31]. Thus, the introduction of a substance like glucose that displays known absorption features in VIS range in a transparent biodegradable polymer enables to tune the release abilities of the system as a function of loaded drug or irradiation times [30, 31].

1.3.3 Mechanical and Thermal Properties

In many engineering applications, bioplastics are forced to meet several criteria in terms of mechanical resistance. In an attempt to replace the conventional petroleum-based plastics, many researchers are facing problems concerning mechanical characteristics. Therefore, many investigations are devoted to the improvement of these features by inserting in the biodegradable matrix many kinds of fillers, such as microcrystalline cellulose [32], epoxidized natural rubber [33], glycerol [34], Sorghum Stalks [35], sorbitol [36], barley straw and grape waste [37], maleic anhydride [38], silicone rubber and/or talc [39, 40], titanium dioxide [41], Miscanthus Biocarbon [42], zinc oxide [43], and so on. The reinforcement of bioplastics is meant to conduct to the significant improvements of the mechanical resistance similar to their synthetic analogs. This kind of approach also determines modification of the thermal behavior of the bioplastics. A relevant example can be found in the study of Zárate-Ramírez and co-workers [44]. They prepared some gluten-based biodegradable polymeric systems where they included distinct types of polysaccharides (0–4.5%). The samples were processed by thermo-molding at high temperature (over 100°C) and at 9 MPa. The dynamic mechanical thermal and the tensile strength experiments revealed that there is a moderate improvement of the network structure after incorporation of the natural polymer at a pH near to the protein isoelectric point, where rendered an additional thermosetting feature to the sample. Furthermore, the system loading with methyl cellulose and carboxymethyl cellulose is producing an enhancement of the sample elongation properties. In any case, the existence of the carriers generated by pH determined a large mismatch among the cellulosic counterpart and protein regions from the multicomponent system. Another important investigation was made by Barkhad and collaborators [45]. Their study was focused on the heat propagation and mechanical characteristics of PLA. This bioplastic was processed in specific conditions, namely melt extrusion, after which the system was subjected to compression molding and finally annealing treatment was applied. As the annealing time is larger (0–24 h), the crystallinity degree of the sample is enhanced. The latter is considerably enhancing the heat conduction abilities of the samples, along with their density, and vitrification temperature. The quickly-cooled materials have lower conductivity (0.0643 $W/m^{-1} K^{-1}$) than those annealed for 24 h (0.0904 $W/m^{-1} K^{-1}$), whereas the vitrification temperature was raised by 11.33%. Another effect of annealing resided in better compressive strength and stiffness noted for bioplastic, while diminishing its ductility. The obtained data showed that optimum annealing conditions for best thermal insulation are 1 to 3 h at 90 °C. In this interval, the sample display insignificant heat conductivity (~0.08 $W/m^{-1} K^{-1}$), smaller density (~1.23 $g\ cm^{-3}$), diminished water absorption (< 0.19%), and big compressive strength (>95 MPa). These aspects recommend these materials as being suitable for green insulation applications. The future of this field is still in development once the renewable materials have proved their efficiency from these points of view.

1.3.4 Electrical Properties

Electrical properties of bioplastics are relatively similar to common polymers, which often present insulation abilities. As in the case of mechanical properties, the main route to enhance the charge transport capacity is to add fillers that have great conductivity properties [46–50]. The work of Huang *et al.* [49] presents recent developments in the area of conductive bioplastics mixed with natural or reusable filling compounds. Several categories of biobased conductive polymer systems are discussed as a function of the filler type, such as CNTs (in bacterial cellulose, pectin, chitosan, or PLA), graphene (in cellulose derivatives, polyvinyl alcohol (PVA), collagen), metallic particles like Ag, Cu, Au, ZrO, WO_3, TiO_2, or ZnO (in glucose oxidase, cellulose, cellulose/polypyrrole, chitosan). Guo *et al.* [46] have synthetized conductive bioplastics by employing as main rules the controlled nanoparticle distribution principle and the work of adhesion. They have employed specific preparation techniques depending on the system components performance. They used graphene nanoplatelets (GNP) and embedded them in a ductile biodegradable matrix, namely poly(butylene adipate-co-butylene terephthalate) (PBAT). Adding in the system of a low-affinity component for the filler (i.e., PLA), their dispersion is limited within PBAT. In this manner, it is easier to form a percolated structure, which at 40 wt% loading raised the electrical and thermal conductivities up to 338 S/m and 3.15 $W/m^{-1} K^{-1}$, respectively. Mergen and collaborators [47] have introduced GNP or carbon nanotubes (CNT) in chitosan using an eco-friendly and not expensive procedure. With the increase of the loading of the bioplastic, the measured surface conduction characteristics, the scattered light intensity, and the tensile parameters were considerably improved. The value of the electrical percolation point is found to be around 24 wt% for graphene-based systems and 10 wt% for the systems with nanotubes. The optical and mechanical percolation points are lower than the electrical ones. The properties of these biocomposites are useful for food packaging or for ultraviolet protection purposes.

1.4 Applications of Bioplastics in Engineering

Practical uses of bioplastics are relying on creating a balance of physical properties to ensure the material's utility in a specific situation. In the following paragraphs, the most relevant examples of engineering applications of bioplastics are reviewed. Their schematic representation is depicted in Figure 1.1.

1.4.1 Bioplastics Applications in Sensors

Biobased sensors are constructed to contain at least one component that has been taken from bioderived sources. The biocomponent could be either the substrate or the sensing element [51]. Many sensors of this sort are linked to biobased products in food monitoring, which can be focused on colorimetric recognition of the analytes. The main idea is to use as natural pH indicators the dyes extracted from fruits, vegetables, berries, and flowers that are covering almost the whole visible wavelengths [51]. After performing the protonation/deprotonation of the dye elements, the delocalized electronic structure is modified, and this is reflected in the quantity of negative charges, but also their restraining is determining the variation in the substance color. A variety of sensors have been reported using as matrix, starch-PVA, cassava starch—glycerol, starch-cellulose, tara gum—PVA, cellulose, chitosan, or carrageenan [52–60]. The principal functions fulfilled by the sensors are colorimetric sensing, antioxidant, or antibacterial, while in some cases, NH_3 or H_2S detection is made. Among the sensing materials in a wide pH range, it is worth mentioning

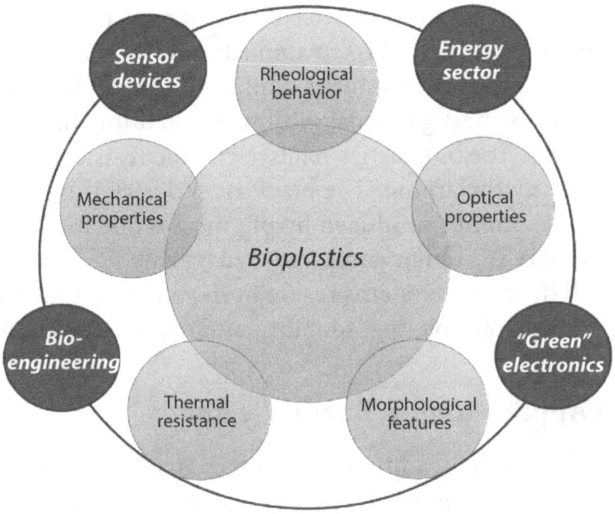

Figure 1.1 The main applications of bioplastics materials in engineering fields.

anthocyanin, limonene, curcumin, alizarin, chlorophyll, and carotenoids. For example, Choi and co-workers [54] obtained a pH detection device constructed from agar and starch having anthocyanin achieved by processing purple sweet potato, which displayed color changes in the pH interval of 2 to 10. Ezati *et al.* [61] have embedded alizarin in a starch-cellulose matrix to make a pH-sensitive device that is able to analyze the quality of rainbow trout. Spectral and morphological examinations revealed that alizarin was extremely compatible with the attained system and interacts with it by hydrogen bonding. The sensor presents color variation with efficiency for the nuances of yellow to violet in the pH interval of 2 to 11. Remarkable color stability was noticed even beyond 8 weeks of depositing of the system under 5 °C. It was demonstrated that the color modification, from orange to reddish brown, is a good indicative of studied substance to recognize the onset of damaging of the cold fish food products. Yoshida *et al.* [62] used chitosan to prepare pH-sensitive systems useful to study shelf-life product. The quality of products can be examined through the pH modifications that are triggered by microorganism growth and chemical reactions, which are affecting the flavor and consistency of products. They manufactured a fast pH-colorimetric indicator by mixing chitosan and anthocyanin. At the beginning, deep purple could be remarked for chitosan sample. Upon introducing it in several environments of specific pH, the nuance of the samples ranged from rosy (acid pH) to blue (neutral pH) and to purple (basic pH). The tensile resistance and stiffness displayed few changes in regard to the control layer (lacking indicator), while the elongation at break was lowered over 40%. Ma *et al.* [59] inserted curcumin in tara gum and PVA blend. The introduced sensing material interacts by hydrogen bonding with the matrix and determines a loosened the film structure. As the curcumin content was higher, the oxygen barrier characteristics diminished, whereas the surface wettability was enlarged. The reaction of the intelligent material in an NH_3 medium with variable relative humidities was analyzed and the color shifts could be remarked within 60 to 180 s. In addition to this, an elevated relative humidity could be suitable for a color response. These systems were tested for shrimp spoilage and led to good sensing results. Yanxiao *et al.* [63] made a gas sensor by utilizing pigment-sensitized TiO_2

as sensing element. Natural pigment-sensitized TiO_2 sensor resulted after sinking it in an ethanol solution of pigments extracted from spinach, radish, black rice and jasmine. These hybrid materials were placed to atmospheres containing methylamine vapors (2–10 ppm) at constant temperature. The best gas sensitivity was noted for the samples with black rice owing to the anthocyanins. The bioplastics relative change resistance was enhanced by augmenting the amounts of methylamine. The black rice–derived material was employed for evaluation of the biogenic amines produced by pig meat while it is in storage. The samples present adequate sensitivity to similar gases, like cadaverine.

The advantages of all described systems reside in the fact that they mainly require simple preparation process, they display biodegradability, and involve natural and safe compounds.

1.4.2 Bioplastics Applications in Energy Sector

The energy problems caused by traditional resources depletion have drawn the attention of researchers toward formulation of alternative solutions for production and storage of energy devices. For instance, in the past decades, the companies manufacturing solar cells have been using bioplastics made from plants to substitute petroleum-derived components in photovoltaics used for the back-sheet [64]. The new bottom layer of the solar module renders environmental protection and also adequate electrical environment so that the device can work properly. It also gives mechanical integrity for the whole module. More recent trends show that biocomposites can be used for making solar panel facades [65]. For this purpose, a biocomposite obtained from fibers and resin, which is durable and highly flexible, was patented. These premises led to development of the researches in this field. Das and collaborators [66] used biocomposites for solar thermal energy storage uses. They attempted to solve the issues related to reduced thermal conduction and form steadiness by using aquatic invasive weed plants. The heat conduction in the phase change material is enlarged of almost 14 times with the incorporation of water hyacinth biochar having the role of matrix. Further embedding of Al powder generates the enhancement of the heat transport by 17 times larger in regard to initial material. Alaaeddin *et al.* [67] studied new backsheet-based biocomposite for solar cells, prepared from polyvinylidene fluoride in which was inserted sugar palm fiber (PVDF-SSPF). The tested module presented acceptable electrical efficiency, attaining adequate and balanced current-voltage plots. The bioplastic composites showed that they are thermally stable since they are exhibiting low heat absorbance and superior temperature shifts, aspects that are promising for solar cells applications. Li and co-workers [68] exploited the advantages arising from optical and thermal properties of the transparent wood for making substrates for perovskite solar cells. The transmittance of this biomaterial (having a thickness of 1 mm) is around 86% at a wavelength of 550 nm, and the haze reaches almost 70% for optical radiations. The latter means that diffused transmittance prevails despite large optical transmittance. Since transparent wood is nonconductive, a layer of ITO is deposited on it. A perovskite photovoltaic system could be made on the prepared substrate. The device has long-term stability and broad photoelectric reaction to the radiations emitted by the sun having the long wavelength edge slightly exceeding 790 nm. The conversion efficiency reached 16.8% when using optically transparent wood substrate during the relatively small temperature stage (under 155°C).

Another application of bioplastics is their use for battery production [69]. They are productive enough to enhance via negative charges exchange, generating great electrochemical movement in the time of their recycling. The main goal in this domain is to improve the electrochemical property during the storing of energy. Knoche *et al.* [70] made a biobattery and supercapacitor based on hybrid glucose/O2. The oxygen-reducing electrode of bilirubin oxidase linked on anthracene-functionalized CNTs and tetrabutylammonium bromide-functionalized Nafion was combined with the other bioelectrode. Dimethylferrocene-modified linear poly(ethylenimine) is utilized not only as bioanode (can interact with enzymes and interferes with negative charges movement) but also as pseudocapacitor (storing feature of bioanode is influenced by high polymer loading). The CNTs and carbon felt embedded into the bioelectrode enhance the material's conductivity, thus determining further increase in the anodic capacitance. The supercapacitor/biobattery was proved to have a capacitance of about 3×10^2 F/g, the value being over four times larger in regard to that of other bioinspired supercapacitors. Baptista *et al.* [71] made a biobattery by using cellulose acetate electrospun fibers. These were chemically modified with polypyrrole and polyaniline by *in situ* chemical oxidation. This approach introduces the benefit of large surface to volume ratio attained by fibrous matrix, where the fibers have the roles of separator and the backing of the electrodes. The cellulose-based fibers display high conductivity (~ 10^{-2} S/m), while the cyclic voltammetry measurements indicated that the battery having the samples as electrodes displayed a power density above 1.5 mW/g under the interaction of the considered biological substances. This is also relevant for medical batteries purposes. Zhu *et al.* [72] employed synthetic enzymatic pathway to produce a large energy-density sugar biobattery. It was proved that 24 electrons in each glucose segment of maltodextrin can be attained via catabolic pathway, which involves 13 enzymes in an air-breathing enzymatic fuel cell. The latter is using nonimmobilized enzymes, which determine storage abilities of almost 600 Ah/kg. This result is one order larger in comparison with that reported for lithium-ion batteries. Lizundia and collaborators [73] have reported the role of cellulose-based compounds as separators in lithium ion batteries. The main advantages introduced by this sort of biopolymers are arising from variable porosity, adequate mechanical and thermal features, reduced toxicity, and good hydrophilicity. The cellulosic membrane separator renders mechanical backing, thus avoids electrical short circuit, but is should allow the flow of ions, while impeding the contact of the electrodes. Moreover, the separator must be electrolyte reservoir of charges movement in the process of either charge or discharge periods. Porous separator layers derived from cellulose have been studied as single polymer materials [74], composites [75], and membranes with adapted surface via polydopamine coatings [76], with the purpose to substitute polyolefin separator in lithium ion devices. Liu *et al.* [77] used starch-based material for Li-S storage purposes. Their solid state electrolyte led to remarkable charges movement that render conduction features of around 4×10^{-4} S/cm. Tests made on starch hosted electrolyte proved that the battery presents mean discharge capacity of slightly above 860 mAh/g after 100 cycles. Conder *et al.* [78] showed that chitosan and chitin can be regarded as precursors of distinguished hard carbons destined to make Na+ containing storage devices. Regardless their similar d-spacings and crystal particle dimensions, pyrolysis of the chitin results in micromesoporous architecture, whereas chitosan leads to nonporous carbon. However, such materials render almost the same initial specific charge and their electrochemical properties begin to depart after sufficient cycling at large rate. The infiltration of the contaminants in chitosan system is believed to produce lagging of the diffusion of the positive charges and diminish the rate of electrochemical processes. Supplementary improvement of the hard carbons from biopolymer via acid exposure allows loosening of the micropores while raising the carbon amount from such samples, hence not only enlarged the active surface area, but also hindered the fading of the some charges. Nowak *et al.* [79] also used chitosan for making electrodes for lithium ion batteries. It is known that tin oxide is ideal for negative electrode owing to its larger specific capacity than graphite. The drawback arises from the unsatisfactory stability caused by volume alteration together with small conduction properties. To address the raised problems, Nowak *et al.* [79] prepared a composite of SnO2 and carbonaceous matrix, where chitosan was the carbon source. The obtained conductive element had slightly above 460 mAh/g. The capacity fade of few percent upon 70 cycles was reported. The attained electrochemical performance is the result of strong interaction among the functional parts of chitosan and those from SnO2. In any case, the charge storage can be principally caused via diffusion-determined procedure revealing that the greatest accomplishments can be attained at small current rates.

Based on these concepts, it can be considered that biodegradable polymer or sugar-powered batteries can be viewed as future-generation green power sources, of great interest in portable electronics. Also, in the context of the environmental safety, the cellulosic separators emerge as an outstanding solution to develop green batteries.

1.4.3 Bioplastics Applications in Bioengineering

Bioplastics can be used in bioengineering field where they introduce several benefits, like flexibility, weight, ease of processing, biocompatibility, cheapness, disposability, and facile integration. In a recent review [80], it was explained that the role of ecofriendly polymer in fabrication of supports for electrochemical biosensing platforms. Chitosan-CNT composites were used to adapt pencil graphite electrode (PGE) with the purpose of electrochemical analysis of vitamin B12 [81]. This substance can be monitored via electrochemical methods owing to the redox chemistry focused on Co element. The composite based modified PGE with chitosan was attained by embedding first the carbon filler in chitosan matrix, and then PGE was inserted in the system. The electrochemical signal of the system was studied in regard to the references of bioplastic sample used for B12 molecule detection. A signal improvement could be noted for diminishment of Co redox couples from monitored substance, owing to catalytic activity of CNTs. Devnani and collaborators [82] have prepared an electrode based on graphene-chitosan to electrochemically examine noradrenaline (NA), a substance with similar redox potential with that of uric acid. To enhance the selectivity, the electrodes were chemically adapted via nanomaterials for mediation of electron transfer. A variety of multicomponent systems (including metal doped polymers) was reported for electrochemical sensing uses [83]. The work of Devnani [82] was aiming to improve current response and to diminish charge transfer resistance of obtained sensing device, which was examined in blood serum. Experiments indicated a good limit of detection of almost 20 nM. No interference was revealed to be caused by ascorbic acid and uric acid regarding the NA detection. Another investigation made by Xia *et al.* [84] pursued fabrication of an electrode modified by using molecularly engineered chitosan/ionic liquid–graphene. The reactions involved bovine serum albumin with the role of template and polypyrrole (formed on top of the electrode) with the role of substrate material. This was a premiere for imprinting a protein using molecularly imprinted electrochemical sensor. Chitosan introduced the advantage of interfacial biocompatibility, whereas graphene helped to attain the sensing interface. In addition, ionic liquids are able to interact with protein and favor its transfer in the liquid balancing the imprinted amount. It also optimized dispersion of graphene, which is reflected on better electrochemical performance [85]. PLA filled with metallic particles is also suitable for electrodes purposes [86]. The carboxylic groups of the PLA are suitable for stabilization of inorganic filler (like gold nanoparticles). Starting from this, a disposable sensor based on printed carbon electrode was made for monitoring the DNA hybridization. Self-assembled monolayer approach relying on thiol-gold of nanodimensions bonding was the route to achieve bonding probe stranded DNA at the outer part of the obtained electrode. The biosensor led to promising results in environments containing methylene blue redox, placed on the biopolymer. Lad and co-workers [87] made a glucose enzyme electrode via embedding glucose oxidase inside the system made of PVA and incompletely prehydrolyzed tetraethyl orthosilicate. The limit of detection for the enzyme prepared electrode was 0.35 mM in conditions of a signal-to-noise ratio of 3. Another report [88] aimed to use graphene-iron oxide-PVA for adapting the properties of PGE useful to record the level of vitamin C. The composites were prepared by one-pot hydrothermal strategy. The utilization of the biodegradable macromolecular compound introduces the advantage of better electrocatalytic activity. The large amount of -OH groups renders numerous electrons during the process. Moreover, PVA enables uniform coating of the electrode and is suitable for mediation of negative charges transfer. The monitoring of the targeted biocompound relies in the following stages: (a) the oxidation of active substance via the deliverance of the negative charges to the new electrode, (b) a current is travelling in the outer circuit—related to oxidation processes, (c) the augmentation of the quantity of the analyte which progressively enhances the current recorded for each spike. The newly designed electrode presents excellent limit of sensing around 0.2 μM, combined with good stability.

Other studies [89, 90] discuss the importance of bioplastics for tissue engineering applications. In a recent study, Wu *et al.* [89] obtained a biocomposite for bone reconstruction uses. The material resulted upon application of cryogelation procedure to chitosan-gelatin loaded with cerium–zinc doped hydroxyapatite. The presence of inorganic phase in the chitosan-derived biosystems led to large growing, and significant protein sorption, concomitant with considerable bactericidal movement. The new chitosan biocomposite presented nontoxic properties in regard to rodent osteoblast cells. Therefore, the reinforced bioplastic material is very promising for applications that involve bone tissue engineering. Neifar *et al.* [90] used thermo-bioplastics, like polyhydroxyalkanoates, and described their use in bioengineering for regeneration of cartilage tissue, bone tissue or for ameliorating the central nervous system problems.

1.4.4 Bioplastics Applications in "Green" Electronics

The accelerated evolution of electronic gadgets shortens the life of many products and this produces enormous quantities of wastes, which produces risks of ambient contamination, while alters health of many beings. For alleviating these undesired aspects, conventional of electronic components made from polymers are nowadays gradually substituted by bioplastics, wherever possible. In the same time, aspects related to functionality, manufacturability, and cost must be taken into account during formulation of the alternative materials [91–101].

A relevant work in the field of "green" electronics was published by Bozó *et al.* [91]. They have reported some materials based on PLA and characterized them to check their suitability for electronic applications. They blended PLA with polyhydroxybutyrate (PHB), but also materials resulted after doping with pyrolyzed lignin (PL) or CNT. These materials were subjected to processes utilized for designing of electronic elements, such as plasma exposure, dip coating, inkjet printing, extrusion, and molding. The flexible films relying on polymer mixtures (in the presence/absence of plasticizer) were prepared by extrusion blowing. The materials exhibit excellent mechanical properties, especially in the presence of the plasticizer, the elongation at break reaches 14 %. The exposure of PLA-PHB to argon plasma creates a greater surface hydrophilicity, which is ideal for further processing with aqueous dispersions of conductive particles. Thus, upon dip coating of the bioplastic in the CNT dispersion, one may notice at the surface the appearance of a homogenous random conductive network. By performing this procedure for several times, the sparse network is changes into a denser one and in this way the conductive features are enhanced. This indicates that such bioplastic films are able to fulfill the demands for a capacitor plate electrode for touch sensor application. The study of Bozó *et al.* [91] continued by depositing conductive micropatterns of Ag and CNT on the obtained flexible supports. This was accomplished by inkjet and screen printing techniques on previously plasma etched surfaces of PLA-PHB supports. Given the conduction abilities of created patterns, the materials can be used to make contact pads and interconnects involved in large impedance circuits. Aside from substrate applications, these bioplastics might be ideal for 3D parts of passive counterparts. The presence of CNTs or metallic particles in bioplastics renders conductive composites, where the properties are controlled by quantity and typical characteristics of the embedded particles. A feasible application in this case is electromagnetic interference (EMI) shielding. The elevated conductivity of a material induces better EMI shielding effectiveness (denoted S), which denotes power dissipation of the radiations within a polymer system. Sufficiently good EMI SE levels (exceeding 30 dB) are failing to impose electrical conductivities above 10^{-1} S/m, presuming the thickness corresponding to shielding layer is adequate. In addition, when the conductive phase has mainly carbon nature (not being metallic) the material can be considered more eco-compatible. The data extracted from current–voltage characteristics and measured DC conductivity for composite of PLA filled with PL and CNTs show that they are fitted for EMI shielding within K-band of frequencies interval of 18.0–26.5 GHz with S values comprised between 40 and 46 dB.

Other relatively recent works [92–94] have described several types of biodegradable polymer materials used in "green" electronics. Bioplastics can be successfully used for integration in electronic devices as substrates and encapsulates. The main issues that must be addressed are linked to mechanical instability of the active layers under external forces and also to creation of good permeation barrier to enable proper operability and endurance to degradation. A general approach to these problems is based on using a multilayered structure, where each component gives the desired barrier features to oxygen and moisture. Also, reduction of the support thickness is employed for diminishing stress–strain ascribed to active elements of the device. Paper is one of the most used semi-natural substrates for modern electronics. It can be employed as flexible support for electronic displays that work on the electrowetting (EW) effect. Kim and Steckl [95] published an interesting study that proved that possibility to make "e-paper" displays straight on paper. In this way, they created an interesting alternative to the conventional e-reader devices. They have tested many sorts of commercially available paper supports that might differ in terms of surface roughness, coating ability, water uptake and thickness. The proposed methodology relies on glassine paper support, ground electrode, an insulating sheet, and an outer substrate of fluoropolymer. Under an applied voltage the EW effect is proved and visualized via the fluctuation of the contact angle of the drop of deionized water containing dye molecules. The EW devices presented large contact angle modulation and remarkable fast switching velocity of about few tens of milliseconds. So, this is an important breakthrough towards cheap and eco-compatible substrates that can be implemented in environmentally friendly electronic devices.

Some of the synthetic polymers offer the possibility to include them as biocompatible substrates for electronics in contact with living tissues. For making implantable devices a variety of polymers are studied, namely PVA, polylactic-co-glycolic acid (PLGA), polyurethane (PU), polyethylene glycol (PEG), polydimethylsiloxane (PDMS) and poly(2-hydroxyethyl methacrylate) (pHEMA) [94, 96]. For example, PVA was integrated in a special device examined for physiological evaluation of the processes occurring at the interfaces to the human epidermis [97]. The benefits of the PVA arise from the controlled transport of aqueous soluble substances by disrupting the crosslinking density of the macromolecules and the resulting swelling ratio. On the other hand, fully biocompatible bioplastics can be investigated as substrates for electronics. For instance, natural silk has a chemical structure derived from the recurrent amino acid part of glycine, serine and alanine. Given its mechanical properties, this biomaterial is used in implantable optical systems for diagnosis [98].

Bioplastic materials that are also biocompatible can be regarded as an ideal solution for contact electrodes and interconnects [94]. Some polymer composites are good candidates for electronic devices which are in contact with biological matter. Their role is to identify events or to modulate them [99]. Conducting electrodes are used for interfacing electronics with the neurons. External linkage with the nervous system is paramount for neuroprosthetics area, including the cochlear implant or for brain machine interfaces and artificial vision. The signal capturing from biological medium implies both electrical and chemical components and their interaction gives the perspective artificial interfacing electronics. An example of bioelectronics device is consisting in metallic electrode arrays placed between polyimide layer and afterwards casted on sacrificial silk support for brain monitoring [100]. The obtained Au electrodes can be encased on the neuron soft layer by flushing with saline liquid to remove the part of silk layer. The electrodes of lowest thickness (under 3 mm) conducted to excellent results concerning average response which is collected from all electrodes.

Another engineering application of the bioplastics is that of active layers. Here, insulators and semiconductors can be regarded as main elements of the electronic products. A relevant example here is represented by the work of Asanduelsa *et al.* [101], which prepared high permittivity dielectrics from a cellulose derivative and barium titanate. It was shown that these dielectrics display high electrical capacity as the ceramic filler amount in the polymer is increased. The obtained composite is highly flexible and displays good mechanical resistance, providing perspectives in development of eco-compatible capacitors for "green" electronics.

Physical properties of bioplastics can also be explored to attain satisfactory progress of thermoelectric devices [94]. To acquire the best results regarding thermoelectric performance, materials with small heat conduction, controlled charge transport and Seebeck coefficient are desirable. Generally, for filled polymers, heat conduction demands almost ten times the percolation levels, including highly closer contact among the loading phase and the continuous phase comparatively to electrons for electrical conductivity.

PLA-carbon fillers composites have high electrical conductivity, while heat conduction is not significantly improved. Introduction of graphene and CNT leads to a thermal conductivity of 5.5 W/m K and 0.41 W/m K, respectively. It was noted that graphite foils are not capable to produce union with CNT bundles during preparation of the biocomposite, while CNT favors crystallization since its displays better nucleating features over graphite. The incorporated filler in the bioplastic are making a layer that impedes travelling of the phonons, making them to scatter. The Seebeck coefficient or thermoelectric power for these materials is around 17 μV/K. The presence of CNTs in the system determines an improvement of the charge conduction, low heat conductivity with appropriate Seebeck coefficient, all these upgrading the thermoelectric properties. Another example is that of bioplastics made from carboxymethyl cellulose mixed with chitosan and then mixed with dodecyltrimethyl ammonium bromide (DTAB). The incorporation of this dopant generates increase of the charge carrier mobility owing to group ions generated by uninterrupted increment in the inserted salt in the polymer matrix. Conduction is increased until the protonium ion present in the salt fills the free position to produce linking to the bioplastic matrix. After attaining the threshold point for DTAB amount, the accumulation of additional ions generates impeding the travelling of the charges and thus inducing sudden lowering of the conduction level.

1.5 Conclusions

The research field of bioplastics is in a continuous development and there are still relevant issues that must be improved in terms of physicochemical properties. The main types of bioplastics and their properties of importance in most applications are briefly reviewed. To solve the drawbacks of this category of materials, such as poor mechanical and/or conductive characteristics, various filling agents and biodegradable polymers were combined over the last years. Such natural or reusable loading particles, macromolecular compounds, and their reinforced systems can be believed to significantly diminish the huge pressure of technological progress on the nature and its life forms, whereas rendering suitable properties. The unique features of the renewable composites allowed the implementation of novel and interesting applications including components for "green" electronics, bioengineering, energy production/storage, and sensor devices. Several relevant case studies are described emphasizing the utility and suitability of bioplastics in these engineering fields and the urgent need to utilize these materials in the current markets for keeping a clean environment.

Acknowledgement

This work was supported by a grant of the Ministry of Research, Innovation and Digitization, CNCS - UEFISCDI, project number PN-III-P1-1.1-TE-2021-0762, within PNCDI III.

Dedication

This work is dedicated in the memory of my mentor Dr. Silvia IOAN (1943-2021), whose achievements, dedication and passion for polymer science have been imprinted on me and guide me throughout my career.

References

1. Dilshad, E., Waheed, H., Ali, U., Amin, A., Ahmed, I., General structure and classification of bioplastics and biodegradable plastics, in: *Bioplastics for sustainable development*, M.K. Roohi (Ed.), pp. 61–82, Springer, Singapore, 2021.
2. Narancic, T., Cerrone, F., Beagan, N., O'Connor, K.E., Recent advances in bioplastics: Application and biodegradation. *Polym. (Basel)*, 12, 920, 2020.
3. Pilla, S., *Handbook of bioplastics and biocomposites engineering applications*, Wiley, New Jersey, 2011.
4. Yang, J., Ching, Y.C., Chuah, C.H., Applications of lignocellulosic fibers and lignin in bioplastics: A review. *Polymers*, 11, 751, 2019.
5. Peelman, N., Ragaert, P., De Meulenaer, B., Adons, D., Peeters, R., Cardon, L., Application of bioplastics for food packaging. *Trends Food Sci. & Technol.*, 32, 128, 2013.
6. Tsiamis, D.A., Torres, M., Castaldi, M.J., Role of plastics in decoupling municipal solid waste and economic growth in the U.S. *Waste Manage.*, 77, 147, 2018.
7. Atiwesh, G., Mikhael, A., Parrish, C.C., Banoub, J., Le, T.-A.T., Environmental impact of bioplastic use: A review. *Heliyon*, 7, e07918, 2021.
8. Lamberti, F.M., Román-Ramírez, L.A., Wood, J., Recycling of bioplastics: Routes and benefits. *J. Polym. Environ.*, 28, 2551, 2020.
9. Rattana, S. and Gheewala, S.H., Environment impacts assessment of petroleum plastic and bioplastic carrier bags in Thailand. *J. Sustain. Ener. Environ.*, 10, 9, 2019.
10. Kumar, S. and Thakur, K.S., Bioplastics—Classification, production and their potential food applications. *J. Hill. Agric.*, 8, 118, 2017.
11. Tonini, D., Schrijvers, D., Nessi, S., Garcia-Gutierrez, P., Giuntoli, J., Carbon footprint of plastic from biomass and recycled feedstock: Methodological insights. *Int. J. Life Cycle Assess.*, 26, 221, 2021.
12. Abdul-Latif, N.-I. S., Ong, M. Y., Nomanbhay, S., Salman, B., Show, P. L, Estimation of carbon dioxide (CO2) reduction by utilization of algal biomass bioplastic in Malaysia using carbon emission pinch analysis (CEPA). *Bioengineered*, 11, 154, 2020.
13. Abe, M.M., Martins, J.R., Sanvezzo, P.B., Advantages and disadvantages of bioplastics production from starch and lignocellulosic components. *Polym. (Basel)*, 13, 2484, 2021.
14. Souza, D.H.S., Andrade, C.T., Dias, M.L., Effect of synthetic mica on the thermal properties of poly(lactic acid). *Polímeros Ciência e Tecnologia*, 24, ESP, 20, 2014.
15. Hemsri, S., Thongpin, C., Somkid, P., Sae-arma, S., Paiykaew, A., Improvement of toughness and water resistance of bioplastic based on wheat gluten using epoxidized natural rubber. *IOP Conference Series: Materials Science and Engineering*, vol. 87, p. 012049, 2015.

16. Shah, M., Rajhans, S., Pandya, H.A., Mankad, A.U., Bioplastic for future: A review then and now. *World J. Adv. Res. Rev.*, 09, 056, 2021.
17. Song, J., Kay, M., Coles, R., Bioplastics, in: *Food and beverage packaging technology*, 2nd, R. Coles and M. Kirwan (Eds.), pp. 295–319, Wiley, Oxford, 2011.
18. Jerez, A., Partal, P., Martínez, I., Gallegos, C., Guerrero, A., Rheology and processing of gluten based bioplastics. *Biochem. Eng. J.*, 26, 131, 2005.
19. Nagy, E.M., Todica, M., Ştefan, R., Coţa, C., Cioca, N., Pop, V.C., Rheological behavior of some starch-based bioplastics. *Bull. Univ. Agric. Sci. Veterinary Med. Cluj-Napoca. Agric.*, 73, 59, 2016.
20. Mazzanti, V., Mollica, F., A review of wood polymer composites rheology and its implications for processing. *Polymers*, 12, 2304, 2020.
21. Cindradewi, A.W., Bandi, R., Park, C.-W., Park, J.-S., Lee, E.-A., Kim, J.-K., Kwon, G.-J., Han, S.-Y., Lee, S.-H., Preparation and characterization of cellulose acetate film reinforced with cellulose nanofibrils. *Polymers*, 13, 2990, 2021.
22. Cosutchi, A.I., Hulubei, C., Stoica, I., Ioan, S., Morphological and structural-rheological relationship in epiclon-based polyimide/hydroxypropylcellulose blend systems. *J. Polym. Res.*, 17, 541, 2009.
23. Stoica, I., Barzic, A.I., Hulubei, C., Polyimide embedding in lyotropic polymer matrix for surface-related applications: Rheological and microscopy investigations. *Rev. Roum. Chim.*, 61, 575, 2016.
24. Barzic, A.I., Soroceanu, M., Rotaru, R., Harabagiu, V., Ciobanu, R.C., Optical dispersion characteristics of polyvinyl alcohol reinforced with a nanoceramic filler. *Mater. Plast.*, 57, 1, 2020.
25. Barzic, A.I. and Albu, R.M., Optical properties and biointerface interactions of chitin. *Polym. Bull.*, 78, 6635, 2021.
26. Albu, R.M., Stoica, I., Barzic, A.I., Postolache, M., Angheluta, M., Dorohoi, D.O., Effect of mechanical treatments on orientation behavior and spectral properties of azoderivative dyes incorporated in poly(vinyl alcohol) films. *Polym. Eng. Sci.*, 61, 2453, 2021.
27. Shan, D., Gerhard, E., Zhang, C., Tierney, J.W., Xie, D., Liu, Z., Yang, J., Polymeric biomaterials for biophotonic applications. *Bioactive Mater.*, 3, 434, 2018.
28. Prasad, P.N., *Introduction to biophotonics*, John Wiley & Sons, New Jersey, USA, 2004.
29. Fan, X., White, I.M., Shopova, S.I., Sensitive optical biosensors for unlabeled targets: A review. *Analytica Chimica Acta*, 620, 8, 2008.
30. Barzic, A.I., Nechifor, C.-D., Stoica, I., Dorohoi, D.-O., On the effects of UV radiation on the release ability of glucose embedded in hydroxypropyl cellulose films. *J. Macromol. Sci. Part B*, 55, 575, 2016.
31. Nechifor, C.-D., Barzic, A.I., Stoica, I., Cloşca, V., Dorohoi, D.-O., Study on glucose release ability from hydroxypropyl cellulose films. *Polym. Bull.*, 72, 549, 2015.
32. Abdullah, A.H.D., Putri, O.D., Fikriyyah, A.K., Nissa, R.C., Intadiana, S., Effect of microcrystalline cellulose on characteristics of cassava starch-based bioplastic. *Polym. Plast. Technol. Mater.*, 59, 1250, 2020.
33. Hemsri, S., Thongpin, C., Somkid, P., Sae-arma, S., Paiykaew, A., Improvement of toughness and water resistance of bioplastic based on wheat gluten using epoxidized natural rubber. *IOP Conf. Series: Mater. Sci. Eng.*, 87, 012049, 2015.
34. Inayati, Pamungkas, D.J. and Matovanni, M.P.N., Effect of glycerol concentration on mechanical characteristics of biodegradable plastic from rice straw cellulose. *AIP Conference Proceedings*, vol. 2097, p. 030110, 2019.
35. Darni, Y., Dewi, F.Y., Lismeri, L., Modification of sorghum starch-cellulose bioplastics with sorghum stalks filler. *Jurnal Rekayasa Kimia dan Lingkungan*, 12, 22, 2017.

36. Chadijah, S., Rustiah, W.O., Munir, M.I.D., Determination of the optimum concentration cellulose baggase in making film bioplastic. *IOP Conf. Series: Journal of Physics: Conf. Series*, vol. 979, p. 012026, 2018.
37. Lopez-Gil, A., Rodriguez-Perez, M.A., De Saja, J.A., Bellucci, F.S., Ardanuy, M., Strategies to improve the mechanical properties of starch-based materials: Plasticization and natural fibers reinforcement. *Polímeros*, 24, 36, 2014.
38. Jang, W.Y., Shin, B.Y., Lee, T.J., Narayan, R., Thermal properties and morphology of biodegradable pla/starch compatibilized blends. *J. Ind. Eng. Chem.*, 13, 457, 2007.
39. Khuenkeao, T., Petchwattana, N., Covavisaruch, S., Thermal and mechanical properties of bioplastic poly(lactic acid) compounded with silicone rubber and talc. *AIP Conference Proceedings*, vol. 1713, p. 080005, 2016.
40. Lee, C., Pang, M.M., Koay, S.C., Choo, H.L., Tshai, K.Y., Talc filled polylacticacid biobased polymer composites: Tensile, thermal and morphological properties. *SN Appl. Sci.*, 2, 354, 2020.
41. Amin, M.R., Chowdhury, M.A., Kowser, M.A., Characterization and performance analysis of composite bioplastics synthesized using titanium dioxide nanoparticles with corn starch. *Heliyon*, 5, e02009, 2019.
42. Li, Z., Reimer, C., Wang, T., Mohanty, A.K., Misra, M., Thermal and mechanical properties of the biocomposites of Miscanthus biocarbon and poly (3-hydroxybutyrate-co-3-hydroxyvalerate)(PHBV). *Polymers*, 12, 1300, 2020.
43. Harunsyah, Yunus, M. and Fauzan, R., Mechanical properties of bioplastics cassava starch film with Zinc Oxide nanofiller as reinforcement. *IOP Conf. Ser.: Mater. Sci. Eng.*, 210, 012015, 2017.
44. Zárate-Ramírez, L.S., Romero, A., Bengoechea, C., Partal, P., Guerrero, A., Thermo-mechanical and hydrophilic properties of polysaccharide/gluten-based bioplastics. *Carbohydr. Polym.*, 112, 24, 2014.
45. Barkhad, M.S., Abu-Jdayil, B., Mourad, A.H.I., Iqbal, M.Z., Thermal insulation and mechanical properties of polylactic acid (PLA) at different processing conditions. *Polymers*, 12, 2091, 2020.
46. Guo, Y., Zuo, X., Xue, Y., Tang, J., Gouzman, M., Fang, Y. *et al.*, Engineering thermally and electrically conductive biodegradable polymer nanocomposites. *Composit. Part B*, 189, 107905, 2020.
47. Mergen, Ö.B., Arda, E., Evingür, G.A., Electrical, optical and mechanical properties of chitosan biocomposites. *J. Compos. Mater.*, 54, 1497, 2020.
48. Abdo, H.S., Elzatahry, A.A., Alharbi, H.F., Khalil, K.A., Electrical conductivity behavior of biopolymer composites, in: *Biopolymer composites in electronics*, K.K. Sadasivuni, D. Ponnamma, J. Kim, J.-J. Cabibihan, M.A. AlMaadeed (Eds.), pp. 13–25, Elsevier, Amsterdam, Netherlands, 2017.
49. Huang, Y., Kormakov, S., He, X., Gao, X., Zheng, X., Liu, Y. *et al.*, Conductive polymer composites from renewable resources: An overview of preparation, properties, and applications. *Polymers*, 11, 187, 2019.
50. Amri, A., Rahmana, H., Utami, S.P., Iriyanti, R.S., Jiang, Z.T., Rahman, M.M., Conductive composites of tapioca based bioplastic and electrochemical-mechanical liquid exfoliation (emle) grapheme. *IOP Conf. Series: Mater. Sci. Eng.*, 345, 012026, 2018.
51. Halonen, N., Pálvölgyi, P.S., Bassani, A., Fiorentini, C., Nair, R., Spigno, G., Kordas, K., Bio-Based smart materials for food packaging and sensors—A review. *Front. Mater.*, 7, 2296, 2020.
52. Yoshida, K., Mori, M., Kondo, T., Blue flower color development by anthocyanins: From chemical structure to cell physiology. *Nat. Prod. Rep.*, 26, 884, 2009.
53. Dudnyk, I., Janeèek, E.-R., Vaucher-Joset, J., Stellacci, F., Edible sensors for meat and seafood freshness. *Sens. Actuat. B*, 259, 1108, 2018.
54. Choi, I., Lee, Y.J., Lacroix, M., Han, J., Intelligent pH indicator film composed of agar/potato starch and anthocyanin extracts from purple sweet potato. *Food Chem.*, 218, 122, 2017.

55. Zhai, X., Li, Zhang, J., Shi, J., Zou, X., Huang, X. *et al.*, Natural biomaterial based edible and pH-sensitive films combined with electrochemical writing for intelligent food packaging. *J. Agric. Food Chem.*, 66, 2836, 2018.
56. Luchese, C.L., Sperotto, N., Spada, J.C., Tessaro, I.C., Effect of blueberry agro-industrial waste addition to corn starch-based films for the production of a pH-indicator film. *Int. J. Biol. Macromol.*, 104, 11, 2017.
57. Chayavanich, K., Thiraphibundet, P., Imyim, A., Biocompatible film sensors containing red radish extract for meat spoilage observation. *Spectrochim. Acta A*, 226, 117601, 2020.
58. Narwade, V.N., Anjum, S.R., Kokol, V., Khairnar, R.S., Ammonia sensing ability of differently structured hydroxyapatite blended cellulose nanofibril composite films. *Cellulose*, 26, 3325, 2019.
59. Ma, Q., Du, L., Wang, L., Tara gum/polyvinyl alcohol-based colorimetric NH3 indicator films incorporating curcumin for intelligent packaging. *Sens. Actuat. B*, 244, 759, 2017.
60. Ahmad, N.A., Heng, L.Y., Salam, F., Zaid, M.H.M., Hanifah, S.A., A colorimetric pH sensor based on Clitoria sp and Brassica sp for monitoring of food spoilage using chromametry. *Sensors*, 19, 4813, 2019.
61. Ezati, P., Tajik, H., Moradi, M., Molaei, R., Intelligent pH-sensitive indicator based on starch-cellulose and alizarin dye to track freshness of rainbow trout fillet. *Int. J. Biol. Macromol.*, 132, 157, 2019.
62. Yoshida, C.M.P., Maciel, V.B.V., Mendonça, M.E.D., Franco, T.T., Chitosan biobased and intelligent films: Monitoring pH variations. *LWT- Food Sci. Technol.*, 55, 83, 2014.
63. Yanxiao, L., Xiao-bo, Z., Xiao-wei, H., Ji-yong, S., Jie-wen, Z., Holmes, M. *et al.*, A new room temperature gas sensor based on pigment-sensitized TiO2 thin film for amines determination. *Biosens. Bioelectron.*, 67, 35, 2015.
64. Sadasivuni, K. K., Deshmukh, K., Ahipa, T. N., Muzaffar, A., Ahamed, M. B., Pasha, S. K. K., Al-Maadeed, M. A.-A., Flexible, biodegradable and recyclable solar cells: A review. *J. Mater. Sci.: Mater. El.*, 30, 951, 2019.
65. Astudillo, J., García González, M., Sacristán, J., Uranga, N., *et al.*, New biocomposites for innovative construction facades and interior partitions. *J. Facade Design Engin.*, 6, 065, 2018.
66. Das, D., Bordoloi, U., Muigai, H.H., Kalita, P., A novel form stable PCM based bio composite material for solar thermal energy storage applications. *J. Energy Storage*, 30, 101403, 2020.
67. Alaaeddin, M.H., Sapuan, S.M., Zuhri, M., Zainudin, E.S., M Al-Oqla, F., Development of photovoltaic module with fabricated and evaluated novel backsheet-based biocomposite materials. *Mater. (Basel, Switzerland)*, 12, 3007, 2019.
68. Li, Y., Cheng, M., Jungstedt, E., Xu, B., Sun, L., Berglund, L.A., Optically transparent wood substrate for perovskite solar cells. *ACS Sustain. Chem. Eng.*, 7, 6061, 2019.
69. Ramesh, M., Maniraj, J., Rajesh Kumar, L., Biocomposites for energy storage, in: *Biobased composites: Processing, characterization, properties, and applications*, 1st, A. Khan, S.M. Rangappa, S. Siengchin, A.M. Asiri (Eds.), pp. 123–142, Wiley, New Jersey, 2021.
70. Knoche, K.L., Hickey, D.P., Milton, R.D., Curchoe, C.L., Minteer, S.D., Hybrid glucose/O2 biobattery and supercapacitor utilizing a pseudocapacitive dimethylferrocene redox polymer at the bioanode. *ACS Energy Lett.*, 1, 380, 2016.
71. Baptista, A.C., Ropio, I., Romba, B., Nobre, J.P., Henriques, C., Silva, J.C., Martins, J.Y., Borges, J.P., Ferreira, I., Cellulose-based electrospun fibers functionalized with polypyrrole and polyaniline for fully organic batteries. *J. Mater. Chem. A*, 6, 256, 2018.
72. Zhu, Z., Kin Tam, T., Sun, F. *et al.*, A high-energy-density sugar biobattery based on a synthetic enzymatic pathway. *Nat. Commun.*, 5, 3026, 2014.

73. Lizundia, E., Costa, C.M., Alves, R., Lanceros-Méndez, S., Cellulose and its derivatives for lithium ion battery separators: A review on the processing methods and properties. *Carbohydr. Polym. Technol. Appl.*, 1, 100001, 2020.
74. Gonçalves, R.F., Lizundia, E., Silva, M.M., Costa, C.M., Lanceros-Mendez, S., Mesoporous cellulose nanocrystal membranes as battery separators for environmentally-safer lithium-ion batteries. *ACS Appl. Energy Mater.*, 2, 3749, 2019.
75. Xu, Q., Wei, C., Fan, L., Peng, S., Xu, W., Xu, J., A bacterial cellulose/Al2O3 nanofibrous composite membrane for a lithium-ion battery separator. *Cellulose*, 24, 1889, 2017.
76. Pan, R., Xu, X., Sun, R., Wang, Z., Lindh, J., Edström, K., Strømme, M., Nyholm, L., Nanocellulose modified polyethylene separators for lithium metal batteries. *Small*, 14, 1704371, 2018.
77. Lin, Y., Li, J., Liu, K., Liu, Y., Liu, J., Wang, X., Unique starch polymer electrolyte for high capacity all-solid-state lithium sulfur battery. *Green Chem.*, 18, 3796, 2016.
78. Conder, J., Vaulot, C., Marino, C., Villevieille, C., Ghimbeu, C.M., Chitin and chitosan—Structurally related precursors of dissimilar hard carbons for Na-Ion battery. *ACS Appl. Energy Mater.*, 2, 4841, 2019.
79. Nowak, A.P., Gazda, M., Łapinski, M., Zarach, Z., Trzcinski, K., Szkoda, M., Mania, S., Li, J., Tylingo, R., Tin oxide encapsulated into pyrolyzed chitosan as a negative electrode for lithium ion batteries. *Materials*, 14, 1156, 2021.
80. Lanzalaco, S. and Molina, B.G., Polymers and plastics modified electrodes for biosensors: A review. *Molecules*, 25, 2446, 2020.
81. Kuralay, F., Vural, T., Bayram, C., Denkbas, E.B., Abaci, S., Carbon nanotube-chitosan modified disposable pencil graphite electrode for Vitamin B 12 analysis. *Coll. Surfaces B*, 87, 18, 2011.
82. Devnani, H., Satsangee, S.P., Jain, R., Nanocomposite modified electrochemical sensor for sensitive and selective determination of noradrenaline. *Mater. Today Proc.*, 3, 1854, 2016.
83. Yang, D., Advances in nanocomposites-synthesis, in: *Characterization and industrial applications*, B. Reddy (Ed.), InTechOPen: London, UK, 2011.
84. Xia, J., Cao, X., Wang, Z., Yang, M., Zhang, F., Lu, B., Li, F., Xia, L., Li, Y., Xia, Y., Molecularly imprinted electrochemical biosensor based on chitosan/ionic liquid-graphene composites modified electrode for determination of bovine serum albumin. *Sens. Actuators B*, 225, 305, 2016.
85. Shu, Y., Liu, M., Chen, S., Chen, X., Wang, J., New insight into molecular interactions of imidazolium ionic liquids with bovine serum albumin. *J. Phys. Chem. B*, 115, 12306, 2011.
86. Nordin, N., Yusof, N.A., Abdullah, J., Radu, S., Hajian, R., Characterization of polylactide-stabilized gold nanoparticles and its application in the fabrication of electrochemical DNA biosensors. *J. Braz. Chem. Soc*, 27, 1679, 2016.
87. Lad, U., Kale, G.M., Bryaskova, R., Glucose oxidase encapsulated polyvinyl alcohol-silica hybrid films for an electrochemical glucose sensing electrode. *Anal. Chem.*, 85, 6349, 2013.
88. Das, T.R., Jena, S.K., Madhuri, R., Sharma, P.K., Polymeric iron oxide-graphene nanocomposite as a trace level sensor of vitamin C. *Appl. Surf. Sci.*, 449, 304, 2018.
89. Wu, S., Ma, S., Zhang, C., Cao, G., Wu, D., Gao, C., Lakshmanan, S., Cryogel biocomposite containing chitosan-gelatin/cerium–zinc doped hydroxyapatite for bone tissue engineering. *Saudi J. Biol. Sci.*, 27, 2638, 2020.
90. Neifar, M., Naili, F., Chouchane, H., Potential use of microbial thermo-bioplastic polyhydroxyalkanoate as promising tissue engineering biomaterial in biomedicine. *Adv. Tissue Eng. Regen. Med.*, 4, 34, 2018.
91. Bozó, É., Ervasti, H., Halonen, N., Shokouh, S.H.H., Tolvanen, J., Pitkanen, O., Jarvinen, T., Palvolgyi, P.S., lgyi, A., Sápi, A., Konya, Z., Bioplastics and carbon-based sustainable materials, components, and devices: Toward green electronics. *ACS Appl. Mater. Interf.*, 13, 49301, 2021.

92. Irimia-Vladu, M., "Green" electronics: Biodegradable and biocompatible materials and devices for sustainable future. *Chem. Soc Rev.*, 43, 588, 2014.
93. Irimia-Vladu, M., Glowacki, E.D., Sariciftci, N.S., Bauer, S., *Green materials for electronics*, Wiley, Weinheim, Germany, 2018.
94. Hou, X., Liu, S., He, C., Designing ultratough, malleable and foldable biocomposites for robust green electronic devices. *J. Mater. Chem. A*, 10, 1497, 2022.
95. Kim, D.Y. and Steckl, A.J., Electrowetting on paper for electronic paper display. *ACS Appl. Mater. Interfaces*, 2, 3318, 2010.
96. Serrano, M.C., Chung, E.J., Ameer, G.A., Advances and applications of biodegradable elastomers in regenerative medicine. *Adv. Funct. Mater.*, 20, 192, 2010.
97. Kim, D.H., Lu, N., Ma, R., Kim, Y.S., Kim, R.H., Wang, S., Wu, J., Won, S.M., Tao, H., Islam, A., Yu, K.J., Kim, T., Chowdhury, R., Ying, M., Xu, L., Li, M., Chung, H.J., Keum, H., McCormick, M., Liu, P., Zhang, Y.W., Omenetto, F.G., Huang, Y., Coleman, T., Rogers, J.A., Epidermal electronics. *Science*, 333, 838, 2011.
98. Guidetti, G., Wang, Y., Omenetto, F.G., Active optics with silk. *Nanophotonics*, 10, 137, 2020.
99. Owens, R.M. and Malliaras, G.G., Organic electronics at the interface with biology. *Mater. Today*, 35, 449, 2010.
100. Kim, D.-H., Viventi, J., Amsden, J.J., Xiao, J., Vigeland, L., Kim, Y.-S., Blanco, J.A., Panilaitis, B., Frechette, E.S., Contreras, D., Kaplan, D.L., Omenetto, F.G., Huang, Y., Hwang, K.-C., Zakin, M.R., Litt, B., Rogers, J.A., Dissolvable films of silk fibroin for ultrathin conformal bio-integrated electronics. *Nat. Mater.*, 9, 511, 2010.
101. Barzic, A.I., Soroceanu, M., Rotaru, R., Doroftei, F., Asandulesa, M., Tugui, C., Dascalu, I.A., Harabagiu, V., Cellulose derivative/barium titanate composites with high refractive index, conductivity and energy density. *Cellulose*, 29, 863, 2022.

2

Biobased Materials: Types and Sources

Kushairi Mohd Salleh[1,2*], Amalia Zulkifli[3], Nyak Syazwani Nyak Mazlan[3] and Sarani Zakaria[3]

[1]School of Industrial Technology, Universiti Sains Malaysia, Penang, Malaysia
[2]Renewable Biomass Transformation Cluster, School of Industrial Technology, Universiti Sains Malaysia, Penang, Malaysia
[3]Bioresources and Biorefinery Laboratory, Department of Applied Physics, Faculty of Science and Technology, Universiti Kebangsaan Malaysia, Bangi, Selangor, Malaysia

Abstract

Knowing the types and sources of the materials is essential. For biobased materials, particularly in bioplastic technology, their classifications can be branched into degradable and nondegradable. The materials' degradability relies on the substituents formed through blending or mixing technology, whether the portions are wholly or partly made of biobased or fossil-based materials. The utilization of biobased materials in many applications is due to the ability to instill numerous value-added properties and be molded into limitless shapes and functions. Furthermore, the variety of biobased materials as a renewable feedstock has delimited complete reliance on nonrenewable materials, fossil-based, and its by-products, for instance. In addition, their biodegradability, dissolvability, and compatibility make them irrefutably attractive in vast applications. Nonetheless, a clear difference between biobased materials, biomaterial, and biodegradable materials has not been well distinguished. As for bioplastics and biocomposite applications, engineered biobased materials are not foreign and have been accepted for delivering remarkable performance. Hence, this book chapter outlines the most well-studied biodegradable and nonbiodegradable biobased materials, specifically for bioplastics products.

***Keywords*:** Biodegradability, biomass sciences, green product, production, polysaccharide, protein

2.1 Introduction

Biobased materials can be defined as naturally occurring or synthesized substances derived from any living matter. "Bio" is a Greek word that carries the meaning of "life." Despite that, the term "biobased" does not decisively indicate that the product is fully derived after biomass but rather a collection of biomasses and fossil-based materials. According to the International Union of Pure and Applied Chemistry (IUPAC), biobased material cannot be deemed environmentally friendly nor biocompatible or biodegradable, mainly whether or not akin to an oil-based polymer [1]. Meanwhile, according to the U.S. Department of Agriculture (USDA), biobased material or product must possess at least 99% materials from

**Corresponding author*: kmsalleh@usm.my

Inamuddin and Tariq Altalhi (eds.) Handbook of Bioplastics and Biocomposites Engineering Applications: 2nd Edition, (23–48) © 2023 Scrivener Publishing LLC

organic sources. The basis of the varied definitions of biomass materials is that the biobased product must contain at least two materials, whether it is mutually immiscible or not, either fully organic, biomass-based, or a combination of organic materials with inorganic, where inorganic is the least portion. From this understanding, it is manifested most in IUPAC definitions. Regardless, scientifically, biobased materials terminology is used explicitly to a material that has been greatly processed with value-added properties. A distinct difference between "biobased materials" and "biomaterials" should also be elucidated. "Biomaterials" are any biocompatible materials to be adapted in a living system, particularly in medical applications, without producing an adverse effect. Howbeit, not all "biobased materials" are thought to be biocompatible and biodegradable. Synonymously, not all biobased materials are bioplastics, as "bioplastics" are defined as "biobased polymers," whereas not all are polymers. Therefore, a highlight is given to biobased materials that are functioned and are engineered as a bioplastic material.

Delineated terminology of bioplastic by IUPAC is expressed as a biobased polymer resulting from the biomass or proceeded from monomers resulting from the biomass. At a certain point of the processing process, flow can shape it [1]. It is discouraged to use the term "bioplastic" as it deludes by suggesting that every polymer resulting from the biomass is environmentally friendly. In addition, the interpretation of bioplastic is used to contrast polymers resulting from fossil resources (monomers). Therefore, the disparity of meanings and expressions between bioplastics and biobased polymers must be distinguished in evading misinterpretation. Correspondingly, with the biobased material, a biobased polymer cannot be deemed environmentally friendly nor biocompatible or biodegradable. Therefore, branching them into biodegradable and nonbiodegradable is much wieldier.

The biobased polymer must comprise carbon, hydrogen, oxygen, etc. If the carbon source is ultimately/partly fossil-based, the biobased polymer is considered nonbiodegradable. Polylactic acid (PLA) and polyhydroxyalkanoates (PHA) are acclaimed biodegradable biobased polymers. Concurrently, biobased polyethylene (PE), biobased polyethylene terephthalate (PET), and biobased polyamide (PA) are among the notable nonbiodegradable biobased polymer. The utilization of biobased polymers with varying biodegradability properties has different trajectories. For instance, the biodegradable biobased polymer is mainly used to substitute synthetic plastic cease as litter and intendedly short-lived. Meanwhile, the partially biobased polymer with a near-pristine quality supersedes oil-based plastics, where durability and robustness are vital. The nonbiodegradable biobased polymer typically has better mechanical and physical stability than biodegradable and long-lived. Due to this, both biodegradable and nonbiodegradable are neck and neck in plastic applications, offering different qualities and needs. Nevertheless, it is important to note that not all biobased polymers are biobased materials. Based on the biobased materials denotations, these biodegradable and nonbiodegradable biobased polymers must be mixed or blended with other materials (not necessarily polymer) to be regarded as biobased materials.

The physical and chemical attributes differentiate biobased material types and sources. These days, new biobased materials are everywhere. Classification of these materials is sometimes referred to as their sources, processing techniques, chemical nature, and physical attributes. Nonetheless, our focus is two large groups of biobased materials according to their decomposition nature: biodegradable and nonbiodegradable. As for biodegradable biobased materials, three major precursors have often been highlighted: polysaccharides, proteins, and lipids. These three materials groups are the largest and the most exploitable biomass materials to be formed into biobased materials products. This is due to the availability and massive range of materials classes, where most bioplastic materials are derived. Meanwhile, for nonbiodegradable biobased materials, there is no specific group to segregate the material, as long as the source of carbon is completely/partly derived from fossil-based materials. Therefore, the utilization of biomass and fossil-based materials together are attractive and sensible. However, even though we fully utilize virgin materials and recycle 100% of used materials, at our consumption rate, these materials are finite. Therefore, in sustaining the existing levels of manufacturing and living, the mounting needs and appetite driven by demand must be fostered intelligently and cooperatively. Henceforth, to solve this, closed-loop recycling and implementation of the collaborative system by encouraging users to consume differently likened to the light of a more sustainable future. Due to this, biobased materials have expanded into incalculable types of products. For instance, casein protein found in milk can be extruded into a fiber with a multitude of uses. Pineapple fibers as an artificial leather were used as a cheaper alternative to leather, and sorghum starch was formed into various edible plastic utensils with added flavor. While recently, filamentous fungi as a new source of biobased materials are excellently used as packaging, beneficial in the automotive industry, textile, and electrical circuit board. These products were made to delimit materials shortage and expand materials variety and quality. Nevertheless, from the preceding discussion, sole materials forming into a new product should not be known as biobased materials. Hereafter, "biobased material" terminology should be used in care, and misunderstanding must be avoided.

This chapter will outline biodegradable and nonbiodegradable biobased materials, specifically the biobased polymer. Only selected biobased materials are chosen according to the availability, potential, vast applications, and implications to the development of biobased polymer and biocomposite engineering applications. New biobased materials are everywhere; distinguishing them according to their degradation behavior helps sustain materials supply and governance.

2.2 Biodegradable Biobased Material

2.2.1 Polysaccharides

Polysaccharides or polycarbohydrates (also called glycan) are the most abundant biopolymer on Earth, with enormous structural diversity. They can be found exceedingly in many organisms, as a structure formation via its supramolecular interactions of myriad interactions. Polysaccharides are biosynthetically built up by a long polymeric chain composing tens of thousands of monosaccharide units, where their glycosidic linkages define their mode, characteristics, functions, and properties. Polysaccharides specifically can be divided into storage-type (starch, glycogen, and galactogen) and structural-type (cellulose and chitin). They may be amorphous but not necessarily edible and dissolve in water. For instance, both starch amylose and cellulose are homopolymers, but due to their different linkages, starch amylose is water-soluble and edible, not cellulose. The repeating unit of polysaccharides' monomers gives distinctive properties among other polysaccharides. If the building block of polysaccharides is made of a similar monomer or monomer derivative, it is called homopolysaccharide or homoglycan. Glycogen, starch, cellulose, dextran, pentosan, and fructan are fallen under this category. While more than one type of monosaccharide they are called heteropolysaccharides or heteroglycan. Pectin, peptidoglycan, and agar are fallen under this category.

To date, polysaccharides have been progressively accepted as key substances in myriad applications due to their activity and acuity. Even if polysaccharides' properties are antecedently remarkable, further modification and compositing with other materials is inevitable and ought to be done. It is a norm to derive polysaccharide materials chemically, physically, thermally, or biologically to improve the given features and adapt to superior materials. Regarding chemical structure modifications, etherification and esterification on polysaccharides provide multidimensional improvements and yield a broad spectrum of polysaccharides derivatives. These derivatives usually show different properties than their precursor, usually more enhanced. For instance, sodium carboxymethyl cellulose (NaCMC) is dissolvable in water and edible, not cellulose. Regardless of treatments, the intention to modify must not defeat the best quality of the materials, where preservation and improvement must be complemented and not overemphasized. Apart from polysaccharides modifications, advancements in new types of solvents, catalysts, and crosslinkers have also improved the quality of polysaccharide-based products. These processes enable polysaccharides to be mixed and blended with ease and sound. Nevertheless, modifications are sometimes insufficient to provide the targeted properties for specific applications. Hence, biomass or fossil-based blend with other polymers or materials is necessary to boost the final properties. Via blending/mixing technology, a new material class is formed, known as biobased materials.

Polysaccharides are constantly misunderstood as biobased material. They can be extracted from biomass but cannot be called biobased materials. Based on terminology definitions outlined by IUPAC and ISO, biobased materials are a combination of different polysaccharides or fossil-based materials forming into new materials. However, if the product is made of one type of material, cellulose, it cannot be called biobased materials. Therefore, the delineated definitions must be well understood to avoid confusion and misinterpretation. The most common polysaccharides used in biobased materials, specifically biobased polymer, are starch, cellulose, chitosan, and alginate.

2.2.2 Starch

Starch is a concoction of two polymers: amylose and amylopectin. Amylose (α-(1→4)-D-glucose) and amylopectin (α-(1→4)- and α-(1→4)-D-glucose) are starch molecules that appear as linear and highly branched polymer, respectively. Starch with basic chemical formula $(C_6H_{10}O_5)_n$ is a storage type of polysaccharide as an energy reservoir, usually found in many diverse crops of higher plants. It is formed by dehydration synthesis, where plants store glucose that is not used as starch inside the amyloplasts. As a storage type of carbohydrate, it will break down to glucose at a reduced photosynthetic activity to nourish the plant as a reserve food supply. The breakdown process is known as the saccharification process. It is an essential source of carbohydrates to the human diet, which occurs naturally as a granular form in plants, particularly abundant in seeds, roots, tubers, and stem pith. Starch molecules molecularly organize themselves as semicrystalline granules. Starch can be isolated by smashing, crushing, or grinding and then dispersed in water, where the resulting paste is sedimented to remove the remaining impurities and then dried. Starch is usually sourced from wheat, tapioca, and corn. They are differentiated with different granular sizes and amylose and amylopectin percentage portions.

Pure amylopectin starch without amylose can also be found in rare circumstances, called waxy starches. While high amylose starch is called amylomaize, it has higher retrogradation and is used as resistant starch. Retrogradation or reprecipitation is a typical process of dissolved starch gradually aggregating to develop a gel when starch is subjected to a freezing or cooling environment for an extended period. Association during aggregation may involve amylose-amylose, amylopectin-amylopectin, and amylose-amylopectin. Consequently, the digestibility of starch deteriorated as it resisted the hydrolytic action of α-amylase. The resistance of starch toward hydrolytic action is called resistant starch, where resistant starch usually acts as dietary fiber. Nevertheless, colonic microbiota can ferment resistant starch, generating metabolic by-products, such as gases and short-chain fatty acids. Fermentation of resistant starch is also beneficial to promote good bacteria growth. Starch is one of the three most common carbohydrates (sucrose and lactose) taken up by the human diet. When humans and other animals consume starch, it is broken down into its constituent sugar molecules. However, if starch is overconsumed, it may lead to dental caries, obesity, and diabetes.

In understanding the behavior of starch, amylose and amylopectin must be differentiated. Typically, starch comprises 70 to 80 wt.% amylopectin and 20 to 30 wt.% amylose. Meyer and workfellow discovered that the amylose in 1940 is not a straight chain of glucose unit but coiled like spring, with six glucose monomers per turn [2]. In consequence of its coiled nature, it can accommodate iodine molecule, where when it is treated with iodine, characteristic blue-violet color is formed caused by amylose-iodine complex formation. Amylose is structurally smaller than amylopectin, linear, branchless, and more soluble in water. Due to their helical structure, they are tightly packed, which aids their resistance from being degraded. Its helical structure accommodates lesser space, thus making them a more preferred starch for storage. In addition, amylose can be dissolved in hot water without being swelled. When the amylose content is higher than usual, starch expansion potential decreases and the starch's gel strength. In addition, starch with higher amylose content induces better film barrier properties and mechanical strength. Meanwhile, amylopectin glucose units are arranged in both linear and branched chains, holding more than 1,000 glucose units, where the branching of the chain disrupts its helical structure. Due to this, when treated with iodine, amylopectin produces reddish-brown instead of blue-violet color. Unlike amylose, amylopectin is easily digestible thus can provide instant energy to its consumers. Since amylopectin is branching, it has many endpoints, making it more soluble in water forming hydrogen bonds with water. Hence, amylopectin is swelled in hot water and developed into a gel. Altogether, the differences between amylose and amylopectin make starch a unique polysaccharide, offering various functionalities.

Starch-based products have been experimented with and developed in various applications. Starch has been consumed as food for ages, works as a structural enhancer, and evolved as self-reinforced materials. So-called self-reinforced materials, modifying starch particles, have improved their plasticizer property and been adapted in starch-based polymeric materials. Starch is proven to adapt various processing techniques, such as injection, extrusion, compression molding, casting, and foaming, reactive extrusion, etc. [3]. Due to this, starch is not foreign as a reliable biobased polymer. As a well-known biobased polymer, starch is often served in paper manufacturing as tensile strength and surface sizing enhancer, also beneficial in the textile industry as warp sizing to impart strength to the thread during weaving. Although classified as biobased polymer, starch is also characterized as a biomaterial. Its biocompatibility has helped medicine technology in accelerating healing and others. They are proven an excellent pharmaceutical mobilizer of controlled release drugs and other bioactive agents. This is due to the starch's unique physicochemical and functional characteristics, good swelling behavior, solubility, gelatinization, and enzymatic digestibility.

The chemically modified starch provides more reactive sites for biologically active compounds also works well as a binder, disintegrant, diluent, absorbents, and glidant/lubricant. Starch granular is usually white, soft, and tasteless that is insoluble in cold water, alcohol but dissolvable in normal temperature or heated water, boiled alkaline solution, and some deep eutectic solvents (DES). Dissolving the starch without degradation with true molecular dispersion can be achieved by using dimethylsulfoxide with a small amount of lithium chloride (LiCl) [4]. Due to the dissolvability of starch, it is often used partly in biobased materials. For instance, starch is composited with natural fiber as a natural adhesive [5]. As a result, the nonpermanent linkages were formed between starch and oil palm strand, resulting in a significantly improved mechanical strength but no resistance to water. Starch's thermoplastic property also lets them melt blending with a natural and synthetic polymer (e.g., polyethylene) for partially biodegradable food packaging applications. Starch is undeniably natural and biodegradable. Via starch blending technology, they accelerate biodegradability as their monomer is a food source for bacterial activity. Starch blending is a way to improve the own starch properties and improve other materials' deficiency.

Granular size is no exception besides the influence of amylose amylopectin ratios on starch characteristics. In general, starch showed polymodal features, where no exact granular size exists for one type of starch. In addition, the granular size influenced the synergistic effect between other materials or polymers. The enzymatic susceptibility and pasting properties are also affected by the granular size and shape. Collectively, starch promises excellent materials properties, coupled with its biodegradability property. They are fond and praised as exquisite biobased materials offering innumerable advantages and serviceability.

2.2.3 Polylactic Acid

Biodegradable polylactic acid (PLA) is a thermoplastic aliphatic polyester resultant from lactic acid, mainly from starch (e.g., tapioca roots, cassava, maize) or sugar cane. They are also called poly(lactic acid) or polylactide, frequently described as poly-α-hydroxy acids, poly-α-esters, or aliphatic polyesters, with the backbone formula $(C3H4O2)_n$ or $[-C(CH3)HC(=O)O-]_n$. Lactic acid, $CH_3CH(OH)COOH$, is an organic acid, white and miscible in water, is derived from: i) polycondensation of 2-hydroxy-propanoic acid (lactic acid), and ii) ring-opening chain polymerization of 3,6-dimethyl-1,4-dioxane-2,5-dione (lactide) [6]. Lactic acid is primarily formed via fermentation, mainly starch. After the starch is sedimented from its source, it is mixed and heated with enzymes to release dextrose (D-glucose). Following this, fermentation ensues, producing lactic acid monomers, mainly lactic acid or lactide. The PLA commercially used is commonly manufactured via ring-opening polymerization of lactide with various metal catalysts. Meanwhile, lactic acid direct polycondensation is similar, but the principal difference is the temperature must be controlled below 200°C, and the by-products (condensates) are released under the circumstances of the reaction.

After thermoplastic starch, PLA is the second most-produced bioplastic that has comparable properties to polypropylene (PP), polyethylene (PE), or polystyrene (PS), as well as being embraced for its eco-friendliness, biodegradability, and many other characteristics. Due to its flexibility, PLA is comparatively easy to work with, requiring minimal effort to produce quality products. Due to their high surface energy, it is often used in 3D printing applications. PLA is undeniably environmentally friendly as it requires lesser energy to produce, is fumeless, also no toxins. Even though PLA has good chemical resistance, it can be degraded via biological activity, thermal, photo, and hydrolysis. Owing to this, PLA has been adapted in many exertions, for example, film, fiber, foam, sutures, food packaging/containers, medical devices (e.g., anchors, screws, plates, mesh), etc. However, PLA's inherent brittleness, low heat resistance, weak UV resistance, comparatively low mechanical strength, hydrophobic, slow degradation rate, lacking reactive side chains, and sometimes demanding processability have limited its widespread use in high-end applications. The advantages and weaknesses showed by PLA, and good compatibility are suitable for synergistically blending and compositing with other materials, offering more remarkable benefits.

PLA is entirely soluble in dioxane, acetonitrile, chloroform, etc., but partially soluble in ethylbenzene, toluene, and acetone [7]. Due to its dissolvability and compatibility, PLA can be mutually blended with polymers or composited with any materials. PLA had been composited with starch (e.g., thermoplastic acetylated starch (TPAS)) to improve toughness, elongation at break, thermal stability, and also its degradability rate [8]. The weaknesses shown by PLA are compensated by TPAS, while PLA also relinquishes good chemical resistance, thermoforming ability, and adaptability of the TPAS. Meanwhile, to prolong its uses in bioplastic application, it is known to be composited with nonbiodegradable polymers or materials. Metal reinforcements, such as bronze and copper on PLA matrix in 3D printing, can boost the coefficient of friction of the composite but impart higher porosity and reduce the mechanical strength. The use of nanosize materials in the PLA matrix is the current approach, particularly nanoadditives and nanofibers. PLA blending is academically and commercially alluring, where it either improves or impairing the PLA properties according to end-use. One of the PLA end-uses that is commercially known is antibacterial packaging. Antibacterial property is incorporated by blending with several substances, for example, bacteriocins, plant extract, essential oils, enzymes, metals, such as silver, which have been attempted [9]. PLA had shown no invertible side effect when antibacterial agents were added into its polymeric system, and the dissipation rate of the antibacterial agents can be accurately monitored and performed. Inclusively, PLA is undoubtedly a promising biodegradable biobased material, where the ingrained weaknesses are tunable, and its compatibility with other materials is inarguably remarkable.

2.2.4 Cellulose

Cellulose is a renewable polymer overtly used as a biobased material and is widely available worthy since it is abundant. The cellulose production was estimated to be around 10^{11} to 10^{12} tons from various sources, such as from seed hair of cotton plants and other polysaccharides from woody plants. Nonetheless, cellulose exists in agricultural waste, grasses, and other nonwoody plants. However, cellulose may also be obtained from other sources, such as bacterial cellulose and algae. The cellulose extraction from these sources may hold other compounds, such as lignin, hemicellulose, and extractive. The most convenient way to obtain cellulose from woody plants is using chemical or mechanical treatments to produce cellulose pulp.

Besides native cellulose, cellulose can undergo some treatment to form regenerated cellulose and cellulose derivatives. Regeneration of cellulose occurs when there are physiochemical changes in cellulose structure. For example, native cellulose can efficiently react with an alkaline solvent like sodium hydroxide. In an alkaline-based solvent, the cellulose will swell, and the crystalline structure of the cellulose is disturbed. The regeneration process occurs when treated cellulose is neutralized with an acidic medium, and a new crystalline cellulose lattice structure is formed, also known as cellulose II. On the other hand, cellulose can be derived into countless derivations of different functionalities. Commercially available cellulose derivatives are cellulose ester and ether. Cellulose ether and ester are widely known in films, bioplastic, thickening agents, explosive powder, and more. An example of cellulose ether and ester is as illustrated in Figure 2.1. These cellulose derivatives are distinguished by the process and chemical use prior treatment of cellulose to produce cellulose derivatives.

2.2.5 Esters

Cellulose ester is a thermoplastic molding compound formed with anhydroglucose units with cellulose's three reactive hydroxyl groups. Cellulose ester can be grouped into two categories: organic and inorganic cellulose ester. The most common esters used in industry are cellulose acetate (CA) (organic cellulose ester) and cellulose nitrate (CN) (inorganic cellulose ester).

CA is the most important organic ester. CA is widely available in various applications, such as cigarette filters, textile fiber, plastic, films, separation membranes, coating, etc. CA price is reasonable with good hydrophilicity, stability in organic and organic chemicals, resistance toward alkaline medium, and more. To obtain CA, highly pure cellulose is reacted

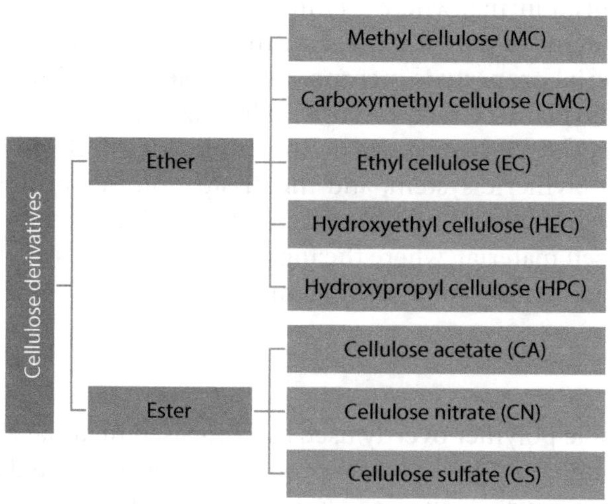

Figure 2.1 Example of cellulose ether and ester.

with glacial acetic acid, acetic anhydride, and a catalyst, such as sulphuric acid. Reactions are allowed to occur for the time needed to substitute virtually all the three hydroxyl groups on each anhydrous glucose unit with the acetyl group. Cellulose triacetate describes fiber form where 92% of the hydroxyl group is acetylated. Plentiful researchers manage to produce cellulose acetate bioplastic with different characteristics—for instance, CA bioplastic modified with TiO_2-organoclay with improved ability to self-photodegrade [10].

CA is used as a coating material for solid drugs, granules, pallets, tablets, and capsules in the medical field. CA is preferred for these applications due to its good pH sensitivity. Other types of commercially available CA are cellulose acetate propionate (CAP) and cellulose acetate butyrate (CAB), often used as inks and coating materials. They are used to form sheeting, molding plastic, film products, lacquer coating, and melt dip coating. CA is suitable to be used as a filter tow as it can remove small particles, such as phenols, nitrosamines, and quinolone. Plastic product produced from CA is mainly carried out by extrusion and blow molding. These products can be seen as jewelry, combs, buttons, packing materials, fashion accessories, and costumes.

CN, known as nitrocellulose or celluloid, is the first cellulose derivative discovered and acknowledged for its application in military and civilian fields. The nitration process is usually carried out by treating cellulose for nitration process where cellulose reacts with a mixture of nitric acid and sulphuric acid solution. Neutralization of treated pulp with water aids in controlling the degree of substitution of CN. This reaction is known as a typical esterification equilibrium. The nitration process is forced to complete upon water and sulfuric acid removal. This process is crucial for producing CN, which is important in producing CN bioplastic materials [11]. CN can be found commercially in a product, such as knife handles, especially for a pocket knife or specially made knife due to its appearance and ability to shrink and grip firmly to the knife extension upon [12]. Other than that, CN membranes that are low cost and have good shelf life are used as hydrophilic membranes to separate and treat various organic-water systems involving industrial usage [13].

2.2.6 Ether

Cellulose ether is a pivotal polymer in the construction and building industries due to its water-solubility properties. Cellulose ether is used as an additive in numerous fields, such as in food production, paint, adhesive, construction, pharmaceutical, paper industry, detergent, ceramic, and more. The addition of cellulose ether improves product quality by acting as a thickener, water retention agent, suspending aid, protecting colloid, film formation, wallpaper paste, and tile adhesive. The viscosity of cellulose ether can be increased with the right molecular weight, chemical substituents attached to it, and conformational characteristics. For instance, cellulose ether with higher molecular weight produces higher media viscosity and vice versa. Commonly available cellulose ether, which are either being used alone or blended with other cellulose ether in the industry, are methyl cellulose (MC), methyl hydroxyethyl cellulose (MHEC), ethyl hydroxyethyl cellulose (EHEC), methyl hydroxypropyl cellulose (HEC).

Cellulose ether is used as a building material as tile adhesive, air-place concrete material, flowable flooring, cement extrudate, thickener, and water retention agents. Recently, Espert and co-workers conducted research on MC and hydroxypropyl methylcellulose (HPMC) oleogels obtained by emulsion template [14]. Investigating the type of cellulose ether, cellulose ether concentration, and initial oil and hydrocolloid concentration had led to impactful findings. It was highlighted that the concentration of cellulose ether decides the physical properties of the oleogels. A higher concentration of cellulose ether resulted in oleogels with higher viscosity (thicken) [14]. As a biobased food material, research on edible sorghum starch files with CMC has improved the composite films' mechanical strength [15]. On the other hand, MC is used as a thin film coating material to encapsulate fruits. MC coating on fruits reduced weight loss in nectarines, increased customer acceptability, and quality, and extended the fruits' shelf-life. In addition, MC as a coating for deep-fried potato strips and dough discs can improve the properties of the fried food by reducing the oil uptake. The potatoes coated with MC reduced the oil uptake up to 40.6% and 35.2% for potato strips and dough, respectively [16].

2.2.7 Chitosan

Chitosan is a polycationic biopolymer derived from chitin that has both deacetylated and acetylated units. Its functional group of -NH_2 and -OH is responsible for its properties and characteristics, giving it unlimited potential application. The history of chitosan can be traced to when Rouget was studying deacetylated from chitin in 1859 [17]. From that moment onward, the knowledge of chitosan had risen scientists' and industries' interest in chitin and chitosan. The main chitosan source is insects, fungi, crustaceans, and mollusk.

The main source of chitin and chitosan is crustaceans. The shrimp and crab waste produced by the seafood industry has become the most demanding source of chitin since its always available, renewable, and at zero cost. In the industry, chitosan is obtained from crab, and shrimp shells, which are also waste from the food industry. Chitin can be obtained by submitting crustaceans through demineralization, deproteinization, deodorization, and drying. Then, chitin is deacetylated, purified, and dry to obtain pure chitosan. The chitin content from the exoskeleton varies from 5% to 42%, depending on the species of the crustaceans. Insects are another alternative to chitosan. Insects, such as the beetle, blowflies, silkworm chrysalides, and houseflies, are some of the insects used to obtain chitosan. Using insects gives the advantage in processing the chitin because the insect has a lower level of inorganic material, resulting in less demineralization treatment.

The most crucial characteristic of chitosan is its molecular weight and deacetylation degree. Other features, such as crystallinity, chitosan source, thermal stability, particle size, and surface area, play a vital role in the further process and final application. These characteristics have a massive impact on chitosan properties, particularly its solubility, polycationic character, antimicrobial property, biocompatibility, and bioadhesive. The characteristics and properties mentioned above are essential for selecting the suitable material to form fiber, membranes, nanoparticles, nanocapsule, hydrogel, etc. These materials and products can be found in various industries, including biomedicine, food, cosmetics, and more.

Based on literature and publish work, chitin and chitosan are widely used and applied in the pharmaceutical and biomedical as drug delivery systems throughout the human body including through colon, nasal, and oral drug delivery. Besides that, chitin and chitosan also had immense potential and were used in tissue engineering. Much research reported chitin/chitosan-based scaffolds and gel in tissue engineering applications. For instance, chitosan is blended with collagen and utilized to design polymetric scaffolds for *in vitro* human cell culture. Chitin/chitosan is also an antimicrobial agent because chitosan-based polymer prevents bacterial growth, accelerating wound healing. Furthermore, chitosan is widely used in pharmaceuticals as an effective drug delivery system, also being able to chelate heavy metal ion binders has become beneficial in environmental fields. Many forms of chitosan-based products can be produced, such as chitosan-based powder, films, membranes, beads, hydrogel, and more, making it useful in many fields and applications.

Chitosan composite can be formed by blending chitosan with other polymers. For instance, to enhance synthetic polymer's biodegradability and water absorbency, blending the synthetic polymer with chitin and chitosan can activate these properties. Polyvinyl alcohol (PVA) and polyethylene glycol (PEG) is commonly blended with chitosan and/or chitin to augment their biological properties and make them environmentally friendly. In addition, the blended polymer enhances its mechanical properties, hydrophilicity, and biocompatibility. Most of these composite polymers are formed via the film casting method. Other polymers, such as PLA, polyvinyl pyrrolidone, and polycaprolactone, are successfully blended with chitosan for diverse applications in the biomedical field. Additionally, chitosan can be melted blended with synthetic and natural polymers, such as polycaprolactone (PCL) and PLA. These blended polymers have shown improvement in their biodegradable properties and thermal stability.

The polymer blending method gained great attention in industries to acquire specific properties and characteristics to particular applications. Many publish research on chitosan blended with a natural polymer, such as gelatin, cellulose, *Antheraea pernyi* silk fibroin, starch, PLA, and more. For instance, the development of porous chitosan and alginate blend with polyelectrolyte by using surfactant Pluronic F68 is useful for making biomedical grade wound dressing and bioscaffolds products [18]. In addition, nanocomposite of chitosan hydrogel with carbon dots can improve pH sensitivity of wound dressing [19]. The presence of chitosan in this composite contributed to antibacterial properties, while carbon dot acts as the pH-sensitive indicator. Moreover, the chitosan reinforced with 1 wt.% carbon dots is reported to have good optical properties and higher mechanical strength.

A bioplastic composite from durian seed starch with sorbitol as a plasticizer was added with chitosan as the filler. The presence of chitosan in the composite improves the bioplastic composite's mechanical strength and biological properties. The mechanical strength of chitosan composite is mainly contributed by new hydrogen bonds between the polymers [20]. However, the degradability of the low-density polyethylene (LDPE) blend is a concern. Therefore, LDPE/chitosan polymer blend was formed to increase bioplastic's degradability. The addition of polymer compatibilizers, such as maleic anhydride (MA) and tert-butyl preoxygenate (TBPB), had resulted in a homogeneous blend and improvement of 40% of the tensile strength of the bioplastic [21].

2.2.8 Alginate

Alginate is a naturally occurring polymer derived from brown marine algae's cell walls. Alginates are a linear anionic polysaccharide polymer of β-(1-4)-D-mannuronic (M-block) and α-L-guluronic acid (G-blocks). Alginates contains two type of linear anionic polysaccharide polymer, which are β-(1-4)-D-mannuronic (M-block) and α-L-guluronic acid (G-blocks). These M-block and G-block can be found in various arrangements that give different properties and characteristics of the alginate polysaccharide. These G and M-block ratio has essential roles in determining the properties of the alginate products. For instance, alginate with higher M-block content is effective for chronic wound healing application as it can produce cytokine production via human monocytes.

There are two main sources of alginate used to prepare alginate: algal and bacterial sources based on microorganisms. The bacterial source of alginate includes two kinds of bacteria: *Pseudomonas* and *Azotobavter*. Others include *Ascophyllus nodosum, Laminaria hyperborean, Macrocytstis pyrifera, Ecklonia maxima, Laminaria digitata,* and *Laminaria japonica*. Alginate obtained from bacterial sources commonly extracted from *Azotobacter* sp. and *Pseudomonas* sp. A different strain of bacterial source and synthesis mechanism will affect the final characteristic and properties of the alginate obtained. For example, *Azotobacter vinelandii* and *P. aeruginosa* are used to produce alginate at a molecular level. Different types of bacterial affected the quality and quantity of alginate gained from the biosynthesis process. For example, *A. vinelandii* strain produces comparatively hard alginate with a high concentration of G-blocks in the alginate molecules, while *P. aeruginosa* is known to produce a higher alginate yield, which has thicker nature.

Extraction of alginate from brown marine algae comprises several steps. First, most brown algae are harvested and dried except for *M. pyrifera*, which must be processed in wet conditions. Dried algae are treated with a mineral acid to remove counter ions by a proton exchange reaction. Afterward, the neutralization process solubilizes the insoluble fraction of alginic acid. Alkali agents are used to neutralizing alginic acid. Next, the precipitate is separated by centrifuge or filtering the solution. By using mineral acid or alcohol, the formed precipitate is treated, sodium alginates are produced as a by-product of this reaction. This treatment eliminates homopolysaccharides, such as laminarin, fucoidan, and other impurities. However, sodium alginate may contain some cytotoxic impurities at this stage and is not suitable for biomedical applications. Therefore, further purification is needed to obtain the purest form of sodium alginate. In pharmaceutical applications, sodium alginates are most widely used as a derive of alginate, while calcium alginates are the other derivatives used as a drug delivery system.

The properties of alginate mainly depend on the arrangement of G- and M-blocks structure. Other than that, alginate is from different sources, and its molecular weight influences the quality and applicability of the alginate. Alginate with higher molecular weight and structural arrangement of G-block components tend to have more robust mechanical properties and are brittle. The solubility of alginate depends on the aqueous solvent's alkalinity or acidity. Alginates are more likely to dissolve in the alkaline solvent. However, as the pH increased to more than 10, the alginate started to depolymerize.

Alginate bioplastic polymers are widely used as biomedical material in various applications, especially for alginate wound dressing, drug delivery systems, and tissue engineering. Properties of alginate are nontoxic, biocompatible, biodegradable, biostable, and hydrophilic, making it suitable for medical application. For example, the alginate wound dressing is useful due to its hydrophilic nature, keeping wound moisture environment for wound healing, absorbing wound exudate, thus increasing wound recovery rate. Alginate bioscaffolds are widely used for tissue engineering applications because it is biocompatible and medically proven their functionality as a biomedical material. Therefore, alginates are used for tissue repair and regeneration in the form of films [22, 23], gels, hydrogels, fibers [24, 25], gauze, foams [26], 3D bioprinting, and wafers [27–29]. The material created an environment necessary for retaining and providing moisture for a faster wound healing process.

In addition, alginate can be blended with antibiotics or other drugs, making it suitable for controlled or slow-release drug delivery systems.

Alginate can be blended with other natural or synthetic polymers to enhance the character and properties of the composite. Common polymer blended with alginate includes chitosan, cellulose, collagen, bacterial cellulose, aloe vera, starch, PVA, pectin, CMC, to name a few. These copolymers are mainly used for the production of alginate-based wound dressing material. Hydrogel-base scaffolds have emerged as an alternative treatment for skin injuries, where alginate-based materials are part of this development. The research on alginate-based scaffolds shows vast potential, as proven by the formation of chitosan/alginate hydrogel containing vitamin E (alpha-tocopherol) for potential wound healing material [30]. The composite has a highly porous structure with intersected pores in the report. The prepared hydrogels are biodegradable and show great wound healing than traditional gauze-treated wounds. Referring to the current development and achievement of alginate-based medical appliances, the application of alginate has ripened in the biomedical field.

2.2.9 Proteins

Proteins are one of the biobased materials known as macromolecules like a string found by Gerardus Johannes Mulder, whereas Jöns Jacob Berzelius proposed the name in 1838. Proteins comprise amino acids that act as the basic building blocks, and covalent peptide bonds bond each amino acid. The proteins can be bonded linearly, branched, or crosslinked with other chains [31]. Twenty different monomers can be found in proteins and offer an exceptional structure of proteins. Natural polymers like proteins are more heterogeneous compared to synthetic polymers. Furthermore, different types of proteins have diverse molecular weights and compositions of amino acids. Protein structures can be divided into four different classes: primary, secondary, tertiary, and quaternary. The primary structure has the basic structure of the protein as amino acids are linked to each other by covalent peptide bonds. α-helix and β-helix are involved in secondary structure. As for tertiary structure, this structure has three-dimensional conformation contributed by interactions of electrostatic, hydrophobic, and hydrogen bonds. In contrast, a quaternary structure involves the interaction of one or more subunits to produce a functional protein.

Proteins can be classified according to the composition of chemicals, functions, structure, and solubility of specific proteins in different solvents. Proteins are sorted into three classes according to their chemical composition, namely simple proteins, conjugated proteins, and derived proteins. Homoproteins are known as simple proteins, generally consist of amino acids only. Second, the conjugated proteins are heteroproteins that comprise nonprotein portions. Third, derived proteins are made up of simple and conjugated proteins. There are two shape classes according to shape, like globular and fibrous proteins. Predominant proteins are globular proteins with compact, spherical structures and soluble in water. Fibrous proteins provide mechanical and structural strength and are insoluble in water. The solubility of proteins in different solvents like alcohol, water, and salt can differentiate protein properties. As for biological functions, proteins can act as enzymes, storage protein, transport protein, generation of movement (muscle), mechanical support, hormones, energy storage, and transmission of nerves.

Proteins have high hydrophilicity, low cost, bounteous feedstock, are biodegradable, and can be utilized as materials in various packaging, agriculture, medical, and food applications. Biobased materials derived from proteins are used in composites predominantly as matrices. Therefore, they can be produced in various products, such as thin films, sponge, fiber, nanofiber, hydrogel, graft, particle, skin, and patch [32]. However, due to the development of science and technology, electro-gel, optics, nanotech, electronics, mechanics, and micro-molding have been synthesized from proteins. As for thin film making, proteins improve the gas barrier and mechanical properties due to assorted amino acid functional groups. Besides, certain polymers can relay antimicrobial agents due to amphiphilic properties. Consequently, protein thin films can preserve the quality of the food, which includes nutrients. In this part, gluten, gelatin, and casein will be discussed in this section. Even though these materials are categorized under proteins, but there are extracted from different sources like animals and plants.

2.2.10 Gluten

The main storage protein of wheat grains is called gluten. The function of gluten is to aid in germination and seedling growth. Gluten can be obtained as an agricultural by-product, categorized as a vegetable protein (plant proteins). By washing the dough with water, gluten is simply produced. T.B. Osborne developed a broad classification of wheat protein chemistry, and he found that gluten consists of two groups called gliadins and glutenins according to their solubility in the alcohol [33]. Gliadins are classified as alcohol soluble, while glutenins are alcohol insoluble. Gluten is known as "two-component glue" because gliadins behave like "plasticizers" or "solvent" for glutenins. The composition of gliadins is 50% to 60%, while glutenins are approximately around 40% to 50%.

Gliadins comprise the mixture of monomeric or single-chained gluten involved by hydrophobic interactions and hydrogen bonding. In addition, gliadins are a heterogenous mixture found in grain or flour. As for glutenin, these fraction protein polymers have more peptide chains linked with interchain disulfide (SS) bonds. The molecular weight of monomeric gliadins in the gluten ranges between 30000 and 60000 while glutenin polymer consists of 80000 too few millions of molecular weight. There are four classes in gliadins, such as α-gliadins, β-gliadins, γ-gliadins, and ω-gliadins, according to the electrophoresis method at low pH [34]. α-, β-, and γ-gliadins have a high content of sulfur, α-helices, and few β-sheets. While sulfur in ω-gliadins is poor, rich in β-turns and few β-sheets. Nevertheless, glutenins have poor solubility because they are a large protein, comprise multiple peptide chains that bond with disulfide (SS) bonds, and cannot be separated by electrophoresis. Hence, the subunits can be divided by employing sodium dodecyl sulfate-polyacrylamide gel electrophoresis (SDS-PAGE) and result in two bands like low molecular weight (LMW), and high molecular weight (HMW) subunits. LMW subunits have 80%, while HMW subunits have 20% in the glutenins. The difference between gliadins and glutenins are gliadins have less cohesiveness and elasticity than glutenins [35].

Complex solvent systems are required to dissolve gluten because of the insolubility characteristic of gluten in water. The medium-like basic or acidic conditions with alcohol solution are needed to dissolve gluten. When the pH of solvents varies, a disturbance occurs on hydrogen and ionic interactions. Hydrophobic interaction was also disrupted due to the presence of ethanol. Besides, gluten is commonly employed in making films because of the great oxygen barrier and mechanical properties. However, a few issues arise from gluten usage as high degree crosslinking and ductility. The heat stability of gluten is immense, and it plays a huge role in extending and binding agents. This vegetable protein is predominantly used as a preservative in manufacturing processed food. This is due to enhancing the flavor and texture of the food. On the other hand, gluten has been utilized extensively as a matrix material in manufacturing composite due to viscoelasticity and thermoplastic properties. Temperature is the crucial parameter that should be optimized when incorporating gluten as one of the raw materials due to its high sensitivity to temperature. Furthermore, gluten has good fiber-matrix interactions and is biodegradable because it does not release toxic gases.

2.2.11 Gelatine

Collagen can be found abundantly on the skin, bones, tendons, and connective tissues. Denatured triple-helix-structure collagen from partial alkaline or acidic hydrolysis is called gelatine [36]. Gelatine is a fibrous and high molecular weight protein resulting from animal-based. The utilization of gelatine in applications begin during the late Ming and earliest Qing Dynasties in China since the 16th century. There are random rearrangements chains during the conversion of collagen into gelatine as a mixture of compounds with various molecular weights like α-chains, β-chains, and γ-chains. A high number of α-chains can be found in gelatine indicates that gel strength is high. Gelatine formation comprises amino acids glycine, hydroxyproline, and proline and has an exceptional sequence of amino acids. The major component in gelatine is a protein (85–92%), then the remaining components are mineral salts (2–4%) and water (8–12%). On the other hand, the extraction of gelatines has different setting temperatures, gel strength, and melting points based on different raw materials. This is because the different content of amino acids and the structure of peptides affect the variables mentioned above.

Gelatine can be obtained from various sources, such as bovine, poultry, marine species, stromal fraction, myofibrillar fraction, ovine and caprine, equine, edible insects, and porcine [37]. This gelatine is commonly utilized in food, cosmetics, and pharmaceutical applications. Conjointly, animal gelatine is economical due to being the simplest, cheapest, and easiest to obtain. Nevertheless, each source of gelatine gives different properties to the end products. Chicken bones, chicken skin, chicken feet, and fish skin are the best alternatives for gelatine production as they are derivative products from industrial processing. These raw materials are safe, biodegradable, and a great solution for religious issues that have been arising due to gelatine derived from porcine being inappropriate to certain religions [31]. Besides, the derivation of gelatine from porcine and bovine are not preferable because of the rising cases of foot or mouth diseases that affect human health. In addition, they are chosen as possible materials to substitute gelatine from the mammalian.

The most common application of gelatine has been used in making the film. Film-based gelatine has a thermoreversible property where the melting point is near the human body temperature. Gelatine is known to be biocompatible and capable of absorbing more water. The produced film can secure the food from oxygen and light as gelatine acts as a UV barrier. Additionally, the lifespan of food can be increased and minimized food loss. Fish skin gelatine-based film has inferior properties compared to bovine and porcine skins as it displays low mechanical properties due to low concentration hydroxyproline and proline.

However, gelatine extracted from fish skin still received great attention due to porcine- and bovine-based gelatine drawbacks.

Due to the high nutritional value of protein, gelatine has been used extensively in food applications. The consumption of gelatine in daily life benefits human health. As for the cosmetics field, gelatine contributes to helping to rejuvenate the skin and improving anti-aging properties. Moreover, it can secure the strength and thickness of hair. Besides, pharmaceuticals also use gelatine as a viscosity enhancer and act as a coating agent to enclose the dry powders. The existence of RGD domains in the structure aids the cells in the gelatine to adhere to tissue regeneration or engineering [38]. RGD is a tripeptide that comprises arginine, glycine, and aspartate. Gelatine is used as biobased material for manufacturing artificial tissue regeneration, bones, and valves.

2.2.12 Casein

Casein is a protein type commonly found in nearly 80% of bovine milk, and this milk protein has great nutritional value. Casein has a porous, tubular arrangement, spherical, and aggregates in the form of micelles structure. There are four diverse proteins, such as (α_{S1}-, α_{S2}-, β-, and κ- caseins) and colloidal calcium phosphate (CCP), with a variety of sizes ranging from 50 to 400 nm aids the formation of colloids with amphiphilic properties (hydrophilic and hydrophobic domains) [39]. CCP is present in the colloidal solution in a nanometre-sized ion. The content of proteins in casein micelle is up to 94%, while 6% is CCP. The casein micelles have high hydration and 3.5 kg of water per kg of protein. These proteins consist of approximately 19 to 25 kDa of molecular mass, known as phosphoproteins. Phosphoproteins are securely held by each other due to hydrophobic force and colloidal CCP that make casein become stable micellar arrangements when disseminated in the aqueous solutions. In the internal micelles, α- and β-caseins can be found, and on the outer surface of micelles is covered with hairy layers of κ-caseins via a steric stabilization mechanism. The presence of κ-caseins made the micelles stable without aggregate because they cannot come closer to each other. The casein micelle size decreases when the proportion of κ-caseins is present in the proteins. On the other hand, the characteristics of α-casein have poor hydrophobicity and low proline content while β-casein is high hydrophobicity.

On top of that, precipitating the milk at low pH, such as pH 4.6 (isoelectric point), could extract casein; meanwhile, the supernatant part is called whey protein. The precipitation of casein occurs when the isoelectric pH is achieved. The range of common isoelectric points is from 4.1 to 5.3. Nevertheless, the isoelectric points of every casein are different because of the difference of amino acid sequences even though the molecular weight and composition are alike. This illustrated that caseins are easily altered by changing pH and ion strength, and these factors affect micelles' integrity, including the size of casein micelles. However, the temperature does not affect the structure of micelles due to the insensitive properties of casein toward heat [40]. This is due to hairy κ-casein layers, making the micelles stable in heat treatment. Furthermore, casein is a rheomorphic protein that obtains a secondary or tertiary structure. This shows that casein has open, many side-chain degrees, and flexible conformation [41].

Rennet casein, caseinate and acid casein are list of casein products. Rennet or acid can be added to skim milk or membrane filtration to separate the casein protein from milk serum. Meanwhile, the mineral acids like sulphuric acid or hydrochloric acid and biological acids like lactic acid can be added to obtain an acidification reaction. Furthermore, when alkalines like potassium hydroxide, sodium hydroxide, or calcium hydroxide are added to the acid casein, they will help caseinates synthesis (neutralization) [42]. Casein is usually used in food, leather, plastics, cement, textile, and especially adhesives. Low cost, nontoxic, high stability, and acquirable properties are the main reasons why casein is utilized in diverse applications, especially as biobased materials. Furthermore, glues and adhesives are the typical application of casein because of the desirable properties like water resistance, ease of use, and excellent bond strength.

2.2.13 Lipid

Lipids received great attention among researchers, and these materials have been employed extensively in various applications. Annual growth of lipids is emerging from 2018 with \$13.5 billion, and by 2023, the growth of lipids is predicted to achieve \$17.5 billion. In the early 1800s, the comprehensive classification of lipids was made by Michel Eugène Chevreul, including the saponification process [43]. Lipids are biobased materials consisting of carbon, hydrogen, oxygen, and a few lipids that contain sulfur, nitrogen, and phosphorus. Lipids are insoluble in water but soluble in organic solvent compounds because of their polarity. Lipids have chains of hydrocarbons along with steroids and fatty acids. Fatty acids comprise aliphatic monocarboxylic acids and a long linear chain with even carbon numbers. On the other hand, lipids can be obtained from the food industry, such as mono- and diglycerides, emulsifiers, and dried fats [44].

The melting point and solubility of lipids in nonpolar solvents are assorted due to the variety of carbon chain lengths in fatty acids [45]. Moreover, there are two classes of fatty acids as saturated and unsaturated fatty acids. The carbon chains with single bonds are categorized under saturated fatty acids, while the double bonds linked to the carbon chains are considered unsaturated fatty acids. Monounsaturated fatty acids are chains bonded with a double bond, while polyunsaturated fatty acids are greater than two double bonds. The presence of double bond and ester or acid groups acting as reactive sites enables optimizing the lipids for specific applications.

There are three different classes of lipids, such as simple, derived, and compound lipids. Simple lipids consist of waxes, fats, and oils. Waxes are esters of alcohols in a solid state at ambient temperature. There is high solubility of waxes in the organic solvents, but they are insoluble in water. Generally, simple lipids like fats and oils can be discovered in food derived from plants (80% from vegetable oil) and animal sources (20%). These lipids are usually applied as filler, matrices, and plasticizers. Fats and oils comprise triglycerides, and the hydrolysis process can alter triglycerides into free fatty acids and glycerol. Phospholipids and glycolipids are under compound lipids, while steroids, terpenes, and carotenoids are derived lipids.

Hydrophobicity is the appealing properties in lipids enhancing the water vapor barrier, antibacterial and antioxidant of products fabricated, such as thin films [46]. This is due to the hygroscopic properties of other biodegradable resources that limit the thin film to function well as it allows high absorption water and destroys the structural integrity. Incorporating lipids can provide the best products with a hydrophobic state and separate the moisture environment. Microbial growth also can be avoided by using lipids like essential oils. Essential oils can be obtained by extracting plants resources, and volatile compounds acquire more than nonvolatile compounds. Hence, the presence of volatile compounds which consist of phenolic or alcoholic components can prevent microbial growth. Thermal stability of products is vital in food packaging applications to ensure the employment of lipids as a product does not easily degrade and avoid food spoilage. The common lipids that have been employed in the previous studies are mostly food waste that contains lipids, phosphate, sugars, and amino acids. This section will discuss polyhydroxyalkanoates (PHAs) that can be synthesized from diverse bacterium via the bacterial fermentation of lipids or sugars.

2.2.14 Polyhydroxyalkanoates (PHA)

Polyhydroxyalkanoates (PHAs) are renewable, biocompatible, and biodegradable macromolecules. PHAs are thermoplastic polyesters synthesized by microorganisms in the state of excessive carbon but the inadequate supply of nutrients like nitrogen, phosphorus, oxygen, and sulfur. The structure of PHAs is isotactic with (R)-configuration in the main chains. When the surrounding conditions of organisms become severe, they can assimilate the carbon source, and hydroxy alkanoates (HA) can be stored [47]. Subsequently, HA underwent polymerization into PHAs. PHAs are accumulated intracellularly in polymer granules that act as carbon and energy sources. The organism uses the energy source whenever the carbon source is reduced in the locality. Microorganisms that accumulate the PHAs have high resistance toward heat, osmotic shock, and ultraviolet irradiation. To synthesize PHAs, a few parameters must be considered, such as microorganisms, pH, temperature, and culture types [48]. Production of PHA can occur by pure and mixed microbial cultures. Mixed microbial cultures are preferred to pure culture because of the effective system for various fermentation techniques and are more economical. Besides, waste sources can synthesize PHAs, such as food waste, coffee waste, lignocellulosic biomass, molasses, and others. Likewise, PHAs can be obtained from agricultural residues like vegetable oils.

PHAs are the largest natural polyester class because approximately 150 PHA monomers have been discovered [49]. There are three different chain lengths of PHAs Firstly; there are three to five carbon monomers (C3-C5) in the short-chain-length (SCL). The characteristics of SCL are low melting temperature with a temperature range from 173°C to 180°C. Besides, the crystallinity of SCL is high (55–80%), which indicates the brittleness of SCL. For instance, poly(3-hydroxybutyrate), poly(4-hydroxybutyrate), and poly(3-hydroxyvalerate) or the copolymer. Next, medium-chain-length (MCL) comprises 6 to 14 carbon monomers in the 3-hydroxyalkanoate unit. The great thermomechanical properties of MCL PHAs made the material more appealing than SCL PHAs with high elasticity and lower melting temperature (39–61°C). MCL PHAs resemble elastomer characteristics. For instance, poly(3-hydroxyoctanoate) (PHO), and poly(3-hydroxyno-nanoate) (PHN), that mainly made from copolymers like 3-hydroxyhexanoate (HHx), 3-hydroxyheptanoate (HH), and/or 3-hydroxydecanoate(HD). Third, the carbon atoms in long-chain-length (LCL) have greater than fourteen HA monomers. However, the existence of LCL is peculiar. Poly(3-hydroxypentadecanoate) and poly(3-hydroxyhexadecanoate) are examples of LCL.

The common properties of PHAs are crystalline, flexible, and elastic, which become the best alternative for replacing synthetic polymer. However, the high crystallinity of PHAs becomes an obstacle for film packaging and biomedical applications as the brittleness of materials is not suitable for that kind of utilization. To enhance the physical properties of materials, PHAs should be blended with other materials like natural resources like lignin, cellulose derivatives, and starch. The viscosity of polymers can be improved by adding the side chain length of MCL PHAs. Additionally, the solubility of PHAs in water is poor, but PHAs are soluble in chloroforms. Besides, the nontoxic properties make the materials biocompatible for any usage. PHAs have been extensively used in diverse industries, such as bottles, containers, films, and biomedical. Bone tissue engineering and drug delivery are the most popular application in the biomedical field. As for bone tissue, when PHAs are degraded, 3-hydroxybutyrate is released as degraded compounds, and they are compatible with an element of the blood composed when the fatty acids are in the liver. Furthermore, the emerging use of PHAs in nanotechnology has been used for hepatitis B virus diagnosis. The formation of films-based PHAs able to use as a mulching film for the agriculture field due to the biodegradability of PHAs.

2.3 Nonbiodegradable Biobased Material

2.3.1 Polyethylene (PE)

Polyethylene (PE) is a thermoplastic polymer with ethylene as a monomer, and hydrogen bonds hold the monomer chains. When heat is applied above a certain temperature, PE will be melted and become solid when the cooling process occurs. The process is reversible due to the thermoplastic properties of PE, and it does not have any significant changes in the aspect of degradation. However, its symmetric molecular structure makes PE become a semi-crystalline polymer as it can develop crystallization and affect the density and chemical stability of PE. Furthermore, PE is insusceptible to various chemicals. However, it is responsive to certain formulations, such as toluene, trichloroethane, xylene, or trichlorobenzene at elevated temperatures.

The commercial PE used vigorously in the industry is mostly made from petrochemical ethylene. Nevertheless, there is a green alternative to produce PE, which is more environmentally friendly and safe to use. Biobased PE is developed from ethanol from processing food crops, namely sugarcane, maize, and wheat. Ethanol has been synthesized from sugarcane through the anaerobic fermentation process [50]. Later on, the catalytic dehydration process was implemented on ethanol to synthesize ethylene [51]. Then, polymerization of ethylene is conducted to produce PE with different types depending on their structure. The standard density of PE is approximately between 0.88 and 0.96 g/cm^3. PE can be divided into different classes of polymers according to their molecular weight and density, namely low-density PE (LDPE), linear low-density PE (LLDPE), medium-density PE (MDPE), high-density PE (HDPE), and ultra-high-molecular-weight PE (UHMWPE).

LDPE has weak intermolecular interactions in the polymer with a density between 0.88 and 0.96 g/cm^3 due to a high degree of chain branching approximately there are sixty branches in one-thousandth carbon atoms. There is long and short-chain branching exists in the LDPE. The amorphous region of LDPE consists of short branches that have more than one co-monomers like 1-hexene, 1-octene, and 1-butene [52]. Therefore, the polymer has flexible and ductile characteristics because the crystalline structure cannot be developed easily. Moreover, the mechanical strength of LDPE is the lowest compared to other PE types. LLDPE is a transparent, excellent toughness and flexibility polymer with a linear structure. This type of polymer is popular among researchers, and it is common for being employed in film applications due to its attractive characteristics [53]. HDPE is considered a versatile and the cheapest polymer. Compared with LDPE, LDPE degradation rate is rapid due to the LPDE closely packed chain's structure. HDPE has the highest degree of crystallinity. Due to fewer short branches packed in the crystal structure very well, HDPE is regarded as the most stable PE. HDPE can be UHMWPE has long polymer chains that give this type of polymer great mechanical strength. By transferring the load to the polymer backbone via overlapping bonds, the long chain of polymer structure can enhance strength. Moreover, UHMWPE's excellent properties benefit from being used in high abrasion resistant and corrosive chemical resistant applications like structural materials [54]. Commercial PEs are commonly utilized in packaging, medical devices, automotive, piping, and coating applications. It can be concluded that each type of PE has its specialties to be applied in specific applications.

2.3.2 Polyethylene Terephthalate (PET)

Polyethylene terephthalate (PET) is an aromatic thermoplastic macromolecule with two monomers: terephthalic acid and ethylene glycol. The usage of PET emerged drastically around the world. The durability, versatility, and water-proof materials have received enormous attention in the industrial sector. It is a semi-crystalline polymer and has a complex structure as it comprises lamellar crystals encapsulated in the amorphous matrix. The crystalline region has a predominant impact on the thermal and mechanical performance of PET [55, 56]. On the other hand, the degree and arrangement of crystallization can be enhanced to improve the effectiveness of PET. Meanwhile, the amorphous part commonly benefits manufacturing bottles and packaging due to its transparency properties. Contrastingly, the opaque and milky white color of PET is contributed by a semi-crystalline structure [57]. Most of the bottles are made of PET due to the high resistance of chemicals, which make this type of polymer are not toxic. Consequently, PET is the best alternative for a soft drinks bottle.

The sustainable PET can be obtained by combining mono ethylene glycol (MEG) and terephthalic acid (TPA). The production of ethanol is aided by the fermentation of sugarcane and undergo various processes like dehydration, catalytic oxidation, and omega process that led to the formation of MEG. Meanwhile, TPA is synthesized by oxidation of *para-xylene* that goes through multiple processes like fermentation, dehydration, dimerization, and dehydrogenation. Later on, MEG and TPA are merged via the transesterification process to form PET. PET is considered partially biobased conventional biopolymers. Hence, the amount of aromatic contents is high and minimizes the mobility of chains. This results in PET being highly resistant to degradation. Therefore, the stockpile of commercial PET is in the environment. PET might be biodegraded, but it took many times to convert into small fragments. This is because the enzymes cannot attack the crystalline region, but they will tend to attack the amorphous region. High crystallinity might cause the degradation process to be lowered [58]. Nonetheless, the glycolysis method can depolymerize PET, and it requires a high temperature, expensive and time-consuming process, which is not economical [59].

Due to its extraordinary properties are primarily utilized in fibers making, packaging, and construction components. As for packaging applications, bottles, containers, and films are the most common materials that have been made from PET. Bottles with 76% become the most dominant packaging application compared to containers (11%) and films (13%) [60]. Besides, PET is used to make composites, and this polymer can be either matrices or reinforcements in any composites. Hybridizing the PET fiber with any fiber also gives advantages to composites' physical and mechanical properties. Besides, the enhancement of barrier properties is also occurring to build up the shelf life of materials and promote the utilization of PET in various industries. For instance, PET and hemp fiber were hybrid, and the composites produced can be further utilized as construction components with extraordinary thermal and mechanical properties [61]. Other than that, the reinforcement of graphene into the PET matrix had been optimized, and improved tensile strength, thermal stability, and crystallinity of nanocomposites were displayed [62].

2.3.3 Polyamide (PA)

Polyamide (PA) is a polymer with a chain recurring amide group, at least 85% of which are attached to the aliphatic or cycloaliphatic group. PA, commercially known as nylon, has been used in numerous applications since the 1930s thanks to its excellent thermal, mechanical, and physical properties, which are contributed by hydrogen bonds in PA chemical structure. In 2019, PA will take up 12% of global production and usage of bioplastic [63]. PA sources are divided into 1) fossil-based and 2) biobased. PA is considered a biobased polymer because it can be derived entirely or partially from renewable sources, such as natural oil and biomass.

Castor oil is the primary source of production of PA. Castor beans contain 40% to 60% of both saturated and unsaturated fatty acids. The main component of the fatty acid is C18 ricinoleic acid. Ricinoleic acid is transformed to form monomers that are useful in producing bio-PA. Functional groups of hydroxyl, carbonyl, and unsaturated bonds are crucial for the chemical modification of ricinoleic acid. Therefore, biomass-derived monomers have high demands from the industry to synthesize bio-PA. Those monomers are 11-aminoundecanoic acid, 1,8-octanedicaboxylic acid, 1,0-decanediamine, adipic acid, caprolactam, 1,4-butanediamine, and sebacic acid.

PA can be yielded by polycondensation process of its diamines functional group with a dicarboxylic acid, amino acids (e.g., PA 11), or ring-opening polymerization of lactams (e.g., PA 4, PA 6, PA 6,6). Synthetic polyamides are classified as 1) aliphatic polyamide, 2) polyphthalamides, and 3) aromatic polyamide (or aramids). Bio-PA composites are produced by physical and chemical modification to reflect industry-specific demand to apply as engineering materials. As a result, Bio-PA composite emerges as attractive material, especially for machinery, automotive, building, electric, electronic, and consumer goods industries. Table 2.1 shows bio-PA and its commercially available products in the market. Bio-PA composites are formed by modifying PA by blending or mixing natural fiber or other additives via chemical or physical modification. For instance, bio-PA/fiber composite is formed by mixing bio-PA with natural and synthetic fiber. Some researchers use modified and unmodified natural fiber as reinforcement for bio-PA, also additives, such as carbon fibers, boron nitride microparticles, modified nutshell powder, lignin, rice husk, and nanoclay.

PA was traditionally used as fibers, injection-molded, and extruded parts in different automotive and textile industries. PA is well known as a polymer nanocomposite material. They are considered as thermoplastic material and various grades of PA glass

Table 2.1 Classification of synthetic bio-PA and products available commercially.

Type	Bio-PA	Product
Aliphatic PA	PA 6	Zytel from DuPont
	PA 6,6	Technyl from Solvay
		Rilsan and Rilsamid from Arkema
		Radipol from Radici Group
Polyphthalamides	PA 6T	Trogamid T from Evonik Industries
		Amodel from Solvay
Aromatic PA	Paraphenylenediamine + terephthalic acid	Kevlar and Nomex from DuPont
		Teijinconex
		Twaron and Technora from Teijin
		Kermel from Kermel

fiber and other filler material available in the market. 10% of plastic in modern cars is made from PA, and 41% of the PAs market is channelled to the automotive industry. Biobased PAs are used in the automotive industry, especially in a well-known automotive company.

2.4 Conclusion

This chapter briefly explains the types and sources of biobased materials: biodegradable and nonbiodegradable types, most used in biobased polymer applications. Generally, biobased materials are from animals and plants that make the finishing products biodegradable, biocompatible, and environmentally friendly. On the other hand, the nonbiodegradable part is usually synthesized from fossil-based materials. The bountiful biomass resources and low cost have made biodegradable materials are the best alternatives to substitute the synthetic polymers for minimizing the depletion of fossil fuel and saving the Earth. The utilization of biobased materials is receiving huge attention from researchers due to various benefits to the performance of end products. This is because products' physical, chemical, thermal, and mechanical properties are improved whenever biobased materials are incorporated. Nevertheless, different biobased materials offer different properties and can be applied to appropriate applications. Moreover, biobased materials can be blended to form composites with great mechanical strength. Reaching a broad consensus on the practicality of using biobased materials would provide a platform to influence consumer attitudes and love to a greener product. Hence, future research endeavors to optimize the parameters are required to manufacture the outstanding products from laboratory scale into the industry sector that derived biobased materials for sustainability.

Acknowledgment

Thanks to the Ministry of Higher Education (MOHE) of Malaysia for providing research grant LRGS/1/2019/UKM-UKM/5/1. Also, thanks to Universiti Kebangsaan Malaysia and Universiti Sains Malaysia for the facilities and supports.

References

1. Vert, M., Doi, Y., Hellwich, K.H., Hess, M., Hodge, P., Kubisa, P., Rinaudo, M., Schué, F., Terminology for biorelated polymers and applications (IUPAC Recommendations 2012). *Pure Appl. Chem.*, 84, 377, 2012.
2. Meyer, K.H., Bretano, W., Bernfeld, P., Starch., I.I., Nonhomogeneity of starch. *Helv. Chim. Acta*, 23, 845, 1940.
3. Jiang, T., Duan, Q., Zhu, J., Liu, H., Yu, L., Starch-based biodegradable materials: Challenges and opportunities. *Adv. Ind. Eng. Polym. Res.*, 3, 8, 2020.
4. Cave, R.A., Seabrook, S.A., Gidley, M.J., Gilbert, R.G., Characterization of starch by size-exclusion chromatography: The limitations imposed by shear scission. *Biomacromolecules.*, 10, 2245, 2009.
5. Salleh, K.M., Hashim, R., Sulaiman, O., Hiziroglu, S., Wan Nadhari, W.N.A., Abd Karim, N., Jumhuri, N., Ang, L.Z.P., Evaluation of properties of starch-based adhesives and particleboard manufactured from them. *J. Adhes. Sci. Technol.*, 29, 319, 2015.
6. Vert, M., Chen, J., Hellwich, K.H., Hodge, P., Nakano, T., Scholz, C., Slomkowski, S., Vohlidal, J., Nomenclature and terminology for linear lactic acid-based polymers (IUPAC Recommendations 2019). *Pure Appl. Chem.*, 92, 193, 2020.
7. Casalini., T., Rossi, F., Castrovinci, A., Perale, G., A Perspective on polylactic acid-based polymers use for nanoparticles synthesis and applications. *Front. Bioeng Biotechnol.*, 7, 1, 2019.
8. Yu, M., Zheng, Y., Tian, J., Study on the biodegradability of modified starch/polylactic acid (PLA) composite materials. *RSC Adv.*, 10, 26298, 2020.
9. Tawakkal, I.S.M.A., Cran, M.J., Miltz, J., Bigger, S.W., A review of poly(lactic acid)-based materials for antimicrobial packaging. *J. Food. Sci.*, 79, R1477–R1490, 2014.
10. Krisnandi, Y.K., Rasanji, D.G.W.K.D., Luthfiyah, S.Z., Sihombing, Z., Photodegradation study of TiO2-organoclay modified acetate cellulose bioplastic. *IOP Conf. Ser. Mater. Sci. Eng.*, 188, 012059, 2017.

11. George, N., Debroy, A., Bhat, S., Singh, S., Bindal, S., Biowaste to bioplastics: An ecofriendly approach for a sustainable future. *J. Appl. Biotechnol. Rep.*, 8, 221, 2021.
12. Gilbert, M., Cellulose plastics, in: *Brydson's plastics materials*, 8th, M. Gilbert (Ed.), pp. 617–630, Elsevier Ltd., Butterworth-Heinemann, 2017.
13. Singh, D.P., Kothari, R., Tyagi, V.V., *Emerging energy alternatives for sustainable environment*, pp. 153–158, CRC Press, Broken Sound Parkway, New York, 2019.
14. Espert, M., Salvador, A., Sanz, T., Cellulose ether oleogels obtained by emulsion-templated approach without additional thickeners. *Food Hydrocolloid*, 109, 106085, 2020.
15. Putri, D.A., Setiawan, A., Anggraini, P.D., Physical properties of edible sorgum starch film added with carboxymethyl cellulose. *J. Phys. Sci.*, 29, 185, 2018.
16. García, M.A., Ferrero, C., Campana, A., Bértola, N., Martino, M., Zaritzky, N., Edible coatings from cellulose derivatives to reduce oil uptake in fried products. *Food Sci. Technol. Int.*, 3, 391, 2002.
17. Candy, T. and Sharma, C.P., Chitosan - As a biomaterial. *Biomat. Art. Cells. Art. Org.*, 18, 1, 1990.
18. Bueno, C.Z., Dias, A.M.A., Sousa, H.J.C.D., Braga, M.E.M., Moraes, Â.M., Control of the properties of porous chitosan-alginate membranes through the addition of different proportions of Pluronic F68. *Mater. Sci. Eng. C*, 44, 117, 2014.
19. Omidi, M., Yadegari, A., Tayebi, L., Wound dressing application of pH-sensitive carbon dots/chitosan hydrogel. *RSC Adv.*, 7, 10638, 2017.
20. Ginting, M.H.S., Kristiani, M., Amelia, Y., Hasibuan, R., The effect of chitosan, sorbitol, and heating temperature bioplastic solution on mechanical properties of bioplastic from durian seed starch (*Durio zibehinus*). *Int. J. Eng. Res. Appl.*, 6, 33, 2016.
21. Kusumastuti, Y., Putri, N.R.E., Timotius, D., Syabani, M.W., Rochmad, Effect of chitosan addition on the properties of low-density polyethylene blend as potential bioplastic. *Heliyon*, 6, e05280, 2020.
22. Cheng, F., Liu, C., Wei, X., Yan, T., Li, H., He, J., Huang, Y., Preparation and characterization of 2,2,6,6-tetramethylpiperidine-1-oxyl (TEMPO)-oxidized cellulose nanocrystal/alginate biodegradable composite dressing for hemostasis applications. *ACS Sustain. Chem. Eng.*, 5, 3819, 2017.
23. Caridade, S.G., Monge, C., Gilde, F., Boudou, T., Mano, J.F., Picart, C., Free-standing polyelectrolyte membranes made of chitosan and alginate. *Biomacromolecules*, 14, 1653, 2013.
24. Chen, S., Cui, S., Hu, J., Zhou, Y., Liu, Y., Pectinate nanofiber mat with high absorbency and antibacterial activity: A potential superior wound dressing to alginate and chitosan nanofiber mats. *Carbohydr. Polym.*, 174, 591, 2017.
25. Chang, J.J., Lee, Y.H., Wu, M.H., Yang, M.C., Chien, C.T., Preparation of electrospun alginate fibers with chitosan sheath. *Carbohydr. Polym.*, 87, 2357, 2012.
26. Catanzano, O., Soriente, A., Gatta, A.L., Cammarota, M., Ricci, G., Fasolino, I., Schiraldi, C., Ambrosio, L., Malinconico, M., Laurienzo, P., Raucci, M.G., d'Ayala, G.G., Macroporous alginate foams crosslinked with strontium for bone tissue engineering. *Carbohydr. Polym.*, 202, 73, 2018.
27. Ahmed, A., Getti, G., Boateng, J., Ciprofloxacin-loaded calcium alginate wafers prepared by freeze-drying technique for potential healing of chronic diabetic foot ulcers. *Drug Deliv. Transl. Res.*, 8, 1751, 2018.
28. Boateng, J., Burgos-Amador, R., Okeke, O., Pawar, H., Composite alginate and gelatin based bio-polymeric wafers containing silver sulfadiazine for wound healing. *Int. J. Biol. Macromol.*, 79, 63, 2015.
29. Atia, N.M., Hazzah, H.A., Gaafar, P.M.E., Abdallah, O.Y., Diosmin nanocrystal–loaded wafers for treatment of diabetic ulcer: *In vitro* and *in vivo* evaluation. *J. Pharm. Sci.*, 108, 1857, 2019.

30. Ehterami, A., Salehi, M., Farzamfar, S., Samadian, H., Vaez, A., Ghorbani, S., Ai, J., Sahrapeyma, H., Chitosan/alginate hydrogels containing alpha-tocopherol for wound healing in rat model. *J. Drug Deliv. Sci. Technol.*, 51, 204, 2019.
31. Abdullah, M.S.P., Noordin, M.I., Mohd Ismail, S.I., Mustapha, N.M., Jasamai, M., Danik, M.F., Wan Ismail, W.A., Shamsuddin, A.F., Recent advances in the use of animal-sourced gelatine as natural polymers for food, cosmetics and pharmaceutical applications. *Sains Malays.*, 47, 323, 2018.
32. Hu, X., Cebe, P., Weiss, A.S., Omenetto, F., Kaplan, D.L., Protein-based composite materials. *Mater. Today*, 15, 208, 2012.
33. Shewry, P.R., Tatham, A.S., Forde, J., Kreis, M., Miflin, B.J., The classification and nomenclature of wheat gluten proteins: A reassessment. *J. Cereal Sci.*, 4, 97, 1986.
34. Shewry, P., What is gluten—Why is it special? *Front. Nutr.*, 6, 101, 2019.
35. Wieser, H., Chemistry of gluten proteins. *Food Microbiol.*, 24, 115, 2007.
36. Sompie, M., Tinangon, R.M., Surtijono, S.E., Said, M.I., Effect of long-time immersion in edible film solution from local chicken claw on the physical and chemical properties of chicken meat. *IOP Conf. Ser. Earth Environ. Sci.*, 492, 012056, 2020.
37. Murrieta-Martínez, C.L., Soto-Valdez, H., Pacheco-Aguilar, R., Torres-Arreola, W., Rodríguez-Felix, F., Márquez Ríos, E., Edible protein films: Sources and behavior. *Packag. Technol. Sci.*, 31, 113, 2018.
38. Bertassoni, L.E., Cardoso, J.C., Manoharan, V., Cristino, A.L., Bhise, N.S., Araujo, W.A., Zorlutuna, P., Vrana, N.E., Ghaemmaghami, A.M., Dokmeci, M.R., Khademhosseini, A., Direct-write bioprinting of cell-laden methacrylated gelatin hydrogels. *Biofabrication.*, 6, 024105, 2014.
39. Horne, D.S., Casein micelle structure: Models and muddles. *Curr. Opin. Colloid Interface Sci.*, 11, 148, 2006.
40. Sinaga, H., Bansal, N., Bhandari, B., Effects of milk pH alteration on casein micelle size and gelation properties of milk. *Int. J. Food Prop.*, 20, 179, 2017.
41. Holt, C. and Sawyer, L., Caseins as rheomorphic proteins: Interpretation of primary and secondary structures of the αs_1-, β- and κ-caseins. *J. Chem. Soc Faraday Trans.*, 89, 2683, 1993.
42. Badem, A. and Uçar, G., Production of caseins and their usages. *Int. J. Food Sci. Nutr.*, 2, 04, 2017.
43. Henderson, J. and Osborne, D.J., The oil palm in all our lives: How this came about. *Endeavour.*, 24, 63, 2000.
44. Kerr, B.J., Kellner, T.A., Shurson, G.C., Characteristics of lipids and their feeding value in swine diets. *J. Anim. Sci. Biotechnol.*, 6, 1, 2015.
45. Fahy, E., Subramaniam, S., Brown, H.A., Glass, C.K., Merrill, A.H., Murphy, R.C., Raetz, C.R.H., Russell, D.W., Seyama, Y., Shaw, W., Shimizu, T., Spener, F., Van Meer, G., VanNieuwenhze, M.S., White, S.H., Witztum, J.L., Dennis, E.A., A comprehensive classification system for lipids. *J. Lipid Res.*, 46, 839, 2005.
46. Zubair, M., Pradhan, R.A., Arshad, M., Ullah, A., Recent advances in lipid derived bio-based materials for food packaging applications. *Macromol. Mater. Eng.*, 26, 306, 2000799, 2021.
47. Anjum, A., Zuber, M., Zia, K.M., Noreen, A., Anjum, M.N., Tabasum, S., Microbial production of polyhydroxyalkanoates (PHAs) and its copolymers: A review of recent advancements. *Int. J. Biol. Macromol.*, 89, 161, 2016.
48. Amaro, T.M.M.M., Rosa, D., Comi, G., Iacumin, L., Prospects for the use of whey for polyhydroxyalkanoate (PHA) production. *Front. Microbiol.*, 10, 1, 2019.
49. Akinmulewo, A.B. and Nwinyi, O.C., Polyhydroxyalkanoate: A biodegradable polymer (a mini review). *J. Phys. Conf. Ser.*, 1378, 042007, 2019.

50. Goldemberg, J., Coelho, S.T., Guardabassi, P., The sustainability of ethanol production from sugarcane. *Energy Policy.*, 36, 2086, 2018.
51. Chen, G., Li, S., Jiao, F., Yuan, Q., Catalytic dehydration of bioethanol to ethylene over TiO_2/γ-Al_2O_3 catalysts in microchannel reactors. *Catal. Today*, 125, 111, 2007.
52. Montazer, Z., Habibi Najafi, M.B., Levin, D.B., Challenges with verifying microbial degradation of polyethylene. *Polymers.*, 12, 123, 2020.
53. Graziano, A., Jaffer, S., Sain, M., Review on modification strategies of polyethylene/polypropylene immiscible thermoplastic polymer blends for enhancing their mechanical behavior. *J. Elastomers Plast.*, 51, 291, 2019.
54. Paxton, N.C., Allenby, M.C., Lewis, P.M., Woodruff, M.A., Biomedical applications of polyethylene. *Eur. Polym. J.*, 118, 412, 2019.
55. Bandla, S., Allahkarami, M. and Hanan, J.C., Thermal crystallinity and mechanical behavior of polyethylene terephthalate, in: *Challenges in Mechanics of Time-Dependent Materials*, vol. 2, pp. 141–146, Springer, Cham, 2015.
56. Zekriardehani, S., Jabarin, S.A., Gidley, D.R., Coleman, M.R., Effect of chain dynamics, crystallinity, and free volume on the barrier properties of poly(ethylene terephthalate) biaxially oriented films. *Macromolecules.*, 50, 2845, 2017.
57. Panowicz, R., Konarzewski, M., Durejko, T., Szala, M., Łazińska, M., Czerwińska, M., Prasuła, P., Properties of polyethylene terephthalate (PET) after Thermo-Oxidative Aging. *Materials*, 14, 3833, 2021.
58. Taniguchi, I., Yoshida, S., Hiraga, K., Miyamoto, K., Kimura, Y., Oda, K., Biodegradation of PET: Current Status and Application Aspects. *ACS Catal.*, 9, 4089, 2019.
59. Salvador, M., Abdulmutalib, U., Gonzalez, J., Kim, J., Smith, A.A., Faulon, J.L., Wei, R., Zimmermann, W., Jimenez, J.I., Microbial genes for a circular and sustainable bio-PET economy. *Genes.*, 10, 1, 2019.
60. Siracusa, V. and Blanco, I., Bio-Polyethylene (Bio-PE), Bio-Polypropylene (Bio-PP) and Bio-Poly(ethylene terephthalate) (Bio-PET): Recent Developments in Bio-Based Polymers Analogous to Petroleum-Derived Ones for Packaging and Engineering Applications. *Polymers.*, 12, 1641, 2020.
61. Ahmad, M.A.A., Abdul Majid, M.S., Ridzuan, M.J.M., Mazlee, M.N., Gibson, A.G., Dynamic mechanical analysis and effects of moisture on mechanical properties of interwoven hemp/polyethylene terephthalate (PET) hybrid composites. *Constr. Build. Mater.*, 179, 265, 2018.
62. Shabafrooz, V., Bandla, S., Allahkarami, M., Hanan, J.C., Graphene/polyethylene terephthalate nanocomposites with enhanced mechanical and thermal properties. *J. Polym. Res.*, 25, 256, 2018.
63. Rahman, M.H. and Bhoi, P.R., An overview of non-biodegradable bioplastics. *J. Clean. Prod.*, 294, 126218, 2021.

3

Bioplastic From Renewable Biomass

N.B. Singh[1,2*], Anindita De[1], Saroj K. Shukla[3] and Mridula Guin[1]

[1]*Department of Chemistry and Biochemistry, Sharda University, Greater Noida, India*
[2]*Research Development Cell, Sharda University, Greater Noida, India*
[3]*Department of Polymer Science, Bhaskaracharya College of Applied Sciences, University of Delhi, Delhi, India*

Abstract

Due to considerable increase in world population and use of different type of materials, several negative impacts have been seen on the environment. One of the most used materials is plastic. These plastics are mostly of petrochemical origin, which are difficult to degrade, and as a result, these plastics accumulate in the environment, making the whole environment polluted. An alternative to petrochemical-based plastics may be biodegradable and bioplastic. However, all the bioplastics do not degrade easily. Scientists are working hard to find bioplastics from wastes, such as biomass. This may limit environmental pollution and manage waste problems. In this chapter, types of bioplastics, their production by different methods, characterizations, and applications have been discussed in detail. Bioplastic waste management strategies have also been highlighted.

Keywords: Bioplastic, biomass, biowaste, polylactic acid, agro-forestry, food packaging, biomedical, anaerobic digestion

3.1 Introduction

The terms polymers and plastics are often used interchangeably but they are not always the same. Polymers may be synthetic or of organic origin and are formed by individual molecules or monomers. Plastics are type of polymers but all polymers are not plastics. The word polymer is obtained from the Greek word: "*polys* meaning *many*, and *meros* meaning *parts*." Polymers are also known as macromolecules as they are made up of small repeating units, or "monomers," linked through covalent bonding. Due to large size, polymers possess unusual properties. In a typical polymer, the structural units, known as monomers, are linked by primary bonds only. Secondary valence forces, e.g., "van der Waals, London dispersion, hydrogen bond," etc. does not form a true polymer *but forms aggregates* and *micelles*. Plastics, in general, are synthetic or semiorganic polymers obtained from petroleum. Plastics are entirely man-made, although polymers can occur

**Corresponding author*: n.b.singh@sharda.ac.in

Inamuddin and Tariq Altalhi (eds.) Handbook of Bioplastics and Biocomposites Engineering Applications: 2nd Edition, (49–80)

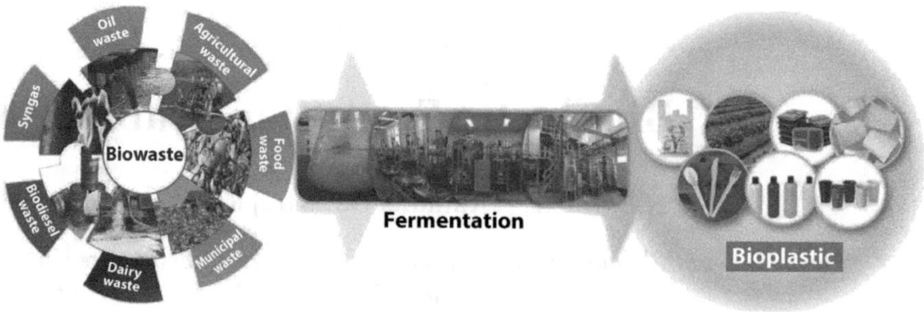

Figure 3.1 Conversion of wastes to bioplastics [2].

naturally. Plastics display properties similar to polymers and used in different sectors for different applications [1].

For industrial economy, polymers are very important materials because of their applications in number of areas. However, most of the plastics are derived from petrochemical industry and are nondegradable. These pollute the environment and are harmful to ecosystem. An alternatives of chemically synthesized plastics are bioplastics, similar to synthetic plastics and by and large environment friendly. It is a myth that all bioplastics are harmless for the environment and biodegradable. However, some of the bioplastics may also contribute to pollution. Now manufactures of bioplastics are being encouraged via wastes. Figure 3.1 [2] shows the conversion of wastes to bioplastics.

In this chapter, attempts have been made to discuss various aspects of bioplastics derived from biomass.

3.2 Plastics and Bioplastics

3.2.1 Plastics

Based on physical properties, there are two types of plastics: thermoplastic and thermosetting [3].

Thermoplastics: Polymers which softens on heat and can be molded into particular shape of choice are called "thermoplastics" (Figure 3.2). Thermoplastics are mainly of four types—polypropylene, polyethylene, poly (vinyl chloride), and polystyrene—together with some more specialized engineering polymers.

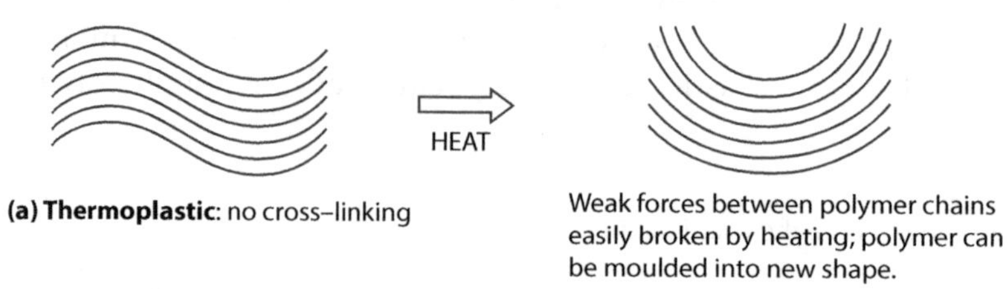

Figure 3.2 (a) Thermoplastics [3].

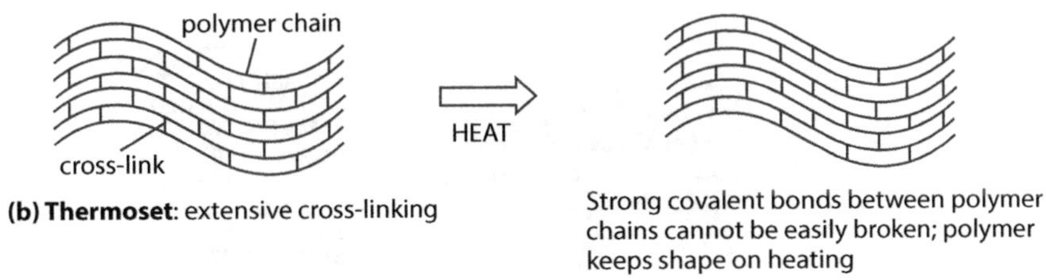

Figure 3.3 (b) Thermoset plastic [3].

Characterization of thermoplastics are made by: "(i) no cross links between chains (ii) weak attractive forces between chains, which are broken by *warming* (iii) change shape and can be remoulded and (iv) weak forces reform in new shape when cooled".

Thermoset polymers: These polymers are liquids in initial stages but on heating are converted to solid polymers exhibiting high cross linking (Figure 3.3). These polymers cannot be softened again by heat and pressure. Once they are formed, they cannot be reprocessed. Important thermosets include unsaturated polymers, alkyds, phenolic resins, amino and epoxies, and polyurethanes. Thermoset polymer is given in Figure 3.3.

3.2.2 Bioplastics

Conventional plastics are light weight with a property, which can be molded into different shapes and a variety of products can be made [4]. In recent years, a large amount of plastics is being used in industry, as well as for household applications, but problems come because of their nondegradability. For this reason, degradable particularly biodegradable plastics are being investigated. The most common is bioplastic. Bioplastics refer to the types of plastics that (1) are biodegradable or (2) may or may not be degradable. However, such plastics are produced from renewable feedstock or naturally occurring materials, such as vegetable oils, vegetable fats, starch, and cellulose. The degradability of bioplastics like any other polymeric materials depends on degree of crystallinity of the polymer, their composition and other environmental factors. The focus on bioplastic research is because of two main reasons: the first is the huge production of plastic materials and the second is problem associated with their disposal. There is great concern for their disposal due to toxicity toward the ecosystem, environmental contamination, and adverse effect on human health [4]. Industrially manufactured plastics are in general nondegradable and hence their continuous use is of great environmental concern. Nondegradable plastics take lot of time, may be tens or hundreds of years to break down [5]. It is found that chemical structure of nondegradable plastics cannot be appropriately altered by microorganisms or other chemical substance, such as water, methane, or carbon dioxide. In order to minimize the problems, degradable bioplastics are needed.

3.3 Classification of Bioplastics

In general plastics are fossil-based and biobased (Figure 3.4) [6].

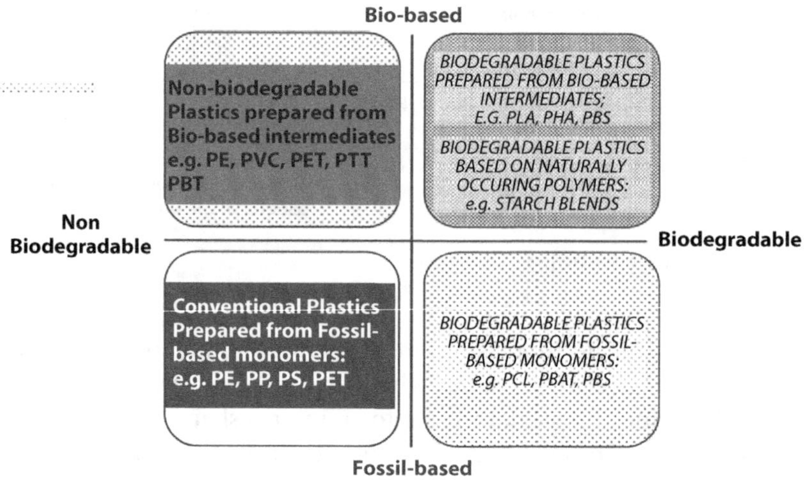

Figure 3.4 Fossil and biobased plastics.

Three groups as given in Figure 3.5 can be considered as bioplastics:

1. Nonbiodegradable plastics, such as poly(ethylene terephthalate) (PET), polyethylene (PE), polypropylene (PP), or poly(trimethylene terephthalate) (PTT), derived partly or completely from biobased intermediates (upper left quadrant).
2. Some of the examples of plastics, which are biobased and biodegradable, are poly(hydroxyalkanoate) (PHA), polylactide (PLA), and or poly(butylene succinate) (PBS) (upper right quadrant).
3. Fossil-based intermediates resources derived from Biodegradable plastics (Lower right quadrant).

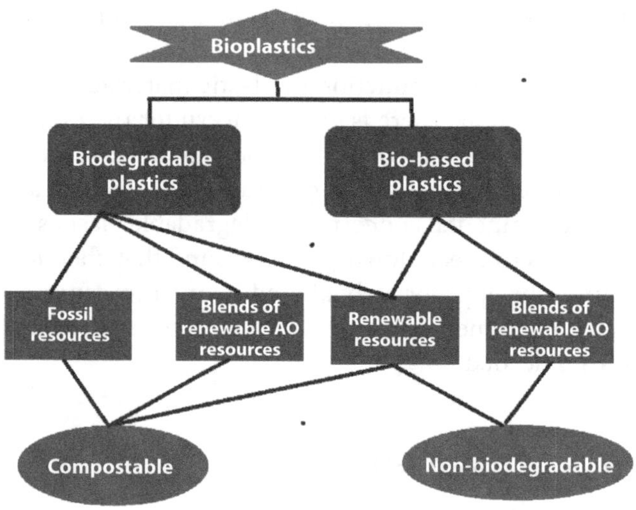

Figure 3.5 Types of bioplastics.

Bioplastics may be biobased/biodegradable or have both the properties. If it is mostly from a natural source, it is called as biobased bioplastics [5]. These can be either nonbiodegradable or biodegradable or both (Figure 3.5) [7]. Different methods have been used for making bioplastics. These reduce energy, nonrenewable raw materials, carbon footprint, nonbiodegradable waste, etc.

3.4 Bioplastic Production

3.4.1 Biowaste to Bioplastic

The high nutrient content in biowaste is the primary cause of serious impact on health and environment. Various conventional methods of managing biowaste includes composting, land filling, incineration, etc. Converting these biowaste into some valuable commercial product, e.g., biopolymers is considered as the best technique to handle the issue of biowaste. The biopolymers are synthesized from biosources and thus are biodegradable and safe to the environment. Some important bioplastics are polyesters, polylactic acid (PLA), and polyhydroxyalkanoates (PHAs). Among them, PHAs are the most important one and received a lot of attention amongst research community. The physicochemical, thermal, and mechanical properties of these biopolymers are similar to that of the petrochemical-based polymers [8]. In spite of high likelihood to take the place of petrochemical-based plastics, the production of PHA is limited due to high cost of production. The industrial level applications are restrained because of complicated synthetic methods and costly feedstocks. Production of PHA from biowastes will not only help to manage wastes but also turn these waste into useful material and improve economic growth. Sources of biowastes for the production of PHA include agro-forestry waste, lignocellulosic biomass, waste cooking oils, wastes from biodiesel industry, municipality wastes, dairy waste, food wastes, etc. The production of PHA depends on concentration of dissolved oxygen and presence of elements in nutrients, such as carbon, nitrogen, and phosphorous.

3.4.1.1 Lipid Rich Waste

3.4.1.1.1 Plant Oils

These feedstocks are economical and provide better yield of PHA. The wastes from various food industries, slaughter houses, oil mills, and edible oil industries are the main source of lipid-rich wastes. The microorganism secret lipase for hydrolyzing the triglyceride (TAG) present in the lipid rich wastes to convert into PHA [9]. There are many reports of using leftover frying oils for production of PHA without any pretreatment. Tufail *et al.* demonstrated 53% accumulation of PHA by cultivating *P. aeruginosa* on leftover frying oils [10]. In a recent work Pernicova *et al.* reported Halomonas strains for synthesis of PHA from waste rapeseed and sunflower frying oil [11]. Further, they investigated *H. hydrothermalis* for the synthesis of PHB to 62% DCW. Effluents from olive oil mill and palm oil mill are also investigated as a potential substrates for production of PHA. Various vegetable oils, such as cottonseed oil, jatropha oil, soybean oil, corn oil, etc., are investigated as carbon source for synthesis of bioplastic.

3.4.1.1.2 Waste From Animal Fats

Waste from the slaughter house in the form of animal fats are investigated as possible substrate for PHA synthesis. Because of the presence of large amount of saturated fatty acids, it remains in solid form. Microbial strains, such as *P. resinovorans*, *C. necator*, etc. are reported for the synthesis of PHA from animal fat waste. The process involves hydrolysis of animal fats to produce two different qualities of biodiesel. The high-quality biodiesel find its application as fuel whereas the low quality biodiesel is used for the synthesis of PHA [12].

3.4.1.1.3 Glycerol/Methanol Waste From Biodiesel Industry

During the transesterification of oils and fats for biodiesel production, glycerol is generated as a byproduct. Production of PHA from this glycerol is a good strategy to reduce the cost of biodiesel production. Recently Meneses *et al.* used *P. chlororaphis* for simultaneous production of mcl-PHA and extracellular polysaccharides using glycerol [13]. Many other microbial species, such as *R. eutropha* and *P. cepacia*, are also studied for PHA and PHB production from crude glycerol. The crude glycerol contains methanol among many other impurities present. Methanol has to be removed as it inhibits the process of fermentation and bacterial growth. Some strain of methylotrophs are reported for PHA production using both glycerol and methanol [14].

3.4.2 Milk Industry Waste

Production of cheese and other milk products generates a large amount of whey. Generally, whey turns out to be 95% of the total processed milk. Whey is composed of proteins, lactic acid, and lactose. To utilize whey as carbon source for the production of PHA, the first step is to convert lactose into glucose and galactose. Halophilic organism, such as *Haloferax mediterranei*, is used for conversion of whey into PHA of 50% to 65% DCW [15]. It is observed that, depending on the composition of cheese, whey PHB or copolymer of 3HV and 3HB can be formed. The PHB production can be improved by using 1% ethanol as stressing agent [16]. Israni *et al.* have synthesized PHA using *B. megaterium Ti3* from whey [17]. The resultant PHA is blended with polyethylene glycol (PEG) to produce a polymer composite with remarkable properties for application in biomedical field. Effective utilization of whey for PHA production needs costly preprocessing. Thus, development of process at ambient temperature and pH is the primary concern now in the field of PHA production from whey [18].

3.4.3 Sugar Industry Waste

It is reported that per year, globally 50 million tons of molasses are generated as waste from sugar industries. The chemical constituents of molasses are primarily sucrose and small amount of glucose and fructose [19]. The earliest report of production of PHB from sugarcane molasses was performed using *Azotobacter vinelandii*. Liquid biofuels and other important commercial products are generated from waste molasses. Park *et al.* investigated a variety of *Bacillus* sp. for production of PHA from molasses [20]. The main issue of using molasses as an effective substrate for PHA production is that removal of various contaminants are extremely important. Impurities, such as polyphenols and inorganic salts, hinder the growth of the microorganisms and thus hampers PHA production [21].

3.4.4 Spent Coffee Beans Waste

The processing of green coffee beans generates a large quantity of nonedible spent coffee grounds. Approximately 650 g wastes are generated as spent coffee grounds to produce 1 kg green coffee beans. The chemical components in the waste is lipids, proteins, and sugars, such as mannose and galactose [22]. To reduce the production cost of PHA, waste oils from spent coffee beans can be a good option. Research results indicate spent coffee oil is better carbon source and more effective for production of PHB than other frying oils [23, 24]. The acidic hydrolysis of the oil from spent coffee beans are investigated for the prepara tion of PHA using various strains of organisms [25, 26]. However, more research is required to understand the process for setting up biorefinary of spent coffee beans. Table 3.1 lists some chemicals in organic waste that are used for bioplastic production.

3.4.5 Bioplastic Agro-Forestry Residue

The agro industry is undergoing a revolution and therefore large amount of wastes are generating in a continuous manner. Thus, agriculture sector residues are gaining importance in bioplastic generation in recent times. The cell walls of plants are made up of cellulose, hemicellulose, and lignin. Lignin is abundantly available feedstock with phenyl-propane monomeric units [27]. The chemical composition of lignin depends on the source of the plant biomass. Depolymerization of lignin is performed by pyrolysis resulting in *o*-vaniline, guaiacol, pyrogallol, *o*-quinone methide, and other products. A variety of pseudomonas species are investigated for valorization of lignin to produce PHA [28]. Moreover, Bacillus and Ralstonia species are also used for the synthesis of PHA from lignin [29].

Among various forms of agricultural residues, straws containing 95 Mt of lignocellulosic feedstock are the most important [30]. Thorenz *et al.* investigated the role of agroforestry residue in improving the bioeconomy in European union [31]. In the agriculture sector,

Table 3.1 Organic wastes for bioplastic production.

Types of organic wastes	Chemicals in the residues
Vegetables and fruits	Flavonoids, carotenoids, antioxidants, lipids, phenols, phytochemicals
Meats and animal waste	Proteins, enzymes, gelatin, collagen
Waste oil	Fatty acids, glycerol, methyl esters, erucic acid
Dairy product	As carbon and nitrogen source
Cellulose	Polypropylene, acrylic acid, ester, thioester, isobutanol, phytosterols
Lignin	PHA, adipic acid

their results revealed wheat straw as the best source for biopolymer production. Further, maize strove, barley straw, and rape straw also exhibit high potential as they contain ample amount (more than 80%) of lignocellulose. In the forestry sector, conifers species outline broad leaf species for potential in bioplastic production. bark residue of spruce and pine shows excellent results. Along with ~ 70% lignocellulose, adequate amount of tannin is also present in coniferous bark, which enhances its industrial application.

Microbial fermentation or enzymatic saccharification of various forms of polysaccharides in agricultural crops is an important method for bioplastic production [32]. Starch, cellulose, chitosan, chitin, proteins, such as casein, gluten are found to be good carbon source. The starch present in the crops are sometimes modified with plasticizers. The starch source, for example, cassava, corn, wheat, potato, tapioca, etc. is investigated by various groups. The bioplastic generated from tapioca starch is flexible, smooth and strong [33]. However, potato-starch–based bioplastics have good texture and are easy to prepare [34]. Asrofi *et al.* have investigated ultrasound treatment on a composite bioplastic made from tapioca starch and sugarcane bagasse [35]. It has been observed that the tensile strength of the polymer is enhanced, while the rate of moisture absorption decreased significantly. Gluten from wheat protein shows promising results in bioplastic production [36]. Moreover, lignocellulosic biomass from food crops are also good resource for bioplastic production [37–39]. A detailed discussion of various waste feedstocks for production of PHA is documented by Bhatia *et al.* [40].

3.4.6 Bioplastic from Microorganism

Polyhydroxyalkanoates (PHAs) are produced mainly by three techniques: enzymatic catalysis, genetically modified synthesis and microbial synthesis [41]. Production of bioplastics from microorganism is promising route among all available processes. PHAs are secondary metabolites of microorganism produced in stressful conditions. Various waste organic sources are used for synthesis of PHA from microorganism, e.g., Bacillus, Pseudomonas, Ralstonia, Methylobacterium, Allochromatium, etc. Fermentable sugars and fatty acids are used for the growth of microorganisms. Thus, sugar and fatty acid containing wastes are considered excellent source for the synthesis of PHA. The yield of PHA is 0.6 to 0.8 g/g from fatty acid substrates, whereas from sugar sources, it is found to be 0.3 to 0.4 g/g [42]. The rate of formation of PHA depends on the source of carbon, concentration of nutrients, and fermentation pathways. The production of PHA is enhanced by excess carbon content and controlling the amount of nitrogen, sulphur, phosphorous, iron, magnesium, etc. nutrients. According to carbon chain length of the monomer, PHAs can be categorized into three types [43].

(i) Short chain length PHA: Monomeric unit contains three to five carbon atoms and prepared from the organism *Cupriavidus necator* and *Alcaligenes latus*. They are highly crystalline in nature and they have similar properties as polypropylene and exhibit thermoplastic behavior. Examples includes P3HB, P4HB, P3HV, etc.

(ii) Medium chain length PHA: For this type of PHA, the monomeric unit carries 6-14 carbon atoms. They are synthesized mostly from *Pseudomonas* sp.

bacteria. Their characteristics are low crystallinity, low tensile strength, low melting point, and high glass transition temperature. Their flexibility and elasticity match with rubber. Examples are P3HHx, P3HO, etc.

(iii) Long chain length PHA: In this case, more than 14 carbon atoms in the monomeric unit are present. Bacteria, such as *Shewanella oneidensis* and *Aureispira marina*, are used for their production.

Presently more than 150 different monomer units are used for PHA production. Scientists are looking for new monomeric units from recombinant microbial strains.

PHAs are prepared under unbalanced growth condition. Although more than 250 different microbial species are used for PHA production, only a handful of bacteria is accepted for commercial production. Among them, *Pseudomonas oleovorans*, *Bacillus megaterium*, *Cupriavidus necator*, and *Alcaligens latus* are important in converting carbon sources to PHA. In addition to these, marine bacteria also have high potential in bioplastic production but only limited research has been performed on them [44]. The cultivation strategy of bacteria can adopt culture methodology, which can be categorized into unadulterated culture and mixed culture.

Moreover, microalgae are recently counted as potentially superior source for bioplastic production due to their rapid growth, high lipid accumulation and no competition with food sources [45]. Several research works have been published recently in this field [46–48]. Different synthetic strategy, e.g., solvent casting, hot molding, injection molding, melt mixing, compression, etc., are applied for synthesis of microalgae polymer blends. Chlorella and spirulina are the widely investigated microalgae species. Bioplastics generated from Chlorella have better properties, whereas spirulina-derived bioplastics exhibit better blend performance. Moreover, other than microalgae, seaweeds and cyanobacteria are also showing promising results as bioplastic sources [49].

3.4.7 Biomass-Based Polymers

Biomass-based polymers are sustainable and recyclable polymers, mostly generated from plant-based materials. As these polymers are obtained from sustainable resources they are of great advantage toward environment. Thus, biomass-based polymers are developed and investigated for their ecological, as well as economical potential [50].

3.4.7.1 Biomass-Based Monomers for Polymerization Process

The primary source of biomass-based monomers is plants, animals, and microbial wastes [51]. The plant sources are from agricultural feedstock and forestry. On the other side, animal sources are mostly from fishery and animal culture. Residual wastes from food processing units and garbage have also been investigated. Among them, the plant-based biomass monomers received higher attention for developing environment benign structural materials [52]. In the next section, three important biomass-based monomers, e.g., trans-anethole (ANE), lignin, and vanilline that are used for different polymer synthesis in recent time, are discussed.

3.4.7.1.1 Biomass ANE

Biomass ANE is used as flavoring and seasoning agent. It is present in anise, fennel, and Chinese star anise as major component. It is synthesized from phenyl propene. Because of its environment benign nature, low cost and easy availability, it is widely studied as biomass monomer. Due to the steric hindrance from the substitution at vinyl positions, homopolymerization of ANE is very tough. However, it can be copolymerized with maleic anhydride, where ANE acts as electron donor and maleic anhydride as electron acceptor [53]. A large number of promising polymers are investigated for various applications. For example, biobased polymer microspheres made from ANE and N-phenylmaleimide co-monomers exhibit remarkable thermal stability as required in high-temperature materials [54]. Biobased hollow polymer particles are synthesized using ANE for the removal of metals and dyes through adsorption [55]. The surface of this polymer can also be modified for enhancing the performance of adsorption. Yuan *et al.* investigated polymer microspheres derived from ANE and maleic anhydride co-monomers for adsorption of methyl red dye and Cr(III) metal ion [53].

3.4.7.1.2 Biomass Lignin

Lignin is the most preferred renewable biomass for production of biobased polymers. Lignin is composed of aromatic phenylpropane. Hydrogels synthesized from hardwood Kraft lignin by Zerpa *et al.* showed improved mechanical and thermal properties than synthetic polymer hydrogels [56]. Lignin-based polymeric adsorbents and zeolitic adsorbents are studied by Nitzsche *et al.* for adsorption of phenol and xylose [57]. Various research groups have investigated lignin-based highly porous polymeric adsorbents for the removal of synthetic dyes and toxic heavy metal ions [58–61]. Wu *et al.* used aerogel made from lignin for adsorption of methylene blue dye [62]. A crosslinked material of lignin with large surface area and adequate functional groups was prepared by Chen *et al.* for removal of rhodamine B dye and Cu(II) metal ion [63]. Moreover, adsorption of organic pollutants is reported by Guo *et al.* using lignin-based nanocomposite of Fe_3O_4/SiO_2/polydopamine [64].

3.4.7.1.3 Biomass Vanillin

Vanillin is used in the synthesis of unique polymers with special properties [65, 66]. Vanillin is phenolic aldehyde with distinct aroma. Polymer based on vanillin are used for the synthesis of polybenzoxazines, polyesters, benzoxazine, and composites. Vanillin-based polymers found application as oxidant, antibacterial agent and adsorbent [67–69]. *Ortho*-vanilline functionalized mesoporous silica coated magnetite is found to exhibit excellent adsorption of Pb(II) metal ion [70]. A polymeric microsphere made from vanilline methacrylate is reported by Zhang *et al.* for adsorption of heavy metals and dyes [71]. Further, magnetic microspheres are synthesized using methacrylated Fe_3O_4 nanoparticles and vanillin methacrylate for magnetic adsorption ability [72].

3.5 Characterization of Bioplastics

Characterization of bioplastic is performed for chemical structure, functionality, morphology, and thermal stability. The techniques used for them are spectroscopy, microscopy, thermal analytical technique, and instruments for evaluation of mechanical properties. The basic principle and use of these techniques in brief are given in Table 3.2.

Some other characteristics of bioplastic are biocompatibility and compost ability. Biocompatibility is an important property of bioplastic to be used in biomedical devices, which comes into direct contact of humans and patients. The standard used by major regulatory bodies for testing of biocompatibility are prescribed in ISO 10993 standards. Further, the standard also varies on the basis of nature indention in body part like membrane, blood, bone, and mucosal membrane. For example ISO 10993-3 is recommended for genotoxicity, carcinogenicity, and reproductive toxicity due to bacterial reverse mutation study using polar and nonpolar extracts. The basic objective of this test is to collect information about the information on risk assessments of a bioplastic during its applications along with mechanical and functionality test [73].

The objective of compost ability is to ensure environmental effect after the use of plastics after monitoring the degradation protocol and products under aquatic, aerial and microbial treatments. It is performed on different scale, i.e., home, industry, water, air, and soil using certified standard by different institutions.

The evaluation is performed under dynamic, as well as static conditions after using batches with defined constitutions [74]. The availability of oxygen is another parameter during compost ability. In general, the samples are treated in specified batch and the products formed after treatment with time are monitored using appropriate techniques. The control in compost ability is an important aspect in designing advanced material using bioplastic after employing molecular level chemical engineering. On this basis, Wan *et al.* have described the importance of molecular engineering, blending and incorporating different additives in synthetic biopolyester to modify their compostability. The basic properties like presence of amorphous character improve water diffusion in polymer along with hydrolytic degradability, and it further improves the compost ability. Thus, the copolymerization of monomer manipulates the crystallinity in polymer matrix, and it further yields tunable compostability in polymer matrix [75].

Table 3.2 Structural information derived from characterization techniques.

S no.	Techniques	Information's
1	Thermal analysis	Stability, thermal transition and degradation kinetics
2	Spectroscopy	Chemical functionality, Bonding, molecular structure and unsaturation
3	Microscopy	Size, morphology and homogeneity
4	X-rays techniques	Crystallinity, size, composition and lattice structure
5	Electroanalytical techniques	Oxidation states
6	Universal testing machine	Strength, plasticity, load bearing capacity and dimensional stability

3.6 Applications of Bioplastics

3.6.1 Food Packaging

The packaging is an important aspect of plastic and creates huge pollution problem. Thus, the substitution of bioplastic in packaging applications promises to control the environmental issues along with holistic use of resources. Bioplastics are effectively used as both hard, soft and flexible packaging applications after due optimization in different properties *viz.* gloss, strength, flexibility, dimensional stability, and load bearing capacity. In this context, different biopolymers, i.e., polylactic acid, cellulose, chitosan, starch, and polyhydroxyalkanoates, are used after adding different fillers, plasticizers and stabilizers. The chemical modifications occurring during developments are another important step in optimization of bioplastics in packaging applications [76].

Polylactic acid (PLA) is complete bioplastic and currently used for different types of packaging applications due to optical transparency, gloss, printability, process ability, stiffness, and excellent barrier properties [77]. PLA exists in two steroisomers, i.e., levo and dextro lactic acid, and their relative individual presence control different properties like molecular weight and thermal stability. Further, change in molecular weight also changes the melting points, glass transition temperature, and stability, which also explores its suitability in packaging in term of hardness and brittleness [78]. Incorporation of bioactive compounds is another area in improving the effectiveness of packaging materials and called as active packaging. Roy *et al.* have developed PLA and curcumin-based active packaging film by solution casting methods. The film exhibits high transparency, flexibility and mechanical strength without compromising thermal stability, high transparency with bright yellow color along with uniform dispersion. Further, the developed film also yields antimicrobial activity to use as an active packaging for sfood items sensitive to oxidative stress to prevent oxidative degradation due to sustain release of curcumin [79].

The control in size of biopolymer like chitosan adds several advantageous features such as surface charge, zeta potential and functionality. These features also control the processability and mechanical strength and functionality. Thus, the nano sized chitosan has been explored by several scientists for use in several types of packaging materials with advanced properties [80]. Furthermore, Chen *et al.* have developed hydrogel film of CHIT containing silicon and β-acid with enhanced antimicrobial activities. The presence of β-acid in hydrogel film also exhibits UV absorbing properties along with increased tensile strength of 7.99 MPa for use in packaging of wide range of food with antimicrobial activity against two microbial strains, *S. aureus and E. coli* [81]. The optimization in barrier of bioplastics is another strategy to modify the bioplastic for use in different packaging applications. The improved barrier properties in film used for storage of edible item with enhanced shelf life for arial and microbial degradation. Some more significant use of bioplastic in packaging are listed in Table 3.3.

3.6.2 Agricultural Applications

The applications of bioplastic in agriculture are in seed coating, soil conditioner, controlled release of fertilizer and micronutrients. Seed coating is current innovation of agriculture scientists to improve seed appearance, preservation, handling, providing active ingredients

Table 3.3 Bioplastic-based packaging materials.

S no.	Composition	Properties	Applications	Ref.
1	Poly(lactic acid) and poly (butylene adipate terephthalate)	Dense matrices with improved crystallinity and reduced permeation to vapor and oxygen	Bakery packaging	[82]
2	Poly(butylene succinate), Poly(butylene adipate-co-terephthalate), and linear low-density polyethylene	Oriented fibrous film with controlled mechanical and barrier properties	Bread packaging	[83]
3	Poly(butylene adipate, terephthalate poly(lactic acid)	Micro structured barrier properties against water vapor and oxygen	Active packaging	[84]
4	Polybutylene adipate-co-terephthalate	Laser induced processing	Packaging for food	[85]
5	Chitosan, polyethylene glycol and methyl ether methacrylate	Antioxidant and antimicrobial nature with mechanical strength	Active food packaging	[86]
6	Poly(butylene adipate terephthalate), poly(lactic acid	Improved compatibility, oxygen permeability and antimicrobial nature	Active packaging for seafood	[87]
7	Poly(lactic acid)	Protected from the environmental factors	Bottles for alcoholic beverages	[88]
8	Thermoplastic starch	Improved polymer compatibility, reduced tensile strength and antimicrobial nature	Fresh noodles	[89]
9	Cassava starch	Optimized tensile strength and elastic modulus	Packaging	[90]
10	Poly(butylene adipate-co-terephthalate) and thermoplastic starch	Good mechanical strength with improved moisture and oxygen barrier	Multipurpose	[91]
11	Starch and citrus peel	Optimized mechanical strength and antioxidant behaviour	Fish	[92]

and improve germination rate [93]. The seed coating is used by horticulture and crop industries with objective to apply colors, i.e., fluorescent dyes, protectants, i.e., pesticides, soil conditioner, i.e., hydro-absorbers, germination stimulants, micronutrient and stress resistance, i.e., salicylic acid. In this regard different biopolymers, i.e., cellulose, guar gum, chitosan after using other additives like plasticizers. The seed coating is further classified in terms of weight, size, and grouping properties of coated seed [94]. Shi *et al.* have developed the seed coating film from bioplastic, pesticide and plant growth promoting agent through chemical process. The coating of composite over corn seed resulted greater elongation of corn than untreated seed in terms of 18 % stem and 21.4 % root along with reduced seed dust. Further, the film also reveals sustained long term release of agrochemicals [95, 96].

The degradation of bioplastic leads to change in the size, porosity and carbon contents, which also change the water holding and retaining capacity of soil. The enhanced porosity

Table 3.4 Bioplastic for agriculture practices.

S no.	Composition	Properties	Applications	Ref.
1	Soya protein	Super absorbent and water absorption capacity	Zinc release	[101]
2	Chitosan and alginate	Safeguard of pesticide from UV light, direct on-field application and up scalability	Sustained release of imidacloprid	[102]
3	Chitosan	Complexing behavior with heavy metals	Abiotic responses	[103]
4	Poly (vinyl alcohol)/ glutaraldehyde	Long-term huge water absorption capacity up to 550%	Controlled release	[104]
5	Cu, Chitosan and PVA	Utilization of water and nutrients	Watermelon	[105]
6	Polyvinyl Alcohol/ Polylactic Acid	Higher encapsulation efficiency and stability	Controlled-Release of Fertilizers	[106]
7	Gelatinized starch	higher swelling degree with delayed release of nutrient	Sustained release of micronutrient	[107]
8	Starch and poly(sodium acrylate)	Outstanding liquid uptake capacity, i.e., 42,000%.	Soil conditioner	[108]
9	Whey proteins and alginic acid	pH conditioned repeated swelling-drying cycles,	Sustained release of urea fertilize	[109]
10	seaweed extract and calcium alginate	pH responsive water holding capacity	Improved soil hydro physical and fertility	[110]

and carbon content also increases oxygen availability as well as humous content. These properties are explored for soil conditioning after composting different biopolymer, which were even used during seed coating. For example, the degradation of three-layered composite structure of gelatin and polylactic acid after buried under indoor soils in 120 days. The abiotic degradation of composite reveals the significant increments in water uptake by 30% [97].

The degradation of developed film supports the condition of soil in terms of water uptake and effective germinations for greater length of radicals under similar time span [98]. The use of bioplastic-based vehicle for sustainable release of different micro nutrients are also explored in agricultural practice due to their microbial degradation. The basic need of bioplastic to be used in release of soil nutrients are responsive degradation along with loosening of matrix. Ghosh *et al.* have designed a hybrid polymer matrix from chitosan, methacrylic acid and potassium. The composite, the potassium encapsulated polymer matrix with pristine structure of chitosan with sustain release of potassium as nano carrier fertilizer delivery system for longer time along with change in porosity and water retaining capacity has been reported [99]. Hydrogel is another form of bioplastic with composites with heterogeneity and ability to absorb huge amount of water. Jha *et al.* have developed guar gum-based super absorbent hydrogel with biodegradable nature along with huge absorption capacity by 356 g/g. The hydrogel bears long term boron sustainable release up to 96.25 days with ability to use in agriculture sector [100].

Some more important applications of bioplastic in agriculture are listed in Table 3.4, along with properties and potentials.

3.6.3 Biomedical Applications

Biomedical applications of bioplastics are explored as organ transplants, suture, scaffolds, tissue engineering, therapeutic devices, wound management, and drug delivery. Biocompatibility, hemocompatibility along with functionality are the essential requirements and is possible through different types of chemical and physical treatments. The grafting of chitosan and alginate has yielded suitable properties for use in liver cell transplant after encapsulation of liver cell. The grafted biopolymer matrix exhibits better immune protection system with a potential to use in xenogenic transplantation [111]. The chemically modified chitosan with biological activities has been explored. The designing of scaffold is another important area for tissue engineering and regenerative medicine [112, 113]. 3 D printing is another development for personalized therapy and diagnostic medical devices suing different bioplastics due to advantageous features of excellent reproducibility, control and manufacturing [114]. Thus, 3 D printed polymeric structures are well matched with natural tissues in terms of shape and structure without any negative immune response or toxicity and also used in cardiovascular stents. These actually degrade with time to reinstate the activity of vessel along with permitted remodeling to maintain recoiling feature [115].

The drug delivery is another important use of bioplastic in medical field. Chitosan grafted polyaniline has been used in drug delivery of Ketoprofen, a model drug with pH responsive nature and it follows anomalous non-Fickian diffusion [116].

Some other applications of bioplastic in biomedical sector are listed in Table 3.5, along with brief description.

Table 3.5 Applications of bioplastic in biomedical applications.

S no.	Composition	Properties	Applications	Ref.
1	Poly vinyl alcohol and polylactic acid	Compressibility, mineralization behavior, and biocompatibility of the viscoelastic characteristics to natural cartilage	Tissue engineering	[117]
2	Chitosan and strontium	Positive effects on downregulation of inflammation	Bone tissue engineering	[118]
3	Chitosan, poly3-hydroxybutyrate-co-3-hydroxyvalerate and calcium sulfate hemihydrate	Porous three-dimensional structure, with an average pore size of 400 μm and swelling rate of 56 %	Tissue engineering	[119]
4	Chitosan and biguanidine	Sequential release behavior of the vascular endothelial growth factors.	Injectable bone tissue engineering	[120]
5	chitosan and polycaprolactone	Transparency, biocompatibility and c	Corneal endothelial regeneration	[121]
6	Starch/hydroxyapatite	Low cost, wide availability, and biocompatibility	Biomedical device	122]
7	Rice starch extract coated iron oxide nanoparticles	Biologica; active and Photoacoustic imaging	Chemo-photothermal therapy	[123]
8	Microporous starch and Fe_3O_4	Excellent hemolysis, cytotoxicity and biodegradation	Hemorrhage control of complex bleeding wounds	[124]
9	Starch and polyurethan	pH-dependent burst release	Drug delivery	[125]
10	Polysaccharides	Tunable properties, low immunogenicity and low toxicity	Micro- and nano-devices for drug delivery	[126]

Another, polysaccharide-based bioplastic is agarose, which is soluble in luke warm water along with thermally induced liquid to gel transition with characteristic hysteresis. The biocompatibility and hemocompatibility of agarose-based film has been explored to use in drug delivery for antibiotic and antiseptic with their potency. Verma *et al.* have prepared a film of agarose using variable concentration of glycerol, i.e., 0% to 50%. Finally, the obtained film of optimized concentration was incorporated with antibiotic and antipyretic drugs, i.e., ampicillin, tetracycline, dettol, and savlon during solution casting of film. Final film showed consistent swelling, water uptake behavior and mechanical strength. Further, the obtained drug encapsulated film was used for sustained release of drug for use in surgical and wound dressings [127].

3.7 Bioplastic Waste Management Strategies

3.7.1 Recycling of Poly(Lactic Acid) (PLA)

There are number of methods for recycling as discussed below.

3.7.1.1 Mechanical Recycling of PLA

In case of PLA, mechanical recycling process is considered to be better than chemical recycling in terms of environmental and human benefits as well as for efficient use of resources. However, the recycled material is of inferior quality compared to the one produced by chemical recycling [128]. Addition of a chain extender while recycling PLA mechanically can compensate the loss of mechanical properties and the obtained material is comparable to virgin PLA. Additionally, once PLA becomes of low quality due to mechanical recycling, it should be recycled chemically to obtain the monomers. It is estimated that for a profitable biopolymer, mechanical recycling plant, at least ~200 Kilo Tonnes of biopolymer production is required and the recycling facility should have the capacity of at least 5–18 Kilo Tonnes annually [129].

3.7.1.2 Chemical Recycling of PLA

Chemical recycling involves either formation of individual monomeric units or conversion of the plastic to other useful products. Both these processes require single or multi-step chemical reactions such as pyrolysis and/or use of catalyst. This technique is more advantageous compared to mechanical recycling as the basic mechanical properties of the starting material are retained. Moreover, potential value added product is obtained from waste plastic material which contributes to a circular economy. There are two distinct methodology available now-a-days for chemical recycling of PLA namely depolymerization and degradation.

3.7.1.2.1 Depolymerization of PLA

In recent times, researches on depolymerization of Polylactic acid have generated interest due to their versatile application as precursor material for various other industrially important chemicals such as acrylic acid, pyruvic acid, 1,2 propanediol, 2,3 pentanedione and many more. There are number of depolymerization methods available in literature. The most common method is hydrolysis at 100-250°C in presence of strong mineral acids such as H_2SO_4 or HNO_3 as catalyst [130]. Hydrolysis of PLA has been performed at around

80-180°C under 10 bar pressures in presence and in absence of NaOH. Use of high temperature and pressure can ensure absence of harsh acid/base catalysts [131]. Similar observations were also reported by Hirao *et al.* who performed the hydrolysis reaction at 250°C under high pressure and demonstrated 90% conversion of PLA to L-lactic acid within 10 to 20 mins [132]. Microwave irradiations are also reported to be highly useful and requires less time compared to the conventional heating [133]. Additionally, depolymerization of polylactic acid scaffolds or films by using NaOH or HCl as catalyst has been tried by various research groups [134]. However, there is still concerns regarding the process optimization and environmental issues related to these methods. Thus, much focus has been placed to make the depolymerization process more efficient as well as environment friendly. One very good alternative for the harsh catalyst use could be enzymes [135–137] although their industrial application still remains a challenge due to difficulty in scaling up the catalytic process [138, 139].

3.7.1.2.2 Degradation of PLA

Thermal degradation of polylactic acid is possible at very high temperature (200-400°C) to produce various different products depending on the system used. Thus, this process is neither cost-effective nor energy efficient. Hence, development of robust and efficient catalyst has been a focal point of research these days. Synthesis of lactate ester is a very potential way of PLA degradation as lactate esters are green substituent for commonly used hydrocarbon solvents due to their low vapor pressure, low toxicity and biodegradability [140].

Typically, lactate ester synthesis involves functional group protection strategies to stop the production of oligolactate side products. The process also requires a catalyst as well as long reaction time. In recent times, metal mediated alcoholysis of lactide have also shown very good results [141]. DuPont company has patented a process where polylactic acid is converted to various lactate ester using 2-3 equivalents of alcohol per polymeric unit in presence of H_2SO_4 as catalyst to yield 69-87% yield in 2h at 150-190°C and reactant autogenic pressure [142]. Instead of using strong, corrosive chemicals such as acids, ionic liquids have also been utilized as catalyst by Song *et al.* [143, 144]. These catalysts are environment friendly as well as possess better catalytic activity compared to the conventional catalyst. However, high cost and high viscosity are some of the demerits which makes them unsuitable for large scale use. Organic molecules such as 1,5,7-triazabicyclo [4.4.0]dec-5-ene (TBD) and 4-(dimethylamino)pyridine (DMAP) have also been used as catalyst with good yield [145, 146]. TBD can catalyze the degradation reaction at room temperature using three equivalents of ethanol per ester group and the yield of lactate ester is greater than 90% [146].

Metal complex-based catalyst for PLA degradation are also known and the mechanism for the same is depicted in Scheme 3.1. Two Zr(IV)/Hf(IV)-Salalen complexes have been used by Jones and coworkers to perform methanolysis of commercially available high manganese methyl lactate at two different temperatures [147] with 75% yield. However, difficult ligand preparation is one of the drawbacks of this type of catalyst. Simple metal salts such as $FeCl_3$ has been used as catalyst for conversion to methyl lactic acid within 4 h at 130°C using 1 mol% of $FeCl_3$ and alcohol/polymer ratio of 5:1. This catalyst is recyclable and can be used up to six times without significant loss of catalytic activity [148]. Other notable

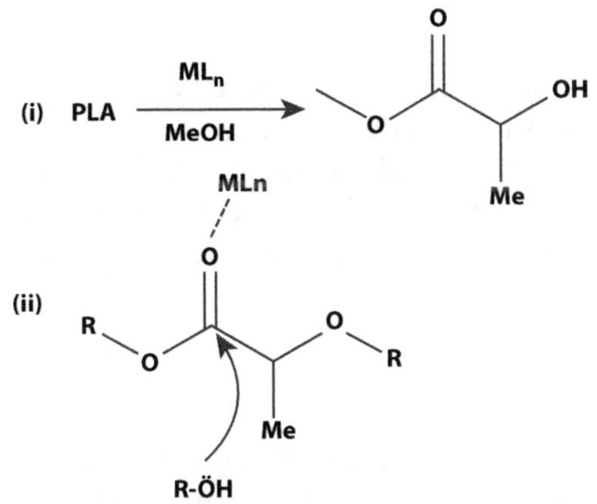

Scheme 3.1 (i) PLA degradation to methyl lactate (ii) Acid catalysed transesterification reaction mechanism.

examples of metal-based catalyst include a Zn-Imino monophenolate complex [149] and Zn(II) complex with N-Heterocyclic carbene-based ligands [24]. Metal-based catalyst can reduce the reaction temperature considerably [150].

3.7.2 Recycling of Poly Hydroxy Alkanoates (PHAs)

The studies on chemical and mechanical recycling of PHAs are limited because of its high cost and less use. Most well-known type of PHA is Poly(3-hydroxy-butyrate) (P3HB) which is synthesized naturally by a microbe called *Alcaligenes eutrophus*. On mechanical recycling of homopolymer of P3HB, the tensile strength is decreased by 50%. On the other hand, copolymer of PHBV can undergo mechanical recycling more efficiently with only 8% loss in tensile strength. Lower grade PHAs can be biodegraded to CO_2 and absorbed by plants which could then be utilized in fermentation or they can be converted to other valuable chemicals also.

Pyrolysis is another method to recycle PHAs. The thermal degradation of P3HB occurs through polymer chain scission reaction at random position to yield cis/trans-crotonic acid (CA) as major product and its oligomers [151]. Crotonic acids can be converted to esters which can again polymerize to give optically active polymers [152, 153].

PHAs can biodegrade in compost, soil and marine environments in presence of some microbes. For example, ~63% conversion of P3HB-co-3HH to CO_2 has been achieved under anaerobic conditions and the same degradation reaction can be carried out under aerobic, seawater conditions at room temperature with a 83% loss of gaseous carbon in 195 days. Microbes that can biodegrade PHAs are mostly found in soil, marine water, compost and sewage sludge.

The last type of recycling method of PHAs is chemical recycling in which PHAs undergo depolymerization to produce monomers or other low molecular weight molecules. These small molecules can be further reused as raw material for synthesis of corresponding biopolymers [154].

3.7.3 Landfill

This is the most conventional approach for solid waste management and constitutes 60% of all the plastic consumed. Although short term impacts are limited, it can have long term impact on the local ground water and soil due to discharge of pollutants and additives discharged upon waste degradation. Thus, in many countries such as UK, landfill taxes have been increased to restrict the use of landfill sites and encourage the alternative waste disposal strategies [148].

3.7.4 Incineration

Incineration is the combustion of substances contained in the waste material. This is a very good alternative to landfills, as with proper designing, for example, use of a boiler, they can lead to energy recovery in form of either thermal or electrical energy. Biodegradable plastics are mostly produced from sustainable resources. For example, PLA consist of common atoms such as carbon, oxygen, and hydrogen atoms without any chlorine atom. As chlorine atoms are absent in PLA, production of dioxins does not happen during burning/incineration. Conventionally, bioplastics are devoid of heavy-metal additives. Thus, commonly incineration can be safely applied for them without releasing harmful dioxins or heavy metals. However, this method is not suitable for mixed plastics which is commonly found due to the risk of releasing chlorinates pollutants in the environment. This problem can be tackled by using filters which on the other hand will increase the capital cost [149]. Many a times, incineration facilities are disapproved by the local communities due to being aesthetically unpleasant and creation of local pollution. In view of these limitations, incineration remains less preferred compared to landfills.

3.7.5 Composting

This is the most appropriate method for bioplastic waste valorization as biodegradability is the inherent property of the bioplastics. This process has minimum environmental impact when operating conditions are carefully monitored and adjusted. However, composting conditions are complex and depends on many factors such as pH, humidity, ventilation etc. which limits its widespread use. Moreover, composting is also sensitive to complex mixtures in the waste. For example, if PET is present in the waste product due to sorting mistake, then it will never degrade leading to low quality compost. The additives used to improve the quality of the bioplastics for example PLA often has a negative effect on the compostability [148].

3.7.6 Anaerobic Digestion

Anaerobic digestion occurs in mesophilic (37°C) or thermophilic (55°C) biogas plants in absence of oxygen. The organic matter is converted to small, harmless molecules such as methane gas, carbon dioxide, water, ammonia, hydrogen and hydrogen sulphide. The residual part is known as digestate residue. As oxygen is not present, less amount of heat is produced. The overall reaction is as follows:

$$\textbf{Organic matter} + H_2O + \textbf{Nutrients} \rightarrow \textbf{Digestate residue} + CO_2 + CH_4 + NH_3 + H_2S + \textbf{Less heat}$$

Bioplastics can be biodegraded by this method. Under ambient condition, polylactic acid is found to be resistant to degradation and hence first the material has to be hydrolyzed at suitable temperature (~58°C) to decrease the molecular weight of the polymer prior to biodegradation. The efficiency of degradation depends on nature of the polymeric substance and on type of microorganism involved. Anaerobic digestion consists of four stages: hydrolysis, acidogenesis, acetogenesis and methanogenesis. In the initial step, the complex organic polymers are converted to simple, soluble monomers by the action of extracellular enzymes, such as cellulase, amylase, protease, etc., secreted by the hydrolytic microbes. After that, the product of hydrolysis is converted into volatile fatty acids such as acetic acid, propanoic acid and butyric acid by fermentative bacteria. Theacetogenic bacteria convert the volatile fatty acid into other products like acetate, hydrogen and/or carbon dioxide. Finally, in the last step, the products of third stage are converted to methane. Brief about various anaerobic degradation is provided in the subsequent sections.

3.7.6.1 Anaerobic Digestion of Poly(Hydroxyalkanoates)

PHAs which are accumulated inside the microbial cells can be degraded by intracellular depolymerase enzymes. However, extracellular degradations are also reported by several research groups using different types of microorganism such *Pseudomonas stutzeri*, a bacteria found in soil and *Alcaligenes faecalis*. Some other species where depolymerization enzyme has been identified includes *Bacillus megaterium*, *Rhodospirillum rubrum*, *Pseudomonas lemoignei* and *Acinetobacter beijerinckii*. The primary depolymerization enzyme responsible for degradation is PHB depolymerase which is an extracellular ento-type hydrolase [155]. According to studies done on degradation of PHAs revealed that complete degradation to carbondioxide and water is possible within 5 to 6 weeks under aerobic conditions. However, when done in anaerobic condition the process is much faster and the degradation product (methane gas) is much more valuable [156]. Almost 85% degradation of PHBV present in anaerobic sludge is observed within 20days [157]. In another report by Verlinden *et al.*, PHB can take months to degrade on anaerobic sewage and can require years in seawater. However, Ultra violet light, can speed up the process [158].

3.7.6.2 Anaerobic Digestion of Poly(Lactic Acid)

Polylactic acid does not biodegrade easily under ambient condition in the environment using microbes. First it has to be hydrolyzed at higher temperature to molecules of smaller molecular weight before the biodegradation process [155]. Yagi *et al.* used a biogas plant sludge to study the biodegradation of PLA under anaerobic aqueous condition. The authors reported ~60% biodegradation of PLA powder in 1 month, ~80% in 40 days and ~90% in 2 months under thermophilic conditions and the same reaction did not even start after 55 days [159]. The same authors in another report predicted the bacterial strain present to be eubacteria between 8-14 days in the test. The authors have also studied the effect of film size on the degradation rate and found that smaller pieces stuck together due to static electricity

Table 3.6 Various types of microorganism used for anaerobic digestion.

Bioplastic type	Microorganism's type	Name of the microorganism	Source
PLA- based	Bacteria	*Amycolatopsis* sp., Lentzea sp, Saccharaothrix sp.	--
	Fungi	*Aspergillus fumigatus, Thermomyces lanuginosus,*	Compost/ Soil
		Fusarium solani, Fennellomyces linderi,	Soil
		Penicillium verrucosum, Aspergillus ustus,	soil
PHB	Bacteria	*Burkholderia capacia, Streptomyces* sp.,	River sediment
PHA		Pseudomonas aerogusina, Streptomyces bangladeshensis,	Soil
		Pseudomonas putida, Leptothrix sp., *Variovorax* sp.	River water
PHA		*Pseudomonas putida, Pseudomonas fluorescens,*	Soil
PHA	Fungi	*Fusarium oxysporum, Candida albicans,*	soil
PHB		*Penicillium* sp., *Aspergillus* sp.	Soil/compost
PHB	Bacteria	*Bacillus* sp., *Gracilibacillus* sp.	Marine water
PHBV		*Actinomadura* sp.	Soil
PHBV		*Micrococssus* sp., *Bacillus* sp.	Active sludge

at the onset of the degradation and thus the total contact area between the sludge and the polymer film becomes less, and the degradation rate decreases [160]. In Table 3.6, various microorganisms utilized for degradation of bioplastics have been elaborated.

3.8 Conclusions and Future Prospects

Bioplastics have functional properties similar to plastics but are obtained partially or completely from biomass resources. This chapter has addressed various aspects of bioplastics manufactured from biomass resources. Methods have been developed to produce bioplastics from different resources of biomass. To minimize plastics/bioplastics in the environment different techniques have been discussed. Bioplastics may play an important role in waste management, energy, and economy. However, bioplastics be evaluated case by case basis. In order to minimize the use of plastics/bioplastics, it is essential to change the mind set of people. For encouraging the production and applications of bioplastics from bioresources, incentives be given to both the manufacturer and users.

References

1. Sperling, L.H., *Introduction to physical polymer science*, John Wiley & Sons, Inc, Hoboken, New Jersey, 2006.
2. Bhatia, S.K., Otari, S.V., Jeon, J.-M., Gurav, R., Choi, Y.-K., Bhatia, R.K., Pugazhendhi, A., Kumar, V., Banu, J.R., Yoon, J.-J., Choi, K.-Y., Yang, Y.-H., Biowaste-to-bioplastic (polyhydroxyalkanoates): Conversion technologies, strategies, challenges, and perspective. *Bioresour. Technol.*, 326, 124733, 2021.
3. Singh, N.B. and Das, S.S., *Introduction to polymer science and technology*, New Age Publishers, New Delhi, 2017.
4. Saratale, R.G., Cho, S.-K., Saratale, G.D., Kadam, A.A., Ghodake, G.S., Kumar, M., Bharagava, R.N., Kumar, G., Kim, D.S., Mulla, S.I., Shin, H.S., A comprehensive overview and recent advances on polyhydroxyalkanoates (PHA) production using various organic waste streams. *Bioresour. Technol.*, 325, 124685, 2021.
5. Atiwesh, G., Mikhael, A., Parrish, C.C., Banoub, J., Le Tuyet-Anh, T., Environmental impact of bioplastic use: A review. *Heliyon*, 7, e07918, 2021.
6. Ross, G., Ross, S., Tighe, B.J., Bioplastics: New Routes, in: *New products, brydson's plastics materials*, 8th, pp. 631–652, 2017.
7. Coles, R. and Kay., M., Food and Beverage Packaging Technology (Chapter 11), in: *Bioplastics*, 2010.
8. Kourmentza, C., Plácido, J., Venetsaneas, N., Burniol-Figols, A., Varrone, C., Gavala, H.N., Reis, M.A., Recent advances and challenges towards sustainable polyhydroxyalkanoate (PHA) production. *Bioengineering*, 4, 55, 2017.
9. Talan, A., Kaur, R., Tyagi, R.D., Drogui, P., Bioconversion of oily waste to polyhydroxyalkanoates: Sustainable technology with circular bioeconomy approach and multidimensional impacts. *Bioresour. Technol. Rep.*, 11, 100496, 2020.
10. Tufail, S., Munir, S., Jamil, N., Variation analysis of bacterial polyhydroxyalkanoates production using saturated and unsaturated hydrocarbons. *Braz. J. Microbiol.*, 48, 629, 2017.
11. Pernicova, I., Kucera, D., Nebesarova, J., Kalina, M., Novackova, I., Koller, M., Obruca, S., Production of polyhydroxyalkanoates on waste frying oil employing selected. *Halomonas strains. Bioresour. Technol.*, 292, 122028, 2019.
12. Titz, M., Kettl, K.H., Shahzad, K., Process optimization for efficient biomediated PHA production from animal-based waste streams. *Clean Techn. Environ. Policy*, 14, 495, 2012.
13. de Meneses, L., Pereira, J.R., Sevrin, C., Grandfils, C., Paiva, A., Reis, M.A.M., Freitas, F., Pseudomonas chlororaphis as a multiproduct platform: Conversion of glycerol into high-value biopolymers and phenazines. *N. Biotechnol.*, 55, 84, 2020.
14. Koller, M. and Braunegg, G., Advanced approaches to produce polyhydroxyalkanoate (PHA) biopolyesters in a sustainable and economic fashion. *Euro Biotech. J.*, 2, 89, 2018.
15. Pais, J., Serafim, L.S., Freitas, F., Reis, M.A.M., Conversion of cheese whey into poly(3-hydroxybutyrate-co-3-hydroxyvalerate) by. *Haloferax mediterranei. N. Biotechnol.*, 33, 224, 2016.
16. Colombo, B., Pepè Sciarria, T., Reis, M., Scaglia, B., Adani, F., Polyhydroxyalkanoates (PHAs) production from fermented cheese whey by using a mixed microbial culture. *Bioresour. Technol.*, 218, 692, 2016.

17. Israni, N., Venkatachalam, P., Gajaraj, B., Varalakshmi, K.N., Shivakumar, S., Whey valorization for sustainable polyhydroxyalkanoate production by Bacillus megaterium: Production, characterization and *in vitro* biocompatibility evaluation. *J. Environ. Manage.*, 255, 109884, 2020.
18. Koller, M., Puppi, D., Chiellini, F., Braunegg, G., Comparing chemical and enzymatic hydrolysis of whey lactose to generate feedstocks for haloarchaeal poly (3-hydroxybutyrate-co-3-hydroxyvalerate) biosynthesis. *Int. J. Pharm. Sci. Res.*, 3, 112, 2016.
19. Raza, Z.A., Abid, S., Banat, I.M., Polyhydroxyalkanoates: Characteristics, production, recent developments and applications. *Int. Biodeterior. Biodegrad.*, 126, 45, 2018.
20. Park, S.J., Jang, Y.-A., Lee, H., Park, A.-R., Yang, J.E., Shin, J., Metab Eng of *Ralstonia eutropha* for the biosynthesis of 2-hydroxyacid-containing polyhydroxyalkanoates. *Metab. Eng.*, 20, 20, 2013.
21. Sjolin, M., Thuvander, J., Wallberg, O., Lipnizki, F., Purification of sucrose in sugar beet molasses by utilizing ceramic nanofiltration and ultrafiltration membranes. *Membranes*, 10, 5, 2020.
22. Saratale, G.D., Bhosale, R., Shobana, S., Banu, J.R., Pugazhendhi, A., Mahmoud, E., Sirohi, R., Bhatia, S.K., Atabani, A.E., Mulone, V., Yoon, J.J., Shin, H.S., Kumar, G., A review on valorization of spent coffee grounds (SCG) towards biopolymers and biocatalysts production. *Bioresour. Technol.*, 314, 123800, 2020.
23. Obruca, S., Benesova, P., Petrik, S., Oborna, J., Prikryl, R., Marova, I., Production of polyhydroxyalkanoates using hydrolysate of spent coffee grounds. *Process Biochem.*, 49, 1409, 2014.
24. Obruca, S., Petrik, S., Benesova, P., Svoboda, Z., Eremka, L., Marova, I., Utilization of oil extracted from spent coffee grounds for sustainable production of polyhydroxyalkanoates. *Appl. Microbiol. Biotechnol.*, 98, 5883, 2014.
25. Obruca, S., Benesova, P., Kucera, D., Petrik, S., Marova, I., Biotechnological conversion of spent coffee grounds into polyhydroxyalkanoates and carotenoids. *New Biotechnol.*, 32, 569, 2015.
26. Bhatia, S.K., Kim, J.-H., Kim, M.-S., Kim, J., Hong, J.W., Hong, Y.G., Kim, H.-J., Jeon, J.-, Kim, S.-H., Ahn, J., Production of (3-hydroxybutyrate-co-3-hydroxyhexanoate) copolymer from coffee waste oil using engineered. *Ralstonia eutropha. Bioprocess Biosyst. Eng.*, 41, 229, 2018.
27. Saratale, G.D., Saratale, R.G., Varjani, S., Cho, S.K., Ghodake, G.S., Kadam, A., Mulla, S.I., Bhargava, R.N., Kim, D.S., Shin, H.S., Development of ultrasound aided chemical pretreatment methods to enrich saccharification of wheat waste biomass for polyhydroxybutyrate prodction and its characterization. *Ind. Crop Prod.*, 150, 112425, 2020.
28. Chio, C., Sain, M., Qin, W., Lignin utilization: A review of lignin depolymerization from various aspects. *Renew. Sustain. Energy Rev.*, 107, 232, 2019.
29. Chen, Z. and Wan, C., Biological valorization strategies for converting lignin into fuels and chemicals. *Renew. Sustain. Energy Rev.*, 73, 610, 2017.
30. Bilo, F., Pandini, S., Sartore, L., Depero, L.E., Gargiulo, G., Bonassi, A., Federici, S., Bontempi, E., A sustainable bioplastic obtained from rice straw. *J. Clean. Prod.*, 200, 357, 2018.
31. Thorenz, A., Wietschel, L., Stindt, D., Tuma, A., Assessment of agroforestry residue potentials for the bioeconomy in the European Union. *J. Clean. Prod*, 176, 348, 2018.
32. Mojibayo, I., Samson, A.O., Johnson, O.Y., Joshusa, I.O.A.S.A., A preliminary investigation of cassava starch potentials as natural polymer in bioplastic production. *Am. J. Interdiscip Innov. Res.*, 02, 31, 2020.
33. Özdamar, E.G. and Murat, A.T.E.Ş., Rethinking sustainability: A research on starch based bioplastic. *J. Sustain. Constr. Mater. Technol.*, 3, 249, 2018.
34. Hamidon, N., Hamidun, M.B., Arish, N.A., Sunar, N.M., Ali, R., Hamid, H.A., Harun, H., Muhamad, M.S., Potential of production bioplastic from potato starch. *Sustain. Environ. Technol.*, 1, 115, 2018.

35. Asrofi, M., Sapuan, S.M., Ilyas, R.A., Ramesh, M., Characteristic of composite bioplastics from tapioca starch and sugarcane bagasse fiber: Effect of time duration of ultrasonication (Bath-Type). *Mater. Today Proc.*, 07, 254, 2020.
36. Jiménez-Rosado, M., Zarate-Ramírez, L.S., Romero, A., Bengoechea, C., Partal, P., Guerrero, A., Bioplastics based on wheat gluten processed by extrusion. *J. Clean. Prod.*, 239, 117994, 2019.
37. Govil, T., Wang, J., Samanta, D., David, A., Tripathi, A., Rauniyar, S., Salem, D.R., Sani, R.K., Lignocellulosic feedstock: A review of a sustainable platform for cleaner production of nature's plastics. *J. Clean. Prod.*, 270, 122521, 2020.
38. Brodin, M., Vallejos, M., Opedal, M.T., Area, M.C., Chinga-Carrasco, G., Lignocellulosics as sustainable resources for production of bioplastics—A review. *J. Clean. Prod.*, 162, 646, 2017.
39. Rahman, M.H. and Bhoi, P.R., An overview of non-biodegradable bioplastics. *J. Clean. Prod.*, 294, 126218, 2021.
40. Bhatia, S.K., Otari, S.V., Jeon, J.M., Gurav, R., Choi, Y.K., Bhatia, R.K., Pugazhendhi, A., Kumar, V., Banu, J.R., Yoon, J.J., Choi, K.Y., Biowaste-to-bioplastic (polyhydroxyalkanoates): Conversion technologies, strategies, challenges, and perspective. *Bioresour. Technol.*, 326, 124733, 2021.
41. Raza, Z.A., Abid, S., Banat, I.M., Polyhydroxyalkanoates: Characteristics, production, recent developments and applications. *Int. Biodeterior. Biodegrad.*, 126, 45, 2018.
42. Choi, S.Y., Rhie, M.N., Kim, H.T., Joo, J.C., Cho, I.J., Son, J., Jo, S.Y., Sohn, Y.J., Baritugo, K.-A., Pyo, J., Lee, Y., Lee, S.Y., Park, S.J., Metabolic engineering for the synthesis of polyesters: A 100-year journey from polyhydroxyalkanoates to non-natural microbial polyesters. *Metab. Eng.*, 58, 47, 2020.
43. Anjum, A., Zuber, M., Zia, K.M., Noreen, A., Anjum, M.N., Tabasum, S., Microbial production of polyhydroxyalkanoates (PHAs) and its copolymers: A review of recent advancements. *Int. J. Biol. Macromol.*, 89, 161, 2016.
44. Takahashi, R., Castilho, N., Silva, M., Miotto, M., Lima, A., Prospecting for marine bacteria for polyhydroxyalkanoate production on low-cost substrates. *Bioeng.*, 4, 60, 2017.
45. Beckstrom, B.D., Wilson, M.H., Crocker, M., Quinn, J.C., Bioplastic feedstock production from microalgae with fuel co-products: A techno-economic and life cycle impact assessment. *Algal Res.*, 46, 101769, 2020.
46. Rahman, A. and Miller, C.D., Microalgae as a source of bioplastics, in: *Algal green chemistry,* R.P. Rastogi, D. Madamwar, A. Pandey (Eds.), pp. 121–138, Elsevier, Oxford, United Kingdom, 2017.
47. Simonič, M. and Zemljič, F.L., Production of bioplastic material from algal biomass. *Chem. Ind. Chem. Eng. Q*, 27, 79, 2021.
48. Onen Cinar, S., Chong, Z.K., Kucuker, M.A., Wieczorek, N., Cengiz, U., Kuchta, K., Bioplastic production from microalgae: A review. *Int. J. Environ. Res. Public Health*, 17, 3842, 2020.
49. Thiruchelvi, R., Das, A., Sikdar, E., Bioplastics as better alternative to petro plastic. *Mater. Today: Proc.*, 37, 1634, 2021.
50. Iwata, T., Biodegradable and bio-based polymers: Future prospects of eco-friendly plastics. *Angew. Chem. Int. Ed.*, 54, 3210, 2015.
51. Llevot, A., Dannecker, P.K., von Czapiewski, M., Over, L.C., Söyler, Z., Meier, M.A., Renewability is not enough: Recent advances in the sustainable synthesis of biomass-derived monomers and polymers. *Chem. Eur. J.*, 22, 11510, 2016.
52. Zhang, D., del Rio-Chanona, E.A., Shah, N., Screening synthesis pathways for biomass-derived sustainable polymer production. *ACS Sustain. Chem. Eng.*, 5, 4388, 2017.
53. Yuan, Y., Yong, X., Zhang, H., Deng, J., Biobased microspheres consisting of poly (trans-anethole-co-maleic anhydride) prepared by precipitation polymerization and adsorption performance. *Eng. ACS Sustain. Chem.*, 4, 1446, 2016.

54. Yong, X. and Deng, J., Biomass trans-anethole-based heat-resistant copolymer microspheres: Preparation and thermostability. *Mater. Today Commun.*, 9, 60, 2016.
55. Raza, S., Yong, X., Yang, B., Xu, R., Deng, J., Biomass trans-anethole-based hollow polymer particles: Preparation and application as sustainable absorbent. *ACS Sustain. Chem. Eng.*, 5, 10011, 2017.
56. Zerpa, A., Pakzad, L., Fatehi, P., Hardwood kraft lignin-based hydrogels: Production and performance. *ACS omega*, 3, 8233, 2018.
57. Nitzsche, R., Gröngröft, A., Kraume, M., Separation of lignin from beech wood hydrolysate using polymeric resins and zeolites–Determination and application of adsorption isotherms. *Sep. Purif. Technol.*, 209, 491, 2019.
58. Ge, Y., Qin, L., Li, Z., Lignin microspheres: An effective and recyclable natural polymer-based adsorbent for lead ion removal. *Mater. Des.*, 95, 141, 2016.
59. González-López, M.E., Robledo-Ortíz, J.R., Rodrigue, D., Pérez-Fonseca, A.A., Highly porous lignin composites for dye removal in batch and continuous-flow systems. *Mater. Lett.*, 263, 127289, 2020.
60. Kwak, H.W., Lee, H., Lee, K.H., Surface-modified spherical lignin particles with superior Cr (VI) removal efficiency. *Chemosphere*, 239, 124733, 2020.
61. Zhang, Y., Ni, S., Wang, X., Zhang, W., Lagerquist, L., Qin, M., Willför, S., Xu, C., Fatehi, P., Ultrafast adsorption of heavy metal ions onto functionalized lignin-based hybrid magnetic nanoparticles. *Chem. Eng. J.*, 372, 82, 2019.
62. Wu, Z., Huang, W., Shan, X., Li, Z., Preparation of a porous graphene oxide/alkali lignin aerogel composite and its adsorption properties for methylene blue. *Int. J. Biol. Macromol.*, 143, 325, 2020.
63. Chen, F., Shahabadi, S.I.S., Zhou, D., Liu, W., Kong, J., Xu, J., Lu, X., Facile preparation of cross-linked lignin for efficient adsorption of dyes and heavy metal ions. *React. Funct. Polym.*, 143, 104336, 2019.
64. Guo, J., Liu, X., Zhang, X., Wu, J., Chai, C., Ma, D., Chen, Q., Xiang, D., Ge, W., Immobilized lignin peroxidase on Fe_3O_4@ SiO_2@ polydopamine nanoparticles for degradation of organic pollutants. *Int. J. Biol. Macromol.*, 138, 433, 2019.
65. Stanzione, M., Gargiulo, N., Caputo, D., Liguori, B., Cerruti, P., Amendola, E., Lavorgna, M., Buonocore, G.G., Peculiarities of vanillin release from amino-functionalized mesoporous silica embedded into biodegradable composites. *Eur. Polym. J.*, 89, 88, 2017.
66. Niradha Sachinthani, K.A., Eumelanin-inspired core derived from vanillin: A new building block for organic semiconductors. *Chem. Commun.*, 51, 2957, 2015.
67. Salmi-Mani, H., Terreros, G., Barroca-Aubry, N., Aymes-Chodur, C., Regeard, C., Roger, P., Poly (ethylene terephthalate) films modified by UV-induced surface graft polymerization of vanillin derived monomer for antibacterial activity. *Eur. Polym. J.*, 103, 51, 2018.
68. Zhao, D., Sun, J., Sun, B., Zhao, M., Zheng, F., Huang, M., Sun, X., Li, H., Intracellular antioxidant effect of vanillin, 4-methylguaiacol and 4-ethylguaiacol: Three components in Chinese Baijiu. *RSC Adv.*, 7, 46395, 2017.
69. Zhang, H., Zhao, R., Pan, M., Deng, J., Wu, Y., Biobased, porous poly (high internal phase emulsions): Prepared from biomass-derived vanillin and laurinol and applied as an oil adsorbent. *Ind. Eng. Chem. Res.*, 58, 5533, 2019.
70. Culita, D.C., Simonescu, C.M., Patescu, R.E., Dragne, M., Stanica, N., Oprea, O., Vanillin functionalized mesoporous silica coated magnetite nanoparticles for efficient removal of Pb (II) from water. *J. Solid State Chem.*, 238, 311, 2016.
71. Zhang, H., Yong, X., Zhou, J., Deng, J., Wu, Y., Biomass vanillin-derived polymeric microspheres containing functional aldehyde groups: Preparation, characterization, and application as adsorbent. *ACS Appl. Mater. Interfaces*, 8, 2753, 2016.

72. Zhang, H., Deng, J., Wu, Y., Biobased magnetic microspheres containing aldehyde groups: Constructed by vanillin-derived polymethacrylate/Fe_3O_4 and recycled in adsorbing amine. *ACS Sustain. Chem. Eng.*, 5, 658, 2017.
73. Sastri, V.R., Material requirements for plastics used in medical devices, in: *Plastics in medical devices*, 2nd, pp. 33–54, William Andrew Publishing, New York, USA, 2014, https://doi.org/10.1016/B978-1-4557-3201-2.00004-5.
74. Pagga, U., Beimborn, D.B., Yamamoto, M., Biodegradability and compostability of polymers-test methods and criteria for evaluation. *J. Environ. Polym. Degrad.*, 4, 173, 1996.
75. Samantaray, P.K., Little, A., Wemyss, A.M., Iacovidou, E., Wan, C., Design and control of compostability in synthetic biopolyesters. *ACS Sustain. Chem. Eng.*, 9, 9151, 2021.
76. Lawal, U. and Valapa, R.B., Bioplastics: An introduction to the role of eco-friendly alternative plastics in sustainable packaging, in: *Bio-Based Packaging: Material, Environmental and Economic Aspects*, Sapuan, S.M., Ilyas, R.A. (Eds.), pp. 319–334, John Wiley and Sons Ltd., USA, 2021.
77. Gadhave, R.V., Das, A., Mahanwar, P.A., Gadekar, P.T., Starch based bio-plastics: The future of sustainable packaging. *Open. J. Polym. Chem.*, 8, 21, 2018.
78. Jiménez, L., Mena, M.J., Prendiz, J., Salas, L., Vega-Baudrit, J., Poly-lactic acid (PLA) as a bioplastic and its possible applications in the food industry. *J. Food Sci. Nut*, 5, 048, 2019.
79. Roy, S. and Rhim, J.W., Preparation of bioactive functional poly (lactic acid)/curcumin composite film for food packaging application. *Int. J. Biol. Macromol.*, 162, 1780, 2020.
80. Mujtaba, M., Morsi, R.E., Kerch, G., Elsabee, M.Z., Kaya, M., Labidi, J., Khawar, K.M., Current advancements in chitosan-based film production for food technology: A review. *Int. J. Biol. Macromol.*, 121, 889, 2019.
81. Tian, B., Wang, J., Liu, Q., Liu, Y., Chen, D., Formation chitosan-based hydrogel film containing silicon for hops β-acids release as potential food packaging material. *Int. J. Biol. Macromol.*, 191, 288, 2021.
82. Klinmalai, P., Srisa, A., Laorenza, Y., Katekhong, W., Harnkarnsujarit, N., Antifungal and plasticization effects of carvacrol in biodegradable poly (lactic acid) and poly (butylene adipate terephthalate) blend films for bakery packaging. *LWT*, 152, 112356, 2021.
83. Bumbudsanpharoke, N., Wongphan, P., Promhuad, K., Leelaphiwat, P., Harnkarnsujarit, N., Morphology and permeability of bio-based poly (butylene adipate-co-terephthalate)(PBAT), poly (butylene succinate)(PBS) and linear low-density polyethylene (LLDPE) blend films control shelf-life of packaged bread. *Food Control*, 132, 108541, 2022.
84. Laorenza, Y. and Harnkarnsujarit, N., Carvacrol, citral and α-terpineol essential oil incorporated biodegradable films for functional active packaging of Pacific white shrimp. *Food Chem.*, 363, 130252, 2021.
85. Gisario, A., Aversa, C., Barletta, M., Natali, S., Veniali, F., Laser joining of aluminum film coated with vinylic resin and plastic/bioplastic films for applications in food packaging. *Opt. Laser Technol.*, 142, 107237, 2021.
86. Kongkaoroptham, P., Piroonpan, T., Pasanphan, W., Chitosan nanoparticles based on their derivatives as antioxidant and antibacterial additives for active bioplastic packaging. *Carbohydr. Polym.*, 257, 117610, 2021.
87. Laorenza, Y. and Harnkarnsujarit, N., Carvacrol, citral and α-terpineol essential oil incorporated biodegradable films for functional active packaging of Pacific white shrimp. *Food Chem.*, 363, 130252, 2021.
88. Aversa, C., Barletta, M., Gisario, A., Pizzi, E., Prati, R., Vesco, S., Corotating twin-screw extrusion of poly (lactic acid) PLA/poly (butylene succinate) PBS/micro-lamellar talc blends for extrusion blow molding of biobased bottles for alcoholic beverages. *J. Appl. Polym. Sci.*, 138, 5129, 2021.

89. Wangprasertkul, J., Siriwattanapong, R., Harnkarnsujarit, N., Antifungal packaging of sorbate and benzoate incorporated biodegradable films for fresh noodles. *Food Control*, 123, 107763, 2021.
90. Luchese, C.L., Rodrigues, R.B., Tessaro, I.C., Cassava starch-processing residue utilization for packaging development. *Int. J. Biol. Macromol.*, 183, 2238, 2021.
91. Chang, C.C., Trinh, B.M., Mekonnen, T.H., Robust multiphase and multilayer starch/polymer (TPS/PBAT) film with simultaneous oxygen/moisture barrier properties. *J. Colloid Interface Sci.*, 593, 290, 2021.
92. Oluwasina, O.O. and Awonyemi, I.O., Citrus Peel Extract Starch-Based Bioplastic: Effect of Extract Concentration on Packed Fish and Bioplastic Properties. *J. Polym. Environ.*, 29, 1706, 2021.
93. Iftime, M.M., Ailiesei, G.L., Ungureanu, E., Marin, L., Designing chitosan based eco-friendly multifunctional soil conditioner systems with urea controlled release and water retention. *Carbohydr. Polym.*, 223, 115040, 2019.
94. Rocha, I., Ma, Y., Souza-Alonso, P., Vosátka, M., Freitas, H., Oliveira, R.S., Seed coating: A tool for delivering beneficial microbes to agricultural crops. *Front. Plant Sci.*, 10, 1357, 2019.
95. De Gregorio, P.R., Michavila, G., Ricciardi Muller, L., de Souza Borges, C., Pomares, M.F., Saccol de Sá, E.L., Pereira, C., Vincent, P.A., Beneficial rhizobacteria immobilized in nanofibers for potential application as soybean seed bioinoculants. *PloS One*, 12, e0176930, 2017.
96. Jiménez-Rosado, M., Perez-Puyana, V., Guerrero, A., Romero, A., Bioplastic Matrices for Sustainable Agricultural and Horticultural Applications, in: *Bioplastics for sustainable development*, K.M. Roohi (Ed.), Springer, Singapore, 2021, https://doi.org/10.1007/978-981-16-1823-9-16.
97. Martucci, J.F. and Ruseckaite, R.A., Biodegradation behavior of three-layer sheets based on gelatin and poly (lactic acid) buried under indoor soil conditions. *Polym. Degrad. Stab.*, 116, 36, 2015.
98. Gull, N., Islam, A., Ghaffar, A., Jabeen, S., Iqbal, S.S., Khan, S.M., Khan, R.U., Hussain, N., Bilal, M., Development and characterization of chitosan and acrylic acid-based novel biodegradable polymeric films for soil conditioning. *Int. J. Biol. Macromol.*, 182, 950, 2021.
99. Kubavat, D., Trivedi, K., Vaghela, P., Prasad, K., Vijay Anand, G.K., Trivedi, H., Patidar, R., Chaudhari, J., Andhariya, B., Ghosh, A., Characterization of a chitosan-based sustained release nanofertilizer formulation used as a soil conditioner while simultaneously improving biomass production of Zea mays L. *Land Degrad. Dev.*, 31, 2734, 2020.
100. Thombare, N., Mishra, S., Shinde, R., Siddiqui, M.Z., Jha, U., Guar gum based hydrogel as controlled micronutrient delivery system: Mechanism and kinetics of boron release for agricultural applications. *Biopolymers*, 112, e23418, 2021.
101. Jiménez-Rosado, M., Perez-Puyana, V., Rubio-Valle, J.F., Guerrero, A., Romero, A., Evaluation of superabsorbent capacity of soy protein-based bioplastic matrices with incorporated fertilizer for crops. *J. Polym. Environ.*, 28, 2661, 2020.
102. Singh, A., Kar, A.K., Singh, D., Verma, R., Shraogi, N., Zehra, A., Gautam, K., Anbumani, S., Ghosh, D., Patnaik, S., pH-responsive eco-friendly chitosan modified cenosphere/alginate composite hydrogel beads as carrier for controlled release of Imidacloprid towards sustainable pest control. *Chem. Eng. Sci.*, 427, 131215, 2022.
103. Hidangmayum, A., Dwivedi, P., Katiyar, D., Hemantaranjan, A., Application of chitosan on plant responses with special reference to abiotic stress. *Physiol. Mol. Biol. Plants*, 25, 313, 2019.
104. Hendrawan, H., Khoerunnisa, F., Sonjaya, Y., Putri, A.D., Poly (vinyl alcohol)/glutaraldehyde/Premna oblongifolia merr extract hydrogel for controlled-release and water absorption application, in: *IOP Conference Series: Materials Science and Engineering*, vol. 509, p. 012048, 2019, April 2019.

105. Gonzalez Gomez, H., Ramirez Godina, F., Ortega Ortiz, H., Benavides Mendoza, A., Robledo Torres, V., Cabrera De la Fuente, M., Use of chitosan-PVA hydrogels with copper nanoparticles to improve the growth of grafted watermelon. *Molecules*, 22, 1031, 2017.
106. Nooeaid, P., Chuysinuan, P., Pitakdantham, W., Aryuwananon, D., Techasakul, S., Dechtrirat, D., Eco-friendly polyvinyl alcohol/polylactic acid core/shell structured fibers as controlled-release fertilizers for sustainable agriculture. *J. Polym. Environ.*, 29, 552, 2021.
107. Chiaregato, C.G. and Faez, R., Micronutrients encapsulation by starch as an enhanced efficiency fertilizer. *Carbohydr. Polym.*, 271, 118419, 2021.
108. de Vasconcelos, M.C., Gomes, R.F., Sousa, A.A., Moreira, F.J., Rodrigues, F.H., Fajardo, A.R., Neto, L.G.P., Superabsorbent hydrogel composite based on starch/rice husk ash as a soil conditioner in melon (cucumis melo L.) seedling culture. *J. Polym. Environ.*, 28, 131, 2020.
109. Di Martino, A., Khan, Y.A., Durpekova, S., Sedlarik, V., Elich, O., Cechmankova, J., Ecofriendly renewable hydrogels based on whey protein and for slow release of fertilizers and soil conditioning. *J. Clean. Prod.*, 285, 124848, 2021.
110. Abdel-Razek, M.K., Hemeid, N.M., Nader, H.R., Beneficial effect of some organic soil-conditioning agents for improving sandy soil productivity under sprinkler irrigation system. *Aust. J. Basic & Appl. Sci.*, 5, 12, 2011.
111. Haque, T., Chen, H., Ouyang, W., Martoni, C., Lawuyi, B., Urbanska, A.M., Prakash, S., *In vitro* study of alginate–chitosan microcapsules: An alternative to liver cell transplants for the treatment of liver failure. *Biotechnol. Lett.*, 27, 317, 2005.
112. Laurencin, C.T., Jiang, T., Kumbar, S.G., Nair, L.S., Biologically active chitosan systems for tissue engineering and regenerative medicine. *Curr. Top. Med. Chem.*, 8, 354, 2008.
113. Rezaei, F.S., Khorshidian, A., Beram, F.M., Derakhshani, A., Esmaeili, J., Barati, A., 3D printed chitosan/polycaprolactone scaffold for lung tissue engineering: Hope to be useful for COVID-19 studies. *RSC Adv.*, 11, 19508, 2021.
114. Ghilan, A., Chiriac, A.P., Nita, L.E., Rusu, A.G., Neamtu, I., Chiriac, V.M., Trends in 3D printing processes for biomedical field: Opportunities and challenges. *J. Polym. Environ.*, 28, 1345, 2020.
115. Jongsung, P., 3D-printed biodegradable polymeric stent integrated with a battery-less pressure sensor for biomedical applications. *19th International Conference on Solid-State Sensors. Actuat Microsyst (Transducers)*, pp. 47–50, 2017.
116. Minisy, I.M., Salahuddin, N.A., Ayad, M.M., *In vitro* release study of ketoprofen-loaded chitosan/polyaniline nanofibers. *Polym. Bull.*, 78, 5609, 2021.
117. Peng, B., Li, J., Wu, J., Chen, N., Chen, G., Preparation and performance of poly (vinyl alcohol)/ polylactic acid/hydroxyapatite composite scaffolds based on 3D printing. *J. Appl. Polym. Sci.*, 139, 5153, 2021.
118. Xu, L., Ma, F., Leung, F.K., Qin, C., Lu, W.W., Tang, B., Chitosan-strontium chondroitin sulfate scaffolds for reconstruction of bone defects in aged rats. *Carbohydr. Polym.*, 273, 118532, 2021.
119. Ye, X., Xia, Y., Wang, B., Kang, Z., Wu, B., Function on 3d printing poly(3 hydroxybutyrate-co-3-hydroxyvalerate)/calcium sulfate hemihydrate scaffold integrated chitosan hydrogel coating. *Chin. J. Tissue Eng. Res.*, 26, 1574, 2022.
120. Divband, B., Aghazadeh, M., Al-Qaim, Z.H., Samiei, M., Hussein, F.H., Shaabani, A., Sedghi, R., Bioactive chitosan biguanidine-based injectable hydrogels as a novel BMP-2 and VEGF carrier for osteogenesis of dental pulp stem cells. *Carbohydr. Polym.*, 273, 118589, 2021.
121. Mohd Roslan, M.R., Mohd Kamal, N.L., Abdul Khalid, M.F., Mohd Nasir, N.F., Cheng, E.M., Beh, C.Y., Tan, J.S., Mohamed, M.S., The state of starch/hydroxyapatite composite scaffold in bone tissue engineering with consideration for dielectric measurement as an alternative characterization technique. *Materials*, 14, 1960, 2021.

122. Vo, T.M.T., Mondal, S., Park, S., Choi, J., Bui, N.T., Oh, J., Rice starch coated iron oxide nanoparticles: A theranostic probe for photoacoustic imaging-guided photothermal cancer therapy. *Int. J. Biol. Macromol.*, 183, 55, 2021.
123. Shi, Z., Lan, G., Hu, E., Lu, F., Qian, P., Liu, J., Dai, F., Xie, R., Targeted delivery of hemostats to complex bleeding wounds with magnetic guidance for instant hemostasis. *Chem. Eng. Sci.*, 427, 130916, 2022.
124. Desai, S.K., Mondal, D., Bera, S., Polyurethane-functionalized starch nanocrystals as anti-tuberculosis drug carrier. *Sci. Rep.*, 11, 1, 2021.
125. Katsarov, J., Shindova, M., Lukova, P., Belcheva, A., Delattre, C., Pilicheva, B., Polysaccharide-based micro- and nanosized drug delivery systems for potential application in the pediatric dentistry. *Polymers*, 13, 3342, 2021.
126. Awadhiya, A., Tyeb, S., Rathore, K., Verma, V., Agarose bioplastic-based drug delivery system for surgical and wound dressings. *Eng. Life Sci.*, 17, 204, 2014.
127. Cosate de Andrade, M.F., Souza, P.M.S., Cavalett, O., Morales, A.R., Life cycle assessment of poly(lactic acid) (PLA): Comparison between chemical recycling, mechanical recycling and composting. *J. Polym. Environ.*, 24, 372, 2016.
128. Niaounakis, M., Chapter 16—Recycling, in: *Biopolymers: Processing and products*, vol. 481, Elsevier, Amsterdam, 2015.
129. Vink, E.T., Rábago, K.R., Glassner, D.A., Springs, B., O'Connor, R.P., Kolstad, J., Gruber, P.R., The sustainability of NatureWorks™ polylactide polymers and Ingeo™ polylactide fibers: An update of the future. *Macromol. Biosci.*, 4, 564, 2004.
130. P. Coszach, J.-C. Bogaert, J. Wilocq, Chemical Recycling of PLA by Hydrolysis. US 8431683 B2, 2013.
131. Hirao, K. and Ohara, H., Synthesis and recycle of poly (L-lactic acid) using microwave irradiation. *Polym. Rev. (Philadelphia, PA, U. S.)*, 51, 1–22, 2011.
132. Hirao, K., Nakatsuchi, Y., Ohara, H., Alcoholysis of Poly (l-lactic acid) under microwave irradiation. *Polym. Degrad. Stab.*, 95, 928, 2010.
133. Odelius, K., Kumar, S., Hakkarainen, M., Ghosh, A.K., Bhatnagar, N., Albertsson, A.C., Porosity and pore size regulate the degradation product profile of polylactide. *Biomacromolecules*, 12, 1258, 2011.
134. MacDonald, R.T., McCarthy, S.P., Gross, R.A., Enzymatic degradability of poly (lactide): Effects of chain stereochemistry and material crystallinity. *Macromolecules*, 29, 7356, 1996.
135. Li, S., Girard, A., Garreau, H., Vert, M., Enzymatic degradation of polylactide stereocopolymers with predominant D-lactyl contents. *Polym. Degrad. Stab.*, 71, 61, 2000.
136. Cigana, P., Favis, B.D., Jérôme, R., Diblock copolymers as emulsifying agents in polymer blends: Influence of molecular weight, architecture, and chemical composition. *J. Polym. Sci., Part B: Polym. Phys.*, 34, 1691, 1996.
137. Tokiwa, Y., Calabia, B.P., Ugwu, C.U., Aiba, S., Biodegradability of plastics. *Int. J. Mol. Sci.*, 10, 3722, 2009.
138. Chapman, J., Ismail, A.E., Dinu, C.Z., Industrial applications of enzymes: Recent advances, techniques, and outlooks. *Catalysts*, 8, 238, 2018.
139. Phomphrai, K., Pracha, S., Phonjanthuek, P., Pohmakotr, M., Facile alcoholysis of L-lactide catalysed by Group 1 and 2 metal complexes. *Dalton Trans.*, 23, 3048, 2008.
140. Bowmer, C.T., Hooftman, R.N., Hanstveit, A.O., Venderbosch, P.W.M., Van der Hoeven, N., The ecotoxicity and the biodegradability of lactic acid, alkyl lactate esters and lactate salts. *Chemosphere*, 37, 1317, 1998.
141. Bykowski, D., Grala, A., Sobota, P., Conversion of lactides into ethyl lactates and value-added products. *Tetrahedron Lett.*, 55, 5286, 2014.
142. L.D. Brake, Preparation of alkyl esters by depolymerization. US Patent US, 5264617, 1993.

143. Song, X., Zhang, X., Wang, H., Liu, F., Yu, S., Liu, S., Methanolysis of poly (lactic acid)(PLA) catalyzed by ionic liquids. *Polym. Degrad. Stab.*, 98, 2760, 2013.
144. Song, X., Wang, H., Zheng, X., Liu, F., Yu, S., Methanolysis of poly (lactic acid) using acidic functionalized ionic liquids as catalysts. *J. Appl. Polym. Sci.*, 131, 40817, 2014.
145. Nederberg, F., Connor, E.F., Glausser, T., Hedrick, J.L., Organocatalytic chain scission of poly (lactides): A general route to controlled molecular weight, functionality and macromolecular architecture. *Chem. Commun.*, 20, 2066, 2001.
146. Leibfarth, F.A., Moreno, N., Hawker, A.P., Shand, J.D., Transforming polylactide into value-added materials. *J. Polym. Sci., Part A: Polym. Chem.*, 50, 4814, 2012.
147. Hong, M. and Chen, E.Y.X., Chemically recyclable polymers: A circular economy approach to sustainability. *Green Chem.*, *19*, 3692, 2017.
148. Payne, J., McKeown, P., Jones, M.D., A circular economy approach to plastic waste. *Polym. Degrad. Stab.*, 165, 170, 2019.
149. Román-Ramírez, L.A., Mckeown, P., Jones, M.D., Wood, J., Poly (lactic acid) degradation into methyl lactate catalyzed by a well-defined Zn(II) complex. *ACS Catalysis*, 9, 409, 2018.
150. Fliedel, C., Vila-Viçosa, D., Calhorda, M.J., Dagorne, S., Avilés, T., Dinuclear zinc–N-heterocyclic carbene complexes for either the controlled ring-opening polymerization of lactide or the controlled degradation of polylactide under mild conditions. *Chem.Cat.Chem.*, 6, 1357, 2014.
151. Aoyagi, Y., Yamashita, K., Doi, Y., Thermal degradation of poly [(R)-3-hydroxybutyrate], poly [ε-caprolactone], and poly [(S)-lactide]. *Polym. Degrad. Stab.*, 76, 53, 2002.
152. Kulawska, M., Grzesik, M., Skrzypek, J., Kinetics of the esterification of crotonic acid with octyl, decyl, or dodecyl alcohol. *Chem. papers-slovak Acad. Sci.*, 57, 424, 2004.
153. Ute, K., Tarao, T., Nakao, S., Kitayama, T., Preparation and properties of disyndiotactic poly (alkyl crotonate)s. *Polymer*, 44, 7869, 2003.
154. Vu, D.H., Åkesson, D., Taherzadeh, M.J., Ferreira, J.A., Recycling strategies for polyhydroxyalkanoate-based waste materials: An overview. *Bioresour. Technol.*, 298, 122393, 2020.
155. Bátori, V., Åkesson, D., Zamani, A., Taherzadeh, M.J., Horváth, I.S., Anaerobic degradation of bioplastics: A review. *Waste Manage.*, 80, 406, 2018.
156. Siracusa, V., Rocculi, P., Romani, S., Dalla Rosa, M., Biodegradable polymers for food packaging: A review. *Trends Food Sci. Technol.*, *19*, 634, 2008.
157. Noda, I., Lindsey, S.B., Caraway, D., Nodax™ class PHA copolymers: Their properties and applications, in: *Plastics from Bacteria*, pp. 237–255, Springer, Berlin, Heidelberg, 2010.
158. Verlinden, R.A., Hill, D.J., Kenward, M.A., Williams, C.D., Radecka, I., Bacterial synthesis of biodegradable polyhydroxyalkanoates. *J. App. Microbiol.*, *102*, 1437, 2007.
159. Yagi, H., Ninomiya, F., Funabashi, M., Kunioka, M., Anaerobic biodegradation tests of poly (lactic acid) under mesophilic and thermophilic conditions using a new evaluation system for methane fermentation in anaerobic sludge. *Int. J. Mol. Sci.*, 10, 3824, 2009.
160. Yagi, H., Ninomiya, F., Funabashi, M., Kunioka, M., Anaerobic biodegradation of poly (lactic acid) film in anaerobic sludge. *J. Polym. Environ.*, 20, 673, 2012.

4

Modeling of Natural Fiber-Based Biocomposites

Fatima-Zahra Semlali Aouragh Hassani[1], Mounir El Achaby[1], Abou el Kacem Qaiss[2] and Rachid Bouhfid[2]*

[1]Materials Science, Energy and Nano-Engineering (MSN) Department, Mohammed VI Polytechnic University (UM6P), Hay Moulay Rachid, Benguerir, Morocco
[2]Composites and Nanocomposites Center (CNC), Moroccan Foundation for Advanced Science, Innovation and Research (MAScIR), Rabat Design Center, Rue Mohamed El Jazouli, Madinat El Irfane, Rabat, Morocco

Abstract
The globalization of markets, along with environmental concerns, has prompted several academic and private laboratories to undertake work on the development of innovative materials. Among these materials, we find "greener" and lighter materials, such as biocomposite materials, which are primarily made up of biopolymer matrices or biobased reinforcements (natural fibers) produced from renewable resources. To extend their use to some structural applications, it is important to predict their mechanical behavior before the production. However, the biocomposite final performance will depend on the choice of the matrix, the shaping, the interfacial adhesion, the fibers concentration, and aspect ratio, and above all on the fiber's orientation in the matrix. This chapter will first introduce the biocomposite materials and their different processing techniques. The parameters that determine their mechanical behavior will next be detailed, followed by the presentation of the main models for the biocomposites mechanical prediction based on the relationship between these parameters.

Keywords: Biocomposites, biopolymer, natural fibers, processing, mechanical properties, fiber orientation, modeling

4.1 Introduction

Thanks to their good mechanical properties and their lightness, composite materials are considered essential materials in various fields, such as the automotive, aerospace, aeronautics, naval, and sports industries [1, 2]. However, environmental issues contribute to a decrease in the use of materials derived from fossil resources. In addition, the movement toward reducing greenhouse gases and the regulations that are imposed to improve the efficiency of fuel exploitation, require lighter materials with a lower carbon footprint. So,

**Corresponding author*: r.bouhfid@mascir.ma

Inamuddin and Tariq Altalhi (eds.) Handbook of Bioplastics and Biocomposites Engineering Applications: 2nd Edition, (81–102) © 2023 Scrivener Publishing LLC

the usual composite materials tend to be replaced by "greener" and lighter materials, such as biocomposite materials [1, 2].

Biocomposite materials are composite materials made from animal- or plant-based reinforcements and resins. To be considered a biocomposite, one of the main elements of the composite material must be a renewable resource [3]. These materials are increasing market share in the composite market, not only because the price of oil has risen and various legislation have been enacted to encourage the synthetic composites recycling but also because the specific mechanical properties of these biocomposites are quite important [3, 4]. Much research focuses on improving the performance of these biocomposites to extend their use to applications where the constraints are critical. The final performance will depend on the choice of the matrix, the reinforcements concentration and orientation, the shaping and above all on their interaction, or the interface's resistance. The major obstacle to their use in some structural applications is the lack of their mechanical behavior knowledge [5].

In order to overcome this problem, it is imperative to understand and visualize the main physical mechanisms (damage, microstructural interaction, viscoelastic, and plastic deformation ...) in order to propose a modeling strategy, making it possible to predict the performance of these composites in order to optimize their future performance [4, 5].

According to the literature, numerous theoretical models have been established and optimized to predict the biocomposites mechanical properties depending on the interaction between parameters that influence their performance, such as fiber orientation [5, 6]. Two forms of orientation state can be identified: unidirectional and random. The simplest technique to predict the total material mechanical properties is to postulate that the fibers (Long or Short) are completely aligned in the parallel or perpendicular direction as done by Voigt, Reuss, Halpin–Tsai, and Hui–Shia. Contrariwise, others consider the relative contributions of both parallel and perpendicular models for their mechanical prediction, among them are the Hirsch model, the self-consistent approach based on the modified Hirsch model and Tsai–Pagano model [4–7]. This chapter will first introduce the biocomposites materials, in addition to detailing the main parameters affecting their mechanical performance. Also, different processing techniques will be presented according to the biocomposites nature. Finally, the natural fiber orientation (unidirectional or random) effect on the biocomposites mechanical properties is investigated by presenting various models that allow it.

4.2 Generality of Biocomposites

The environmental movement has been gaining traction for several decades. Indeed, the rise in pollution concerns of all kinds (ocean pollution, accumulation of solid waste, etc.) is forcing the scientific and industrial communities to find solutions [8]. The field of materials is not spared, and the development of these from renewable resources is the approach adopted. Then there is the topic of biocomposites. The latter are composites in which at least one of the phases is of biological origin. In this regard, three categories of biocomposites should be distinguished [9]:

- Biocomposites with synthetic matrix and natural reinforcement.
- Biocomposites with natural matrix and synthetic reinforcement.
- Biocomposites with natural matrix and reinforcement.

The last category is also called "Green biocomposites." Indeed, it makes it possible to avoid the use of fossil fuels while avoiding the accumulation of waste through its biodegradable nature. This type of material is, therefore, completely environmentally friendly. Furthermore, when the biocomposite is obtained by mixing several biopolymers or types of fibers, it is qualified as being hybrid [9, 10].

4.2.1 Natural Matrix

Due to the various oil crises, environmental awareness, and thanks to innovations in the field of biomaterials and their processability, biopolymers experienced a strong emergence. The latter being defined as any polymer directly synthesized by living organisms or whose monomer comes from renewable resources [11]. Figure 4.1 proposes a classification of biopolymers according to their production process.

Another classification is often encountered in the literature. This is done according to the source of the polymer. Therefore, biopolymers would fall into three categories [8]:

- Natural biopolymers, found directly in nature (Starch, protein, and collagen).
- Synthesized biopolymers, the monomer is obtained from biomass and then subjected to a polymerization reaction by conventional methods (poly lactic acid PLA).
- Microbial biopolymers, synthesized by microorganisms from different sources of sugar (polyhydroxyalkanoate (PHA) family).

In addition, resins derived from vegetable oils, such as soybean oil, are also classified as biopolymers. However, their production generally requires the presence of catalysts or crosslinking agents, such as anhydrides (synthetic additives). This could, therefore, alter the "organic" character of the resulting composite. For this purpose, the matrices are usually made of biothermoplastics.

The most common in the literature are starch, PLA, and PHA [12].

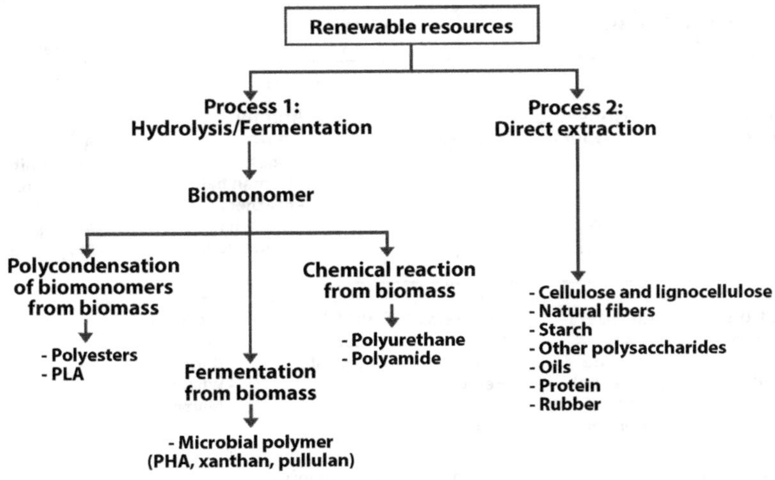

Figure 4.1 Classification of biopolymers according to the production process.

4.2.2 Natural Reinforcement

Fiber is an elongated object with an aspect ratio (L/D) greater than one. Although the term "fiber" has a lot of use, natural fibers can be thought of as fibrous materials, which are extracted from the environment. Thus, while the fibrous materials made from petrochemicals and carbon are definitely made from components found in the environment, they are not termed natural fibers because the basic elements are not fibers in their natural condition and are not collected directly from living organisms [13].

4.2.3 Natural Fiber Classification

Natural fibers can be handled by hand, with basic equipment, or through sophisticated industrial processes to make them useful for a specific purpose, but they are distinct from synthetic fibers. Natural fibers can be classified according to their origin: animal fibers, mineral fibers, and vegetable fibers (Figure 4.2) [14–16]. The first two are, respectively, harmful to health and limited from a performance point of view. It is the vegetable fibers that arouse strong enthusiasm [17]. These in turn are subdivided into two categories, depending on their origin (Figure 4.2) [15, 18]:

- Woody fibers, obtained from wood, whether coniferous or deciduous. They give the composite very good mechanical properties. However, their use would amount to causing a phenomenon of deforestation, which would go against the concept of biodesign.
- Nonwoody fibers, obtained from herbaceous plants. They represent an excellent alternative to synthetic fibers. Fibers can be distinguished from roots, leaves, seeds, stems, fruits, or even herbs, depending on the organ of the plant from which they are derived.

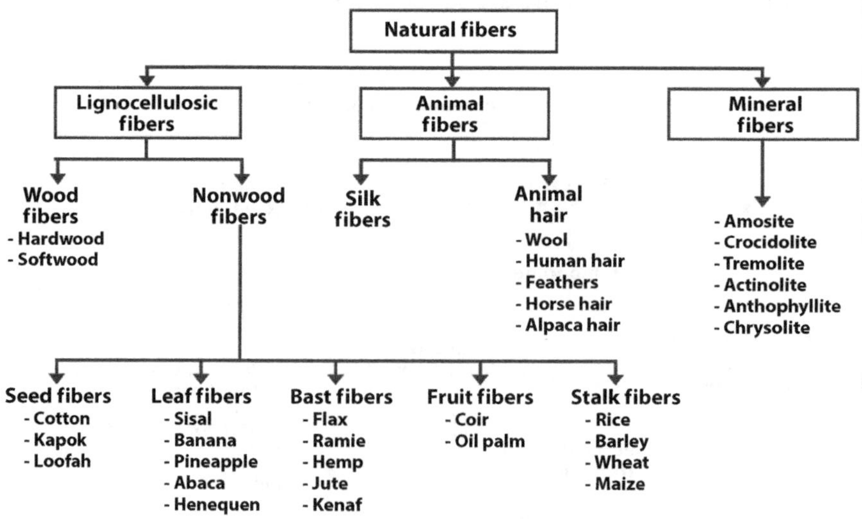

Figure 4.2 Classification of natural fibers according to their origin.

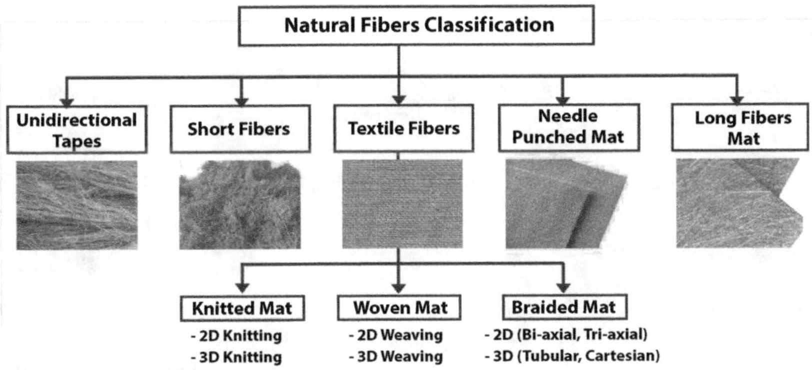

Figure 4.3 Classification of fibers architecture.

The fibers architecture describes the fibers arrangement in the matrix, which can be linear, two-dimensional, or three-dimensional. In linear reinforcements, the fibers can be continuous or discontinuous [16, 19–22]. Continuous fibers are used in wound, pultruded, or laminated composite structures in which the fibers orientation can be precisely controlled. While, discontinuous fibers are mixed directly with the matrix, for example in injection molded structures, or in combination with a binder to form a flat mat. In this case, the orientation of the fibers is little controlled [19–21, 23]. The two- and three-dimensional architectures are produced using textile processes, which allow good control of the fiber's orientation. Therefore, the fibers can be used to produce a variety of the complex shaped preforms in a relatively short time. Two-dimensional architecture is the most widely used mainly in layered structures. In such arrangements, the fibers disposition can be unidirectional, bidirectional, multidirectional, or random (Figure 4.3). However, random discontinuous tissues are the most developed [19–21].

4.2.4 Biocomposites Processing

Different process can be employed to manufacture biocomposite materials. These operations will allow the assembly of the different phases, to obtain a homogeneous material. The limited fiber dispersion within the matrix prevents fully homogeneous materials when producing biocomposite materials based on natural fibers [24–26]. As a result, the formulation channel and operating conditions will have a direct impact on the elaborate biocomposite's mechanical properties.

4.2.4.1 Extrusion and Injection

The most common plastics procedures are extrusion and injection molding (Figure 4.4). The extrusion is defined as the thermoplastic material transformation (powder or granulated) into a continuous product having a given profile, by forcing through a die the material previously softened by heat. The different constituents are introduced at one end of the extruder, they are then kneaded and forced by one or more screws in rotation, through

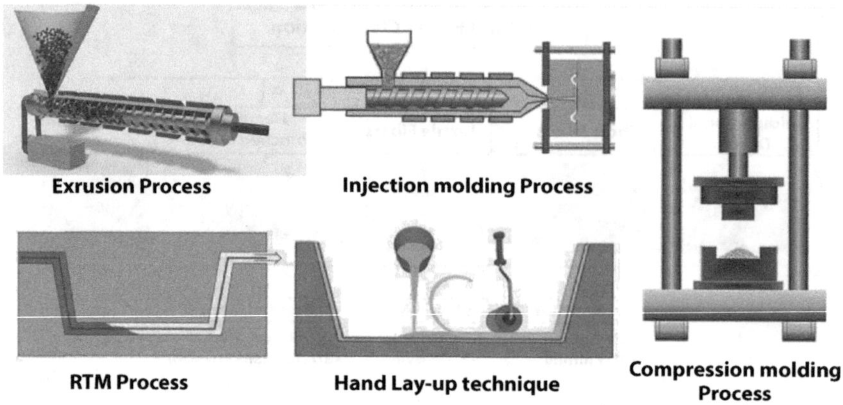

Figure 4.4 Biocomposites processing.

the heated tools gradually [27–31]. It comes out at the other end a warm and deformable extrude that must be maintained and cooled during its final formatting. The extrusion of biocomposites based on natural fiber optimizes the mixture of the reinforcement within the matrix while predominantly orienting the fibers in the flow direction. Injection molding allows for its part to directly obtain the material in the desired form. The assembly can be attached to the extruder's outlet to inject hot extruded directly into the heated mold without having to cool [27–31].

4.2.4.2 Compression Molding

In the compression molding process, a polymer solid mass is heated up or melted and forced to flow through hot mold surfaces that close to form a final shape (Figure 4.4). This method is widely used on an industrial scale, due to its high reproducibility and the short time required for its implementation. A mixing step can be carried out beforehand to optimize the homogeneity of the composite. To do this, the use of an extruder or a heated roller mixer is required [30, 31].

4.2.5 RTM-Resin Transfer Molding

The classic RTM process corresponds to resin injection molding in a rigid and closed mold. This process is relatively like standard plastic part injection processes except that the mold is not initially empty but contains a dry fiber reinforcement (Figure 4.4). In this process, the flow of the resin through the fibrous reinforcement is generally oriented in the plane of the layers of yarns (reinforcement). After complete crosslinking of the resin, the final part can be demolded [30, 31].

4.2.6 Hand Lay-Up Technique

The hand lay-up process uses a dry reinforcement and a liquid matrix (Figure 4.4). This is a very basic process, which does not allow parts with very high mechanical performance to be obtained. It is also subject to the experience of the operator who will determine the quality of the product. This will be explained by the description of the operating procedure. The shaping consists of possibly having a gel coat film, which will give the part the desired surface state but also its color. After solidification of the gel coat, the dry fibers are added. They can be preceded by a surface film to avoid marking the gel coat and transmitting the shape of the reinforcement to the surface. After each reinforcement fold, the resin is placed with a brush or roller. These steps are repeated until the final thickness of the part is reached. The part is polymerized at room temperature [30, 31].

4.3 Parameters Affecting the Biocomposites Properties

The biocomposites performance is governed by the synergy of several factors. Indeed, the natural fibers constituting the framework of the biocomposite, their aspect ratio, orientation, and dispersion will govern the material behavior during various stresses. Furthermore, the hydrophilic nature of these fibers versus the hydrophobic nature of the matrix will result in adhesion issues at the fiber-matrix interface, which must be considered during the design process [5, 32, 33].

4.3.1 Fiber's Aspect Ratio

The natural fibers aspect ratio depends directly on the extraction technique as well as the processing method. This improves the mechanical properties of the biocomposites the larger it is. As for the orientation, it will generate anisotropy of the material. This makes it possible to determine strong axes, for the material, along which the fibers will be stressed longitudinally (the most resistant direction of the fiber). Finally, a good dispersion of the fibers ensures a more homogeneous mixture and, therefore, a continuity in the distribution of the stress [27, 29, 34].

Also, it has been shown that the longer the fibers, the more they are able to absorb energy before breaking, and thus resulting to superior mechanical properties. In the majority of biocomposites, it is advantageous to have the maximum number of longer fibers to obtain the best possible mechanical properties, but this generally decreases the ease of implementation. Thwe *et al.* [35] have shown that in PP/bamboo-glass composites, increasing the length of bamboo fibers causes a slight decrease in tensile stress but an increase in Young's modulus. The interface between bamboo fiber and a PP matrix decreases as the diameter of the fiber bundles increases. The shape of the section of the latter becomes more and more complex as the diameter increases leading to imperfect contact between the matrix and the fiber in several places.

For polyester polymer reinforced with short sisal fibers Joseph *et al.* [36] demonstrated the existence of a critical length of these fibers required for good stress transfer. Below 55 mm, the composites tensile strength strongly increases with fiber's lengths, while it decreases when the length is between 55 and 75 mm. Concerning the modulus and elongation at break, they were unaffected by fibers length.

González *et al.* [37] studied, for PLA/bamboo composites, the influence of fiber size not only on mechanical properties but also on water absorption. Three sizes of fibers were used: bamboo wood flour (<150 μm), short bamboo fibers (between 150 and 500 μm) and large bamboo fibers (between 500 and 1000 μm). No significant change in modulus is observed, on the other hand, the stress and the strain at break are increased but only when switching from bamboo wood flour to short fibers, while the impact resistance increases by increasing fibers length. With regard to water absorption, an increase in the length of the fibers leads to a higher water uptake. As this is governed by a diffusion process, a greater length of fibers facilitates the penetration of water into the composite.

Soykeabkaew *et al.* [38] developed biocomposites based on starch reinforced with jute or flax by varying the fibers aspect ratio (L/D) using the same amount of fibers (10wt.%) in order to estimate the variation in physical and mechanical properties. It was found that the mechanical properties of the biocomposites strongly increase with increasing the fibers aspect ratio. This can be explained by the fact that fibers with a higher aspect ratio have a larger contact surface with the matrix. In addition, these authors also found a strong interaction between fiber and starch.

In summary of those works, there is, therefore, a critical length of fibers below which the mechanical performance of the material is not optimal. It depends a priori on the stress at break of the fiber and on the interfacial shear strength. However, the dimensions of the parts should also be considered with a view to optimizing the mechanical performance of staple fiber materials. The optimum fiber length in a staple fiber composite part, that is, that allowing maximum load transfer between fibers by shear, therefore, depends on both the properties of the fiber and the dimensions of the manufactured part.

4.3.2 Fiber/Matrix Interfacial Adhesion

The fiber-matrix interface is used to transfer the loads applied to the material to the reinforcement. Thus, to impart good mechanical strength, it is necessary to control the fiber/matrix adhesion. Poor adhesion is due to weak interactions between the two phases. To remedy this, various compatibilization methods, aimed at reducing the hydrophilic nature of the fiber, have been listed in the literature [22, 23, 39].

The incorporation of plant fibers into a matrix requires prior knowledge of the reaction of fibers in the polymer and also to optimize the matrix/fiber interfacial adhesion [16]. The strength of a fiber reinforced biocomposite depends on the degree to which an applied load is transmitted to the fibers. This is a function of the magnitude of the fiber/matrix interfacial bond [33]. The importance of the interactions at the interface between the polymer matrix and the fibers has been widely recognized in improving the hygrothermal stability and mechanical properties of natural composites and fiber-reinforced polymer. For this, different treatments have been studied to improve this bond [33]. A force applied directly to the matrix on the composite's surface is transferred to the adjacent fibers and propagates through the matrix and interface from fiber to fiber. If the adhesion is poor, forces are not distributed efficiently, which compromise the biocomposites mechanical properties [30, 33]. On the other hand, a good adhesion can enable the composite to endure stress even after some fibers have been cut since the force can be transmitted to the unbroken fibers. Several studies in the past have already investigated changes in the surface properties of cellulose fibers to improve their adhesion with the matrix and reduce moisture absorption. It has been demonstrated that correct fiber treatment can result in good compatibility with the polymer matrix, which increases composite performance [16, 33]. Various methods, such as plasma, alkaline, thermal treatments, grafting by copolymerization, treatment with silanes or even with other chemicals have made it possible to improve the fiber/matrix compatibility. All these studies show that the adhesion between the fibers and the matrix plays a key role in the biocomposites mechanical properties. Before performing a mechanical characterization of a composite, it is necessary to understand the chemical surface interactions between the various constituents [16, 33].

4.3.3 Fibers Orientation and Dispersion

4.3.3.1 Short Fibers Orientation

The fiber orientation in a coordinate system (X,Y,Z) is defined by the Euler angles θ and Ø, or by the Cartesian coordinates of a unit vector parallel to the fiber (Figure 4.5).

The angle θ, called angle of inclination, is the angle formed between the Z axis and the direction of the fiber, while the angle Ø, or azimuthal angle, is defined in the plane (X,Y), between the projection of the unit vector parallel to the fiber and the X axis [5]. The values of θ, therefore, vary between 0° and 180°, those of Ø between −90° and 90°. To characterize the fibers orientation in injected parts, the Z axis is generally defined normal to the material flow plane. Several mathematical representations making it possible to describe the orientation distribution and can be easily integrated into the equations of mechanics, used during modeling. The most widely used are orientation functions [5, 40], orientation distribution functions [41, 42] and orientation tensors [43]. When the orientation of the fibers is taken into consideration, the latter have the advantage of having the same tensor shape as representations of the elastic behavior of composites, which explain their use in micromechanical models [29, 44].

The filling phase is the main phase influencing the fibers orientation, there is, therefore, a coupling between the rheological behavior of the material during injection and the orientation of the reinforcements during the filling flow. During this phase, the polymer acts as a carrier fluid that transmits to the fibers a state of stress deformation, which is at the origin of orientation movements [5, 29]. A deformation applied to a fluid volume around a fiber, transmits forces to the contour of the fiber. This takes the form of a torque distributed along the length of the fiber. If the volume is stretched, contracted, or even sheared, then the fiber follows the strain and orients itself in the strain volume as shown in Figure 4.5.

Shear and elongation are the two types of stresses that will preferentially orient the fibers during the filling step, and there is competition between these effects [5].

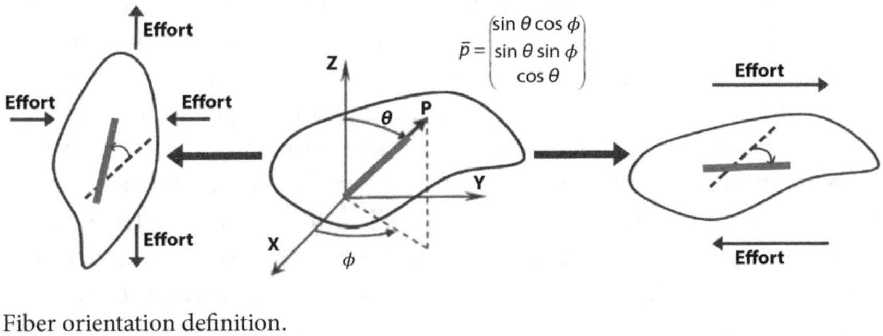

Figure 4.5 Fiber orientation definition.

4.3.3.2 *Fiber's Orientation in Simple Shear Flow*

The fibers in shear flow, in a Newtonian fluid, are animated by a movement of rotation, or of rocking on themselves, periodic out of the shear plane in which they find a new position of equilibrium between each "half-turn." The effects have been demonstrated experimentally and theoretically [5, 40]. The rotating speed of the fiber is not constant, it reaches its maximum when the fiber is perpendicular to the flow direction (Figure 4.6a). The fiber also rotates when it is immersed in a non-Newtonian fluid. However, the rotation period increases sharply with the pseudoplasticity of the fluid [5, 40].

Theoretically, it is possible to show that the frequency of this rotation decreases with the aspect ratio of the fiber but increases as a function of the shear rate in the flow. The expression of the rotation period, denoted T, as a function of the aspect ratio of the fibers and the shear rate in the carrier fluid is given in (Figure 4.6a). This ability of shear flow to orient fibers parallel to the direction of flow is observed for long staple fiber and thermosetting matrix materials. However, due to the fibers interactions, the fibers have little possibility to turn, and some of them bend [5].

Gibson and Williamson [45] associated the rheological behavior of long fibers in a thermosetting polymer with shear-like flow near the walls (inducing orientation of the fibers parallel to the direction of flow). In general, shear flows take place far from the injection threshold and upstream from the material front, thus promoting orientation of the fibers in the direction of the flow. Flows in thin parts also create high shear rates, which are the site of a quasi-unidirectional orientation in the flow direction [5]. At the scale of an injected part, most of the flow occurs through thin sections (in the sense that one dimension is much smaller than the other two). The flows are thus plane and more particularly generate shear stresses, the preferred orientation thus follows the direction of flow [5].

4.3.3.3 *Fiber's Orientation in Elongational Flow*

In elongational flow, the short fibers orient perpendicular or parallel to the direction of flow depending on whether the rate of elongation is negative or positive. Unlike shear flow, in elongation, there is a stable equilibrium position of the fiber, which orients in the direction of elongation as shown in Figure 4.6b [5]. The behavior of long fibers in elongational flow is more complicated. Some orient themselves perpendicular to the flow by bending slightly, while others fold over the same several times in order to orient themselves perpendicular to the flow in pieces. They then relax abruptly to form a helix (Figure 4.6b) [5].

In parts made by injecting short fibers into a thermoplastic polymer, the shear is practically zero at heart and the fibers there retain their initial orientation, transverse to the direction of flow [5]. Conversely, Gibson and Williamson [45] associated with the rheological behavior of long fibers in a thermosetting polymer a flow of elongation type at the heart of the parts (the direction of orientation of the fibers depends on the rate of the elongation in fluid). Near the injection threshold, the material front adopts a radially extending flow, which is akin to an elongational flow of the divergent type and causes a transverse orientation of the fibers, with respect to the direction of flow. This type of flow can also be linked to the presence of contractions, bifurcations, or extra thicknesses in the mold. One of the peculiarities of these geometric singularities, and the kind of flow they generate, is that they often result in a completely three-dimensional orientation. However, these are still localized effects on the scale of a room [5, 45].

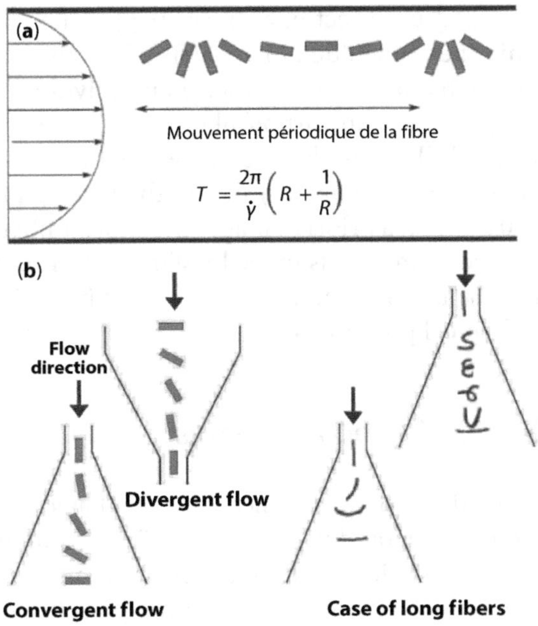

Figure 4.6 Isolated fiber motion in a Newtonian fluid in shear (a) and elongation flow (b).

Little work has been done on the observation of the long fiber's orientation in composite materials with a thermosetting matrix produced by injection. Blanc [46] nevertheless observed the orientation of the fibers in injected parallelepipedal pieces of unsaturated polyester reinforced with 20% by mass of glass fibers (between 13 and 25 mm long). Two types of population are thus observed: fibers, formed of filaments parallel to each other, of apparent lengths of up to 10 mm and independent unit filaments, in large quantity, not or only slightly curved, and of length lower because they break under the effect of the high stresses applied to them during flow.

The skin zone (core zone) of these materials is devoid of fibers, only unitary filaments are observed over a thickness representing 5% to 10% of the half-thickness of the part. This explains the good surface appearance of the parts [5, 46]. However, both fibers and filaments are very deformed and often loop, so it is impossible to assign them a direction of orientation that would correspond, in terms of mechanical property, to a preferential direction of reinforcement. It is also evidenced that the fibers are not strictly contained in planes parallel to the mean plane although they do not stray too far from it. On the other hand, Blanc does not manage to distinguish as clearly as in the case of thermoplastics the different orientation layers in the thickness [5, 46]. He nevertheless managed to demonstrate an organization in five layers of different orientation within the parallelepipedal specimens. The surface layers have no fibers. The layer at the heart of the part occupies about 80% of the thickness and corresponds to an orientation perpendicular to the direction of flow. Between the two, no privileged orientation is distinguished [5, 46].

In the case of parts produced by compression, the distribution of unidirectional fiber bands at the mesoscopic scale affects the value of the macroscopic modulus of elasticity more than the fiber/matrix properties at the microscopic scale [47]. The substructure responsible for the mechanical properties of the material is, therefore, the mesostructure (scale structure of the matrix coated unidirectional fiber strip) rather than the microstructure (scale structure of the unit fiber embedded in a matrix). It is, therefore, advisable to reason in terms of the orientation of bands of fibers rather than individual fibers. In discontinuous carbon fiber materials, the decrease in carbon fiber length results in a greater amount of insulated filaments and small filament clusters in the final part [47]. The explanation for these microstructural changes is undoubtedly due to the cutting of the fibers, prior to the implementation of the part, which a priori damages the shorter fibers more. Many fiber orientation defects are observed in materials made by stamping processes [5, 47]. Indeed, the latter generate folds undulations and crumpling of layers of fibers inside the spokes, locally leading to defects in mechanical properties.

4.4 Process Molding of Biocomposites

Studies of the fiber-reinforced biocomposite mechanical behavior mainly focus on the simplest structures: unidirectional or random (isotropic). These studies generally involve the use of the concept of homogenization [48]. This concept makes it possible to circumvent the problems associated with heterogeneities due to the presence of different constituents arranged quite often in a random manner. It is, therefore, essential to idealize the material by considering it as continuous (homogeneous) and, therefore, by averaging its properties at a certain scale, depending on the microstructure [5, 49].

Many micromechanical approaches have been developed to predict the mechanical properties of composite materials and their dependence on the properties of their constituents and their microstructures (Figure 4.7). In most cases, these approaches are reduced only to the calculation of the modulus of elasticity, assuming a perfect fiber/matrix adhesion and a regular arrangement of continuous or parallel fibers [27, 48].

The mechanical elasticity properties of a biocomposite depend essentially on the intrinsic properties of the constituents, the architecture of the reinforcement, the fiber/matrix interface, and the fiber aspect ratio. There are several theoretical approaches in the literature for determining the elastic modulus of the composite with varying precision [50]. All models consider the volume composition of the mixture as well as the fibers and matrix elastic modulus. These models are built based on the law of mixtures, applied in the case of unidirectional UD fibers.

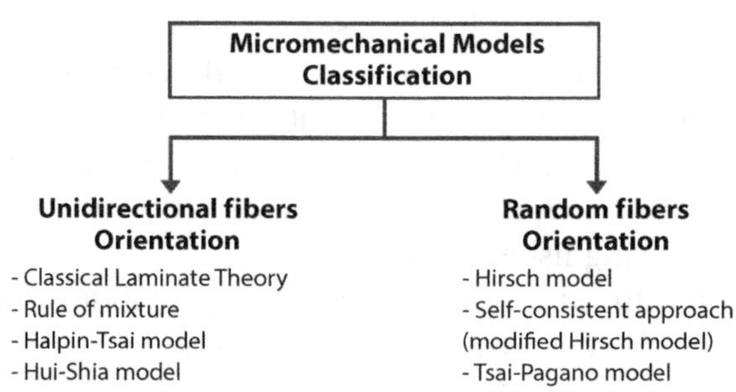

Figure 4.7 Micromechanical molding of biocomposites.

4.4.1 Unidirectional Fibers

4.4.1.1 Classical Laminate Theory

Using the classical laminate theory, the elastic behavior of a multilayer orthotropic material can be measured using the same constants that describe the mechanical behavior of a single layer. The latter is based on the kinematic assumptions of Kirchhoff-Love, employed for the study of structures of the type plate and curved shell. According to this theory, a straight line and perpendicular to the undeformed mean surface of the structure remains straight and perpendicular to this surface; not changing its shape in the direction of the thickness, it remains inextensible in this direction. The CLT theory do not consider the effects of transverse shear strains (γ_{xz} and γ_{yz}) and the transverse normal strain (ε_{zz}), and assumes that the folds are perfectly linked to each other, that the displacements according to the folds are continuous, that each ply's material is elastic, that the laminate is thin, and that deformations, displacements, and rotations are small [49, 51–54].

The assumed displacement field can be written as (Eq. 4.1):

$$u(x,y,z,t) = u_0(x,y,t) - zw_{0,x}(x,y,t)$$

$$v(x,y,z,t) = v_0(x,y,t) - zw_{0,y}(x,y,t)$$

$$w(x,y,t) = w_0(x,y,t) \tag{4.1}$$

where (x,y,z) are the Cartesian displacements, (x,y,0) is the undistorted medium plane, and t is the time. While u(x,y,z,t), v(x,y,z,t) and w(x,y,t) are, respectively, the displacements according to the directions of the x, y, and z axes of the laminate. u_0, v_0 and w_0 are, respectively, the displacements of a material in the reference surface. $w_{0,x}$ and $w_{0,y}$ are, respectively, the rotations around x and y.

4.4.1.2 Rule of Mixture

Several models have been created and optimized to predict the mechanical properties of biocomposite materials, according to the literature, but the simplest and most evident method is to assume that fiber orientations are ordered totally in parallel or in series as follows:

Voigt model (Rule of mixture)

This model was originally developed for composites that were reinforced with fibers. It applies in the case where the applied stress is parallel to the fibers. It also presupposes that the fibers' and matrix's deformations are the same [55, 56]. The value obtained is an excess value.

It expresses the Young's modulus of the biocomposite, by the following relation (Eq. 4.2):

$$E_V = V_f E_f + (1 - V_f) E_m \tag{4.2}$$

where E_V is the Voigt elastic modulus, E_f and E_m represent the fiber and the matrix elastic modulus, respectively, and V_f is the fiber volume fraction.

Reuss model (Inverse rule of mixture)

The series model (or Reuss model) gives the modulus of elasticity where the stresses in the two phases are equal. The Young's modulus in the transverse direction "E_R" is defined by (Eq. 4.3) [55, 56]:

$$E_R = \frac{E_f\ E_m}{V_f E_m + (1 - V_f) E_f} \tag{4.3}$$

Cox-Krenschel model (Modified rule of mixture)

Cox-Krenschel's modified the rule of mixture by proposing a new model defined by Equation 4.4. It considers the fibers contribution in the law of mixtures and the transfer of stress between the fibers and the matrix. The stress transfer is defined via the ratio G_m/E_f between the matrix shear modulus and the fiber modulus [57, 58].

$$E_c = \eta_l V_f E_f + (1 - V_f) E_m \tag{4.4}$$

The term η_l defined the fiber length efficiency factor or factor of the fiber length distribution and corresponds to the fibers contribution. It is usually between 0 and 1 and defined by Equation 4.5 [57, 58]:

$$\eta_l = 1 - \frac{\tanh(\beta.l/2)}{(\beta.l/2)} \quad \text{with } \beta = \frac{1}{r}\sqrt{\frac{E_m}{E_f.(1-v).\ln.\sqrt{\pi/4.V_f}}} \tag{4.5}$$

where β is the coefficient of stress concentration rate at the ends of the fibers, r is the mean radius of fiber, l is the fiber's length, and v is the Poison's ratio.

Fukuda and Kawata model (Modified rule of mixture)

Fukuda and Kawata developed a new method for calculating the efficiency factor of orientation η_o based on the fiber distribution in a matrix by employing a limit angle (α_o) (Eq. 4.6) [57, 58]:

Rectangular distribution:

$$\eta_o = \frac{\sin(\alpha_0)}{(\beta.l/2)} \cdot \left(\frac{3.v.\sin(\alpha_0)}{4.\alpha_0} + \frac{(1-v).\sin(3.\alpha_0)}{4.3.\alpha_0} \right) \tag{4.6a}$$

Sinusoidal distribution:

$$\eta_o = \frac{\pi^2}{16} \cdot \left(\frac{1}{\frac{\pi}{2}+\alpha_0} + \frac{1}{\frac{\pi}{2}-\alpha_0} \right) \cos(\alpha_0) . \left[\frac{3-v}{4} \cdot \left(\frac{1}{\frac{\pi}{2}+\alpha_0} + \frac{1}{\frac{\pi}{2}-\alpha_0} \right) \cos(\alpha_0) + \frac{1+v}{4} \cdot \left(\frac{1}{\frac{\pi}{2}+3\alpha_0} + \frac{1}{\frac{\pi}{2}-3\alpha_0} \right) \cos(3\alpha_0) \right] \tag{4.6b}$$

Triangular distribution:

$$\eta_o = 4.\frac{1-\cos(\alpha_0)}{\alpha_0^2}.\left(\frac{3-v}{4}.\frac{1-\cos(\alpha_0)}{\alpha_0^2}+\frac{1+v}{4}.\frac{1-\cos(3\alpha_0)}{9.\alpha_0^2}\right) \tag{4.6c}$$

4.4.1.3 Halpin-Tsai Model

It is employed in polymeric blend systems for both unidirectional continuous and discontinuous phases. It also takes into account fiber morphology by introducing an adjustable parameter denoted ξ that is expressed as follows (Eq. 4.7) [55, 57–59]:

$$E_{H-T/L} = E_m\left(\frac{1+\xi.\eta_L.V_f}{1-\eta_L.V_f}\right) \text{ with } \eta_L = \frac{(E_f/E_m)-1}{(E_f/E_m)+\xi};\ \xi = 2L/T \text{ or } \xi = 2L/D \tag{4.7a}$$

$$E_{H-T/T} = E_m\left(\frac{1+2.\eta_T.V_f}{1-\eta_T.V_f}\right) \text{ with } \eta_T = \frac{(E_f/E_m)-1}{\left(\frac{E_f}{E_m}\right)+2};\ \xi = 2L/T \text{ or } \xi = 2L/D \tag{4.7b}$$

It is worth noting that for $L\rightarrow 0$ and $\xi\rightarrow 0$, the H-T model is reduced Reuss model. Contrariwise, For $L\rightarrow\infty$ and $\xi\rightarrow\infty$ it is reduced to Voigt model [55, 57–59].

Where $E_{H-T/L}$ and $E_{H-T/T}$ denotes, respectively, the Halpin-Tsai longitudinal and transversal elastic modulus. L, T, and D are the fibers length, thickness, and diameter, respectively. ξ represents the fiber geometry, fiber distribution, and fiber loading conditions.

Modified Halpin-Tsai model

The particle packing fraction, which is the true (apparent) volume occupied by filler noted φ_{max}, was added to the Halpin–Tsai equation by Nielson. For the configuration of square, hexagonal, and random fibers, it can theoretically have a maximum value of 0.785, 0.907, and 0.82. Following Equation 4.8 [55]:

$$E_{m-H-T/L} = E_m\left(\frac{1+\zeta.\eta_L.V_f}{1-\eta_L.\Psi.V_f}\right) \text{ with } \Psi = 1+1\left(-\frac{\phi_{max}}{\phi^2_{max}}\right)V_f \tag{4.8}$$

where $E_{m-H-T/L}$ is the modified Halpin-Tsai longitudinal elastic modulus.

4.4.1.4 Hui-Shia Model

It was created for parallel and perpendicular fibers orientation based on their aspect ratio and is represented as follows (Eq. 4.9) [55, 56]:

$$E_{H-S/L} = E_m \left(1 - \frac{V_f}{\zeta}\right)^{-1} \tag{4.9a}$$

$$E_{H-S/T} = E_m \left(1 - \frac{V_f}{4}\left(\frac{1}{\zeta} + \frac{3}{\zeta + J}\right)\right)^{-1} \tag{4.9b}$$

where

$$\xi = V_f + \frac{E_m}{E_f + E_m} + 3(1 - V_f)\left[\frac{(1-g)\alpha^2 - \frac{g}{2}}{\alpha^2 - 1}\right] \tag{4.9c}$$

$$g = 1 - \frac{\ln(2\alpha) - 1}{\alpha^2} \text{ with } \alpha \geq 1 \tag{4.9d}$$

$$J = (1 - V_f)\left[\frac{3(\alpha^2 + 0.25)g - 2\alpha^2}{\alpha^2 - 1}\right] \tag{4.9e}$$

$E_{H\text{-}S/L}$ and $E_{H\text{-}S/T}$ are, respectively, the Hui-Shia longitudinal and transversal elastic modulus.

4.4.2 Random Fibers

Biocomposites with random reinforcements are composites reinforced with fibers (long or short) oriented randomly, and the mechanical properties are predicted by accounting for the relative contributions of both longitudinal and transversal approaches. In the case of a mat fold, it is considered as a stack made up of an infinite number of unidirectional folds, which can take all possible directions. This analogy can only be made if the fibers are long compared to the critical length (length necessary to transfer the charges between the matrix and the fibers without their release).

4.4.2.1 Hirsch Model

It accounts for the impacts of fiber orientation and stress concentration (fiber ends) by including into its equation (Eq. 4.10) an empirical parameter "x" ($0 < x < 1$) which is also the combination of Voigt and Reuss models. A value of $x = 0.5$ has been proposed for a random fiber orientation [58–60].

$$E_H = x\,E_V + (1 - x)E_R \tag{4.10}$$

where E_H is the Hirsch elastic modulus.

4.4.2.2 *Self-Consistent Approach (Modified Hirsch Model)*

To estimate the mechanical characteristics of biocomposites while taking reinforcement content and orientation into account, Semlali *et al.* [29] proposed a self-consistent technique based on a modified version of Hirsch model (Eq. 4.11). Thus, the empirical parameter (x) in the Hirsch model was taken proportional to the fiber orientation factors (α) and ranging from 0 to 1. The self-consistent method is based on the iterative insertion of a fiber orientation family; as a result, the first category of fibers was included in the matrix to form an equivalent homogeneous material, which was then integrated into an infinite composite.

$$E_h = \alpha\,E_V + (1 - \alpha)E_R \tag{4.11}$$

where E_h is the modified Hirsch elastic modulus.

4.4.2.3 *Tsai-Pagano Model*

It assumes that the matrix filler orientation is random, that the dispersion is good, and that the fiber-matrix interfacial adhesion is perfect. The length and diameter of the reinforcing fibers are also provided. It uses the Voigt and Reuss models' contributions of 3/8 and 5/8, respectively, according to (Eq. 4.12) [57, 58, 61].

$$E_{T-P} = \frac{3}{8}E_V + \frac{5}{8}E_R \tag{4.12a}$$

$$E_{T-P} = \frac{3}{8}E_{H-T/L} + \frac{5}{8}E_{H-T/T} \tag{4.12b}$$

where E_{T-P}, E_V, and E_R are the Tsai-pagan, Voigt and Reuss elastic modulus, respectively.

4.5 Conclusion

Environmental awareness of plastics' permanence in nature, as well as the limitation of fossil fuels on which they depend, has led researchers to seek other materials that are more environmentally friendly. These are from natural and biodegradable origins. Nonetheless, the hydrophobic polymers and hydrophilic reinforcements have distinct characteristics, resulting in surfaces with reduced properties. Thus, to develop biocomposites with improved performance, it is imperative to understand the main parameters affecting the materials mechanical properties. Indeed, increasing the fiber length and aspect ratio to an optimum value has been found to significantly improve mechanical properties. Furthermore, fiber/

matrix interfacial adhesion is important for biocomposites' performance. As a result, it is essential to consider the chemical surface interactions among the various ingredients. Above all, the fibers' orientation (unidirectional and random) field during processing greatly impacts the composite performances. By evaluating the interaction between all those parameters, theoretical models have been developed and optimized to predict the mechanical properties of biocomposites before manufacture. This will allow the production of biocomposites with high quality, reduce the production cost, and avoid the lengthy experimental testing.

References

1. Mahmud, S., Hasan, K.M.F., Jahid, M.A., Mohiuddin, K., Zhang, R., Zhu, J., *Comprehensive review on plant fiber-reinforced polymeric biocomposites*, Springer, US, 2021.
2. Essabir, H., Bensalah, M.O., Rodrigue, D., Bouhfid, R., Qaiss, A.E.K., Biocomposites based on Argan nut shell and a polymer matrix: Effect of filler content and coupling agent. *Carbohydr. Polym.*, 143, 70, 2016.
3. Avérous, L. and Le Digabel, F., Properties of biocomposites based on lignocellulosic fillers. *Carbohydr. Polym.*, 66, 480, 2006.
4. Liu, B., Feng, X., Zhang, S.M., The effective Young's modulus of composites beyond the Voigt estimation due to the Poisson effect. *Compos. Sci. Technol.*, 69, 2198, 2009.
5. Semlali Aouragh Hassani, F., Ouarhim, W., Zari, N., Bouhfid, R., Qaiss, A., Kumar, K., Paulo Davim, J., Natural fiber-based biocomposites, biodegradable composites materials, in: *Manufacturing and Engineering. De Gruyter*, 1st, pp. 49–79, 2019.
6. Sun, C.T. and Vaidya, R.S., Prediction of composite properties from a representative volume element. *Compos. Sci. Technol.*, 56, 171, 1996.
7. Chivrac, F., Gueguen, O., Pollet, E., Ahzi, S., Makradi, A., Averous, L., Micromechanical modeling and characterization of the effective properties in starch-based nano-biocomposites. *Acta Biomater.*, 4, 1707, 2008.
8. Gurunathan, T., Mohanty, S., Nayak, S.K., A review of the recent developments in biocomposites based on natural fibres and their application perspectives. *Compos. Part A*, 77, 1, 2015.
9. Akampumuza, O., Wambua, P.M., Ahmed, A., Li, W., Qin, X., Review of the applications of biocomposites in the automotive industry. *Polym. Polym. Compos.*, 38, 101, 2016.
10. John, M.J. and Thomas, S., Biofibres and biocomposites. *Carbohydr. Polym.*, 71, 343, 2008.
11. Smith, A.M., Moxon, S., Morris, G. Biopolymers as wound healing materials, in: *Wound Healing Biomaterials: Volume 2: Functional Biomaterials*, Ågren, M. (Ed.), Woodhead Publishing, 2016.
12. Koronis, G., Silva, A., Fontul, M., Green composites: A review of adequate materials for automotive applications. *Compos. Part B Eng.*, 44, 120, 2013.
13. Stokke, D.D., Wu, Q., Han, G., *Wood and natural fibers composites: An overview*, 1st, John Wiley & Sons, 2013.
14. Wallenberger, F.T. and Weston, N.E., *Natural fibers, plastics and composites*, Springer Science & Business Media, New York, 2004.
15. Ramamoorthy, S.K., Skrifvars, M., Persson, A., A review of natural fibers used in biocomposites: Plant, animal and regenerated cellulose fibers. *Polym. Rev.*, 55, 107, 2015.
16. Semlali Aouragh Hassani, F., Chakchak, H., El Achaby, M., Bouhfid, R., Qaiss, A.E.K., Date palm fiber composites, in: *Date palm fiber composites: Processing, properties and applications*, 1st, pp. 75–91, Springer, Singapore, 2020.

17. Raza, Z.A., Abid, S., Banat, I.M., Polyhydroxyalkanoates: Characteristics, production, recent developments and applications. *Int. Biodeterior. Biodegrad.*, 126, 45, 2018.
18. Fuqua, M.A., Huo, S., Ulven, C.A., Natural fiber reinforced composites. *Polym. Rev.*, 52, 259, 2012.
19. Ahmad, F., Yuvaraj, N., Bajpai, P.K., Effect of reinforcement architecture on the macroscopic mechanical properties of fiberous polymer composites: A review. *Polym. Compos.*, 41, 2518, 2020.
20. Goutianos, S., Peijs, T., Nystrom, B., Skrifvars, M., Development of flax fibre based textile reinforcements for composite applications. *Appl. Compos. Mater.*, 13, 199, 2006.
21. Mallick, P.K., *Fibre-reinforced composites: Materials, manufacturing and design*, 3rd, CRC Press, United States of America, 2007.
22. Semlali Aouragh Hassani, F.Z., El Bourakadi, K., Merghoub, N., Qaiss, A., Bouhfid, R., Effect of chitosan/modified montmorillonite coating on the antibacterial and mechanical properties of date palm fiber trays. *Int. J. Biol. Macromol.*, 148, 316, 2020.
23. Semlali Aouragh Hassani, F.Z., Ouarhim, W., Raji, M., Mekhzoum, M.E.M., Bensalah, M.O., Essabir, H., Rodrigue, D., Bouhfid, R., Qaiss, A., N-silylated benzothiazolium dye as a coupling agent for polylactic acid/date palm fiber bio-composites. *J. Polym. Environ.*, 27, 2974, 2019.
24. Sengupta, R., Chakraborty, S., Bandyopadhyay, S., Dasgupta, S., Mukhopadhyay, R., Auddy, K., Deuri, A.S., A short review on rubber/clay nanocomposites with emphasis on mechanical properties. *Engineering*, 47, 21, 2007.
25. Ouarhim, W., Semlali Aouragh Hassani, F., Qaiss, A., Bouhfid, R., Sabu Thomas, S., Nithin Chandran, C., Rheology of polymer nanocomposites, in: *Rheology of polymer blends and nanocomposites theory, modelling and applications*, 1st, pp. 73–96, Elsevier, 2019.
26. Semlali Aouragh Hassani, F.Z., Ouarhim, W., Bouhfid, R., Qaiss, A. el kacem, Effect of mold design on the fiber orientation in the case of injection molding: Experiment and simulation, in: *Polym. compos. manuf. gruyter*, 1st, vol. 25, Kaushik Kumar, J. and Davim, P. (Eds.), 2020.
27. Semlali Aouragh Hassani, F., Achaby, M., Bensalah, M.O., Rodrigue, D., Bouhfid, R., Qaiss, A., Injection molding of short fiber thermoplastic bio-composites: Prediction of the fiber orientation. *J. Compos. Mater.*, 54, 1, 2020.
28. Ouarhim, W., Semlali Aouragh Hassani, F-Z., Bouhfid, R., Qaiss, A., Numerical, experimental and simulation study of natural fiber-based composites on injection molding, in: *Polymers and Composites Manufacturing*, De Gruyter, Boston, 2020.
29. Semlali Aouragh Hassani, F., Ouarhim, W., Zari, N., Bensalah, M.O., Rodrigue, D., Bouhid, R., Qaiss, A., Injection molding of short coir fiber polypropylene biocomposites: Prediction of the mold filling Phase. *Polym. Compos.*, 40, 4042, 2019.
30. Semlali Aouragh Hassani, F., Bouhfid, R., Qaiss, A., Recent advances in the fabrication of hybrid natural fiber composites, in: *Hybrid natural fiber composites: Material formulations, processing, characterization, properties and engineering applications*, 1st, pp. 113–129, Woodhead Publishing, 2021.
31. Semlali Aouragh Hassani, F., El Achaby, M., Bouhfid, R., Qaiss, A.E.K., Modeling for the process and the prediction of the thermal and mechanical behavior for the biopolymers and bio-composites, in: *Biopolymers and biocomposites from agro-waste for packaging applications*, 1st, p. 271, Woodhead Publishing, 2021.
32. Bhat, A.H., Dasan, Y.K., Khan, I., Jawaid, M., M.J.M.S.S., and Alothman, O.Y., Green Biocomposites: Design and applications, pp. 69–100, Springer, Cham, 2017.
33. Semlali Aouragh Hassani, F., Ouarhim, W., Kassab, Z., Achaby, M.E.L., Bouhfid, R., Qaiss, A., Mechanical characterization of cellulose nanoparticle-based advanced materials, in: *Cellulose nanoparticles: Synthesis and manufacturing*, Kumar Thakur, V., Frollini, E., Scott, J. (Eds.), pp. 298–322, 2021.

34. Kabir, M.M., Wang, H., Lau, K.T., Cardona, F., Chemical treatments on plant-based natural fibre reinforced polymer composites: An overview. *Compos. Part B Eng.*, 43, 2883, 2012.
35. Thwe, M.M. and Liao, K., Effects of environmental aging on the mechanical properties of bamboo-glass fiber reinforced polymer matrix hybrid composites. *Compos. - Part A Appl. Sci. Manuf.*, 33, 43, 2002.
36. Joseph, K., Medeiros, E.S., Carvalho, L.H., Compósitos de matriz poliéster reforçados por fibras curtas de sisal. *Polímeros*, 9, 136, 1999.
37. González, D., Santos, V., Carlos, J., Manufacture of fibrous reinforcements for biocomposites and hemicellulosic oligomers from bamboo. *Chem. Eng. J.*, 167, 278, 2011.
38. Soykeabkaew, N., Supaphol, P., Rujiravanit, R., Preparation and characterization of jute- and flax-reinforced starch-based composite foams. *Carbohydr. Polym.*, 58, 53, 2004.
39. Fowler, P.A., Hughes, J.M., Elias, R.M., Biocomposites: Technology, environmental credentials and market forces. *J. Sci. Food Agric.*, 86, 1781, 2006.
40. Vincent, M. and Agassant, J.F., Experimental study and calculations of short glass fiber orientation in a center gated molded disc. *Polym. Compos.*, 7, 76, 1986.
41. Advani, S.G., The use of tensors to describe and predict fiber orientation in short fiber composites. *J. Rheol. (N. Y. N. Y)*, 31, 751, 1987.
42. Verleye, V. and Couniot, F.D., A., Numerical prediction of fibre orientation in complex injection-moulded parts. *Trans. Eng. Sci.*, 4, 303, 1994.
43. Hand, G.L., A theory of anisotropic fluids. *J. Fluid Mech.*, 13, 33, 1961.
44. Semlali Aouragh Hassani, F., Ouarhim, W., Bensalah, M.O., Essabir, H., Rodrigue, D., Bouhfid, R., Qaiss, A., Mechanical properties prediction of polypropylene/short coir fibers composites using a self-consistent approach. *Polym. Compos.*, 40, 1919, 2019.
45. Gibson, A.G. and Williamson, G.A., Shear and extensional flow of reinforced plastics in injection molding. I. Effects of temperature and shear rate with bulk molding compound. *Polym. Eng. Sci.*, 25, 968, 1985.
46. Blanc, R., Agassant, J.F., Vincent, M., Injection molding of unsaturated polyester compounds. *Polym. Eng. Sci.*, 32, 1440, 1993.
47. Feraboli, P., Peitso, E., Deleo, F., Cleveland, T., Stickler, P.B., Characterization of prepreg-based discontinuous carbon fiber/epoxy systems. *J. Reinf. Plast. Compos.*, 28, 1191, 2009.
48. Cox, H.L., The elasticity and strength of paper and other fibrous materials. *Br. J. Appl. Phys.*, 72, 72, 1952.
49. Semlali Aouragh Hassani, F., Bouhfid, R., Qaiss, A.E.K., Jawaid, M., Hamdan, A., Thariq Hameed Sultan, M., Modeling of damage evaluation and failure of laminated composite materials, in: *Structural Health Monitoring System for Synthetic, Hybrid and Natural Fiber Composites*, 1st, pp. 101–125, Springer, Singapore, 2021.
50. Kalaprasad, G., Joseph, K., Thomas, S., Pavithran, C., Theoretical modelling of tensile properties of short sisal fibre-reinforced low-density polyethylene composites. *J. Mater. Sci.*, 32, 4261, 1997.
51. Di Sciuva, M., An improved shear-deformation theory for moderately thick multilayered anisotropic shells and plates. *J. Appl. Mech. Trans. ASME*, 54, 589, 1987.
52. Robbins, D.H. and Reddy, J.N., Modelling of thick composites using a layerwise laminate theory. *Int. J. Numer. Methods Eng.*, 36, 655, 1993.
53. Casavola, C., Cazzato, A., Moramarco, V., Pappalettere, C., Orthotropic mechanical properties of fused deposition modelling parts described by classical laminate theory. *Mater. Des.*, 90, 453, 2016.
54. Wei-zang, C., Qian, H., Wei, F., 3-D Numerical study on the bending of symmetric composite laminates. *Appl. Math. Mech.*, 15, 1, 1994.

55. Krishnan, K.A., Anjana, R., George, K.E., Effect of alkali-resistant glass fiber on polypropylene/polystyrene blends: Modeling and characterization. *Polym. Compos.*, 37, 2, 398–406, 2016.
56. Hu, H. and Onyebueke, L., and Abatan, A., Characterizing and modeling mechanical properties of nanocomposites - Review and evaluation. *J. Miner. Mater. Charact. Eng.*, 9, 275, 2010.
57. Delgado-Aguilar, M., Julián, F., Tarrés, Q., Méndez, J.A., Mutjé, F.X.E., P., Bio composite from bleached pine fibers reinforced polylactic acid as a replacement of glass fiber reinforced polypropylene, macro and micro-mechanics of the Young's modulus. *Compos. Part B Eng.*, 125, 203, 2017.
58. Oliver-Ortega, H., Granda, L.A., Espinach b, F.X., Delgado-Aguilar a, M., Duranc, J., P.M. a, A., Stiffness of bio-based polyamide 11 reinforced with softwood stone ground-wood fibres as an alternative to polypropylene- glass fibre composites. *Eur. Polym. J.*, 84, 481, 2016.
59. Gupta, R., Sulaiman, N., Gupta, A., An improved algorithm for prediction of Young's modulus of wood plastic composites. *Sci. Res. Essays*, 8, 649, 2013.
60. Sawpan, M.A., Pickering, K.L., Fernyhough, A., Hemp fibre reinforced poly(lactic acid) composites. *Adv. Mater. Process. IV*, 29, 337, 2007.
61. López, J.P., Mutjé, P., Pèlach, M.À., Mansouri, N.-E. El, Boufi, S., Vilaseca, F., Analysis of the tensile modulus of polypropylene composites reinforced with stone groundwood fibers. *BioResources*, 7, 1310, 2012.

5

Process Modeling in Biocomposites

Joy Hoskeri H.[1*], Nivedita Pujari S.[1] and Arun K. Shettar[2]

[1]*Department of Bioinformatics, Karnataka State Akkamahadevi Women's University, Jnana Shakti Campus, Vijayapur, Karnataka, India*

[2]*Department of Applied Genetics, Karnatak University, Dharwad, Karnataka, India*

Abstract

Biocomposites are biodegradable substrate materials made up of natural fibers. Fibers obtained from plant and animals are termed as natural fibers. Biocomposites are manufactured by reinforcing two different natural fibers for developing potential composite material. Natural fibers used in reinforcement mainly involve feathers, silk, and wool, which are known to be animal fibers, whereas plant fibers include jute, kenaf, bamboo fiber, and coconut fiber. These natural fibers are reinforced with biopolymers like polylactic acid (PLA), polyglycolic acid (PGA), and polyvinyl alcohol (PVA). Biocomposites reinforced with various natural fibers are modeled using different moulding techniques, such as injection moulding, compression moulding, and extrusion method. Each of modeling techniques possesses unique mechanical and physical abilities, such as biocompatibility, moisture absorbing, and biodegradability. Also, different formulations of biocomposites can be manufactured using natural fibers for development of biocomposites with improved composite properties. Formulations may be prepared by adding different agents like additive agent and coupling agent. Potential biocomposites have showed promising implications in pharmaceuticals, cosmetic, food packaging, tissue engineering, and agriculture. Currently, various types of biocomposite models can be developed through altering the formulation, composite modeling technique and based on use of natural fiber reinforced. Current chapter provides detail insight on different process modeling technique that can be used in the development of potential biocomposite.

***Keywords*:** Natural fibers, biocomposites, biocomposite models, moulding, and formulations

5.1 Introduction

Biopolymers are one among the most widely used composite materials in the development of different substrate materials with diversified applications. Biocomposites are made by reinforcing two or more different natural polymers for improved composite properties. One of the

**Corresponding author*: joybioinfo@gmail.com

Inamuddin and Tariq Altalhi (eds.) Handbook of Bioplastics and Biocomposites Engineering Applications: 2nd Edition, (103–120) © 2023 Scrivener Publishing LLC

constituent sources of natural polymers or biopolymers is fibers. Fibers extracted from animals and plants are referred to as natural fibers. With growing concern about environment, eco-friendly alternatives for synthetic polymers are being studied for sustainable usage. Therefore biocomposites are biodegradable natural polymers isolated from natural fibers. Plant fibers mainly include cellulose, lignin, hemicelluloses, and pectin, which can be easily isolated from natural fibers like cotton, jute, bamboo, palm, and kneaf. Animal fibers includes chitin, chitosan, gelatin, collagen, keratin, silk sourced from insect shell, crustacean shell, muscle tissue, wool, body hair, and silk worm cocoons. Biocomposites produced by such conventional natural fibers possess several applications; for example, animal fibers like silk are huge source of fiber for textile industry. Likewise, plant fibers like jute, kneaf, and bamboo contribute in food packaging industries. Apart from this natural fibers are also extensively used as drug delivery systems, in cosmetic industries and also in agriculture. Biocomposite are manufactured by reinforcing two different natural fibers into a polymer matrix. Polymer matrix used for producing biocomposites mainly involves polylactic acid, polyvinyl alcohol, polyglycolic acid etc. Natural fibers are used as fillers for development of potential biocomposite. Biocomposites are produced by three widely used methods namely compression moulding, where two or more natural fibers are compressed at different temperatures to produce biocomposite sheets. Injection moulding is carried out by filling natural fibers as micro or nanofillers in polymer matrix to produce biocomposites. Extrusion method is usually coupled with injection moulding, where a desired natural fiber is reinforced into the polymer matrix. Further, various formulations of biocomposites can be produced through these methods by changing composite ratios or by adding certain additives to produce biocomposites with improved composite properties. Depending on various parameters, such as type of natural fiber used, method of processing technique employed or depending on type of formulations, biocomposites can be classified into various types. A detail information of biocomposites and also their varied form with different applications are discussed in this chapter.

5.2 Biopolymer Composites

With increase in global pollution and impact on environment, synthetic polymers are appearing to pose threat in future years. Composites are materials made with composition of two elements and biocomposites are made up of materials that are resourced from natural living systems [1]. Biopolymer composites are made up of natural, biodegradable materials that are either synthesized chemically or produced biologically by living organisms [2]. Therefore Biopolymers are combination of one or two polymers synthesized naturally by reinforcement of natural fibers obtained from plants and animals. Some of the most widely used biopolymers include polylactic acid, polyhydroxy alkanotes, and cellulose esters [3]. Biopolymer composites also exhibit advantages due to their excellent physiochemical properties (discussed in later part of the chapter). Biopolymer composites are synthesized by various reinforcement techniques where natural fibers are reinforced into a polymer matrix to obtain their respective composites.

5.2.1 Natural Fiber-Based Biopolymer Composites

Natural fibers are extensively used biopolymers as polymer matrices for reinforcement purposes in order to provide stable dimensional binding materials used to immobilize the fibers in correct position [4]. Classification of natural fibers can include animal fibers and plant fibers. Animal fibers material mainly include bird feather, silk and animal hair (Figure 5.1). Due to excellent their physiochemical properties animal fibers are considered to be a potential reinforcement material in biopolymer composites [5]. Plant fibers include cellulose, lignin, pectin, and hemicelluloses. In comparison with synthetic fibers, natural plant fibers are of greater demand, since fibers obtained from plants are both renewable and eco-friendly. Thus, plant fibers are considered to be potential reinforcing materials for the development of biopolymer composites [6]. A variety of natural fibers and their sources are enlisted in Table 5.1.

5.2.2 Applications of Biopolymer Composites

Biopolymer composites developed by reinforcement of natural fibers have gained much interest in recent years due to wide range of applications. In comparison to synthetic or

Figure 5.1 Natural fibers from animal and plant origin.

Table 5.1 Different types of natural fibers and their sources.

Natural fiber	Fibers	Animal/Plant	Application	Reference
Animal	Wool	Sheep, rabbit, goats, and alpaca	Textile industries	[4, 5]
	Body hair	Horse, camel and goat	Fabrication	
	Feathers	Birds	Microbial corrosion-resistant	
	Silk	Spider, butterfly, and silkworm	Textile industries fabrication, tissue engineering	
Plant	Kenaf fiber	*Hibiscus cannabinus* L.	Application in injection moulding	[6]
	Flax fiber	Flax (*Linum usitatissimum*)	Alternatives for reinforcement used for packaging, and in concretes and polymers.	[7]
	Bast fiber	Dicotyledonous plants	Reinforcement materials for paper and textile industries. And also in construction materials.	[8]
	Jute fiber	*Corchorus capsularis*	Reinforcement Materials for polymers	[9]
	Ramie fiber	*Boehmeria nivea*	Textile industries	[9]
	Abaca fiber	Banana plant (*Musa textilis*)	Best alternative for glass fibers in automobile industries.	[8]

(*Continued*)

Table 5.1 Different types of natural fibers and their sources. (*Continued*)

Natural fiber	Fibers	Animal/Plant	Application	Reference
	Coconut fiber	Coconut (*Cocus nucifera*)	Fibers produced with reinforcement of coir are used in manufacture of cushions.	[10]
	Bamboo fiber	Plants belonging to Bambusoideae family	Employed in production of hand-made paper and carpentry works.	[6]

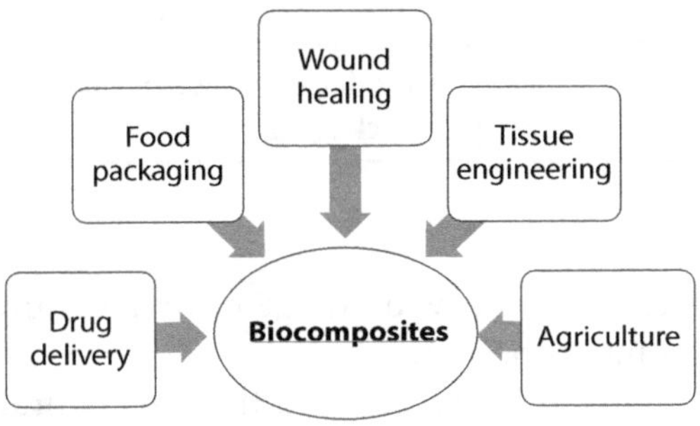

Figure 5.2 Applications of biocomposites.

artificial fibers, naturally obtained fiber-based reinforced composites are known to avail lower density and therefore exhibit varied applications, including medical implants, tissue engineering, drug deliver, textile, automobiles, packaging etc [11] (Figure 5.2). Some of the well-known applications of biopolymer composites are mentioned below in Table 5.2.

5.2.3 Properties of Biopolymer Composites

Biopolymer composites are influenced by various environmental factors, which include temperature, humidity, moisture, and radiation [20]. Therefore to understand the efficiency of biopolymer composites depends on their durability studies. Biodegradability is the outmost important feature of biopolymer composites. Microbial degradation determines biodegradability of biocomposites, natural fiber reinforced biocomposite polymers with polylactic acid and polyhdroxubutaroyl is biodegradable by anerobic bacteria and are also capable of undergoing enzymatic degradation [21]. Biodegradability also varies depending on the method employed for processing and modeling the composites, hence, it is important to select right method for manufacturing biocomposites [20]. Temperature stability is also important aspect while designing biopolymer composites. Since biocomposites are

Table 5.2 Biopolymers and applications.

Biopolymer composite	**Application**	**Reference**
Polylactic acid (PLA) biopolymers with kenaf and jute	Good substitute for polymers used in tissue engineering, agriculture, personal care and for petroleum based polymers.	[12–14]
Poly-3-hydroxy butyrate (PHB)	Food packaging	[15]
Bamboo fiber reinforced PLA	Replacement for aluminum alloys and epoxy composites.	[16]
Nano fibers	Wound healing and drug delivery systems, release of vitamins/drug to target tissues.	[17]
Potato starch-wheat gluten composite	Development of bio-based plastics	[18]
Polyvalanic acid–chitosan	Drug delivery systems and food packaging	[19]

used in a variety of applications with wide range of temperature, thus, thermal stability becomes one of the important properties.

Investigations on biocomposites have revealed that the natural fiber reinforcement can improve the thermal insulation properties of biopolymers where the ability to withstand higher temperatures is exhibited. A polylactic acid composite reinforced with natural fibers, such as wheat or rice husk or wood fibers, have shown decreased initial degradation temperature [21]. Exposure to moisture can negatively affect the thermal properties of biocomposites where it can cause reduced durability, and exposure to moisture leads to delamination [22]. However, polymers developed by reinforcement of natural polymers like starch and cellulose are known to exhibit potential moisture absorption in comparison with synthetic polymers [20]. Resistance to radiation is another important aspect for developing biocomposites. Degradation by UV light can be avoided by binding chemicals to plant cell wall polymers, which reduces degradation of lignin, also addition of polymers helps in keeping polymer matrix structure intact [23].

5.3 Classification of Biocomposites

Biocomposites have broad range classification based of methods employed in manufacturing, source of origin, and also based on their applications. Biocomposites, as discussed above, are biodegradable and hence are eco-friendly. As they are made by reinforcing two materials they are classified into different types. Some of the well-studied biocomposites are discussed here.

5.3.1 PLA Biocomposites

Polylactic acid (PLA) is a renewable polymer found in human body. Because of its biocidal activity and its ability to hydrolyze different surface by producing lactic acid makes its best alternative for petroleum-based polymers that are used in biomedical, tissue engineering, personal care and cosmetic industries [24]. Studies are done for reinforcement of PLA with different natural fibers like cellulose and chitin for developing biocomposites with excellent physiochemical properties [25]. For example, PLA reinforced with potato pulp biocomposites is a potential composite for the use in food packaging applications [26]. Reinforcement of chicken feather fiber with polylactic acid is known to be a potential biocomposite used in biomedical applications [27]. Therefore, PLA-based biocomposites can serve as a potential biocomposite in comparison with other synthetic composites.

5.3.2 Nanobiocomposites

Nanocomposites are made by incorporating small amount of fillers into the matrix of polymer and the fillers used are often nanosized, which provides improved polymeric properties. For biomedical and tissue engineering applications nanofillers are added into biopolymers for development of nanobiocomposites, which can be biocompatible to form stiffer material. Natural fibers derived from animal and plants are also used as natural fillers [28]. Some of the examples include cellulose nanofillers [29], nanofillers extracted from banana [30], and lignin nanofillers.

5.3.3 Hybrid Biocomposites

These composites are produced through mixing more than two or two different natural fibers in to the matrix of polymers. Studies are done to figure out best combination of natural fibers to obtain maximum benefits and to reduce negative aspects [31]. Combining different natural fibers is reported to enhance tensile properties of composites [32]. For example, PLA reinforced with cellulose, abaca, and jute fibers showed improved rigidity and strength [33]. Also, PLA composites developed by reinforcing sisal and banana fibers using chemical modifications showed improved physiochemical properties mainly, including high tensile strength and flexibility [34].

5.3.4 Natural Fiber-Based Composites

In recent years, there is increase employment of natural fibers mainly jute, flax, hemp, and sisal for producing eco-friendly biocomposites. A material that can be used as a composite in any medical application has to meet several important criteria of an ideal biocomposite. Among which, one of the fundamental property is biocompatibility, so that the composite used will not have negative effect on the host tissue. Therefore, nonbiocompatible matrix can easily be substituted by traditional biocompatible matrix or composites. Natural fibers are classified based on their origin, such as plant or animal, or type of plant or animal. Plant-based natural fibers generally include lignocelluloses and mainly include cellulose, lignin, and hemicelluloses sourced from jute, flax, and sisal. On the other focus, animal fibers include silk, spider, and wool. A detailed information on the different types of natural fibers is given in Table 5.1.

5.4 Process Modeling of Biocomposite Models

Biopolymers are reinforced with two or more natural polymers based on the need and applications. The reinforcement process is done using several process modeling techniques to obtain improved properties of biopolymer composites. Compression moulding, injection moulding, and extrusion are the three widely used process modeling techniques.

5.4.1 Compression Moulding

Among the several modeling techniques, compression moulding is considered to be most frequently used method. In this method, specific design of the natural fiber that will be used for reinforcement is required. In order to obtain optimum composite products, here it is important to maintain suitable parameters, such as moulding temperature, heating time, and pressure, which has great influence on its mechanical properties [35]. Compression moulding is performed in a processing machine, which fabricates natural fibers into different shapes. Several layers of fabrics are prepared to achieve a minimum composite with thickness of 3 mm. Further, these composites are processed by pressing into a stainless steel container or also in a teflon-coated container to produce a composite sheet of desired dimension. The composites are then moulded with a hot pressing machine with a moulding temperature ranging from 180°C to 200°C and a moulding pressure between 30 and 70 bars for about 5 to 10 minutes [36]. Demonstration of compressing moulding technique is described in Figure 5.3. Compress modeled biocomposites based with flax and PLA are known to be used for several applications [36]. PLA biocomposites produced by compress moulding using jute, flax, or cotton fibers resulted in potential composites with significant reduction of production time, which in return reduces production cost of biocomposites [37].

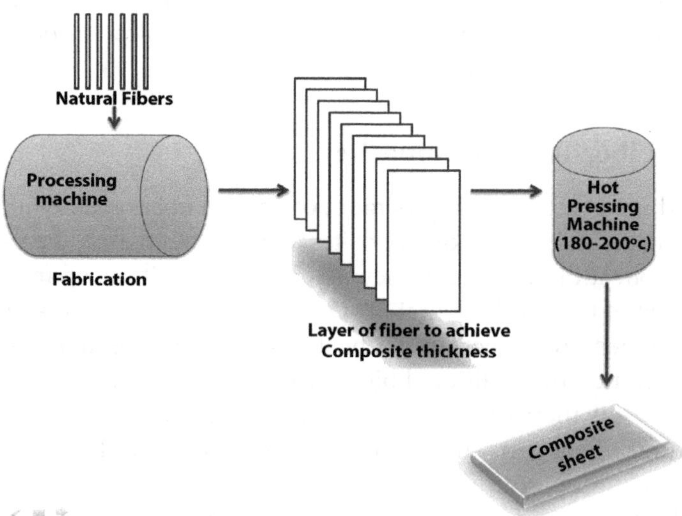

Figure 5.3 Compression modeling of natural fibers.

5.4.2 Injection Moulding

Injection moulding is mainly recognized for rapid production rates. It is known that operational conditions used and quantity of fiber used in injection moulding will affect properties of composite materials [38, 39]. However, research on addressing the limitations of injection moulding parameters has lead to production of enhanced properties in biocomposites. Injection moulding of biocomposites is carried out using specialized machines called screw injection moulding machines, which is used to develop specimens for testing strength. Specimens produced can be obtained of desired dimensions and thickness. Due to the sensitivity of biocomposite components to higher temperatures, usually low temperatures are preferred for processing. An injection moulding system has different units, such as feed zone (30°C), heating zone (variable temperature, 125–160°C), and injection nozzle (155°C). Also an optimum injection pressure 120 MPa and flow rate of polymer of 20 cm^3/s is maintained to ensure proper moulding. Packing pressure up to 120 to 80 MPa is maintained to prevent uneven filling of mould spaces occurring at the lower values of these parameters [40]. A detailed description of injection moulding is described in Figure 5.4. Several studies are reported on injection modeled biocomposites for their efficient composite properties. Injection moulded biocomposites from PLA and wheat bran are reported to show significant physiochemical properties that are efficient in their use as biocomposites [40]. Starch-based biopolymer composites developed via injection moulding with cellulose fibers are also studied to be efficient biocomposites [41]. Flax a natural fiber is reinforced with polyethylene via injection moulding to produce biocomposites. The biocomposite is known to show good thermostable properties at different temperatures testing [42]. Cross linked biocomposites made up of gelatin, produced by injection moulding caused crosslinking, thus, changing protein structure, which further showed decrease in solubility, density and swelling. These biocomposites also reported with good mechanical properties [43]. Biocomposites made up of albumin produced by injection moulding using chitosan served as physiochemical modifier, which helped in improving physiological properties of biocomposite [44]. A combination of crayfish flour and polycaprolactone were efficiently used to develop a two step process: mixing with glycerol as plasticizer and injection moulding of glycerol, crayfish flour, and polycaprolactone blends. Remarkable enhancement in mechanical properties was found in modified biocomposites [45].

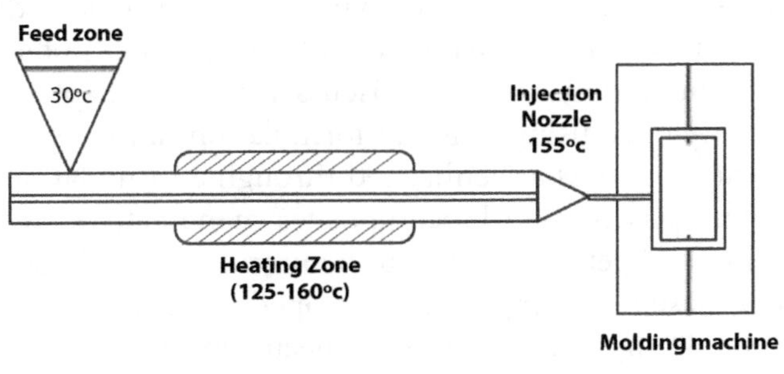

Figure 5.4 Industrial setup of injection moulding.

5.4.3 Extrusion Method

Extrusion methods are widely used for reinforcement of biopolymers in manufacturing of biocomposites. This method is usually coupled with injection moulding or any other method for increasing the potential of the product. Twin-screw extrusion method is a high shear process that provides better fiber dispersion. Improved mechanical properties were observed when polypropylene and cellulose of wood were reinforced together through the twin-screw extruder method [46]. However, extrusion method is also known as a process with certain drawbacks, such as fiber breakage [46]. Some studies have also been reported to show reduction in the diameter and length of bamboo fibers developed via extrusion and injection molding [47]. Extremely shortened flax-based biodegradable composites can also be produced by extrusion method [48]. Extrusion moulding is executed by dissolving fibers extracted from biological sources in organic solvents, such as chloroform using soxhlet extraction. Later images of fibers can be captured with binocular magnifiers to determine fiber size. Images can be analyzed using special software, such as Archimed 4.0 software, such software determines number of fibers with the help of high end resolution camera. Length and diameter of each of the fibers can later be measured manually with imaging software like Image J software [49]. A detailed description of extrusion method is described below in Figure 5.5. Other examples of biocomposites developed using extrusion molding involves PLA and sisal fiber biocomposites, these composites are developed by direct and indirect extrusion moulding for improved mechanical properties, which reported to show enhanced mechanical properties when compared to injection moulded biocomposites [50]. Further, twin screw extrusion moulding is also efficient in reducing cellulose fibers into micro or nanosize materials, thus, capable of developing bionanocomposites. Biocomposites developed using extrusion process with addition of cellulose fillers using twin screw moulding are reported to show improved mechanical properties and dispersion properties [51]. Polylactic acid–based *Paulownia elongata* biocomposites produced via extrusion molding reported to show PLA as an efficient biofiller, thus, reporting to have improved physical and mechanical properties [52].

5.5 Formulation of Models

Biocomposites are developed by reinforcement of two or more natural polymers. For a biocomposite to be able to achieve standard parameters of conventional composite different formulations are used. Use of different formulations makes biocomposites to meet standard physical and mechanical properties and makes them efficient for their use in different applications. Various formulations are studied so far to develop potential biocomposites. Some of the well-known formulations are discussed in this chapter. Natural fibers like flax and epoxy were developed using different formulation ratios reported variable tensile properties. Mechanical properties are enhanced through a formulation that is rich in flax [53]. Calcium orthophosphate-based biocomposites manufactured with varied formulations of collagen, chitin and gelatin appears to be promising candidates for clinical applications [54]. Chitosan-based biocomposites developed by formulation of 2% chitosan, 4% calcium carbonate and 4% hydroxyapatite have been reported to act as a potential bone adhesive [55]. Similarly, wood fiber biocomposites prepared with different formulations of

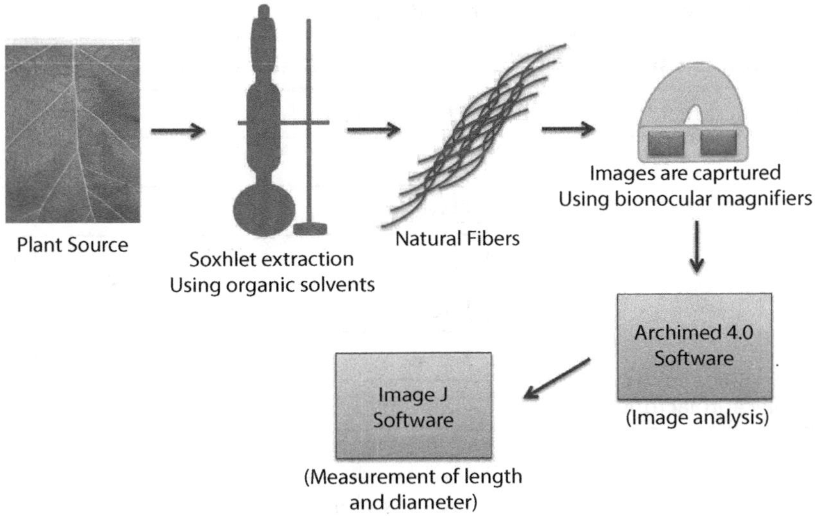

Figure 5.5 Extrusion moulding of natural fibers.

polylactide, thermoplastic starch blends, and two different variety of wood fiber (hard and soft) showed varied composite properties. Among which, polylactide and wood fiber formulation is reported to exhibit improved tensile strength and compatibility [56]. Addition of coupling agent like maleic acid in the formulation of polylactide-sisal–based biocomposites are known to have lesser hydrothermal aging then compared to other formulations that were prepared without a coupling agent [57]. Nanobiocomposite films prepared by reinforcement of polylactic acid with different formulation containing different concentration of montomorillonite and thymol show variation in its mechanical properties (Figure 5.6). Where formulation containing montomorillonite show better mechanical abilities and lower oxygen transmission, therefore biocomposites made by using different combinations of formulations could be potential packaging agents that can also be used as antioxidants [58].

5.5.1 Types of Model

As discussed above, biocomposites models can be classified based on different parameters, which mainly involves type of natural fiber used, method of processing and also depending on the type of formulation employed. Biocomposite models are developed for their benefits in various fields, which include pharmaceuticals, agriculture, food packaging, and cosmetic industries. Comprehensive information of different types of biocomposite models are discussed in Table 5.3.

5.6 Conclusion

Biocomposites have received considerable interest in recent years with increase in global pollution rate due to synthetic composite polymers. With the search of an alternative over the synthetic polymers, biopolymers have made their way into different fields of research.

Table 5.3 Different types of biocomposite models.

SL. no.	Biocomposite	Formulation/Reinforcement	Reference
1.	Polylactic acid	Chicken feather	[59]
		Silk	[60]
		Flax	[61]
		Sisal	[50]
		Jute	[62]
		Kenaf	[63]
2.	Polycaprolactone	Chitosan	[64]
		Hemp	[65]
		Brewer's spent grain	[66]
		Gelatin	[67]
		Palm fiber	[68]
		Fucoidan	[69]
		Oil palm empty fruit bunch fiber	[70]
		Alfa fiber	[71]
3.	Polybutylene succinate	silk	[72]
		Wood fiber and rice husk	[73]
		Red algae fiber	[74]
		Jute fiber	[75]
		Henequen fiber	[76]
		Basalt fiber	[77]
4.	Polyglycolic acid	Hydroxyapatite	[78]
5.	Polyvinyl alcohol	collagen	[79]
		Palm leave fiber	[80]
		Jute microfiber	[81]
		starch	[82]
		Microalgae	[83]

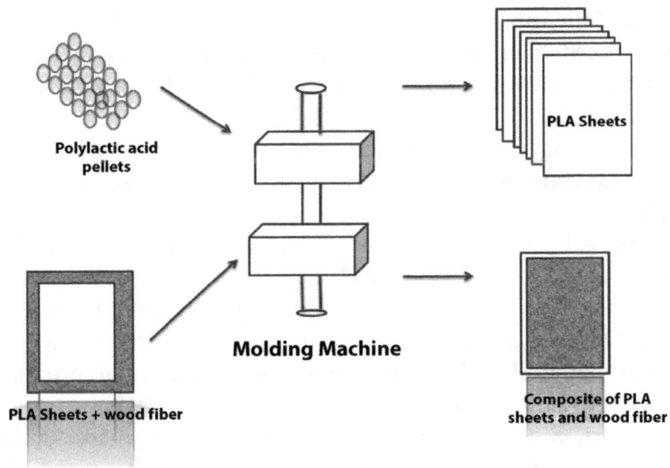

Figure 5.6 Overview of biocomposite moulding.

Use of natural fibers is the key for development of potential biocomposites. Biocomposites produced by reinforcement of natural fibers with various polymer matrix have proved to be potential in providing improved composite properties, such as improved tensile strength, thermostability, and also are capable to withstand high pressure. Biocomposites are mainly biodegradable and biocompatible, which makes them efficient in tissue engineering applications. Process modeling techniques, such as injecting molding, compression moulding, and extrusion methods used for the manufacture of biocomposites, help in the development of desired composite materials with improved mechanical properties, such biocomposites are, thus, used for different applications due to their exhibition of desired properties, similar to that of an ideal composite.

References

1. Christian, S.J., Natural fibre-reinforced noncementitious composites (biocomposites), in: *Nonconventional and Vernacular Construction Materials*, pp. 169–187, Woodhead Publishing, Sawston, United Kingdom, 2020.
2. Azammi, A.N., Ilyas, R.A., Sapuan, S.M., Ibrahim, R., Atikah, M.S.N., Asrofi, M., Atiqah, A., Characterization studies of biopolymeric matrix and cellulose fibres based composites related to functionalized fibre-matrix interface, in: *Interfaces in Particle and Fibre Reinforced Composites*, pp. 29–93, Woodhead Publishing, Sawston, United Kingdom, 2020.
3. Díez-Pascual, A.M. and Diez-Vicente, A.L., ZnO-reinforced poly (3-hydroxybutyrate-co-3-hydroxyvalerate) bionanocomposites with antimicrobial function for food packaging. *ACS Appl. Mater. Interfaces*, 6, 9822, 2014.
4. Asim, M., Saba, N., Jawaid, M., Nasir, M., Potential of natural fiber/biomass filler-reinforced polymer composites in aerospace applications, in: *Sustainable Composites For Aerospace Applications*, pp. 253–268, Woodhead Publishing, Sawston, United Kingdom, 2018.
5. Sanjay, M.R. and Yogesha, B., Studies on natural/glass fiber reinforced polymer hybrid composites: An evolution. *Mater. Today: Proc.* Elsevier, 4, 2, 2739–2747, 2017.

6. Mohamed, S.A.N., Zainudin, E.S., Sapuan, S.M., Azaman, M.D., Arifin, A.M.T., Introduction to natural fiber reinforced vinyl ester and vinyl polymer composites, in: *Natural Fibre Reinforced Vinyl Ester and Vinyl Polymer Composites*, pp. 1–25, Woodhead Publishing, Sawston, United Kingdom, 2018.
7. Terzopoulou, Z.N., Papageorgiou, G.Z., Papadopoulou, E., Athanassiadou, E., Alexopoulou, E., Bikiaris, D.N., Green composites prepared from aliphatic polyesters and bast fibers. *Ind. Crops Prod.*, 68, 60, 2015.
8. Faruk, O., Bledzki, A.K., Fink, H.P., Sain, M., Biocomposites reinforced with natural fibers: 2000–2010. *Prog. Polym. Sci.*, 37, 1552, 2012.
9. Varghese, A.M. and Mittal, V., Polymer composites with functionalized natural fibers, in: *Biodegradable and Biocompatible Polymer Composites*, p. 157, 2018.
10. Sadeghi, S., Dadashian, F., Eslahi, N., Recycling chicken feathers to produce adsorbent porous keratin-based sponge. *Int. J. Environ. Sci. Technol.*, 16, 1119, 2019.
11. Abidin, N.I.Z., Sabri, M.F.M., Kalantari, K., Afifi, A.M., Ahmad, R., Corrosion detection for natural/synthetic/textiles fiber polymer composites, in: *Structural Health Monitoring of Biocomposites, Fibre-Reinforced Composites and Hybrid Composites*, pp. 93–112, 2019.
12. Bharath, K.N. and Basavarajappa, S., Applications of biocomposite materials based on natural fibers from renewable resources: A review. *Sci. Eng. Compos. Mater.*, 23, 123, 2016.
13. Yu, L., Dean, K., Li, L., Polymer blends and composites from renewable resources. *Prog. Polym. Sci.*, 31, 576, 2006.
14. Naffakh, M., Díez-Pascual, A.M., Marco, C., Polymer blend nanocomposites based on poly (l-lactic acid), polypropylene and WS 2 inorganic nanotubes. *RSC Adv.*, 6, 40033, 2016.
15. Díez-Pascual, A.M. and Díez-Vicente, A.L., Poly (3-hydroxybutyrate)/ZnO bionanocomposites with improved mechanical, barrier and antibacterial properties. *Int. J. Mol. Sci.*, 15, 10950, 2014.
16. Pozo Morales, A., Güemes, A., Fernandez-Lopez, A., Carcelen Valero, V., De La Rosa Llano, S., Bamboo–polylactic acid (PLA) composite material for structural applications. *Materials*, 10, 1286, 2017.
17. Tekade, R.K., Maheshwari, R., Tekade, M., Biopolymer-based nanocomposites for transdermal drug delivery, in: *Biopolymer-Based Composites*, pp. 81–106, Woodhead Publishing, Sawston, United Kingdom, 2017.
18. Muneer, F., Johansson, E., Hedenqvist, M.S., Plivelic, T.S., Kuktaite, R., Impact of pH modification on protein polymerization and structure–function relationships in potato protein and wheat gluten composites. *Int. J. Mol. Sci.*, 20, 58, 2019.
19. Sanchez-Alvarado, D.I., Guzmán-Pantoja, J., Páramo-García, U., Maciel-Cerda, A., Martínez-Orozco, R.D., Vera-Graziano, R., Morphological study of chitosan/poly (vinyl alcohol) nanofibers prepared by electrospinning, collected on reticulated vitreous carbon. *Int. J. Mol. Sci.*, 19, 1718, 2018.
20. Christian, S.J., Natural fibre-reinforced noncementitious composites (biocomposites), in: *Nonconventional and Vernacular Construction Materials*, pp. 169–187, Woodhead Publishing, Sawston, United Kingdom, 2020.
21. Lee, S.H. and Wang, S., Biodegradable polymers/bamboo fiber biocomposite with bio-based coupling agent. *Compos. Part A Appl. Sci. Manuf.*, 37, 80, 2006.
22. Azwa, Z.N., Yousif, B.F., Manalo, A.C., Karunasena, W., A review on the degradability of polymeric composites based on natural fibres. *Mater. Des.*, 47, 424, 2013.
23. Faruk, O. and Ain, M.S., Biofiber reinforced polymer composites for structural applications, in: *Developments in Fiber-Reinforced Polymer (FRP) Composites for Civil Engineering*, Woodhead Publishing, Sawston, United Kingdom, 2013.

24. Saini, P., Arora, M., Kumar, M.R., Poly (lactic acid) blends in biomedical applications. *Adv. Drug Deliv. Rev.*, 107, 47, 2016.
25. Jonoobi, M., Harun, J., Mathew, A.P., Oksman, K., Mechanical properties of cellulose nanofiber (CNF) reinforced polylactic acid (PLA) prepared by twin screw extrusion. *Compos. Sci. Technol.*, 70, 1742, 2010.
26. Zhao, Y., Liu, B., Bi, H., Yang, J., Li, W., Liang, H., Liang, Y., Jia, Z., Shi, S., Chen, M., The degradation properties of MgO Whiskers/PLLA composite in vitro. *Int. J. Mol. Sci.*, 19, 2740, 2018.
27. O'Brien, F.J., Harley, B.A., Yannas, I.V., Gibson, L.J., The effect of pore size on cell adhesion in collagen-GAG scaffolds. *Biomaterials*, 26, 433, 2005.
28. John, M.J. and Thomas, S., Biofibres and biocomposites. *Carbohydr. Polym.*, 71, 343, 20082008.
29. Prakobna, K., Galland, S., Berglund, L.A., High-performance and moisture-stable cellulose–starch nanocomposites based on bioinspired core–shell nanofibers. *Biomacromolecules*, 16, 904, 2015.
30. Tibolla, H., Pelissari, F.M., Menegalli, F.C., Cellulose nanofibers produced from banana peel by chemical and enzymatic treatment. *LWT Food Sci. Technol.*, 59, 1311, 2014.
31. Asim, M., Saba, N., Jawaid, M., Nasir, M., Potential of natural fiber/biomass filler-reinforced polymer composites in aerospace applications, in: *Sustainable Composites For Aerospace Applications*, pp. 253–268, Woodhead Publishing, Sawston, United Kingdom, 2018.
32. Ramamoorthy, S.K., Åkesson, D., Rajan, R., Periyasamy, A.P., Skrifvars, M., Mechanical performance of biofibers and their corresponding composites, in: *Mechanical and Physical Testing of Biocomposites, Fibre-Reinforced Composites and Hybrid Composites*, pp. 259–292, Woodhead Publishing, Sawston, United Kingdom, 2019.
33. Bax, B. and Müssig, J., Impact and tensile properties of PLA/Cordenka and PLA/flax composites. *Compos. Sci. Technol.*, 68, 1601, 2008.
34. Asaithambi, B., Ganesan, G., Kumar, S.A., Bio-composites: Development and mechanical characterization of banana/sisal fibre reinforced poly lactic acid (PLA) hybrid composites. *Fiber. Polym.*, 15, 847, 2014.
35. Saravana Bavan, D. and Mohan Kumar, G.C., Examination of Tensile strength and FTIR features of maize fibers reinforced polymer composites. *Int. J. Recent Trends Sci. Technol.*, 5, 27, 2011.
36. Kandar, M.M. and Akil, H.M., Application of design of experiment (DoE) for parameters optimization in compression moulding for flax reinforced biocomposites. *Procedia Chem.*, 19, 433, 2016.
37. Rubio-López, A., Olmedo, A., Díaz-Álvarez, A., Santiuste, C., Manufacture of compression moulded PLA based biocomposites: A parametric study. *Compos. Struct.*, 131, 995, 2015.
38. Saint-Martin, G., Schmidt, F., Devos, P., Levaillant, C., Voids in short fibre-reinforced injection-moulded parts: Density control vs. mass control. *Polym. Test.*, 22, 947, 2003.
39. Mohanty, A.K., Wibowo, A., Misra, M., Drzal, L.T., Effect of process engineering on the performance of natural fiber reinforced cellulose acetate biocomposites. *Compos. Part A Appl. Sci. Manuf.*, 35, 363, 2004.
40. Sasimowski, E., Majewski, Ł., Grochowicz, M., Analysis of selected properties of injection moulded sustainable biocomposites from poly (butylene succinate) and wheat bran. *Materials*, 14, 7049, 2021.
41. Belhassen, R., Boufi, S., Vilaseca, F., Lopez, J.P., Méndez, J.A., Franco, E., Pelach, M.A., Mutje, P., Biocomposites based on alfa fibers and starch-based biopolymer. *Polym. Adv. Technol.*, 20, 1068, 2009.
42. Panigrahi, S., Li, X., Tabil, L., Injection moulding processing of flax fibre-reinforced polyethylene biocomposites, in: *International Conference on Flax and Other Bast Plants*, p. 399, 2008.

43. Etxabide, A., de la Caba, K., Guerrero, P., A novel approach to manufacture porous biocomposites using extrusion and injection moulding. *Eur. Polym. J.*, 82, 324, 2016.
44. Martín-Alfonso, J.E., Félix, M., Romero, A., Guerrero, A., Development of new albumen based biocomposites formulations by injection moulding using chitosan as physicochemical modifier additive. *Compos. B Eng.*, 61, 275, 2014.
45. Felix, M., Romero, A., Martín-Alfonso, J.E., Guerrero, A., Development of crayfish protein-PCL biocomposite material processed by injection moulding. *Compos. B Eng.*, 78, 291, 2015.
46. Bledzki, A.K., Letman, M., Viksne, A., Rence, L., A comparison of compounding processes and wood type for wood fibre—PP composites. *Compos. Part A Appl. Sci. Manuf.*, 36, 789, 2005.
47. Tokoro, R., Vu, D.M., Okubo, K., Tanaka, T., Fujii, T., Fujiura, T., How to improve mechanical properties of polylactic acid with bamboo fibers. *J. Mater. Sci.*, 43, 775, 2008.
48. Wollerdorfer, M. and Bader, H., Influence of natural fibres on the mechanical properties of biodegradable polymers. *Ind. Crops Prod.*, 8, 105, 1998.
49. Gamon, G., Evon, P., Rigal, L., Twin-screw extrusion impact on natural fibre morphology and material properties in poly (lactic acid) based biocomposites. *Ind. Crops Prod.*, 46, 173, 2013.
50. Chaitanya, S. and Singh, I., Processing of PLA/sisal fiber biocomposites using direct-and extrusion-injection molding. *Mater. Manuf. Process.*, 32, 468, 2017.
51. Hietala, M., Rollo, P., Kekäläinen, K., Oksman, K., Extrusion processing of green biocomposites: Compounding, fibrillation efficiency, and fiber dispersion. *J. Appl. Polym. Sci.*, 131, 39981, 2014.
52. Tisserat, B., Joshee, N., Mahapatra, A.K., Selling, G.W., Finkenstadt, V.L., Physical and mechanical properties of extruded poly (lactic acid)-based Paulownia elongata biocomposites. *Ind. Crops Prod.*, 44, 88, 2013.
53. Amroune, S., Belaadi, A., Dalmis, R., Seki, Y., Makhlouf, A., Satha, H., Quantitatively investigating the effects of fiber parameters on tensile and flexural response of flax/epoxy biocomposites. *J. Nat. Fibers*, 1, 2366–2381, 2020.
54. Dorozhkin, S.V., Calcium orthophosphate-based biocomposites and hybrid biomaterials. *J. Mater. Sci.*, 44, 2343, 2009.
55. Pinzón, L.M., Cedano, F.J., Castro, C.I., Briceño, J.C., Casas, J.P., Tabima, D.M., Salcedo, F., Formulation and characterization of chitosan-based biocomposites with potential use for bone adhesion. *Int. J. Polym. Mater. Polym. Biomater.*, 66, 697, 2017.
56. Mihai, M., Legros, N., Alemdar, A., Formulation-properties versatility of wood fiber biocomposites based on polylactide and polylactide/thermoplastic starch blends. *Polym. Eng. Sci.*, 54, 1325, 2014.
57. Gil-Castell, O., Badia, J.D., Kittikorn, T., Strömberg, E., Ek, M., Karlsson, S., Ribes-Greus, A., Impact of hydrothermal ageing on the thermal stability, morphology and viscoelastic performance of PLA/sisal biocomposites. *Polym. Degrad. Stab.*, 132, 87, 2016.
58. Dairi, N., Ferfera-Harrar, H., Ramos, M., Garrigós, M.C., Cellulose acetate/AgNPs-organoclay and/or thymol nano-biocomposite films with combined antimicrobial/antioxidant properties for active food packaging use. *Int. J. Biol. Macromol.*, 121, 508, 2019.
59. Cheng, S., Lau, K.T., Liu, T., Zhao, Y., Lam, P.M., Yin, Y., Mechanical and thermal properties of chicken feather fiber/PLA green composites. *Compos. B Eng.*, 40, 650, 2009.
60. Cheung, H.Y., Lau, K.T., Tao, X.M., Hui, D., A potential material for tissue engineering: Silkworm silk/PLA biocomposite. *Compos. B Eng.*, 39, 1026, 2008.
61. Wróbel-Kwiatkowska, M., Czemplik, M., Kulma, A., Żuk, M., Kaczmar, J., Dymińska, L., Hanuza, J., Ptak, M., Szopa, J., New biocomposites based on bioplastic flax fibers and biodegradable polymers. *Biotechnol. Prog.*, 28, 1336, 2012.
62. Yang, Y., Murakami, M., Hamada, H., Molding method, thermal and mechanical properties of jute/PLA injection molding. *J. Polym. Environ.*, 20, 1124, 2012.

63. Graupner, N. and Müssig, J., A comparison of the mechanical characteristics of kenaf and lyocell fibre reinforced poly (lactic acid)(PLA) and poly (3-hydroxybutyrate)(PHB) composites. *Compos. Part A Appl. Sci. Manuf.*, 42, 2010, 2011.
64. Hong, S. and Kim, G., Fabrication of electrospun polycaprolactone biocomposites reinforced with chitosan for the proliferation of mesenchymal stem cells. *Carbohydr. Polym.*, 83, 940, 2011.
65. Dhakal, H.N., Ismail, S.O., Zhang, Z., Barber, A., Welsh, E., Maigret, J.E., Beaugrand, J., Development of sustainable biodegradable lignocellulosic hemp fiber/polycaprolactone biocomposites for light weight applications. *Compos. Part A Appl. Sci. Manuf.*, 113, 350, 2018.
66. Hejna, A., Formela, K., Saeb, M.R., Processing, mechanical and thermal behavior assessments of polycaprolactone/agricultural wastes biocomposites. *Ind. Crops Prod.*, 76, 725, 2015.
67. Wang, Y., Chang, H.I., Wertheim, D.F., Jones, A.S., Jackson, C., Coombes, A.G., Characterisation of the macroporosity of polycaprolactone-based biocomposites and release kinetics for drug delivery. *Biomaterials*, 28, 4619, 2007.
68. Dhakal, H., Bourmaud, A., Berzin, F., Almansour, F., Zhang, Z., Shah, D.U., Beaugrand, J., Mechanical properties of leaf sheath date palm fibre waste biomass reinforced polycaprolactone (PCL) biocomposites. *Ind. Crops Prod.*, 126, 394, 2018.
69. Lee, J.S., Jin, G.H., Yeo, M.G., Jang, C.H., Lee, H., Kim, G.H., Fabrication of electrospun biocomposites comprising polycaprolactone/fucoidan for tissue regeneration. *Carbohydr. Polym.*, 90, 181, 2012.
70. Hamid, M.Z.A., Ibrahim, N.A., Yunus, W.M.Z.W., Zaman, K., Dahlan, M., Effect of grafting on properties of oil palm empty fruit bunch fiber reinforced polycaprolactone biocomposites. *J. Reinf. Plast. Compos.*, 29, 2723, 2010.
71. Marrakchi, Z., Oueslati, H., Belgacem, M.N., Mhenni, F., Mauret, E., Biocomposites based on polycaprolactone reinforced with alfa fibre mats. *Compos. Part A Appl. Sci. Manuf.*, 43, 742, 2012.
72. Lee, S.M., Cho, D., Park, W.H., Lee, S.G., Han, S.O., Drzal, L.T., Novel silk/poly (butylene succinate) biocomposites: The effect of short fibre content on their mechanical and thermal properties. *Compos. Sci. Technol.*, 65, 647, 2005.
73. Kim, H.S., Yang, H.S., Kim, H.J., Biodegradability and mechanical properties of agro-flour–filled polybutylene succinate biocomposites. *J. Appl. Polym. Sci.*, 97, 1513, 2005.
74. Lee, M.W., Han, S.O., Seo, Y.B., Red algae fibre/poly (butylene succinate) biocomposites: The effect of fibre content on their mechanical and thermal properties. *Compos. Sci. Technol.*, 68, 1266, 2008.
75. Liu, L., Yu, J., Cheng, L., Qu, W., Mechanical properties of poly (butylene succinate) (PBS) biocomposites reinforced with surface modified jute fibre. *Compos. Part A Appl. Sci. Manuf.*, 40, 669, 2009.
76. Han, S.O., Cho, D., Park, W.H., Drzal, L.T., Henequen/poly (butylene succinate) biocomposites: Electron beam irradiation effects on henequen fiber and the interfacial properties of biocomposites. *Compos. Interfaces*, 13, 231, 2006.
77. Zhang, Y., Yu, C., Chu, P.K., Lv, F., Zhang, C., Ji, J., Zhang, R., Wang, H., Mechanical and thermal properties of basalt fiber reinforced poly (butylene succinate) composites. *Mater. Chem. Phys.*, 133, 845, 2012.
78. Ko, H.S., Lee, S., Lee, D., Jho, J.Y., Mechanical properties and bioactivity of poly (lactic acid) composites containing poly (glycolic acid) fiber and hydroxyapatite particles. *Nanomaterials*, 11, 249, 2021.
79. Degirmenbasi, N., Kalyon, D.M., Birinci, E., Biocomposites of nanohydroxyapatite with collagen and poly (vinyl alcohol). *Colloids Surf. B Biointerfaces*, 48, 42, 2006.
80. Mohanty, J.R., Das, S.N., Das, H.C., Swain, S.K., Effective mechanical properties of polyvinylalcohol biocomposites with reinforcement of date palm leaf fibers. *Polym. Compos.*, 34, 959, 2013.

81. Das, K., Ray, D., Bandyopadhyay, N.R., Sahoo, S., Mohanty, A.K., Misra, M., Physico-mechanical properties of the jute micro/nanofibril reinforced starch/polyvinyl alcohol biocomposite films. *Compos. B Eng.*, 42, 376, 2011.
82. Sofyane, A., Ablouh, E., Lahcini, M., Elmeziane, A., Khouloud, M., Kaddami, H., Raihane, M., Slow-release fertilizers based on starch acetate/glycerol/polyvinyl alcohol biocomposites for sustained nutrient release. *Mater. Today: Proc.*, 36, 74, 2021.
83. Tran, D.T., Lee, H.R., Jung, S., Park, M.S., Yang, J.W., Lipid-extracted algal biomass based biocomposites fabrication with poly (vinyl alcohol). *Algal Res.*, 31, 525, 2018.

6

Microbial Technology in Bioplastic Production and Engineering

Dileep Francis* and Deepu Joy Parayil

Department of Life Sciences, Kristu Jayanti College (Autonomous), Bengaluru, India

Abstract

Plastics have become an indispensable part of human culture. Plastics are conventionally derived from petro-based feedstock, a nonrenewable resource. The chemical nature of plastics renders them non-biodegradable. Plastic production and use contribute to the depletion of fossil reserves. Further, plastic waste accumulates over time and causes large-scale pollution. Technologies that seek to use renewable, carbon-neutral resources to derive chemicals and fuels for human consumption are a priority concern for environmental sustainability. While biofuels are the most crucial innovation in this line, bioplastics are emerging as equally important. Bioplastics, also called biobased polymers, are derived from biobased feedstock such as plant polysaccharides, starch, vegetable oils, and lignocellulosic biomass. Lignocellulose is the most suitable starting material for bioplastic synthesis as it can be easily sourced from agricultural waste, which is abundant and cheap. Biomass can be converted to polymers either chemically or through microbial technology. Microbial technology for bioplastic production utilizes the capacity of microorganisms to naturally synthesize plastics or their monomeric building blocks using biomass as the substrate. Advances in microbial technology fostered by the advent of metabolic engineering have opened new avenues for bioplastic production. The present chapter summarizes the significant advancements in the production of bioplastics using microbes

***Keywords*:** Bioplastics, microbial technology, fermentation, lignocellulosic biomass polyhydroxyalkanoates, polylactic acid, polyamides

6.1 Introduction

Plastics, giant organic polymers built by linking together a large number of organic monomers, have become an essential and indispensable commodity in the day-to-day life of humans. Conventional polymers like polyethylene (PE), polypropylene (PP), polyethylene terephthalate (PET), polystyrene (PS), and polyvinylchloride (PVC) that constitute the lion's share of the global production by weight are produced from petrochemical feedstock derived from fossil fuels [1]. Petro-based plastics find applications as packaging materials, in fabrics and textiles, for building and construction activities, automotive, electronics,

**Corresponding author*: dileep@kristujayanti.com

Inamuddin and Tariq Altalhi (eds.) Handbook of Bioplastics and Biocomposites Engineering Applications: 2nd Edition, (121–148)

and electrical devices, among others [2]. The global production of plastics in 2019 was approximately 370 million metric tonnes [3]. Global production of plastics is anticipated to grow exponentially in the coming years as there is an increasing demand for plastic-based commodities across the globe [4]. Increasing production and use of plastics is a threat to environmental sustainability for three significant reasons. First, it fast depletes the valuable fossil reserve, a nonrenewable source of energy and carbon. Second, plastic refuse accumulates in the biosphere at alarming rates because of the increased usage and the non-biodegradable nature of plastics [5–7]. Third, the petrochemical processes involved in the production of plastics and the current methods for decomposition (including land-filling and incineration) release significant amounts of greenhouse gases [8]. The growing concern for environmental sustainability in industrial manufacturing practices has accelerated the search for alternatives to petro-based plastics. An ideal replacement should be derived from renewable resources, be entirely or partially biodegradable, possess physicochemical, mechanical, and thermal properties similar to petro-based plastics, and be amenable to cost-effective, industrial-scale production.

Bioplastics, technically designated as biobased polymers by the International Union for Pure and Applied Chemistry (IUPAC), are viable replacements for petro-based plastics. They are, by definition, polymers with properties similar to conventional plastics derived entirely or partially from biomass [9]. This definition, however, does not imply that they are biodegradable. Some bioplastics are biodegradable or biocompatible, while others are not [10]. The most significant environmental impact of bioplastics is that, since they are derived from biomass, they help protect the earth's valuable fossil reserves.

Further, the biodegradation of bioplastics is a carbon-neutral process [11, 12]. The carbon dioxide released to the biosphere on the degradation of bioplastics originates from the biomass used in its production. A living organism fixed this *via* photosynthesis in the recent past. Thus, there is no net increase in greenhouse gases in the atmosphere [12, 13]. In addition, fermentation-based bioplastics production enables low-temperature processes that save energy [11]. Despite all these benefits, the current market share of bioplastics is just around 1% of the global output, majorly due to technological and cost-based constraints in industrial manufacturing and various other challenges [5]. However, the annual growth rate is estimated to be around 30% by 2025, making bioplastic manufacturing an up-and-coming industry [5].

Direct chemical conversion and microbial fermentation can be employed for the conversion of biomass into bioplastics. In the chemical synthesis route, biomass is directly processed into bioplastics using either mechanical or chemical processes. Additives, both biobased and fossil-based, may be added to the processed product. For example, thermoplastic starch (TPS), perhaps the most abundant type of bioplastic in the market, is processed directly from starch by adding flexibilizers and plasticizers [14, 15]. Native starch derived from cereal grains and tubers is devoid of thermoplastic properties. After including suitable additives, applying thermal processes and sheer stress converts natural starch into a thermoplastic polymer amenable to molding into a wide range of products. Another biopolymer, sold under the brand name Rilsan, is a polyamide synthesized by the chemical polymerization of 11-aminoundecanoic acid, a major component of castor bean oil [16].

Microbial fermentation involves the conversion of organic substrates into valuable products by microorganisms grown in controlled growth conditions. The process exploits the highly specialized and unique enzymes present in microorganisms—as a part of their metabolic pathways—to convert a suitable substrate into a commercial product. Recent developments in metabolic engineering have furthered the potential of microbial technology to derive a diverse array of commercially relevant chemicals. Despite the apparent benefits of fermentation, conversion of biomass into bioplastics using this route is in its infancy except for polymers like polyhydroxyalkanoates (PHA) and poly (lactic acid) (PLA). Nevertheless, the environmental advantages of using microbial technology for bioplastics have led to many developments in the field. The present chapter summarizes the significant advancements in using microorganisms as agents to produce bioplastics from renewable biomass.

6.2 Fundamental Principles of Microbial Bioplastic Production

Humans have relied on microbial metabolites for various applications from time immemorial. Natural microbial processes result in the production of many beneficial chemicals. Tapping this natural reservoir to synthesize valuable substances has been a priority of industrial microbiology. Antibiotics, ethanol, organic acids, solvents, amino acids, enzymes, polysaccharides, and a diverse array of drug molecules are obtained from microorganisms and find applications in medicine, agriculture, food, feed, etc. [17–19]. Microorganisms produce these molecules through diverse metabolic pathways by employing enzymes as catalysts. This process has been referred to as microbial fermentation [20]. These molecules serve as a source of carbon or energy (primary metabolites) or perform specialized functions that help adapt to challenging environments (secondary metabolites). Some of these metabolites are used directly for human use, while others undergo biological or synthetic modification to enhance utility. It is pertinent to state that many of these molecules can also be synthesized using chemical methods. However, the fermentation-based process offers the advantage of being environmentally sustainable. Microbial fermentation utilizes cheap, renewable raw materials, employs low-temperature production methods, and generates byproducts less harmful to the environment [21, 22]. The major downside to the fermentative route is the difficulties in scaling up the bioprocess to meet commercial demands [22].

Microorganisms capable of producing commercially beneficial molecules are sometimes referred to as microbial cell factories [23]. Cell factories producing bioplastics utilize crude or refined forms of biomass as substrates for the synthesis of biopolymers that get molded into various commodities. The production host employed may be natural or genetically engineered. The fermentative route for bioplastic production can adopt direct (one-step) and indirect (two-step) methods [24]. The direct production of bioplastics, which is preferred over the latter, utilizes microbial fermentation to convert biomass into bioplastics directly. However, this method warrants the use of metabolic engineering in most cases because of the absence of microorganisms with a known metabolic pathway that leads to the synthesis of the desired polymer. In the indirect method, biobased monomers produced through fermentation are used to synthesize plastics using chemical or enzymatic methods. The advent of these methods can be attributed to the search for a biobased alternative to synthesizing the monomeric building blocks currently produced through the petrochemical route.

The general steps involved in devising a microbial process for synthesizing bioplastics are as follows: (1) Identification of a production host through biochemical screening or, development of a metabolically engineered strain, (2) developing a pilot-scale bioprocess based on a suitable substrate and production media, (3) optimizing the fermentation conditions for enhancing yield and reducing the cost of production, (4) scaling up the production process to meet industry demands, (5) establishment of efficient processes for the recovery of the desired biopolymer. Figure 6.1 depicts the general process involved in bioplastic production using microbial technology

One of the significant challenges in bioplastic production is the development of suitable production hosts [25]. As pointed out earlier, few microorganisms directly synthesize bioplastics as a byproduct of central metabolic pathways. In some instances, the capability to produce a polymer directly results from stressful or special growth conditions. Optimization of the growth conditions to maximize production in such hosts is a challenging feat to be accomplished. Similarly, microorganisms identified to possess metabolic pathways that lead to monomeric building blocks for plastics or intermediates channeled to bioplastic production via genetic engineering might not be readily suitable for growth and maximal yields in the regulated environment used in bioprocesses. The advent of metabolic engineering that enables directed modifications of metabolic and regulatory pathways has helped exploit the natural capabilities of microorganisms in better, previously unimaginable ways. Metabolic engineering utilizes the principles of recombinant DNA (rDNA) technology to manipulate microbial metabolic pathways for rerouting the metabolic flux in favor of the synthesis of the desired product with high yields. Approaches used for engineering hosts include the expression of heterologous enzyme/s in the natural producers to facilitate production, knockout of undesirable genes that divert the metabolic flux, cause feedback regulation, etc. High throughput data from the emerging field of metabolomics is expected to complement metabolic engineering in industrial fermentation. Recently, synthetic bioengineering approaches employing computational simulation to develop theoretical/artificial metabolic pathways to synthesize the desired product from a substrate of choice have been used to make rational decisions in formulating metabolic engineering strategies for production hosts [26, 27].

Another consideration in the use of microbial technology in bioplastic production is the choice of substrate. Crude or refined forms of biomass are utilized for production. The amount of biomass produced in the biosphere through photosynthesis amounts to around 170 billion metric tonnes, out of which humans consume only a meager 4%

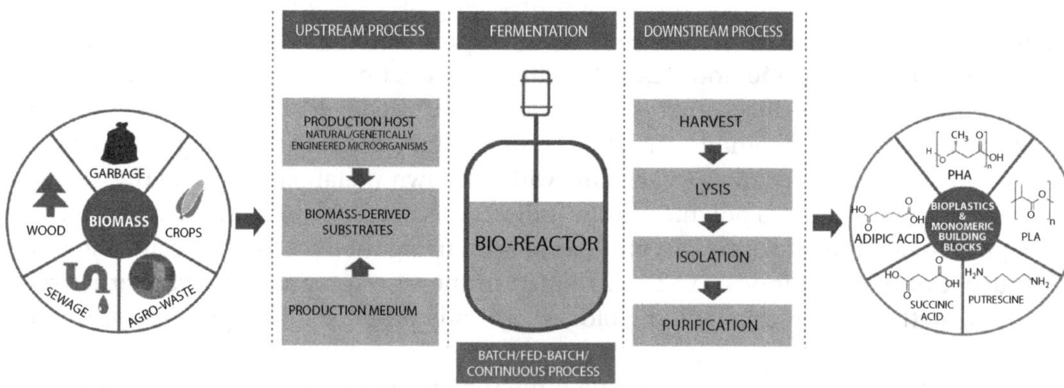

Figure 6.1 Overview of bioplastic production using microbial technology.

as food [28, 29]. This biomass is majorly contributed by plants and, to some extent, by microalgae. Biomass is considered to be an abounding renewable resource for carbon and energy to synthesize valuable chemicals, including bioplastics. The predominant chemical components of plant biomass include carbohydrates, vegetable oils, proteins, lignins, terpenes, etc. Carbohydrates constitute about 75% of the biomass composition and serve as the predominant substrate for bioplastics [30]. Carbohydrates, such as starch, cellulose, and glucose, are used widely for production. These substrates are mainly obtained from edible feedstock, such as corn, sugar cane, cassava, malt, beet, etc. In addition to plant-based biomass, animal-based biomass, such as oils and proteins, are also used as feedstock for bioplastic production. A major technological challenge in utilizing biomass for plastic production stems from the need to distinguish and prioritize the development of technologies and processes that use nonedible biomass instead of edible substrates. Lignocellulosic biomass, which is present abundantly in agricultural waste and wood, is the top candidate for the green synthesis of fuels and plastics. It is sought as a replacement for biomass derived from corn and sugarcane. Advancements in the process of converting this abundant resource, which usually goes as waste, into fermentable substrates are expected to push forward the biofuel and bioplastic industry by leaps and bounds.

6.3 Bioplastics Obtained Directly from Microorganisms

6.3.1 PHA

Poly (3-hydroxybutyrate) (PHB), a short-chain PHA, is the first bioplastic to be identified from a microorganism. Lemoigne, in 1923, described the presence of this biopolymer in *Bacillus megaterium* [31]. After about five decades of the initial description, PHAs gained popularity as an alternative to conventional plastics in the 1970s, triggered by the petroleum crisis. Today PHAs are the most widely produced group of microbial bioplastics. PHAs are naturally occurring polyesters stored in intracellular granules by various microorganisms when grown under unbalanced nutrient conditions. They serve as energy and carbon reserves and help survive stress [32, 33]. More than 90 genera of microorganisms produce and catabolize these polymers [34]. PHAs exhibit remarkable monomer composition variability, but these biopolymers invariably show physicochemical, thermal, and mechanical properties similar to petro-based plastics. They are amenable to thermal molding and, with specific additives, display properties similar to polypropylene. Importantly, bioplastics derived from PHAs are biocompatible and biodegradable [35]. PHAs find applications as packaging materials, such as refuse sacks, compost bin liners, carriers, gift bags, and shrink wrap; in food service as foam packaging, coffee cup liners, disposable gloves, and aprons; in agriculture as plant pots and mulching film; in medicine as absorbable sutures, mesh; surgical films, and composite mesh; in consumer products as furniture and phone case and in chemicals as plastic additives [36, 37].

More than 150 PHAs made up of hydroxy fatty acid monomers with the hydroxyl group attached to the C-3, C-4, C-5, and C-6 positions and diverse R-side chains are known [38]. The R-side chains can be linear, branched, saturated, unsaturated, or aromatic. Figure 6.2 shows the general structure of PHAs.

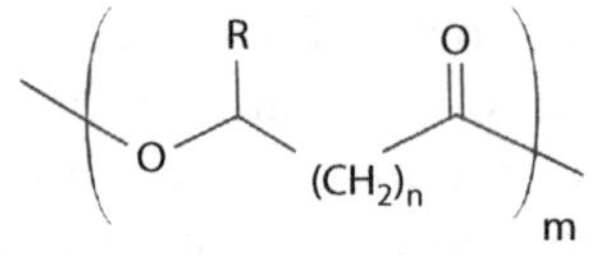

PHB: R=CH_3; PHV: R=CH_2CH_3; PHH: R=CH_2CH_2Ch

Figure 6.2 General structure of PHAs. The substituent groups for PHB, poly 3-hydroxyvalerate (PHV), and poly 3-hydroxyhexanoate (PHH), are also shown.

PHAs are classified into three groups based on the number of carbon atoms in their monomeric building blocks [39]. They are (1) short-chain-length PHAs (scl-PHAs) with the number of carbon atoms in the monomer ranging from 3 to 5. They display properties similar to PP. PHB, poly (4-hydroxybutyrate) [P(4HB)], and PHV belongs to this class. Although PHB has properties identical to PP, its brittleness hampers its applications. The brittleness is overcome by copolymerization with other scl-PHAs. (2) Medium-chain-length PHAs (mcl-PHAs) with the number of carbon atoms in the monomer ranging from 6 to 14. They possess flexibility and elasticity similar to rubber. (3) Long-chain-length PHAs (lcl-PHAs) with the number of carbon atoms in the monomer more than 14.

Microorganisms accumulate PHAs as amorphous polymers in intracellular granules surrounded by a lipid monolayer. The granules also incorporate phasins, a group of proteins that modulate the size and volume of the granules, polymerases, depolymerases, and other nonspecific cytosolic proteins [40]. PHAs are produced as energy reserves in conditions of nutritional stress. Biosynthesis of PHAs involves the production of monomeric intermediates, such as the R-enantiomers of 3-hydroxybutyryl–CoA and 3-hydroxyacyl-CoA, through diverse metabolic pathways using different carbon sources and specific enzymes. The enzymes are encoded by genes present in biosynthetic operons. The organization of the operons and the characteristics of the enzymes encoded vary according to the producer species. The monomers are linked together to derive polymeric PHAs with the help of Pha synthases (also called polymerases) [40, 41]. The type of PHA produced by a particular microbe depends on the Pha synthase present. Depolymerases facilitating hydroxyacyl–CoA's catabolic release from PHAs help mobilize the energy reserves. The biodegradability of PHA-based plastics stems from depolymerase enzymes in the soil microflora [42]. Biosynthesis of PHB and other PHAs utilizes intermediates derived from metabolic processes, including glucose break down, alkane oxidation, fatty acid biosynthesis, fatty acid breakdown (β-oxidation), along with other pathways [40, 43]. Figures 6.3 (a) and (b) summarize the metabolic routes that lead to the synthesis of PHAs. The addition of specific substrates in the fermentation medium may produce copolymers with superior properties. For example, the inclusion of propionic acid together with glucose as a substrate for *Ralstonia eutropha* results in the synthesis of a copolymer composed of 3-hydroxybutyrate (HB) and 3-hydroxyvalerate (HV) [44]. Thus, the monomeric composition of PHAs may be engineered according to specific needs by altering the media composition.

The microbial production of PHAs involves culturing a producer strain on a medium containing a fermentable carbon source, such as glucose, dextrose, or triglycerides, and the critical macronutrients and micronutrients for the growth and multiplication of the microorganism. The culture media composition varies with the organism used and the process employed. Pure-culture fermentations and mixed-culture fermentations are utilized

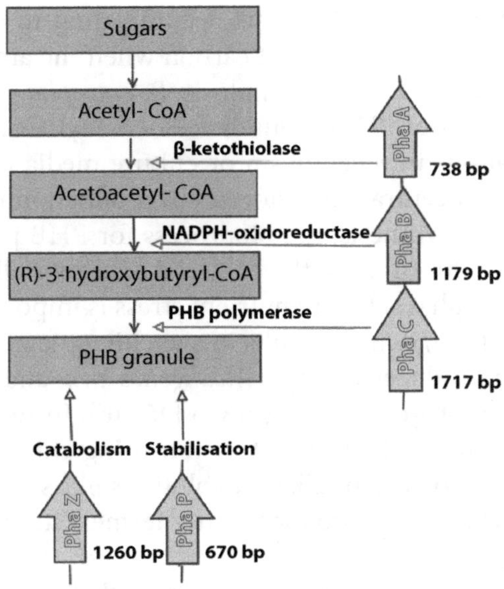

Figure 6.3 **(a)** Biosynthesis of PHB in *R. eutropha*. The gene cluster (phaCAB operon) responsible for the metabolism of sugars into PHB granules is shown in boxes with arrowheads. Two additional genes, which are not a part of the operon, *viz*, Pha Z and Pha P, are also involved in the metabolism of PHBs

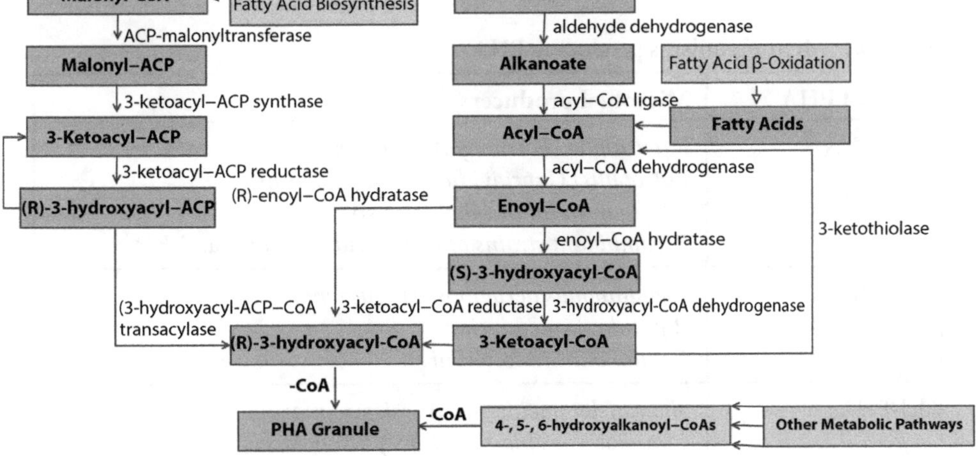

Figure 6.3 **(b)** Biosynthetic pathways in the synthesis of PHAs.

for production. In some bacteria, PHAs start accumulating in conditions of nutritional imbalance, such as the presence of abundant carbon when the amounts of nitrogen, phosphorus, or oxygen are limited [37]. For example, in *R. eutropha*, a commercially employed bacterium for the production of PHBs, abundance of acetyl-CoA leads to a reduction in PHB biosynthesis. Alterations in metabolism or culture media composition that leads to a decrease in acetyl-CoA concentration restore PHB production [40]. By deriving insights from the regulation of PHA synthesis, the bioprocess for PHB production in *R. eutropha* is divided into two stages [45]. Initially, the bacterial culture is allowed to reach a suitable population density (growth phase) before nutrient stress is imposed by reducing the nutrient content (production phase), which results in switching metabolic pathways in favor of PHA production. In other producers, such as *Alcaligenes latus* and recombinant *Escherichia coli*, PHA production is an integral part of growth [45, 46]. In such organisms, the growth and synthesis of PHAs coincide. Solvent extraction using chloroform, dichloromethane, dichloroethane, and chloropropane, or alkaline solutions like sodium hypochlorite, is used for the extraction of PHAs from the producers after fermentation. Enzymatic methods are also employed [47].

A significant number of microorganisms produce PHAs. More than 300 species are known producers of the polymer. Members of both Eubacteria and Archaea are included in the group of producers. Natural producers have shown excellent yield of PHAs in fermentation conditions. *R. eutropha* cells, for example, can store PHAs such that the granules contribute 90% of their dry weight [48]. However, the growth rate of these organisms in fermentation conditions is slow, and product recovery is challenging. Recombinant DNA technology allows transferring protein-coding genes from one organism to another efficiently. Microbial technology for the production of commercially important products has used the potential of rDNA technology to improve processes and yields. For example, *E. coli*, an organism that grows fast and is amenable to convenient lysis and product recovery, is an ideal choice for producing many commercial microbial products [49]. Several transgenes involved in the production of PHAs have been incorporated into *E. coli*—starting with the introduction of *pha* genes in 1988—to enable the heterologous production of PHAs. Engineered strains of bacteria have been used to increase yield, facilitate efficient

Table 6.1 Microorganisms producing PHAs.

Type of PHA	Microbial producers
scl-PHAs	*R. eutropha (Formerly Alcaligenes Eutrophus) Cupriavidus necator, A. latus, B. megaterium, Burkholderia* sp. *Recombinant strains of Cupriavidus necator* and *E. coli*
mcl-PHAs	*Pseudomonas aeruginosa, P. guezennei* *P. putida, P. mendocina.* *Comamonas testosteroni*
lcl-PHAs	*Shewanella oneidensis, Aureispira marina* *Chromatium vinosum, Thiocystis violacea, Thiocapsa pfennigii* and *Synechocystis* sp.

recovery, tailor monomer compositions/ratios, and regulate the molecular mass of the polymer [50–52]. Table 6.1 shows the major microbial producers of PHAs. Currently, corn, cassava, and vegetable oil are used mainly as substrates for PHA production [53]. Research and development efforts are underway to develop methods to utilize lignocellulosic biomass for PHA production

6.3.2 Poly (γ-Glutamic Acid) (PGA)

Another emerging class of biopolymer obtained through direct fermentation is PGA, with demonstrated potential to be developed into biodegradable bioplastics [54]. PGA-based materials are currently used as functional membranes for biochemical applications, drug delivery agents in pharmaceutics, and humectants and dispersants in health care products [55]. PGA possesses a backbone similar to nylon and is structurally analogous to polyacrylic acid [56]. PGA is a polyamide with only glutamic acids as its building block (Figure 6.4a). Unlike gene-encoded proteins, in which amino acids are held together by covalent bonds between the α-amino and α-carboxylic groups (α-amide linkage) formed using the peptidyl transferase activity of the ribosomes during translation, poly amino acids, such as PGA, are synthesized by microorganisms through metabolic reactions [57]. They possess amide linkages involving side-chain functional groups. Glutamic acids in PGA, for example, are covalently connected through the α-amino group and the γ-carboxylic acid group. Ivanovic and Bruckner first identified PGA as the capsule material of *B. anthracis* [54].

Under ambient humidity, PGA is not a thermoplastic. Lyophilized PGA (with lower than 5% moisture content) acquires thermoplastic properties and is considered a promising polyamide (nylon) plastic [58]. However, the plasticization of PGA has been largely unsuccessful due to its hygroscopic nature and the presence of multiple carboxylic acid groups [59]. Esterification and polymer γ-irradiation techniques are strategies for converting PGA into bioplastics [60, 61]. Recently, plasticization of PGA was achieved by mixing it with hexadecylpyridinium cation (HDP^+) to yield a complex called poly-γ-glutamate ion-complex (PGAIC) (Figure 6.4b), which is amenable to molding *via* simple pressurization. An ethanol solution of PGAIC was used to synthesize a stable PGA-based nanofiber [58]. Thus, with further advances in the chemical reformation strategies of PGA into thermoplastic material amenable to industrial molding, PGA may find applications as a high-performance, biodegradable plastic in a large number of commercial applications.

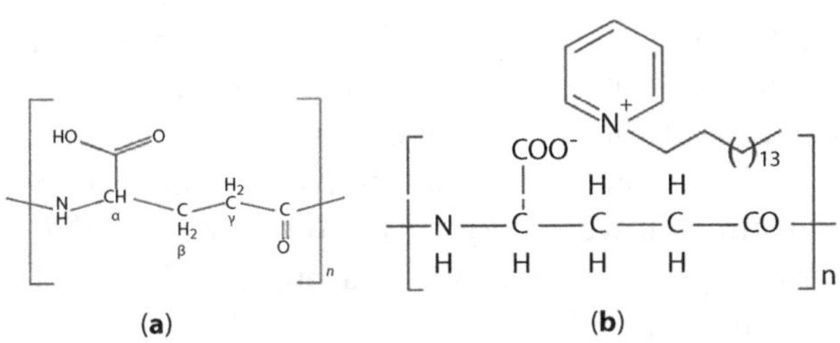

Figure 6.4 Structure of (a) PGA and (b) PGAIC.

PGA is produced naturally by many bacteria and possesses diverse physiological roles based on the producer and growth conditions. It can help tide over adverse conditions. In some pathogenic microorganisms, PGA mediates immune evasion. Bacteria that produce PGA include *Bacillus* species, such as *B. amyloliquefaciens, B. atrophaeus, B. cereus, B. licheniformis, B. megaterium, B. mojavensis,* and *B. pumilis.* It is also produced by *Staphylococcus aureus, Fusobacterium nucleatum, Lysinibacillus sphaericus,* and *Natrialba aegyptiaca,* a halotolerant archaebacterium. PGA is produced in three forms: (1) D-PGA, a polymer made only of D-glutamate; (2) L-PGA, a polymer made only of L-glutamate; and (3) DL-PGA, a heteropolymer composed of randomly incorporated D- and L-glutamate. Biosynthesis involves the synthesis of L- and D-glutamic acids and linking them together to yield PGA. Bioprocess for the production of PGA employs *B. licheniformis and B. subtilis* and utilise substrates, such as casein hydrolysate, maltose, soy sauce, and sucrose. Addition of glutamic acid and metabolic intermediates in PGA synthesis, such as pyruvic acid, citric acid, malic acid, α-ketoglutaric acid, and succinic acid, directly in the production media results in improved yield [57]. Further, the addition of glycerol, Tween-80, or DMSO is advantageous because it improves cellular permeability and increases the uptake of extracellular substrates and secretion of PGA, thereby improving the overall yield [62–64]. The same additives also help redistribute the metabolic flux favoring glutamate production. Metabolically modified strains of *E. coli* carrying genes for PGA production have also been successfully used for its synthesis [65].

6.4 Bioplastics from Microbial Monomers

Although constrained by many technological bottlenecks, biobased thermoplastic polyesters and polyamides can also be derived from monomers obtained from bacterial fermentation of renewable biomass. These molecules are sometimes called biobased platform chemicals. Both aliphatic and aromatic monomers can be produced from microorganisms.

6.4.1 Bioplastics from Aliphatic Monomers

6.4.1.1 PLA

Lactic acid (LA) obtained by bacterial fermentation of a variety of edible and nonedible biomass is the building block for PLA, a biodegradable bioplastic. PLA is a thermoplastic polyester that finds application in healthcare as suture threads, scaffolds in regenerative medicine, cardiovascular implants, drug carriers, orthopaedic screws, cancer therapy, skin and tendon healing, and medical equipment, in food packaging as flexible films, containers, extruded packages, and lunch box, and agriculture as mulching films [66]. It is also used widely in 3D printing applications. In the chemical method for synthesis, PLA is derived through the polycondensation of LA. LA is obtained through petrochemical, as well as microbial processes. The microbial route is attractive because it allows renewable low-cost raw materials and consumes lower energy. Moreover, microbial fermentation produces pure isomers, L(+)- or D(−)-lactic acid. Globally, around 90% of the LA requirement is met through microbial fermentation [67].

LA, first identified from soured milk by the Swedish chemist Carl Wilhelm Scheele in 1780, has a long history of applications in the food, chemical, textile, and pharmaceutical industries. The microbial production of LAs mainly employs lactic acid bacteria (LAB). LAB are mostly facultative anaerobic cocci with high acid tolerance. *Lactobacilli* and *Carnobacteria* are exceptions because they are rod-shaped. LAB is divided into two groups based on the biosynthetic mechanism involved in LA production [68]. Homofermentative LAB converts sugars exclusively into LA through the Embden-Meyerhof pathway (glycolysis).

On the other hand, heterofermentative LAB utilizes the pentose phosphate pathway for the fermentation of sugars and yields ethanol/acetic acid and CO_2 in addition to LA. While obligatory heterofermentative LAB utilizes only the pentose phosphate pathway, facultative heterofermenters use glycolysis and pentose phosphate pathways. Some filamentous fungi, such as *Rhizopus,* are employed for the aerobic fermentation of LA [69]. They possess amylolytic enzymes that convert starch directly into L(+)-lactic acid. Corn, potato, rice, maize, wheat, rye, barley, and cassava rich in starch are potential substrates for the production of LA. More than 90% of commercial LA comes from the submerged fermentation of corn [70].

LA production by bacteria has gained renewed interest due to its application in PLA production. Currently, LA monomers are incorporated into PLA through chemical processes. Two methods are adopted for the polymerization: (1) solvent-based condensation that produces polymers with low to medium molecular weight, and (2) ring open polymerization (ROP) through a cyclic dimer intermediate (called lactide) [71–73]. Solvent-based condensation is challenging due to the need for large reactors and energy-intensive steps for evaporation and solvent recovery [74, 75]. ROP is preferred because it can produce high molecular weight polymers using direct esterification in an azeotropic distillation process. But the process employs metal catalysis for polymerization [76, 77]. Although the advancements in the fermentative production of LA, the advent of ROP-based polymerization, and methods to co-polymerase PLA with other monomers have facilitated PLA production and application, it still involves complex chemical processes. Thus, there is high demand for the direct synthesis of PLA using fermentation.

The rationale for the direct production of PLA in microbial systems stemmed from the presence of similar functional groups (hydroxyl and carboxyl groups) in the monomers of PHA and PLA [78]. Thus there seemed to be a theoretical possibility of adding LA monomers to a growing PHA chain in a metabolically engineered strain of bacteria expressing the correct set of heterologous enzymes. Incorporating LA monomers into PHAs would result in biopolymers with properties partially similar to the PLA produced through the chemobioprocess. Such as process would forego the need for chemical processing and hence result in reduced expense and lower energy consumption. In 2008, a direct biobased process for LA-based PHA production was reported [79]. Poly (lactate-co-3-hydroxybutyrate) [(P (LA-co-3HB)] was synthesized in genetically modified *E. coli* by utilizing the PHB biosynthetic pathway supplemented with an LA polymerizing enzyme (LPE). P(LA-co-3HB) incorporated 6 mol% of LA units in the copolymer. The LPE used in the genetically modified bacteria is a mutant form of the PHA synthase (Ser325Thr/Glu481Lys) [the mutant gene is designated *phaC1*(STQK)] from *Pseudomonas* sp. 61-3 with LA polymerizing activity. Another heterologous enzyme, propionate CoA-transferase (PCT), obtained from *Megasphaera elsdenii,* which converts lactate into lactyl-CoA, was also used in the engineered biosynthetic pathway to derive the building block monomer. However, 6 mol%

of LA incorporation proved to be a low monomer ratio to derive copolymers with thermal, mechanical, and chemical properties similar to PHAs or homopolymeric PLA.

Nevertheless, this pioneering study established the fundamental metabolic engineering strategy to produce PLA in a one-step process. Several approaches have been employed to derive PHB-based copolymers of LA with higher proportions of LA. Use of *E. coli* overproducing LA along with the genes for the copolymer synthesis, use of anaerobic fermentation, and the use of better evolved LPE with more beneficial mutations were some of the strategies that enabled the synthesis of the PHB-LA copolymer with higher proportions of LA (up to 45 mol% of LA). These modifications also resulted in high polymer accumulation (up to 65% of dried cell weight, DCW) [80–83]. Another strategy involved the knockout of the genes supplying the PHB-CoA monomers, PhaA and PhaB, from the biosynthetic construct for the copolymer in the production host. Instead, the gene *phaJ*, which can produce the valerate monomer HV-CoA using valeric acid supplied in the medium, was expressed. When grown in a medium containing glucose and sodium valerate, *E. coli* harboring these modifications produced a terpolymer of LA, HB, and HV with 90 to 96 mol% of LA, thus proving that elimination of HB-CoA monomers is a feasible approach for synthesizing LA-rich polymers [79, 84].

In addition to *E. coli*, other microbes, including *C. glutamicum, Sinorhizobium meliloti*, and *P. putida*, have also been tried as producer strains for PLA production. *E. coli* produces endotoxins, such as lipopolysaccharides, and hence is not suitable for producing PLA utilized to produce food-grade or pharmaceutical-grade plastics [85–87]. *C. glutamicum* is considered a superior alternative to *E. coli* for commercial production and has shown promise in pilot-scale studies. Genetically engineered *C. glutamicum* with the genes required for the synthesis of the LA-HB copolymer could produce bioplastics with greater than 90 mol% LA, albeit with significantly lower rates of polymer accumulation than in *E. coli* (2.4% of DCW). Quantitative metabolite analysis revealed that polymerization was the rate-limiting step for polymer accumulation. Using a codon-optimized LPE, the polymer content could be increased four-fold [88–91]. Genetically engineered *Sinorhizobium meliloti* could produce homopolymeric PLA, but the polymer yield was low (3.2% of DCW). Engineered *P. putida* produced a quaterpolymer comprised of HB, LA, HH, and hydroxyoctanoate (HO) (42% of DCW) when grown in the presence of sodium octanoate [38]. Other

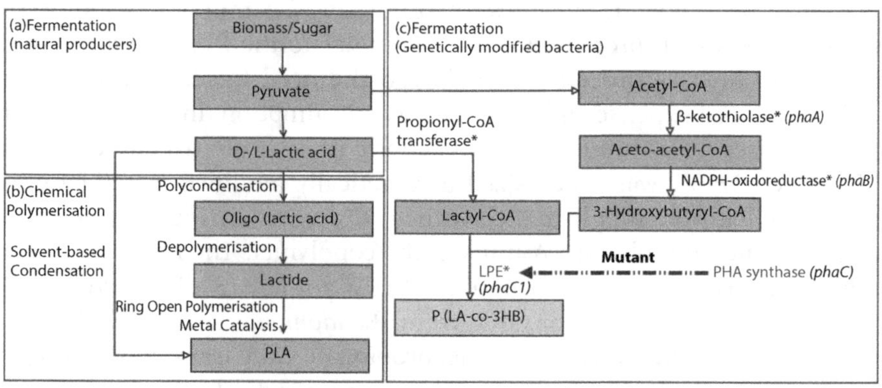

Figure 6.5 Methods for production of PLA.

major metabolic engineering strategies employed to enhance PLA production include genome-wide transposon mutagenesis to select for better producers, disrupting σ factors that regulate global transcription levels, heterologous expression of D-lactate dehydrogenase (ldhD) gene in *E. coli* along with the genes for LA-HB copolymer synthesis, and the use of copolymerizing monomers other than HB [38]. Identification of rate-limiting steps in the biosynthesis of PLA and exploring better production hosts is the way forward for a feasible commercial strategy for the direct synthesis of PLA in microorganisms. Figure 6.5 summarizes the different methods for the production of PLA.

6.4.1.2 *Poly (Butylene Succinate)*

Poly (1, 4, butylene succinate) (PBS) is an emerging bioplastic produced through the condensation polymerization of succinic acid (SA) and 1, 4 butanediol (BDO) (Figure 6.6). It is a biodegradable, semicrystalline thermosetting plastic similar to PET. Among the properties that make PBS an attractive bioplastic are its relatively lower melting points (115°C, lower than PLA), flexibility, and mechanical and physical proteins at par with conventional plastics [92]. However, slow crystallization rates, low melt viscosity, and softness limit its use in processed products, especially those requiring injection molding [93]. PBS is amenable to copolymerization with other dicarboxylic acids and diols, yielding polymers with a diverse range of properties, such as superior tensile strength and crystallinity. The major copolymers of PBS are poly (butylene succinate-co-butylene adipate) (PBSA), poly(butylene succinate-co-butylene terephthalate) (PBST), and poly(butylene succinate-co-propylene succinate) (PBSPS) [94]. SA and BDO, used for the synthesis of PBS, were exclusively obtained from petrochemical feedstock until recently. However, these chemicals can also be obtained through microbial technology, using biomass as a substrate.

Chemically, the hydrogenation of maleic anhydride to succinic anhydride and subsequent hydration yields SA. Replacement of chemically derived succinic acid with biobased SA in PBS has obvious environmental benefits. In aerobic organisms, SA is a metabolic intermediate of the tricarboxylic acid (TCA) cycle, and in anaerobic metabolism, it is one of the end products. A large number of bacteria and many eukaryotes produce SA. Fungal/yeast

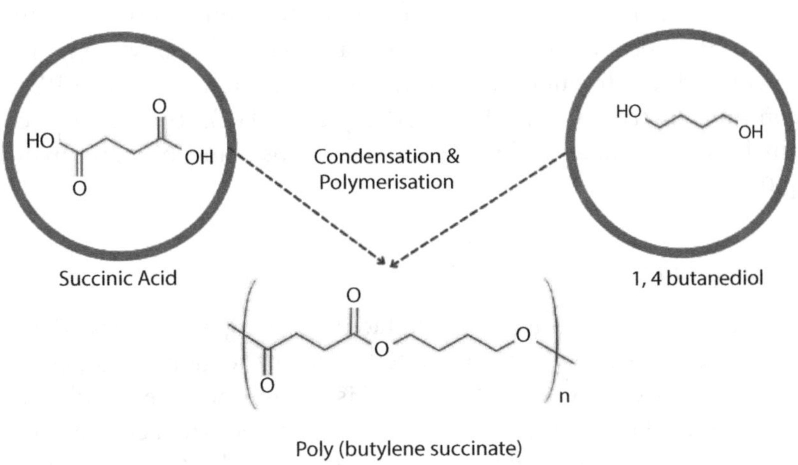

Figure 6.6 Synthesis of PBS.

species employed to synthesize SA include *Aspergillus niger, A. fumigatus, Byssochlamys nivea, Candida tropicalis, Lentinus degener, Paecilomyces varioti, Penicillium viniferum, Saccharomyces cerevisiae*, and *Pichia kudriavzevii* [95]. Bacteria, such as *Actinobacillus succinogenes, Anaerobiospirillum succiniciproducens, C. glutamicum, E. coli, Mannheimia succiniciproducens*, and *Basfia succiniciproducens*, are also used for the production of SA [95]. Among these organisms, *S. cerevisiae, E. coli and C. glutamicum, P. kudriavzevii, A. succinogenes, M. succiniciproducens*, and *B. succiniciproducens*, have been genetically modified to improve the production of SA [95]. A general strategy for metabolic engineering involves the knockout of genes that code for enzymes that compete with SA production. *E. coli* produces SA as a minor product under anaerobic fermentation conditions along with the major products, acetate, formate, and ethanol. In order to redirect the metabolic flux toward the production of SA, *E. coli* was genetically modified by deleting the genes responsible for producing formic acid (*pfl*, pyruvate formate lyase) and LA (*ldhA*, lactate dehydrogenase) [95]. However, the production host was not able to grow appropriately under anaerobic conditions because the deletion of NADH dependant lactate dehydrogenase resulted in the limited regeneration of NAD^+. Overexpression of *E. coli* malate dehydrogenase gene (*mdh*) helped overcome the hurdle. The resultant *E. coli* strain yielded 1.19 mol of SA per mol of glucose using fed-batch fermentation [96]. Several other genetic modifications to increase yield, overcome growth deficiencies, and facilitate the efficient utilization of substrates have been employed [95]. *E. coli* was engineered to produce SA aerobically utilizing the glyoxylate shunt pathway. The gene which codes for the glyoxylate cycle transcription repressor, *iclR*, was knocked out in the producer strain [97]. The world's first commercial biobased SA plant was built in France in 2008 by Bioamber, a start-up jointly launched by companies in the US and France. Many more companies, including Myriant, Succinity, and Reverdia, have successfully ventured into producing biobased SA [98].

BDO is industrially derived from petrochemicals, such as acetylene, butane, propylene, and butadiene [99]. A partially biobased approach to the production of BDO involves the microbial conversion of glucose to succinic acid followed by its catalytic reduction to BDO [100]. A significant bottleneck in the direct synthesis of BDO through microbial fermentation is that it is a nonnatural compound in the microbial world. However, metabolic engineering has enabled the direct production of BDO in *E. coli*. A cohort of heterologous genes was introduced into *E. coli*, such that the intermediates from central metabolic pathways are rerouted into BDO production [101]. In one approach, *E. coli* was engineered to channel succinic acid and α-ketoglutarate derived from the TCA cycle into BDO production through a common intermediate, 4-hydroxybutyrate (4HB). The bacteria could ferment many carbohydrate substrates, including sucrose, xylose, and biomass hydrolysates, and yield BDO [102].

6.4.1.3 Biopolyamides (Nylons)

Thermoplastic polyamides (nylons) are manufactured using precursors obtained from petrochemical feedstock through chemical routes. 90% of the global markets for nylons are met by nylon 6, 6, and nylon 6 [103]. Nylon 6, 6 is a heteropolymer synthesized using petrochemically derived adipic acid (AA) (1, 4-butanedicarboxylic acid) and hexamethylene diamine [104]. Although challenged by many technological bottlenecks, AA can be derived by fermenting renewable biomass. The International Energy Agency (IEA) considers AA a priority platform chemical for the biobased production of polyamide plastics. The annual global production of AA is nearly 3 million tonnes, with a 3% to 5% yearly growth rate [105]. The majority of the AA is utilized to produce nylons, while a minor fraction goes into the production of other commodities, such as polyurethrenes and plasticizers [106]. The petrochemical route for producing AA from benzene produces nitrous oxide (N_2O), a potent greenhouse gas as a byproduct in the final synthesis stage [104]. Thus the synthesis of biobased AA becomes a priority. A nonmicrobial, chemical route for synthesizing AA employs the chemocatalytic conversion of glucose. In this method, glucaric acid derived from the aerobic oxidation of glucose is hydrodeoxygenated using metal catalysts to form AA [107]. In an approach that partially relies on microbial technology, precursor chemicals, glucaric acid, or muconic acid, is derived through fermentation and converted to AA acid using chemical methods [108]. Direct production of AA in bacteria is not straightforward since no central metabolic pathways that lead to its production are known except for an endogenous pathway reported in *Thermobifida fusca* [109]. However, pathways that could theoretically lead to AA production have been identified [105]. Some of the pathways with demonstrated potential to be harnessed for direct production are (1) reverse AA degradation, (2) ω-oxidation in combination with β-, or reverse β-oxidation, and (3) 2-oxopimelic acid pathway. Figure 6.7 summarizes the major pathways employed for the direct synthesis of AA by using engineered microorganisms. For the development of production hosts, microorganisms are metabolically engineered by incorporating a combination of heterologous enzymes so as to execute these pathways. The reverse AA degradation route relies on the flipping of the metabolic process involved in the breakdown of AA naturally found in *P. chrysogenum* [110]. The pathway was successfully deployed in metabolically engineered *E. coli* [111–113] and *T. fusca* [109]. A process utilizing the modified β-oxidation pathway engineered in recombinant yeast *(Candida spp.)* was used by Verdezyne to produce biobased AA from vegetable oils. The pathway starts with the conversion of palmitic acid, a 16-carbon fatty acid, into hexadecanedioic acid (a dicarboxylic acid) through the ω-oxidation pathway. In the β-oxidation pathway, in each cycle, the dicarboxylic acid would

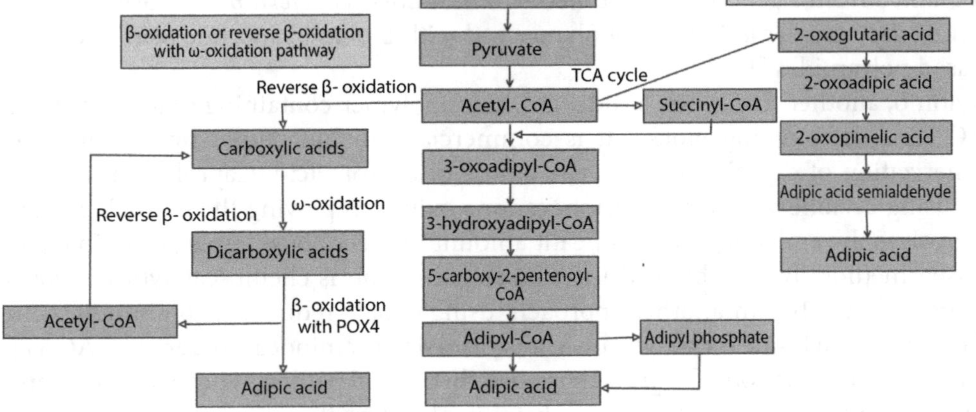

Figure 6.7 Metabolic routes employed for the synthesis of AA in engineered microorganisms.

undergo the removal of two carbon atoms and result in the formation of acetyl-CoA until the fatty acid chain is completely degraded. In the engineered strain, however, the composition of acyl-CoA oxidase, the first and the rate-limiting enzyme in the β-oxidation, was manipulated to alter substrate specificity. While the natural strain harbors two isoforms of this enzyme, one which identifies and acts on a wide range of chain lengths (4–20 carbons) (POX5 in *Candida* spp.) and another one which acts only on chain lengths above six carbons (POX4), the production host contained only the latter. Thus, once the dicarboxylic acid is oxidized and shortened to 6 carbon length forming AA, it no longer enters the oxidation cycle because the acyl-CoA oxidase shows little affinity to the substrate. AA accumulates and is secreted extracellularly [114]. However, it is not economically and energetically favorable to lose carbon atoms in each round of β-oxidation. Thus, approaches that channel acetyl-coA back into AA production are required. In engineered *E. coli,* reverse β-oxidation was employed to synthesize carboxylic acids from acetyl-CoA, which gets converted to AA via the ω-oxidation pathway and β-oxidation. 3-ketoacyl-CoA thiolase from *R. eutropha* was used for condensation because it synthesizes short chain length dicarboxylic acids (6–10 carbons). Thioesterase, responsible for removing coenzyme A, was sourced from *E. coli* (*ydil* gene) because it shows little affinity to acetyl- and butyryl-CoA and high specificity for C-6, C-8, and C-10 acyl-CoA substrates, thus favoring AA production [115]. In the 2-oxopimelic acid pathway, AA was obtained as a byproduct of an engineered metabolic route fine-tuned to convert glucose to aminocaproic acid (the building block of nylon-6) in *E. coli* [116]. Although several other attempts to ferment sugars and cheap renewable raw materials directly into AA using genetically modified bacteria have been reported in the literature, biobased synthesis of AA is still in the research and development phase.

Putrescine (1, 4-diaminobutane), a structural analog of hexamethylenediamine, is a four-carbon building block for Nylon 4, 6 (manufactured by Dutch State Mines (DSM) under the trade name Stanyl) along with AA. Industrially, putrescine is produced through the hydrogenation of succinonitrile. Metabolic engineering has been successfully employed to obtain putrescine through microbial fermentation. Engineered strains of microorganisms convert L-ornithine or its derivative L-arginine into putrescine using ornithine decarboxylase and arginine decarboxylase enzymes, respectively [117]. Another important monomer for biobased polyamides is cadaverine (1,5-diaminopentane). Decarboxylation of L-lysine derived from oxaloacetatic acid leads to the biosynthesis of cadverine. Biobased cadaverine obtained from a genetically modified strain of *Corynebacterium* harboring the L-lysine decarboxylase gene (*cadA*) was copolymerized with biobased succinic acid to derive fully biobased polyamides [118].

Nylon 6, another major polyamide, is a homopolymer containing 6-aminocaproic acid (6-ACA) as the building blocks. It is commercially manufactured by the ring-opening polymerization of caprolactam, a cyclic amide of caproic acid. Caprolactam is manufactured using cyclohexanone and cyclohexanone oxime employing the petrochemical route and involves the emission of a significant amount of greenhouse gases [116]. In a partially biobased method, lysine obtained through fermentation is chemo-catalytically converted to caprolactam [116]. In another approach, using enoate reductases, 6-aminohex-2-enoic acid derived from lysine is catalytically transformed to 6-aminocaproic acid (6-ACA) [119]. For the direct production of caprolactam, two theoretical pathways have been explored, (1) the adipyl-CoA pathway and (2) the α-ketopimelate (AKP) pathway. In the first method, 3-oxoadipyl-CoA derived through the 3-oxoadipyl-CoA thiolase mediated condensation of succinyl-CoA, and acetyl-CoA is reduced in a series of reactions to adipyl-CoA. A CoA-dependant aldehyde dehydrogenase then catalyzes the synthesis of adipate semialdehyde from adipyl-CoA. Adipate semialdehyde undergoes transamination with the help of an aminotransferase to yield 6-ACA. The second pathway starts with AKP, an intermediate chemical involved in the synthesis of coenzyme B in methanogenic archaea. Enzymatic decarboxylation and transamination of α-ketopimelate results in the formation of 6-ACA. Six-ACA can be derived using the left-AKP pathway or the right-AKP pathway. The left-AKP route starts with the decarboxylation of AKP to adipate semialdehyde and subsequent transamination to yield 6-ACA. The right-AKP route involves the synthesis of amino α-aminopimelate (AAP) by the transamination of AKP followed by decarboxylation to yield 6-ACA. The AKP pathway was deployed in *E. coli* using metabolic engineering to produce 6-ACA directly [116].

6.4.1.4 1, 3-Propanediol (PDO)

PDO is a platform chemical in synthesizing polyesters and polyurethanes. Owing to its wide applications in the synthesis of fibers used in floor coverings and sportswear, laminates, and adhesives, it has a global market demand of over 50 000 tons per year [120, 121]. PDO can be procured from petrochemical feedstock (ethylene dioxide or acrolein) using chemical catalysis in a relatively expensive process than similar diols. The method also yields toxic byproducts. PDO can also be synthesized using microbial fermentation. It is one of the oldest known products of anaerobic fermentation. Bacterial genera, such as *Klebsiella, Enterobacter, Citrobacter, Clostridium*, and *Lactobacillus*, produce PDO from glycerol. Fermentation of glycerol by wild-type strains also produces 2,3-butanediol, lactic acid, acetic acid, formic acid, and ethanol as byproducts [122]. Moreover, most of the wild-type producers are opportunistic pathogens limiting their use in industrial production. Genetically modified *E. coli*, producing PDO from glycerol through anaerobic fermentation, was obtained by transferring the genes for glycerol dehydratase (*dhaB*) and 1,3-propanediol oxidoreductase (*dhaT*) from *Klebsiella pneumoniae* [123]. Glycerol, however, is not an appropriate substrate for the commercial manufacturing of PDO due to cost considerations. Glycerol is still costlier than glucose despite being a byproduct of biodiesel production. Since all the natural producers of PDO employ glycerol as the substrate, genetically engineered strains have been used to enable PDO production from glucose [124]. One strategy is to incorporate genes that would convert glucose into glycerol in wild-type PDO producers such that glucose in the production media is converted to PDO. Another approach is integrating PDO-producing genes in bacteria that can produce glycerol from glucose [125]. There is an increased interest in optimizing the microbial process for PDO production to meet commercial demands, especially after its applications in producing biobased polyesters. In the year 2000, DuPont commercialized Sorona, a polytrimethylene terephthalate (PTT), by copolymerizing biobased PDO with petrochemically derived terephthalic acid, thereby creating a partial bioplastic with wide applications [126].

6.4.2 Bioplastics from Aromatic Monomers

Aromatic monomers are used for producing high-performance plastics with superior thermal and mechanical properties compared to polyesters or polyamides. They can be used in performance fabrics, electronics, automobiles, etc. Production of aromatic monomers through microbial fermentation to synthesize next-generation, high-performance bioplastics is an active field of research. Aromatic monomers, such as D-phenyllactic acid (D-PhLA), phenylpropanoid derivatives of cinnamic acid, 4-hydroxycinnamic acid (*p*-coumaric acid), and caffeic acid (3,4-dihydroxybenzeneacrylic acid), have been produced using engineered microorganisms through the fermentation of glucose [127–129]. 3-Amino-4-hydroxybenzoic acid (3, 4-AHBA), the building block for the textile polymer polybenzoxazole, is naturally produced by specific classes of microorganisms. Genetically modified strains of bacteria capable of converting biomass into 3, 4-AHBA have also been developed. Monomers for the synthesis of biobased polyimides, a high-performance plastic owing to superior chemical stability, and thermal and mechanical properties, can also be derived using microbial technology. Fermentation-based systems for synthesizing 4-aminophenylalanine (4APhe) from biomass have been developed. Enzymatic deamination of 4APhe yields 4-aminocinnamic acid (4ACA). In a fermentation-based approach to synthesize polyimide plastics, an engineered strain of *E. coli* harboring a phenyl alanine ammonia lyase (*AtPal4*) gene from *Arabidopsis* was used to convert 4APhe into 4ACA [130]. The aromatic monoamine was then converted to a diamine through the photocycloaddition of 4ACA and subsequently into polyimide films through chemo catalytic conversion. These films exhibited a glass transition temperature of 250 degrees, the best thermal resistance profile ever reported for a bioplastic [130]. 2,5-furandicarboxylic acid (FDCA), considered to be a potential alternative to petroleum-based terephthalic acid (TPA) in the polymer industry, was also produced through microbial fermentation [131]. FDCA was obtained from cellulose through a two-step process. Cellulose was first dehydrated to yield 5-hydroxymethylfurfural (HMF), which was then oxidized to FDCA [132, 133]. In a promising development, a direct approach for the synthesis of aromatic polyesters from glucose using genetically modified *E. coli* was recently reported. A copolymer of 3HB and D-PhLA (P(3HB)-co-D-phenyllactate) with 47.7 mol% of D-PhLA was produced by transferring the isocaprenoyl-CoA:2-hydroxyisocaproate-CoA transferase (HadA) from *Clostridium difficile* and an evolved polyhydroxyalkanoate (PHA) synthase gene into a D-PhLA producing strain. Further, by expressing *R. eutropha* β-ketothiolase (*phaA*) and reductase (*phaB*) genes, it was possible to obtain copolymers with different mol% of D-PhLA [134].

6.5 Lignocellulosic Biomass for Bioplastic Production

Green synthesis platforms focus on the sustainable production of bioplastics. Sustainability is partly achieved through the use of renewable biomass instead of petrochemicals. However, long-term sustainability can only be achieved by developing technologies that use inedible biomass as substrates instead of edible ones. In recent years, lignocellulose biomass has been recognized as a highly potent source for fermentable sugars [135]. Lignocellulose is the recalcitrant, non-starch, fibrous material found in plant cell walls. It is made up of cellulose (30–47%), hemicellulose (20–40%), and lignin (15–25%) [136, 137]. Hydrolysis of lignocellulose yields a mixture of fermentable carbohydrates, including glucose. Lignocellulose is abundant, cheap, and underutilized. Sources of lignocellulosic biomass include agricultural and agroindustrial (bagasse, rice husk, coconut shells and husk, areca nut husk, etc.) waste, forest biomass, and energy crops. Glucose in cellulose is held together by a relatively recalcitrant 1, 4-β glycosidic bond, unlike the 1, 4-α glycosidic bond in starch, making the hydrolytic release of fermentable sugars from lignocellulose a daunting task. The major challenge stems from the natural host's incapability to secrete hydrolytic enzymes that can convert cellulose into fermentable sugars. Physicochemical pretreatment and enzymatic hydrolysis are used to generate fermentable sugars so that lignocellulosic biomass can be used as a substrate to synthesize biobased products, including bioplastics.

The bioprocess approaches for producing biobased material from pretreated lignocellulose are classified into three groups [138]. In separate hydrolysis and fermentation (SHF), lignocellulose is hydrolyzed using exogenous or endogenous enzymes under optimal conditions to yield fermentable sugars (saccharification). Subsequently, in a separate process, the sugars are fermented to yield the desired product. In this approach, hydrolysis and saccharification can be carried out under optimal conditions without striking a balance because they are carried out in separate vessels. However, feedback inhibition of cellulolytic enzymes is a significant challenge. In simultaneous saccharification and fermentation (SSF), the breakdown of lignocellulose and fermentation is carried out in a single vessel by adding exogenous enzymes. Although the approach requires balancing the optimal reaction conditions, feedback inhibition of hydrolytic enzymes is less pronounced. In the third approach, called consolidated bioprocessing (CBP), the microorganism that carries out the fermentation secretes or surface-displays the hydrolytic enzymes [139].

Most of the commercial production of bioplastics relies on edible and refined biomass. A significant concern in the commercialization of the processes is the cost of sugars and fatty acids. The use of edible biomass for the production of nonfood commodities, including bioplastics, will result in a global increase in food price indices and worsen the food crisis. Bioplastic production from lignocellulose biomass is an active field of research. Lignocellulose has been successfully employed to produce PHB by using both SHF and CBP [140]. Alkali pretreated and enzymatic hydrolysates of softwood yielded good titers of PHB [141, 142]. Xylan from beechwood was used as a substrate to derive PHB, employing an engineered strain of *E. coli* capable of producing xylanase [143]. CBP was also employed to derive PHB from alkaline-pretreated lignin using *Rhodococcus jostii* [139]. CBP, however, yielded lower titers compared to SHF. Of note, neither CBP nor SHF could induce PHB titers comparable to glucose-based production. Aliphatic monomers like LA and SA have been successfully produced from lignocellulosic biomass. Hydrolysates of wood and sugarcane bagasse were used to make LA by SHF in *L. paracasei* and a heat-tolerant *Bacillus* sp. SSF was effectively employed for the production of LA from lignocellulosic biomass by the inclusion of exogenous cellulase enzyme in the fermentation media [144, 145]. CBP was also employed by using *L. plantarum*, an endoglucanase-producing strain [146]. However, SSF- and CBP-based processes yielded low titers of the product. Lignocellulose hydrolysates were also employed for the production of SA. SSF using cassava pulp yielded high titers of SA compared to SHF [147]. However, the yield was lower compared with the SHF of cassava flour. By employing an engineered strain of *E. coli* capable of producing endoxylanase and xylosidase, beechwood xylan was converted to SA with yields comparable to CBP of starch but significantly lower than an SSF-based approach [148]. Other aliphatic monomers produced by the fermentation of lignocellulosic biomass include cis, cis-muconic acid, and PDO [149, 150]. Production of aromatic monomers, such as cinnamic acid, *p*-coumaric acid, D-phenyllactic acid, using lignocellulosic biomass has also been reported [139].

6.6 Conclusion

Bioplastics are gaining popularity because they offer viable solutions to some of the most critical environmental issues. Bioplastics reduce the reliance on fossil fuels, decrease greenhouse emissions, check the accumulation of nonbiodegradable plastic waste, and reduce the synthesis of toxic byproducts. Employing microbial technology for bioplastic production furthers all these advantages significantly. Biomass that goes into the production of bioplastics is abundant, cheap, and renewable. However, optimization and scaling-up of the bioprocess is challenging and contributes to the overall cost of production. Laboratory-scale production parameters and yield are not always replicated in industry-scale operations. While most of the bioplastic production across the globe has relied on edible biomass and vegetable oil as substrates, there is an urgent need to switch to nonedible biomass, such as lignocellulose. Extracting glucose and other fermentable sugars from lignocellulose is not straightforward. The single most giant leap for the bioplastic industry would emerge from a commercially viable process for the saccharification and fermentation of lignocellulose. Advancements in microbiology, metabolomics, and metabolic engineering offer the possibility of developing optimal production hosts with rationally designed metabolic pathways and minimal interference from endogenous inhibition mechanisms. In spite of all the technological constraints, research efforts in favor of bioplastic production have to be pursued with rigor because of its environmental benefits.

References

1. Coppola, G., Gaudio, M.T., Lopresto, C.G., Calabro, V., Curcio, S., Chakraborty, S., Bioplastic from renewable biomass: A facile solution for a greener environment. *Earth Syst. Environ.*, 5, 231, 2021.
2. Sharma, B. and Jain, P., Deciphering the advances in bioaugmentation of plastic wastes. *J. Clean. Prod.*, 275, 123241, 2020.
3. Kumar, R., Verma, A., Shome, A., Sinha, R., Sinha, S., Jha, P.K., Kumar, R., Kumar, P., Shubham Impacts of plastic pollution on ecosystem services, sustainable development goals, and need to focus on circular economy and policy interventions. *Sustainability*, 13, 9963, 2021.
4. Geyer, R., Jambeck, J.R., Law, K.L., Production, use, and fate of all plastics ever made. *Sci. Adv.*, 3, e1700782, 2017.
5. Di Bartolo, A., Infurna, G., Dintcheva, N.T., A review of bioplastics and their adoption in the circular economy. *Polymers*, 13, 1229, 2021.
6. Gallo, F., Fossi, C., Weber, R., Santillo, D., Sousa, J., Ingram, I., Nadal, A., Romano, D., Marine litter plastics and microplastics and their toxic chemicals components: The need for urgent preventive measures. *Environ. Sci. Eur.*, 30, 13, 2018.
7. Lebreton, L., Slat, B., Ferrari, F., Sainte-Rose, B., Aitken, J., Marthouse, R., Hajbane, S., Cunsolo, S., Schwarz, A., Levivier, A., Noble, K., Debeljak, P., Maral, H., Schoeneich-Argent, R., Brambini, R., Reisser, J., Evidence that the Great Pacific Garbage Patch is rapidly accumulating plastic. *Sci. Rep.*, 8, 4666, 2018.
8. Royer, S.J., Ferrón, S., Wilson, S.T., Karl, D.M., Production of methane and ethylene from plastic in the environment. *PloS One*, 13, e0200574, 2018.

9. García-Depraect, O., Bordel, S., Lebrero, R., Santos-Beneit, F., Börner, R.A., Börner, T., Muñoz, R., Inspired by nature: Microbial production, degradation and valorization of biodegradable bioplastics for life-cycle-engineered products. *Biotechnol. Adv.*, 53, 107772, 2021.
10. Niaounakis, M., Recycling of biopolymers—The patent perspective. *Eur. Polym. J.*, 114, 464–475, 2019.
11. Atiwesh, G., Mikhael, A., Parrish, C.C., Banoub, J., Le, T.A.T., Environmental impact of bioplastic use: A review. *Heliyon*, 7, e07918, 2021.
12. Elsawy, M.A., Kim, K.H., Park, J.W., Deep, A., Hydrolytic degradation of polylactic acid (PLA) and its composites. *Renew. Sust. Energ. Rev.*, 79, 1346, 2017.
13. Janssen, P.J.D., Lambreva, M.D., Plumeré, N., Bartolucci, C., Antonacci, A., Buonasera, K., Frese, R.N., Scognamiglio, V., Rea, G., Photosynthesis at the forefront of a sustainable life. *Front. Chem.*, 2, 36, 2014.
14. Martinez Villadiego, K., Arias Tapia, M.J., Useche, J., Escobar Macías, D., Thermoplastic starch (TPS)/polylactic acid (PLA) blending methodologies: A review. *J. Polym. Environ.*, 30, 75–91, 2022.
15. Nazrin, A., Sapuan, S.M., Zuhri, M.Y.M., Ilyas, R.A., Syafiq, R., Sherwani, S.F.K., Nanocellulose reinforced thermoplastic starch (TPS), polylactic acid (PLA), and polybutylene succinate (PBS) for food packaging applications. *Front. Chem.*, 8, 213, 2020.
16. Radzik, P., Leszczyńska, A., Pielichowski, K., Modern biopolyamide-based materials: Synthesis and modification. *Polym. Bull.*, 77, 501, 2020.
17. Vitorino, L.C. and Bessa, L.A., Technological microbiology: Development and applications. *Front. Microbiol.*, 8, 827, 2017.
18. Pham, J.V., Yilma, M.A., Feliz, A., Majid, M.T., Maffetone, N., Walker, J.R., Kim, E., Cho, H.J., Reynolds, J.M., Song, M.C., Park, S.R., Yoon, Y.J., A review of the microbial production of bioactive natural products and biologics. *Front. Microbiol.*, 10, 3389, 2019.
19. Francis, D., Antimicrobials from microbes, in: *Bioresources and Bioprocess in Biotechnology: Volume 2: Exploring Potential Biomolecules*, S. Sugathan, N.S. Pradeep, S. Abdulhameed (Eds.), pp. 291–326, Springer, Singapore, 2017.
20. Feng, R., Chen, L., Chen, K., Fermentation trip: Amazing microbes, amazing metabolisms. *Ann. Microbiol.*, 68, 717, 2018.
21. Hara, K.Y., Araki, M., Okai, N., Wakai, S., Hasunuma, T., Kondo, A., Development of bio-based fine chemical production through synthetic bioengineering. *Microb. Cell Fact.*, 13, 173, 2014.
22. Chen, G.Q., New challenges and opportunities for industrial biotechnology. *Microb. Cell Fact.*, 11, 111, 2012.
23. Gohil, N., Bhattacharjee, G. *et al.*, An introduction to microbial cell factories for production of biomolecules, in: *Microbial Cell Factories Engineering for Production of Biomolecules*, V. Singh (Ed.), pp. 1–19, Academic Press, 2021.
24. Jung, Y.K. and Lee, S.Y., Efficient production of polylactic acid and its copolymers by metabolically engineered *Escherichia coli*. *J. Biotechnol.*, 151, 94, 2011.
25. Khatami, K., Perez-Zabaleta, M., Owusu-Agyeman, I., Cetecioglu, Z., Waste to bioplastics: How close are we to sustainable polyhydroxyalkanoates production? *Waste Manage.*, 119, 374, 2021.
26. Choi, S.Y., Rhie, M.N., Kim, H.T., Joo, J.C., Cho, I.J., Son, J., Jo, S.Y., Sohn, Y.J., Baritugo, K.-A., Pyo, J., Lee, Y., Lee, S.Y., Park, S.J., Metabolic engineering for the synthesis of polyesters: A 100-year journey from polyhydroxyalkanoates to non-natural microbial polyesters. *Metab. Eng.*, 58, 47, 2020.
27. Chen, J., Li, W., Zhang, Z.-Z., Tan, T.W., Li, Z.J., Metabolic engineering of *Escherichia coli* for the synthesis of polyhydroxyalkanoates using acetate as a main carbon source. *Microb. Cell Fact.*, 17, 102, 2018.

28. Corma, A., Iborra, S., Velty, A., Chemical routes for the transformation of biomass into chemicals. *Chem. Rev.*, 107, 2411, 2007.
29. Röper, H. and Koch, H., New carbohydrate derivatives from biotechnical and chemical processes. *Starch - Stärke*, 40, 453, 1988.
30. Tursi, A., A review on biomass: Importance, chemistry, classification, and conversion. *Biofuel Res. J.*, 6, 962, 2019.
31. Lemoigne, M., Produit de deshydratation et de polymerisation de l'acide β-oxybutyrique. *Bull. Soc. Chim. Biol.*, 8, 770, 1926.
32. Anderson, A.J. and Dawes, E.A., Occurrence, metabolism, metabolic role, and industrial uses of bacterial polyhydroxyalkanoates. *Microbiol. Rev.*, 54, 450–472, 1990.
33. Sudesh, K., Abe, H., Doi, Y., Synthesis, structure and properties of polyhydroxyalkanoates: Biological polyesters. *Prog. Polym. Sci.*, 25, 1503, 2000.
34. Raza, Z.A., Abid, S., Banat, I.M., Polyhydroxyalkanoates: Characteristics, production, recent developments and applications. *Int. Biodeterior. Biodegradation*, 126, 45, 2018.
35. Anjum, A., Zuber, M., Zia, K.M., Noreen, A., Anjum, M.N., Tabasum, S., Microbial production of polyhydroxyalkanoates (PHAs) and its copolymers: A review of recent advancements. *Int. J. Biol. Macromol.*, 89, 161, 2016.
36. Ray, S. and Kalia, V.C., Biomedical applications of polyhydroxyalkanoates. *Indian J. Microbiol.*, 57, 261, 2017.
37. Poltronieri, P. and Kumar, P., Polyhydroxyalcanoates (PHAs) in industrial applications, in: *Handbook of Ecomaterials*, L.M.T. Martínez, O.V. Kharissova, B.I. Kharisov (Eds.), pp. 1–30, Springer, Cham, 2017.
38. Nduko, J.M. and Taguchi, S., Microbial production of biodegradable lactate-based polymers and oligomeric building blocks from renewable and waste resources. *Front. Bioeng. Biotechnol.*, 8, 618077, 2021.
39. Carpine, R., Olivieri, G., Hellingwerf, K.J., Pollio, A., Marzocchella, A., Industrial production of poly-β-hydroxybutyrate from CO2: Can cyanobacteria meet this challenge? *Processes*, 8, 323, 2020.
40. Luengo, J.M., Garcia, B., Sandoval, A., Naharro, G., Olivera, E.R., Bioplastics from microorganisms. *Curr. Opin. Microbiol.*, 6, 251, 2003.
41. Verlinden, R.A.J., Hill, D.J., Kenward, M.A., Williams, C.D., Radecka, I., Bacterial synthesis of biodegradable polyhydroxyalkanoates. *J. Appl. Microbiol.*, 102, 1437, 2007.
42. Mergaert, J., Webb, A., Anderson, C., Wouters, A., Swings, J., Microbial degradation of poly(3-hydroxybutyrate) and poly(3-hydroxybutyrate-co-3-hydroxyvalerate) in soils. *Appl. Environ. Microbiol.*, 59, 3233, 1993.
43. Lageveen, R.G., Huisman, G.W., Preusting, H., Ketelaar, P., Eggink, G., Witholt, B., Formation of polyesters by *pseudomonas oleovorans*: Effect of substrates on formation and composition of poly-(R)-3-hydroxyalkanoates and poly-(R)-3-hydroxyalkenoates. *Appl. Environ. Microbiol.*, 54, 2924, 1998.
44. Du, G., Chen, J., Yu, J., Lun, S., Feeding strategy of propionic acid for production of poly (3-hydroxybutyrate-co-3-hydroxyvalerate) with *Ralstonia eutropha*. *Biochem. Eng. J.*, 8, 103, 2001.
45. Kessler, B., Weusthuis, R., Witholt, B., Eggink, G., Production of microbial polyesters: Fermentation and downstream processes. *Adv. Biochem. Eng. Biotechnol.*, 71, 159, 2001.
46. Hrabak, O., Industrial production of poly-β-hydroxybutyrate. *FEMS Microbiol. Lett.*, 103, 251, 1992.
47. Kaur, L., Khajuria, R., Parihar, L., Singh, G.D., Polyhydroxyalkanoates: Biosynthesis to commercial production—A review. *J. Microbiol. Biotechnol. Food Sci.*, 10, 1098, 2021.

48. Riedel, S.L., Jahns, S., Koenig, S., Bock, M.C.E., Brigham, C.J., Bader, J., Stahl, U., Polyhydroxyalkanoates production with *Ralstonia eutropha* from low quality waste animal fats. *J. Biotechnol.*, 214, 119, 2015.
49. Rosano, G.L. and Ceccarelli, E.A., Recombinant protein expression in *Escherichia coli:* Advances and challenges. *Front. Microbiol.*, 5, 172, 2014.
50. Schubert, P., Steinbüchel, A., Schlegel, H.G., Cloning of the *Alcaligenes eutrophus* genes for synthesis of poly-beta-hydroxybutyric acid (PHB) and synthesis of PHB in *Escherichia coli. J. Bacteriol.*, 170, 5837, 1988.
51. Slater, S.C., Voige, W.H., Dennis, D.E., Cloning and expression in *Escherichia coli* of the *Alcaligenes eutrophus* H16 poly-beta-hydroxybutyrate biosynthetic pathway. *J. Bacteriol.*, 170, 4431, 1998.
52. Lee, S.Y. and Lee, Y., Metabolic Engineering of *Escherichia coli* for production of enantiomerically pure (R)-(−)-hydroxycarboxylic acids. *Appl. Environ. Microbiol.*, 69, 3421, 2003.
53. Surendran, A., Lakshmanan, M., Chee, J.Y., Sulaiman, A.M., Thuoc, D.V., Sudesh, K., Can polyhydroxyalkanoates be produced efficiently from waste plant and animal oils? *Front. Bioeng. Biotechnol.*, 8, 169, 2020.
54. Bajaj, I. and Singhal, R., Poly (glutamic acid)–an emerging biopolymer of commercial interest. *Bioresour. Technol.*, 102, 5551, 2011.
55. Nair, P., Navale, G.R., Dharne, M.S., Poly-gamma-glutamic acid biopolymer: A sleeping giant with diverse applications and unique opportunities for commercialization. *Biomass Convers. Biorefin.*, 2021.
56. Ashiuchi, M., Microbial production and chemical transformation of poly-γ-glutamate. *Microb. Biotechnol.*, 6, 664, 2013.
57. Kunioka, M., Biosynthesis and chemical reactions of poly (amino acid) s from microorganisms. *Appl. Microbiol. Biotechnol.*, 47, 469, 1997.
58. Ashiuchi, M., Fukushima, K., Oya, H., Hiraoki, T., Shibatani, S., Oka, N., Nishimura, H., Hakuba, H., Nakamori, M., Kitagawa, M., Development of antimicrobial thermoplastic material from archaeal poly-γ-L-glutamate and its nanofabrication. *ACS Appl. Mater. Interfaces*, 5, 1619, 2013.
59. Sabbah, M., Di Pierro, P., Ruffo, F., Schiraldi, C., Alfano, A., Cammarota, M., Porta, R., Glutamic acid as repeating building block for bio-based films. *Polymers*, 12, 1613, 2020.
60. Choi, H.J. and Kunioka, M., Preparation conditions and swelling equilibria of hydrogel prepared by γ-irradiation from microbial poly (γ-glutamic acid). *Radiat. Phys. Chem.*, 46, 175, 1995.
61. Biagiotti, M., Borghese, G., Francescato, P., Morelli, C.F., Albertini, A.M., Bavaro, T., Ubiali, D., Mendichi, R., Speranza, G., Esterification of poly (γ-glutamic acid) (γ-PGA) mediated by its tetrabutylammonium salt. *RSC Adv.*, 6, 43954, 2016.
62. Du, G., Yang, G., Qu, Y., Chen, J.C., Lun, S., Effects of glycerol on the production of poly(gamma-glutamic acid) by *Bacillus licheniformis. Process Biochem.*, 40, 2143, 2005.
63. Wu, Q., Xu, H., Shi, N., Yao, J., Li, S., Ouyang, P., Improvement of poly(gamma-glutamic acid) biosynthesis and redistribution of metabolic flux with the presence of different additives in *Bacillus subtilis* CGMCC 0833. *Appl. Microbiol. Biotechnol.*, 79, 527, 2008.
64. Jian, X., Chen, S., Yu, Z., Optimization of process parameters for poly γ-glutamate production under solid state fermentation from *Bacillus subtilis* CCTCC202048. *Process Biochem.*, 40, 3075, 2005.
65. Jiang, H., Shang, L., Yoon, S.H., Lee, S.Y., Yu, Z., Optimal production of poly-gamma-glutamic acid by metabolically engineered *Escherichia coli. Biotechnol. Lett.*, 28, 1241, 2006.
66. Jamshidian, M., Tehrany, E.A., Imran, M., Jacquot, M., Desobry, S., Poly-lactic acid: Production, applications, nanocomposites, and release studies. *Compr. Rev. Food Sci. Food Saf.*, 9, 552, 2010.

67. Alexandri, M., Schneider, R., Mehlmann, K., Venus, J., Recent advances in D-lactic acid production from renewable resources: Case studies on agro-industrial waste streams. *Food Technol. Biotechnol.*, 57, 293, 2019.
68. Müller, T., Comparison of methods for differentiation between homofermentative and heterofermentative lactic acid bacteria. *Zentralbl. Mikrobiol.*, 145, 363, 1990.
69. Jin, B., Yin, P., Ma, Y., Zhao, L., Production of lactic acid and fungal biomass by Rhizopus fungi from food processing waste streams. *J. Ind. Microbiol. Biotechnol.*, 32, 678, 2005.
70. Abedi, E. and Hashemi, S.M.B., Lactic acid production—producing microorganisms and substrates sources-state of art. *Heliyon*, 6, e04974, 2020.
71. Kim, S.H. and Kim, Y.H., Direct condensation polymerization of lactic acid. *Macromol. Symp.*, 144, 277, 1999.
72. Ajioka, M., Enomoto, K., Suzuki, K., Yamaguchi, A., The basic properties of poly(lactic acid) produced by the direct condensation polymerization of lactic acid. *J. Environ. Polym. Degrad.*, 3, 225, 1995.
73. Robert, J.L. and Aubrecht, K.B., Ring-opening polymerization of lactide to form a biodegradable polymer. *J. Chem. Educ.*, 85, 258, 2008.
74. Auras, R., Harte, B., Selke, S., An overview of polylactides as packaging materials. *Macromol. Biosci.*, 4, 835, 2004.
75. Mehta, R., Kumar, V., Bhunia, H., Upadhyay, S.N., Synthesis of poly(lactic acid): A review. *J. Macromol. Sci. Part C*, 45, 325, 2004.
76. Pretula, J., Slomkowski, S., Penczek, S., Polylactides—methods of synthesis and characterization. *Adv. Drug Deliv. Rev.*, 107, 3, 2016.
77. Singhvi, M.S., Zinjarde, S.S., Gokhale, D.V., Polylactic acid: Synthesis and biomedical applications. *J. Appl. Microbiol.*, 127, 1612, 2019.
78. Steinbüchel, A. and Valentin, H.E., Diversity of bacterial polyhydroxyalkanoic acids. *FEMS Microbiol. Lett.*, 128, 219, 1995.
79. Taguchi, S., Yamada, M., Matsumoto, K., Tajima, K., Satoh, Y., Munekata, M., Ohno, K., Kohda, K., Shimamura, T., Kambe, H., Obata, S., A microbial factory for lactate-based polyesters using a lactate-polymerizing enzyme. *Proc. Natl. Acad. Sci. U. S. A.*, 105, 17323, 2008.
80. Shozui, F., Matsumoto, K., Motohashi, R., Yamada, M., Taguchi, S., Establishment of a metabolic pathway to introduce the 3-hydroxyhexanoate unit into LA-based polyesters via a reverse reaction of β-oxidation in *Escherichia coli* LS5218. *Polym. Degrad. Stab.*, 95, 1340, 2010.
81. Yamada, M., Matsumoto, K., Nakai, T., Taguchi, S., Microbial production of lactate-enriched poly[(R)-lactate-co-(R)-3-hydroxybutyrate] with novel thermal properties. *Biomacromolecules*, 10, 677, 2009.
82. Chek, M.F., Hiroe, A., Hakoshima, T., Sudesh, K., Taguchi, S., PHA synthase (PhaC): Interpreting the functions of bioplastic-producing enzyme from a structural perspective. *Appl. Microbiol. Biotechnol.*, 103, 1131, 2019.
83. Yamada, M., Matsumoto, K., Shimizu, K., Uramoto, S., Nakai, T., Shozui, F., Taguchi, S., Adjustable mutations in lactate (LA)-polymerizing enzyme for the microbial production of LA-based polyesters with tailor-made monomer composition. *Biomacromolecules*, 11, 815, 2010.
84. Shozui, F., Matsumoto, K., Nakai, T., Yamada, M., Taguchi, S., Biosynthesis of novel terpolymers poly(lactate-co-3-hydroxybutyrate-co-3-hydroxyvalerate)s in lactate-overproducing mutant *Escherichia coli* JW0885 by feeding propionate as a precursor of 3-hydroxyvalerate. *Appl. Microbiol. Biotechnol.*, 85, 949, 2010.
85. Mamat, U., Wilke, K., Bramhill, D., Schromm, A.B., Lindner, B., Kohl, T.A., Corchero, J.L., Villaverde, A., Schaffer, L., Head, S.R., Souvignier, C., Meredith, T.C., Woodard, R.W.,

Detoxifying *Escherichia coli* for endotoxin-free production of recombinant proteins. *Microb. Cell Fact.*, 14, 57, 2015.
86. Grage, K., McDermott, P., Rehm, B.H.A., Engineering *Bacillus megaterium* for production of functional intracellular materials. *Microb. Cell Fact.*, 16, 211, 2017.
87. Taguchi, S., Ooi, T., Mizuno, K., Matsusaki, H., Advances and needs for endotoxin-free production strains. *Appl. Microbiol. Biotechnol.*, 99, 9349, 2015.
88. Jo, S.J., Maeda, M., Ooi, T., Taguchi, S., Production system for biodegradable polyester polyhydroxybutyrate by *Corynebacterium glutamicum*. *J. Biosci. Bioeng.*, 102, 233, 2006.
89. Jo, S.-J., Matsumoto, K., Leong, C.R., Ooi, T., Taguchi, S., Improvement of poly(3-hydroxybutyrate) [P(3HB)] production in *Corynebacterium glutamicum* by codon optimization, point mutation and gene dosage of P(3HB) biosynthetic genes. *J. Biosci. Bioeng.*, 104, 457, 2007.
90. Song, Y., Matsumoto, K., Yamada, M., Gohda, A., Brigham, C.J., Sinskey, A.J., Taguchi, S., Engineered *Corynebacterium glutamicum* as an endotoxin-free platform strain for lactate-based polyester production. *Appl. Microbiol. Biotechnol.*, 93, 1917, 2012.
91. Matsumoto, K., Tobitani, K., Aoki, S., Song, Y., Ooi, T., Taguchi, S., Improved production of poly(lactic acid)-like polyester based on metabolite analysis to address the rate-limiting step. *AMB Express*, 4, 83, 2014.
92. Zeng, J.B., Li, Y.D., Zhu, Q.Y., Yang, K.K., Wang, X.L., Wang, Y.Z., A novel biodegradable multiblock poly(ester urethane) containing poly(l-lactic acid) and poly(butylene succinate) blocks. *Polymer*, 50, 1178, 2009.
93. Rafiqah, S.A., Khalina, A., Harmaen, A.S., Tawakkal, I.A., Zaman, K., Asim, M., Nurrazi, M.N., Lee, C.H., A review on properties and application of bio-based poly(butylene succinate). *Polym. (Basel)*, 13, 1436, 2021.
94. Xu, J. and Guo, B.-H., Poly (butylene succinate) and its copolymers: Research, development and industrialization. *Biotechnol. J.*, 5, 1149, 2010.
95. Ahn, J.H., Jang, Y.-S., Lee, S.Y., Production of succinic acid by metabolically engineered microorganisms. *Curr. Opin. Biotechnol.*, 42, 54, 2016.
96. Singh, A., Lynch, M.D., Gill, R.T., Genes restoring redox balance in fermentation-deficient *E. coli* NZN111. *Metab. Eng.*, 11, 347, 2009.
97. Balzer, G.J., Thakker, C., Bennett, G.N., San, K.Y., Metabolic engineering of *Escherichia coli to* minimize byproduct formate and improving succinate productivity through increasing NADH availability by heterologous expression of NAD^+-dependent formate dehydrogenase. *Metab. Eng.*, 20, 1, 2013.
98. Choi, S., Song, C.W., Shin, J.H., Lee, S.Y., Biorefineries for the production of top building block chemicals and their derivatives. *Metab. Eng.*, 28, 223, 2015.
99. Cukalovic, A. and Stevens, C.V., Feasibility of production methods for succinic acid derivatives: A marriage of renewable resources and chemical technology. *Biofuel. Bioprod. Biorefin.*, 2, 505, 2008.
100. Xu, J. and Guo, B.H., Poly (butylene succinate) and its copolymers: Research, development and industrialization. *Biotechnol. J.*, 5, 1149, 2010.
101. Cheng, J., Li, J., Zheng, L., Achievements and perspectives in 1, 4-butanediol production from engineered microorganisms. *J. Agric. Food Chem.*, 69, 10480, 2021.
102. Yim, H., Haselbeck, R., Niu, W., Pujol-Baxley, C., Burgard, A., Boldt, J., Khandurina, J., Trawick, J.D., Osterhout, R.E., Stephen, R., Estadilla, J., Teisan, S., Schreyer, H.B., Andrae, S., Yang, T.H., Lee, S.Y., Burk, M.J., Van Dien, S., Metabolic engineering of *Escherichia coli* for direct production of 1,4-butanediol. *Nat. Chem. Biol.*, 7, 445, 2011.
103. Tsuge, Y., Kawaguchi, H., Sasaki, K., Kondo, A., Engineering cell factories for producing building block chemicals for bio-polymer synthesis. *Microb. Cell Fact.*, 15, 19, 2016.

104. Niu, W., Draths, K.M., Frost, J.W., Benzene-free synthesis of adipic acid. *Biotechnol. Prog.*, 18, 201, 2002.
105. Skoog, E., Shin, J.H., Saez-Jimenez, V., Mapelli, V., Olsson, L., Biobased adipic acid—The challenge of developing the production host. *Biotechnol. Adv.*, 36, 2248, 2018.
106. Polen, T., Spelberg, M., Bott, M., Toward biotechnological production of adipic acid and precursors from biorenewables. *J. Biotechnol.*, 167, 75, 2013.
107. T.R. Boussie, E.L. Dias, Z.M. Fresco, V.J. Murphy, Production of adipic acid and derivatives from carbohydrate-containing materials. US Patent 8501989B2, assigned to Archer Daniels Midland Co, 2013.
108. Weber, C., Brückner, C., Weinreb, S., Lehr, C., Essl, C., Boles, E., Biosynthesis of cis,cis-muconic acid and its aromatic precursors, catechol and protocatechuic acid, from renewable feedstocks by *Saccharomyces cerevisiae*. *Appl. Environ. Microbiol.*, 78, 8421, 2012.
109. Deng, Y., Ma, L., Mao, Y., Biological production of adipic acid from renewable substrates: Current and future methods. *Biochem. Eng. J.*, 105, 16, 2016.
110. Thykaer, J., Christensen, B., Nielsen, J., Metabolic network analysis of an adipoyl-7-ADCA-producing strain of *Penicillium chrysogenum*: Elucidation of adipate degradation. *Metab. Eng.*, 4, 151, 2002.
111. Yu, J.L., Xia, X.-X., Zhong, J.J., Qian, Z.G., Direct biosynthesis of adipic acid from a synthetic pathway in recombinant *Escherichia coli*. *Biotechnol. Bioeng.*, 111, 2580, 2014.
112. Babu, T., Yun, E.J., Kim, S., Kim, D.H., Liu, K.H., Kim, S.R., Kim, K.H., Engineering *Escherichia coli* for the production of adipic acid through the reversed β-oxidation pathway. *Process Biochem.*, 50, 2066, 2015.
113. Kallscheuer, N., Gätgens, J., Lübcke, M., Pietruszka, J., Bott, M., Polen, T., Improved production of adipate with *Escherichia coli* by reversal of β-oxidation. *Appl. Microbiol. Biotechnol.*, 101, 2371, 2017.
114. Beardslee, T. and Picataggio, S., Bio-based adipic acid from renewable oils. *Lipid Technol.*, 24, 223, 2012.
115. Clomburg, J.M., Blankschien, M.D., Vick, J.E., Chou, A., Kim, S., Gonzalez, R., Integrated engineering of β-oxidation reversal and ω-oxidation pathways for the synthesis of medium chain ω-functionalized carboxylic acids. *Metab. Eng.*, 28, 202, 2015.
116. Turk, S.C.H.J., Kloosterman, W.P., Ninaber, D.K., Kolen, K.P.A.M., Knutova, J., Suir, E., Schürmann, M., Raemakers-Franken, P.C., Müller, M., de Wildeman, S.M.A., Raamsdonk, L.M., van der Pol, R., Wu, L., Temudo, M.F., van der Hoeven, R.A.M., Akeroyd, M., van der Stoel, R.E., Noorman, H.J., Bovenberg, R.A.L., Trefzer, A.C., Metabolic engineering toward sustainable production of nylon-6. *ACS Synth. Biol.*, 5, 65, 2016.
117. Qian, Z.-G., Xia, X.-X., Lee, S.Y., Metabolic engineering of *Escherichia coli* for the production of putrescine: A four carbon diamine. *Biotechnol. Bioeng.*, 104, 651, 2009.
118. Ma, W., Chen, K., Li, Y., Hao, N., Wang, X., Ouyang, P., Advances in cadaverine bacterial production and its applications. *Engineering*, 3, 308, 2017.
119. P.C. Raemakers-Franken, P.M.M. Nossin, P.M. Brandts, M.G. Wubbolts, W.P.H. Peeters, S. Ernste, S.M.A.D. Wildeman, M. Schuermann, Biochemical synthesis of 6-amino caproic acid. European Union Patent EP1706501B1, 2010.
120. Zhou, M., Chen, J., Freguia, S., Rabaey, K., Keller, J., Carbon and electron fluxes during the electricity driven 1, 3-propanediol biosynthesis from glycerol. *Environ. Sci. Technol.*, 47, 11199, 2013.
121. Saxena, R.K., Anand, P., Saran, S., Isar, J., Microbial production of 1, 3-propanediol: Recent developments and emerging opportunities. *Biotechnol. Adv.*, 27, 895, 2009.
122. Biebl, H., Menzel, K., Zeng, A.P., Deckwer, W.D., Microbial production of 1, 3-propanediol. *Appl. Microbiol. Biotechnol.*, 52, 289, 1999.

123. Stephanopoulos, G.N., Aristidou, A.A. *et al.*, Examples of pathway manipulations: Metabolic engineering in practice, in: *Metabolic Engineering*, G.N. Stephanopoulos, A.A. Aristidou, J. Nielsen (Eds.), pp. 203–283, Academic Press, 1998.
124. Vivek, N., Hazeena, S.H., Alphy, M.P., Kumar, V., Magdouli, S., Sindhu, R., Pandey, A., Binod, P., Recent advances in microbial biosynthesis of C3–C5 diols: Genetics and process engineering approaches. *Bioresour. Technol.*, 322, 124527, 2021.
125. Nakamura, C.E. and Whited, G.M., Metabolic engineering for the microbial production of 1,3-propanediol. *Curr. Opin. Biotechnol.*, 14, 454, 2003.
126. Bhatia, S.K. and Kurian, J.V., Biological characterization of Sorona polymer from corn-derived 1,3-propanediol. *Biotechnol. Lett.*, 30, 619, 2008.
127. Kawai, Y., Noda, S., Ogino, C., Takeshima, Y., Okai, N., Tanaka, T., Kondo, A., p-Hydroxycinnamic acid production directly from cellulose using endoglucanase- and tyrosine ammonia lyase-expressing *Streptomyces lividans*. *Microb. Cell Fact.*, 12, 45, 2013.
128. Noda, S., Miyazaki, T., Miyoshi, T., Miyake, M., Okai, N., Tanaka, T., Ogino, C., Kondo, A., Cinnamic acid production using Streptomyces lividans expressing phenylalanine ammonia lyase. *J. Ind. Microbiol. Biotechnol.*, 38, 643, 2011.
129. Rodriguez, A., Kildegaard, K.R., Li, M., Borodina, I., Nielsen, J., Establishment of a yeast platform strain for production of p-coumaric acid through metabolic engineering of aromatic amino acid biosynthesis. *Metab. Eng.*, 31, 181, 2015.
130. Suvannasara, P., Tateyama, S., Miyasato, A., Matsumura, K., Shimoda, T., Ito, T., Yamagata, Y., Fujita, T., Takaya, N., Kaneko, T., Biobased polyimides from 4-aminocinnamic acid photodimer. *Macromolecules*, 47, 1586, 2014.
131. Luo, K., Wang, Y., Yu, J., Zhu, J., Hu, Z., Semi-bio-based aromatic polyamides from 2,5-furandicarboxylic acid: Toward high-performance polymers from renewable resources. *RSC Adv.*, 6, 87013, 2016.
132. Sajid, M., Zhao, X., Liu, D., Production of 2,5-furandicarboxylic acid (FDCA) from 5-hydroxymethylfurfural (HMF): Recent progress focusing on the chemical-catalytic routes. *Green Chem.*, 20, 5427, 2018.
133. Kim, H., Lee, S., Ahn, Y., Lee, J., Won, W., Sustainable production of bioplastics from lignocellulosic biomass: Technoeconomic analysis and life-cycle assessment. *ACS Sustain. Chem. Eng.*, 8, 12419, 2020.
134. Yang, J.E., Park, S.J., Kim, W.J., Kim, H.J., Kim, B.J., Lee, H., Shin, J., Lee, S.Y., One-step fermentative production of aromatic polyesters from glucose by metabolically engineered *Escherichia coli* strains. *Nat. Commun.*, 9, 79, 2018.
135. Fatma, S., Hameed, A., Noman, M., Ahmed, T., Shahid, M., Tariq, M., Sohail, I., Tabassum, R., Lignocellulosic biomass: A sustainable bioenergy source for the future. *Protein Pept. Lett.*, 25, 148, 2018.
136. Mosier, N., Wyman, C., Dale, B., Elander, R., Lee, Y.Y., Holtzapple, M., Ladisch, M., Features of promising technologies for pretreatment of lignocellulosic biomass. *Bioresour. Technol.*, 96, 673, 2005.
137. Menon, V. and Rao, M., Trends in bioconversion of lignocellulose: Biofuels, platform chemicals & biorefinery concept. *Prog. Energy Combust. Sci.*, 38, 522, 2012.
138. den Haan, R., van Rensburg, E., Rose, S.H., Görgens, J.F., van Zyl, W.H., Progress and challenges in the engineering of non-cellulolytic microorganisms for consolidated bioprocessing. *Curr. Opin. Biotechnol.*, 33, 32, 2015.
139. Kawaguchi, H., Hasunuma, T., Ogino, C., Kondo, A., Bioprocessing of bio-based chemicals produced from lignocellulosic feedstocks. *Curr. Opin. Biotechnol.*, 42, 30, 2016.

140. Cesário, M.T., Raposo, R.S., de Almeida, M.C.M.D., van Keulen, F., Ferreira, B.S., da Fonseca, M.M.R., Enhanced bioproduction of poly-3-hydroxybutyrate from wheat straw lignocellulosic hydrolysates. *N. Biotechnol.*, 31, 104, 2014.
141. Saratale, G.D. and Oh, M.K., Characterization of poly-3-hydroxybutyrate (PHB) produced from *Ralstonia eutropha* using an alkali-pretreated biomass feedstock. *Int. J. Biol. Macromol.*, 80, 627, 2015.
142. Bowers, T., Vaidya, A., Smith, D.A., Lloyd-Jones, G., Softwood hydrolysate as a carbon source for polyhydroxyalkanoate production. *J. Chem. Technol. Biotechnol.*, 89, 1030, 2014.
143. Salamanca-Cardona, L., Scheel, R.A., Lundgren, B.R., Stipanovic, A.J., Matsumoto, K., Taguchi, S., Nomura, C.T., Deletion of the *pflA* gene in *Escherichia coli* LS5218 and its effects on the production of polyhydroxyalkanoates using beechwood xylan as a feedstock. *Bioengineered*, 5, 284, 2014.
144. Kuo, Y.-C., Yuan, S.-F., Wang, C.-A., Huang, Y.-J., Guo, G.-L., Hwang, W.-S., Production of optically pure l-lactic acid from lignocellulosic hydrolysate by using a newly isolated and d-lactate dehydrogenase gene-deficient *Lactobacillus paracasei* strain. *Bioresour. Technol.*, 198, 651, 2015.
145. Peng, L., Xie, N., Guo, L., Wang, L., Yu, B., Ma, Y., Efficient open fermentative production of polymer-grade L-lactate from sugarcane bagasse hydrolysate by thermotolerant *Bacillus* sp. strain P38. *PloS One*, 9, e107143, 2014.
146. Okano, K., Zhang, Q., Yoshida, S., Tanaka, T., Ogino, C., Fukuda, H., Kondo, A., D-lactic acid production from cellooligosaccharides and β-glucan using l-LDH gene-deficient and endoglucanase-secreting *Lactobacillus plantarum*. *Appl. Microbiol. Biotechnol.*, 85, 643, 2010.
147. Sawisit, A., Jantama, S.S., Kanchanatawee, S., Jantama, K., Efficient utilization of cassava pulp for succinate production by metabolically engineered *Escherichia coli* KJ122. *Bioprocess Biosyst. Eng.*, 38, 175, 2015.
148. Zheng, Z., Chen, T., Zhao, M., Wang, Z., Zhao, X., Engineering *Escherichia coli* for succinate production from hemicellulose via consolidated bioprocessing. *Microb. Cell Fact.*, 11, 37, 2012.
149. Vardon, D.R., Franden, M.A., Johnson, C.W., Karp, E.M., Guarnieri, M.T., Linger, J.G., Salm, M.J., Strathmann, T.J., Beckham, G.T., Adipic acid production from lignin. *Energy Environ. Sci.*, 8, 617, 2015.
150. Przystałowska, H., Zeyland, J., Szymanowska-Powałowska, D., Szalata, M., Słomski, R., Lipiński, D., 1, 3-Propanediol production by new recombinant *Escherichia coli* containing genes from pathogenic bacteria. *Microbiol. Res.*, 171, 1, 2015.

7

Synthesis of Green Bioplastics

J.E. Castanheiro[1]*, P.A. Mourão[1] and I. Cansado[1,2]

[1]*MED, D.Q.B., ECT, Universidade de Évora, Évora, Portugal*
[2]*LAQV-RQUIMTE, D.Q.B., ECT, Universidade de Évora, Évora, Portugal*

Abstract

Conventional plastics are used in the last years/decades. However, these materials have environmental problems. After its utilization, they are not biodegradable and generate an increase of greenhouse gases. Thus, it is important to replace the traditional plastics by a new material: bioplastics. Bioplastics are biodegradable, and some of them are biocompatible. Starch, cellulose, polyhydroxyalkanoates, poly(lactic acid), and cellulose are bioplastics, which can be used in agriculture, medicine, packaging, structural, and automotive engineering. Bioplastics are produced from polysaccharide, protein, and other glyceride resources. Starch is a thermoplastic, and it can be used for different applications. Also, cellulose and polyhydroxyalkanoates have different applications.

In this work, it made a review in relation to bioplastics (most common) as green plastic and its production.

***Keywords*:** Bioplastics, biopolymer, starch, PHA, cellulose, chitosan

7.1 Introduction

Traditionally, plastics are materials derived from the petrochemical industries. These materials are composed of a molecular unit (a monomer) that are repeated in the chains. Polymers have chains with different molecular weights. Therefore, polymers have a molecular distribution of weight. These traditional polymers are prepared by chemical synthesis. Different synthesis methods can be used to prepare polymers from monomers: polycondensations, polyadditions. There are three groups of polymers: thermoplastics, thermosets, and elastomers. Some polymers are produced in large amounts like polyethylene, polypropylene, polystyrene, polyethylene terephthalate [1–5].

However, the petrochemical-based polymers have a high environmental impact due to the increase of the greenhouse gases amount and energy consumption. Therefore, bioplastics can reduce these environmental problems. Bioplastics can replace the traditional polymers. They are produced from renewable raw material [6]. Different raw materials, like polysaccharides (chitosan, cellulose, starch), proteins (gluten), organic waste, wastewater, algae, are used as raw material to produce a new class of polymers: biopolymers [7].

**Corresponding author*: jefc@uevora.pt

Inamuddin and Tariq Altalhi (eds.) Handbook of Bioplastics and Biocomposites Engineering Applications: 2nd Edition, (149–160)

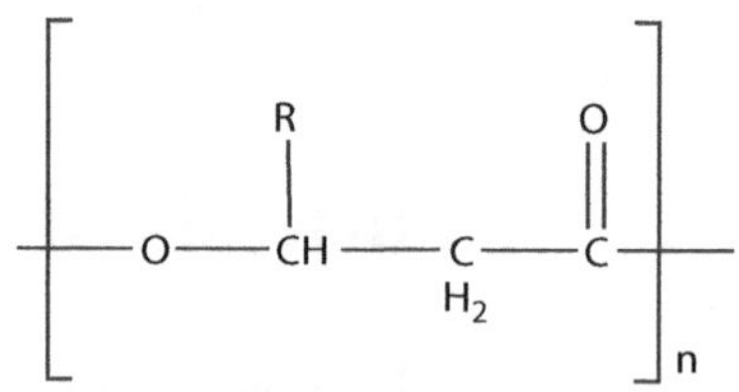

Figure 7.1 Monomeric structure of PHA.

Bioplastics are biodegradable. In this class, the starch, polyhydroxyalkanoates (PHA), and polylactide (PLA) bioplastics are included. Figure 7.1 shows the monomeric unit of the PHA.

PHAs have several characteristics, such as completely biodegradable in water and in soil [8]; good resistance at temperature (until 120°C) [9].

This chapter focuses on bioplastics synthesis and applications.

7.2 Bioplastic

7.2.1 Polyhydroxyalkanoates (PHAs)

Some bacteria (about 250 types) can produce PHAs from different renewable raw material, such as lipids and sugar. These polymers are intracellular products of these microorganisms. To recover the PHAs, the cell wall of bacteria needs to be destroyed [10–12]. This material is thermoplastic and biodegradable. There are several PHA composites (about 100). However, polyhydroxybutyrate (PHB) is thermoplastic common. The PHAs can be an alternative to traditional polymers, due to their physicochemical properties [1, 12]. Table 7.1 shows some characteristics of PHAs [12].

Table 7.1 Characteristics of PHAs.

Properties	**Uses**	**Biodegradable**
Gas barrier ☺	Packing ☺	In water
UV resistant ☺	Adhesives ☺	In soil
Biocompatible ☺	Fibers ☺	
Thermostable ☹	Biomedical applications ☺	
Elastic ☺		
Rigid ☺		
Hydrophobic ☺		
Thermoplastic ☺		

7.2.2 Poly(lactic acid) (PLA)

Poly(lactic acid) (PLA) is produced from lactic acid, by a polymerization process, which is obtained from sugarcane, chips, and starch. This polymer is a polyester [13, 14].

Figure 7.2 shows the monomeric unit of polylactic acid.

PLA is a thermoplastic polyester, which is biodegradable by hydrolysis. The main application of PLA is the food industry. It is used to produce cups, food plates, and trays. However, it is important to note that the PLA, due to its fragility, is not recommended to others packing processes The PLA can be also used for agriculture films, shopping bags, and hygiene products [15, 16].

Different PLA transformation processes, like extrusion, injection molding, can be applied to this polymer [16]. PLA is a biopolymer that can be used to replace traditional polymers, like polyethylene.

Table 7.2 shows some properties and applications of PLA [12].

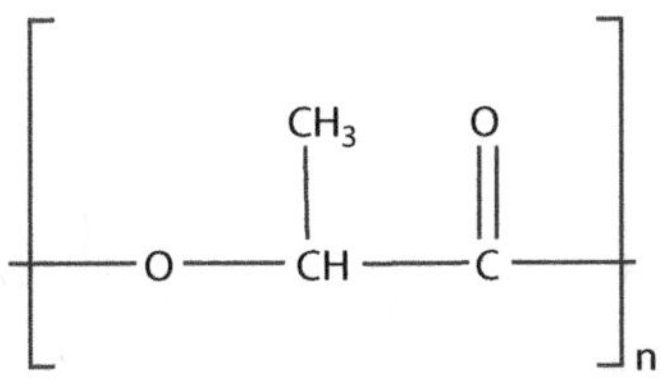

Figure 7.2 Monomeric structure of PLA.

Table 7.2 Properties of PLA.

Properties	**Uses**	**Biodegradable**
Gas barrier ☺	Packing	In water
UV resistant ☺	Adhesives	In soil
Biocompatible ☺	Fibers	
Thermostable ☹	Biomedical applications	
Elastic ☺	Films	
Rigid ☺		
Hydrophobic ☺		
Thermoplastic ☺		

7.2.3 Cellulose

Cellulose is a biopolymer present in plant cell walls, and it is obtained from wood. It is a linear chain which is linked by units that give structure to plants. Due to its structure, the cellulose has crystalline areas (corresponding to some order of polymer chains), and it does not have crystalline areas (corresponding to some disorder of polymer chains). The last areas are known as amorphous polymeric areas. Due to the crystallinity of cellulose, the polymer is insoluble in water. This polymer is biodegradable, nontoxic, and it can be used in the food, pharmaceutical industry [17, 18]. The cellulose can be derivatized by reaction of OH groups with alcohols or an acid. The ester of cellulose is a biopolymer that can be used to manufacture parts by injection and extrusion. This new material has the properties of a thermoplastic polymer. Also, from this polymer, it can be obtained ethers of cellulose by reaction of OH groups of cellulose with methanol or ethanol. The polymer obtained by reaction with methanol is soluble in water. However, the product obtained by reaction with ethanol is not soluble in water. These materials can be used as films [12, 18].

Figure 7.3 shows the monomeric unit of cellulose. Table 7.3 shows some characteristics of cellulose [12].

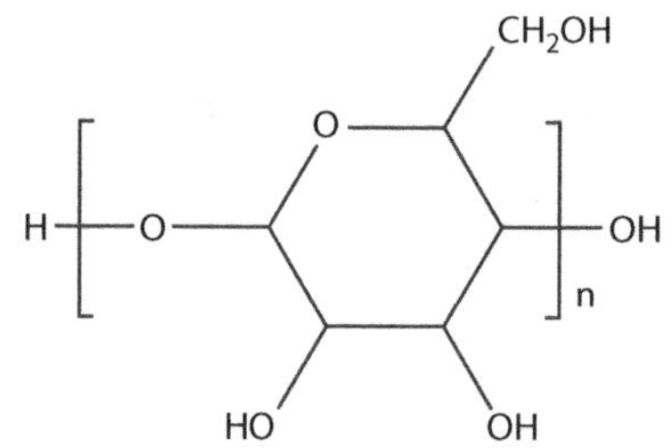

Figure 7.3 Monomeric unit of cellulose.

Table 7.3 Characteristics of cellulose.

Properties	**Uses**	**Biodegradable**
Gas barrier ☹	Textile	In water
UV resistant ☹	Air filters	In soil
Biocompatible ☺	Wound dress	
Thermostable ☺	Coatings	
Elastic ☹		
Rigid ☺		
Hydrophobic ☹		
Thermoplastic ☹		

7.2.4 Starch

This material is a biopolymer and is recyclable. It is obtained from natural sources. To transform this polymer into a plastic used in film production, it is necessary to introduce water or glycerol (a plasticizer). This material is called a starch with thermoplastic properties. This polymer can be used to replace polyethylene. These biopolymer materials can be applied for different applications like films. Different industrial processes can be used, such as injection and extrusion. However, the starch needs to be blended with other polymers [1, 7].

7.3 Renewable Raw Material to Produce Bioplastic

7.3.1 Raw Material from Agriculture

Starch is a polysaccharide and can be classified as a thermoplastic. There are different resources of starch, such as potatoes, wheat, and tapioca [19]. However, the starch obtained from potatoes exhibited the greatest properties [20]. It is obtained by enzymatic hydrolysis followed by a fermentation with microorganisms. Also, this polymer can be modified to improved hydrophilic characteristics [21]. However, the mechanical characteristics of starch-biobased plastic are reduced, which can overcome added nanoparticles. This bioplastic can be used in the car components, drug delivery, and food industry [22]. To improve its mechanical properties, starch bioplastic obtained from tapioca was treated under ultrasound technique [23].

Wheat gluten (a protein) can be used to make bioplastics [24]. Also, sugarcane can be also used to produce a bioplastic [25]. To replace the food harvests, lignocellulosic biomass can be a source to production of bioplastic. However, this renewable raw material needs pretreatment, becoming an increase of process cost [26, 27].

7.3.2 Organic Waste as Resources for Bioplastic Production

The waste of the food industry can be used as a recourse of bioplastics [28, 29]. Some waste of carrots, parsley, and cauliflowers are used to produce bioplastic films [30]. A waste of bioethanol production, rice straw can be applied to produce starch-biobased plastics and cellulose-biobased plastics [31, 32]. The waste of chocolate industry (cocoa pod husk), and the waste of sugar industry (sugarcane bagasse) can be used to produce bioplastic films [33]. Also, potato peels were used to prepare a biopolymer [34, 35].

PHAs are polymers produced from renewable raw material, which is transformed by microorganisms [36]. As example, *Bacillus cereus* [37] or *Staphylococcus epidermidis* [38] can be used to transform sugarcane molasses into PHA. Also, *Bacillus* sp. can be used to transform low-cost agro-wastes [39], and *Pseudodonghicola xiamenensis* can be applied to convert glycerol (a by-product of biodiesel industry) into PHAs [40].

7.3.3 Algae as Resources for Bioplastic Production

Microalgae can be also used to produce bioplastics. The great advantage of microalgae is that bioplastic production becomes independent of food [41, 42].

Microalgae, *Chlorella* and *Spirulina*, have been used in the polymerization of protein biomass to produce bioplastic and thermoplastic blends. The biomass from algal grows on wastewater from farm animals, municipal, or industrial effluent resources, which in its composition have high amounts of nitrogen and phosphorus. This bioplastic can be used for agriculture products. Different variables, such as temperature, pressure, content of plasticizer, can influence the polymerization process. Microalgae *Chlorella* exhibited a biopolymer with improved properties then *Spirulina* algae [43].

Chlorella pyrenoidosa produced a biopolymer (poly(hydroxybutyrate), PHB) under optimized conditions. After 10 days, the growth is maximum. After the biodiesel production, it was obtained as a waste. Biomass was applied to generate biopolymers. The poly(hydroxybutyrate) is a biodegradable material. After 14 days, about 27% of poly(hydroxybutyrate) was obtained in this culture [44].

Chlorella sorokiniana is a microalga, which can produce starch with the shape of granules. The polymer was characterized by size distribution. It was observed that the molecular weight of starch (obtained by this process) is like the molecular weight starch present on plants. Also, the amount of amylose and crystallinity of starch (microalgal) is like the starch present on the cereal. This bioplastic exhibited high temperature of gelatinization (110°C) [45].

Microalgae consortium, *Scenedesmus* sp and *Desmodesmus* sp, (about 80%), were applied to eliminate compounds from wastewater. This consortium can produce bioplastics, due to the high protein amount (about 48%). The bioplastics produced, after adding glycerin, can be obtained by injection molding. These materials demonstrated a lower capability to absorb water. Some mechanical bioplastic properties, such as tensile, rheological, and water uptake, were evaluated [46].

Bioplastic production was performed using the seaweed *Kappaphycus alvarezii*, which has high amount of polysaccharides. This plastic was used as biofilm. The amount of polyethylene glycol (plasticizer) with seaweed biomass was adjusted to obtain a thin film with good tensile strength. The bioplastic was characterized by different techniques, such as thickness, ductile strength, extension at break, and O_2 transference rate. The seaweed can be an alternate resource for bioplastic production [47].

A bioplastic produced from seaweed (*Eucheuma cottonii*) was studied. Different amounts of glycerol (10, 20, and 30 mL), which was used as a plasticizer, was applied to estimate the ductile strength. It used H_2O (solvent), corn-starch (thickener), and vinegar (to dissolve the starch). It was concluded that when the glycerol amount increases, a weaker bioplastic is produced. However, the biodegradability of plastic was not affected by the amount of glycerol [48].

7.3.4 Wastewater as Resources for Bioplastic Production

The amount of organic compounds exists in wastewaters, these sources can be used to produce bioplastics. An example is the wastewater from dairy industries. This waste has soluble organic compounds, suspended solids and small amounts of fats, oils, phosphates and grease. Bioplastics can be produced from this wastewater in the presence of microorganisms [49].

From this dairy waste, it can be observed that the PHB is a bioplastic synthesized with high amounts. Under optimized conditions, the buttermilk is the carbon source at pH = 7, at T = 37° C and at a speed of 150 rpm. With these conditions, a high amount of PHB is produced [50].

Mixed microbial cultures were used to produce polyhydroxyalkanoates (PHA) [51]. This biopolymer was accumulated in microbial cultures. It was used as a batch reactor, using an effluent with a composition like oil mill wastewater. A yield of 74 % and a recovery purity of 100 % was observed when it was using NH_4-Laurate for extraction. Under optimized conditions, it is necessary a previous process with sodium hypochlorite at T = 85 °C during 60 min. Also, a treatment with lauric acid during 3 h, is necessary. However, the polymers have a molecular weight low [51].

From the cheese production, it is obtained as a waste, which can be used, as raw material, to produce a bioplastic [52]. This process has two steps: a fermentative production and poly(3-hydroxybutyrate) production. The Lactobacillus sp. transforms lactose into lactic acid. In PHB production, the *Rhodovulum sulfidophilum* transforms lactic acid into bioplastic. It was observed that the lactic acid production revealed a lactic acid amount of 29.5 g/L. It was also observed, in the presence of *Rhodovulum sulfidophilum*, high amount of PHB (about 67 %) [52].

The production of PHA was performed using a microorganism on wastewater (paper mill) [53]. The production of this biopolymer has three steps: in a batch process, occurs an acidogenic fermentation of wastewater; in second steps, occurs the enhancement of a PHA, with feast-famine regime; the third step, consist in the increases of PHAs amount on cellular microbial community, using a reactor fed-batch. After 5 h, a maximum of 77 % of PHA of biomass dry weight can be obtained. The biomass for this production included *Plasticicumulans acidivorans*. The PHA production process can be integrated with wastewater treatment [53].

7.4 Bioplastics Applications

7.4.1 Food Industry

Due to the environmental problems, the food industries need bioplastics to packaging with high economic and low environmental impact. It is imperative to replace the petrochemical-based packaging by bio-based polymers [6].

The use of bioplastics for packaging does not need exceptional O_2 and/or H_2O barrier properties [54].

The bioplastic used in the food industry is an O_2 and H_2O protector. It can be developed as a bio-based polymer with multicomponent coatings, which can act as a barrier. However, this option has a disadvantage due to the difficulty for a recycle [6]. The starch/polyvinyl alcohol, proteins, cellulose/cellophane show high, high-medium to moisture permeability. However, starch/polyvinyl alcohol and proteins show low oxygen permeability, while the cellulose/cellophane shows very high oxygen permeability. The polyhydroxyalkanoates (PHA) shows low moisture permeability and low oxygen permeability [6, 7].

Starch-based bioplastic is used for food packaging. The advantage of this material over traditional polymers, in special packaging food, is food respiration (fruit). Another example, PLA can be used for yoghurt boxes, with high mechanical properties [55].

7.4.2 Agricultural Applications

Traditionally, the petrochemical-based polymers, like polyethylene and polypropylene are used as films for agricultural applications. However, these polymers have some problems associated with the environment. To overcome these problems, the PHAs can be an alternative to traditional polymers. The PHAs can be used as films to protect the cultures from insects, weather, and birds. Also, due their structure, the PHA polymer can be used to prevent contamination of soil, to control weeds and retention of water in soil [56].

7.4.3 Medical Applications

Biopolymers have a particularly important play in different areas of medicine and biomedicine [57].

Cellulose is a biopolymer, not toxic, and biocompatible, which can be used for tissue engineering, neural and pharmaceutical fields [58]. Nanocellulose, which is polymer extracted from cellulose, has achieved much importance due to its application as a biomedical raw material. This polymer has physical and chemical properties and becomes the cellulose as a promisor material. The nanocellulose can be nanocrystals, nanofibrils and bacterial cellulose. Nanocellulose can be used in biomedicine. It is used for drug delivery and skin tissue repair materials, as example [58].

Polyhydroxyalkanoates (PHAs) is a biopolymer, biocompatible, and biodegradable. PHAs can be used for cancer detection, treating ulcers, drug delivery matrices, and therapeutic applications [56].

7.4.4 Other Applications

Bioplastics can be used in different areas, such as structure, electric, and chemical engineering. Composite polymers, which combine the properties of polymer and the properties of fillers (clays, zeolite, graphene, and structured silica), can be applied on different fields [59, 60].

Chitosan is a biopolymer with different applications. Chitosan has been used as a catalyst in different reactions, like ethoxylation, hydrogenation [61, 62]. As examples, sulfonated chitosan was used in the reaction between fatty acid and alcohol [63, 64]. Also, composites, chitosan/MCM-41-SO_3H, were applied as acid material in alkoxylation of limonene [65].

7.5 Conclusions

Traditional plastics, which are petroleum-based materials, are produced from compounds obtained from distillation of crude oil. From crude oil distillation different compounds are obtained, which are used to produce other compounds (monomers) to produce plastics. These materials are prepared by polyaddition and/or polycondensation reactions in the presence of catalysts. However, the conversion of these products after its utilization, creates an increase of greenhouse gases. It is imperative to try to replace the traditional plastics, which are not biodegradable and persistent on the environment, by bioplastics. Bioplastics are materials with excellent properties for the environment. Starch, cellulose, PHAs can be applied in different fields, like the cooking industry, medical specialty area. Bioplastics are materials with low environmental impact, which possibility an intensive production of the bioplastics. The PHA polymers are the materials that can be a substitute of traditional plastic in different areas, such as the engineering field. However, PHAs have high production costs. It is important to realize further research to decrease the production costs and make PHAs more attractive for different applications.

Replacing traditional plastics with bioplastics materials becomes a very important goal.

References

1. Huang, J., Veksha, A., Chan, W.P., Giannis, A., Lisak, G., Chemical recycling of plastic waste for sustainable material management: A prospective review on catalysts and processes. *Renew. Sust. Energ. Rev.*, 154, 111866, 2022.
2. Alshehrei, F., Biodegradation of synthetic and natural plastic by microorganisms. *J. Appl. Environ. Microbiol.*, 5, 8, 2017.
3. So, W.-M.W., Cheng, N.-Y.I., Chow, C.-F., Zhan, Y., Learning about the types of plastic wastes: Effectiveness of inquiry learning strategies. *Education*, 44, 311, 2016.
4. Tokiwa, Y., Calabia, B.P., Ugwu, C.U., Aiba, S., Biodegradability of plastics. *Int. J. Mol. Sci.*, 10, 3722, 2009.
5. Chia, W.Y., Tang, D.Y.Y., Khoo, K.S., Lup, A.N.K., Chew, K.W., Nature's fight against plastic pollution: Algae for plastic biodegradation and bioplastics production. *Environ. Sci. Ecotechnol.*, 4, 100065, 2020.
6. Coppola, G., Gaudio, M.T., Lopresto, C.G., Calabro, V., Curcio, S., Chakraborty, S., Bioplastic from renewable biomass: A facile solution for a greener environment. *Earth Syst. Environ.*, 5, 231, 2021.
7. Atiwesh, G., Mikhael, A., Parrish, C.C., Banoub, J., Le, T.-A.T., Environmental impact of bioplastic use: A review. *Heliyon*, 7, e07918, 2021.
8. Meereboer, K.W., Misra, M., Mohanty, A.K., Review of recent advances in the biodegradability of polyhydroxyalkanoate (PHA) bioplastics and their composites. *Green Chem.*, 22, 5519, 2020.
9. Philp, J.C., Bartsev, A., Ritchie, R.J., Baucher, M.-A., Guy, K., Bioplastics science from a policy vantage point. *N. Biotechnol.*, 30, 635, 2013.
10. Chen, G.-Q., Amicrobial polyhydroxyalkanoates (PHA) based bio- and materials industry. *Chem. Soc. Rev.*, 38, 2434, 2009.
11. Singh, M., Kumar, P., Ray, S., Kalia, V.C., Challenges and opportunities for customizing polyhydroxyalkanoates. *Indian J. Microbiol.*, 55, 235, 2015.
12. Karan, H., Funk, C., Grabert, M., Oey, M., Hankame, B., Green bioplastics as part of a circular bioeconomy. *Trends Plant Sci.*, 24, 237, 2019.
13. Semprini, L., Hopkins, G.D., Roberts, P.V., Grbic-Galic, D., McCarty, P.L., A field evaluation of *in-situ* biodegradation of chlorinated ethenes: Part 3, studies of competitive inhibition. *Ground Water*, 29, 239, 1991.
14. Sin, L.T., Rahmat, A.R., Rahman, W.A.W.A., *Polylactic Acid: PLA Biopolymer Technology and Applications*, pp. 1–70, Elsevier, Oxford, 2012.
15. Nagarajan, V., Mohanty, A.K., Misra, M., Perspective on polylactic acid (PLA) based sustainable materials for durable applications: Focus on toughness and heat resistance. *ACS Sustain. Chem. Eng.*, 4, 2899, 2016.
16. Elsawy, M.A., Kim, K.-H., Park, J.-W., Deep, A., Hydrolytic degradation of polylactic acid (PLA) and its composites. *Renew. Sust. Energ. Rev.*, 79, 1346, 2017.

17. Verma, V., Verma, P., Ray, P., Ray, A.R., 2, 3-Dihydrazone cellulose: Prospective material for tissue engineering scafolds. *Mater. Sci. Eng. C*, 28, 1441, 2008.
18. Li, J., Wan, Y., Li, L., Liang, H., Wang, J., Preparation and characterization of 2,3-dialdehyde bacterial cellulose for potential biodegradable tissue engineering scafolds. *Mater. Sci. Eng. C*, 29, 1635, 2009.
19. Gökçe, E., Rethinking sustainability: A research on starch based bioplastic. *J. Sustain. Construct. Mater. Technol.*, 3, 249, 2018.
20. Hamidon, N., Hamidun, M.B., Arish, N.A., Sunar, N.M., Ali, R., Hamid, H.A., Harun, H., Muhamad, M.S., Potential of production bioplastic from potato starch. *Sustain. Environ. Technol.*, 1, 115, 2018.
21. Mojibayo, I., Samson, A.O., Johnson, O.Y., Joshusa, I.O.A.S.A., A preliminary investigation of cassava starch potentials as natural polymer in bioplastic production. *Am. J. Interdiscip. Innov. Res.*, 02, 31, 2020.
22. Mose, B.R. and Maranga, S.M., A Review on starch-based nanocomposites for bioplastic materials. *J. Mater. Sci. Eng. B*, 1, 239, 2011.
23. Asrof, M., Sapuan, S.M., Ilyas, R.A., Ramesh, M., Characteristic of composite bioplastics from tapioca starch and sugarcane bagasse fiber: Effect of time duration of ultrasonication (Bath-Type). *Mater. Today Proc.*, 46, 1626, 2021.
24. Jiménez-Rosado, M., Zarate-Ramírez, L.S., Romero, A., Bengoechea, C., Partal, P., Guerrero, A., Bioplastics based on wheat gluten processed by extrusion. *J. Clean. Prod.*, 239, 117994, 2019.
25. Pohare, M.B., Bhor, S.A., Patil, P.K., Sugarcane for economical bioplastic production. *Pop. Kheti*, 5, 20, 2017.
26. Brodin, M., Vallejos, M., Opedal, M.T., Area, M.C., Chinga-Carrasco, G., Lignocellulosics as sustainable resources for production of bioplastics—A review. *J. Clean. Prod.*, 162, 646, 2017.
27. Govil, T., Lignocellulosic feedstock: A review of a sustainable platform for cleaner production of nature's plastics. *J. Clean. Prod.*, 270, 122521, 2020.
28. Tsang, Y.F., Kumar, V., Samadar, P., Yanga, Y., Leed, J., Ok, Y.S., Song, H., Kim, K.-H., Kwon, E.E., Jeon, Y.J., Production of bioplastic through food waste valorization. *Environ. Int.*, 127, 625, 2019.
29. Jõgi, K. and Bhat, R., Valorization of food processing wastes and by-products for bioplastic production. *Sustain. Chem. Pharm.*, 18, 100326, 2020.
30. Perotto, G., Ceseracciu, L., Simonutti, R., Paul, U.C., Guzman-Puyol, S., Tran, T.-N., Bayer, I.S., Athanassiou, A., Bioplastics from vegetable waste: Via an ecofriendly water-based process. *Green Chem.*, 20, 894, 2018.
31. Agustin, M.B., Ahmmad, B., Alonzo, S.M.M., Patriana, F.M., Bioplastic based on starch and cellulose nanocrystals from rice straw. *J. Reinf. Plast. Compos.*, 33, 2205, 2014.
32. Bilo, F., Pandini, S., Sartore, L., Depero, L.E., Gargiulo, G., Bonassi, A., Federici, S., Bontempi, E., A sustainable bioplastic obtained from rice straw. *J. Clean. Prod.*, 200, 357, 2018.
33. Azmin, S.N.H.M., Binti, N.A., Hayat, M., Nor, M.S.M., Development and characterization of food packaging bioplastic film from cocoa pod husk cellulose incorporated with sugarcane bagasse fiber. *J. Bioresour. Bioprod.*, 5, 255, 2020.
34. Sultan, N.F.K. and Johari, W.L.W., The development of banana peel/corn starch bioplastic film: A preliminary study. *Bioremediation Sci. Technol. Res.*, 5, 12, 2017.
35. Azieyanti, N.A., Amirul, A., Othman, S.Z., Misran, H., Mechanical and morphology studies of bioplastic-based banana peels. *J. Phys. Conf. Ser.*, 1529, 032091, 2020.
36. Shraddha, G., Yogita, R., Simanta, S., Aparna, S., Kamlesh, S., Screening and production of bioplastic (PHAs) from sugarcane rhizospheric bacteria. *Int. Multidiscip. Res. J.*, 1, 30, 2011.

37. Suryawanshi, S.S., Sarje, S.S., Loni, P.C., Bhujbal, S., Kamble, P.P., Bioconversion of sugarcane molasses into bioplastic (Polyhydroxybutyrate) using *Bacillus cereus* 2156 under statistically optimized culture conditions. *Anal. Chem. Lett.*, 10, 80, 2020.
38. Sarkar, K., Ray, B., Banerjee, R., Saha, S., Roy, S., Chatterjee, S., Poly-β-hydroxybutyrate (Bioplastic) production utilizing waste effluent of a sugar industry. *IOSR J. Environ. Sci. Toxicol. Food Technol.*, 8, 26, 2014.
39. Getachew, A. and Woldesenbet, F., Production of biodegradable plastic by polyhydroxybutyrate (PHB) accumulating bacteria using low-cost agricultural waste material. *BMC Res. Notes*, 9, 509, 2016.
40. García, I.L., López, J.A., Dorado, M.P., Kopsahelis, N., Alexandri, M., Papanikolaou, S., Villar, M.A., Koutinas, A.A., Evaluation of by-products from the biodiesel industry as fermentation feedstock for poly(3-hydroxybutyrate-co-3-hydroxyvalerate) production by *Cupriavidus necator*. *Bioresour. Technol.*, 130, 16, 2013.
41. Beckstrom, B.D., Wilson, M.H., Crocker, M., Quinn, J.C., Bioplastic feedstock production from microalgae with fuel co-products: A techno-economic and life cycle impact assessment. *Algal Res.*, 46, 101769, 2020.
42. Rajendran, N., Puppala, S., Sneha, R.M., Ruth, A.B., Rajam, C., Seaweeds can be a new source for bioplastics. *J. Pharm. Res.*, 5, 1476, 2012.
43. Zeller, M.A., Hunt, R., Jones, A., Sharma, S., Bioplastics and their thermoplastic blends from *Spirulina* and *Chlorella* microalgae. *J. Appl. Polym. Sci.*, 130, 3263, 2013.
44. Das, S.K., Sathish, A., Stanley, J., Production of biofuel and bioplastic from *chlorella pyrenoidosa*. *Mater. Today Proc.*, 5, 16774, 2018.
45. Gifuni, I., Olivieri, G., Krauss, I.R., D'Errico, G., Pollio, A., Marzocchella, A., Microalgae as new sources of starch: Isolation and characterization of microalgal starch granules. *Chem. Eng. Trans.*, 57, 1423, 2017.
46. López-Rocha, C.J., Álvarez-Castillo, E., Estrada-Yáñez, M.R., Bengoechea, C., Guerrero, A., Orta-Ledesma, M.T., Development of bioplastics from a microalgae consortium from wastewater. *J. Environ. Manage.*, 263, 110353, 2020.
47. Sudhakar, M.P., Magesh, P.D., Dharani, G., Studies on the development and characterization of bioplastic film from the red seaweed (*Kappaphycus alvarezii*). *Environ. Sci. Pollut. Res.*, 28, 33899, 2021.
48. Rajendran, N., Puppala, S., Sneha, R.M., Ruth, A.B., Rajam, C., Seaweeds can be a new source for bioplastics. *J. Pharm. Res.*, 5, 1476, 2012.
49. Bosco, F. and Chiampo, F., Production of polyhydroxyalcanoates (PHAs) using milk whey and dairy wastewater activated sludge. Production of bioplastics using dairy residues. *J. Biosci. Bioeng.*, 109, 418, 2010.
50. Mehta, V., Patel, E., Vaghela, K., Marjadi, D., Dharaiya, N., Production of biopolymer from dairy waste: An approach to alternate synthetic plastic. *Int. J. Res. Biosci.*, 6, 1, 2017.
51. Mannina, G., Presti, D., Montiel-Jarillo, G., Suárez-Ojeda, M.E., Bioplastic recovery from wastewater: A new protocol for polyhydroxyalkanoates (PHA) extraction from mixed microbial cultures. *Bioresour. Technol.*, 282, 361, 2019.
52. Carlozzi, P., Giovannelli, A., Traversi, M.L., Touloupakis, E., Poly(3-hydroxybutyrate) bioproduction in a two-step sequential process using wastewater. *J. Water Process Eng.*, 39, 101700, 2021.
53. Jiang, Y., Marang, L., Tamis, J., van Loosdrecht, M.C.M., Dijkman, H., Kleerebezem, R., Waste to resource: Converting paper mill wastewater to bioplastic. *Water Res.*, 46, 5517, 2012.
54. Jabeen, N., Majid, I., Nayik, G.A., Bioplastics and food packaging: A review. *Cogent Food Agric.*, 1, 1117749, 2015.

55. Bastioli, C., Global status of the production of biobased packaging materials. *Starch/Stärke*, 53, 351, 2001.
56. El-malek, F.A., Khairy, H., Farag, A., Omar, S., The sustainability of microbial bioplastics, production and applications. *Int. J. Biol. Macromol.*, 157, 319, 2020.
57. Parisi, O.I., Scrivano, L., Candamano, S., Ruffo, M., Vattimo, A.F., Spanedda, M.V., Puoci, F., Molecularly imprinted microrods via mesophase polymerization. *Molecules*, 23, 1, 63, 2018.
58. Lin, N. and Dufresne, A., Nanocellulose in biomedicine: Current status and future prospect. *Eur. Polym. J.*, 59, 302, 2014.
59. Mohammed, L., Ansari, M.N.M., Pua, G., Jawaid, M., Islam, M.S., A review on natural fiber reinforced polymer composite and its applications. *Int. J. Polym. Sci.*, 243947, 2015.
60. Candamano, S., Crea, F., Iorfda, A., Mechanical characterization of basalt fabric-reinforced alkali-activated matrix composite: A preliminary investigation. *Appl. Sci.*, 10, 2865, 2020.
61. Guibal, E., Heterogeneous catalysis on chitosan-based materials: A review. *Prog. Polym. Sci.*, 30, 71, 2005.
62. Kadib, A.E., Chitosan as a sustainable organocatalyst: A concise overview. *ChemSusChem*, 8, 217, 2015.
63. Caetano, C.S., Caiado, M., Farinha, J., Fonseca, I.M., Ramos, A.M., Vital, J., Castanheiro, J.E., Esterification of free fatty acids over chitosan with sulfonic acid groups. *Chem. Eng. J.*, 230, 567, 2013.
64. Castanheiro, J., Chitosan with sulfonic groups: A catalyst for the esterification of caprylic acid with methanol. *Polymers*, 13, 3924, 2021.
65. Castanheiro, J.E., Chitosan/MCM-41-SO_3H composites as catalyst of the etherification of limonene. *Results Mater.*, 12, 100233, 2021.

8

Natural Oil-Based Sustainable Materials for a Green Strategy

Figen Balo[1]*, Berrak Aksakal[2], Lutfu S. Sua[3] and Zeliha Mahmat[4]

[1]*Department of Industrial Engineering, Firat University, Elazig, Turkey*
[2]*Department of Public Health Services, Elazig Provincial Health Directorate, Elazig, Turkey*
[3]*Department of Management and Marketing, Southern University and A&M College, Baton Rouge, USA*
[4]*Independent Researcher, Firat University, Elazig, Turkey*

Abstract

Utilization of sustainable resources is significant, considering the effects of fossil-sourced plastics. Using sustainable sources, such as plant oils, for biobased plastic fabrication proposes the benefits of lesser dependency on petroleum-based energy resources and sustainability. Natural epoxidized oils are additives to replace and strengthen fossil plastics. Epoxidized vegetable oils attracts more interest recently, particularly the plastic sector as they become available, economicaly feasible, environmentally friendly, renewable, and nonnoxious. Renewable components for biobased plastics corresponding to a large share of the plastics market and epoxidized vegetable oil products can replace many of these. The fatty acids and epoxidized natural-based oils are broadly utilized in manufacturing glycols, polyols, carbonyl combinations, plastic additives for polymers, lubricants, adhesives and cast resins, coatings, and anticorrosion inhibitors, and so on. Currently, there are important search endeavors being undertaken to prepare epoxidized vegetable oil-based plastics in a profitable manner.

The literature has numerous scientific publications focusing on the testing and preparation of epoxidized vegetable oil-based plastics. In this paper, we concentrate on epoxidized vegetable oils produced as green plasticizer that have an actual high probability of entering the market or have already accessed the market.

Keywords: Green material, epoxidized plant oils, sustainable material, renewable material, bioplastic, multicriteria decision making

8.1 Introduction

The polymer sector, as a huge chemical end-user, contributes significantly to the consumption of fossil fuels. This requisition will occur more and more crucial as the population grows, and endeavors must be sustained. For many decades, the way of renewable feedstock's utilization in the chemical sector is actively improved because of anticipated decline of the petroleum resources and rise of the environmental problems. Global concerns have

**Corresponding author*: figenbalo@gmail.com

Inamuddin and Tariq Altalhi (eds.) Handbook of Bioplastics and Biocomposites Engineering Applications: 2nd Edition, (161–174) © 2023 Scrivener Publishing LLC

made the valorization of renewable feed stocks a prominent focus of industrial and scientific research. The need to free the polymer sector from its reliance on finite sources is a serious concern, prompting researchers to look for industrially viable sustainable options. The plant-based oils are promising types of resources, which in combination with low toxicity and biodegradability made them an appealing feed stock for chemical sector [1, 2]. Plant-based oils indicate a renewable source that can be utilized as trustworthy products to acquire novel materials with a wide range of functional and structural differences. Plant-based oils are an industrially appealing feed stock for the plastics sector because of their inexpensive cost and abundant availability [3, 4]. Plant-based oils and derivatives have been employed through polymeric chemists for a long time because of their global availability, renewable nature at an inexpensive cost, and diverse implementation eventualities. Plant-based oils have been in high demand as an option source for the manufacturing of nanocomposites and adhesives in recent years [5–7].

Plant-based oils, in this scenario, provide a number of benefits in addition to their renewable nature. As technological oleo chemical demonstrates on a daily basis, their low prices and relatively worldwide availability make them feasible and appealing. The greatest resources of plant oils are yearly vegetables like corn, soybean, linseed, peanuts, or cottonseed. Nonetheless, other resources are oil bearing longevity like the olive, coconut or palm [8].

The biobased plastic definition is general. The biobased plastic involves plastics, which are nondegradable (partial or neat mix) and durable acquired from a biobased resource or plastics that are biodegradable [9]. The biobased bio-degradable plastics can contain bio-degradable biological-sourced plastics, but also contain petrochemical-based biodegradable plastics, like polybutylene-succinate and polybutylene-adipate-terephthalate [10].

Because of their high double bond concentration, plant-based oils appear to be one of the most encouraging feedstocks. Furthermore, they are plentiful, inexpensive, nondepletable, and nontoxic. These oils are the mix of glycerin triesters and triglycerides with fatty acids, which are around 95% from triglyceride molecules' total mass, their chain width

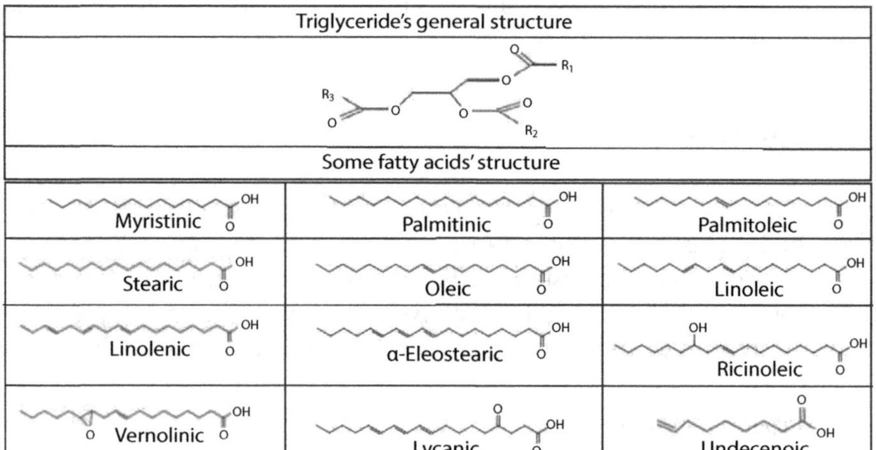

Figure 8.1 Triglyrcesides' and some fatty acids' structure.

is vibrated between 14 and 22 C atoms between zero and three pair linkages per chain. More than thousand fatty acids have been determined, but solely approximately twenty are exist in vegetable oil in appreciable amounts. The fatty acid composition changes relying on growing conditions, season, crop, and plant species [11–18]. Triglycerides' and some fatty acids' structure are given in Figure 8.1 [11–18].

Table 8.1 presents fatty acid distribution of the plant-based oils.

A few types of fatty acids have been poly-saturated (linolenic and linoleic acids) or mono-saturated (erucic and oleic acids), while the others are saturated (myristinic, lauric, palmitinic, arachic, stearic, lignoceric, and behenic acids) [22]. The plant oils with lycanic, vernolic, and ricinoleic fatty acids can comprise also the other characteristic classes, like carbonyl, epoxide and hydroxyl classes, respectively [23, 24]. Active characteristic classes in the triglyceride structure for vegetable oil chemical modification is given in Figure 8.2.

The chemical and physical features of vegetable oils can be changed via diverse chemical processes. Essentially, the chemical alteration of vegetable oils is the significant way for improvement of the environmentally friendly and novel high-efficient ways of processes primary to novel produces and for research of novel implementations to already present oleochemicals [19–21, 25].

Table 8.1 The fatty acid distribution in the plant-based oils.

	Fatty acid (percent)					
Oil	**Linolenic**	**Palmitinic**	**Linoleic**	**Oleic**	**Stearic**	**Other**
Cotton	0.7	21.6	54.4	18.6	2.6	–
Rapeseed	9.0	5.0	14.0	15.0	14.0	43.0
Castor	0.5	1.5	4.0	5.0	0.5	87.5
Sunflower	1.0	6.0	47.0	42.0	4.0	–
Olive	0.6	13.7	10.0	71.1	2.5	–
Soya bean	7.8	11.0	53.3	23.4	4.0	–
Palm	–	39.0	9.0	45.0	5.0	–
Linseed	56.6	5.5	15.3	19.1	3.5	–
Corn	1.2	10.9	59.6	25.4	2.0	–

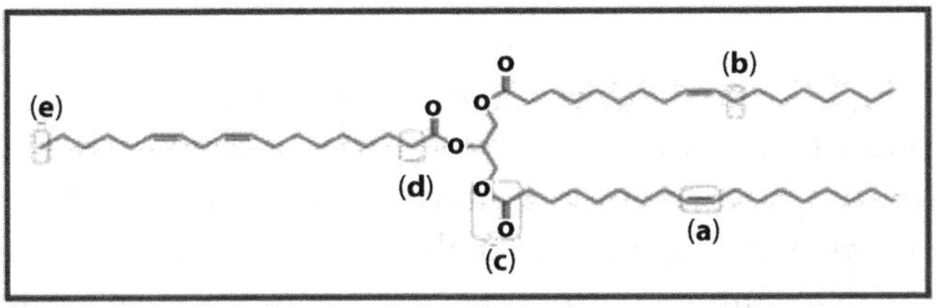

Figure 8.2 Active characteristic classes in the triglyceride structure. (a) double chain, (b) allyl C atoms, (c) ester classes, (d) α carbonyl C, (e) ω-C atoms.

A profusion of the reactive effective classes in vegetable oils makes them handy oleo-chemical feed stocks [25]. The molecules of triglyceride in the vegetable oil structure have some active effective classes, which can be utilized for their alteration [19–21, 25]. The chemical alteration technics on double chain, as displayed in Figure 8.2, contain the epoxidation process, oxidation, hydroformylation, carbonylation, hydrogenation, and so on.

The most significant physical features of the epoxidized vegetable oils are the transparency and viscosity. The chemically epoxidized vegetable oils are fundamentally defined through epoxide equivalent weight or through oxirane oxygen content, through unreacted double chain's quantity obtained on value of iodine number value, through water quantity in the resin after drying, through acidic number's values for hydroxyl number and residual acid [26]. The iodine number's low values, water content, and hydroxyl number are preferred for epoxide resin based on the vegetable oil. The lower values signify to higher proportion of unsaturated chains' conversion in the fatty acids' bond. The less hydroxyl number signifies to the least opening of epoxy ring to hydroxyl class. It is continually essential to sustain a lower water ingredient, as the epoxy ring's water favors beginning. In the polyols' direct synthesis from epoxidized vegetable oil, the humidity ingredient is not taken into consideration.

The fundamental chemical surds' permissible values for trading epoxidized soybean oil are: water – lesser 300 ppm, ring of oxirane—more 6.3% and number value of iodine—lesser 0.05 [27]. Nevertheless, the oxirane ingredient is diversified relying on the oil variety type and epoxidation degree. A step of conversion or epoxidation of double chains can be computed on value of iodine number after and before epoxidizing. A viscosity of epoxidized vegetable oil is substantially greater than the corresponding plant oils due to creating oxirane characteristic. The maximum viscosity limitation established is 500 mPas [28, 29].

The vegetable oils' other significant index– this is number of an acidic, which detects the condensation of the acidic classes in the oils. Its valuation is found through potassium-hydroxide mass measurement in milligram, which is desired for acid's neutralization in the specimen's 1 gr [30, 31].

In this study, epoxidized plant oils, which have an important place among the most widely used biobased plasticizers in the industry were investigated. The general properties of epoxidized plant oils, which were determined as the most common and most suitable in scientific research, were determined. Then, the important properties of these epoxidized plant oils were analyzed with a quantitative decision-making technique, taking into account the properties that are important in the production of suitable bioplastics. Entropy and Copras were used as analysis methods.

8.2 Methodology

For the purpose of this analysis, a hybrid quantitiative analysis method is utilized in order to evaluate a a number of alternatives based on a series of factors and choose the best available alternative. The proposed method involves a combination of Entropy and Copras Methodologies as a two-step algorithm. Analytical results indicate the efficiency of the developed analysis in determining optimum choice among a selection of alternatives.

8.2.1 Entropy Methodology

Entropy weight method measures value dispersion when making a selection among various alternatives. Higher levels of dispersion indicate higher differentiation. This method starts with ranking the factors based on their relative importance when making a decision. Weights are assigned to each factor for this purpose. The entropy method process steps are explained below [32].

Step 1: Normalization of the values in the initial decision matrix is performed. Its formulation is as shown in Equation 8.1.

$$P_{ij} = \frac{r_{ij}}{\sum_{i=1}^{m} r_{ij}} \quad \forall j \tag{8.1}$$

i = Variable showing alternatives
j = Variable showing criteria
r_{ij} = Indicates importance of the ith alternative according to jth criterion
P_{ij} = Normalized value

Step 2: Entropy value and entropy coefficient are calculated by formulating as shown in Equation 8.2 and Equation 8.3.

$$E_j = -k * \sum_{i=1}^{m} [P_{ij} * \ln(P_{ij})] \quad \forall j \tag{8.2}$$

$$k = (\ln (m))^{-1} \tag{8.3}$$

E_J = j. entropy value of the value
k = Entropy coefficient

Step 3: The weights of each criterion are calculated with the formulation expressed in Equation 8.6.

$$W_J = \frac{1 - E_J}{\sum_{j=1}^{n} (1 - E_j)} \tag{8.4}$$

W_j = The weight value of jth criterion

Initial decision matrix is given below in Table 8.2. This matrix was created as a result of evaluating each alternative according to the pre-determined set of criteria. Epoxidized palm oil (EPO), linseed oil (ELO), castor oil (ECO), tall oil (ETO), soybean oil (ESO), olive oil (EZO), sunflower oil (ESFO), and canola oil (EKO) were used for comparison.

Table 8.2 Initial decision matrix.

	Brilliance	Iodine value	Boiling point	Melting point in water (at 25°C)	Oxirane value	Resin	Acid value	Ignition point	Cross-linker	Thermal conductivity coefficient	Saponification number	Viscosity	Density (25 °C)	Refractive index(at 25°C)	Loss on heating
EPO	391	0.44	145	0.011	3,15	42	1,14	304	6,7	0.144	125	390	0.919	1,398	0.51
ELO	402	0.62	163	0.032	9,4	131,5	1,32	321	11	0.163	141	430	0.9965	2,153	0.71
EHO	372	1,44	155	0.0273	9,93	76	3,01	316	9,89	0.151	192,5	400	0.9865	2,07	0.6
ETO	388	1,61	136	0.01	4,895	127,3	1,545	212	7,9	0.133	164,15	510	0.9405	1,315	0.43
ESO	400	3	150	0.01	6,4	145	2	310	15	0.156	184	325	0.99	1,472	0.5
EZO	412	0.85	167	0.035	4,14	56	1,21	359	7,1	0.14	192	276	0.9145	1,865	0.65
ESFO	400	3	150	0.01	6,4	136,2	2	310	10.8	0.156	184	325	0.99	1,472	0.5
EKO	365	131,4	153	0.0265	9,96	131,9	2,97	311	9,77	0.147	189,5	370	0.9755	2,01	0.58
Total	3130	142,36	1219	0.1618	54,275	845,9	15,195	2443	78,16	1,19	1372,15	3026	7,7125	13,755	4,48

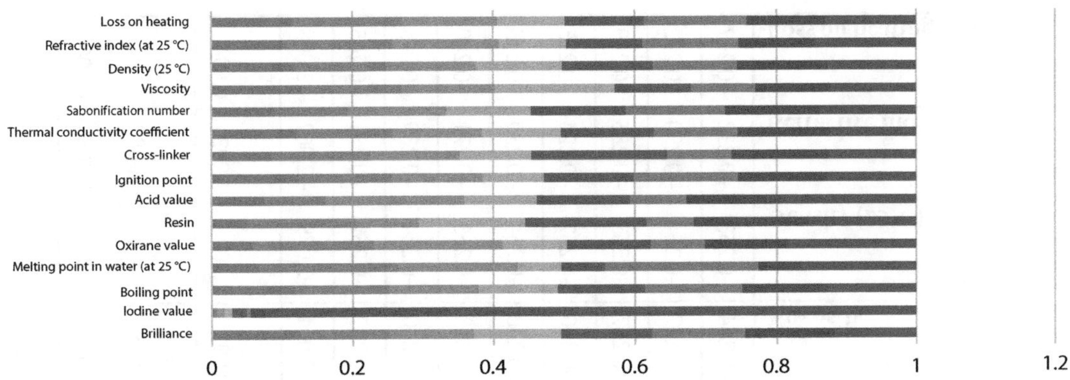

	Brilliance	Iodine value	Boiling point	Melting point in water (at 25°C)	Oxirane value	Resin	Acid value	Ignition point	Cross-linker	Thermal conductivity coefficient	Sabonification number	Viscosity	Density (25 °C)	Refractive index (at 25°C)	Loss on heating
EPO	0.12492	0.003091	0.11895	0.067985	0.058038	0.049651	0.075025	0.124437	0.085722	0.121008	0.091098	0.128883	0.119157	0.101636	0.113839
ELO	0.128435	0.004355	0.133716	0.197775	0.173192	0.155456	0.086871	0.131396	0.140737	0.136975	0.102758	0.142102	0.129206	0.156525	0.158482
EHO	0.11885	0.010115	0.127153	0.168727	0.182957	0.089845	0.198091	0.129349	0.126535	0.126891	0.140291	0.132188	0.127909	0.150491	0.133929
ETO	0.123962	0.011309	0.111567	0.061805	0.090189	0.150491	0.101678	0.086779	0.101075	0.111765	0.11963	0.168539	0.121945	0.095602	0.095982
ESO	0.127796	0.021073	0.123052	0.061805	0.117918	0.171415	0.131622	0.126893	0.191914	0.131092	0.134096	0.107403	0.128363	0.107016	0.111607
EZO	0.131629	0.005971	0.136998	0.216316	0.076278	0.066202	0.079631	0.14695	0.090839	0.117647	0.139926	0.09121	0.118574	0.135587	0.145089
ESFO	0.127796	0.021073	0.123052	0.061805	0.117918	0.161012	0.131622	0.126893	0.138178	0.131092	0.134096	0.107403	0.128363	0.107016	0.111607
EKO	0.116613	0.923012	0.125513	0.163782	0.18351	0.155929	0.195459	0.127302	0.125	0.123529	0.138104	0.122274	0.126483	0.146129	0.129464

EPO ELO EHO ETO ESO EZO ESFO EKO

Figure 8.3 Normalized values by entropy methodology.

After the initial decision matrix, the normalization process is applied. The entropy method process steps are explained below. Figure 8.3 illustrates the results obtained by the utilization of the entropy methodology.

After the normalization process, entropy values and the weight of each criterion are shown in Table 8.3 which provides the entropy values along with the criteria weights.

8.2.2 Copras Methodology

The operation of the COPRAS technique is explained below [33];

Step 1: Decision matrix is created. In the matrix expressed in Equation 8.5, i are symbols showing alternatives and criteria.

$$X_{ij} = \begin{bmatrix} x_{11} & \cdots & x_{1n} \\ \vdots & \ddots & \vdots \\ x_{m1} & \cdots & x_{mn} \end{bmatrix} \tag{8.5}$$

X_{ij} = Decision matrix

x_{ij} = value of j. criterion and for alternative i.

Table 8.3 Entropy values and criteria weights.

Pij*Ln (Pij)	Brilliance	Iodine value	Boiling point	Melting point in water	Oxirane value	Resin	Acid value	Ignition point	Cross-linker	Thermal conduct. coeff.	Sabonification no.	Viscosity	Density (25°C)	Refractive index	Loss on heating
EPO	−0.260	−0.018	−0.253	−0.183	−0.165	−0.149	−0.194	−0.259	−0.211	−0.256	−0.218	−0.264	−0.253	−0.232	−0.247
ELO	−0.264	−0.024	−0.269	−0.321	−0.304	−0.289	−0.212	−0.267	−0.276	−0.272	−0.234	−0.277	−0.264	−0.290	−0.292
EHO	−0.253	−0.046	−0.262	−0.300	−0.311	−0.216	−0.321	−0.265	−0.262	−0.262	−0.276	−0.267	−0.263	−0.285	−0.269
ETO	−0.259	−0.051	−0.245	−0.172	−0.217	−0.285	−0.232	−0.212	−0.232	−0.245	−0.254	−0.300	−0.257	−0.224	−0.225
ESO	−0.263	−0.081	−0.258	−0.172	−0.252	−0.302	−0.267	−0.262	−0.317	−0.266	−0.269	−0.240	−0.264	−0.239	−0.245
EZO	−0.267	−0.031	−0.272	−0.331	−0.196	−0.180	−0.201	−0.282	−0.218	−0.252	−0.275	−0.218	−0.253	−0.271	−0.280
ESFO	−0.263	−0.081	−0.258	−0.172	−0.252	−0.294	−0.267	−0.262	−0.273	−0.266	−0.269	−0.240	−0.264	−0.239	−0.245
EKO	−0.251	−0.074	−0.260	−0.296	−0.311	−0.290	−0.319	−0.262	−0.260	−0.258	−0.273	−0.257	−0.262	−0.281	−0.265
Total	−2.079	−0.406	−2,078	−1,947	−2,008	−2,006	−2,014	−2,071	−2,048	−2,078	−2,069	−2,064	−2,079	−2,062	−2,068
Entropi Values	1,000	0.195	0.999	0.936	0.966	0.965	0.969	0.996	0.985	0.999	0.995	0.992	1,000	0.992	0.994
1-Entropi Value	0.000	0.805	0.001	0.064	0.034	0.035	0.031	0.004	0.015	0.001	0.005	0.008	0.000	0.008	0.006
Wj Values	0.0003	0.7908	0.0009	0.0624	0.0336	0.0347	0.0309	0.0040	0.0149	0.0009	0.0049	0.0075	0.00026	0.0080	0.0055

Step 2: The decision matrix is normalized and weighted as expressed in Equation 8.6.

$$d_{ij} = \frac{x_{ij} * w_j}{\sum_{i=1}^{m} x_{ij}} \tag{8.6}$$

d_{ij} = Weighted normalized values
w_j = The previously found weight value of each criterion
Here, the sum of the d_ij values is equal to the previously found weight value of the relevant criterion. So $w_j = \sum_{i=1}^{m} d_{ij}$.

Step 3: Normalized weighted values are summarized as expressed in Equation 8.7.

$$S_{+i} = \sum_{j=1}^{n} d_{+ij}; \quad S_{-i} = \sum_{j=1}^{n} d_{-ij}, \tag{8.7}$$

S_{+i} = Expresses sum of normalized weighted values(maximization directional)
S_{-i} = Expresses sum of normalized values (minimization directional)

Here, it is aimed that the sum of the maximization-oriented criteria is large and the sum of the minimization-oriented criteria is small.

Step 4: The relative importance values of the alternatives are calculated as expressed in Equation 8.8. This value is ordered from largest to smallest, and the larger it is, the greater its relative importance.

$$Q_i = S_{+i} + \frac{S_{-min} * \sum_{i=1}^{m} S_{-i}}{S_{-i} * \sum_{i=1}^{m} \frac{S_{-min}}{S_{-i}}}, \tag{8.8}$$

Q_i = Expresses the relative importance of alternatives
S_{-min} = It refers to the smallest of the values collected in the minimization direction.

Step 5: Each alternative's utility degree is calculated as expressed in Equation 8.9. The degree of utility is ranked in terms of closeness to 100 and the most suitable alternative is found.

$$N_i = \left(\frac{Q_0}{Q_{max}} \right) * 100 \tag{8.9}$$

N_i = Expresses the degree of utility of alternatives
Q_{max} = Expresses the highest relative importance

Table 8.4 Weighted normalized values.

	Brilliance	Iodine value	Boiling point	Melting point in water	Oxirane value	Resin	Acid value	Ignition point	Cross-linker	Thermal conduct. coeff.	Saponific. no.	Viscosity	Density (25°C)	Refractive index	Loss on heating
EPO	0.000	0.002	0.000	0.004	0.002	0.002	0.002	0.001	0.001	0.000	0.000	0.001	0.000	0.001	0.001
ELO	0.000	0.003	0.000	0.012	0.006	0.005	0.003	0.001	0.002	0.000	0.001	0.001	0.000	0.001	0.001
EHO	0.000	0.008	0.000	0.011	0.006	0.003	0.006	0.001	0.002	0.000	0.001	0.001	0.000	0.001	0.001
ETO	0.000	0.009	0.000	0.004	0.003	0.005	0.003	0.000	0.002	0.000	0.001	0.001	0.000	0.001	0.001
ESO	0.000	0.017	0.000	0.004	0.004	0.006	0.004	0.001	0.003	0.000	0.001	0.001	0.000	0.001	0.001
EZO	0.000	0.005	0.000	0.014	0.003	0.002	0.002	0.001	0.001	0.000	0.001	0.001	0.000	0.001	0.001
ESFO	0.000	0.017	0.000	0.004	0.004	0.006	0.004	0.001	0.002	0.000	0.001	0.001	0.000	0.001	0.001
EKO	0.000	0.730	0.000	0.010	0.006	0.005	0.006	0.001	0.002	0.000	0.001	0.001	0.000	0.001	0.001

Table 8.5 Performance values.

	S+i	S-i	S-min	S-i-top	(S-min)/ (S-i)	Total[(S-min)/ (S-i)]	Qi	Ni	Rank
EPO	0.015	0.003	0.003	0.027	1,000	7,134	0.018	2,414	8
ELO	0.033	0.004			0.777		0.035	4,646	5
EHO	0.037	0.004			0.794		0.040	5,178	4
ETO	0.026	0.003			0.914		0.030	3,889	7
ESO	0.038	0.003			0.970		0.042	5,468	2
EZO	0.028	0.003			0.882		0.031	4,066	6
ESFO	0.037	0.003			0.970		0.041	5,316	3
EKO	0.760	0.004			0.827		0.763	100	1

The weighted normalized values and the performance values obtained by Copras methodology are provided in Table 8.4 and Table 8.5, respectively.

The resulting performance values indicate that Epoxidized Canola Oil with a performance value of 0.763 is the highest ranked epoxidized plant oil based on the factors used for within study.

8.3 Conclusions

In the last 20 years, most plastics utilized are reproduced petro chemically materials that are not biodegradable. The raise in worldwide ecological troubles like decreasing petroleum sources and greenhouse gas emission has concentrated more carefulness on the green plastic composite improvement. Bio-sourced plastic composites are one type of composite that is biodegradable, compostable, environmentally friendly and because they are made from sustainable and renewable sources. They degrade our dependence on consuming petrochemical-based energy resources and the hazardous substance generation.

One of the most optimum products for biodegradable fabrication, environmentally friendly, high performing plastics is epoxidized vegetable oil-based plastics. Thus, remarkable endeavors have been made as an effort to improve the characteristic properties of epoxidized vegetable oil-based oils by employing plasticizer. Fossil sourced plasticizer are normal compounding contents, nevertheless epoxidized plant oil-sourced plasticizer is improved as a practicable option [34]. The vegetal-based oils are sourced from biomass and are formed chemically of diverse triglycerols, in another saying, fatty acids and glycerol esters [35]. The vegetal-based oils are interesting feed stocks for many technological implementations as they are obtained from sustainable sources, easily available, environment-friendly, biodegradable and obtained in big amounts at a competitive expense [36].

A quantitative decision-making model is proposed within the scope of this study, which involves a hybrid multifactor decision-making technique to evaluate a number of alternatives based on a series of factors and obtain the best one. The proposed method involves a combination of Entropy and Copras Methodologies as a two-step algorithm. Analytical results indicate the efficiency of the developed method in determining an optimum choice among a selection of alternatives. This paper promises to inspire the replacement of present fossil-sourced plastics with sustainable plant oil-based materials for a variety of implementations as part of a movement in the direction of a more circular financial organization.

References

1. Biermann, U., Friedt, W., Lang, S., Lühs, W., Machmüller, G., Metzger, J.O., Rüsch gen. Klaas, M., Schäfer, H.J., Schneider, M.P., New syntheses with oils and fats as renewable raw materials for the chemical industry. *Angew. Chem. Int. Edit.*, 39, 2206, 2000.
2. Aluyor, E.O., Obahiagbon, K.O., Ori-jesu, M., Biodegradation of vegetable oils: A review. *Sci. Res. Essays*, 4, 543, 2009.
3. Belgacem, M.N. and Gandini, A., Chapter 3 Materials from vegetable oils: Major sources, properties and applications, in: *Monomers, Polymers and Composites from Renewable Resources*, pp. 39–66, 2008.
4. Wool, R.P. and Sun, X.S., Polymers and composite resins from plant oils, in: *Bio-Based Polymers and Composites*, pp. 6–113, Elsevier Academic Press, Burlington, 2005.
5. Khot, S.N., Lascala, J.J., Can, E., Morye, S.S., Williams, G.I., Palmese, G.R., Kusefoglu, S.H., Wool, R.P., Development and application of triglyceride-based polymers and composites. *J. Appl. Polym. Sci.*, 82, 703, 2001.
6. Xia, Y. and Larock, R.C., Vegetable oil-based polymeric materials: Synthesis, properties, and applications. *Green Chem.*, 12, 1893, 2010.
7. Lu., Y. and Larock, R.C., Novel polymeric materials from vegetable oils and vinyl monomers: Preparation, properties and applications. *ChemSusChem*, 2, 136, 2009.
8. Hui, Y.H., *Bailey's Industrial Oil and Fats Products, Edible Oil and Fat Products: General Application Vol. 1*, 5th Edition, pp. 19–44, Wiley, Blackwell, NJ, USA, 1995.
9. Soroudi, A. and Jakubowicz, I., Recycling of bioplastics, their blends and biocomposites: A review. *Eur. Polym. J.*, 49, 2839, 2013.
10. Spierling, S., Knupffer, E., Behnsen, H., Mudersbach, M., Krieg, H., Springer, S., Albrecht, S., Herrmann, C., Endres, H.J., Knüpffer, E., Behnsen, H., Mudersbach, M., Krieg, H., Springer, S., Albrecht, S., Herrmann, C., Endres, H.J., Bio-based plastics—A review of environmental, social and economic impact assessments. *J. Clean. Prod.*, 185, 476, 2018.
11. Aluyor, E.O., Ozigagu, C.E., Oboh, O.I., Aluyor, P.P., Chromatographic analysis of vegetable oils: A review. *Sci. Res. Essays*, 4, 191, 2009.
12. Aluyor, E.O., Obahiagbon, K.O., Ori-jesu, M., Biodegradation of vegetable oils: A review. *Sci. Res. Essays*, 4, 543, 2009.
13. Shahidi, F. (Ed.), *Bailey's Industrial Oil and Fat Products Six Volume Set*, Sixth Edition, p. 3616, John Wiley & Sons Inc, NJ, USA, 2005.
14. Barnwal, B.K. and Sharma, M.P., Prospects of biodiesel production from vegetable oils in India. *Renew. Sust. Energ. Rev.*, 9, 363–378, 2005.
15. Karaosmanoglu, F., Vegetable oil fuels: A Review. *Energy Sources*, 21, 221, 1999.
16. Lathi, P.P. and Mattiasson, B., Green approach for the preparation of biodegradable lubricant base stock from epoxidized vegetable oil. *Appl. Catal. B Environ.*, 69, 207, 2007.
17. Goering, C.E., Schwab, A.W., Daugherty, M.J., Pryde, E.H., Heakin, A.J., Fuel properties of eleven vegetable oils. *Trans. ASAE*, 25:6, 1472, 1982.

18. Abbasov, V.M., Nasirov, F.A., Rzayeva, N.S., Nasirova, L., II, Musayeva, K.Z., Epoxidated vegetable oils: Preparation, properties and application. *PPOR*, 19, 4, 427–449, 2018.
19. Gunstone, F.D., Supplies of vegetable oils for non-food purposes. *Eur. J. Lipid Sci. Technol.*, 113, 3, 2011.
20. Corma, A., Iborra, S., Velty, A., Chemical routes for the transformation of biomass into chemicals. *Chem. Rev.*, 107, 2411, 2007.
21. Cahoon, E.B., Shockey, J.M., Dietrich, C.R., Gidda, S.K., Mullen, R.T., Dyer, J.M., Engineering oilseeds for sustainable production of industrial and nutritional feedstocks: Solving bottlenecks in fatty acid flux. *Curr. Opin. Plant Biol.*, 10, 236–244, 2007.
22. Stauffer, C.E., *Fats & oils: Practical Guides for the Food Industry, Eagan Press Handbook Series*, WILEY-VCH Verlag GmbH & Co. KGaA, Weinheim, p. 149, 1996.
23. Belgacem, M.N. and Gandini, A. (Eds.), *Monomers, Polymers and Composites From Renewable Resources*, p. 553, Elsevier, Amsterdam, 2008.
24. Sharma, V. and Kundu, P.P., Addition polymers from natural oils – A review. *Prog. Polym. Sci.*, 31, 983, 2006.
25. Hill, K., Fats and oils as oleochemical raw materials. *Pure Appl. Chem.*, 72, 1255, 2000.
26. Parreira, T.F., Ferreira, M.M.C., Sales, H.J.S., de Almeida, W.B., Quantitative determination of epoxidized soybean oil using near-infrared spectroscopy and multivariate calibration. *Appl. Spectrosc.*, 56, 1607, 2002.
27. Wakeman, R.J., *Liquid-solid separation. A to Z Guide to Thermodynamics, Heat & Mass Transfer and Fluids Engineering*, Begell House Inc. Connecticut, USA, 2011.
28. La Scala, J. and Wool, R.P., Rheology of chemically modified triglycerides. *J. Appl. Polym. Sci.*, 95, 774, 2005.
29. Espinoza-Perez, J.D., Haagenson, D.M., Pryor, S.W., Ulven, C.A., Wiesenborn, D.P., Production and characterization of epoxidized canola oil. *T. ASABE*, 52, 1289, 2009.
30. Mahajan, S., Konar, S., Boocock, D.B., Determining the acid number of biodiesel. *J. Am. Oil Chem. Soc.*, 83, 567, 2006.
31. Wang, H., Tang, H., Wilson, J., Salley, S.O., Ng, K.Y., Total acid number determination of biodiesel and biodiesel blends. *J. Am. Oil Chem. Soc.*, 85, 1083, 2008.
32. Tunca, M.Z., Ömürbek, N., Cömert, H.G., Aksoy, E., Evaluation of the performances of Opec countries with Entropy and Maut, which are multi-criteria decision making methods. *Süleyman Demirel Univ. Visionary J.*, 7, 1, 2016.
33. Zavadskas, E.K., Kaklauskas, A., Peldschus, F., Turskis, Z., Multi-attribute assessment of road design solutions by using the COPRAS method. *Balt. J. Road Bridge Eng.*, 2, 195, 2007.
34. Cataldo, F., Ursini, O., Angelini, G., Biodiesel as a plasticizer of a SBR-based tire tread formulation. *ISRN Polym. Sci.*, 2013, 340426, 1, 2013.
35. Tan, C.P. and Man, Y.B.C., Comparative differential scanning calorimetric analysis of vegetable oils: I. Effects of heating rate variation. *Phytochem. Anal.*, 13, 129, 2002.
36. Lligadas, G., Ronda, J.C., Galia, M., Cadiz, V., Renewable polymeric materials from vegetable oils: A perspective. *Mater. Today*, 16, 337, 2013.

Part II

APPLICATIONS OF BIOPLASTICS IN HEALTH AND HYGIENE

9

Biomedical Applications of Bioplastics

Syed Tareq[1], Jaison Jeevanandam[2], Caleb Acquah[3] and Michael K. Danquah[1]*

[1]Department of Chemical Engineering, University of Tennessee, Chattanooga TN, USA
[2]CQM - Centro de Química da Madeira, MMRG, Universidade da Madeira, Campus da Penteada, Funchal, Portugal
[3]School of Nutrition Sciences, Faculty of Health Sciences, University of Ottawa, Ottawa, Ontario, Canada

Abstract

Bioplastics are derived from natural and renewable sources with enhanced biodegradability. Distinct bioplastics, including polylactic acid (PLA), and polyhydroxyalkanoate (PHA), are extensively proposed to be beneficial in the biomedical industries. Some notable raw materials used in the formation of biodegradable and biocompatible plastics include cellulose, chitosan, dietary protein, and starch. This chapter highlights the properties of these bioplastic materials. Bioplastics can offer desired mechanical characteristics, including tensile strength and break elongation. However, bioplastics have their flaws. They are fragile and have a low thermal distortion temperature. Hence, the mechanical characteristics of bioplastic are considered significant for most applications. Bioplastics can increase the eutrophication tendency compared to petrochemical plastics. This chapter highlights the multi-functional uses of bioplastics in tissue engineering, fabrication of implants, as an antimicrobial agent, memory-enhancing, and biological control of pests. In addition, the different properties of bioplastics, including mechanical, electrical, magnetic, optical, and biological, are also discussed.

Keywords: Bioplastic, biomedical applications, natural polymers, implants, wound dressing

9.1 Introduction

The word "plastic" originates from the Greek word "plastikos," which implies "capable of being shaped or molded" about petrochemically derived solid products [1]. With the advancement in technology and a continual increase in the global population, the use of plastic materials has also increased significantly for commercial and domestic applications. Numerous scientists, particularly Herman Mark, "the father of polymer chemistry" have provided immense background to the material science of plastics [2]. Synthetic plastic materials which include polyethylene, polystyrene, polybutylene, polyvinyl chloride, and polyethylene terephthalate, possess harmful effects on the environment, as they are non-biodegradable. Synthetic plastics are resistant to various natural degradation processes due

**Corresponding author*: michael-danquah@utc.edu

Inamuddin and Tariq Altalhi (eds.) Handbook of Bioplastics and Biocomposites Engineering Applications: 2nd Edition, (177–198) © 2023 Scrivener Publishing LLC

to their chemical structure and durability and can stay in the environment for centuries or longer. This is due to their chemical structures, which enhance their stability [3].

Recent data show that about eight million plastic tons are dumped each year into the sea [4]. According to recent European Commission estimates, 1.5% to 4% of worldwide plastic production escapes into the ocean per annum [5]. There have also been reports of harmful chemical spillage, plastic waste, and microplastics wreaking havoc on the terrestrial and aquatic ecosystem [6–8]. Moreover, the United Nations Development Program (UNDP) emphasized the significance of striving toward a balanced financial system that uses natural resources more efficiently and produces almost no waste or pollution [9]. The chances of survival for numerous species are jeopardized by the presence of nonbiodegradable remnants. As a result, researchers are developing technologies to advance the commercial use and production of biodegradable polymers instead of petroleum-based polymers.

Biodegradable plastics have gained traction as a suitable alternative to conventional plastics to overcome the main challenge of nonbiodegradability [10]. Biodegradable plastics can be designed to easily decompose naturally upon contact with environmental microorganisms from the three domains of cellular life, such as Bacteria, Eukarya, and Archaea [11]. This occurs by metabolizing the chemical bonds in the structure of the polymer. Certain plants, such as the genetically engineered plant *Arabidopsis thaliana*, help in producing biodegradable plastics. The plant utilizes its enzymes with the help of soil microorganisms, such as *Pseudomonas* and *Moraxella* species to produce plastic [12]. The microorganisms can form plastics by consuming carbon sources along with sunlight and converting them into energy. Several researchers have replicated this natural phenomenon by transferring the respective genes into plants for the production of plastics [13, 14]. Plastics derived from plants can be extracted using a solvent via the distillation process. Hence, the plants serve as a plastic renewable feedstock for the extraction of biodegradable plastics to potentially reduce greenhouse gas emissions compared to conventional synthetic plastics.

Bioplastics are made primarily from polysaccharides with starch, proteins, and other sources of carbon, including residues of wastewater management or algae growth [15]. For instance, bioplastics can be synthesized by isolating carbohydrates from sources, such as starch and sugar, to convert them into polylactic acids (PLAs), or by utilizing engineered microorganisms to fabricate poly-hydroxy-alkanoates (PHAs) [16]. Recently, PHA was widely proposed to be useful in the manufacturing of clinical equipment, such as sutures and cardiac stents, whereas PLA is extensively used in the packaging of products [17]. Bioplastics are grouped into three categories, namely (i) partly biobased, nonbiodegradable plastics, (ii) both biobased and biodegradable, (iii) biodegradable plastics from fossil sources [18]. Polymers, such as poly butyrate adipate terephthalate derived from fossil fuels, have been used to produce biodegradable plastics. Bioplastics that are not biodegradable include polyol-polyurethane, biopolyethylene terephthalate, and biopolyethylene (bio-PE). Polybutylene succinate (bio-PBS) [19], PLA [20, 21], and PHA [21] are biodegradable and biobased plastics. Remarkably, polybutylene succinate (PBS) from corn, sugarcane, and cassava (bio-PBS) has been in production since 2017 [22]. This industrial shift to biobased and biodegradable plastic certainly highlights the target towards ultimate sustainability. Biological plastics with biodegradability also include polymers produced from cornstarch, cellulose, and lignin [23].

Bioplastics are currently viewed as a new polymer class due to their minimal carbon footprint and the use of recyclable raw materials. Deriving maximal benefits from the promising resource in pharmaceutical applications is driving the explosive growth of biomedical technology. Components of scaffolds for bone reconstruction, artificial circulatory systems, transient skin substitutes (Biofills), renal dialysis membranes, and controlled medication release systems are just a few of the applications for cellulose-based polymers [24–26]. These biopolymers offer a humid environment for maximum healing in chronic wound dressing application due to their potential to directly absorb water effectively [27]. It has been suggested that cellulose-based polymers can be employed as a dialysis membrane for bioartificial pancreas and islet transplant immune isolation [28]. The cellulose-based materials are also expected to be utilized in orthopedics, bone substitutions, and systematic delivery of drugs [29, 30]. Starch and cellulose composites are more biodegradable, and polymers intended for bone cement and controlled medication delivery have taken use of this potential [31]. Consequently, the biodegradability of bioplastics combined with the sustainability of their resources, end up making each other a prime fit in the health care field [32]. Bioplastics generated from clean energy sources may be reused spontaneously by biological systems, decreasing the dependence on fossil fuels and protecting the environment [10]. Figure 9.1 shows the global production capacity of bioplastics in 2018.

Biofriendly materials need to be addressed as suitable alternatives for widely used synthetic plastics. The current reduction techniques for polymer waste are not best suited due to the requirement for incineration and therefore do not reach the stated objectives of enhancing recycling and reusing [33]. Bioplastics can aid in the prosperous development of society and reduce CO_2 emissions emitted by synthetic plastics through the utilization of renewable resources [34]. Recycling plastics is challenged with issues in mixed plastic pollutants, and the outcome of wastes may not be acceptable for obtaining postconsumer items [35]. The nontoxicity, biocompatibility, renewability, and biodegradability characteristics of bioplastics have increased its global demand as an alternative to traditional plastics in industries, which includes packaging of food, horticulture, agriculture, bags for composting, and hygiene [36]. Packaging has been and continues to be one of the most popular uses of bioplastics, but other areas, including biomedical, textiles, and automotive applications, are also expected to expand [37]. Thus, this chapter aims to elaborate on distinct aspects of bioplastics in biomedical

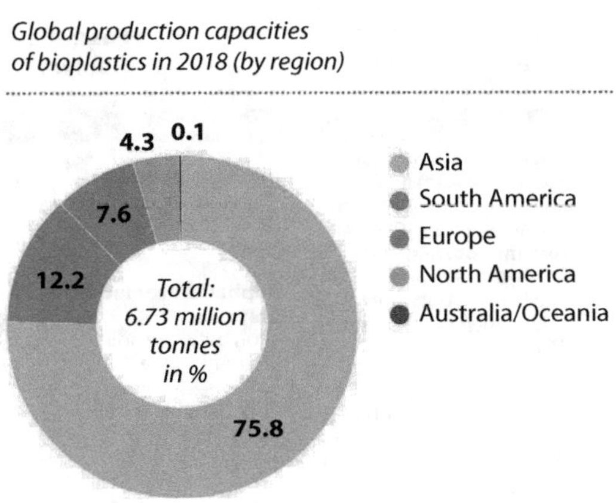

Figure 9.1 Global production capacities of bioplastics in 2018. Source: European bioplastics ©Nova institute [118].

applications. In addition, the different properties of bioplastics, including mechanical, electrical, magnetic, optical, and biological, are also discussed.

9.2 Synthesis of Bioplastics

Bioplastics can be synthesized from a wide range of affordable, abundant, renewable, and organic raw materials that are derived from fauna, floral, algal, and microbial sources as shown in Table 9.1. The synthesis of monomers derived from natural sources yields a

Table 9.1 Definition of different types of bioplastics.

Types of bioplastic	Definition
Bio-based bioplastics	Produced partially or entirely with biological polymer sources
Bioplastics-synthetic polymers	Derived from synthetic polymers
Oxo-degradable plastics	Conventional plastics with additives for rapid break down
Photo-biodegradable plastics	Degradation by ultraviolet rays facilitated by oxo-degradation
Hydro-biodegradable plastics	Plant source-based plastics with initial degradation initiated by hydrolysis
Compostable plastics	Nonbiodegradable, differs in setting and breakdown of the materials

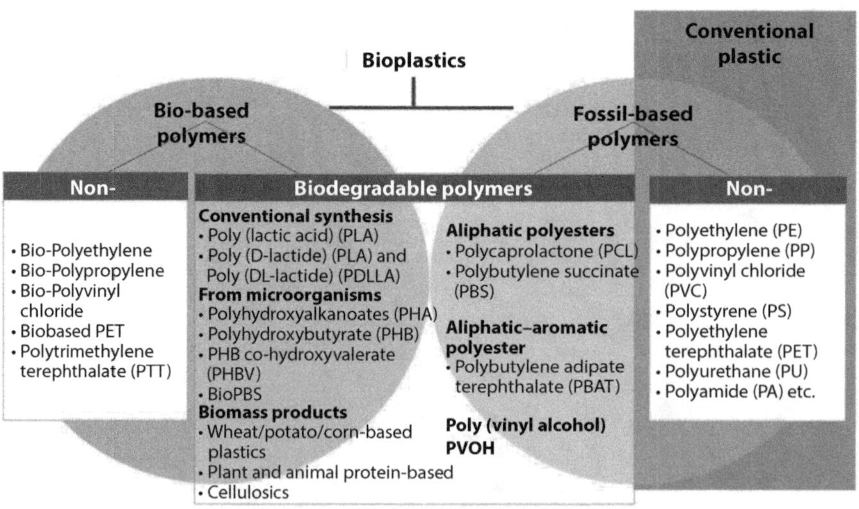

Figure 9.2 Classification of bioplastics depending on their biodegradability and source. Reproduced with permission from [119], © Springer Nature, 2020.

variety of biobased polymers as shown in Figure 9.2. Polysaccharides, which are the most common macromolecules in both flora and fauna, are an excellent source of bioplastics in the form of starch.

9.2.1 Starch-Based Bioplastics

This type of bioplastic is fabricated by the extraction of starch, which begins with the selection of tubers, specifically with high content of starch such as potato and yam. Samples are ground and labored to yield raw starch which is purified via centrifugation to obtain absolute starch [38, 39]. Later, pure starch is allowed to react with hydrochloric acid for the conversion of amylopectin into bioplastic. Finally, plasticizers are added to enhance the elasticity of the product [40]. Examples of plasticizers used for starch-based bioplastics include formamide, glycerol, ethylene glycol, urea, sorbitol, and propane-1,2,3-triol [38].

9.2.2 Cellulose-Based Bioplastics

The most extensively distributed natural polymer is cellulose, which is produced by the removal of lignin from timber pulp or textile fibers, according to Cyras *et al.* (2009) [41] and Liu (2006) [42]. Cellulose is a polysaccharide with biologically degradable ability that is solubilized in a solution of carbon disulfide and hydroxide of sodium to produce xanthate of cellulose. The cellulose xanthate solution can then be cast in an acid solution to produce cellophane films [41].

9.2.3 Chitin and Chitosan

Apart from cellulose, chitin is the most frequent agriculture-based polymer generated in nature as organized crystalline microfibrils. It forms a structural framework in arthropod exoskeletons or fungus and yeast cell walls [22]. However, it is compostable, safe, and biocompatible [43]. Flieger *et al.* (2003) [44] observed that chitin can be typically removed from crab crumb by adding sodium hydroxide (NaOH) solution (diluted, pH 13.5) to dissolve residual tissues and prevent additional microbial activity or deterioration of shells during a pretreatment phase. The crushed shells are introduced into a reaction chamber and treated with hydrochloric acid to volatilize the materials. Later, the resultant material is rinsed before the subsequent addition of the sodium hydroxide solution at a mildly increased heat for protein liquefication and chitin production. A yield of 12%, dehydrated weight basis (DWB) can be obtained during the extraction of chitin from the crab crumb. Chitin must be cleaned and then immersed in steamed lye for the elimination of acetate to form chitosan from the molecule. The resultant chitosan must be cleaned, dried, crushed, weighed, and packed for sale after the hydrolyzation process [42, 45].

9.2.4 Polyhydroxyalkanoates (PHA)

Polyhydroxyalkanoates are linear polyesters inherently formed via the bacterial fermentation of sugars or lipids. PHA is generated from the bacterial fermentation process to store carbon and energy. The fermentation process can be optimized to enable commercial production.

9.2.5 Polylactic Acid (PLA)

Polylactic acid (PLA) is a visible plastic that is typically dextrose or maize extracted [46]. Lactic acid is produced during carbohydrate fermentation and polymerizes to form low molecular weight PLA. This breaks down into lactide, which subsequently polymerizes into high-molecular-weight PLA. Fermentation of agricultural feedstocks, which includes sugar cane or maize can lead to the synthesis of polyethylene. Although nonbiodegradable, bio-derived polyethylene is identified to be recyclable and is chemically as well as physically equivalent to regular polyethylene. Further, castor oil acids can be reacted with a diamine to produce PLA while manufacturing polyamides.

9.2.6 Bioplastics from Microalgae

Algal species are broadly grouped into macroalgae and microalgae with more than one million species en masse. Notably, microalgal species make use of solar energy to generate adenosine triphosphate (ATP). They are ubiquitous, do not require the use of arable land, and can be cultivated in freshwater or marine environments [16]. Microalgae can be used as supplements in pharmaceutical formulations. Generally, microalgae have a high content of lipid which is also regarded as a good biomass candidate to produce bioplastics as shown in Figure 9.3. Numerous approaches to form bioplastics from microalgae exist and they include (i) biomass of microalgae mediated starch and cellulose extraction; (ii) conventional plastics or chemical-based polymers mixed with bioplastics; (iii) PHA or PHB generation via natural organisms or with genetic modification; and (iv) biological monomer synthesis (example: polylactic acid) [47, 48].

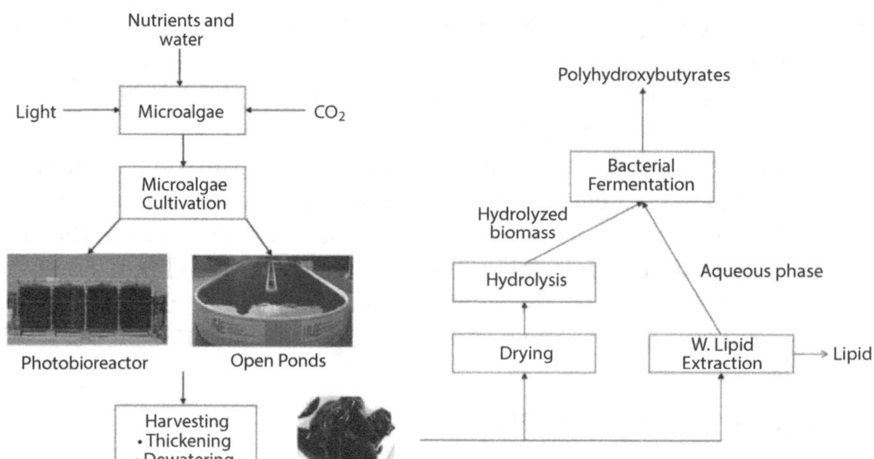

Figure 9.3 Bioplastic fabrication via microalgae. Reproduced with permission from [120], ©MDPI, 2020 [Open access].

9.3 Properties of Bioplastics

Bioplastics are made from renewable sources and undergo polymerization similar to crude oil derivatives. However, bioplastics possess good mechanical properties that make them ideal replacements for conventional plastic. They can also be processed by traditional procedures, such as melt extrusion and injection molding processes.

9.3.1 Material Strength

The tensile strength and percentage elongation can be measured using thin sheets of bioplastic. Distinct concentrations of sorbitol and glycerol plasticizers had a substantial effect on the mechanical properties of the bioplastics. For instance, an increment in the composition of plasticizer with sorbitol (5 ml) and glycerol (3 ml) could lead to a decline in the strength of tensile and enhance the percentage of elongation [49]. Further, the tensile strength decreases at higher concentrations of plasticizer due to the weakening of hydrogen bonds which eventually increases the flexibility [50]. The resultant product becomes soft and flexible which decreases tensile strength. The viscoelastic responsiveness and molecular mobility of the plastic polymer chain will improve with an increment in the concentration of plasticizers [51, 52]. Further, it is noteworthy that the polymer chains become more mobile to make the bioplastic more elastic with an increment in their elongation values [53].

In another study, the biological degradation of five distinct bioplastic types was investigated [54]. The bioplastics were tested in three distinct soil types to determine the best option for the production of biodegradable drip tubes. It was observed that the weight reduction, the strength of tensile (TS), and elongation in percentage (percent E) were resistant to the different soil types for 0 to 5 months. The resultant Bi-OPL materials showed minimal degeneration with reduced tensile strength. Moreover, the subsurface decomposition of all biopolymer samples was quicker than surface biodegradation [54].

Recently, the bioplastic composite performance of nanosized oxide particles of titanium with starch extracted from corn was also analyzed [55]. Enhanced strength of tensile and reduction in the boosted elongation of the resultant bioplastics' mechanical properties,

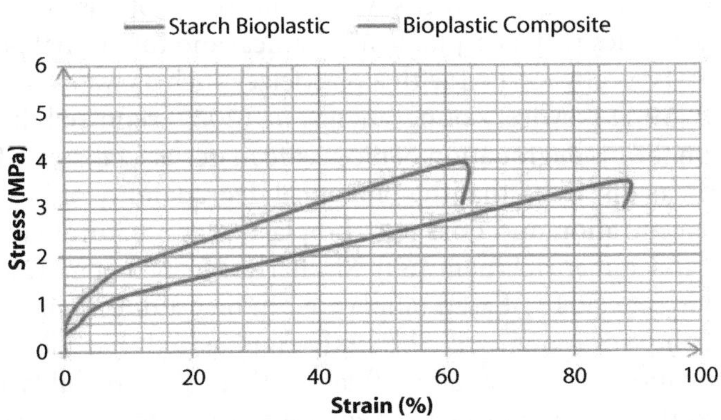

Figure 9.4 Comparison of Stress vs Strain curves for starch-based bioplastic and composite bioplastic. Reproduced with permission from [121], ©Elsevier, 2019 [Open access].

relative to the starch bioplastic. Fiber reinforcement is also significant in altering the tensile strength and flexibility of bioplastic made from corn starch. The mechanical strength of the bioplastic samples was tested after reinforcement with fibers of natural origin, such as wool, cotton, and jute. Although cotton and wool have poor mechanical strength, their samples had a greater mean tensile strength than jute, due to the fineness of cotton, wool, and hair, which allows for optimal blending and dispersion of natural fibers inside the specimens. Figure 9.4 compares the stress and strain curves of starch-based bioplastic and composite bioplastic.

9.3.2 Electrical, Mechanical, and Optical Behavior of Bioplastic

A bioplastic made from tapioca was tested using graphene along with electrochemical-mechanical liquid exfoliation (EMLE) [56]. The results emphasized that utilization of greater graphene volume can improve the conductivity of bioplastic. The optimum conductivity value of $1.57\ \text{x}10^{-1}$/cm or 6.34 cm of resistivity was observed in a specimen with a graphene concentration of 9% v/v and 50 minutes of mixing duration. The magnetic behavior of bioplastics was analyzed in another recent investigation where a smart bioplastic with anocomposites was fabricated using cellulose sheets. Cotton and sugarcane bagasse were dignified to obtain the biopolymers. Magnetic studies on nanoparticles revealed a superparamagnetic hysteresis structure, which is preserved in magnetic bioplastics. Magnetic force microscopy displayed the nanoparticle distribution over the bioplastics as well as its single domain characteristics. The study implied that the adsorption of the nanoparticles does not possess a significant impact on the optical transmittance of the composite material [56].

9.4 Biological Properties of Bioplastics

Products derived from starch, such as Polyhydroxybutyrate (PHB) and Polylactic acid (PLA) possess qualities that make them enhanced antimicrobial agents. PHB belongs to the class of Polyhydroxyalkanoates (PHA) [57]. PHB films containing different antimicrobial agents exhibited high efficacy in preventing the growth of microorganisms and controlled discharge of disinfectants [58].

Dietary proteins, such as egg albumens, whey, and soy proteins, are also being used in the formation of bioplastics with acceptable mechanical and functional properties. Protein-based bioplastics are typically formed with a plasticizer such as water, glycerol, and natural rubber latex (NRL) [59]. Likewise, Zoungranan *et al.* (2020) reported the potential natural influential factors on the biodegradation of bioplastic materials, fabricated via corn and cassava starch. The results disclosed that the biological degradability of bioplastic with starch depends on the concentration and dosage of bioplastics and natural factors, such as heat, relative dampness, and soil enrichment [60]. Similarly, Mostafa *et al.* (2020) demonstrated that polyhydroxybutyrate (PHB)-based bioplastic can be produced via *Erythrobacter aquimaris* in the rhizosphere of Mangrove. The study emphasized that the initial pH is critical in the PHB accumulation and eventually influences their bioavailability [61]. Further, several reports showed that bioplastics possess enhanced drug delivery properties. Kim *et al.* (2014) fabricated novel polyethylene glycol/poly-hydroxy octanoate (PEG/PHO) copolymer nanoparticles for the formulation of an anticancer drug named paclitaxel to inhibit

Table 9.2 Number of days for the decomposition of certain bioplastics [123]. This table is adapted with permission from © Journal of Mechanical Engineering Science and Technology, 2018 (Open access).

Bioplastic	Number of days to decompose
Bioplastic from cassava starch with 5% of nanoclay filler	6
Bioplastic from potato starch	5
Bioplastic from cassava starch	12
Gluten-based bioplastic	50
Bioplastic from corn starch	7
Bioplastic from cassava starch with chitosan filler	8
Synthetic plastic	> 50

colon carcinoma cells in mice model [62]. Furthermore, He *et al.* (2017) demonstrated an environment-friendly pathway to construct bioplastic from chitin for bone repair application. The study showed that the chitin/hydroxyapatite composite plastics possess good hemocompatibility, histocompatibility, and *in vivo* biodegradability towards osteoblast cells [63]. The decomposition of certain bioplastic from plant-based polymers was listed in Table 9.2.

9.5 Biomedical Applications of Bioplastics

A long chain polysaccharide, chitin, is crucial as a functional unit for various living organisms. It forms the cell wall in fungus, the exoskeletons in arthropods, and has widespread applications in the medical and biotechnology industry. They also provide mechanical support to the cuticles in insects [64]. Chitosan, on the other hand, is made when the chitin is treated with alkali-like sodium hydroxide. It is a linear polysaccharide and is frequently used to increase farm yields [65]. Cellulose, chitin, and chitosan are the major natural polymers that are used to fabricate bioplastics for biomedical applications [66].

9.5.1 Antimicrobial Property

Yilmaz *et al.* (2020) analyzed the antiseptic properties of chitosan nanoparticles. In this study, chitosan was used to transform an acrylic resin into an antimicrobial covering material. The solid-state (powders) and colloid phase of chitosan were introduced separately into the matrix. Later, chitosan was crushed in a mechanical mixture at room temperature for 5 h in the presence of air to obtain the powders. The goal of this mixture was to achieve a uniform dispersion of particles and to maximize the impact of the particles by raising the number of contact sites. The study proposed that chitosan possessed a positive-charged amine group that binds with the oppositely charged membranes of microorganisms. This led to damage and release of the proteinaceous bacteria and its other intracellular components

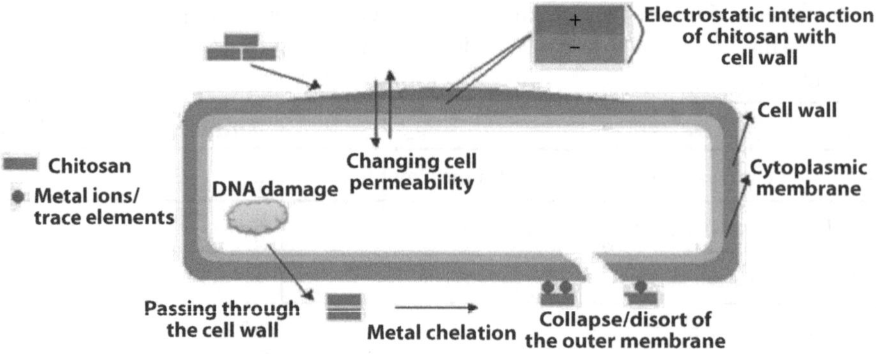

Figure 9.5 Proposed antibacterial mechanism of cellulose nanoparticles. Reproduced with permission from [122], © Springer (Singapore), 2019.

as shown in Figure 9.5. The study also emphasized that the chitosan's bactericidal ability is influenced by pH [67].

Esmaeili and Ghobadianpour (2016) utilized $MnFe_2O_4$ superparamagnetic nanomaterials as the core and chitosan crosslinked with glutaraldehyde as the coating, which was subsequently treated with PEG to improve particle stability against the reticuloendothelial system (RES). The study showed that chitosan coating offers antimicrobial effects by improving the characteristics of ferrite nanoparticles [68]. When developing a wound healing material, several aspects must be addressed to maintain an appropriately damp condition and facilitate the gas exchange to minimize water loss and exudate buildup. Medical dressings with absorbable and nonabsorbable properties were manufactured into sponges using a phase separation approach facilitated by thermal induction technique (CS-g-PHEA) in a recent study [56]. Fourier transform infrared (FTIR) spectroscopy and nuclear magnetic resonance (^{1}H-NMR) spectroscopies were used to characterize the material, which was fabricated via a free-radical polymerization process. *In vitro* experiments were used to assess the biocompatibility, swelling ability, and biodegradability of the resultant bandages, while antimicrobial activity investigations were carried out using three distinct strains of bacteria. The antiseptic ability of the specimen and loaded sponges was demonstrated in the resultant dressings by a substantial inhibition zone of bacteria [69].

Recently, the relationship between molecular weight, chitosan deacetylation level, and its respective antimicrobial property was also investigated [70]. It was identified that microbe inhibition activity rises with an increment in their molecular weight, particularly when dealing with the bacterium *E. coli*. An increment in the chitosan's molecular weight beyond 100 kDa, on the other hand, has been demonstrated to decrease antimicrobial effectiveness against *E. coli*, and not against gram-positive bacteria [70]. Chitosan is also effective against fungi; however, this area has not been studied extensively. The association of chitosan with the cell's wall or membrane is primarily responsible for its antifungal effects [71]. Despite this, minimum inhibitory concentrations (MICs) of fungi mediated by chitosan fluctuate and are strongly linked to their molecular weight, deacetylation level (DDA), and the target fungus [71, 72].

The fluidity of the membrane is significant in influencing the chitosan's degree of antifungal activity. Chitosan-sensitive fungal species are identified to possess more polyunsaturated fatty acids in their cytoplasmic membrane than non–chitosan-sensitive fungi, indicating that chitosan permeability is reliant on membrane fluidity. The tolerance to chitosan was enhanced in *Neurospora crassa*, which is a fatty acid desaturated mutant fungal variety with decreased membrane fluidity. Chitosan tolerance might have been developed in nematophagous and entomopathogenic fungi via infection of arthropods and nematodes, according to a phylogenetic study of fungi with confirmed vulnerability to chitosan [73].

9.5.2 Biocontrol Agents

Bioplastics are useful as biocontrol agents as they can be developed to prevent the growth of pathogens and pests. The impact of the biodegradable plastic utilization on the pest population of *Helopeltis* sp. and *Conopomorpha cramerella* at cocoa plantations in West Java, Indonesia, has been recently studied [74, 75]. Thick and thin bioplastics were used, and the former absorbed more water causing the bioplastic sleeve to become more humid. This enabled fungal growth to thrive in humid conditions. Bioplastic sleeve acts as a physical barrier like a bioplastic sleeve for the protection and prevention of adult *Helopeltis* sp from producing eggs, which reduces the degree of infection [76, 77]. Bioplastics from starch have a variety of unique characteristics that make them ideal for biocontrol formulations. They are fabricated in a variety of formats for enhanced delivery and effectiveness. The starch component of the bioplastic plays an important function in each form. The starch-based bioplastics component in granule formulations allows adsorption of spore suspensions with high viability retention while providing a nutrition source for the fungus [78].

9.5.3 Pharmaceutical Applications of Bioplastics

Bioplastics are considered harmless to be utilized in pharmaceutical applications when they degrade in the environment and are also biocompatible. Bioplastics are commonly used in the pharmaceutical industry as multifunctional ingredients. In this section, the widespread applications of cellulose bioplastics in pharmaceutical engineering have been discussed. Cellulose, due to its ubiquitous nature, is one of the most utilized raw materials in the development of bioplastics. The usage of cellulose varies with the size, shape, and extent of crystallization in the linkage [79]. One of the important applications of cellulose bioplastic is its use as fillers for tablets and capsules. Products that originate from cellulose have several benefits over other excipients in multiple criteria, including compatibility, pharmacological inertness, and indigestibility by enzymes in the gastrointestinal tract in humans. On the protective mucosa of the stomach and esophagus, these polymers have no irritant potential. These formulations can employ a variety of pure cellulose and cellulose ether derivatives as fillers. Current developments have shown that advantages of using cellulose-based polymers in filler modification include cheap cost, widespread accessibility, eco-friendliness, and higher filler loading levels [79].

Tablet binders act as adhesives and help other ingredients to fuse. In a process known as granulation, they convert the powder into larger granules. Microcrystalline cellulose (MCC), marketed as Avicel, is one of the most widely used direct compression binders. Direct compression is a simple, efficient production process that produces a consistent output, and it is growing in popularity. MCC's unusual binding characteristics are explained by its flexibility, as well as its limited bulk density, large surface area, and greater hygroscopicity [80]. In a recent study, the paracetamol tablets were bound with MCC, which is extracted from *Sachharum officinarum*, to determine their mechanical properties [81]. The oral administration of these tablets showed low toxicity in higher doses in mice models. The extract was given at a rate of 5000 mg/kg to the animals, which resulted in their mortality. As a result, the oral LD50 of these drugs range from 2000 to 5000 mg/kg. Nevertheless, normalcy and negligible changes in body weight and health indicators demonstrated the extract's tolerability at 300 mg/kg of doses per kg of body weight [81].

Cellulose nanocrystals, also known as CNCs, are advanced nanomaterials made of cellulose and possess various uses in the medical, aviation, and petrochemical industries. They are used as drug carriers for the delivery of drugs. These are useful because of their exclusive properties, such as recyclability, biocompatibility, minimal neurotoxic effects, and low coefficient of thermal expansion. Limited loading capacity, hydrophilic inclinations, and weak emulsification performance are the limitations of CNCs that can be easily fixed by the preparation of cellulose composites [82]. CNCs are better suited to transport hydrophilic medicines due to their enhanced biological properties. In another study, CNCs were successfully modified using rarasaponins (RS) from bamboos to increase their hydrophobicity. The CNCs-RSs could be diffused in a nonpolar solvent, such as tetrahydrofuran, indicating a significant rise in hydrophobicity. These are innovative materials that could be used to make hydrophobic drug carriers. Moreover, these are harmless for human health and the environment because they are made from cellulose and organic surfactants [83, 84].

9.5.4 Implantation

A medical procedure where an implant is inserted to replace natural organs is known as implantation. The implants are artificial devices and are often made of biomedical materials, such as titanium and silicon, which are used in dental, cardiovascular, neurological, contraceptive, or cosmetic applications. Instead of typical metal alloy and ceramic, green polymers of bioplastic are employed for soft transplanted tissue as well as implantations. This includes catheters, a prosthesis for the ureter, ligaments, tendons, and grafts of vascular tissues [85, 86]. Bacterial cellulose has become increasingly important in the biomedical industry in recent years. Microbial cellulose is a natural bioplastic contender for a variety of medicinal and tissue-engineered applications due to its unique supramolecular nanostructure and valuable characteristics. It is also highly recommended for wound-dressing materials [87, 88].

Further, *Gluconacetobacter xylinus* bacterium produces microbial cellulose that is utilized to rebuild human dental tissues [89]. The biologically synthesized membranes in veterinary medicine are well tolerated by the body, safeguarding, and assisting in the regeneration of injured parts of the body by selective permeability, and preventing microbial contamination [90, 91]. In tracheloplasty tests in dogs, a biosynthetic membrane cellulosic bioplastic was used and was found not to interfere with biomechanics or even intra-articular movement. It also aided the early restoration of the tracheloplasty area with the newly produced tissue integrating well with the neighboring cartilage. However, it is ineffective in promoting full articular cartilage restoration as a single component [92].

9.5.5 Tissue Engineering Applications

The fundamental principle of tissue engineering is to create tissue substitutes that reinstate or ameliorate the function of diseased or damaged human tissues. Several types of scaffolds are now manufactured using a variety of manufacturing techniques; however, the primary difficulty for tissue engineering is selecting the suitable materials for scaffold synthesis. Scaffolds have been made from a variety of biological polymers or bioplastics, including alginate, collagen, proteoglycans, chitin, and chitosan. They are biodegradable and highly bioavailable on the surface of damaged tissues, encouraging cellular attachment and development. Cellulose is an attractive choice for biomaterial production [93, 94], due to its adjustable chemical, physical, and mechanical characteristics as a polymer of glucose subunits [95]. The use of cellulose-based material provides a low-cost foundation for tissue engineering applications. However, they frequently exhibit weak mechanical characteristics and rapid biodegradability, limiting their application. Biopolymers, such as PHA, have flexible mechanical characteristics, have little tissue toxicity, and are easily degradable *in vivo* [96, 97]. Chitosan promotes cell attachment and can engage with glycosaminoglycans in body tissues due to its positive surface charge. Costa-Pinto *et al.* (2009) developed mesenchymal stem cells from human bone marrow on permeable chitosan bioplastic frameworks via a distinguishable osteogenic medium [98, 99].

Osteochondral abnormalities need the use of a prosthesis, such as a bone transplant or bone replacement. Poly-(3-hydroxybutyric acid-co-3-hydroxy valeric acid) (PHBV) has been utilized as bound with collagen loaded with phosphate of calcium (CaP-Gelfix) to offer great mechanical strength to bone tissues [100]. In addition, Puppi *et al.* (2010) explored and briefly explained the application of PHA in tissue engineering applications [101]. Further, porous poly-(3-hydroxybutyrate-co-3-hydroxy hexanoate) (PHBHHx)/polyhydroxybutyrate (PHB)-based bioplastic ad scaffolds have been employed as a development of articular cartilage chondrocyte cells (RAC) matrix in a rabbit. When compared to plain PHB, the cells grew and proliferated better on these scaffolds [102].

In the engineering of cardiac tissue applications, polyhydroxyalkanoates (mcl-PHA) are demonstrated to have unique features with the chain length of the medium. They are proposed to be integrated into the cardiac system and conjugate with biologically active compounds, which proceed to elevate adhesion of cells, proliferation, and survival [103, 104]. Cardiac engineering applications utilize biomaterials arranged in the shape of a heart valve, followed by cell seeding, to assist regenerate functioning valve replacements in a live heart [105]. Similarly, PHB and 3-hydroxybutyrate-co-4-hydroxybutyrate (PHB4HB) have been investigated as potential blood circulatory components [106]. Likewise, PHA, as well as PHB4HB, have been utilized to fabricate wound dressing and closure materials with enhanced mechanical strength for the treatment of muscular wounds and skin tissue healing applications [107]. Moreover, PHA is under extensive research to be used in the healing of the nerves. It has been identified that PHA conduits have good mechanical characteristics, which may be altered via various manufacturing processes [108].

9.5.6 Memory Enhancer

The brain is composed of interconnected neurons that communicate with each other. Progressive loss of memory caused by illnesses, such as Alzheimer's disease might be perilous. Alzheimer's disease is a degenerative neurologic disease that involves the death of brain cells and atrophy of the brain. In addition, the connections between the brain cells are destroyed in Alzheimer's disease. This is attributable to the accumulation of protein, which leads to the development of aberrant structures known as "plaques" and "tangles." Hydroxybutyrate methyl ester (HBME), derived from bacterial PHAs, defends mitochondria by rescuing mitochondrial respiratory complexes, maintaining membrane integrity, and decreasing oxidative damage [109]. Again, PHA-derived hydroxyapatite (HA) monomers can activate calcium ion (Ca^{2+}) channels, which can aid in memory enhancement [110, 111].

9.6 Limitations

Despite several beneficial aspects of biopolymers, they also possess certain limitations to be used in biomedical applications. Biopolymers are fragile, have a low thermal distortion temperature, and have low resilience to long-term processes. For example, PLA, which is one of the most popular synthetic biopolymers, performs poorly as a food packaging material, due to its low thermal distortion temperature, limited resilience to high heat and humidity, and poor flexibility. The main limitation of the current biopolymers is that they need more energy to produce than petro-plastic, and it is also more expensive than petroleum-based plastics. However, bioplastics made of nonfinite sources are presently being produced and require further study to address performance constraints [112].

The total amount of greenhouse gases produced by human actions (CO_2 equivalent) is known as carbon footprint. The total human demand for natural capital is called ecological footprint and this is widely used in sustainability measures [19]. In a previous publication, the carbon footprints of fibers incorporated into chemical synthesized plastic polypropylene (PP) were examined. It was observed that the jute, cotton, and kenaf fibers had a lower carbon footprint when compared to polypropylene. This metric refers to the energy quantity and product of petroleum used in polymerized propylene manufacture. However, the jute and kenaf fibers had a lower environmental effect [113].

Relative to petrochemical plastics, bioplastics can result in a potential increment in eutrophication (up to 400%) and land usage [114]. One of the difficulties in developing bioplastics is that their mechanical characteristics, such as barrier properties, are inferior to petro-plastics. Researchers have created numerous prototypes based on bioplastics and nanoparticles to be beneficial in the packaging industry in recent years, with impressive outcomes [115, 116]. One of the most intriguing elements of using nanofillers is the ability to dramatically reduce the amount that must be introduced to the polymer, therefore eliminating the negative effects caused by using traditional inorganic additives [117]. Apart from a few large-scale applications, the area of nanocomposites is still in its early stages of development, which can be used to enhance bioplastics. Even though these polymers have been the focus of study for over two decades, polymer manufacturers are only now taking the initial steps toward significant commercial advancement of these technologies. The production of these novel nanocomposite bioplastic materials is important for numerous reasons, such as ecological benefits to be beneficial in improving packing performance of bioplastics and reducing their cost in the future [84].

9.7 Conclusion

Recent advancements of bioplastic may result in increased manufacturing efficiency, as well as the fabrication of novel bioplastics with enhanced uses and possibilities in the future. Biobased plastics provide major benefits, which include a reduction in the dependency on fossil fuels and biocompatibility. As such, biobased polymers have been deployed in numerous applications with a huge potential in the biomedical industry. The number of effective nanomedicine and tissue regeneration technologies employing biobased polymers might rise due to competent and creative research that molds device design for the desired application. Though biobased plastics can provide certain technical advancements, they are not the comprehensive technological panacea to our ecosystem issues. All bioplastics are made from natural elements; however, this does not imply that all bioplastics are compostable. Bioplastics must, therefore, be built on an integrated ecologically friendly system to enhance the material's sustainability to meet the ultimate goals of being biodegradable and biocompatible while having good mechanical characteristics analogous to conventional plastics based on the application.

References

1. Sangawar, V. and Gadre, A., A review on biodegradation of polymer and its applications. *Inventi Rapid: Chem. Eng.*, 2, 1, 2013.
2. Teegarden, D.M., *Polymer Chemistry: Introduction To An Indispensable Science*, NSTA Press, United States, 2004.
3. Kabir, E., Kaur, R., Lee, J., Kim, K.-H., Kwon, E.E., Prospects of biopolymer technology as an alternative option for non-degradable plastics and sustainable management of plastic wastes. *J. Clean. Prod.*, 258, 120536, 2020.
4. Neufeld, L., Stassen, F., Sheppard, R., Gilman, T., The new plastics economy: Rethinking the future of plastics, in: *World Economic Forum*, 2016.
5. Narancic, T. and O'Connor, K.E., Microbial biotechnology addressing the plastic waste disaster. *Microb. Biotechnol.*, 10, 1232, 2017.
6. Lebreton, L.C., Van Der Zwet, J., Damsteeg, J.-W., Slat, B., Andrady, A., Reisser, J., River plastic emissions to the world's oceans. *Nat. Commun.*, 8, 1, 1–10, 2017.
7. Andrady, A.L., Microplastics in the marine environment. *Mar. Pollut. Bull.*, 62, 1596, 2011.
8. Lamb, J.B., Willis, B.L., Fiorenza, E.A., Couch, C.S., Howard, R., Rader, D.N., True, J.D., Kelly, L.A., Ahmad, A., Jompa, J.J.S., Plastic waste associated with disease on coral reefs. *Science*, 359, 460, 2018.
9. SDGs, U.N., *Sustainable Development Knowledge Platform*, United Nations, United States, 2017.
10. Bano, K., Pandey, R., Jamal-e-Fatima, Roohi, New advancements of bioplastics in medical applications. *Int. J. Pharm. Sci. Rev. Res.*, 9, 402, 2018.
11. García-Depraect, O., Bordel, S., Lebrero, R., Santos-Beneit, F., Börner, R.A., Börner, T., Muñoz, R., Inspired by nature: Microbial production, degradation and valorization of biodegradable bioplastics for life-cycle-engineered products. *Biotechnol. Adv.*, 53, 107772, 2021.
12. Kathiresan, K., Polythene and plastics-degrading microbes from the mangrove soil. *Rev. Biol. Trop.*, 51, 629, 2003.
13. Gill, M., Bioplastic: A better alternative to plastics. *Int. J. Res. Appl. Nat. Soc. Sci.*, 2, 115, 2014.

14. Khan, A.K., Anjum, I., Hano, C., Abbasi, B.H., Anjum, S., An overview on feasible production of bioplastic polyhydroxyalkanoate (PHA) in transgenic plants, in: *Bioplastics for Sustainable Development*, p. 555, 2021.
15. Coppola, G., Gaudio, M.T., Lopresto, C.G., Calabro, V., Curcio, S., Chakraborty, S., Bioplastic from renewable biomass: A facile solution for a greener environment. *Earth Syst. Environ.*, 5, 1, 2021.
16. Kumar, S. and Thakur, K.S., Bioplastics-classification, production and their potential food applications. *J. Hill Agric.*, 8, 118, 2017.
17. Philip, S., Keshavarz, T., Roy, I., Polyhydroxyalkanoates: Biodegradable polymers with a range of applications. *J. Chem. Technol. Biotechnol.*, 82, 233, 2007.
18. Gironi, F. and Piemonte, V., Bioplastics and petroleum-based plastics: Strengths and weaknesses. *Energ. Sources Part A*, 33, 1949, 2011.
19. Xu, J. and Guo, B.H., Poly (butylene succinate) and its copolymers: Research, development and industrialization. *Biotechnol. J.*, 5, 1149, 2010.
20. Garlotta, D., A literature review of poly (lactic acid). *J. Polym. Environ.*, 9, 63–84, 2001.
21. Nampoothiri, K.M., Nair, N.R., John, R.P., An overview of the recent developments in polylactide (PLA) research. *Bioresour. Technol.*, 101, 8493, 2010.
22. Thakur, S., Chaudhary, J., Singh, P., Alsanie, W.F., Grammatikos, S.A., Thakur, V.K., Synthesis of bio-based monomers and polymers using microbes for a sustainable bioeconomy. *Bioresour. Technol.*, 344, 126156, 2021.
23. Iwata, T., Biodegradable and bio-based polymers: Future prospects of eco-friendly plastics. *Angew. Chem. Int. Ed.*, 54, 3210, 2015.
24. Mohan, D., Teong, Z.K., Bakir, A.N., Sajab, M.S., Kaco, H., Extending cellulose-based polymers application in additive manufacturing technology: A review of recent approaches. *Polymers*, 12, 1876, 2020.
25. Oprea, M. and Voicu, S.I., Recent advances in composites based on cellulose derivatives for biomedical applications. *Carbohydr. Polym.*, 247, 116683, 2020.
26. Ullah, H., Wahid, F., Santos, H.A., Khan, T., Advances in biomedical and pharmaceutical applications of functional bacterial cellulose-based nanocomposites. *Carbohydr. Polym.*, 150, 330, 2016.
27. Kamoun, E.A., Kenawy, E.-R.S., Chen, X., A review on polymeric hydrogel membranes for wound dressing applications: PVA-based hydrogel dressings. *J. Adv. Res.*, 8, 217, 2017.
28. Waheed, H. and Hussain, A., Effect of polyvinyl pyrolidone on morphology and performance of cellulose acetate based dialysis membrane. *Eng. Technol. Appl. Sci. Res.*, 9, 3744, 2019.
29. Andra, S., Kumar Balu, S., Jeevanandam, J., Muthalagu, M., Danquah, M.K., Surface cationization of cellulose to enhance durable antibacterial finish in phytosynthesized silver nanoparticle treated cotton fabric. *Cellulose*, 1, 5895–5910, 2021.
30. Barhoum, A., Jeevanandam, J., Rastogi, A., Boluk, Y., Dufresne, A., Danquah, M.K., Bechelany, M., Plant celluloses, hemicelluloses, lignins, and volatile oils for the synthesis of nanoparticles and nanostructured materials. *Nanoscale*, 12, 22845, 2020.
31. Sharma, S., Sudhakara, P., Singh, J., Ilyas, R.A., Asyraf, M.R.M., Razman, M.R., Critical review of biodegradable and bioactive polymer composites for bone tissue engineering and drug delivery applications. *Polymers*, 13, 2623, 2021.
32. Meereboer, K.W., Misra, M., Mohanty, A.K., Review of recent advances in the biodegradability of polyhydroxyalkanoate (PHA) bioplastics and their composites. *Green Chem.*, 22, 5519, 2020.
33. Windfeld, E.S. and Brooks, M.S.-L., Medical waste management–A review. *J. Environ. Manage.*, 163, 98, 2015.

34. Abdul-Latif, N.I.S., Ong, M.Y., Nomanbhay, S., Salman, B., Show, P.L., Estimation of carbon dioxide (CO2) reduction by utilization of algal biomass bioplastic in Malaysia using carbon emission pinch analysis (CEPA). *Bioengineered*, 11, 154, 2020.
35. Hahladakis, J.N. and Iacovidou, E., An overview of the challenges and trade-offs in closing the loop of post-consumer plastic waste (PCPW): Focus on recycling. *J. Hazard. Mater.*, 380, 120887, 2019.
36. Degli Esposti, M., Morselli, D., Fava, F., Bertin, L., Cavani, F., Viaggi, D., Fabbri, P., The role of biotechnology in the transition from plastics to bioplastics: An opportunity to reconnect global growth with sustainability. *FEBS Open Bio*, 11, 967, 2021.
37. Marichelvam, M.K., Jawaid, M., Asim, M., Corn and rice starch-based bio-plastics as alternative packaging materials. *Fibers*, 7, 32, 2019.
38. Shafqat, A., Al-Zaqri, N., Tahir, A., Alsalme, A., Synthesis and characterization of starch based bioplatics using varying plant-based ingredients, plasticizers and natural fillers. *Saudi J. Biol. Sci.*, 28, 1739, 2021.
39. Rao, L.S., Naidu, C.D., Tiwari, S., Investigation on synthesis, structure and degradability of starch based bioplastics. *Mater. Today: Proc.*, 49, 257–261, 2021.
40. Ismail, N.A., Mohd Tahir, S., Yahya, N., Abdul Wahid, M.F., Khairuddin, N.E., Hashim, I., Rosli, N., Abdullah, M.A., Synthesis and characterization of biodegradable starch-based bioplastics, in: *Materials Science Forum*, Trans Tech Publications, 2016.
41. Cyras, V.P., Soledad, C.M., Analía, V., Biocomposites based on renewable resource: Acetylated and non acetylated cellulose cardboard coated with polyhydroxybutyrate. *Polymer*, 50, 6274, 2009.
42. Liu, L., Bioplastics in food packaging: Innovative technologies for biodegradable packaging. *San Jose State University Packaging Eng.*, 13, 1348, 2006.
43. Itoh, Y., Kawase, T., Nikaidou, N., Fukada, H., Mitsutomi, M., Watanabe, T., Itoh, Y., Functional analysis of the chitin-binding domain of a family 19 chitinase from Streptomyces griseus HUT6037: Substrate-binding affinity and cis-dominant increase of antifungal function. *Biosci. Biotechnol. Biochem.*, 66, 1084, 2002.
44. Flieger, M., Kantorova, M., Prell, A., Řezanka, T., Votruba, J., Biodegradable plastics from renewable sources. *Folia Microbiol.*, 48, 27, 2003.
45. Rinaudo, M., Chitin and chitosan: Properties and applications. *Prog. Polym. Sci.*, 31, 603–632, 2006.
46. Sorrentino, A., Gorrasi, G., Vittoria, V., Potential perspectives of bio-nanocomposites for food packaging applications. *Trends Food Sci. Technol.*, 18, 84, 2007.
47. Onen Cinar, S., Chong, Z.K., Kucuker, M.A., Wieczorek, N., Cengiz, U., Kuchta, K., Bioplastic production from microalgae: A review. *Int. J. Environ. Res. Public Health*, 17, 3842, 2020.
48. Hempel, F., Bozarth, A.S., Lindenkamp, N., Klingl, A., Zauner, S., Linne, U., Steinbüchel, A., Maier, U.G., Microalgae as bioreactors for bioplastic production. *Microb. Cell Fact.*, 10, 1, 1–6, 2011.
49. Aznury, M., Mechanical properties of bioplastics product from Musa paradisica formatypica concentrate with plasticizer variables, IOP Publishing, England, 2019.
50. Tarique, J., Sapuan, S.M., Khalina, A., Effect of glycerol plasticizer loading on the physical, mechanical, thermal, and barrier properties of arrowroot (Maranta arundinacea) starch biopolymers. *Sci. Rep.*, 11, 1, 1–17, 2021.
51. Hammer, C.F., Polymeric plasticizers, in: *Polymer Blends*, pp. 219–241, Elsevier, United States, 1978.
52. Antoniou, J., Liu, F., Majeed, H., Qazi, H.J., Zhong, F., Physicochemical and thermomechanical characterization of tara gum edible films: Effect of polyols as plasticizers. *Carbohydr. Polym.*, 111, 359, 2014.

53. Aznury, M., Mechanical properties of bioplastics product from Musa paradisica formatypica concentrate with plasticizer variables, in: *Journal of Physics*, IOP Publishing, 1167, 012048, 2019.
54. Mostafa, H. and Sourell, H., Characteristics of some biochemical materials used for agricultural foil mulch. *J. Agric. Res.*, 49, 539, 2011.
55. Amin, M.R., Chowdhury, M.A., Kowser, M.A., Characterization and performance analysis of composite bioplastics synthesized using titanium dioxide nanoparticles with corn starch. *Heliyon*, 5, e02009, 2019.
56. Amri, A., Rahmana, H., Utami, S.P., Iriyanti, R.S., Jiang, Z.T., Rahman, M.M., Conductive composites of tapioca based bioplastic and electrochemical-mechanical liquid exfoliation (emle) graphene. IOP Publishing, England, 2018.
57. Aydemir, D. and Gardner, D.J., Biopolymer blends of polyhydroxybutyrate and polylactic acid reinforced with cellulose nanofibrils. *Carbohydr. Polym.*, 250, 116867, 2020.
58. Muñoz-Bonilla, A., Echeverria, C., Sonseca, Á., Arrieta, M.P., Fernández-García, M., Bio-based polymers with antimicrobial properties towards sustainable development. *Materials*, 12, 641, 2019.
59. Xing, Z.C., Chae, W.-P., Baek, J.-Y., Choi, M.-J., Jung, Y., Kang, I.-K., *In vitro* assessment of antibacterial activity and cytocompatibility of silver-containing PHBV nanofibrous scaffolds for tissue engineering. *Biomacromolecules*, 11, 1248, 179, 2010.
60. Zoungranan, Y., Lynda, E., Dobi-Brice, K.K., Tchirioua, E., Bakary, C., Yannick, D.D., Influence of natural factors on the biodegradation of simple and composite bioplastics based on cassava starch and corn starch. *J. Environ. Chem. Eng.*, 8, 104396, 2020.
61. Mostafa, Y.S., Alrumman, S.A., Otaif, K.A., Alamri, S.A., Mostafa, M.S., Sahlabji, T., Production and characterization of bioplastic by polyhydroxybutyrate accumulating erythrobacter aquimaris isolated from mangrove rhizosphere. *Molecules*, 25, 179, 2020.
62. Kim, H.Y., Ryu, J.H., Chu, C.W., Son, G.M., Jeong, Y.-I.L., Kwak, T.-W., Kim, D.H., Chung, C.-W., Rhee, Y.H., Kang, D.H., Paclitaxel-incorporated nanoparticles using block copolymers composed of poly (ethylene glycol)/poly (3-hydroxyoctanoate). *Nanoscale Res. Lett.*, 9, 1, 1–10 2014.
63. He, M., Wang, X., Wang, Z., Chen, L., Lu, Y., Zhang, X., Li, M., Liu, Z., Zhang, Y., Xia, H., Zhang, L., Biocompatible and biodegradable bioplastics constructed from chitin via a "Green" pathway for bone repair. *ACS Sustain. Chem. Eng.*, 5, 9126, 2017.
64. Pilla, S., *Handbook of Bioplastics and Biocomposites Engineering Applications*, vol. 81, John Wiley & Sons, United States, 2011.
65. Fernandez, J.G. and Ingber, D.E., Manufacturing of large-scale functional objects using biodegradable chitosan bioplastic. *Macromol. Mater. Eng.*, 299, 932, 2014.
66. Pilla, S., Engineering applications of bioplastics and biocomposites-An overview, in: *Handbook of Bioplastics and Biocomposites Engineering Applications*, p. 1, 2011.
67. Atay, H.Y., Antibacterial activity of chitosan-based systems, in: *Functional Chitosan*, p. 457, 2019.
68. Esmaeili, A. and Ghobadianpour, S., Vancomycin loaded superparamagnetic MnFe2O4 nanoparticles coated with PEGylated chitosan to enhance antibacterial activity. *Int. J. Pharm.*, 501, 326, 2016.
69. Siafaka, P.I., Zisi, A.P., Exindari, M.K., Karantas, I.D., Bikiaris, D.N., J.C.p., Porous dressings of modified chitosan with poly (2-hydroxyethyl acrylate) for topical wound delivery of levofloxacin. *Carbohydr. Polym.*, 143, 90, 2016.
70. Tachaboonyakiat, W., Antimicrobial applications of chitosan, in: *Chitosan Based Biomaterials Volume 2*, pp. 245–274, Elsevier, United Kingdom, 2017.
71. Kong, M., Chen, X.G., Xing, K., Park, H., Antimicrobial properties of chitosan and mode of action: A state of the art review. *Int. J. Food Microbiol.*, 144, 51, 2010.

72. Raafat, D., Von Bargen, K., Haas, A., Sahl, H.-G., Insights into the mode of action of chitosan as an antibacterial compound. *Appl. Environ. Microbiol.*, 74, 3764, 2008.
73. Aranda-Martinez, A., Ortiz, M.Á.N., García, I.S.A., Zavala-Gonzalez, E.A., Lopez-Llorca, L.V., Ethanol production from chitosan by the nematophagous fungus Pochonia chlamydosporia and the entomopathogenic fungi Metarhizium anisopliae and Beauveria bassiana. *Microbiol. Res.*, 204, 30, 2017.
74. Eris, D., Yusuf, C., Purwantara, A., Isroi, I., *Application of the Biodegradable Bioplastic Sleeves to Control Helopeltis sp in Cocoa Plantation*, 2019.
75. Rosmana, A., Shepard, M., Hebbar, P., Mustari, A., Control of cocoa pod borer and phytophthora pod rot using degradable plastic pod sleeves and a nematode, steinernema carpocapsae. *Indones. J. Agric. Sci.*, 11, 41, 2013.
76. Hosang, M.L.A., Schulze, C.H., Tscharntke, T., Buchori, D., The potential of artificial nesting sites for increasing the population density of the black cacao ants. *Indones. J. Agric.*, 3, 45, 2010.
77. Saripah, B., Azhar, I., Board, M.C., Floor, W.S., *Handling of Cocoa Black Ants As A Biological Control Agent Against Cocoa Pod Borer in Monococoa Ecosystem.*
78. Nishat, N. and Malik, A., Antimicrobial bioplastics: Synthesis and characterization of thermally stable starch and lysine-based polymeric ligand and its transition metals incorporated coordination polymer. *ISRN Inorg. Chem.*, 2013, 538157, 2013.
79. Shokri, J. and Adibkia, K., Application of cellulose and cellulose derivatives in pharmaceutical industries, IntechOpen, United Kingdom, 2013.
80. Mousavipazhouh, H., Azadfallah, M., Jouybari, I.R., Encapsulation of precipitated calcium carbonate fillers using carboxymethyl cellulose/polyaluminium chloride: Preparation and its influence on mechanical and optical properties of paper. *Maderas Cienc. Tecnol.*, 20, 703–714, 2018.
81. Yenon, A.A., Acute toxicity and phytochemical screening of stem bark extracts of Entandrophragma angolense (Welw.) C. DC. (Meliaceae). *Asian J. Biomed. Pharm. Sci.*, 4, 67, 2014.
82. Cherian, B.M., Leão, A.L., de Souza, S.F., de Olyveira, G.M., Costa, L.M.M., Brandão, C.V.S., Narine, S.S., Bacterial nanocellulose for medical implants, in: *Advances in Natural Polymers*, pp. 337–359, Springer, Germany, 2013.
83. Aslan, M., Simsek, G., Dayi, E., Guided bone regeneration (GBR) on healing bone defects: A histological study in rabbits. *J. Contemp. Dent. Pract.*, 2, 114, 2004.
84. Carvalho, R.S., Nelson, D., Kelderman, H., Wise, R., Guided bone regeneration to repair an osseous defect. *Am. J. Orthod. Dentofacial Orthop.*, 123, 455, 2003.
85. Narancic, T., Cerrone, F., Beagan, N., O'Connor, K.E., Recent advances in bioplastics: Application and biodegradation. *Polymers*, 12, 920, 2020.
86. Čolnik, M., Knez-Hrnčič, M., Škerget, M., Knez, Z., Biodegradable polymers, current trends of research and their applications, a review. *Chem. Ind. Chem. Eng. Q.*, 26, 401–418, 2020.
87. Iamaguti, L.S., Brandao, C.V., Pellizzon, C.H., Ranzani, J.J., Minto, B.W., Histological and morphometric analysis for the use of a biosynthetic cellulose membrane in experimental trochleopasty. *Pesqui. Vet. Bras.*, 28, 195, 2008.
88 Nath, A., Kanjilal, T., Bhattacharjee, C., *Application of bioactive composite green polymer for the development of artificial organs.* New Polymeric Composite Materials, United States, MaterialsResearchForum LLC, 2016.
89. Canas-Gutierrez, A., Osorio, M., Molina-Ramirez, C., Arboleda-Toro, D., Castro-Herazo, C., Bacterial cellulose: A biomaterial with high potential in dental and oral applications. *Cellulose*, 27, 1, 2020.
90. Bianchera, A., Catanzano, O., Boateng, J., Elviri, L., The place of biomaterials in wound healing, in: *Therapeutic Dressings and Wound Healing Applications*, p. 337, 2020.

91. Shukla, R., Mishra, P., Ujjwal, R.R., Kesharwani, P., Electrospun nanofibers for wound healing, in: *Theory and Applications of Nonparenteral Nanomedicines*, p. 289, 2021.
92. Robotti, F., Sterner, I., Bottan, S., Rodríguez, J.M.M., Pellegrini, G., Schmidt, T., Falk, V., Poulikakos, D., Ferrari, A., Starck, C., Microengineered biosynthesized cellulose as anti-fibrotic *in vivo* protection for cardiac implantable electronic devices. *Biomaterials*, The Netherlands, 229, 119583, 2020.
93. Domingues, R.M., Gomes, M.E., Reis, R.L., The potential of cellulose nanocrystals in tissue engineering strategies. *Biomacromolecules*, 15, 2327, 2014.
94. Courtenay, J.C., Sharma, R.I., Scott, J.L., Recent advances in modified cellulose for tissue culture applications. *Molecules*, 23, 654, 2018.
95. Haworth, W.N., Hirst, E.L., Webb, J.I., CCCLIV.—Polysaccharides. Part II. The acetylation and methylation of starch. *J. Chem. Soc.*, 2681, 1928.
96. Liu, F., Liu, X., Chen, F., Fu, Q., Mussel-inspired chemistry: A promising strategy for natural polysaccharides in biomedical applications. *Prog. Polym. Sci.*, 123, 101472, 2021.
97. de Souza, R.F.B., de Souza, F.C.B., Rodrigues, C., Drouin, B., Popat, K.C., Mantovani, D., Moraes, A.M., Mechanically-enhanced polysaccharide-based scaffolds for tissue engineering of soft tissues. *Mater. Sci. Eng. C*, 94, 364, 2019.
98. Barbeck, M., Jung, O., Smeets, R., Koržinskas, T., *Biomaterial-Supported Tissue Reconstruction or Regeneration*, BoD–Books on Demand, United Kingdom, 2019.
99. Costa-Pinto, A.R., Correlo, V.M., Sol, P.C., Bhattacharya, M., Charbord, P., Delorme, B., Reis, R.L., Neves, N.M., Osteogenic differentiation of human bone marrow mesenchymal stem cells seeded on melt based chitosan scaffolds for bone tissue engineering applications. *Biomacromolecules*, 10, 2067, 2009.
100. Köse, G.T., Korkusuz, F., Korkusuz, P., Hasirci, V.J., *In vivo* tissue engineering of bone using poly (3-hydroxybutyric acid-co-3-hydroxyvaleric acid) and collagen scaffolds. *Tissue Eng.*, 10, 1234, 2004.
101. Puppi, D., Chiellini, F., Piras, A.M., Chiellini, E., Polymeric materials for bone and cartilage repair. *Prog. Polym. Sci.*, 35, 403, 2010.
102. Deng, Y., Zhao, K., Zhang, X.-F., Hu, P., Chen, G.-Q., Study on the three-dimensional proliferation of rabbit articular cartilage-derived chondrocytes on polyhydroxyalkanoate scaffolds. *Biomaterials*, 23, 4049, 2002.
103. Constantinides, C., Basnett, P., Lukasiewicz, B., Carnicer, R., Swider, E., Majid, Q.A., Srinivas, M., Carr, C.A., Roy, I., *In vivo* tracking and 1H/19F magnetic resonance imaging of biodegradable polyhydroxyalkanoate/polycaprolactone blend scaffolds seeded with labeled cardiac stem cells. *ACS Appl. Mater. Interfaces*, 10, 25056, 2018.
104. Rai, R., Keshavarz, T., Roether, J., Boccaccini, A.R., Roy, I., Medium chain length polyhydroxyalkanoates, promising new biomedical materials for the future. *Mat. Sci. Eng. R.*, 72, 29, 2011.
105. Wu, Q., Wang, Y., Chen, G.-Q., Medical application of microbial biopolyesters polyhydroxyalkanoates. *Artif. Cells Blood Substit. Immobil. Biotechnol.*, 37, 1, 1–12, 2009.
106. Cheng, S.-T., Chen, Z.-F., Chen, G.-Q., The expression of cross-linked elastin by rabbit blood vessel smooth muscle cells cultured in polyhydroxyalkanoate scaffolds. *Biomaterials*, 29, 4187, 2008.
107. Brigham, C.J. and Sinskey, A.J., Applications of polyhydroxyalkanoates in the medical industry. *Int. J. Biotechnol. Wellness Ind.*, 1, 52, 2012.
108. Masaeli, E., Morshed, M., Rasekhian, P., Karbasi, S., Karbalaie, K., Karamali, F., Abedi, D., Razavi, S., Jafarian-Dehkordi, A., Nasr-Esfahani, M.H., Does the tissue engineering architecture of poly (3-hydroxybutyrate) scaffold affects cell–material interactions? *J. Biomed. Mater. Res. A*, 100, 1907, 2012.

109. Zhang, J., Cao, Q., Li, S., Lu, X., Zhao, Y., Guan, J.-S., Chen, J.-C., Wu, Q., Chen, G.-Q.J., 3-Hydroxybutyrate methyl ester as a potential drug against Alzheimer's disease via mitochondria protection mechanism. *Biomaterials*, 34, 7552, 2013.
110. Chen, G.-Q., Biofunctionalization of polymers and their applications, in: *Biofunctionalization of Polymers and Their Applications*, pp. 29–45, Springer, Germany, 2010.
111. Cheng, S., Chen, G.-Q., Leski, M., Zou, B., Wang, Y., Wu, Q., The effect of D, L-β-hydroxybutyric acid on cell death and proliferation in L929 cells. *Biomaterials*, 27, 3758–3765, 2006.
112. Abe, M.M., Martins, J.R., Sanvezzo, P.B., Macedo, J.V., Branciforti, M.C., Halley, P., Botaro, V.R., Brienzo, M., Advantages and disadvantages of bioplastics production from starch and lignocellulosic components. *Polymers*, 13, 2484, 2021.
113. Korol, J., Hejna, A., Burchart-Korol, D., Wachowicz, J., Comparative analysis of carbon, ecological, and water footprints of polypropylene-based composites filled with cotton, jute and kenaf fibers. *Materials*, 13, 3541, 2020.
114. Broeren, M.L., Kuling, L., Worrell, E., Shen, L., Environmental impact assessment of six starch plastics focusing on wastewater-derived starch and additives. *Resour. Conserv. Recy.*, 127, 246, 2017.
115. Aversa, C., Barletta, M., Gisario, A., Pizzi, E., Prati, R., Vesco, S., Design, manufacturing and preliminary assessment of the suitability of bioplastic bottles for wine packaging. *Polym. Test.*, 100, 107227, 2021.
116. Razza, F., Degli Innocenti, F., Dobon, A., Aliaga, C., Sanchez, C., Hortal, M., Environmental profile of a bio-based and biodegradable foamed packaging prototype in comparison with the current benchmark. *J. Clean. Prod.*, 102, 493, 2015.
117. Yunus, M. and Fauzan, R., Mechanical properties of bioplastics cassava starch film with Zinc Oxide nanofiller as reinforcement, IOP Publishing, England, 2017.
118. Ifbb, F.A.O., European Bioplastics, Nova-Institute, Statista, Germany, 2018.
119. Bhagwat, G., Gray, K., Wilson, S.P., Muniyasamy, S., Vincent, S.G.T., Bush, R., Palanisami, T., Benchmarking bioplastics: A natural step towards a sustainable future. *J. Polym. Environ.*, 28, 3055, 2020.
120. Onen Cinar, S., Chong, Z.K., Kucuker, M.A., Wieczorek, N., Cengiz, U., Kuchta, K., Bioplastic production from microalgae: A review. *Int. J. Environ. Res. Public Health*, 17, 3842, 2020.
121. Amin, M.R., Chowdhury, M.A., Kowser, M.A., Characterization and performance analysis of composite bioplastics synthesized using titanium dioxide nanoparticles with corn starch. *Heliyon*, 5, e02009, 2019.
122. Yilmaz Atay, H., Antibacterial activity of chitosan-based systems, in: *Functional Chitosan: Drug Delivery and Biomedical Applications*, S. Jana and S. Jana (Eds.), pp. 457–489, Springer Singapore, Singapore, 2019.
123. Wahyuningtiyas, N.E. and Suryanto, H., Properties of cassava starch based bioplastic reinforced by nanoclay. *JMEST*, 2, 20, 2018.

10

Applications of Bioplastics in Hygiene Cosmetic

Anuradha[1] and Jagvir Singh[2*]

[1]*Department of Zoology, Raghuveer Singh Government P.G. College, Lalitpur, U.P., India*
[2]*Department of Chemistry, Nehru Mahavidyalaya, Lalitpur, U.P., India*

Abstract

The ever-increasing use of plastics has led to its production increased to more than 330 million tonnes almost every year. The main uses are the packaging, building, and automotive industries, with the largest share in packaging accounting for about 45%. Plastic acts as pollutants so these substances have various environmental and health issues. Such type issues can be resolved to produce such materials which have biodegradable properties, and bioplastics may be an example of this. More use of it will reduce environmental problems. Industries like bioplastic packaging, health care, and cosmetics will get a new direction of progress. Human beings can also be saved from the ill effects of chemicals by being more hygienic in cosmetics with more use of bioplastics because bioplastics are biobased and biodegradable plastics that provide excellent biodegradability.

Keywords: Bioplastic, polysaccharides, starch, cellulose, bionanocomposite, hygiene cosmetic packaging

10.1 Introduction

Plastic has come into existence for almost the last decade. It is the most widely used. These materials have been synthesized by organic or inorganic substances and are known as synthetic and if they are synthesized with natural substances are called semisynthetic materials [1, 2]. The monomer unit of such types of compounds is manufactured as plastic materials by the industries. Plastics are usually polymers with high molecular weight and may combine with other materials to improve performance and reduce costs [3, 4].

The word plastic means mold which refers to the damage or formation of them during production, which allows them to be thrown, pressed, or removed into a wide variety of shapes, like films, threads, plates, tubes, bottles, boxes, and much more products that are used in our daily life [5–8]. The common term plastic is not to be confused with the technical marker plastic, which is used for any permanent plastic deformation when exposed to a particular point. Aluminum, for example, is plastic in this sense, but not plastic in the general sense; conversely, in their finished forms, some plastics will break down before they

**Corresponding author*: singhjagvir0143@gmail.com
ORCID: 0000-0001-7493-8243

Inamuddin and Tariq Altalhi (eds.) Handbook of Bioplastics and Biocomposites Engineering Applications: 2nd Edition, (199–210) © 2023 Scrivener Publishing LLC

are damaged, so they are not plastics in the technical sense [9–12]. Plastic is dangerous to the environment because it will never be destroyed, toxic chemicals in the plastic can alter human hormones, damage water, and endanger marine life.

Powders and foundations, in addition to skin cleansers and pores and skincare merchandise [13, 14]. Microbeads are small, stable plastic beads that are deliberately introduced to cosmetic merchandise for exfoliating, smoothing, or sprucing the pores and skin. However, polyethylene is not introduced to cosmetic merchandise totally within the form of microbeads. Polyethylene must not constantly be in a seen or solid shape [15–19]. In truth, a wonderful wide variety of cosmetics encompass Polyethylene of their formulation in other forms to offer the product with extraordinary styles of benefits. Polyethylene is a polymer used in cosmetic products, such as face powder and foundation, as well as skin cleansers and skincare. These cosmetics is in the form of microplastic. The manufacturing of microbeads is also one of its similar uses [20–25]. It is a type of synthetic material in which ethylene remains as a unit and explains its polymerization. Its use is abundant at the industrial level. Microbeads are used in cosmetics to smooth or exfoliate the skin. Small, solid spheres do this work. Microbeads are used in cosmetics for a variety of benefits [26–28].

10.2 The Need to Find an Alternative to Plastic

Along with the nonbiodegradable nature, the second problem is also that petroleum products are used in the production of plastics, whose limited reserves are continuously decreasing, as a result of which the prices are increasing. This is a matter of concern for economists and environmentalists. Scientists are actively engaged in the search for alternatives to plastics for sustainable development [29–31]. For the appropriate replacement of plastic, it is included in the priority of scientists to develop new materials from renewable biomaterials. Taking inspiration from nature, scientists are involved in the research of biodegradable materials and simple methods [32, 33]. Nature has an amazing ability to make complex and diverse structures from the simplest materials. In the produced glucose in plants through photosynthesis, many (thousands) molecules of this glucose join to form cellulose and starch. The properties of these two natural substances are miraculously different due to the difference in the way the glucose molecules interact [34–36].

Biodegradable plastics are plastics that can decompose and do not pose any environmental hazard. Biodegradable plastics are made from natural materials or petrochemical resources, which do not contain chemicals available in ordinary plastics [37–40]. Microorganisms living in their habitat work to decompose the structure of biodegradable plastic, which does not cause any harm to the environment. The enzymes of microbes use plastic as their source of energy [41–44]. However, the biodegradation rate of this plastic depends on many factors, such as its components, weather, oxygen levels, and temperature. The process of biodegradation in some biodegradable plastics requires high temperatures in a controlled environment.

The process of degradation sounds environmentally friendly, but this process has not been proven yet. The main reason for the idea of changing from ordinary plastic to biodegradable plastic is to reduce greenhouse gases. Which is the main cause of global warming. Some biodegradable plastics release two types of gases, methane, and carbon dioxide, during the decomposition process [45]. Both these gases contribute to global warming.

Despite being compostable, composting biodegradable plastics can be a difficult task, as ny biodegradable plastic requires a precise temperature of between 77 degrees and 140 degrees. Composting is a difficult process that requires sorting, shredding, mixing, and precise temperature which is not possible for a layman. Composting at home is not possible as it requires a specific environment, temperature and humidity level to make it. Due to the lack of such facilities in many cities of India, bioplastics are littered on the roads and harm the environment like normal plastics [46].

Most of the time, normal plastic bottles are sold with a bioplastic label. Biodegradable plastic is a great alternative to eco-friendly, but it has different degradation processes due to the components used to make it. Your local waste management system should be ready to accept biodegradable plastics. The use of biodegradable plastics in India will not be appreciable because to biodegrade this plastic, it needs to be treated differently under restricted conditions. Starting the process of collecting biodegradable waste from people and sending it to industries where many things have to be taken care of, such a process seems difficult to hear.

10.3 Bioplastics

Bioplastics are a type of plastic polymeric materials, such as starch, cellulose, lactic acid, fat, vegetable, and animal proteins, among others as seen as Figure 10.1 [47]. The term bioplastic is used to distinguish these materials of biological origin, from petro-plastics, which are synthesized from petroleum derivatives. Plastics are easily moldable materials that can deform without breaking under more or less elaborate conditions. It is for this reason that they are materials of great versatility.

10.3.1 Characteristic of Bioplastics

1) Economic and Environmental Importance of Bioplastics [7]
2) Biodegradability
3) Limitations of Bioplastics
4) Improving the Properties of Bioplastics

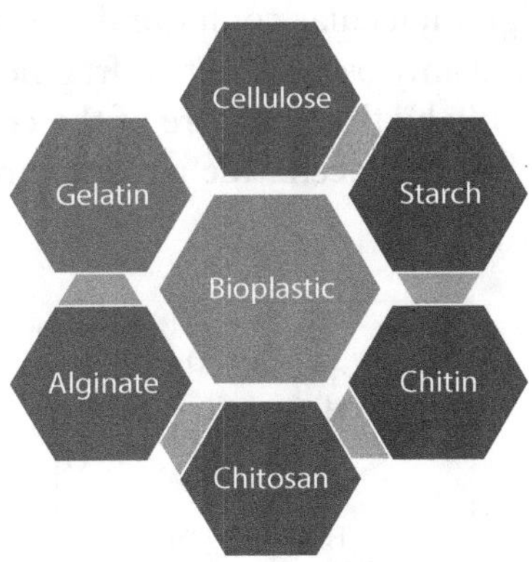

Figure 10.1 Different sources of bioplastic [7].

10.3.2 Types (Classification)

1) Classification according to its preparation
2) Classification according to its raw material
3) Industrial productions of bioplastics

10.3.3 Uses of Bioplastics

1) Disposable articles
2) Construction and civil engineering
3) Pharmaceutical applications
4) Medical applications
5) Air, sea and land transport and industry
6) Agriculture

10.4 Resources of Bioplastic

10.4.1 Polysaccharides

Polysaccharides are complex carbohydrates. They are made up of many small monosugar molecules. These are molecules of huge size. Like other carbohydrates, they are neither soluble in water nor have a sweet taste. Polysaccharides are those carbohydrates whose hydrolysis produces several monosaccharides known as polysaccharides. These are molecules of huge size [48, 49]. Like other carbohydrates, they are neither soluble in water nor have a sweet taste. Here it is formed by the condensation of many monosaccharides, in this reaction, one water molecule is lost. It is a type of polysaccharide [50].

10.4.2 Starch or Amylum

Amylum is a type of polymer broadly found in vegetation. Kind approximately 20% through the weight of the particle is a linear polymer called amylose. It is the most important representative of the group of carbohydrates (Figure 10.2). The human body has to depend on it for energy. These are polysaccharides are a type of polymers that made up of monosaccharide units of carbohydrate. It is a high molecular compound. The composition of starch consists of two fractions: amylopectin and amylose. These two fragments are formed by a-D-glucose.

They are distinguished not only by the structure of the chain but also by the number of glucose residues [51–53]. These monosaccharide residues are linked to oxygen by bridges.

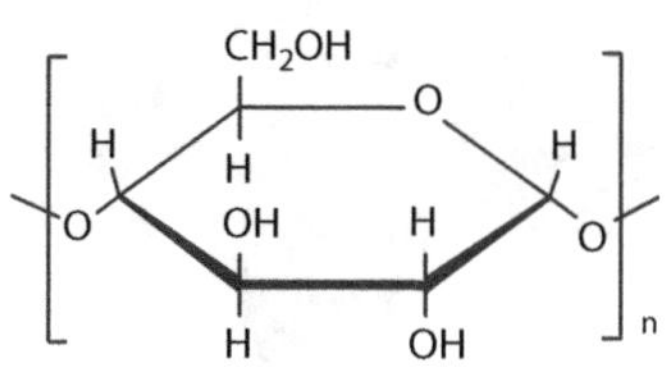

Figure 10.2 Structure of starch [53].

In the process of chain formation, the glycoside hydroxyl of one molecule interacts with the alcohol hydroxyl of another molecule.

10.4.3 Cellulose

Cellulose is a kind of chemical compound that belongs to a type of organic compound that has a molecular weight of 162.1406 g/mol and formula $(C_6H_{10}O_5)n$. Cellulose is the most abundant natural substance. Cellulose belongs to the category of biopolymers, which exist as a type of complex carbohydrate or polysaccharide. Cellulose is abundant in all plants and their related substances [54–57]. Cellulose is broken down into smaller components. The molecule of cellulose contains about hundreds to thousands of small units of glucose molecules, which are linked together to form a long chain and huge molecule. Fauna is not a source of cellulose. In addition to plants, it is present in algae and some bacteria, and other microorganisms.

Cellulose production globally is estimated to be about 1010 to 1011 tonnes annually through biological activities [58]. Out of this, about 6×109 tonnes of it are used for various industries like paper, textile furniture, building use, chemical, and other materials processing. The history of making innovative materials from cellulose is very old. In 1870, the Hyatt Manufacturing Company reacted with nitric acid on cellulose to form cellulose nitrate, which was widely used in the film industry as celluloid [59]. It was the first man-made thermoplastic. Inspired by this effort, a variety of materials were made on an industrial scale, which was made from the chemical transformation of cellulose obtained from wood.

Viscose rayon or synthetic silk is made on a large scale from regenerated cellulose. In addition, cellulose ethers and esters are commercially used in coatings, films, membranes, building materials, pharmaceuticals, and foods. It created what was widely used in the film industry as celluloid [60]. It was the first man-made thermoplastic. Inspired by this effort,

Table 10.1 Commercial sources of cellulose [62].

S. no.	Fiber source	World production (in thousand tons)
1.	Jute	2300
2	Flex	830
3.	Hemp	214
4.	Remy	100
5.	Sugarcane Bagasse	75000
6.	Bamboo	30000
7.	Kenaf	970
8.	Sisal	378
9.	Coconut	100
10.	Ebeca	70
11.	Grass	700

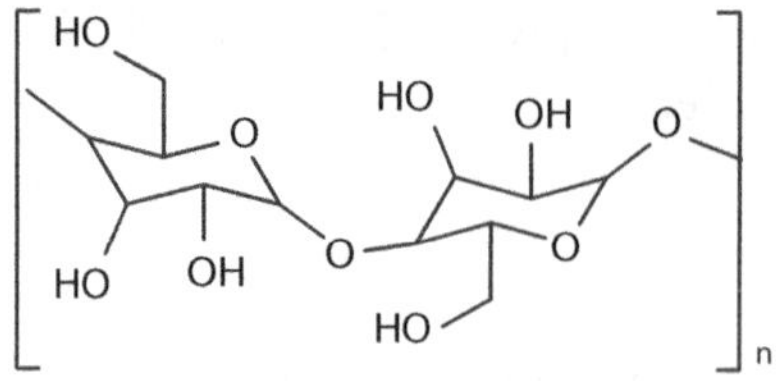

Figure 10.3 Structure of cellulose [54].

a variety of materials were made on an industrial scale, which was made from the chemical transformation of cellulose obtained from wood. Viscose rayon or synthetic silk is made on a large scale from regenerated cellulose. In addition, cellulose ethers and esters are commercially used in coatings, films, membranes, building materials, pharmaceuticals, and foods [61].

10.4.3.1 Source of Cellulose

Cellulose has many uses as a natural fiber [54]. Their major sources and annual production are shown in Table 10.1 and depicted in Figure 10.3.

Apart from these, wastes from agricultural industries such as straw, peels of rice and pulses, kernels and peels of fruits, residues of seasonal crops, etc. are sources of cellulose. Some of it is used as fodder for livestock, some are burnt and the rest is left to rot [63]. If a suitable method of proper use of this waste is developed, then farmers will get an additional source of income, and materials that are harmful to the environment can be manufactured.

10.5 Use of Biodegradable Materials in Packaging

Biopolymers obtained from renewable sources will be more useful for this purpose. According to modern requirements, for the replacement of plastics in packaging, after scraping of biopolymers obtained from agricultural waste, their widespread and commercial use is possible. The term bioplastic is used for substances developed from bio-based materials. For bioplastics to be acceptable, it is necessary to determine the conditions under which they decompose. It is their mechanical strength and immunity against harmful micro-organisms for food items. They can also adsorb different substances like antimicrobial, dyes which help increase the life of food items. In general, the biggest drawback of biopolymers is their moisture-absorbing properties. The mechanical strength and the ability to be molded are also low. Scientists are actively engaged in the development of nanocomposites to address these shortcomings.

10.6 Bionanocomposite

Bionanocomposites are types of composite materials. Nanoparticles are particles whose sizes range from 1 to 100 nm. A nanometer is one-millionth of a millimeter. Biopolymers used in bionanocomposites can be broadly divided into three categories:

A. Natural Biopolymers
Carbohydrates obtained by plants-cellulose, starch, chitin, alginate, agar, carrageenan, and proteins obtained from plants or animals—soy protein, genes from maize, gluten from wheat, gelatin, collagen, protein from buttermilk, casein.
B. Synthesized biodegradable polymers-polylactic acid, polyglycolic acid, polycaprolactone, polybutylene succinate, polyvinyl alcohol.
C. Polyesters are synthesized by fermentation from micro-organisms, such as polyhydroxyalkanoate polysaccharides, such as pullulan.

Due to the low mechanical strength of biopolymers, some inorganic supplements are added to them. The flaky silicates found in the soil are very suitable for this work. Even after adding 5% soil by weight of the biopolymer, the mechanical strength (strength) increases manifold and a barrier property to carbon dioxide, oxygen, and moisture is produced.

It is not possible to describe all the substances used in nanocomposites in this article, so only research related to cellulose is being briefly described. Cellulose nanoparticles are an ideal material for biopolymer composite industries. By nature, the macromolecules of cellulose are joined together by hydrogen bonds to form very strong structures. The cell wall of plants is made up of cellulose. Tissues are formed from many cells, which form different parts of the plant like roots, stems, branches, leaves, etc.

Crystalline cellulose has a coefficient of elasticity (ability to be stretched) higher than that of steel and similar to that of Kevlar (a very strong synthetic filament used in bulletproof jackets). Mechanical strength is similar to that of other reinforcing complements—glass fiber, metal fiber.

The basis of the unique properties of nanocomposites is the homogeneous mixture of the smallest particles of the complementary material in the matrix. The usefulness of the properties of cellulose will be proved only when it is obtained in the form of nanoparticles. The biggest obstacle to synthesizing nanocellulose is the high cost of manufacturing them. Cellulose molecules in natural sources are strongly linked to each other, it is not easy to separate them, so scientists are busy finding low-cost and simple methods, which can be used on an industrial scale.

There are two categories of nitrocellulose or cellulose nanoparticle-nanocrystalline cellulose or cellulose whiskers and microfibrillated cellulose. Wood or plant wood obtained by homogenization (grinding) under high pressure exhibits gel-like properties. Nanocellulose is obtained when the amorphous part of cellulose is separated by acidic hydrolysis. The method of its preparation is shown in the diagram. In addition, the third type of nanocellulose vinegar can be made from compounds like sugars, using the bacteria *Gluconoaceto* bacteria.

Many scientists are also active in making nanocellulose by the action of enzymes on agricultural/forestry residues. The nanocellulose composites made so far are transparent having a greater magnitude of tensile and lower thermal expansion than cast iron, i.e., the size does not change with an increase or decrease in temperature. Some of their applications are barrier films, flexible display boards (such as in foldable TVs, mobile phones), reinforcing fillers in plastics, biomedical implants such as balloons for angioplasty, drug delivery, electronic components that have potential in templates, filter membranes, batteries, supercapacitors, electrically active polymers.

10.7 Hygiene Cosmetic Packaging

Packaging has a very important role in the storage and transportation of food items to meet the needs of the growing population of the world. The invention of synthetic polymer (plastic) proved to be revolutionary in the field of packaging [50]. The popularity of plastics for packaging is unparalleled due to their important properties, such as their ability to mold, transparent properties, moisture, and adsorption, the barrier to air penetration. The cardboard, paper, glass have been replaced by plastics due to the high mechanical strength (load-carrying capacity) of extremely light and thin plastic films. Due to the variety of plastic materials and various organizations, packaging can be designed to suit the specific material.

Plastics in packaging are used in almost all types of consumer goods. About half the amount of plastic used in packaging is used to hold food items. On the one hand, the non-biodegradable property of plastic is useful for storing food items for a long time, on the other hand, this property is very harmful to the environment. There is no mechanism in nature for the disposal of human synthesized plastic materials that are thrown away after use.

The fact to be considered is whether the use of plastic materials for short-term uses like packaging, catering, surgery, hygiene is justified. It is conceivable, whether, for short-term convenience, living things should be thrown into the environment in the form of waste. Food packaging and medical waste constitute a major part of everyday waste, which contains various types of plastics [54, 55]. The greed for immediate convenience and the tendency to ignore the long-term consequences have increasing dependence on plastics.

The habit of going to the market with a bag from home has become extinct. The business instinct to woo the customers with attractive packaging is leading to indiscriminate use of plastic in packaging, resulting in an increasing amount of plastic in the waste. Animals like cows swallow the plastic along with the food items lying in the garbage, which proves painful for them. When plastic material reaches the river, pond, sea through drains, it is swallowed by aquatic organisms, which is harmful to them. It also affects birds dependent on fish for food.

10.8 Conclusion

Bioplastics, compostable, biodegradable is from plastics of today to a sustainable future. Plastic waste is not garbage. If used and recycled properly, it is a social resource. Make bioplastics and recycling and recycling programs for plastics a recycling economy. Even more problematic than traditional polymers are bio/compostable plastics, which, unfortunately, are the new buzzword in cosmetic packaging. If it accidentally mixes with conventional plastic after disposal, it ruins the entire recycling chain.

To overcome this problem, scientists have developed eco-friendly biodegradable plastic as an alternative to plastic, which can be used as packaging material. Scientists have developed an eco-friendly biodegradable polymer using guar gum and chitosan. These polymers are made from polysaccharides extracted from guar beans, crab, and shrimp. It has a very high capacity to tolerate water, that is, it does not get spoiled by water. This Guar Gum-Chitosan

Film is great for packaging use in harsh environmental conditions. Polysaccharide packaging material is one of the high-capacity biopolymers. However, polysaccharides are not preferred due to some drawbacks, such as low mechanical properties, fast soluble in water, and low barrier properties.

References

1. Science matters: The case of plastics, accessed on December 11 2020. Available online: https://www.sciencehistory.org/science-of-plastics.
2. Lestari, R.A.S., Kasmiyatun, M., Dermawan, K., Aini, A.N., Riyati, N., Putri, F.R., Bioplastic from jackfruit seeds and rice. *IOP Conf. Ser. Mater. Sci. Eng.*, 835, 012035, 2020.
3. Coltelli, M.B. and Aglietto, M., *Riutilizzo dei Materiali Polimerici*, Edizioni Nuova Cultura, Rome, Italy, 2011.
4. Ehilen, O., Obadoni, B., Imade, F., Eseigbe, D., Mensah, J.K., The effect of detergents on the germination and growth of *amaranthus hybridus* L. and *solanum lycopersicon* L. *Nigerian Ann. Natural Sci.*, 16, 100–108, 2017.
5. Fatoba, P.O., Olorunmaiye, K.S., Adepoju, A.O., Effects of soaps and detergents wastes on seed germination, flowering and fruiting of tomato (*lycopersico nesculentum*) and okra (*abelmoschus esculentus*) plants. *Ecol. Environ. Conserv.*, 17, 7–11, 2011.
6. Alvarenga, P., Palma, P., Goncalves, A.P., Fernandes, R.M., Cunha-Queda, A.C., Duarte, E., Vallini, G., Evaluation of chemical and ecotoxicological characteristics of biodegradable organic residues for application to agricultural land. *Environ. Int.*, 33, 505–513, 2007.
7. Muthusamy, M.S. and Pramasivam, S., Bioplastics-an eco-friendly alternative to petrochemical plastics. *Curr. World Environ.*, 14, 49, 2019.
8. Jerez, A., Partal, P., Mart, I., Gallegos, C., Guerrero, A., Rheology and processing of gluten-based bioplastics. *Biochem. Eng. J.*, 26, 131, 2005.
9. Jerez, A., Partal, P., Martínez, I., Protein-based bioplastics: Effect of thermo-mechanical processing. *Rheol. Acta*, 46, 711, 2007.
10. Mohammadi, A., Karim, A., Mahmud, S., Robal, M., Antimicrobial, rheological, and physicochemical properties of sago starch films filled with nanorod-rich zinc oxide. *J. Food Eng.*, 113, 511, 2012.
11. Xie, F., Pollet, E., Halley, P., Averous, L., Starch-based nano-biocomposites. *Prog. Polym. Sci.*, 38, 1590, 2013.
12. Gironi, F., Bioplastics and petroleum-based plastics: Strengths and weaknesses. *Energ. Source. Part A*, 33, 1949, 2011.
13. Jabeen, N., Majid, I., Nayik, G.A., Jabeen, N., Majid, I., Nayik, G.A., Bioplastics and food packaging: A review. *Cogent Food Agric.*, 1, 1, 42, 2015.
14. Kalia, S., Dufresne, A., Cherian, B.M., Kaith, B.S., Av, L., Njuguna, J., Cellulose-based bio- and nanocomposites: A review. *Int. J. Polym. Sci.*, 2011, 837875, 2011.
15. Sanyang, M.L., Sapuan, S.M., Jawaid, M., Ishak, M.R., Sahari, J., Recent developments in sugar palm (*Arenga pinnata*) based biocomposites and their potential industrial applications: A review. *Renew Sust. Energ. Rev.*, 54, 533, 2016.
16. Pilla, S., Engineering applications of bioplastics and biocomposites-An overview, in: *Handbook of Bioplastics and Biocomposites Engineering Applications*, pp. 1–15, Wiley, Hoboken, NJ, USA, 2012.
17. Bashir, A.S.M. and Manusamy, Y., Recent developments in biocomposites reinforced with natural biofillers from food waste. *Polym. Plast. Technol. Eng.*, 54, 87, 2015.

18. Suryanto, H., Kharismawan, F., Solichin Rahmawan, A., Sahana, R., Muhajir, M., Influence of nanoclay on thermal decomposition of biocomposite matrix starch/carrageenan blend influence of nanoclay on thermal decomposition of biocomposite matrix starch/carrageenan blend, in: *Proceedings of The 2nd International Conference on Mechanical Engineering Research and Application (Paper Submission System)*, Malang, East Java, Indonesia, October 7–9 2020.
19. Said, P., Pradhan, R., Sharma, N., Naik, B., Protective coatings for shelf-life extension of fruits and vegetables. *J. Bioresour. Eng. Technol.*, 1, 1, 2014.
20. Coltelli, M.B., Coiai, S., Bronco, S., Passaglia, E., Nanocomposites based on phyllosilicates: From petrochemicals to renewable thermoplastic matrices, in: *Advanced Nanomaterials*, K.E. Geckeler, and H. Nishide, (Eds.), pp. 403–458, Wiley-VCH Verlag GmbH & Co. KGaA, Weinheim, Germany, 2010.
21. Rittigstein, P., Priestley, R.D., Broadbelt, L.J., Torkelson, J.M., Model polymer nanocomposites provide an understanding of confinement effects in real nanocomposites. *Nat. Mater.*, 6, 257, 2007.
22. Harunsyah, Sariadi, Raudah, The effect of clay nanoparticles as reinforcement on mechanical properties of bioplastic base on cassava starch the effect of clay nanoparticles as reinforcement on mechanical properties of bioplastic base on cassava starch. *J. Phys. Conf. Ser.*, 953, 012021, 2018.
23. Singha, A.S. and Thakur, V.K., Chemical resistance, mechanical and physical properties of biofibers-based polymer composites. *Polym. Plast. Technol. Eng.*, 48, 736, 2009.
24. Geueke, B., Bioplastics as food contact materials. *Food Packaging Forum*, accessed on December 15 2020, Available online: http://www.foodpackagingforum.org/fpf-2016/wp-content.
25. Brizga, J., Hubacek, K., Feng, K., The unintended side effects of bioplastics: Carbon, land, and water footprints. *One Earth*, 3, 45, 2010.
26. Wellenreuther, C. and Wolf, A., Innovative feedstocks in biodegradable bio-based plastics: A literature review. *HWWI Res. Pap.*, 1, 194, 2020.
27. Muhammad, A., Rashidi, A.R., Roslan, A., Idris, S.A., Development of bio based plastic materials for packaging from soybeans waste. *AIP Conf. Proc.*, 1885, 020230, 2017.
28. Di Bartolo, A., Infurna, G., Dintcheva, N.T., A review of bioplastics and their adoption in the circular economy. *HWWI Res.*, 13, 1229, 2021.
29. Pantani, R. and Sorrentino, A., Influence of crystallinity on the biodegradation rate of injection-moulded poly (lactic acid) samples in controlled composting conditions. *Polym. Degrad. Stab.*, 98, 1089, 2013.
30. Fortunati, E., Armentano, I., Iannoni, A., Barbale, M., Zaccheo, S., Scavone, M., Visai, L., Kenny, J.M., New multifunctional poly (lactide acid) composites: Mechanical, antibacterial, and degradation properties. *J. Appl. Polym. Sci.*, 124, 87, 012.
31. Castro-Aguirre, E., Iniguez-Franco, F., Samsudin, H., Fang, X., Auras, R., Poly (lactic acid)-mass production, processing, industrial applications, and end of life. *Adv. Drug Deliv. Rev.*, 107, 333, 2016.
32. Signori, F., Coltelli, M.B., Bronco, S., Thermal degradation of poly (lactic acid) (PLA) and poly(butylene adipate-*co*-terephthalate) (PBAT) and their blends upon melt processing. *Polym. Degrad. Stab.*, 94, 74, 2009.
33. Zhao, X., Cornish, K., Vodovotz, Y., Narrowing the gap for bioplastic use in food packaging: An update. *Environ. Sci. Technol.*, 54, 4712, 2020.
34. Emadian, S.M., Onay, T.T., Demirel, B., Biodegradation of bioplastics in natural environments. *Waste Manage.*, 59, 526, 2017.
35. van Den, O.M., Molenveld, K., van Der, Z.M., Bos, H., Bio-based and biodegradable plastics-facts and figures. *Wagening. Food Biobased Res.*, 67, 1722, 2017.

36. Fu, Y., Wu, G., Bian, X., Zeng, J., Weng, Y., Biodegradation behavior of poly (butylene adipate-co-terephthalate) (PBAT), poly (lactic acid) (PLA), and their blend in freshwater with sediment. *Molecules*, 25, 3946, 2020.
37. Maisanaba, S., Prieto, A.I., Pichardo, S., Jordá-Beneyto, M., Aucejo, S., Jos, Á., Cytotoxicity and mutagenicity assessment of organomodified clays potentially used in food packaging. *Toxicol. In Vitro*, 29, 1222–1230, 2015.
38. Andrzejewska, A., Mazurkiewicz, A., Ligaj, B., Investigation of mechanical properties the polylactide in function its degradation rate. *IOP Conf. Ser. Mater. Sci. Eng.*, 393, 012033, 2018.
39. Iniguez-Franco, F., Auras, R., Rubino, M., Dolan, K., Soto-Valdez, H., Selke, S., Effect of nanoparticles on the hydrolytic degradation of PLA-nanocomposites by water-ethanol solutions. *Polym. Degrad. Stab.*, 146, 287, 2017.
40. Zhang, P., He, Y.-Y., Gao, D., Cai, Y., Liu, B., Hydrolytic and thermal degradation of polyethylene glycol compatibilized poly (lactic acid)-nanocrystalline cellulose bionanocomposites. *J. Appl. Polym. Sci.*, 136, 46933, 2019.
41. Dilkes-ho, L., Ashworth, P., Laycock, B., Pratt, S., Lant, P., Resources, conservation & recycling public attitudes towards bioplastics-Knowledge, perception and end-of-life management. *Resour. Conserv. Recycl.*, 151, 104479, 2019.
42. Amin, A.M.M., Saiud, S.M., Hamid, K.H.K., Musa, M., Biodegradation behaviour of thermoplastic starch films derived from tacca leontopetaloides starch under controlled composting condition biodegradation behaviour of thermoplastic starch films derived from tacca leontopetaloides starch under controlled comp. *IOP Conf. Ser. Mater. Sci. Eng.*, 358, 012050, 2018.
43. Imam, S.H. and Nadathur, G.S., Degradation of starch–poly(β-Hydroxybutyrate-Co-β-Hydroxyvalerate) bioplastic in tropical coastal waters. *Appl. Environ. Microbiol.*, 65, 2, 431–435, 2014.
44. Puls, J., Wilson, S.A., Ho, D., Degradation of cellulose acetate-based materials: A review. *J. Polym. Environ.*, 19, 152, 2011.
45. Seggiani, M., Cinelli, P., Balestri, E., Mallegni, N., Stefanelli, E., Rossi, A., Lardicci, C., Lazzeri, A., Novel sustainable composites based on poly(hydroxybutyrate-*co*-hydroxyvalerate) and seagrass beach-CAST fibers: Performance and degradability in marine environments. *Materials*, 11, 5, 772, 2018.
46. Mlalila, N., Hilonga, A., Swai, H., Devlieghere, F., Ragaert, P., Antimicrobial packaging based on starch, poly(3-hydroxybutyrate) and poly(lactic-*co*-glycolide) materials and application challenges. *Trends Food Sci. Technol.*, 74, 1, 2018.
47. Nakayama, A., Yamano, N., Kawasaki, N., Biodegradation in seawater of aliphatic polyesters. *Polym. Degrad. Stab.*, 166, 290, 2019.
48. Sashiwa, H., Fukuda, R., Okura, T., Sato, S., Nakayama, A., Microbial degradation behavior in seawater of polyester blends containing poly(3-hydroxybutyrate-co-3-hydroxyhexanoate) (PHBHHx. *Mar. Drugs*, 16, 34, 2018.
49. Kale, G., Kijchavengkul, T., Auras, R., Rubino, M., Selke, S.E., Singh, S.P., Compostability of bioplastic packaging materials: An overview. *Macromol. Biosci.*, 7, 255, 2007.
50. Ruggero, F., Gori, R., Lubello, C., Methodologies to assess biodegradation of bioplastics during aerobic composting and anaerobic digestion: A review. *Waste Manage. Res.*, 37, 959, 2019.
51. Coltelli, M.-B., Gigante, V., Panariello, L., Aliotta, L., Morganti, P., Danti, S., Cinelli, P., Lazzeri, A., Chitin nanofibrils in renewable materials for packaging and personal care applications. *Adv. Mater. Lett.*, 10, 6, 425–430, 2019.
52. Coltelli, M.-B., Cinelli, P., Gigante, V., Aliotta, L., Morganti, P., Panariello, L., Lazzeri, A., Chitin nanofibrils in poly (lactic acid) (PLA) nanocomposites: Dispersion and thermo-mechanical properties. *Int. J. Mol. Sci.*, 20, 504, 2019.

53. Panariello, L., Coltelli, M.-B., Buchignani, M., Lazzeri, A., Chitosan and nano-structured chitin for biobased anti-microbial treatments onto cellulose-based materials. *Eur. Polym. J.*, 113, 328–339, 2019.
54. Othman, S.H., Majid, N.A., Tawakkal, I.S.M.A., Basha, O.K., Nordin, N., Shapi't, R.A., Tapioca starch film reinforced with microcrystalline cellulose for potential food packaging application. *Food Sci. Technol.*, 39, 605, 2019.
55. Jiang, T., Duan, Q., Zhu, J., Liu, H., Yu, L., Starch-based biodegradable materials: Challenges and opportunities. *Adv. Ind. Eng. Polym. Res.*, 3, 8, 2020.
56. Risyon, N.P., Othman, S.H., Basha, R.K., Talib, R.A., Characterization of polylactic acid/halloysite nanotubes bionanocomposite films for food packaging. *Food Packag. Shelf Life*, 23, 100450, 2020.
57. Morganti, P. and Coltelli, M.B., A new carrier for advanced cosmeceuticals. *Cosmetics*, 6, 10, 2019.
58. Swaroop, C. and Shukla, M., Development of blown polylactic acid-MgO nanocomposite films for food packaging. *Compos. Part A Appl. Sci. Manuf.*, 124, 105482, 2019.
59. Mangaraj, S., Yadav, A., Bal, L.M., Dash, S.K., Mahanti, N.K., Application of biodegradable polymers in food packaging industry: A comprehensive review. *J. Packag. Technol. Res.*, 3, 77, 2019.
60. Lazzeri, A., Cosmetic packaging to save the environment: Future perspectives. *Cosmetics*, 6, 26, 2019.
61. Lu, P., Yang, Y., Liu, R., Liu, X., Ma, J., Wu, M., Preparation of sugarcane bagasse nanocellulose hydrogel as a colourimetric freshness indicator for intelligent food packaging. *Carbohydr. Polym.*, 249, 116831, 2020.
62. Ling, H., Tapan, C., Nath, K., Chong, S., Foo, V., Gibbins, C., The plastic waste problem in Malaysia: Management, recycling and disposal of local and global plastic waste. *SN Appl. Sci.*, 3, 1, 2021.
63. Ncube, L.K., Ude, A.U., Ogunmuyiwa, E.N., Zulkifli, R., Beas, I.N., Environmental impact of food packaging materials: A review of contemporary development from conventional plastics to polylactic acid-based materials. *Materials*, 13, 4994, 2020.

11

Biodegradable Polymers in Drug Delivery

Ariane Regina Souza Rossin[1,2], Fabiana Cardoso Lima[1,2], Camila Cassia Cordeiro[2], Erica Fernanda Poruczinski[1,2], Josiane Caetano[2] and Douglas Cardoso Dragunski[1,2*]

[1]*Department of Chemistry, State University of Maringá, Colombo Avenue, Maringá, Paraná, Brazil*
[2]*Center of Engineering and Exact Sciences, State University of West Paraná, Street da Faculdade, Toledo, Paraná, Brazil*

Abstract
This chapter presents applications of biodegradable polymers in drug release systems. An overview of different polymers and their classifications of synthetic and natural. Examples of biodegradable drug systems surrounding the types of devices are discussed in this chapter. The reader will also obtain brief information related to intravenous, transdermal, oral, and ocular applications. Furthermore, products already on the market are also exemplified. Finally, future expectations for the field of biodegradable polymers in drug delivery are discussed.

Keywords: Pharmaceutical applications, biomedical applications, biodegradable polymer devices

11.1 Introduction

Research aimed at discovering and improving polymers for drug delivery (DD) has made significant advances since the 20th century, proposing the synthesis of new polymers and improved conditions for the therapeutic delivery of the drug [1]. The choice of the polymer must take into account its degradation's toxicity and immune response in the organism [2]. However, research still has difficulties in the formulation of materials with DD effectively and with ensured biocompatibility to the patient [1, 3]. Furthermore, characteristics, such as degradation rate, target application, cost, and stability in the biological environment, should be taken as criteria for the selection of the ideal polymer [2, 4].

Biodegradable polymers (BP) have wide applicability because they do not harm the environment (eco-friendly) and still have biomedical applications in implants, scaffolds and drug delivery systems (DDS) [2, 5]. Since the use of the first medical suture in the late 1960s, research aimed at obtaining BP materials became of great scientific interest, highlighting the use in DDS [5]. A striking feature of BP in DDS is its versatility in administering the drug in a controlled manner and degradation into oligomers or monomers by natural metabolism of the human body and subsequent physiological elimination [6].

**Corresponding author*: dcdragunski@gmail.com

Inamuddin and Tariq Altalhi (eds.) Handbook of Bioplastics and Biocomposites Engineering Applications: 2nd Edition, (211–228) © 2023 Scrivener Publishing LLC

When it comes to implantable DDS devices, a BP has the advantage of not requiring surgical removal after full delivery of the drug [7].

In general, BPs can be categorized taking into account the feedstock that originates them, and can be designated as natural and synthetic. Synthetic polymers (SP) can be molded to provide improved release along with optimal physical and chemical properties. However, the SP may contain primer residues, in addition to impurities, which prevent cell growth. Natural polymers by their unique composition can provide a more desirable cellular response than synthetic polymers [8]. For the versatility of formulating new devices with unique properties, BP blends in the form of polymer blends have high visibility in research to improve many existing biomaterials [4].

A brief overview of BPs and their ratings will be addressed throughout this chapter, highlighting some natural and synthetic BP species with a focus on DDS. Due to their manufacturing versatility, the types of DDS devices from BPs are also handled in this chapter. The number of studies involving BPs in DSS in the medical segments of intravenous, transdermal, oral, and ocular are of great importance to providing well-being to the patient and are exemplified in this chapter. Because there are already successful products in the market, they are also highlighted in this chapter with the use in practical clinics. Finally, BPs' future is discussed briefly.

11.2 Biodegradable Polymer (BP)

BPs have been on the rise as an alternative in the advancement of sustainable materials with less environmental impact and stood a new research in the economy and society due to their potential for applicability in several areas [9], such as DDS [10], eco-friendly packaging [11–13] and tissue engineering [14]. They can come from renewable natural sources, such as grains, vegetables, and sugarcane, be synthesized by bacteria and fungi, and be obtained from animal exoskeletons, as well as from fossil sources, such as oil or the combination of biomass and oil. This variety of sources contributes to greater availability of material and in the future with less dependence on exclusively petrochemical resources [9].

They are classified as biodegradable due to the possibility of being naturally degraded and reverted to small molecules for the microorganisms' action present in the environment, such as bacteria of natural occurrence [15, 16]. The natural degradation time of these polymers is quite variable as it can be influenced by environmental conditions, oxygenation of the medium, availability of organic, and mineral nutrients and by specific factors of each polymer, such as molar mass and physical and chemical structure [16]. According to the sources and the synthesis processes of BPs, they can be classified into natural or synthetic [17].

11.2.1 Natural

Natural BPs have very common properties, originating from microorganisms, plants, and animals, being of great interest for food, cosmetics, and biomedical industries [18] because they present similarities with the natural extracellular matrix [19], are biocompatible, bioactive, bioadhesive, and demonstrate low inflammatory reaction [14, 17]. Polysaccharides and proteins are examples of polymers classified as natural and biodegradable and are briefly commented on below.

11.2.1.1 Polysaccharides

Polysaccharides are mostly polymers consisting of basic units of glucose and functional groups of amines, aldehydes, alcohols, carboxyls, hydroxyls, and sulfates. Some examples of polymers originating from polysaccharides exploited on a larger or smaller scale are cellulose, chitin, chitosan, starch, and alginate [17].

Cellulose is the polymer easily found nature, renewable, crystalline and found in plants, animals, green algae, fungi and minerals. Plant fiber is its main source. It is a fibrous substance, infused and insoluble in water [19] therefore, to increase its applicability and improve its solubility in alkaline solutions, additives such as urea can be used to decrease hydrophobic interactions [20]. Or for cellulose to be used in hydrogels, for example, one can make a chemical modification of their hydroxyl groups in esters or cationic groups [21].

Chitin, on the other hand, is the second polymer easily found in nature. It is linear, renewable, insoluble in common organic solvents and aqueous solvents dilution by the fact that its hydrogen-bond semicrystalline structure gives it high hydrophobicity [22]. It is found in the cell wall of fungi and yeasts, and in the exoskeletons of crustaceans (shrimps, crabs, crabs and lobsters) and some insects and fungi [23]. Despite being abundant, the insoluble characteristic limits its use [16].

Cellulose and chitin are relatively similar, with subunits connected by β-1.4 glycolipid bonds and hydroxyl groups that allow intermolecular and intramolecular hydrogen bonds [20, 21]. However, chitin has an acetamide group in position C2. In addition, both are structural BPs that provide support and protection since they are produced on the cell walls of plants in the case of cellulose and the exoskeleton of insects and crustaceans in the case of chitin [22].

Chitosan is derived from the chitin deacetylation reaction. The degree of deacetylation can range from 30% to 90% and directly influences the molar mass of the polymer. It has a rigid crystalline structure with strong interactions between intra and intermolecular hydrogens, providing a low solubility characteristic at pH less than 6.5 and in commonly used organic solvents [17, 24, 25]. It is soluble in acids, even weak ones, due to the protonation of amine groups and the formation of a positive charge, making chitosan soluble in water [20]. Chitosan has several applications in the production of biomaterials due to its adhesive and antimicrobial properties [26].

Starch is considered a linear polymer, formed by two polysaccharides: amylose and amylopectin [27]. It is low cost, low crystallinity, available in large quantities, water-soluble and produced by green plants, roots (cassava and sweet potatoes) and cereals (wheat and rice) [16, 17]. Its mechanical and biodegradability properties are altered according to the amounts of the polysaccharides amylose and amylopectin present, since the higher amylose content results in elongation and higher resistance to the polymer [27].

Alginic acid is a bioadhesive anionic polymer formed by β-D-mannuronic acid and its α-L-guluronic acid epimer, joined by a 1,4-glycoside bond in variable proportions along the chain [27]. It has good solubility in water, on the other hand, it is insoluble in alcoholic and organic solutions, being able to obtain a gelled form in the presence of tortuous cations [28]. It can be produced by some bacteria and mainly by brown algae cell walls and intracellular spaces, and can be applied to various biomaterials due to the versatility of being processed in different morphologies [26]. The presence of long alkyl groups, aromatic groups and carboxylic groups, which are protonated at a pH of around 3.28-3.65, grant the polymer maximum viscosity values and a decrease in solubility, resulting in its precipitation [29]. Alginate acid has a high swelling capacity and can absorb an amount of water 300 times greater than its weight and excels in biomaterial production studies for controlled release of cationic drugs, enzyme immobilization and cell encapsulation [17].

11.2.1.2 Proteins

Proteins are polypeptides formed by α polar and nonpolar amino acids in different forms, with high insolubility and infusibility, generally used in the form they are found in nature. Some examples proteins are collagen, gelatin and silk [16, 17, 27]

Collagen is an essential protein in the functioning of the human body, as it is present in the structure of the extracellular matrix [21]. It is a biocompatible, bioactive polymer, abundant in bonding surface sites for cells, capable of promoting cell adhesion, proliferation and differentiation [26]. It consists of several polypeptides such as glycine, proline, hydroxyproline, and lysine [27], can be obtained from fish, pigs, cattle, horses and humans, and tendons and skin are the main sources of fibrous collagen [17]. It features three polypeptide chains forming a triple helix structure [28]. And when applied to wounds collagen plays the role of regulating the healing process acting as a natural substrate for fixation, proliferation, cell differentiation and healing [30].

Gelatin is a polymer derived from acid, alkaline or enzymatic collagen hydrolysis [17] with a high percentage of glycine, proline and hydroxyproline, water-soluble, biocompatible, moderate cost, high availability and mechanical properties depending on their source, which is usually pigs or cattle [21]. Its isoelectric point can also vary from its processing method and may result in the production of basic gelatin with a positive surface charge, or acidic surface load with a negative surface charge [26]. Because of breaks in hydrogen bonds and covalent collagen bonds, gelatin has a random spiral structure and is widely employed as an emulsifier and foaming agent due to its viscoelastic property [28].

Silk is a polymer composed of silk fibroin and sericin, naturally produced by insects (silkworms), spiders, mites and flies [28]. Its structure is composed of repetitive segments of hydrophobic peptides and short hydrophilic sequences. One of its main properties is the ability to self-assemble into regular structures because of the amphiphilic nature of its sequences [31]. In addition, the silk stands out concerning the other fibrous polymers for having excellent physical-chemical properties such as mechanical resistance, tenacity, biocompatibility, non-toxicity and low immunogenicity when in contact with the human body, which are determined by its structure with crystalline and amorphous parts and with the presence of nanoconfined crystals in β-sheet [32].

11.2.2 Synthetic

Synthetic polymers play stand out in tissue engineering and are one of the main materials chosen for scaffolding production that requires great processing flexibility [33], there is greater possibility of adjusting their mechanical properties than natural polymers but, in return, they have longer time for the biodegradation process and lower bioactivity and biocompatibility [21]. Examples of synthetic BPs are polyesters, polyanhydrides, polycarbonates, polyphosphazenes, and polyurethane [33].

11.2.2.1 Polyesters

Biodegradable polyester when compared to other classes of synthetic polymers stands out for its biocompatibility and adjustable biodegradability characteristics [14]. Degradation of polymers classified as polyesters may occur under alkaline or acidic conditions and in an environment where microorganisms decrease the mass of polymers, especially water-insoluble to soluble intermediates [34]. Some examples of polyesters of great relevance are poly(lactic acid) (PLA), poly(glycolic acid) (PGA), poly(d,l-lactide-*co*-glycolide) (PLGA) and poly(ε-caprolactone) (PCL) [3, 17].

PLA is an aliphatic, semicrystalline, or amorphous polymer, biocompatible, biodegradable and compostable, obtained from lactic acid present in renewable sources such as maize, wheat, sugarcane, beetroot and potatoes [9]. It has properties such as high modulation of elasticity and tensile strength and is classified as a potential substitute for packaging production and an important biomaterial for medicine in the production of biocompatible devices [35, 36].

PGA is an aliphatic, crystalline polymer with low solubility in organic solvents when presented in high molar mass and widely used in the production of biomedical materials for drug release and tissue regeneration because it is biodegradable, biocompatible and non-toxic [3, 14]. It is of interest to the oil, gas and packaging industry for presenting rigidity, high gas barrier, mechanical resistance, thermal stability, biodegradation and compostability [37]. PGA has a similar structure to PLA and a combination of these two polymers can yield an interesting material to be used in the plastic industry, mainly for low reuse products that require rapid degradation [16].

PLGA is a biocompatible, biodegradable, nontoxic copolymer synthesized from glycolic acid monomers and lactic acid. Its properties, such as crystallinity, degradation rate and hydrophilicity varied depending on the proportions of each component, and a higher amount of glycolic acid results in a more hydrophilic PLGA and faster biodegradation [37]. It is a widely investigated polymer in DDS and tissue engineering [17, 38].

PCL is an aliphatic, half-crystalline, biocompatible and slow-degradation polymer [39]. It can be used in the production of biomaterials, since its mechanical properties can be adjusted because it presents low toxicity [40] and because its degradation time tends to favor tissue regeneration time [41]. And can be used in packaging products due to its processing flexibility, given its ability to be soluble in various organic solvents [42].

11.2.2.2 Polyanhydrides

Polyanhydrides are biodegradable, biocompatible polymers, widely applied in controlled DDS due to their ability to suffer surface erosion processes [43]. This trait and the high hydrophobicity of polyanhydride ensure that the penetration of water inside the polymer only happens after the erosion process and degradation, and drug release is therefore controlled by polymer erosion time [33, 44].

The degradation of polyanhydrides can occur both in acidic conditions and in alkaline conditions. In the latter, the degradation tends to be faster [45]. However, in addition to the conditions of the reaction medium, the polymer structure also influences the degradation rate, as polyanhydrides with an aliphatic structure are soluble in organic solvents and degrade more rapidly, while polyanhydride with aromatic structures are insoluble in these solvents and degrade more slowly [17]. Intermediate polyanhydride degradation products are metabolized and excreted in urine and feces and are therefore considered a safe polymer [46].

11.2.2.3 Polycarbonates

Polycarbonates are an interesting and commercially successful class of polymers. They are resistant polymers, amorphous, transparent and known for their excellent mechanical and thermal properties, high calorific capacity, hydrophobicity, antioxidant properties, transparency and excellent strength [47–49]. They can be divided into aliphatic polycarbonates and aromatic polycarbonates, according to structural class based on their polymer carbon chain [50, 51]. For application in the biomedical field, aliphatic polycarbonates are more desirable.

Aliphatic polycarbonates (APC) are used in the biomedical field for their softness, biocompatibility, biodegradability and mechanical properties very similar to living tissue [52]. The hydrophobic segments of APCs can self-assemble in a micellar structure and interact with hydrophobic drugs for DDS. The DD is governed by the rate of degradation and self-assembled disorganization of APCs [53]. Examples of APCs used in DDS are poly(trimethylene carbonate) (PTMC) and poly(dimethyl trimethylene carbonate) (PDTC), poly (5-methyl-5-allyloxycarbonyl-trimethylene carbonate and poly(5-methyl-5-allyloxycarbonyl)-trimethylene carbonate) [3, 54].

11.2.2.4 Polyphosphazenes

Polyphosphazenes (PPS) are macromolecules with phosphorus and nitrogen, replaced by two organic groups in phosphorus atoms to form poly(organophosphazenes) [55, 56]. Poly(dichlorothane) was the first successfully synthesized polyphosphazene that has been used as an intermediate for the synthesis of different polyphosphazene using replacement reaction [57]. PPSs with amino acid ester and peptide esters that are easily hydrolyzed are considered degradable [58].

Polyphosphazenes that suffer hydrolysis and degradation in an aqueous medium can provide a different range in properties and in your structure so that they can be adjusted in various ways to be applied for DDS [58, 59]. Hydrophilic-lipophilic adjustments within the substitute can control the degradation characteristics of PPS through their physical-chemical properties [1]. The degradation of these polymers produces a product composed of a lateral chain group along with phosphate and ammonia [57].

11.2.2.5 Polyurethanes

Polyurethanes (PU) are polymers that have carbamate and urethane bonds in their chain, which may vary from elastomers and thermoplastics produced by reactions of polyols with diisocyanates (aromatic, aliphatic, or cyclo-aliphatic) [3, 60]. Biodegradable PU (BPU) are synthesized by aliphatic diisocyanates [60]. PU in general is used in the manufacture of blood contact devices, tissue manufacturing, implant coating and DDS [48, 61]. In DDS, the degree of intumescence, percentage of drug loading, degradation rate, pH and enzymes are fundamental to designate the kinetic release profile in BPU [62].

11.3 Device Types

11.3.1 Three-Dimensional Printing Devices

In the last 30 years, the engineering of three-dimensional (3D) devices for biomedical applications has gained prominence, having as main applications the production of 3D prosthetic models [63], biocompatible skins [64], scaffolds for cell growth [65] and devices for controlled drug release. Initially, the most common methods for active principle incorporation in polymeric matrices are the techniques of compression, spraying, dip coating and encapsulation. These systems used water-soluble polymers, such as (polyvinyl pyrrolidone), cellulose derivatives, and PEG (polyethylene glycol) in the structure of hydrogels, capsules, tablets, topical ointments, and others. With the advent of 3D printing, new types of polymeric devices could be designed based on their delivery control mechanism (diffusion, solvent activation, biodegradation, or external triggers such as pH and temperature) [66] and the proper geometry for a given drug load and application [67], becoming important medicinal tools in the treatment of illnesses and diseases. Examples of commonly used 3D printed polymeric devices are implants, tablets, micro/nanoneedles and nanofibers.

11.3.1.1 Implants

Implants represent an important class of DDS. These devices are capable of maintaining a therapeutic dose for a long time without the need for constant ingestion of pills or injections. Most of these implants consist of BP rods whose Implants can be inserted subcutaneously or intramuscularly through a surgical procedure. As the stick degrades, the occluded drug inside the polymer matrix starts to be continuously released in the body, ensuring the therapeutic dose for a long period. In addition, morphological characteristics of implants can increase the drug release kinetics, requiring adjustment of geometry, porosity, and surface area to obtain more controlled zero-order kinetics [68]. This class of device can be produced using the fusion deposition method (FDM) using thermoplastic polymers: PLA, PGA, PCL and PLGA. However, less aggressive printing methods such as stereolithography (SLA) are more suitable for greater control of printing parameters and to avoid thermal degradation of drugs.

11.3.1.2 Tablets

Printed tablets are a more traditional conception of DDS simulating the action of conventional tablets. However, as it is possible to determine the porosity pattern and the solubility of the polymer, there is an solubilization slow, making them more suitable for a long and continuous release of the drug in the individual's body [69]. In addition, this device has advantages due to the possibility of being applied in the flexibility and customization of therapeutic doses for each individual and the development of "polypills," a concept of a single tablet containing a pre-set mixture of drugs according to the individual needs of the user [67].

11.3.1.3 Microneedles

Another class of devices used for this purpose is microneedles for transdermal delivery. The device consists of polymeric structures in sharp shapes with sizes varying from 50 to 150 μm, placed and organized surface of a small coupler device. These structures have dimensions such that they cross the epidermis without reaching the dermis's nerve receptors. With this, the problem of low absorption of the active principle in topical ointments is solved, increasing the bioavailability of the inoculated drug, without causing pain to the patient as in the cases of subcutaneous injections. MNs can be produced by FDM using thermoplastic polymers (PLA, PGA, PCL, and PLGA) [70].

11.3.1.4 Nanofibers

A tool for DDS is quite attractive for its characteristics of large surface area, superior mechanical properties, cell proliferation and mass transfer, and a variety in drug loading are nanofibers [71]. A enough used technique in the obtaining of nanofibers is electrospinning, a methodology that is versatile, easy to operate and with good yield [72]. The advantage of using electrospinning is in the possibility of insertion in the polymer matrix drugs with reduced solubility for bioavailability and controlled release improvement. Electrospinning of chitosan, alginate, silk, PCL gelatin, PGA, PEO and others, have already been authorized by the Food and Drug Administration (FDA) for medical use [73].

11.3.2 Nanocarriers

Nanostructured materials, for example, have a high potential in scientific and technological areas, especially in the controlled release of drugs [74, 75]. A BP at a nanometric size is considered a new class of technology for controlled drug release, especially polysaccharide nanoparticles and nanocapsules [76, 77]. Biopolymers are directly related to macromolecular science, pharmacy, cell biology, pharmacology and medicine for the formulation of ideal nanostructures (nanoparticles, microspheres, microparticles, etc.) for various medical applications [78].

11.3.2.1 Nanoparticles

One of the main characteristics of nanoparticles (NPs) is that they present considerable surface area, which makes them important materials concerning the bioavailability of substances [79]. In addition, they can release compounds, and what will determine this factor are their molecular and physicochemical characteristics [80]. Several researchers present polymeric micelles as another example of nanosphere materials, which are particles dispersed in water on a nanometric scale (10–100 nm) whose preparation is made with synthetic polymers. There is still the possibility of being designed as solid NPs, core-shell, and polyplexes [81]. Studies indicate that polymeric micelles are examples of efficient DDS for intravenous administration, mainly due to their distribution and application, being possible for them to couple with bioactive substances [82, 83]. The liposome is an example of a polymeric micelle, as it has a spherical shape and the presence of an amphiphilic lipid bilayer with dimensions up to 100 nm. Its preparation can be done in basically four ways. The first one is to remove the lipids contained in organic solvents; the second is to spread these lipids in an aqueous solution; the third is to use the resulting liposome to purify it; and finally, note the product formed [3, 4].

11.3.2.2 Dendrimers

Dendrimers are polymers with three-dimensional architecture that have a very versatile molecular structure, having several applications, such as in biomedicine, controlled DD, paint composition [84, 85], nanocomposites [3], among others. The use of dendrimers in DDS is possible due to abundant cavities on their inside, encapsulating a large number of drugs physically or chemically [86]. Examples of BPs used in the formation of dendrimers are polyester, peptides, polyacetal, and others.

11.3.2.3 Hydrogels

Hydrogels are solid forms of polymers with a stable and crosslinked structure formed by three-dimensional networks made of polymers with high water absorption capacity [87]. They have high water content in their structure, which can easily provide biocompatibility characteristics and the encapsulation of hydrophilic drugs. The formed cross-linked network can protect the drug from enzyme attacks and therefore maintain therapeutic bioactivity [88]. In the course of administering DDS hydrogels, application can be: oral, parental, nasal, ocular, topical, and others. Examples of BPs used in the formation of hydrogels are chitosan, hyaluronic acid, alginic acid, collagen, PLGA, PCL, PLA, and others [89].

11.4 Applications

BPs have desirable characteristics for biomedical application, because they can be degraded, excreted, or absorbed by the body in a non-invasive manner, provide sites for cellular growth, biocompatibility and additionally be used as a carrier of bioactive molecules [3]. In DDS, biodegradable polymers may have the characteristics of nano and microspheres, injectable gels or implants, and can be applied to the most varied branches of medicine [2]. Examples of DDS applications (Figure 11.1) are in the intravenous, transdermal, ocular and oral area.

11.4.1 Intravenous

Intravenous DDS (IDDS) is one of the fastest and most common routes and can offer the bioavailability of the drug at 100%. In this type of methodology, the administration of the drug is inserted directly into the blood network, which circulates throughout the body. Intravenous formulations should be produced in aqueous medium or extremely

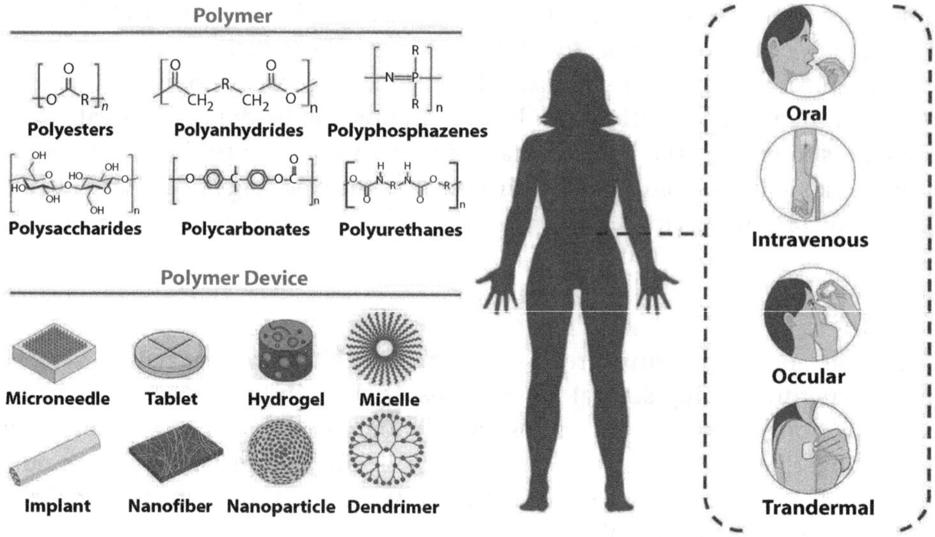

Figure 11.1 Examples of DDS applications.

thin suspensions so that the process of blockage of vessels and arteries does not occur [90]. Characteristics such as size, shape, surface load, and targeting type are key parameters for the overlap of biological barriers imposed by the body. Thus, biopolymeric NPs have great biomedical potential for intravenous applications [91]. Studies of IDDS of PEG-coated PLGA NPs loaded with docetaxel (anticancer agent) in an animal model showed an increase in the length of stay of the drug in the blood when compared to pure PLGA and a more pronounced concentration [92]. Intravenous administration of anti-inflammatory Ibuprofen sodium in PEG-containing gelatin NPs demonstrated tests *in vivo* at sustained release of up to 4 days, superior bioavailability and biocompatibility with the pure drug, what may affect the frequency of administration and side effects to the gastric system [93].

11.4.2 Transdermal

Transdermal DDS (TDDS) can be defined as vehicles for drug delivery using intact skin until it reaches the systemic circulation system. The use of this type of TDDS may decrease absorption effects by the gastric system, reduce dosage and hepatic elimination [94]. Biopolymers in TDDS can be topical, solid, or implantable systems, such as lipidic biopolymer-based nanosystems, patches, nanofibers, film, and microneedles [95]. Nanofibers of PLA and poly(butylene adipate-co-terephthalate) (PBAT) obtained by the electrospinning technique and with the addition of acetophenone indicate a rapid release of 80% of the drug, with subsequent slow and sustained release, relieving pain quickly and continuously without harm to the gastric system [96]. Microneedles composed of alginate and maltose in insulin were studied to reduce injections in patients with diabetes. The material studied *in vivo* provided a sustained hypoglycemic effect in diabetic rats, in addition to pharmacological availability and bioavailability improved when compared to subcutaneous injections [97].

11.4.3 Oral

Oral drug administration is the tool most used due to its ease, patient acceptance and flexibility in formulations, problems such as the degradation of the compound before the target site and bioavailability are common. BPs can be used to maintain chemical stability and targeted delivery of medicines in oral drug delivery systems (OrDDS) [98]. Alginate and chitosan microspheres were loaded with Icariin, an anti-inflammatory for the treatment of inflammatory bowel disease in rats, and demonstrated that the largest amount of the drug is released in colonic fluid, being retained for up to 12 hours in the colon. Furthermore, there is an inflammatory decrease and protection of the colon being indicated for ulcerative colitis. Chitosan and alginate NPS with measles antigen can provide a systemic immune response, low cytotoxicity, and still protect the antigen in gastric pH, this system can be used as an oral vaccine DDS [99].

11.4.4 Ocular

The bioavailability of drugs to the eyes and their innermost cavities have a great challenge penetrating the layers, complex structures, and natural defense mechanisms [100]. Biopolymers in ocular DDS (OcDDS) are used to optimize contact time with the cornea and conjunctival epithelium present in the eyes and for mucoadhesive activity, preventing the fast elimination of the formulation in the natural cleansing process of the eyes. Liposome and dendrimers NPS, hydrogels, and other BPs are being greatly investigated for drug delivery into the ocular cavity and also as an improvement of existing products in the market [101]. Gelatin NPS containing for corneal antiangiogenic effect and showed a 98% drug transport inhibition of vessel growth in the cornea with single and daily *in vivo* treatment tests in rats. Therefore, it is a simple and comfortable treatment method [102]. Microspheres of poly (ethylene glycol)-co-(L-lactic acid) diacrylate/N-isopropylacrylamide (PEG-PLLA-DA/NIPAAm) obtaining satisfactory results in controlled delivery of ranibizumab, with longer bioavailability of the drug, aiming at a better treatment for macular edema. In *in-vitro* testing, the hydrogel achieved easier degradation and a faster, more sustained release [103].

11.5 Existing Materials in the Market

There is an infinite number of materials that use BPs and are available on the market for DDS. The natural BPs in DDS with the most prominence in the market are starch, cellulose, hyaluronic acid, alginic acid, chitosan, collagen, gelatin, and albumin. The most used synthetics are polyesters, polyanhydrides, polycarbonates, polyphosphazene, and polyurethanes [57]. Table 11.1 shows the main BPs available on the market.

Table 11.1 Commercially available formulations of biodegradable polymers.

BP	Commercial name	Application in DDS	Company
Albumin	Abraxane	Metastatic pancreatic adenocarcinoma	Abraxis Bioscience
PLA	Lactel®	Excipient	Evonik Industries
	Cynviloq™	Ovarian cancer and bladder cancer	Sorrento Therapeutics, Inc.
	Resomer®	Pancreatic cancer	Evonik Industries
PLGA	Lactel®	Excipient	Evonik Industries
	Resomer®	Excipient	Evonik Industries
	Purasorb®	Resorbable excipients	Corbion
Poly (ethylene glycol–propylene glycol) copolymer	Pluronic®	Esophageal adenocarcinoma, vascular disease, hypertriglyceridemia	BASF Corporation
	Lutrol®	Gel-transported chemotherapy drugs, local anesthetics and antibiotics	BASF Corporation
PEO	Polyox® Polioxirano	Tablet bindings, tablet coatings and transdermal	Colorcon®
Poly (methacrylate)	Eudragit®	Excipient for oral, vaginal, rectal and topical administration	Evonik
	Acryl-EZE®		Colorcon®
Hyaluronic acid	Monovisc®	Osteoarthritis	Anika Therapeutics, Inc.
	Juvéderm®	Cosmetic	Allergan Laboratories
	MicroHyala®	Microneedles for Wrinkle treatment	CosMED Pharmaceutical Co.

11.6 Conclusions and Future Projections

Degradable materials play a great role in *in-vivo* applications by providing less invasive methods of body excretion for transient implants, tissue engineering, and DD. In this respect, biodegradable polymers suitable for biomedical applications, biocompatibility, mechanical properties, low cytotoxicity, and ease of manipulation. Given these qualities, current research seeks to improve DDS using biodegradable polymeric matrices for application in the delivery of genes, vaccines, anticancer drugs, enzymatic immobilization, and others, in various segments. However, there is still room for more advanced research in the formulation of biomaterials with high potential in drug delivery, addressing mechanical properties, degradation, loading capacity, and surface. The wide use of biodegradable polymers in the market and medical routine also needs more complete research on the safe and effective delivery of the drug. Understanding chemical composition and interactions is also an aspect that should be explored in the improvement of angiogenesis, reactions in biological system, bacterial growth, and others. The approach of responsive materials with multiple stimuli comes in a future projection with great utility for more selective DD in inaccessible environments and complex diseases. Finally, the joining of biodegradable polymers, DD, and applied nanotechnology to nanomedicine will bring benefits to the medical field for more advanced treatment.

References

1. Kwon, G.S. and Furgeson, D.Y., Biodegradable polymers for drug delivery systems, in: *Biomedical polymers*, 1st, pp. 83–110, Woodhead Publishing, New York, 2007.
2. Prakash Jain, J., Ayen, W.Y., Kumar, N., Biodegradable polymers in drug delivery, in: *Biodegradable polymers in clinical use and clinical development*, 1st, pp. 1–58, John Wiley & Sons, Inc, New Jersey, 2011.
3. Prajapati, S.K., Jain, A., Jain, A., Jain, S., Biodegradable polymers and constructs: A novel approach in drug delivery. *Eur. Polym. J.*, 120, 109191, 2019.
4. Pires, A.L.R., Bierhalz, A.C.K., Moraes, Â.M., Biomateriais: Tipos, aplicações e mercado. *Quim. Nova*, 38, 957, 2015.
5. Pappalardo, D., Mathisen, T., Finne-Wistrand, A., Biocompatibility of resorbable polymers: A historical perspective and framework for the future. *Biomacromolecules*, 20, 1465–1477, 2019.
6. Bartzoka, E.D., Crestini, C., Lange, H., Biomass derived and biomass inspired polymers in pharmaceutical applications, in: *Handbook of Polymers for Pharmaceutical Technologies*, vol. 3, 2015.
7. Mody, N., Agrawal, U., Sharma, R., Vyas, S.P., Structured biodegradable polymers for drug delivery, in: *Handbook of polymers for pharmaceutical technologie*, vol. 3, pp. 243–274, John Wiley & Sons, Ltd, Massachusetts, 2015.
8. Shoichet, M.S., Polymer scaffolds for biomaterials applications. *Macromolecules*, 43, 581, 2010.
9. Brito, G.F., Agrawal, P., Araújo, E.M., Mélo, T.J.A., Biopolímeros, Polímeros Biodegradáveis e Polímeros Verdes. *Rev. Eletrônica Mater. e Process*, 6, 127, 2011.
10. George, A., Shah, P.A., Shrivastav, P.S., Natural biodegradable polymers based nano-formulations for drug delivery: A review. *Int. J. Pharm.*, 561, 244, 2019.
11. Siracusa, V., Rocculi, P., Romani, S., Rosa, M.D., Biodegradable polymers for food packaging: A review. *Trends Food Sci. Technol.*, 19, 634, 2019.
12. Zhong, Y., Godwin, P., Jin, Y., Xiao, H., Biodegradable polymers and green-based antimicrobial packaging materials: A mini-review. *Adv. Ind. Eng. Polym. Res.*, 3, 27, 2020.
13. Din, M.I., Ghaffar, T., Najeeb, J., Hussain, Z., Khalid, R., Zahid, H., Potential perspectives of biodegradable plastics for food packaging application-review of properties and recent developments. *Food Addit. Contam. Part A*, 37, 665, 2020.

14. Asghari, F., Samiei, M., Adibkia, K., Akbarzadeh, A., Davaran, S., Biodegradable and biocompatible polymers for tissue engineering application: A review. *Artif. Cells Nanomed. Biotechnol.*, 45, 185, 2016.
15. Elvers, D., Song, C.H., Steinbüchel, A., Leker, J., Technology trends in biodegradable polymers: Evidence from patent analysis. *Polym. Rev.*, 56, 584, 2016.
16. Araújo, B.A., Freitas, L.S., Sarmento, K.K.F., Bezerra, V.R., Lima, C.A.P.de, Medeiros, K.M. de, A aplicação de polímeros biodegradáveis como uma alternativa sustentável. *Res. Soc Dev.*, 10, e49010918248, 2021.
17. Doppalapudi, S., Jain, A., Khan, W., Domb, A.J., Biodegradable polymers—An overview. *Polym. Adv. Technol.*, 25, 427, 2014.
18. Alavi, M. and Nokhodchi, A., An overview on antimicrobial and wound healing properties of ZnO nanobiofilms, hydrogels, and bionanocomposites based on cellulose, chitosan, and alginate polymers. *Carbohydr. Polym.*, 227, 115349, 2020.
19. Dutta, S.D., Patel, D.K., Lim, K.-T., Functional cellulose-based hydrogels as extracellular matrices for tissue engineering. *J. Biol. Eng.*, 13, 1, 2019.
20. Yang, J., Dahlström, C., Edlund, H., Lindman, B., Norgren, M., pH-responsive cellulose–chitosan nanocomposite films with slow release of chitosan. *Cellul.*, 26, 3763, 2019.
21. Hogan, K.J. and Mikos, A.G., Biodegradable thermoresponsive polymers: Applications in drug delivery and tissue engineering. *Polym. (Guildf).*, 211, 123063, 2020.
22. Zargar, V., Asghari, M., Dashti, A., A review on chitin and chitosan polymers: Structure, chemistry, solubility, derivatives, and applications. *Chem. Bio. Eng. Rev.*, 2, 204, 2015.
23. Venkatesan, J., Lowe, B., Pallela, R., Kim, S.-K., Chitosan-based polysaccharide biomaterials. *Polysaccharides Bioactivity Biotechnol.*, 1837, 1837–1850, 2015.
24. Guellis, C., Rossin, A.R.S., Bessegato, G.G., Dragunski, D.C., Bariccatti, R.A., Lindino, C.A., Incorporation of fluorescent dyes in electrospun chitosan/poly(ethylene oxide). *J. Braz. Chem. Soc*, 32, 1040–1050, 2021.
25. Li, L. and Hsieh, Y.L., Chitosan bicomponent nanofibers and nanoporous fibers. *Carbohydr. Res.*, 341, 374, 2006.
26. Filippi, M., Born, G., Chaaban, M., Scherberich, A., Natural polymeric scaffolds in bone regeneration. *Front. Bioeng. Biotechnol.*, 0, 474, 2020.
27. Vroman, I. and Tighzert, L., Biodegradable polymers. *Mater.*, 2, 307, 2020.
28. Guo, L., Liang, Z., Yang, L., Du, W., Yu, T., Tang, H., Li, C., Qiu, H., The role of natural polymers in bone tissue engineering. *J. Control. Release*, 338, 571, 2021.
29. Barbu, A., Neamtu, B., Zăhan, M., Iancu, G.M., Bacila, C., Mireşan, V., Current trends in advanced alginate-based wound dressings for chronic wounds. *J. Pers. Med.*, 11, 890, 2021.
30. Mathew-Steiner, S.S., Roy, S., Sen, C.K., Collagen in wound healing. *Bioeng.*, 8, 63, 2021.
31. Chambre, L., Martín-Moldes, Z., Parker, R.N., Kaplan, D.L., Bioengineered elastin- and silk-biomaterials for drug and gene delivery. *Adv. Drug Deliv. Rev.*, 160, 186, 2020.
32. Saric, M. and Scheibel, T., Engineering of silk proteins for materials applications. *Curr. Opin. Biotechnol.*, 60, 213, 2019.
33. BaoLin, G. and Peter, M.X., Synthetic biodegradable functional polymers for tissue engineering: A brief review. *Sci. China Chem.*, 57, 490, 2014.
34. Rydz, J., Sikorska, W., Kyulavska, M., Christova, D., Polyester-based (bio)degradable polymers as environmentally friendly materials for sustainable development. *Int. J. Mol. Sci.*, 16, 564, 2014.
35. Farah, S., Anderson, D.G., Langer, R., Physical and mechanical properties of PLA, and their functions in widespread applications — A comprehensive review. *Adv. Drug Deliv. Rev.*, 107, 367, 2016.

36. Xiang, S., Feng, L., Bian, X., Li, G., Chen, X., Evaluation of PLA content in PLA/PBAT blends using TGA. *Polym. Test.*, 81, 106211, 2020.
37. Samantaray, P.K., Little, A., Haddleton, D.M., McNally, T., Tan, B., Sun, Z., Huang, W., Ji, Y., Wan, C., Poly(glycolic acid) (PGA): A versatile building block expanding high performance and sustainable bioplastic applications. *Green Chem.*, 22, 4055, 2020.
38. Xu, Y., Kim, C.-S., Saylor, D.M., Koo, D., Polymer degradation and drug delivery in PLGA-based drug–polymer applications: A review of experiments and theories. *J. Biomed. Mater. Res. Part B Appl. Biomater.*, 105, 1692, 2017.
39. Asadian, M., Onyshchenko, I., Thiry, D., Cools, P., Declercq, H., Snyders, R., Morent, R., De Geyter, N., Thiolation of polycaprolactone (PCL) nanofibers by inductively coupled plasma (ICP) polymerization: Physical, chemical and biological properties. *Appl. Surf. Sci.*, 479, 942, 2019.
40. Dong, Q., Zhang, M., Zhou, X., Shao, Y., Li, J., Wang, L., Chu, C., Xue, F., Yao, Q., Bai, J., 3D-printed Mg-incorporated PCL-based scaffolds: A promising approach for bone healing. *Mater. Sci. Eng. C*, 129, 112372, 2021.
41. McNeil, E., S., R., Griffiths, H., Perrie, Y., Polycaprolactone Fibres as a Potential Delivery System for Collagen to Support Bone Regeneration. Current drug delivery, 8, 4, p. 448–455, 2011.
42. Bartnikowski, M., Dargaville, T.R., Ivanovski, S., Hutmacher, D.W., Degradation mechanisms of polycaprolactone in the context of chemistry, geometry and environment. *Prog. Polym. Sci.*, 96, 1, 2019.
43. Trindade, J. dos S., Silva, H. de J.B. da, F., D.R. da S. Sa, M.L. de, Silva, H.D. de A., Oliveira, R.S. de, Matos, J.M.E., de Estudo sobre a utilização de polianidrido como abordagem terapêutica para o tratamento de lesões no tecido ósseo. *Res. Soc Dev.*, 10, e42610414204, 2021.
44. Poetz, K.L., Shipp, D.A., Poetz, K.L., Shipp, D.A., Polyanhydrides: Synthesis, properties, and applications. *Aust. J. Chem.*, 69, 1223, 2016.
45. Domanskyi, S., Poetz, K.L., Shipp, D.A., Privman, V., Reaction-diffusion degradation model for delayed erosion of cross-linked polyanhydride biomaterials. *Phys. Chem. Chem. Phys.*, 17, 13215, 2015.
46. Huntimer, L., Ramer-Tait, A.E., Petersen, L.K., Ross, K.A., Walz, K.A., Wang, C., Hostetter, J., Narasimhan, B., Wannemuehler, M.J., Evaluation of biocompatibility and administration site reactogenicity of polyanhydride-particle-based platform for vaccine delivery. *Adv. Healthc. Mater.*, 2, 369, 2013.
47. Brunelle, D.J., Advances in polycarbonates: An overview. *ACS Symposium Series*, 1–5, 2005.
48. Metters, A.T. and Lin, C.-C., Biodegradable hydrogels: Tailoring properties and function through chemistry and structure, in: *Biomaterials*,1st, pp. 1–5, CRC Pres, New York , 2007.
49. W. Brenna and S.W. F., United States Patent [19]. 57, 3, 1978.
50. Kyriacos, D. *Polycarbonates*. In: *Brydson's Plastics Materials,* 8th edition, pp. 457-48, Elsevier Ltd, Amsterdam, 2017.
51. Artham, T. and Doble, M., Biodegradation of aliphatic and aromatic polycarbonates. *Macromol. Biosci.*, 8, 14, 2008.
52. Guillaume, S.M. and Mespouille, L., Polycarbonates and green chemistry. *J. Appl. Polym. Sci.*, 131, p. 1–1, 2014.
53. Xu, J., Feng, E., Song, J., Renaissance of aliphatic polycarbonates: New techniques and biomedical applications. *J. Appl. Polym. Sci.*, 131, 1–6, 2014.
54. Wang, C.F., Lin, Y.X., Jiang, T., He, F., Zhuo, R.X., Polyethylenimine-grafted polycarbonates as biodegradable polycations for gene delivery. *Biomaterials*, 30, 4824, 2009.
55. Rothemund, S. and Teasdale, I., Preparation of polyphosphazenes: A tutorial review. *Chem. Soc Rev.*, 45, 5200, 2016.
56. Zhou, X., Qiu, S., Mu, X., Zhou, M., Cai, W., Song, L., Xing, W., Hu, Y., Polyphosphazenes-based flame retardants: A review. *Compos. Part B Eng.*, 202, 108397, 2020.

57. Doppalapudi, S., Jain, A., Khan, W., Domb, A.J., Biodegradable polymers-An overview. *Polym. Adv. Technol.*, 25, 427, 2014.
58. Ogueri, K.S., Ogueri, K.S., Ude, C.C., Allcock, H.R., Laurencin, C.T., Biomedical applications of polyphosphazenes. *Med. Devices Sensors*, 3, e10113, 2020.
59. Ullah, R.S., Wang, L., Yu, H., Abbasi, N.M., Akram, M., Ul-Abdin, Z., Saleem, M., Haroon, M., Khan, R.U., Synthesis of polyphosphazenes with different side groups and various tactics for drug delivery. *RSC Adv.*, 7, 23363, 2017.
60. Guelcher, S.A., Biodegradable polyurethanes: Synthesis and applications in regenerative medicine. *Tissue Eng. Part B Rev.*, 14, 3, 2008.
61. Reddy, M.S.B., Ponnamma, D., Choudhary, R., Sadasivuni, K.K., A comparative review of natural and synthetic biopolymer composite scaffolds. *Polym. (Basel).*, 13, p. 1105, 2021.
62. Cherng, J.Y., Hou, T.Y., Shih, M.F., Talsma, H., Hennink, W.E., Polyurethane-based drug delivery systems. *Int. J. Pharm.*, 450, 145, 2013.
63. van der Stelt, M., Grobusch, M.P., Koroma, A.R., Papenburg, M., Kebbie, I., Slump, C.H., Maal, T.J.J., Brouwers, L., Pioneering low-cost 3D-printed transtibial prosthetics to serve a rural population in Sierra Leone – an observational cohort study. *EClinicalMedicine*, 35, 100874, 2021.
64. Augustine, R., Skin bioprinting: A novel approach for creating artificial skin from synthetic and natural building blocks. *Prog. Biomater.*, 7, 77, 2018.
65. Fayyazbakhsh, F. and Leu, M.C., A brief review on 3D bioprinted skin substitutes. *Proc. Manuf.*, 48, 790, 2020.
66. Abuchowski, A., Es, T., Palczuk, N.C., Davis, F.F., Alteration of immunological properties of bovine serum albumin by covalent attachment of polyethylene glycol. *J. Biol. Chem.*, 252, 3578, 1977.
67. Kotta, S., Nair, A., Alsabeelah, N., 3D printing technology in drug delivery: Recent progress and application. *Curr. Pharm. Des.*, 24, 5039, 2018.
68. Prasad, L.K. and Smyth, H., 3D Printing technologies for drug delivery: A review. *Drug Dev. Ind. Pharm.*, 42, 1019, 2015.
69. Kyobula, M., Adedeji, A., Alexander, M.R., Saleh, E., Wildman, R., Ashcroft, I., Gellert, P.R., Roberts, C.J., 3D inkjet printing of tablets exploiting bespoke complex geometries for controlled and tuneable drug release. *J. Control. Release*, 261, 207, 2017.
70. Waghule, T., Singhvi, G., Dubey, S.K., Pandey, M.M., Gupta, G., Singh, M., Dua, K., Microneedles: A smart approach and increasing potential for transdermal drug delivery system. *Biomed. Pharmacother.*, 109, 1249, 2019.
71. Torres-Martinez, E.J., Cornejo Bravo, J.M., Serrano Medina, A., Pérez González, G.L., Villarreal Gómez, L.J., A summary of electrospun nanofibers as drug delivery system: Drugs loaded and biopolymers used as matrices. *Curr. Drug Deliv.*, 15, 1360, 2018.
72. Rossin, A.R.S., de Oliveira, É., de Moraes, F., da S. Júnior, R., Scheidt, D.T., Caetano, W., Hioka, N., Dragunski, D.C., Oliveira, E.L., Moraes, F.A.P., Junior, R.C., Scheidt, D.T., Caetano, W., Hioka, N., Dragunski, D.C., Terapia Fotodinâmica em Eletrofiação: Revisão de técnicas e aplicações. *Quim. Nova*, 43, 613, 2020.
73. Kai, D., Liow, S.S., Loh, X.J., Biodegradable polymers for electrospinning: Towards biomedical applications. *Mater. Sci. Eng. C*, 45, 659, 2014.
74. Mosqueira, V.C.F., Legrand, P., Gulik, A., Bourdon, O., Gref, R., Labarre, D., Barratt, G., Relationship between complement activation, cellular uptake and surface physicochemical aspects of novel PEG-modified nanocapsules. *Biomaterials*, 22, 2967, 2001.
75. Do Rego Monteiro, V.A., De Souza, E.F., De Azevedo, M.M.M., Galembeck, F., Aluminum polyphosphate nanoparticles: Preparation, particle size determination, and microchemistry. *J. Colloid Interface Sci.*, 217, 237, 1999.
76. Young, R.J. and Lovell, P.A., *Introduction to polymers*, 3rd edition, CRC Press, 2011.

77. Gao, Y., Xie, J., Chen, H., Gu, S., Zhao, R., Shao, J., Jia, L., Nanotechnology-based intelligent drug design for cancer metastasis treatment. *Biotechnol. Adv.*, 32, 761, 2014.
78. Shukla, K., S., K., P. ovender, P., Giri, G., N., Biodegradable polymeric nanostructures in therapeutic applications: Opportunities and challenges. *RSC Adv.*, 6, 94325, 2016.
79. Cheong, J.N., Tan, C.P., Man, Y.B.C., Misran, M., α-Tocopherol nanodispersions: Preparation, characterization and stability evaluation. *J. Food Eng.*, 89, 204, 2008.
80. McClements, D.J., Decker, E.A., Park, Y., Weiss, J., Structural design principles for delivery of bioactive components in nutraceuticals and functional foods. *Crit. Rev. Food Sci. Nutr.*, 49, 577, 2009.
81. Su, S. and Kang, P.M., Systemic review of biodegradable nanomaterials in nanomedicine. *Nanomater.*, 10, 656, 2020.
82. Adams, M.L. and Kwon, G.S., Relative aggregation state and hemolytic activity of amphotericin B encapsulated by poly(ethylene oxide)-block–poly(N-hexyl-l-aspartamide)-acyl conjugate micelles: Effects of acyl chain length. *J. Control. Release*, 87, 23, 2003.
83. Scholz, C., Iijima, M., Nagasaki, Y., Kataoka, K., Polymeric micelles as drug delivery systems: A reactive polymeric micelle carrying aldehyde groups, *Polym Adv Technol.*, 9, 10-11, p. 768–776, 1998.
84. Barth, R.F., Adams, D.M., Soloway, A.H., Alam, F., Darbyl, M.V., Starburst dendrimer-monoclonal antibody immunoconjugates: Evaluation as a potential delivery system for neutron capture therapy 1,2. *Bioconjugate Chem.*, 5, 58, 1994.
85. Wiener, E., Brechbiel, M.W., Brothers, H., Magin, R.L., Gansow, O.A., Tomalia, D.A., Lauterbur, P.C., Dendrimer-based metal chelates: A new class of magnetic resonance imaging contrast agents. *Magn. Reson. Med.*, 31, p. 1–8, 1994.
86. Huang, D. and Wu, D., Biodegradable dendrimers for drug delivery. *Mater. Sci. Eng. C*, 90, 713, p. 1–17, 2018.
87. Chirani, N., Yahia, H., Gritsch, L., Motta, F.L., Chirani, S., Faré, S., History and applications of hydrogels. *J. Biomed. Sci.*, 4, p. 1–23, 2014.
88. Li, J. and Mooney, D.J., Designing hydrogels for controlled drug delivery. *Nat. Rev. Mater.*, 1, 1–17, 2016.
89. Ghasemiyeh, P. and Mohammadi-Samani, S., Hydrogels as drug delivery systems; pros and cons. *Trends Pharm. Sci.*, 5, 7, 2019.
90. Sharma, M., Transdermal and intravenous nano drug delivery systems: Present and future. *Appl. Targeting Nano Drugs Deliv. Syst.*, 1st, p. 499–550, Elsevier Inc. All, United Kingdom, 2019.
91. Chevalier, M.T., García, M.C., Alvarez, V.A., Biopolymeric nanoparticles as drug carriers for intravenous administration, in: *Biopolymers and nanocomposites for biomedical and pharmaceutical application*, 2017.
92. Rafiei, P., Haddadi, A., Docetaxel-loaded, P.L.G.A., and PLGA-PEG nanoparticles for intravenous application: Pharmacokinetics and biodistribution profile. *Int. J. Nanomed.*, 12, 935, 2017.
93. Narayanan, D., Geena, M.G., Lakshmi, H., Koyakutty, M., Nair, S., Menon, D., Poly-(ethylene glycol) modified gelatin nanoparticles for sustained delivery of the anti-inflammatory drug Ibuprofen-Sodium: An *in vitro* and *in vivo* analysis. *Nanomed. Nanotechnol. Biol. Med.*, 9, 818, 2013.
94. Basavaraj, K.H., Johnsy, G., Navya, M.A., Rashmi, R., Siddaramaiah, Biopolymers as transdermal drug delivery systems in dermatology therapy. *Crit. Rev. Ther. Drug Carr. Syst.*, 27, 155, 2010.
95. Tekade, R.K., Maheshwari, R., Tekade, M., Biopolymer-based nanocomposites for transdermal drug delivery, in: *Biopolymer-Based Composites: Drug Delivery and Biomedical Applications*, S. Jana, S. Maiti, S. Jana (Eds.), pp. 81–106, Elsevier, 2017.

96. Carvalho, B.M., Pellá, M.C.G., Hardt, J.C., de Souza Rossin, A.R., Tonet, A., Ilipronti, T., Caetano, J., Dragunski, D.C., Ecovio®-based nanofibers as a potential fast transdermal releaser of aceclofenac. *J. Mol. Liq.*, 325, 115206, 2021.
97. Zhang, Y., Jiang, G., Yu, W., Liu, D., Xu, B., Microneedles fabricated from alginate and maltose for transdermal delivery of insulin on diabetic rats. *Mater. Sci. Eng. C*, 85, 18, 2018.
98. Jana, P., Shyam, M., Singh, S., Jayaprakash, V., Dev, A., Biodegradable polymers in drug delivery and oral vaccination. *Eur. Polym. J.*, 142, 110155, 2021.
99. Biswas, S., Chattopadhyay, M., Sen, K.K., Saha, M.K., Development and characterization of alginate coated low molecular weight chitosan nanoparticles as new carriers for oral vaccine delivery in mice. *Carbohydr. Polym.*, 121, 403, 2015.
100. Lynch, C., Kondiah, P.P.D., Choonara, Y.E., Toit, L.C., Ally, N., Pillay, V., Advances in biodegradable nano-sized polymer-based ocular drug delivery. *Polym.* 11, 1371, 2019.
101. Tsai, C.-H., Wang, P.-Y., Lin, I.-C., Huang, H., Liu, G.-S., Tseng, C.-L., Ocular drug delivery: Role of degradable polymeric nanocarriers for ophthalmic application. *Int. J. Mol. Sci.*, 19, 2830, 1–20, 2018.
102. Chuang, Y.-L., Fang, H.-W., Ajitsaria, A., Chen, K.-H., Su, C.-Y., Liu, G.-S., Tseng, C.-L., development of Kaempferol-loaded gelatin nanoparticles for the treatment of corneal neovascularization in mice. *Pharm.*, 11, 635, 2019.
103. Liu, W., Borrell, M.A., Venerus, D.C., Mieler, W.F., Kang-Mieler, J.J., Characterization of biodegradable microsphere-hydrogel ocular drug delivery system for controlled and extended release of ranibizumab. *Transl. Vis. Sci. Technol.*, 8, 12, 2019.

12

Microorganism-Derived Bioplastics for Clinical Applications

Namrata Sangwan, Arushi Chauhan, Jitender Singh and Pramod K. Avti*

Department of Biophysics, Postgraduate Institute of Medical Education and Research (PGIMER), Chandigarh, India

Abstract
The word "bioplastic" refers to chemically unrelated items produced by microorganisms (or parts of them) under various environmental circumstances. They can be described as polyesters having physicochemical qualities similar to conventional plastics. But can be synthesized through microorganisms by accumulating within the cell microbes in the shape of granules for storage. They are generally manufactured from hydroxy-acyl–CoA and derivatives through a number of metabolic processes. Bioplastics vary in monomer concentration, macromolecular structure, and physical qualities based on the microbial species. Because many of these are biodegradable and biocompatible, they might be suitable for clinical applications. Due to their high biocompatibility, and naturally biodegradable nature, polyhydroxyalkanoate (PHA) monopolymers and Poly(3-hydroxybutyrate) (PHB) have captured the clinical applications interest in the recent years. Therefore, bioplastics in the recent times are extensively being research in the biomedical field. They might play an essential role in bone engineering, cardiovascular tissue engineering, drug carrier, drug and gene co-carrier, and antigen carrier. Due to its antibacterial activity, it can be employed in masks, sutures, and wound dressing as a packaging material. This chapter tries to provide an overview of microbially derived bioplastics and highlight their probable potential in the clinical applications.

Keywords: Bioplastics, biomedical applications, microbes, poly(3-hydroxybutyrate), poly (ε-caprolactone, poly(3-hydroxybutyrate), poly-lactic acid, poly lactic-co-glycolic acid

12.1 Introduction

Bioplastics are polymers made from renewable biological sources such as plants, microorganisms, and algae. The widespread use of plastics is causing severe environmental contamination, which is a significant concern. Plastics are seen as a hazard to the environment due to their inability to decompose and accumulate in the environment. The employment of microbes for plastic breakdown, transforming them into low molecular weight molecules that are not hazardous to the ecosystem, is the most cost-effective and environmentally

**Corresponding author*: pramod.avti@gmail.com
ORCID: 0000-0001-5603-4523

Inamuddin and Tariq Altalhi (eds.) Handbook of Bioplastics and Biocomposites Engineering Applications: 2nd Edition, (229–250) © 2023 Scrivener Publishing LLC

benign technique of reducing plastic waste. Bacteria are efficient cell producers, capable of effectively converting carbon and nitrogen supplies to polysaccharides, polyamides, polyesters, and other internal and extracellular bioplastics [1].

Gram stain–positive and Gram stain–negative bacteria can synthesize biodegradable polymers by over 300 different bacterium species. The commonly used microorganisms for the synthesis of bioplastics are strains that are both wild and recombinant, like *Ralstonia eutropha*, also known as *Wautersia eutropha* [2], *Azotobacter* [3], *Bacillus* [4], *Pseudomonas* [5], *Burkholderia* [6], *Halomonas, Rhodobacter sphaeroides* [7], *Zobellella denitrificans* [8], *Chromobacterium* [9], recombinant *E. coli.* [10]. Microbes can thrive in many habitats and infect many living creatures. Secreted bioplastics can operate as virulence factors when generated by bacterial pathogens, but intracellular polymers are mostly reserved resources that improve survival under the deprivation stage. The transition of bacterial pathogens from mobility to sessility is a strategic decision that is frequently associated with the development of polysaccharides. Glycogen is a polymer present inside the cell that helps microbe survivability during this stage of infection. Polyhydroxyalkanoates (PHAs) significantly decrease bioplastics which function as reserve molecules, enhancing microbial viability and perhaps acting as a charge carrier sink in anaerobic zones [11].

Most well-known bioplastics are comprised of monomers produced via beta-oxidation precursors [(R)-3-hydroxy acyl–CoAs], polymerized with enzymes via condensation of a carboxy group, expressed in a monomeric–CoA thioester with the 3-hydroxy moiety of the next unit. Other biopolymers with atypical monomers (e.g., 4-, 5-, or 6-hydroxy alkanoic acids or glutamic acid) are produced by various processes [18–23], implying the presence of biosynthetic enzymes widely. A broad range of physical and chemical characteristics may be accomplished through polymerization dependent on the monomer units and copolymers containing various components in adjustable proportions [14]. Biocompatibility is essential for biomaterials, which means that they must function well without eliciting an immune reaction [15]. A biodegradable biomaterial must meet the required parameters for a safe and convenient product to be used in a medical setting. Among these requirements are nonantigenicity, nonimmunogenicity, and anti-inflammatory nature to the host machinery; if antibacterial qualities exist, they should be convincingly proven by experimental research. Similarly, antibacterial activity should be maintained for a minimum of a year or two; must have the requisite mechanical qualities (suitable permeation, resilience to preoperative and postoperative strain); etc. [16]. The biodegradability time should be short, and degradation intermediates should be less toxic and readily absorbed by the biological system [15, 17, 18].

Most techniques for producing and fabricating intricately designed user-friendly gadgets are well known. These polymers are primarily utilized to meet physical and mechanical requirements. Incorporating the same material's oriented fiber, mechanical self-reinforcement comes from insertion into the matrix [19, 20]. There are also sophisticated automatic features; shape memory polymers may be flexibly deformed and achieve the same shape due to a unique impetus, such as pH, temperature, and magnetic field. In biomedical institutions, drug carrier tools, cardiovascular devices, surgical sutures, clot clean-up devices, aneurysm or ductus arteriosus obstruction, and orthodontic treatment are all examples of their use [21].

This chapter will highlight the use of microorganism-based bioplastics and their role in biomedical applications, focusing on some ambitious work that may be immensely beneficial soon. We shall also estimate future development patterns based on social expectations and previous progress. Moreover, typical themes, such as the production, modification, and applications of bioplastics in the biomedical sector and possible prospects for future development, are also discussed.

12.2 Types of Bioplastics

Bioplastics are natural products generated and catabolized by many organisms and have widespread medical applications. They are metabolized by various organisms and have no adverse effects on the host, which means these are biocompatible.

Bioplastics belong to a large family of different materials possessing various properties and functions. These bioplastics may be derived from renewable sources like poly-lactic acid (PLA), poly-ε-caprolactone (PCL), polylactic-co-glycolic acid (PLGA), and polyhydroxyalkanoate (PHA). Biodegradable plastics are called so because they can be degraded by enzymes derived from living organisms [22]. The properties of bioplastics are crystallinity, storage modulus, transition temperature, and bioresorbability. For a biomedical device, biocompatibility is essential and results from hydrolyzable links in anhydride, esters, orthoesters, amide, urethane, urea, and urea carbonate backbones.

Many recognized bioplastics involve monomers with various beta-oxidation intermediates. They remain biochemically connected via condensation of the carboxy moiety, which is evident in a monomeric-CoA thioester and a 3-hydroxy group (or thiol group) of the next. Lemoigne first described poly (3-hydroxybutyrate) (PHB) bioplastics in *B. megaterium* [23]. Microbes that can amass and catabolize these polyesters include aerobes, anaerobes, cyanobacteria, archaea, and lower eukaryotes. PHB, PHA, and derivatives are the most extensively manufactured microbial bioplastics. Microorganisms can, nevertheless, yield other polyesters. Most of these require analogous biosynthetic enzymes. Because of its biocompatibility and versatility, polyhydroxy butyrate (PHB) is a promising polyhydroxyalkanoates (PHAs). It is generated by bacteria, fungus, molds, and other microorganisms.

12.2.1 Poly(3-hydroxybutyrate) (PHB)

Poly(3-hydroxybutyrate) (PHB) is generated from inclusion bodies by numerous bacterial species to act as a carbon and energy source. Bacteria, fungi, molds, and other microorganisms produce polyhydroxybutyrate, it is an eco-friendly plastic with a broad array of applications. As the bacteria enter the stationary growth phase, these lipid inclusions accumulate. Under nitrogen-limited conditions in terms of a source of carbon, some bacteria generate more than 60% of their mass as PHB. The most researched PHA constituent is firstly identified polyhydroxy butyrate (PHB).

Polyhydroxyalkanoates (PHAs) are polyesters synthesized by numerous microbes, including *Pseudomonas putida, Aeromonas hydrophila, Alcaligenes latus, Ralstonia eutropha*, and *Bacillus subtilis* [3–5]. According to a recent investigation, PHB was synthesized via a marine bacterium *Pseudodonghicola xiamenensis* isolated from the Red Sea. The greatest concentration of this bioplastic was obtained on enhancing the conditions of fermentation time (96 h), NaCl (4% w/v), temperature (35°C), pH (7.5), and peptone as nitrogen (N) source [24].

In *Ralstonia eutropha,* the phaCBA cluster of PHB metabolic genes encodes three main proteins: the enzyme PhaA (β-ketothiolase), which speed up chemical reactions for the formation of acetoacetyl–CoA from acetyl–CoA; PhaB (NADPH-oxidoreductase), stereospecifically leads to the formation of (R)-3-hydroxybutyryl-CoA from acetoacetyl-CoA.

Since it can generate polymers with low amounts of extremely long monomers (C6–C8), PhaC will be most effective with monomers fewer than five carbon atoms. The PHB pathway appears to be intricate in its regulation. PHB synthesis is inhibited by an overabundance of acetyl–CoA, whereas all energy metabolism or surroundings that lead to a decline in the level of acetyl–CoA can result in the revival of PHB synthesis. PHB synthesis is dependent on different signals like metabolic, genetic, and environmental. Besides PhaA, PhaB, and PhaC, other proteins, PhaZ and PhaP, have been produced that help in PHB synthesis and stabilization. PhaZ facilitates the release of Poly(3-hydroxybutyrate) from polymers. It is generally present in an inactive form and requires an activator + PHB to get activated. Therefore, for the degradation of PHBs depolymerases are necessary, i.e., PhaZ1, PhaZ2, and PhaZ3. PhaP, also called phasin, with low molecular weight, plays an essential role in synthesizing Poly(3-hydroxybutyrate) by managing the number, size, and surface: volume PHB inclusions. For the functioning of PhaP presence of PhaB and a repressor is necessary.

12.2.2 Polyhydroxyalkanoate

Polyhydroxyalkanoates (PHAs) are polyesters made up of monomers with mcl-PHAs (medium-chain lengths, C5-C14) or lclPHAs (long-chain lengths >C14). Even though PHAs are highly identical to PHBs (short-chain length), microbes that produce PHBs rarely produce PHAs. Recombinant organisms with mixed biochemical processes, on the other hand, can make polymers or copolymers with short-chain, medium-chain, or both length monomers. Short-chain-length PHAs (scl-PHAs), like polyhydroxybutyrate P(3HB), poly(4-hydroxybutyrate) P(4HB), and poly(3-hydroxyvalerate) P(3HB), are produced by *Cupriavidus necator* and *Alcaligenes latus* (3HV) [1, 25]. *P. putida* and *P. mendocina* produce mcl-PHAs [26, 27]. The lcl-PHAs are derived from *Aureispira marina* and *Shewanella oneidensis* [28].

PHA was identified and reported by Maurice Lemoigne around a generation ago as luminous inclusion body in the Gram-positive bacteria *B. megaterium* [23]. PHAs can be generated via mutant bacteria, mixed bacteria, or engineered organisms such as *Corynebacterium glutamicu, R. eutropha, W. eutropha, Bacillus, Azotobacter, Burkholderia, Pseudomonas, Halomonas, Aeromonas* [27]. On a low nutritional medium, *Ralstonia eutropha* manufactures single-chain length-PHA, poly(3-hydroxybutyrate-co-3-hydroxy valerate) (P3HB3HV), and primarily P(3HB) [28]. Microorganisms typically biosynthesize PHA in response to nutritional (noncarbon) constraints. The polymer exhibits intracellular carbon storage in many species, thus lowering potential in many organisms. PHAs build up inside cells and are extracted using organic solvents like chloroform or alkaline solutions [29]. The biosynthesis of PHA-monomers is typically derived from one of three ways: 1) by *de novo* technique, they are produced from acetyl-CoA and propionyl-CoA; 2) the intracellular intermediates of fatty acid degradation (β-oxidation) provide substrates for PHA, like CoA thioesters, are converted to (R)-3-hydroxy acyl-CoA [27]. For the polymerization of PHA, hydroxy acyl-ACP (acyl carrier protein) conjugates are transformed to (R)-3-hydroxy acyl-CoA molecules [30]. One molecule of acetyl-CoA, two acetyl-CoA, and/or one molecule of propionyl-CoA is used to make short-chain PHA precursors. The enzyme binds them together -ketothiolase, which produces 3-hydroxybutyryl-CoA (3HB-CoA) or 3-hydroxyvaleryl-CoA (3HV-CoA). Intermediates of fatty acid degradation or biosynthesis can be used to synthesize long-chain monomers. Trans2-enoyl-CoA molecules, which are fatty acid beta-oxidation intermediates, are substrates of an (R)-specific enoyl-CoA hydratase enzyme (PhaJ), which transforms substrate into a PHA precursor [27].

The PHA polymerase, also known as PhaC, PhaEC, or sometimes PhaRC, catalyzes the production of hydroxyalkanoate monomers. The above polymerases are grouped as four sets depending on their unit structure and polymer type produced. Class I synthases, including enzymes like *R. eutropha* PhaC, are well investigated. Class II synthases commonly expressed in *Pseudomonas sp* polymerize medium-sized chain length PHAs. The classes I and II PHA synthases constitute the same protein domains, known as PhaC. Class III synthases comprise a heterodimer comprising PhaC and PhaE proteins and enzymes from *Allochromatium vinosum* and *Thiocapsa pfennigii.*

Pseudomonas, particularly those in the rRNA homology group, generate mcl-PHA, *I. P. oleovorans* produces mcl-PHA copolymers from n-alkane, n-alkene, and long-chain fatty acids. *Ralstonia eutropha,* a Gram-negative nonsporulating aerobe, requires mineral salts and simple carbon sources, as well as organic molecules, to develop [31].

12.2.3 Poly-Lactic Acid

Poly-lactic acid (PLA) is hydrophobic, member of the aliphatic polyester family, produced naturally by bacteria, plants, and mammals. It possesses stereoisomerism, i.e., L (+) or D (−) optical activity. The homopolymers are synthesized from lactic acid (L-lactic acid or D-lactic acid) at a melting temperature of 175°C, with the transition temperature of 55°C [32]. And the heteropolymers from DL-lactic acid are unstructured due to disordered polymer chains [33]. The lactic acid dimerization into lactide via polymerization due to ring-opening is a way of polymerization and depolymerization by interchanging esters [34]. The polymer comprises a linear polyester, i.e., a six-membered cyclic ester [35]. Lately, PLA has grown in popularity for medical applications because of its nontoxicity and biodegradability. Polymerization of lactic acid is an essential step in the synthesis of PLA. It is a thermoplastic aliphatic polyester that involves the condensation of 2-hydroxy-propionic acid. LA has both the functional groups, the hydroxyl and carboxyl group, and undergoes self-esterification to form linear or cyclic polymers. The process of condensation involving catalytic hydrolysis leads to the formation of linear polyesters, like lactoyllactic acid and poly (lactic acid). PLA is created by the condensation of two monomeric domains, which is followed by the release of one water molecule. PLA is typically made in three steps: lactic acid fermentation via bacteria, (ii) lactic acid refinement and cyclic dimer (lactide) synthesis, and (iii) cyclic lactide ring-opening polymerization (ROP) [29, 36].

Lactic acid produced by the monosaccharide fermentation has a very high molecular weight (> 90%) [37]. Lactic acid (CH_3CHOH COOH) is obtained from the fermentation of plant waste, roots, sugarcanes (starch and sugar), and other natural sources by the action of anaerobic bacteria. Lactic acid manufacturing microorganisms serve as the generators of lactic acid [38]. Homofermentative and heterofermentative are the two categories of bacteria that lead to fermentation. *Lactococcus, Enterococcus, Streptococcus,* and some *Lactobacilli* are the homofermentative that produce just one product at a time to reduce the number of side products. These usually employ the Embden Meyerhof pathway (utilization of glucose metabolically to generate ATP, NADH, and several biosynthetic precursors). Other bacteria aiding in L-lactic acid production are *Lactobacillus amylophilus,*

L. delbrueckii, *L. leichmannii*, and *L. bulgaricus* [38, 39]. Heterofermentative bacterium, such as *Leuconostoc*, *Weissella*, and *Lactobacillus brevis* generate by-products, such as carbon dioxide and ethanol along with LA [37, 38]. Further, LA polymerization may depend on the 3HB monomer to inspect this concept and the study was conducted by mutating or knocking out two enzymes to exclude the supply of 3HB [40]. PhaA and PhaB were knocked out, and a PhaJ enzyme supplying 3-HV-CoA from valerate was added. When expressed in *E. coli*, this setup produced 90 to 96 mol% of lactic acid units supplemented with glucose. *E. coli* being gram-negative results in the generation of harmful endotoxins [12, 25]. The host, *C. glutamicum* created polymers based on lactic acid because it is an endotoxin-free Gram stain–positive bacterial species used to produce several industrial compounds and PHAs [27]. As a result, different host strains may be employed to produce bioplastics.

12.2.4 Poly Lactic-co-Glycolic Acid (PLGA)

Poly lactic-co-glycolic acid (PLGA) is a non-natural, biomedically important polymer. It is hydrophilic and crystalline and creates composites integrating extracellular matrix elements: polyglycolic acid (PGA) and polylactic acid copolymer (PLA). Polylactic acid comprises an uneven α-carbon which is characterized as the D or L stereochemical form and sometimes as R and S form [35]. The PLGA copolymers have a high degradation rate, biocompatibility, and biodegradation due to such properties and molecular weight, and these are considered safe and compatible for medical use. As a result of their reduced hydrophobicity compared to PLAs, these materials can maintain a high rate of biodegradation and, in the same way, maintain similar mechanical properties and accepted via Food and Drug Administration (FDA) [41]. The mixture of PLGA copolymers may be used for bone replacements and scaffolds; further, by altering the ratio, more excellent safety or compatibility may be attained of its copolymers. PLGA particles play an important role in drug delivery to achieve an intended therapeutic effect.

Different amounts of glycolide and lactide monomers can be used to make various forms of PLGA. On average, 75% lactic acid with 25% glycolic acid makes PLGA. Because the monomer ratios used significantly impact the final physicochemical properties of the material, PLGA is synthesized utilizing several techniques. The low molecular weight of PLGA can be attained by a process, solution poly-condensation of Glycolic acid and Lactic acid at a temperature of 120 °C. This requires continuous removal of water [42]. Enzyme polymerization occurs under favorable reaction conditions, including pressure, pH, temperature, and in the presence of enzyme lipases.

They were inspired by fermentation by the synthesis of natural bioplastics from microorganisms like PHAs, PHBs, and PLAs. The researchers devised a metabolic route to manufacture PLGA in *Escherichia coli* (*E. coli*) strains by microbial fermentation straight from carbohydrates [43]. The team has been working since 2011 and was successful in PLGA production using the glyoxylate shunt route and producing glycolate from glucose with a reduced PLGA polymer composition. The synthesis was modified by adding enzymes for the co-polymerization of PLGA and incorporating different metabolic pathways to synthesize nonnatural polymers like PLGA. *In silico* studies with an insight into the metabolic fluxes and the genomic simulations of *E. coli* were conducted to achieve the maximum yield and efficient polymer and least by-products.

12.2.5 Poly (ε- caprolactone) (PCL)

Poly (ε-caprolactone) (PCL) is a partially crystalline chemically synthesized material with a low melting point of 60°C, whereas transient −60°C glass temperature [44]. Their physical and biochemical characteristics, physio-chemical state, degradation rate, and tensile stability could be modified, and thus they can be used without significantly losing their properties under different circumstances [45]. The degradation rate takes a long time, and readily enzymes, such as lipases and esterases, can degrade it [45]. And can therefore be used to provide stiffness to the body tissues, replace hard tissues with healing the injury, and in tissue engineering. PCL has fewer ester bonds for each monomer, so the degradation takes more time for PCL structural segments to be enzymatically dissolved in the body than polylactides [46]. Therefore, it is more stable than polylactides. Though its low biocompatibility, its rubber like characteristics, extendable biodegradability, and convenience to form, blend, and copolymers render them a choice of material for bio medicinal use. Especially for bone graft, a good cladding material, surgical sutures, microdrug vesicles, and nano-medicinal use [47].

The ring-opening polymerization of circular monomers (-Caprolactone) produces PCL [48]. It may also be made using the free-radical ring-opening polymerization of 2-methylene-1-3-dioxepane with different catalysts (anionic, cationic, and co-ordination) [49]. The optimal weight of PCL varies with its molecular weight and spans from 3000 to 80,000 g/mol. The drug's release rate of microcapsules has been influenced by the combination of other polymers such as cellulose acetate butyrates (CAB), cellulose propionate (CP), polylactic glycolic acid (PLGA) and polylactic acid (PLA) [50]. PCL can dissolve fully in a range of solvents at ambient temperatures, like methylene chloride, carbon tetrachloride, and chloroform. PCL has a reduced solubility in alcohol, diethyl ether, etc. [51]. The hydrolytic breakdown of Poly (-hydroxy) esters might take either on the surface or in bulk [41]. Hydrolytic splitting on the polymer backbone causes surface degradation or erosion. It occurs when the amount of hydrolysing bond breakage exceeds the water infiltration rate into the bulk of the polymer. This permits the polymers to thin over time without changing the molecular weight of the inner dimension [52]. Water entry leads to bulk degradation by hydrolysing the chains, thus leading to the diffusion of monomers and oligomers out of the polymer. Slow erosion and diffusion balance will be attained [41].

12.3 Properties of Bioplastics

Bioplastics are materials that can differ significantly from one another rather than a single polymeric form [53]. The bioplastics family were subcategorised into 4 groups, each with its own individual characteristics: (a) Partially bio- based non-biodegradable plastics (PE, PP or PET), (b) bio-based technical performance polymers (PTT), (c) plastics that are both bio-based and biodegradable (PLA and PHA/PBS) and (d) plastics that are based on fossil resources and are biodegradable (PBAT) [54]. Bio-renewable polymers are made from renewable biomass sources like oils, vegetables lipids, maize and microbes. Plastics made from petroleum, like fossil-fuel plastics, rely on fossil fuels more heavily and emit more greenhouse gases [55]. Bioplastics are a diverse group of materials. They are made up of various materials with different qualities and uses (Figure 12.1). Bioplastics are described as plastics that are either biobased, biodegradable or have both qualities [55].

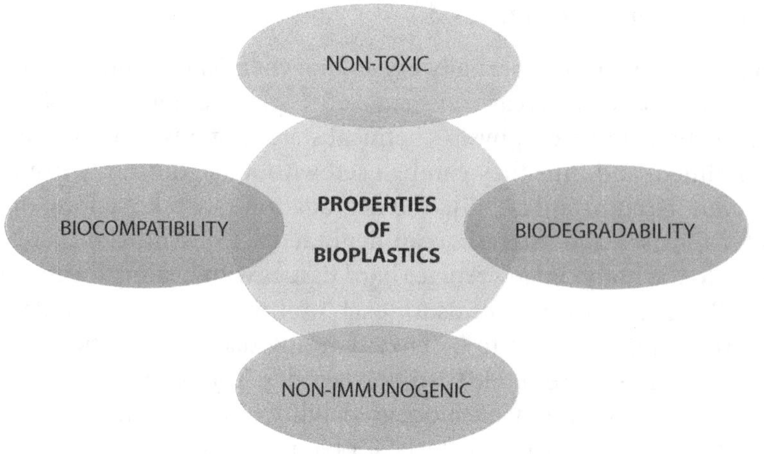

Figure 12.1 Properties of bioplastics.

12.3.1 Physiochemical, Mechanical, and Biological Properties of Bioplastics

12.3.1.1 Polylactic Acid

Young's modulus, tensile strength, and flexural properties are higher in polylactic acid (PLA) [56]. PLA's uses have been widened by a novel technique of synthesis (ring-opening polymerization), which allows for the cost-effective manufacturing of a high molecular weight PLA polymer. PLA seems to be a versatile and biodegradable polysaccharide that could be generated using abundant recycled materials [57].

The well-known biodegradable polyester polylactic acid (PLA) is studied broadly for tissue engineering and drug delivery systems, and it is also widely employed in human medicine [34]. PLA has demonstrated biomaterial promise in various healthcare applications, including tissue regeneration, skin and tendon mending, dental niches, orthopedic therapies, drug carriers, cancer therapy, cardiovascular implants, and medical devices and equipment. PLA has shown to be an excellent three-dimensional (3D) material printed biopolymer, as evidenced by its involvement in the global pandemic of coronavirus disease of 2019 (COVID-19) [58].

12.3.1.2 Poly Lactic-co-Glycolic Acid

These qualities are utilized to modify the rigid, hydrophobic, and slow decomposing lactic acid properties of poly lactic-co-glycolic acid (PLGA) versus the flexible, less hydrophobic, and quicker decomposing glycolic acid properties of PLGA [59]. According to research, the frequency of PLGA breakdown, enhancing the glycolic acid to lactic acid ratio, upsurges the PLGA copolymer's hydrophilicity, resulting in the quicker breakdown [60]. Polyethylene glycol (PEG) is the most often utilized copolymer because of its biocompatible nature and readily adsorbed on the exterior of PLGA. The PLGA carrier is protected from opsonin capture by the hydrophilic PEG [61]. It is employed in delivering drugs as it is biocompatible. PLGA is a commonly utilized synthetic polymer in bone regeneration as it is biocompatible and biodegradable. As a result, numerous PLGA-containing medical devices are authorized for clinical trials in humans via Food and Drug Administration. The copolymer PLGA is significantly more versatile than the homopolymers of lactic acid poly (lactic acid) and Poly (lactic acid) in terms of controlling breakdown rate (glycolic acid). This polymer has been reported widely as a material for bone restoration, and it is employed as frameworks, coatings, fibers, or microspheres and nanospheres to fulfil a range of therapeutic criteria [62].

12.3.1.3 Polycaprolactone

The polycaprolactone (PCL) polyester is biodegradable with a glass transition temperature of about 60°C and a low melting point. Polycaprolactone were most commonly used to make specialized polyurethanes. Water, oil, solvents, and chlorine are all essential components of polyurethane manufactured from polycaprolactones [63]. In physiological circumstances (such as the human body), PCL is damaged by the hydrolysis of its ester bonds. As a result, it has attracted much interest for usage as an implantable biomaterial. It is particularly intriguing to develop long-term implanted devices, as it degrades even more slowly than polylactide [64]. PCL is an artificial polyester polymer that has been extensively studied in the field of bone regeneration. PCL has little bioactivity and degrades rapidly, which can be regulated by altering its molecular weight, crystallinity, or hydrophilic polyethylene structure of glycol, copolymers, or ceramics, with PGA or PLA. The adhesion, migration, and proliferation of osteoblast and endothelial cells are improved when PCL is coated with TCP, HAp, calcium phosphate, gelatin, collagen, or bone sialoprotein. PGA and PCL are considered standard sustained genetically engineered human cells and demonstrated regeneration of cementum-like tissue, ligaments, and bone structures, indicating that they may be helpful in periodontal regeneration [63].

12.3.1.4 Polyhydroxyalkanoates

Polyhydroxyalkanoates (PHAs), or polyhydroxyalkanoates, are created naturally by various microbes and bacteria that digest carbohydrates or lipids [65]. More than 150 distinct monomers may be mixed to form materials with a wide range of characteristics within this family, which serves as a source of energy and carbon storage [66]. In the manufacturing of bioplastics, biodegradable polymers are employed. Altering the surface mingling PHA along with other polymers, enzymes, and inorganic materials can also change its mechanical characteristics and biocompatibility, allowing for a wider range of uses. These are hydroxy-derived alkanoic acid polymers with various chemical compositions that offer useful features, such as biocompatibility and biodegradability. As a result, such bioplastics are potential materials with a multi-source purpose, such as municipal engineering and agriculture, pharmacology, and biomedicine [67]. Polyhydroxyalkanoates (PHAs), biodegradable and biocompatible polymers, are employed in a range of medical devices and dosage formulations [68].

12.3.1.5 Polyethylene Glycol (PEG)

Water, methanol, ethanol, acetonitrile, benzene, and dichloromethane are all soluble in polyethylene glycol (PEG), while diethyl ether and hexane are not. Nonionic surfactants are created by combining it with hydrophobic compounds. It is nontoxic, nonvolatile, colourless, inert, and odourless. It is also very soluble in water and organic solvents, including carbon tetrachloride, benzene, and chloroform. The ability to create PEG compounds with variable molecular weights is contingent on producing PEG with varying chain lengths [69]. Poly (ethylene glycol) has several chemical and biological characteristics that are especially useful in various physical, chemical, and medical applications. Nonimmunogenic and nontoxic can be added to media, bonded to surfaces, and conjugated to molecules without interfering with cellular activities or target immunogenicity. Attachment to proteins and other biomolecules is hydrophilic (aqueous-soluble), reducing aggregation and enhancing solubility. Highly adaptable and allows for surface modification or bioconjugation without steric interference [70]. Polyethylene Glycols (PEGs) are nonimmunogenic, physiologically inert, and hydrophilic molecules that are very intriguing for diagnostics and therapeutics development applications (ex. Antibody-Drug Conjugation, imaging, enzymatic detection).

12.4 Applications

12.4.1 Tissue Engineering

This subfield in biomedical engineering utilizes a mix of cells, engineering, biomaterials technologies, and appropriate biochemical and biophysical parameters to repair, augment, or replace biological tissues that enables them to substitute, preserve, or promote the growth of bone, cartilage, and blood vessels [71]. The ability to regenerate living tissues under physiological conditions by attaching cells to biomaterials that provide a structural framework for three-dimensional multiplication has piqued interest in research, engineering, and medicine as a possible substitute to tissue or organ transplantation. Scaffolds constructed of different materials have already been tested in this discipline to satisfy the varying demands of tissue engineering. On the other hand, plastics provide a lot of design freedom since their structure and composition may be changed to match specific needs. Scaffolding for tissue regeneration is typically composed of polyesters such as PLA and PGA, their copolymers (PLGA). Such polymers are some of the few polymeric materials that the FDA has authorized for use in human clinical trials [71].

Scaffolds manufactured from different biomaterials have been explored in this discipline to satisfy the varying demands of tissue engineering. PLA was used to make tissue engineering scaffold, means of delivery materials, or encapsulating membranes, and other resorbable biomedical implants, including in dermatological, attributable to its bioresorbable and biocompatibility inside the body. For instance, in an *in vivo* rabbit, calvarial defect model 3D electrospun fibrous PLA scaffold has been proposed as a viable means of bone tissue regeneration. The applicability of these 3D scaffolds for osteoblast growth was investigated, resulting in 1.8-fold more osteoblast development than 2D electrospun. Thus, it is indicated that the 3D electrospun microfibrous matrix offered a suitable surface after two and four weeks for cell infiltration and bone growth [71].

PCL has attracted interest as a biomaterial for cardiovascular and bone tissue regeneration because of its low melting temperature and excellent tensile qualities. Likewise, PLA 3D scaffold synthesized 3D nanofiber scaffold utilizing poly-(-caprolactone) (PCL) by employing electrospinning method. The PCL-mediated scaffolds enable cell attachment, growth, and differentiation, enhancing bone regeneration *in vitro* and *in vivo* transplant studies [41].

Furthermore, 40 Sprague Dawley rats were employed in an *in vivo* investigation of rat calvarial lesions to assess bone tissue repair utilizing AL/HA/SF blends as a scaffold. The AL/HA/SF group had considerably more new bone development four weeks after surgery. They led to a significant in novo bone development and were degradable in a stable manner devoid of foreign body inflammatory responses. As a result, it is claimed that AL/HA/SF can be helpful to scaffolding to engineer bone tissue [72].

Similarly, *in vitro* and *in vivo* studies of calvarial deficits revealed that bioplastic poly3-hydroxybutyrate4-hydroxybutyrate (P34HB) electrospun fiber scaffolds exhibited exceptional porosity physicochemical parameters, allowing cell attachment and proliferation without cytotoxicity. Consequently, the P34HB electrospun fiber scaffold features a 3D porous network with high biocompatibility and biosafety and the capacity to grow and mend a bone, making it a viable bone regeneration scaffold [73].

Because of its required properties, P(3HB-co-4HB) is a viable biomaterial choice. Engineered cardiac patches, vascular grafts, and blood vessel replacements were made using P(3HB-co-4HB) [74]. Niu *et al.* discovered that mouse BMSCs were induced to develop into cardiomyocytes when put on the P(3HB-co-4HB) patch, suggesting that this scaffold facilitated the proliferation and adherence of differentiated cardiac cells acted as a cardiac patch in mice [75].

Considering the efficacy of these biomaterials at pre-clinical levels, PLA-based materials are successfully deployed. So far, clinical studies for such benefit have been conducted on some parts like bladder [76], cartilage [77], liver [78], adipose [79], and bone tissues [80]. Only a few scaffolds are clinically used as a copolymer of PGA/PLA and polydioxanone for cartilage regeneration; BioSeed®-B and BioSeed®-C are the brand names for these products (Biotissue Technologies AG, Freiburg, Germany) [81].

Synthetic cardiac replacements have been developed to replace injured tissue and increase circulatory function recovery by giving a constructed version of endogenous heart healing. Creating a patch for the heart is a viable replacement cardiac regeneration technique. GORE-TEX® cardiovascular patch [82] and IMPRA® ePTFE cardiovascular patch [83], both developed as tissue regeneration patches for implant inside the human heart, are available on the present cardiovascular patch market [81].

Furthermore, in clinical research, incorporating PHB synthesizing genes from *Ralstonia eutropha* into the flax genome (*Linum usitatissimum*) improved fibers' biochemical properties and is considered a new dressing way for chronic wounds. Thus, the introduction of novel flax bandages for manufacturing "bioplastic" fibers from *Linum usitatissimum*, a flax-growing plant, might cause stimulation of cell division is required for the recovery process. As a result, the fabric manufactured from PHB-producing plants was tested for wound healing capabilities, it demonstrated a significant reduction in the size of the wound. To summarise, the unprocessed fabric made from flax plants that produce PHB is a highly efficient wound dressing agent and may be used to treat chronic skin ulcers. As a result, the moderate antibacterial and antifungal activity of plant fiber, as well as its high production, make it a viable therapeutic method for clinical therapies, particularly in the long-term treatment of chronic wounds [84].

Bioplastic copolymer-based coatings have been demonstrated in studies to minimize wound discomfort, speed re-epithelization, reduce wound infection and scars. Some commercial products based on bioplastic copolymer for wound dressings include Xcell®, Bioprocess®, and Biofill®. For treating third-degree burns, Legeza *et al.* developed a BC-based wound dressing incorporating superoxide dismutase or poviargol [85]. Few more examples of commercially available different type biomaterial-based product of tissue regeneration are TachoSil®, a cardiac wound sealant; NeuroFlex®, NeuroMatrix®, and NeuroMend® for nerve repair and regrowth; Dynamatrix®, Xelma, Apligraft®, and Dermagraft® by Advanced Tissue Sciences Inc. for skin, ulcer and wound healing; and INFUSE® for bone repairment [81]. In conclusion, tissue engineering is among today's most intriguing multidisciplinary topics, and it is quickly expanding. *In vivo* or clinical investigations have been the only criterion for research on scaffolds suitable for clinical use. Therefore, specific artificially constructed scaffold properties are required for optimum tissue regeneration applications (Figure 12.2).

12.4.2 Drug Delivery System

There has long been a need to distribute bioactive substances to specific body parts to optimize clinical efficacy while minimizing adverse effects. The drug-delivery method is tailored to the bioactive ingredient for maximum bioavailability, resulting in a longer duration of treatment and optimum doses. The establishment of this kind of drug-delivery system is dependent on the polymer's physicochemical and mechanical characteristics, as well as a polymer-based modality procedure that encompasses drug solubilization and preservation from deactivation and deterioration caused by drug interactions with

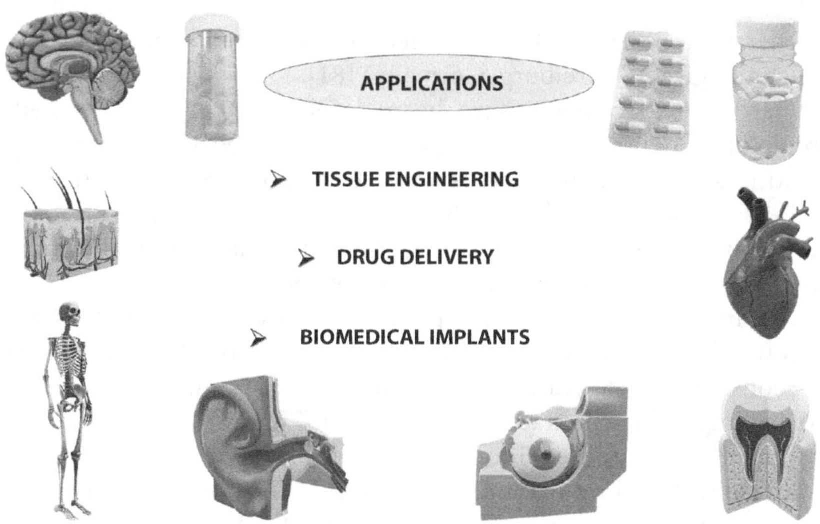

Figure 12.2 Applications of bioplastics.

biological molecules. These drug-delivery techniques are employed to transport drugs towards intended areas while allowing specific drug release [74]. They provide a stable therapeutic concentration of drugs, eliminating the necessity of repeated dosage and the adverse concerns that come with it. Following the emergence of polymers with intrinsic features of release retardation, biodegradability, and biocompatibility, the notion found its way from bench to bedside. Several clinical trials using these bioplastics-based (PLA, PLGA, PHA, or PHBS) commercial goods have been tested with various agents/therapies [86].

The debiopharm company's release of decapeptyl for prostate cancer therapy to the European market in 1986 was the first impactful move. The FDA authorized this technology as the first injectable, biodegradable microparticle deposit method, and it is still on the market today. After that, Abbott Laboratories and Takeda Pharmaceuticals merged to become Takeda-Abbott Products, which advanced a PLGA microsphere formulation to inject an LHRH analogue (leuprolide, Lupron Depot®) authorized via FDA in 1989 [92–94]. In 1996, Zoladex® (goserelin) was authorized as an LHRH analogue [88]. Trelstar® got approved by FDA in 2000 to treat prostate cancer in an advanced stage. Suprecur®, Suprefact®/Profact®, and Enantone® were all authorized in different nations at about the same time. Eligard®, a PLA-LHRH solution in N-methyl pyrrolidone that creates *in situ* implants following intramuscular injections, was licensed to treat prostate cancer in 2002. [86, 89]. Other disorders have also been treated with PLA-based medicines. In 2003, the USFDA authorized Resperidal® Consta® for psychotic illnesses. In 2006 and 2010, Vivitrol® was licensed for alcoholism and opioid addiction therapy, respectively [90]. Since the USFDA authorized Atridox® (1998) and Arestin® (2001), PLA copolymers have been used to treat periodontal disease. Atrisorb®, a periodontal tissue-directed growth biodegradable PLA-based impermeable membrane, is already on the market [87, 91]. Sculptra®, an aesthetic composition to treat facial lipoatrophy, was authorized by the US Food and Drug Administration (USFDA) in 2004. Sandostatin® LAR (approved by the FDA in 1998), Somatuline® (approved by the FDA in 2007) to treat acromegaly, and Nutropin® Depot to treat hormone insufficiency are all medications used to treat growth hormone-related illnesses [87, 92]. The combination of PLA–PEG triblock polymers exhibit temperature-induced gelation. The hydrophobic connections between PLGA segments get stronger as the temperature rises, culminating in gelation. OncoGel™ is a paclitaxel-free, chromophore-free three copolymer (PLGA–PEG–PLGA) composition for regional tumor control. This thermally induced sol-gel transition was termed ReGel® technology, and it was used to develop OncoGel™, paclitaxel-based PLGA–PEG–PLGA copolymer for regional tumor therapy [92]. Medisorb™ is an Alkermes patented technique that encapsulates drugs in PLGA copolymer microspheres. When the microspheres are injected, they quickly absorb moisture, resulting in swelling and polymer degradation with continuous medicine delivery over time. Vivitrol® and Resperidal® Consta®, two effective medications based on this approach, have been approved to treat addictions caused due to alcohol and psychotic disorders, respectively [93].

In phase IV to treat hill and valley pimple lesions with Poly (lactic acid) (Sculptra®), acne vulgaris is treated with injectable poly (lactic acid) to assess the amount of restoration possible (NCT00795327). The phase II trial of Genexol®-PM with gemcitabine (NCT01770795) to treat lung cancer was completed to investigate the efficiency and safety of Genexol®-PM with gemcitabine [86]. Similarly, PHAs are important in drug administration as they demonstrate biocompatibility, biodegradability, and nontoxicity qualities. As per a study by Fisol *et al.*, the biodegradable polymer binds nanomaterials P(3HB-co-4HB). Docetaxel, an anticancer medication, was effectively entrapped in P(3HB-co-4HB) without compromising its physicochemical characteristics [94]. Kawaguchi *et al.* made microspheres using P(3HB) of 3 molecular weights (65,000, 135,000, and 450,000) that contained derivatives of 5-fluoro-2′-deoxyuridine. Microspheres are injected intraperitoneally with a maximum molecular weight polymer enclosing prodrug had a more significant anticancer impact on P388 Leukemia in mice was treated with a free prodrug during five days. The polymer sphere demonstrated inadequate response and high biocompatibility [95]. Yagmurlu *et al.* developed antibiotic-loaded PHBV (22 % HV) rods, while its *in vivo* assessment was conducted in rabbits. Around day 15, radiographic and histologic research revealed the signs of *Staphylococcus aureus* infection in rod-implanted rabbits had considerably improved, even though osteomyelitis had persisted in the untreated controls [96]. In a comparable investigation, antibiotic-loaded P(3HB–4HB) rods were used in a comparative comparison [percent w/w Sulperazone or Duocid (ampicillin-sulbactam)]. These were inserted in rabbit tibias infected with S. aureus to cause implant-related osteomyelitis (IRO). Within a 6-week evaluation period after administering Sulperazone-P(3HB–4HB), infectious mediators cannot be extracted again through infection sites, indicating that the disease had been wholly treated [97]. Because of their superior biocompatibility and customizable physicochemical features, these bioplastics-based therapeutic implications have changed the area of drug delivery. These have aided the healthcare sector by supplying several items to tackle various ailments, ranging from bench to bedside applications.

12.4.3 Implants and Prostheses

Because of their rising popularity and advancements in healthcare, implants are regularly used as a first-line therapy choice. Stents are prostheses implanted in the coronary artery and used to restore constricted or damaged arteries and avoid restenosis by employing an inflating balloon on a catheter. Poly-L-lactic acid was used to create the very first biodegradable polymeric stent. The stent has a 9-month deterioration time. Relatively limited thrombosis, modest neointimal development, and a very minimal inflammatory reaction were reported after stent placement in mice [74]. New Zealand white rabbits had PHB-based biodegradable stents inserted in their iliac arteries over 30 weeks. After insertion, inflammatory responses, stent thrombosis, and lumen constriction were all induced by PHB [74]. Other everolimus-eluting PLA bioabsorbable stents (ABSORB II, Abbott Vascular) are available in almost 60 countries across the globe, namely Korea, the United States, Japan, and China, with encouraging results so far. The ABSORB II stent and the everolimus-eluting metallic stent had comparable one-year composite secondary medical outcomes (Xience, Abbott Vascular). Compared to second-generation drug-eluting stents, the ABSORB II stent demonstrated identical post-procedure area, minimal lumen area, stenosis, and eccentricity index [98]. In contrast, PLA-based stents dominate the marketplace of stents. Kyoto Medical Planning's Igaki-Tamai®, Elixir Medical Corporation's DeSolve®, Manli Cardiology's Mirage BRMS®, Abbott's Absorb GT1®, Meril's MeRes®, Xinsorb®, Arterial Remodelling Technologies' ART 18AZ®, and Terumo's Nobori® are examples of products among the commercially available PLA-based stents. Each of these stents is bioresorbable in 18 to 24 months and are drug-eluting stents (Sirolimus® or Myolimus®) that avoid restenosis [98].

Medical sutures are often used to help close and cure wounds caused by surgery or trauma by binding tissues together throughout the process of healing. Suture textiles for medical use come in a range of types, including absorbable and nonabsorbable; among the many options for clinical use are: Monofilament Glycolide and epsilon-caprolactone copolymer (PGA-PCL) Two-way communication (Quill™ SRS) (PGA-PCL) Barbed Poliglecaprone 25 (Monoderm™) (100). 910 Polyglactin Coated (Vicryl™, Polysorb™) Lactide and glycolide monofilament copolymer Vicryl Plus™ (polyglactin 910) coated braided lactide and glycolide monofilament lactide and glycolide copolymer braided [100, 101].

TephaFLEX solubilized stitches were the first PHA invention to be accepted via FDA for use in clinics. The item comprises P(4HB), a superior material with exceptional properties such as great elasticity, allowing it to grow ten times its original length without splitting [16]. PHA is used in medical devices, such as artery augments, vascular grafts, pericardial patches, stents, dressings, and carriers for drugs and is manufactured in the United States by companies like Tepha [102]. The VICRYL™ material for stitches focused on PGA/PLA copolymers is a marketable product with a proven record of therapeutic efficacy [43]. Several commercially available prostheses for whole or partial disc implants are emerging. For example, two FDA-approved cervical and lumbar disc prosthesis made of polymers are presently extensively used for disc replacement applications [103]. The FDA authorized the first artificial disc (DePuy Inc.), 2004. It was created using a long process that mixed a CoCrMo alloy with a biomaterial with an ultra-high molecular weight. ProDisc-C™ was built on identical properties of the polymer composites and was authorized in 2006 and 2007, including lumbar and cervical implants. Medtronic has released the Bryan™ prostheses made of titanium metals and polyurethane polymer [104].

Various polymeric implantable devices and prostheses are employed in clinical practice for orthobiologic purposes. Artificial menisci are among the most prevalent uses. Menisci are cartilage structures in the knee joint that help to distribute friction [105]. Meniscal replacements are frequently required after traumatic injuries or degenerative rips resulting from tissue degeneration with age. Polymeric PU implants (Actifit®) and Collagen meniscus implants (CMI®) are two partially meniscal replacements now authorized by the FDA [106]. A novel prosthetic device for transradial amputees termed the neural-enabled prosthetic hand (ANS-NEPH) is being tested in a clinical trial. It sends electrical stimulation impulses to electrodes implanted in nerve fascicles in the remnant limb's upper arm to give the patient the feeling of touching, grasp force, and hand movement. The system adjusts the impulses continually based on readings from sensors in the prosthetic hand and wireless communication with the implanted neurostimulator so that the patient feels while doing activities with the prosthesis [107].

Many implanted prostheses contain polymers that are useful in human health. Data from postimplant monitoring of devices in clinical use may only be used to enhance appropriate polymeric biomaterials iteratively. Therefore, the use of biomaterial-based biodegradable and biocompatible stents, sutures, and vivid biomedical implants aid in the prevention of restenosis and other surgical and post-operative complications. The applications based on the types employed are summarized in Table 12.1.

Table 12.1 Types of bioplastics and their associated applications.

Sr. no.	Types	Applications	Brand name	References
1.	Poly(3-hydroxy-butyrate) (PHB)	Wound healing, Cardiovascular patch	TachoSil®; NeuroFlex®, NeuroMatrix®, NeuroMend®, Dynamatrix®, GORE-TEX®, IMPRA® ePTFE Xelma, Apligraft®, Dermagraft® and INFUSE®	[81–84]
2.	Polyhydro-xyalkanoate (PHA)	Monofilaments, Artificial disc	TephaFLEX, DePuy Inc., ProDisc-C™	[100, 103, 104]
3.	Poly-Lactic Acid (PLA)	Stents tissue regeneration, skin and tendon mending, dental niches, orthopaedic prostheses, drug carriers, cancer therapy, cardiovascular implants.	ABSORB II, Abbott Vascular, Igaki-Tamai®, Elixir Medical Corporation's DeSolve®, Manli Cardiology's Mirage BRMS®, Abbott's Absorb GT1®, Meril's MeRes®, Xinsorb®, Arterial Remodelling Technologies' ART 18AZ®, and Terumo's Nobori®, and ANS-NEPH	[71, 98, 107]
4.	Poly lactic-co-glycolic acid (PLGA)	Tissue engineering, Cartilage Regeneration, and drug delivery	BioSeed®-B and BioSeed®-C, OncoGel™, Medisorb™	[71, 81, 92, 93]
5.	Poly (ε-caprolactone) (PCL)	Monofilament	Quill™ SRS, Monoderm™, VicrylTM, Polysorb™ and Vicryl Plus™	[99–101]

12.5 Conclusion

"Biologically safe," microorganism-derived bioplastics poly-lactic acid was derived from renewable sources PLA, PHA, PLGA, and PCL. These bioplastics are being utilized or are a possible option for various consumer and biomedical applications due to their numerous desired qualities, such as renewability, biocompatibility, clarity, and thermo-plasticity. The synthesis and characterization of naturally produced biopolymer-based composites have revealed that they have great promise for tissue regeneration to increase cell adhesion or to effectively drive cell migration, proliferation, and differentiation. In clinical practice, biobased scaffolds have been effectively used to heal epidermis, bones, cartilage, artificial skin, and implants for human body parts. Biopolymer scaffolding is also being utilized to create biocompatible stents and barrier membranes which do not need further surgery. Several of the adverse reactions with synthetic polymers could be alleviated by using bioplastics. As a result, bioplastics offer a viable alternative to synthetic polymers. Preclinically and clinically, the biopolymer-based composite resorbable acts as a potential therapeutic implant because of their high biocompatibility, long disintegration time, and anti-inflammatory characteristics. Moreover, medication and bioactive agent delivery systems based on bioplastics should be created as multipurpose release systems instead of one-way release systems to keep biomolecules functioning. The resemblance between the configuration of the human body and the polymeric materials is crucial for creating biobased composites for medicinal purposes. Therefore, bioplastic-based composites must be extensively explored to develop potential therapeutics from bench to bedside for future biomedical applications.

References

1. Anjum, A., Zuber, M., Zia, K.M., Noreen, A., Anjum, M.N., Tabasum, S., Microbial production of polyhydroxyalkanoates (PHAs) and its copolymers: A review of recent advancements. *Int. J. Biol. Macromol.*, 89, 161, 2016.
2. Budde, C.F., Riedel, S.L., Willis, L.B., Rha, C., Sinskey, A.J., Production of poly(3-hydroxybutyrate-co-3-hydroxyhexanoate) from plant oil by engineered Ralstonia eutropha strains. *Appl. Environ. Microbiol.*, 77, 2847, 2011.
3. Page, W.J., Manchak, J., Rudy, B., 'Formation of poly(hydroxybutyrate-co-hydroxyvalerate) by Azotobacter vinelandii UWD. *Appl. Environ. Microbiol.*, 58, 2866, 1992.
4. Halami, P.M., Production of polyhydroxyalkanoate from starch by the native isolate Bacillus cereus CFR06. *World J. Microbiol. Biotechnol.*, 24, 80512, 2008.
5. Young, F.K., Kastner, J.R., May, S.W., Microbial production of poly-β-hydroxybutyric acid from d-Xylose and Lactose by Pseudomonas cepacia. *Appl. Environ. Microbiol.*, 60, 4195, 1994.
6. Silva, L.F., Taciro, M.K., Michelin Ramos, M.E., Carter, J.M., Pradella, J.G.C., Gomez, J.G.C., 'Poly-3-hydroxybutyrate (P3HB) production by bacteria from xylose, glucose and sugarcane bagasse hydrolysate. *J. Ind. Microbiol. Biotechnol.*, 31, 245, 2004.
7. Sangkharak, K. and Prasertsan, P., Optimization of polyhydroxybutyrate production from a wild type and two mutant strains of Rhodobacter sphaeroides using statistical method. *J. Biotechnol.*, 132, 331, 2007.
8. Ibrahim, M.H.A. and Steinbüchel, A., Zobellella denitrificans strain MW1, a newly isolated bacterium suitable for poly (3-hydroxybutyrate) production from glycerol: PHB accumulation from glycerol by Zobellella denitrificans MW1. *J. Appl. Microbiol.*, 108, 214, 2010.
9. Kimura, H., Takahashi, T., Hiraka, H., Iwama, M., Takeishi, M., Effective biosynthesis of poly (3-hydroxybutyrate) from plant oils by chromobacterium sp. *Polym. J.*, 31, 210, 1999.
10. Andreeßen, B., Lange, A.B., Robenek, H., Steinbuchel, A., Conversion of glycerol to Poly(3-hydroxypropionate) in recombinant Escherichia coli. *Appl. Environ. Microbiol.*, 76, 622, 2010.
11. Moradali, M.F. and Rehm, B.H.A., Bacterial biopolymers: From pathogenesis to advanced materials. *Nat. Rev. Microbiol.*, 18, 195, 2020.
12. Steinbüchel, A. and Füchtenbusch, B., Bacterial and other biological systems for polyester production. *Trends Biotechnol.*, 16, 419, 1998.

13. Madison, L.L. and Huisman, G.W., Metabolic engineering of poly(3-Hydroxyalkanoates): From DNA to Plastic. Microbiol. *Mol. Biol. Rev.*, 63, 21, 1999.
14. Angelova, N. and Hunkeler, D., Rationalizing the design of polymeric biomaterials. *Trends Biotechnol.*, 17, 409, 1999.
15. Ulery, B.D., Nair, L.S., Laurencin, C.T., Biomedical applications of biodegradable polymers. *J. Polym. Sci. B. Polym. Phys.*, 49, 832, 2011.
16. De, Souza, L. and Shivakumar, S., Polyhydroxyalkanoates (PHA)–Applications in wound treatment and as precursors for oral Drugs, in: *Biotechnological Applications of Polyhydroxyalkanoates*, pp. 227–270, Springer,Singapore, 2019.
17. Gallo, J., Holinka, M., Moucha, C., Antibacterial surface treatment for orthopaedic implants. *Int. J. Mol. Sci.*, 15, 13849, 2014.
18. Francis, L., Meng, D., Locke, I.C., Knowles, J.C., Mordan, N., Salih, V., Boccaccini, A.R., Roy, I., Novel poly(3-hydroxybutyrate) composite films containing bioactive glass nanopar ticles for wound healing applications. *Polym. Int.*, 65, 661, 2016.
19. Maitz, M.F., Applications of synthetic polymers in clinical medicine. *Biosurface Biotribology.*, 1, 161, 2015.
20. Törmälä, P., Biodegradable self-reinforced composite materials; Manufacturing structure and mechanical properties. *Clin. Mater.*, 10, 29, 1992.
21. Lendlein, A., Behl, M., Hiebl, B., Wischke, C., Shape-memory polymers as a technology platform for biomedical applications. *Expert. Rev. Med. Devices.*, 7, 357, 2010.
22. Chan, B.P. and Leong, K.W., Scaffolding in tissue engineering: General approaches and tissue-specific considerations. *Eur. Spine. J.*, 4, 467, 2008.
23. Biedendieck, R., Knuuti, T., Moore, S.J., Jahn, D., The "beauty in the beast"—The multiple uses of Priestia megaterium in biotechnology. *Appl. Microbiol. Biotechnol.*, 105, 5719, 2021.
24. Mostafa, Y.S., Alrumman, S.A., Alamri, S.A., Otaif, K.A., Mostafa, M.S., Alfaify, A.M., Bioplastic (poly-3-hydroxybutyrate) production by the marine bacterium Pseudodonghicola xiamenensis through date syrup valorization and structural assessment of the biopolymer. *Sci. Rep.*, 10, 8815, 2020.
25. Raza, Z.A., Abid, S., Banat, I.M., Polyhydroxyalkanoates: Characteristics, production, recent developments and applications. *Int. Biodeterior. Biodegrad.*, 126, 45, 2018.
26. Gopi, S., Kontopoulou, M., Ramsay, B.A., Ramsay, J.A., Manipulating the structure of medium-chain-length polyhydroxyalkanoate (MCL-PHA) to enhance thermal properties and crystallization kinetics. *Int. J. Biol. Macromol.*, 119, 1248, 2018.
27. Suriyamongkol, P., Weselake, R., Narine, S., Moloney, M., Shah, S., Biotechnological approaches for the production of polyhydroxyalkanoates in microorganisms and plants — A review. *Biotechnol. Adv.*, 25, 148, 2007.
28. Koller, M. and Braunegg, G., Advanced approaches to produce polyhydroxyalkanoate (PHA) biopolyesters in a sustainable and economic fashion. *EuroBiotech. J.*, 2, 89, 2018.
29. Lackner, M., Biopolymer, in: *Handbook of climate change mitigation and adaptation*, pp. 3211–3230, Springer International Publishing, New York, 2017.
30. Tsuge, T., Fukui, T., Matsusaki, H., Taguchi, S., Kobayashi, G., Ishizaki, A., Doi, Y., Molecular cloning of two (R)-specific enoyl-CoA hydratase genes from Pseudomonas aeruginosa and their use for polyhydroxyalkanoate synthesis. *FEMS Microbiol. Lett.*, 184, 193, 2000.
31. Hoffmann, N., Steinbüchel, A., Rehm, B.H., The Pseudomonas aeruginosa phaG gene product is involved in the synthesis of polyhydroxyalkanoic acid consisting of medium-chain-length constituents from non-related carbon sources. *FEMS Microbiol. Lett.*, 184, 253, 2000.
32. Luzier, W.D., Materials derived from biomass/biodegradable materials. *Proc. Natl. Acad. Sci. U S A.*, 89, 839, 1992.

33. El-Hadi, A.M., Miscibility of crystalline/amorphous/crystalline biopolymer blends from PLLA/PDLLA/PHB with additives. *Polym.-Plastics Technol. Mater.*, 58, 31, 2019.
34. Li, Y., Xin, S., Bian, Y., Dong, Q., Han, C., Xu, K., Dong, L., Stereocomplex crystallite network in poly(D,L-lactide): Formation, structure and the effect on shape memory behaviors and enzymatic hydrolysis of poly(D,L-lactide). *RSC Adv.*, 5, 24352, 2015.
35. Makadia, H.K. and Siegel, S.J., Poly Lactic-co-Glycolic Acid (PLGA) as biodegradable controlled drug delivery carrier. Polymers. *Polym. (Basel).*, 3, 1377, 2011.
36. Kulkarni, R.K., Polylactic acid for surgical implants. *Arch. Surg.*, 93, 839, 2011.
37. Auras, R., Harte, B., Selke, S., An overview of polylactides as packaging materials. *Macromol. Biosci.*, 4, 835, 2004.
38. Othman, M., Ariff, A.B., Rios-Solis, L., Halim, M., Extractive fermentation of lactic acid in lactic acid bacteria cultivation: A review. *Front. Microbiol.*, 8, 2285, 2017.
39. Jamshidian, M., Tehrany, E.A., Imran, M., Jacquot, M., Desobry, S., Poly-lactic acid: Production, applications, nanocomposites, and release studies. *Compr. Rev. Food Sci. Food Saf.*, 9, 552, 2010.
40. Mehta, R., Kumar, V., Bhunia, H., Upadhyay, S.N., Synthesis of poly (Lactic Acid): A review. *J. Macromol. Sci., Part C: Polym. Rev.*, 45, 325, 2005.
41. Jo, S.-J., Maeda, M., Ooi, T., Taguchi, S., Production system for biodegradable polyester polyhydroxybutyrate by Corynebacterium glutamicum. *J. Biosci. Bioeng.*, 102, 233, 2006.
42. Bano, K. and Pandey, R., Jamal-e-Fatima, Roohi., New advancements of bioplastics in medical applications. *Int. J. Pharm. Sci. Res.*, 9, 402, 2018.
43. Wang, Z.-Y., Zhao, Y.-M., Wang, F., Wang, J., Syntheses of poly(lactic acid-co-glycolic acid) serial biodegradable polymer materials via direct melt polycondensation and their characterization. *J. Appl. Polym. Sci.*, 99, 244, 2006.
44. Choi, S.Y., Park, S.J., Kim, W.J., Yang, J.E., Lee, H., Shin, J., Lee, S.Y., One-step fermentative production of poly(lactate-co-glycolate) from carbohydrates in Escherichia coli. *Nat. Biotechnol.*, 34, 435, 2016.
45. McKeen, L.W., Permeability properties of plastics and elastomers, pp. 21–37, Elsevier, 2016.
46. Pandey, A., Negi, S., Soccol, C.R., Current developments in biotechnology and bioengineering: Production, isolation and purification of industrial products, pp. 25–41, Elsevier, 2016.
47. Malikmammadov, E., Tanir, T.E., Kiziltay, A., Hasirci, V., Hasirci, N., PCL and PCL-based materials in biomedical applications. *J. Biomater. Sci. Polym. Ed.*, 29, 863, 2018.
48. Luciani, A., Coccoli, V., Orsi, S., Ambrosio, L., Netti, P.A., PCL microspheres based functional scaffolds by bottom-up approach with predefined microstructural properties and release profiles. *Biomaterials.*, 29, 4800, 2008.
49. Carothers, W.H., Dorough, G.L., Natta, F.J., van Studies of polymerization and ring formation. X. The reversible polymerization of six-membered cyclic esters. *J. Am. Chem. Soc*, 54, 761, 1932.
50. Sivabalan, A., Subramani, R.H., Meenarathi, B., Palanikumar, S., Anbarasan, R., Synthesis and characterization of poly (ε-caprolactone): A comparative study. *Int. J. Sci. Res. Eng. Technol.*, 1, 9, 2014.
51. Komur, B., Bayrak, F., Ekren, N., Eroglu, M.S., Oktar, F.N., Sinirlioglu, Z.A., Yucel, S., Guler, O., Gunduz, O., Starch/PCL composite nanofibers by co-axial electrospinning technique for biomedical applications. *Biomed. Eng. OnLine.*, 16, 40, 2017.
52. Huang, H., Oizumi, S., Kojima, N., Niino, T., Sakai, Y., Avidin–biotin binding-based cell seeding and perfusion culture of liver-derived cells in a porous scaffold with a three-dimensional interconnected flow-channel network. *Biomaterials.*, 28, 3815, 2007.
53. Göpferich, A., Karydas, D., Langer, R., Predicting drugre lease from cylindric polyanhydride matrix discs. *Eur. J. Pharm. Biopharm.*, 41, 81, 1995.
54. Song, J.H., Murphy, R.J., Narayan, R., Davies, G.B.H., Biodegradable and compostable alternatives to conventional plastics. *Phil. Trans. R. Soc B*, *364*, 2127, 2009.

55. Di Bartolo, A., Infurna, G., Dintcheva, N.T., A review of bioplastics and their adoption in the circular economy. *Polymers*, 13, 1229, 2021.
56. Atiwesh, G., Mikhael, A., Parrish, C.C., Banoub, J., Le, T.-A.T., Environmental impact of bioplastic use: A review. *Heliyon*, 7, e07918, 2021.
57. Zhang, L., Chai, W., Li, W., Semple, K., Yin, N., Zhang, W., Dai, C., Intumescent-grafted bamboo charcoal: A natural nontoxic fire-retardant filler for polylactic acid (PLA) composites. *ACS Omega*, 6, 26990, 2021.
58. Balla, E., Daniilidis, V., Karlioti, G., Kalamas, T., Stefanidou, M., Bikiaris, N.D., Vlachopoulos, A., Koumentakou, I., Bikiaris, D.N., Poly(lactic acid): A versatile biobased polymer for the future with multifunctional properties—From monomer synthesis, polymerization techniques and molecular weight increase to PLA applications. *Polymers*, 13, 1822, 2021.
59. DeStefano, V., Khan, S., Tabada, A., Applications of PLA in modern medicine. *ER*, 1, 76, 2020.
60. Engineer, C., Parikh, J., Raval, A., Review on hydrolytic degradation behavior of biodegradable polymers from controlled drug delivery system. *Trends Biomater. Artif. Organs*, 25, 79–85, 2011.
61. Keles, H., Naylor, A., Clegg, F., Sammon, C., Investigation of factors influencing the hydrolytic degradation of single PLGA microparticles. *Polym. Degrad. Stab.*, 119, 228, 2015.
62. Vllasaliu, D., Fowler, R., Stolnik, S., PEGylated nanomedicines: Recent progress and remaining concerns. *Expert Opin. Drug Deliv.*, 11, 139, 2014.
63. Lanao, R.P.F., Jonker, A.M., Wolke, J.G., Jansen, J.A., van Hest, J.C., Leeuwenburgh, S.C., Physicochemical properties and applications of poly (lactic-co-glycolic acid) for use in bone regeneration. *Tissue Eng. Part B Rev.*, 19, 380, 2013.
64. Labet, M. and Thielemans, W., Synthesis of polycaprolactone: A review. *Chem. Soc Rev.*, 38, 3484, 2009.
65. Li, L. and LaBarbera, D.V., 3D High-content screening of organoids for drug discovery, in: *Comprehensive Medicinal Chemistry III*, 2017.
66. Lu, J., Tappel, R.C., Nomura, C.T., Mini-review: Biosynthesis of poly (hydroxyalkanoates). *J. Macromol. Sci. Polym. Rev., Part C: Polym. Revs.*, 49, 226, 2009.
67. Doi, Y. and Steinbüchel, A., Biopolymers, Polyesters II-Properties and Chemical Synthesis, Wiley-Blackwell, UK, 2002.
68. Zhila, N.O., Sapozhnikova, K.Y., Kiselev, E.G., Vasiliev, A.D., Nemtsev, I.V., Shishatskaya, E.I., Volova, T.G., Properties of degradable polyhydroxyalkanoates (PHAs) synthesized by a new Strain, Cupriavidus necator IBP/SFU-1, from various carbon sources. *Polymers*, 13, 3142, 2021.
69. Bonartsev, A.P., Bonartseva, G.A., Reshetov, I.V., Kirpichnikov, M.P., Shaitan, K.V., Application of polyhydroxyalkanoates in medicine and the biological activity of natural poly (3-hydroxybutyrate). *Acta Naturae.*, 11, 41, 2019.
70. Bhattacharya, S. and Shunmugam, R., Unraveling the effect of PEG chain length on the physical properties and toxicant removal capacities of cross-linked network synthesized by Thiol–Norbornene photoclick chemistry. *ACS omega*, 5, 2800, 2020.
71. D'souza, A.A. and Shegokar, R., Polyethylene glycol (PEG): A versatile polymer for pharmaceutical applications. *Expert Opin. Drug Deliv.*, 13, 1257, 2016.
72. Xiao, L., Wang, B., Yang, G., Gauthier, M., Poly (lactic acid)-based biomaterials: Synthesis, modification and applications. *J. Biomed. Eng. Technol.*, 11, 247, 2012.
73. Jo, Y.Y., Kim, S.G., Kwon, K.J., Kweon, H., Chae, W.S., Yang, W.G., Lee, E.Y., Seok, H., Silk fibroin-alginate-hydroxyapatite composite particles in bone tissue engineering applications *in vivo*. *Int. J. Mol. Sci.*, 18, 858, 2017.
74. Fu, N., Meng, Z., Jiao, T., Luo, X., Tang, Z., Zhu, B., Sui, L., Cai, X., P34HB electrospun fibres promote bone regeneration *in vivo*. *Cell Prolif.*, 52, e12601, 2019.
75. Chai, J.M., Amelia, T.S.M., Mouriya, G.K., Bhubalan, K., Amirul, A.A.A., Vigneswari, S., Ramakrishna, S., Surface-modified highly biocompatible bacterial-poly

(3-hydroxybutyrate-co-4-hydroxybutyrate): A review on the promising next-generation biomaterial. *Polymers.*, 13, 51, 2021.
76. Niu, H., Mu, J., Zhang, J., Hu, P., Bo, P., Wang, Y., Comparative study of three types of polymer materials co-cultured with bone marrow mesenchymal stem cells for use as a myocardial patch in cardiomyocyte regeneration. *J. Mater. Sci. Mater. Med.*, 24, 1535, 2013.
77. Engelhardt, E.M., Micol, L.A., Houis, S., Wurm, F.M., Hilborn, J., Hubbell, J.A., Frey, P., A collagen-poly (lactic acid-co-ε-caprolactone) hybrid scaffold for bladder tissue regeneration. *Biomaterials.*, 32, 3969, 2011.
78. John, A., Katiyar, V., Pang, K., Shaikh, M.M., Nanavati, H., Ghosh, P., Ni (II) and Cu (II) complexes of phenoxy-ketimine ligands: Synthesis, structures and their utility in bulk ring-opening polymerization (ROP) of L-lactide. *Polyhedron.*, 26, 4033, 2007.
79. Lv, Q., Hu, K., Feng, Q., Cui, F., Cao, C., Preparation and characterization of PLA/fibroin composite and culture of HepG2 (human hepatocellular liver carcinoma cell line) cells. *Compos. Sci. Technol.*, 67, 3023, 2007.
80. Mauney, J.R., Nguyen, T., Gillen, K., Kirker-Head, C., Gimble, J.M., Kaplan, D.L., Engineering adipose-like tissue *in vitro* and *in vivo* utilizing human bone marrow and adipose-derived mesenchymal stem cells with silk fibroin 3D scaffolds. *Biomaterials.*, 28, 5280, 2007.
81. Mathieu, L.M., Mueller, T.L., Bourban, P.E., Pioletti, D.P., Müller, R., Månson, J.A., Architecture and properties of anisotropic polymer composite scaffolds for bone tissue engineering. *Biomaterials.*, 27, 905, 2006.
82. Park, S.B., Lih, E., Park, K.S., Joung, Y.K., Han, D.K., Biopolymer-based functional composites for medical applications. *Prog. Polym. Sci.*, 68, 77, 2017.
83. Reconstructions, P.V., The GORE-TEX Cardiovascular Patch: The performance you expec, 2014. https://www.goremedical.com/resource/AT0904-EN2.
84. Cardiovascular Patches—ePTFE, bd.com/en-us/offerings/capabilities/vascular-surgery/surgical-fabric-and-patches/cardiovascular-patches—eptfe.
85. Kulma, A., Skórkowska-Telichowska, K., Kostyn, K., Szatkowski, M., Skała, J., Drulis-Kawa, Z., Preisner, M., Zuk, M., Szperlik, J., Wang, Y.F., Szopa, J., New flax producing bioplastic fibers for medical purposes. *Ind. Crops Prod.*, 80, 80–89, 2015.
86. Torres, F.G., Commeaux, S., Troncoso, O.P., Biocompatibility of bacterial cellulose based biomaterials. *J. Funct. Biomater.*, 3, 864, 2012.
87. Jain, A., Kunduru, K.R., Basu, A., Mizrahi, B., Domb, A.J., Khan, W., Injectable formulations of poly (lactic acid) and its copolymers in clinical use. *Adv. Drug Deliv. Rev.*, 15, 213, 2016.
88. Periti, P., Mazzei, T., Mini, E., Clinical pharmacokinetics of depot leuprorelin. *Clin. Pharmacokinet.*, 41, 485, 2002.
89. Zoladex® (10.8 mg Goserelin Acetate Implant), 2003. https://www.rxlist.com/zoladex-drug/patient-images-side-effects.htm#info.
90. Wright, J.C. and Hoffman, A.S., Historical overview of long acting injections and implants, in: *Springer-Long acting injections and implants*, pp. 11–24, 2012.
91. Shue, L., Yufeng, Z., Mony, U., Biomaterials for periodontal regeneration: A review of ceramics and polymers. *Biomatter.*, 2, 271, 2012.
92. Danny, V.M.D., Facial enhancement and the European experience with Sculptra™ (poly-l-lactic acid). *J. Drugs Dermatol.*, 3, 542, 2004.
93. Dean, R.L., The preclinical development of Medisorb Naltrexone, a once a month long acting injection, for the treatment of alcohol dependence. *Front. Biosci.*, 10, 643, 2005.
94. Bartus, R.T., Emerich, D.F., Hotz, J., Blaustein, M., Dean, R.L., Perdomo, B., Basile, A.S., Vivitrex®, an injectable, extended-release formulation of naltrexone, provides pharmacokinetic and pharmacodynamic evidence of efficacy for 1 month in rats. *Neuropsychopharmacology.*, 28, 1973, 2003.

95. Faisalina, A.F., Sonvico, F., Colombo, P., Amirul, A.A., Wahab, H.A., Majid, M.I.A., Docetaxel-Loaded Poly(3HB-co-4HB) Biodegradable Nanoparticles: Impact of Copolymer Composition. *Nanomaterials.*, 26, E2123, 2020.
96. Kawaguchi, T., Tsugane, A., Higashide, K., Endoh, H., Hasegawa, T., Kanno, H. *et al.*, Control of drug release with a combination of prodrug and polymer matrix: Antitumor activity and release profiles of 2',3'-diacyl-5-fluoro-2'-deoxyuridine from poly(3-hydroxybutyrate) microspheres. *J. Pharm. Sci.*, 81, 508, 1992.
97. Yagmurlu, M.F., Korkusuz, F., Gürsel, I., Korkusuz, P., Örs, Ü., Hasirci, V., Sulbactam-cefoperazone polyhydroxybutyrate-co-hydroxyvalerate (PHBV) local antibiotic delivery system: *In vivo* effectiveness and biocompatibility in the treatment of implant-related experimental osteomyelitis. *J. Biomed. Mater. Res.*, 15, 494, 1999.
98. Gürsel, İ., Korkusuz, F., Türesin, F., Alaeddinoğlu, N.G., Hasırcı, V., *In vivo* application of biodegradable controlled antibiotic release systems for the treatment of implant-related osteomyelitis. *Biomaterials.*, 22, 106, 2000.
99. Narancic, T., Cerrone, F., Beagan, N., O'Connor, K.E., Recent advances in bioplastics: Application and biodegradation. *Polymers.*, 12, 920, 2020.
100. D. Kirsch, S. Marczyk, L.P. Covidien, Multifilament barbed suture. United States patent US 8, 2013.
101. Manual, W.C., Ethicon, Inc. A Johnson and Johnson company, USA, 2005.
102. Greenberg, J.A. and Clark, R.M., Advances in suture material for obstetric and gynecologic surgery. *Obstet. Gynecol.*, 2, 146, 2009.
103. Williams, S.F., Rizk, S., Martin, D.P., Poly-4-hydroxybutyrate (P4HB): A new generation of resorbable medical devices for tissue repair and regeneration. *Biomed. Tech/Biomed. Eng.*, 58, 439, 2013.
104. Dang, T.T., Nikkhah, M., Memic, A., Khademhosseini, A., Polymeric biomaterials for implantable prostheses, in: *Natural and synthetic biomedical polymers*, pp. 309–331, Elsevier, USA, 2014.
105. Link, H.D., McAfee, P.C., Pimenta, L., Choosing a cervical disc replacement. *Spine J.*, 1, S294, 2004.
106. Vrancken, A.C., Buma, P., van, Tienen, T.G., Synthetic meniscus replacement: A review. *Int. Orthop.*, 37, 291, 2013.
107. Spencer, S.J., Saithna, A., Carmont, M.R., Dhillon, M.S., Thompson, P., and Spalding, T., Meniscal scaffolds: Early experience and review of the literature. *Knee*, 19, 760, 2012.

13

Biomedical Applications of Biocomposites Derived From Cellulose

Subhajit Kundu[1], **Debarati Mitra**[1] **and Mahuya Das**[2]*

[1]Department of Chemical Technology, University of Calcutta, Kolkata, India
[2]Greater Kolkata College of Engineering and Management, Baruipur, Kolkata, India

Abstract
This chapter contains a detailed discussion on cellulose-based biocomposites for biomedical application. The synthesis of cellulose by different method and their comparison is included in the chapter along with the different forms of cellulose and the processes by which they are converted into biocomposites. The applications of those biocomposites are also mentioned here. The included applications in the present chapter are tissue engineering, wound dressing, medicinal implants, dental application, etc.

Keywords: Cellulose, cellulose nanofiber, cellulose nanocrystals, bacterial cellulose, biocomposites, applications

13.1 Introduction

The phrase "biomaterials" has a wide range of meanings. Its meanings vary, but it generally refers to a material that may be used to heal, replenish, or rejuvenate organs (in terms of structure and functions) [1]. According to Black and Hastings, a biomaterial is an untenable material utilised in a medical device that is meant to interact with biological processes [2]. Biomaterials may be natural or synthetic materials that are utilised for replacing elements of a living system or function in close proximity to biological tissues, according to another description. Three generations of biomaterial applications have been discovered. The first phase was developed during 1960s in order to create a sufficient mix of functional qualities that could properly resemble those of the replacement tissue without causing the host to react negatively. Artificial tissues made by embedding cells in scaffold materials, for instance, aid in directing cell growth and differentiation. In the 1980s, the second series of biomaterials was designed to cause a regulated response with the tissues into which they were inserted in order to obtain a particular therapeutic effect. The emergence of superabsorbent biomaterials with rates of disintegration could be tuned to the needs of a specific

**Corresponding author*: d_mahuya@yahoo.com

Inamuddin and Tariq Altalhi (eds.) Handbook of Bioplastics and Biocomposites Engineering Applications: 2nd Edition, (251–274) © 2023 Scrivener Publishing LLC

application. The third generation of biomaterials aims to help people regenerate functional tissue in injured bladders, tracheas, skin, corneal epithelium, and cartilage [3].

Because of its low density, gentle exfoliating, flammable, cheap cost, nontoxic and biodegradable features, cellulose-based polymer composites have gotten a lot of attention. Many studies on the use of cellulose fibers as a reinforcing material in the fabrication of several types of composites have been conducted all over the world [4]. Nevertheless, cellulose reinforced composites are less appealing because of lack of strong interfacial adhesion, low sensitivity toward water, as well as low melting point. Because of its great strength and durability paired with biodegradability and low weight, nanocellulose fibers and their use in composite materials have gotten a lot of interest [5]. The use of cellulose nanofibers in polymer reinforcement is a comparatively recent subject of investigation. In the field of biomedical application specially for the applications where the biomaterial is in touch with living tissue, the material must not exhibit any kind of negative effects. For this reason, assessment of material's immunogenicity and verification of their interaction with the cells is mandatory. Because of their unique characteristics cellulose-based composites as bioresorbable polymers are becoming more relevant in biomedical applications [6].

13.2 Importance of Cellulose in the Field of Biocomposite

Biocomposites contains natural fiber in polymeric resin and if the resin is from natural source, i.e., biodegradable then it is called green composite. In this chapter, we are focussing on biocomposites and, obviously, the reinforcement will be natural fiber. Natural fiber–based biocomposites are successfully applicable in various field then why there is increasing importance in cellulose-based biocomposite. Cellulose fibers are highly attractive as a as potential reinforcing materials due to its high abundancy, low density, biodegradability, renewability, low abrasive nature, and many more interesting properties [7]. Being the most important constituent of plant cell wall, it helps plant to be stiff and upright indicating good mechanical property of cellulose. However, the disadvantages of cellulose include its hydrophilicity, low thermal stability, high moisture absorption and inconsistency in its quality depending on its source [8].

13.3 Classification of Cellulose

According to the fibril size, the cellulose can be divided into two types, i) microcellulose/microcrystalline cellulose (MCC) and ii) nanocellulose. Nanomaterials based on cellulose are usually classified in three types: i) micro/nanofibrillated cellulose (NFC)/(MFC), ii) nanocrystalline cellulose (NCC), and iii) bacterial nanocellulose (BNC). The length of NFC/MFC is higher than 1 μm, and the diameter is 20 to 50 nm. NFC and MFC differ in their processing techniques; NFC production involves a high-pressure homogenization followed by chemical treatment, whereas MFC required only chemical treatment. NFC contains both crystalline and amorphous part of cellulose. Meanwhile, nanocrystalline cellulose (NCC) possesses only crystalline zone of cellulose fibril with the average diameter in the range of 5 to 70 nm and length around 100 to 200 nm [9, 10]. MCC is generated by preliminary removal of the amorphous parts giving short rod-like fibrils with high

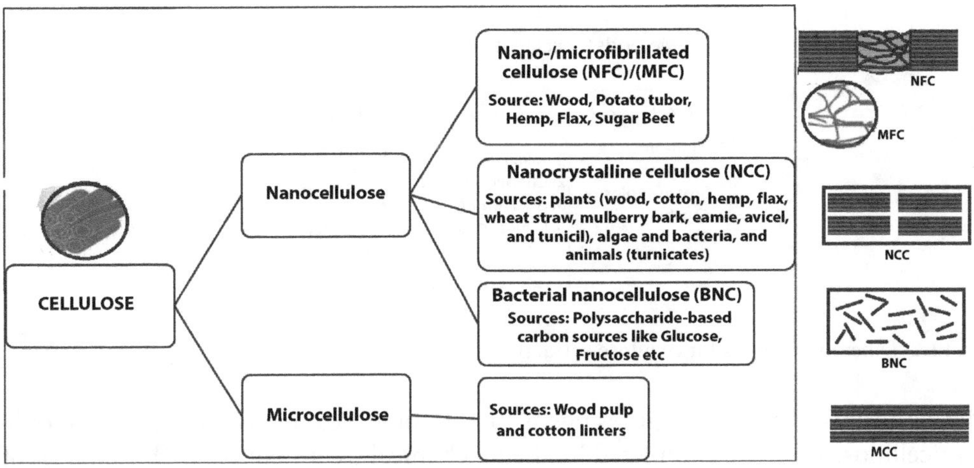

Figure 13.1 Different types of cellulose and their source.

crystallinity while in MFC both the crystalline and amorphous regions present conserving the molecular mass of cellulose resulting into a three-dimensionally network structure consisting of long and thin interconnected fibrils having wide range of distribution in internal dimensions. Bacterial nanocellulose (BNC) is the third kind of cellulose that can be synthesized from carbonaceous material like glucose by bacteria. In Figure 13.1, different source and types of cellulose are represented [11–14].

13.4 Synthesis of Cellulose in Different Form

Cellulose is a macromolecule containing linear polymer chains consisting of β (1,4) D glucopyranose units. In the plant cell wall of natural fibers, cellulose fibers are embedded in the matrix of hemicellulose and lignin forming a natural biocomposite. These cellulose fibers are actually combination of bundle of microfibrils resulting from bunching of elementary cellulose fibrils consisting of the basic β (1,4) D glucopyranose units. These elementary fibrils are nanodimensional with diameter in the range of about 2 to 20 nm and length of few micrometers. Each cellulose fibrils contains about 30 to 100 cellulose chains in total. Those cellulose chains are arranged in two region one is highly ordered, crystalline, and another is disordered or amorphous [15].

Cellulose is to be extracted from its source when to be used for any application. Different techniques are employed for cellulose extraction from its source, which is given in Figure 13.2.

13.4.1 Mechanical Extraction

This process involves inter-fibrillar hydrogen bond breaking by mechanical force leading to reduction of degree of polymerization from 1,200 and 1,400 to 850 and 500. The mechanical methods involve costly tools and materials hence high production costs (tools and materials). This extraction process is less efficient and requires higher energy than the chemical one [16].

Methods for cellulose extraction		
Mechanical	Chemical	Bacterial
1. High-pressure homogenization 2. Steam explosion 3. Crushing 4. Grinding	1. Alkali treatment 2. Chemical retting 3. Acid retting 4. Degumming	Bacteria of Acetobacter genera (Most efficiently Acetobacter xylinum)

Figure 13.2 Different methods for cellulose extraction.

Mechanical treatment sometimes needs further modification as the separation of lignin or hemicellulose is not be completed by the mechanical treatment only. The analysis of different types of mechanical treatment is given in Table 13.1.

13.4.2 Electrochemical Method

There is another electromechanical method, known as electrospinning, which involves extrusion of a cellulose dispersion made by mechanical treatment and electrospinning of cellulose nanofiber under the application of high electrical field. After the evaporation of solvent, it produces a randomly oriented nanocellulose fibers in the collector morphology of which will be controlled by the feeding rate of cellulose dispersion, tip to collector distance, electrical field strength etc. This is an economic and simple method [23].

13.4.3 Chemical Extraction

Chemical extraction involves extract cellulose by using chemical retting, acid retting, alkali hydrolysis, or degumming for the removal of lignin from the fibers. This entire chemical actually reacts with the different component of lignocellulosic fiber other than cellulose leading to isolation of cellulose in different crystalline form. Chemical treatment results in change of fine structure, morphology, and properties of cellulose fiber causing less fiber damage than mechanical treatment. It is also more economic than the mechanical one. All the variety of chemical treatment involves heating, soaking, and swelling of the fiber in the chemical. However, combination of different chemical treatment along with mechanical one is also used for isolation of cellulose effectively.

Mechanical and chemical extractions may be followed by enzymatic hydrolysis method or that is another way to extract cellulose by itself as described below.

13.4.4 Enzymatic Hydrolysis

Enzymatic hydrolysis process involves the use of enzyme to isolate cellulose from the natural source by modifying/degrading the noncellulosic part of plant fiber. The mechanism of conversion starting from the plant fiber to the ultimate nanocellulose structure is quite complex. The overall process includes two steps, the extraction of cellulose and the production of nanocellulose. The raw material may be in the native form without having any

Table 13.1 Different types of mechanical operations for cellulose extraction.

Serial no.	Name of the treatment	Parameter of the process	Nature of the fiber	Remarks
1.	High-Pressure Homogeni-zation (HPH)	High pressure (~8,000 psi)	Reduced DP of cellulose fibers from 2720 to 740, diameters obtained between 20–100 nm and several tens of micrometers length	Homogenizer clogging, consumption of high energy and cost , crystalline microfibril damaged due to high mechanical force [17–19]
2.	Microfluidization	High pressure (~30,000 psi)	Cellulose nanofibers having length of several micrometers and diameter less than 100 nm	Involvement of high energy and high cost
3.	Grinding	1400–1500 rpm	Obtained cellulose microfiber rather than nanofiber	Do [20]
4.	Cryocrushing	Immersion of swelled cellulosic fibers in liquid nitrogen	Cellulose nanofibers having diameters between 30-80 nm	Do [21]
5.	Steam explosion	High steam pressure (14–16 bar)	Cellulose nanofibers	Do
6.	High-intensity ultrasonication	Oscillating power generated by hydrodynamic forces of ultrasound to isolate cellulose nanofibers. Power: 1000 W, frequency: 20–25 kHz and time: ~30 min, high temperatures		Use of higher feed concentration and a greater distance between probe and beaker is not suitable for fibrillation [22]. High energy consumption and high cost

earlier processing treatments or processed material like bleached Kraft pulp [24–28]. The unprocessed raw material may also involve other physicochemical or mechanical pretreatment process to remove the noncellulosic components whereas for the processed cellulose a simple preliminary steps is required to disperse or swell the cellulose followed by enzymatic hydrolysis [29].

The complete extraction of NCs most commonly involves a mechanical separation step like sonication [30, 31] fibrillation, microfluidization [32, 33], etc. followed by the enzymatic hydrolysis, leading to an intensive fragmentation of cellulose particles while reports are there without involving any posttreatment [34–37].

The enzyme used in this treatment method is based on cellobiohydrolases, which comprises of A and B types of cellulase with the ability to attach crystalline zone of cellulose chains and endoglucanases consisting of C and D types having the capacity of affecting amorphous region of cellulose [38, 39]. Cellulases are used to pretreated fibers with a dose of 0.02% concentration and produced from different cellulolytic organisms from the genera *Trichoderma*, *Clostridium*, and *Aspergillus* [40]. Though cellulase are most efficient enzyme for disintegrating cellulose but there are other enzymes like *A. xylinum* and *Trichoderma reesei* for reduction of the size of the microcrystalline cellulose. The process of enzymatic hydrolysis is very costly, including the isolation process of the enzymes and very much time consuming while considering successful hydrolysis [41].

13.4.5 Bacterial Production of Cellulose

Bacterial cellulose has the composition of plant origin cellulose with a crystalline nanofibrillar structure. This cellulose is synthesized extracellularly by some bacterial genera, like Acetobacter (most efficient), Aerobacter, Agrobacterium, Azotobacter, Salmonella, Rhizobium, etc., [42]. Bacteria from these genera can oxidize different polysaccharide-based carbon sources, like glucose, galactose, fructose [43, 44] and alcohols, such as ethanol [45], and produce cellulose pulp at 25°C to 30°C and pH of 3 to 7. The fibrillar arrangement of cellulose maybe parallel via hydrogen bonding or a three-dimensional network, depending on the synthesis environment [46]. Two main methods are used to produce BNC, i) static culture and ii) stirred/shaking culture. The white, thick, and leathery BNC pellicle is accumulated at the air-liquid interface by static culture while in shaking culture dispersed cellulose accumulated in the culture media in the form of suspended fiber or irregular pellet. The static culture is a better way to produce BNC as it leads to higher yield and mechanical properties while agitated culture leads to higher microorganism mutation probability. The main disadvantage of static culture is that it consumes longer time and a larger area for cultivation. Depending on the various method of synthesis like static, bioreactor cultures and agitated or shaking bacterial cellulose may differ in fine structure, morphology, mechanical properties. Bacterial cellulose exhibits higher thermal stability and higher critical surface tension in comparison to the chemically and mechanically extracted cellulose [47–49].

13.5 Formation of Biocomposite Using Different Form of Cellulose

The most important drawback of cellulose-based reinforcement is the presence of large number of hydroxyl groups, leading to high hydrophilicity in turn resulting weak interfacial adhesion with polymers. Hence, for proper application of cellulose fibers in composites, it needs surface modification by different treatment. Pretreatment process during extraction of cellulose in different form by different method also modifies the surface of generated cellulose fibers and also reduces the energy from 20,000 to 30,000 kWh/tonne to 1000 kWh/tonne consumed in mechanical processes. Different types of pretreatments include alkaline treatment, TEMPO (2,2,6,6-tetramethylpiperidine-1-oxyl), enzymatic pretreatment etc. The pretreatment can be either enzymatic pretreatment or chemical pretreatment. Enzymatic pretreatment produces strong synergistic effect involving cellobiohydrolases and endoglucanases. TEMPO-mediated oxidation carried out in aqueous solution of TEMPO (2,2,6,6 tetramethyl-1 piperi-dinyloxy radical)containing catalytic amounts with NaBr as a catalyst at pH ranging from 9 to 11 and at room temperature in presence of NaClO. This oxidation reaction involves conversion of C6 primary hydroxyl group of cellulose into carboxylate groups [50, 51]. Higher amount of NaClO in the reaction medium leads formation of higher number of carboxylic groups at the surface of the cellulose fibrils with higher reduction in DP [52, 53]. It generates negative charges on the surface of fibers [54] preventing postaggregation of nanoparticle during drying and without changing the desired morphology of the fiber. This facilitates separation of cellulose nanofibrils from each other overcoming the hydrogen bonding. Modification of TEMPO oxidation is reported by using 4-hydroxy derivative of TEMPO, which is less costly, and also by employing TEMPO/NaClO/$NaClO_2$ system with slightly acidic or neutral conditions [55] to avoid depolymerization or dis colouration. Other less commonly used oxidation treatments are the periodate treatment followed by chlorite oxidation and with ammonium persulfate at 60°C [56].

Surface treatment also involved alkaline treatments, bleaching for extensive hemicellulose, lignin and pectin solubilization and removal. Different temperature or treatment conditions may be employed for alkaline extraction considering the point of cellulose degradation [57, 58]. Bleaching improves ageing resistance and yellowing imparting brittleness and is carried out by the common bleaching agents like sodium chlorite ($NaClO_2$), hydrogen peroxide (H_2O_2), ozone (O_3), sulfite, chlorine dioxide (ClO_2) and peracetic acid [59]. Esterification, acetylation, halogenation, etherification are the other methods of chemical treatments to modify the hydrophilicity, moisture/water absorptivity, thermal resistance, microbial resistance etc of cellulose fiber along with higher structural stability and adsorptivity. Acetylation of NFC with acetic anhydride for 30 min at 105°C substituted 43% of hydroxyl group, i.e., degree of substitution (DS) is 0.43 resulting in an increased contact angle from 33° to 115°, reduced crystallinity as the crystalline part of cellulose degrades to some extent. Chemical treatment to the surface of BC, NCC, or NFC also prevents the inherent tendency of agglomeration of nanofibers imparting hydrophobic character and better adhesion to nonpolar polymer [60–62]. Carboxymethylation of those nanocellulose renders negatively charged surface giving more stable dispersion forming electrical double layer. This treatment, prior to mechanical treatment, leads more dispersibility and lower crystallinity [61]. Ionic liquids are another suitable agent for surface modification of cellulose fiber without having any corrosive properties, low viscosity, a melting point below 100°C, no flammability and low vapor pressure [63]. The commonly used anionic groups like $(BF_4)^-$, $(PF_6)^-$, $(CF_3CO_2)^-$, $(SbF_6)^-$, $(ClO_4)^-$, $(OTS)^-$,$(Al_2Cl_7)^-$, $(GeCl_3)^-$, and $(AlCl_4)^-$ [63, 64] dissolute cellulose by breaking the intermolecular hydrogen bonding of cellulose while the cation attached with the oxygen atom of the hydroxyl group leading to variety of morphology after precipitation [63]. Besides ionic liquids surfactants (both anionic like fluorosurfactant, e.g., perfluorooctadecanoic acid and cationic) are also used to coat the cellulose surface for improving hydrophobicity [64–68]. Grafting is another technique in which the suitable reactant molecule is covalently attached to the surface of the cellulose chain either by using suitable coupling agent or by modifying the surface reactivity with appropriate treating agent. Grafting agents like polyhydroxybutyrate, vinyl monomer, aminosilane, polyethylene glycol, maleic anhydride grafted polypropylene, etc, can modify the surface morphology, crystallinity, adhesion of fiber to the matrix [69–72]. Silylation of CNF with chlorodimethylisopro-pylsilane obtained from bleached pulp produced fiber changing DS from 0.6 to 1 while using isopropyl dimethylchlorosilane forms a shear thinning CNF suspension [73]. Isocyanates, like octadecyl isocyanate, produce NFC and CNC with degree of substitution of 0.09 and 0.07, respectively [74] via covalent bond formation with hydroxyl groups of nanocellulosics.

13.6 Biocomposites Derived from Cellulose and Their Application

Cellulose is gaining its attraction in biomedical field due to mainly its biocompatibility, low cost and absence of cytotoxicity and along with the other features already have been discussed in the earlier sections [75]. Different derivatives of cellulose are already applicable in biomedical field which is given in Figure 13.3.

Here we are focussing on biocomposites based on nanocellulose itself. Extracted cellulose in nanodimension is more useful for formation of biocomposites targeting biomedical field. Cellulose nanoparticles are mostly obtained as stable aqueous suspensions with which different hydrosoluble or hydrodispersible polymer solution is mixed thoroughly. The composite film is then formed by either solvent casting, freeze drying or hot pressing method. Different works have been reported in literature about formation of cellulose polymer

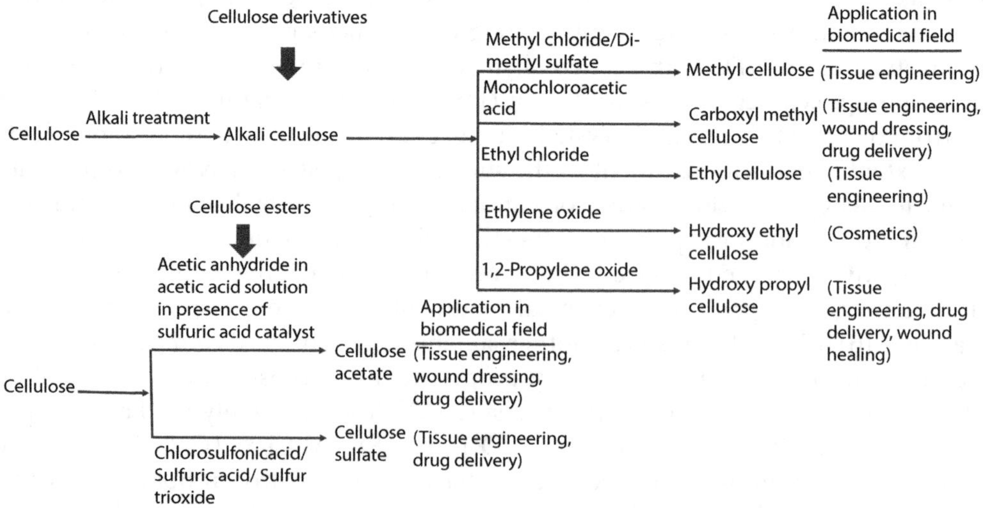

Figure 13.3 Different types of cellulose-based biomaterials and their biomedical applications.

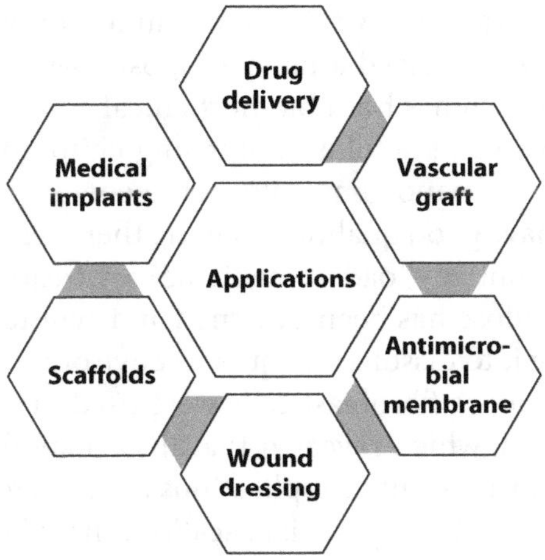

Figure 13.4 Major biomedical applications by nanocellulose derive biocomposites.

nanocomposites following these methods involving polymers like polyvinyl alcohol, starch, hydroxypropyl cellulose etc., [76–82]. On the other hand, for hydrophobic polymer non-aqueous dispersion of nanocellulose is usually prepared using surfactant or chemical grafting by reducing hydrophilicity of nanocellulose increasing interfacial adhesion between cellulose and polymer. Melt extrusion as the processing technique has also been reported when polylactic acid be the matrix [83] and also use of polyvinyl alcohol as a compatibilizer for cellulose whisker-based PLA composite is separately reported [84]. Different biocomposites based on nanocellulose is presented below according to their application as mentioned in Figure 13.4.

13.6.1 Tissue Engineering

Tissue engineering, being a multidisciplinary field, combines engineering and life science concepts to create smart biological replacements that might possibly heal, maintain, and upgrade tissue activities that have been disrupted. Biomaterials are commonly used in the tissue engineering area to create structures for medical procedures [85, 86]. The human body's live biological constituents, based on pharmacological fluids and biomolecules through cells, tissues, and organs, will be subjected to such structures. Regenerative tissue material must have ideal strength in mechanical properties for bone regeneration or elastic modulus for prosthetic blood vessels and other ligamentous repairs. The type and qualities of scaffolds for cell adhesion are also critical in tissue engineering [87]. Electrospinning fiber scaffolds, for example, replicate cytoskeleton and have a higher surface-to-volume ratio, allowing for more surface area for cell adhesion. They are used in neural tissue engineering, cardiac tissue engineering, orthopedic tissue engineering, and stem cell tissue engineering [88, 89].

In order to produce a suitable scaffold for neural tissue engineering applications, gabapentin, an antipsychotic medicine, used by Farzamfar *et al.* as an analgesic to alleviate neuropathic symptoms [90]. It was combined with cellulose acetate and gelatin as gel. The drug-loaded three-dimensional scaffolds were made using the approach of wet-electrospinning with a

1:1 cellulose acetate/gelatine ratio (by wt. %) in a solution of water/ethanol in the ratio of 3:7 (by vol%). Hasan *et al.* synthesized a nanocomposite with silver nanoparticles coated on carboxylated material, which has both biomechanical and antibacterial properties [91]. Using the freeze-drying process, scaffolds comprising chitosan and carboxymethyl cellulose with different percents of nanocomposite were generated. These tests demonstrated that constructed scaffolds have good qualities, making them self-sustaining and prospective antibacterial scaffolds (without any loaded medicine) for treating bone infections, such as osteomyelitis. Bacterial cellulose has been designed and synthesized with sulphate groups by aceto-sulfation, allowing it to dissolve in aqueous environments and be used in a variety of applications [92]. L929 mouse fibroblast cells were used in the biological assessment of BCS/PVA nanofiber scaffolds, which revealed that these nanofibers are good matrices for tissue growth. One of the most prevalent applications of biomaterials for tissue engineering is the creation of a physiologically appropriate scaffold that allows living cells to connect, proliferate, and differentiate in order to stimulate tissue healing *in vitro* as well as *in vivo*. Mammalian cells are unable to attach to the cellulosic surfaces used in simulated tissues due to its hygroscopic nature and low quasi protein adsorption [93]. The addition of matrix ligands to cellulosic materials, on the other hand, can increase cell adherence to substrate surfaces. Sajjad *et al.* developed nontmorillonite-based bacterial cellulosed composites and by showing the zone of inhibition against pathogens, they established that its use as substitute material for the purpose of wound healing, as well as scaffold for tissue engineering [94]. Hydroxyethyl cellulose is a covalent, hydrophilic polymer with a β-glucose bond, making it a good choice for tissue engineering. Tohamy *et al.* showed that this type of cellulose improves cell viability and accelerates cell proliferation significantly [95]. Because of its superior mechanical qualities and biocompatibility, nanocrystalline cellulose might be a suitable medium for cell adhesion strength. One of the benefits of using nanocrystalline cellulose is the micro-fibrils higher aspect ratio essential components that generate natural fiber networks of fibers or nanostructures kept together by hydrogen bonding along with mechanical contact [96]. The particular nanofibers might be cross-linked to strengthen the network even more mechanical means. The porous of bacterial cellulose can be well adapted to host oxygen and nutrients, creating an environment that is favourable to cell development and reproduction. Halib *et al.* reviewed the unique porosity found from its morphology [97]. Luo *et al.* showed how nanocellulose-based scaffolds may be used to repair, improve, or replace damaged tissues and organs, such as the skin, heart, ophthalmology, nerves, skeletal muscle, blood vessels liver. NC-based scaffolds are intriguing materials for tissue and organ regeneration because of their unique characteristics [98]. Because of their unique properties, NC-based scaffolds are appealing materials for tissue and organ regeneration. The importance of different types of cellulose used in tissue engineering as biocomposites are tabulated in Table 13.2.

13.6.2 Wound Dressing

The delayed tissue development and activity of local cellular proteins cause a delay in persistent wound healing. For speedier wound healing, wound dressings that offer an optimal microenvironment for tissue regeneration would be ideal. Cellulose-based composites have the tendency toward water in filtration as well as high absorption capacity of water as wound dressing materials, or encourage porosity and dryness to prevent adhesion of

Table 13.2 Different types of cellulose and their importance as derived biocomposites in tissue engineering.

Types of cellulose	Importance	References
Nanocellulose or microcellulose	Film made of nanocomposites with antibacterial effects.	[99]
Nanocellulose or microcellulose	Exceptionally porous polar and solubilized scaffolds are interlinked.	[100]
Macrofibrils/nanofibrils	To modulate cell response, specialized films with tailored scaffolding capabilities are used.	[101]
Macrofibrils/nanofibrils	Coating cellulose nanofibrils on a 3D-printed scaffold.	[102]
Bacterial cellulose	Bone tissue repair and products produced based on hydrogels.	[103]
Methyl cellulose	As a scaffold material, it was treated as thermo-responsive and has controlled cross-linking kinetics.	[104]
Carboxy-methyl cellulose	Murine bone marrow precursors inhibit osteoclast formation.	[105]
Ethyl cellulose	Nano-fibrous as well as micro-porous scaffold composite materials.	[106]
Cellulose acetate	Tissue engineering scaffolds made on hybrid scaffolds composites.	[107]
Cellulose sulphate	Biobased nanofiber systems that are both benign and bioactive.	[108]

bacterial activity. In this scenario, the cellulose substance must be shaped into sheets and utilized to heal the wounds. Making drug-loaded cellulose-based bandage is also really useful [109].

Wang *et al.* studied that copper-based bioactive glass with mesoporous nature and cellulose containing nano-fibrillation could be a great material for chronic wound [110]. In a 3D spherical culture environment of human uterine vein vascular endothelium, the Cu2+ released from the composite aerogel of nanocellulose and Cu-bioglass had a significant thrombotic impact. Nuutila *et al.* investigated the ability of hydroxymethyl-based cellulose and nano-fibrilled cellulose in the application of wound dressing [111]. They found that the carboxymethyl-cellulose hydrogel was unsuccessful, but the nanofibrillar cellulose hydrogel blocked 70% of punch-biopsy wound contraction. Remarkably, nanofibrillar cellulose did not prevent epithelialization of the intercellular spaces or cell motility from the patch when applied to separated skin grafts. Wutticharoenmongkol *et al.* synthesized nanofiber mats based on cellulose acetate loaded by gallic acid [112]. The fabrication was done by electrospinning. Total immersion, as well as transdermal diffusion method, was applied via a pig skin at slightly basic medium to identify the features of gallic acid's release propensity. The antimicrobial activity of the gallic acid-based cellulose acetate fiber mats over *Staphylococcus aureus* suggested that they may be used as wound dressing materials. Based on examination of skin regeneration and angiogenesis in burn injury skin of Sprague-Dawley rats, the effects of bacteria cellulose membrane treatment for 15 days on burn wound healing were explored by Kwak and co-workers [113]. At all time periods, the skin roughness rating in the rats with burnt skin was lower in the membrane treated group than in the gauze treated group, while the epidermis and dermis thickness and number of blood vessels were higher in the membrane treated group. The bacteria cellulose membrane has been shown to speed wound healing in the rats with burn injuries by regulating angiogenesis and connective tissue development while causing no particular harm to the liver or kidney. Biocompatible pellicle from bacterial cellulose was synthesized with the modification by incorporation of 2,2,6,6-tetrametgylpiperidine-1-oxyl radical at pH 10 in NaClO/NaBr system by Wu *et al.* to increase the biocompatibility of bacterial cellulose [114]. The silver nano-particles was also synthesized here. By in vitro cytotoxicity test with cell viability >95% the composite showed significant biocompatibility and excellent antibacterial properties of 100% and over 99% against *E. coli* and *S. aureus*, respectively. Wound healing and regeneration materials made of bacterial cellulose-montmorillonite-reinforced composites have been produced for therapeutic uses with no negative effects. The bacterial cellulose-chitosan membranes have antibacterial properties and a low cytocompatibility, making them ideal for wound treatment [115]. The different types of cellulose and their advantages in biomaterials for wound healing and skin repairing are given in Table 13.3.

13.6.3 Drug Delivery

The administration of pharmaceuticals in a specified amount to specifically targeted organs at a predetermined time is characterised as a drug delivery system. Drug delivery systems

Table 13.3 Advantages of different types of cellulose in biomaterials for wound healing and skin repairing.

Types of cellulose	Advantages	References
Micro-/nanofibrils	The donor site, which is covered by a dressing, has a higher level of epithelialization.	[116]
Bacterial cellulose	Micro-particles of hydrogel material for wound dressing.	[117]
Methyl cellulose	Hydrogel type wound dressing materials.	[118]
Ethyl cellulose	Prospective wounds healing materials based on nanofibers.	[119]
Hydroxy-propyl methyl cellulose	Hydrogel nanocomposite exhibiting wound-healing properties.	[120]
Cellulose acetate	Wound healing nanofibers made of polymeric nanofibers with antibacterial characteristics.	[121]

transport pharmaceuticals to particular organs, tissues, or cells, with the transfer mechanism adapting to environmental conditions including exposure, chemical activity, magnetic fields, temperature, pH, and fields. The drug delivery system is also favoured by its biocompatibility as well as its biodegradability [122].

Silks are biodegradable, self-assembling proteins, biocompatible that may be genetically altered to include certain chemical properties, making them useful for medication and gene delivery. The biosynthesis and use of silk-based multiblock copolymer systems and related silk protein drug delivery methods were highlighted by Numata *et al.* [123]. By directing antigen to skin-resident dendritic cells and inducing local inflammatory reactions in the skin, microneedle vaccinations imitate numerous elements of cutaneous pathogen invasion, which are linked to increased immune responses. DeMuth *et al.* have studied that microparticles by loading of vaccine composed of a silk and polyacrilic acid composite have improved vaccine release kinetics, resulting in a 10 % increase in antigen-specific T-lymphocytes, leading in greater immunity [124]. Hashemikia *et al.* synthesized silica particles modified with (3-aminopropyl) triethoxysilane for betamethasone sodium phosphate, used as an anti-inflammatory drug [125]. They investigated that even after several washing cycles, drug loaded silica particles blended with textiles treated from chitosen softener, demonstrated excellent drug transport and robust anti-bacterial activity against both *E. coli* and *S. aureus*. Godakanda and team have developed composites of ethyl cellulose and poly-vinyl pyrrolidone with the loading of naproxen [126]. Controllable drug release may be achieved by adjusting the polymer ratios. The formulations with zero-order drug release over 20 and 80 hours have also been developed by them. Jeddiand Mahkam prepared nano-carboxymethyl cellulose by carboxymethylation via nanocrystallization of cellulose, which was used in beads of hydrogels made by bilayer alginate and chitosen [127]. In comparison to other pH media, the beads demonstrated great pH sensitivity for the dexamethasone medication in pH 5.8. The findings suggest that hydrogel beads might be used as a high-potential drug delivery technology. When a precise dose or dosage type is unavailable conventionally, or if a particular dosing schedule is required, compounded medications are utilised on the verge of getting. Thoorens *et al.* monitored the dependency of strain rate design and dosing of micro-crystalline cellulose [128]. Bhandari *et al.* synthesized cellulose nanofiber-based aerogel with the loading ~19% of Bendamustinehydrochlorite drug via physical adsorption [129]. The assessment of drug release was analyzed by dialysis followed by different mathematical models. The results suggested the promising application of cellulose nanofiber-based composites in gastroretentive drug delivery system with enhanced bioavailability. Mahmoudian and Ganji created a new composite vancomycin hydrochloride release method which is depending on hydroxy propyl methyl cellulose microscopic particles with chitosan-glycerophosphate thermoresponsive hydrogel for the treatment of osteomyelitis on the local level [130]. *In vitro* release assays demonstrated that encasing the cellulose mostly in hydrogel reduced the speed but also actual dosage of vancomycin released, suggesting that the hydroxy propyl methyl cellulose microparticles entrapped in the hydrogels had better sustained release kinetics for local antibiotic treatment. The significance of different cellulose as biocomposites in drug delivery applications are focused in Table 13.4.

13.6.4 Dental Applications

Many dental along with oral tissue problems occur, such as bone in apical periodontitis, cartilage in the muscles of mastication, skin and mucosal, enamel, and tooth pulp, as well

Table 13.4 Role of different cellulose in drug delivery applications as biocomposites.

Types of cellulose	Significance	References
Bacterial cellulose	Excipient for tablets	[131]
Carboxy-methyl cellulose	Injectable complex with thermosensitive hydrogel	[132]
Ethyl cellulose	Development of altered medication dose formulations	[133]
Ethyl cellulose	Tunable medication release from a fiber mix	[134]
Cellulose acetate	Nanofibers for medication administration via the skin as patches	[135]
Cellulose sulphate	Fabrication of a sodium cellulose sulphate with chitosan microcapsule structure	[136]

as salivary glands [137]. In dental applications, metals, ceramics, polymers, and composites have been employed as tooth crowns and root restorations. Apart from having nearly identical qualities like dental enamel, pulp, or other tissues, restorative dental materials must be compatible and have a lasting connection with tooth structure, just like other biomedical intended uses. In contrast to bone and cartilage grafting, it is critical for dental nanocomposites or materials to last as long as feasible [138].

13.6.5 Other Applications

- ✓ **Fixation of the bones both inside and externally:** Chandramohan *et al.* compared the predicted flexural capacity of the natural fiber reinforced polymer composites using ANSYS solution [139]. The goal of the study was to take use of the benefits that renewable resources provide in the production of biocomposite materials made of biopolymers and natural fibers. This plate material, which was externally covered with a hybrid composite made of calcium phosphate and hydroxyapatite, employed for both internal and exterior fixing of broken bones.
- ✓ **Sockets for a prosthetic just below knee:** The hot and humid atmosphere produces thermal pain and high residual limb skin temperature owing to such stiffness between the skin as well as the prosthesis socket, so the skin being not easily breathable. Perspiration and sweating on the prosthetic socket are caused by the inadequate environment, which leads to a variety of skin problems, such as blisters, ulcerations, and other skin problems. Natural fiber as a primary prosthetic socket material, when used in place of existing materials, can reduce issues, particularly in the prosthetic and orthopaedic fields. The use of natural fiber-based polymer composites in parallel with the advancement of prosthesis socket design would alleviate the problem of thermal discomfort while also providing an environmentally favourable product alternative to the widely utilised plastic-based materials [140].

✓ **Pharmaceuticals:** A controlled, monitored method is needed to assure the delivery of the appropriate dose at the correct time, as well as to prevent the medicine from being released too early before it reaches the target spot [141]. Coating the narcotic matrix or carrier with a fluid matrix has become one of the most preferred ways in generating prolonged dosage forms, owing to its adaptable and well-designed framework that enables for a consistent sustained release.

✓ **Cardiovascular medical devices:** Heart valve transplant might be used to treat difficulties with the aortic and/or mitral valves, like significant age-related calcific aortic dissection (valve narrowing due to calcium absorption). Stents and vascular grafts are made of a variety of coated and noncoated polymers, as well as coated and noncoated metals like Ti, cellulose, polytetrafluoroethylene [142]. Furthermore, the growth of a cytocompatible cellulose-chitosan hollow tube having diameter of 4 mm with mechanical characteristics, similar with those of a human coronary artery was noted, with mechanical behavior which may be changed by altering the ratio of cellulose as well as chitosan in the composition [143].

✓ **Orthopedics:** Grafting materials help to rebuild damaged bone in addition to replacing missing bones. Bone grafting is traditionally done by transplanting natural bones from a variety of sources, including auto-grafting, allo-grafting, xeno-grafting. Bone grafts are obtained from other parts of the recipient's body, bones from a different member with same kind, then bones from such a various organisms, perhaps from an animal to a human, in that sequence [144].

✓ **Ophthalmologic applications:** Devices or substances are being used to cure visual impairment to low vision, myopia-hyperopia, blindness, astigmatism, cataracts, presbyopia, open-angle glaucoma (primary), macular degeneration relating age, diabetic retinopathy etc. in ophthalmologic treatments [145].

✓ **Oral applications:** Several dental and oral tissue difficulties arose, such as bone in apical periodontitis, cartilage in the temporomandibular joint, skin as well as oral mucosa, dentin and tooth pulp, and salivary glands. Bacterial cellulose has indeed been employed in tissue engineering in dental and oral applications, namely in the renewal of dental pulp tissue, periodontal repair, and the healing of surgical wounds of the mucosal surface [146].

13.7 Conclusion

In this chapter, the use of biocomposites based on different types of cellulose in biomedical applications is addressed. Tissue engineering, wound dressings, and skin replacements, drug delivery, dentistry and others like cardiac applications and applications in orthopaedics, ophthalmologic fields, etc. are just a few of the biomedical applications where cellulose-based biocomposites show promising potential. Various chemical derivatives based on cellulose and other alterations have been used to change certain cellulose characteristics in order to broaden the spectrum of uses, particularly in biomedical field, as discussed above. Hydroxy-ethyl cellulose along with carboxy-methyl cellulose is employed in particular because they are adaptable materials for tissue engineering. Cellulose acetate also increases patient comfort by promoting tissue repair in wound healing formulations. Entrapment is done with bacterial cellulose and cellulose nanoparticles. Regarding sustained release, carboxymethyl cellulose derivatives of ethyl, hydroxy propyl methyl, and also sulphates of cellulose are utilized in drug carriers. Regardless of the fact that cellulose substances are routinely utilized in ordinary household products, such as paper and, perhaps, textiles, they provide the qualities necessary for the growth of new biological materials.

References

1. Todros, S., Todesco, M., Bagno, A., Biomaterials and their biomedical applications: From replacement to regeneration. *Processes*, 9, 1949, 2021.
2. Hench, L.L. and Thompson, I., Twenty-first century challenges for biomaterials. *J. R. Soc. Interface*, 7, S379, 2010.
3. Cheah, C.W., Al-Namnam, N.M., Lau, M.N., Lim, G.S., Raman, R., Fairbairn, P., Ngeow, W.C., Synthetic material for bone, periodontal, and dental tissue regeneration: Where are we now, and where are we heading next? *Mater. (Basel)*, 14, 6123, 2021.
4. Kali, S., Dufresne, A., Charian, B.M., Kaith, B.S. *et al.* Cellulose-based bio- and nanocomposites: A review. *Int. J. Polym. Sci.*, Article ID 837875, 1–35, 2011.
5. Trache, D., Thakur, V.K., Boukherroub, R., Cellulose nanocrystals/graphene hybrids-A promising new class of materials for advanced applications. *Nanomaterials*, 10, 1523, 2020.
6. Salimi, S., Sotudeh-Gharebagh, R., Zarghami, R., Chan, S.Y., Yues, K.H., Production of nanocellulose and its applications in drug delivery: A critical review. *ACS Sustain. Chem.*, 7, 15800, 2019.
7. Sharma, A., Thakur, M., Bhattacharya, M., Mandal, T., Goswami, S., Commercial application of cellulose nano-composites - A review. *Biotechnol. Rep.*, 21, e00316, 2019.
8. Karimah, A., Ridho, M.R., Munawar, S.S., Adi, D.S., Damayanti, R., Subiyanto, B., Fatriasari, W., Fudholi, A., A review on natural fibers for development of eco-friendly bio-composite: Characteristics, and utilizations. *J. Mater. Res. Technol.*, 13, 2442, 2021.
9. Islam, M.T., Alam, M.M., Patrucco, A., Montarsolo, A., Zoccola, M., Preparation of nanocellulose: A review. *AATCC J. Res.*, 1, 17, 2014.
10. Kumar, V., Bollström, R., Yang, A., Chen, Q., Chen, G., Salminen, P., Bousfield, D., Toivakka, M., Comparison of nano and microfibrillated cellulose films. *Cellulose*, 21, 3443, 2014.
11. Siqueira, G., Bras, J., Dufresne, A., Cellulosic bionanocomposites: A review of preparation, properties and applications. *Polymers*, 2, 728, 2010.
12. Lavoine, N., Desloges, I., Sillard, C., Bras, J., Controlled release and long-term antibacterial activity of chlorhexidine digluconate through the nanoporous network of microfibrillated cellulose. *Cellulose*, 21, 4429, 2014.
13. Moon, R.J., Martini, A., Nairn, J., Simonsen, J., Youngblood, J., Cellulose nanomaterials review: Structure, properties and nanocomposites. *Chem. Soc Rev.*, 40, 3941, 2011.
14. Brinchi, L., Cotana, F., Fortunati, E., Kenny, J.M., Production of nanocrystalline cellulose from lignocellulosic biomass: Technology and applications. *Carbohydr. Polym.*, 94, 154, 2013.
15. Li, J., Xu, Q.H., Jin, L.Q., Research development on hydrophobic modification of cellulose nanofibrils. *Adv. Mater. Res.*, 785–786, 440, 2013.
16. Adil, S.F., Bhat, V.S., Batoo, L.M., Imran, A., Assal, M.E., Madhusudhan, B., Khan, M., Al-Warthan, A., Isolation and characterization of nanocrystalline cellulose from flaxseed Hull: A future onco-drug delivery agent. *J. Saudi Chem. Soc*, 24, 374, 2020.

17. Li, J., Wei, X., Wang, Q., Chen, J., Chang, G., Kong, L., Su, J., Liu, Y., Homogeneous isolation of nanocellulose from sugarcane bagasse by high pressure homogenization. *Carbohydr. Polym.*, 90, 1609, 2012.
18. Liu, L.L., Tian, Y.B., Tang, C.C., Zhong, C., Liu, X., Preparation of nanocrystalline cellulose from soybean dregs by acid hydrolysis followed by high-pressure homogenization. *Food Sci.*, 0, 6–10, 2011.
19. Wang, Y., Wei, X., Li, J., Wang, Q., Wang, F., Kong, L.L., Homogeneous isolation of nanocellulose from cotton cellulose by high pressure homogenization. *J. Mater. Chem. Eng.*, 1, 49, 2013.
20. Zheng, H., Production of fibrillated cellulose materials—Effects of pretreatments and refining strategy on pulp properties, vol. 74, Aalto University, PK, 2014.
21. Stelte, W. and Sanadi, A.R., Preparation and characterization of cellulose nanofibers from two commercial hardwood and softwood pulps. *Ind. Eng. Chem. Res.*, 48, 11211, 2009.
22. He, W., Jiang, X., Sun, F., Xu, X., Extraction and characterization of cellulose nanofibers from Phyllostachys nidulariamunro via a combination of acid treatment and ultrasonication. *BioResources.*, 9, 6876, 2014.
23. Frey, M.W., Electrospinning cellulose and cellulose derivatives. *Polym Rev.*, 48, 378, 2008.
24. Squinca, P., Bilatto, S., Badino, A.C., Farinas, C.S., Nanocellulose production in future biorefineries: An integrated approach using tailor-made enzymes. *ACS Sustain. Chem. Eng.*, 8, 2277, 2020.
25. Lourenço, A.F., Gamelas, J.A.F., Sarmento, P., Ferreira, P.J.T., Enzymatic nanocellulose in papermaking–The key role as filler flocculant and strengthening agent. *Carbohydr. Polym.*, 224, 115200, 2019.
26. Siqueira, G.A., Dias, I.K.R., Arantes, V., Exploring the action of endoglucanases on bleached eucalyptus kraft pulp as potential catalyst for isolation of cellulose nanocrystals. *Int. J. Biol. Macromol.*, 133, 1249, 2019.
27. Li, L., Zhuang, J., Zou, H., Pang, J., Yu, S., Partition usage of cellulose by coupling approach of supercritical carbon dioxide and cellulase to reducing sugar and nanocellulose. *Carbohydr. Polym.*, 229, 115533, 2020.
28. Cui, S., Zhang, S., Ge, S., Xiong, L., Sun, Q., Green preparation and characterization of size-controlled nanocrystalline cellulose via ultrasonic-assisted enzymatic hydrolysis. *Ind. Crops Prod.*, 83, 346, 2016.
29. Tong, X., Shen, W., Chen, X., Jia, M., Roux, J.C., Preparation and mechanism analysis of morphology-controlled cellulose nanocrystals via compound enzymatic hydrolysis of eucalyptus pulp. *J. Appl. Polym. Sci.*, 137, 1, 2020.
30. Ghafari, R., Scaffaro, R., Maio, A., Gulino, E.F., Lo Re, G., Jonoobi, M.M., Processing-structure-property relationships of electrospun PLA-PEO membranes reinforced with enzymatic cellulose nanofibers. *Polym. Test.*, 81, 106182, 2020.
31. Sacui, I.A., Nieuwendaal, R.C., Burnett, D.J., Stranick, S.J., Jorfi, M., Weder, C., Foster, E.J., Olsson, R.T., Gilman, J.W., Comparison of the properties of cellulose nanocrystals and cellulose nanofibrils isolated from bacteria, tunicate, and wood processed using acid, enzymatic, mechanical, and oxidative methods. *ACS Appl. Mater. Interfaces.*, 6, 6127, 2014.
32. de Campos, A., Correa, A.C., Cannella, D., de Teixeira, E.M., Marconcini, J.M., Dufresne, A., Mattoso, L.H.C., Cassland, P., Sanadi, A.R., Obtaining nanofibers from curauá and sugarcane bagasse fibers using enzymatic hydrolysis followed by sonication. *Cellulose.*, 20, 1491, 2013.
33. Fortunati, E., Puglia, D., Monti, M., Peponi, L., Santulli, C., Kenny, J.M., Torre, L., Extraction of cellulose nanocrystals from Phormium tenax fibres. *J. Polym. Env.*, 21, 319, 2013.
34. Tibolla, H., Pelissari, F.M., Rodrigues, M.I., Menegalli, F.C., Cellulose nanofibers produced from banana peel by enzymatic treatment: Study of process conditions. *Ind. Crops Prod.*, 95, 664, 2017.

35. Martelli-Tosi, M., Torricillas, M.D.S., Martins, M.A., de Assis, O.B.G., Tapia-Blácido, D.R., Using commercial enzymes to produce cellulose nanofibers from soybean straw. *J. Nanomater.*, 2016, 8106814, 2016.
36. Bondancia, T.J., Mattoso, L.H.C., Marconcini, J.M., Farinas, C.S., A new approach to obtain cellulose nanocrystals and ethanol from eucalyptus cellulose pulp via the biochemical pathway. *Biotechnol. Prog.*, 33, 1085, 2017.
37. Bondancia, T.J., Corrêa, L.J., Cruz, A.J.G., Badino, A.C., Mattoso, L.H.C., Marconcini, J.M., Farinas, C.S., Enzymatic production of cellulose nanofibers and sugars in a stirred-tank reactor: Determination of impeller speed, power consumption, and rheological behavior. *Cellulose.*, 25, 4499, 2018.
38. Anderson, S.R., Esposito, D., Gillete, W., Zhu, J.Y., Baxa, U., McNeil, S.E., Enzymatic preparation of nanocrystalline and microcrystalline cellulose. *Tappi. J.*, 13, 35, 2014.
39. Satyamurthy, P., Jain, P., Balasubramanya, R.H., Vigneshwaran, N., Preparation and characterization of cellulose nanowhiskers from cotton fibres by controlled microbial hydrolysis. *Carbohydr. Polym.*, 83, 122, 2011.
40. Lin, D., Li, R., Lopez-Sanchez, P., Li, Z., Physical properties of bacterial cellulose aqueous suspensions treated by high pressure homogenizer. *Food Hydrocoll.*, 44, 435, 2015.
41. Kalia, S., Boufi, S., Celli, A., Kango, S., Nanofibrillated cellulose: Surface modification and potential applications. *Colloid Polym. Sci.*, 292, 5, 2014.
42. Shoda, M. and Sugano, Y., Recent advances in bacterial cellulose production. *Biotechnol. Bioprocess Eng.*, 10, 1, 2005.
43. Sheykhnazari, S., Tabarsa, T., Ashori, A., Shakeri, A., Golalipour, M., Bacterial synthesized cellulose nanofibers. Effects of growth times and culture mediums on the structural characteristics. *Carbohydr. Polym.*, 86, 1187, 2011.
44. Ruka, D.R., Simon, G.P., Dean, K.M., Altering the growth conditions of Gluconaceto bacterxylinus to maximize the yield of bacterial cellulose. *Carbohydr. Polym.*, 89, 613, 2012.
45. Yamada, Y., Hoshino, K., Ishikawa, T., The phylogeny of acetic acid bacteria based on the partial sequences of 16S ribosomal RNA: The elevation of the subgenus Gluconoa-cetobacter to the generic level. *Biosci. Biotechnol. Biochem.*, 6, 1244, 1997.
46. Donini, ÍAN, De Salvi, D.T.B., Fukumoto, F.K., Lustri, W.R., Barud, H.S., Marchetto, R., Mes-saddeq, Y., Ribeiro, S.J.L., Biosynthesis and recent advances in bacterial cellulose production. *Eclet. Chem.*, 35, 165, 2010.
47. Santos-Ebinuma, V.C., Roberto, I.C., Simas Teixeira, M.F., Pessoa, A., Improving of red colorants production by a new Penicillium purpurogenum strain in submerged culture and the effect of different parameters in their stability. *Biotechnol. Prog.*, 29, 778, 2013.
48. Jozala, A.F., de Lencastre-Novaes, L.C., Lopes, A.M., de Carvalho Santos-Ebinuma, V., Mazzola, P.G., Pessoa, A., Jr., Grotto, D., Gerenutti, M., Chaud, M.V., Bacterial nanocellulose production and application: A 10-year overview. *Appl. Microbiol. Biotechnol.*, 100, 2063, 2016.
49. Krystynowicz, A., Czaja, W., Wiktorowska-Jezierska, A., Gonçalves-Miśkiewicz, M., Turkiewicz, M., Bielecki, S., Factors affecting the yield and properties of bacterial cellulose. *J. Ind. Microbiol. Biotechnol.*, 29, 189, 2002.
50. Fraschini, C., Chauve, G., Bouchard, J., TEMPO-mediated surface oxidation of cellulose nanocrystals (CNCs). *Cellulose*, 24, 2775, 2017.
51. Gamelas, J.A., Pedrosa, J., Lourenço, A.F., Mutjé, P., González, I., Chinga-Carrasco, G., Singh, G., Ferreira, P.J., On the morphology of cellulose nanofibrils obtained by TEMPO-mediated oxidation and mechanical treatment. *Micron*, 72, 28, 2015.
52. Li, J., Xu, Q.H., Jin, L.Q., Research development on hydrophobic modification of cellulose nanofibrils. *Adv. Mater. Res.*, 440, 785–786, 2013.

53. Anderson, S.R., Esposito, D., Gillete, W., Zhu, J.Y., Baxa, U., McNeil, S.E., Enzymatic preparation of nanocrystalline and microcrystalline cellulose. *Tappi J.*, 13, 35, 2014.
54. Kalia, S., Boufi, S., Celli, A., Kango, S., Nanofibrillated cellulose: Surface modification and potential applications. *Colloid Polym. Sci.*, 292, 5, 2014.
55. Lam, E., Male, K.B., Chong, J.H., Leung, A.C.W., Luong, J.H.T., Applications of functionalized and nanoparticle-modified nanocrystalline cellulose. *Trends Biotechnol.*, 30, 283, 2012.
56. Castro-Guerrero, C.F. and Gray, D.G., Chiral nematic phase formation by aqueous suspensions of cellulose nanocrystals prepared by oxidation with ammonium persulfate. *Cellulose*, 21, 2567, 2014.
57. Zheng, H., *Production of fibrillated cellulose materials-effects of pretreatments and refining strategy on pulp properties*, vol. 74, Aalto University, PK, 2014.
58. Siqueira, G., Tapin-Lingua, S., Bras, J., da Silva Perez, D., Dufresne, A., Morphological investigation of nanoparticles obtained from combined mechanical shearing, and enzymatic and acid hydrolysis of sisal fibers. *Cellulose*, 17, 1147, 2010.
59. Kargarzadeh, H., Ahmad, I., Abdullah, I., Dufrense, A., Zainudin, S.Y., Sheltami, R.M., Effects of hydrolysis conditions on the morphology, crystallinity, and thermal stability of cellulose nanocrystals extracted from kenaf bast fibers. *Cellulose*, 19, 855, 2012.
60. Hassan, M.L., Bagasse and rice straw nanocellulosic materials and their applications, in: *Handbook of Polymer Nanocomposites Processing, Performance and application*, vol. 47, 2015.
61. Laurén, P., *Small-animal SPECT/CT and nanofibrillar cellulose hydrogels: A preclinical evaluation of a potential novel biomaterial application*, vol. 71, University of Helsinki, Finland, 2013.
62. Wicaksono, R., Syamsu, K., Yuliasih, I., Nasir, M., Cellulose nanofibers from cassava bagasse: Characterization and application on tapioca-film. *Chem. Mater. Res.*, 3, 79, 2013.
63. Li, J., Wei, X., Wang, Q., Chen, J., Chang, G., Kong, L., Su, J., Liu, Y., Homogeneous isolation of nanocellulose from sugarcane bagasse by high pressure homogenization. *Carbohydr. Polym.*, 90, 1609, 2012.
64. Xiao, Y.T., Chin, W.L., Abd Hamind, S.B., Facile preparation of highly crystalline nanocellulose by using ionic liquid. *Adv. Mater. Res.*, 1087, 106, 2015.
65. Lazko, J., Sénéchal, T., Landercy, N., Dangreau, L., Raquez, J.M., Dubois, P., Well defined thermostable cellulose nanocrystals via two-step ionic liquid swelling-hydrolysis extraction. *Cellulose*, 21, 4195, 2014.
66. Aulin, C., Shchukarev, A., Lindqvist, J., Malmström, E., Wågberg, L., Lindström, T., Wetting kinetics of oil mixtures on fluorinated model cellulose surfaces. *J. Colloid Interface Sci.*, 317, 556, 2007.
67. Kaboorani, A. and Riedl, B., Surface modification of cellulose nanocrystals (CNC) by a cationic surfactant. *Ind. Crops Prod.*, 65, 45, 2015.
68. Eyley, S. and Thielemans, W., Surface modification of cellulose nanocrystals. *Nanoscale*, 6, 7764, 2014.
69. Kalia, S. and Sabaa, M.W., *Polysaccharide based graft copolymers*, Springer, Berlin, German, 2013.
70. Wei, L., McDonald, A.G., Stark, N.M., Grafting of bacterial polyhydroxybutyrate (PHB) onto cellulose via *in situ* reactive extrusion with dicumyl peroxide. *Biomacromolecules*, 16, 1040, 2015.
71. Tee, Y.B., Talib, R.A., Abdan, K., Chin, N.L., Basha, R.K., Md Yunos, K.F., Thermally grafting aminosilane onto kenafderived cellulose and its influence on the thermal properties of poly-(Lactic acid) composites. *BioResources*, 8, 4468, 2013.
72. Cheng, D., Wen, Y., Wang, L., An, X., Zhu, X., Ni, Y., Adsorption of polyethylene glycol (PEG) onto cellulose nanocrystals to improve its dispersity. *Carbohydr. Polym.*, 123, 157, 2015.

73. Cheng, Z.L., Xu, Q.H., Gao, Y., Research progress in nano-cellulose modification. *Adv. Mater Res.*, 627, 859, 2012.
74. Missoum, K., Modification chimique de surface de NanoFibrilles de Cellulose (NFC), 2012. available from: http://www.theses.fr./2012GRENI105/document
75. Ates, B., Koytepe, S., Ulu, A., Gurses, C., Thakur, V.K., Chemistry, structures, and advanced applications of nanocomposites from biorenewable resources. *Chem. Rev.*, 120, 9304, 2020.
76. Roohani, M., Habibi, Y., Belgacem, N.M., Ebrahim, G., Karimi, A.N., Dufresne, A., Cellulose whiskers reinforced polyvinyl alcohol copolymers nanocomposites. *Eur. Polym. J.*, 44, 2489, 2008.
77. Paralikar, S.A., Simonsen, J., Lombardi, J., Poly(vinyl alcohol)/cellulose nanocrystal barrier membranes. *J. Membr. Sci.*, 320, 248, 2008.
78. Lu, J., Wang, T., Drzal, L.T., Preparation and properties of microfibrillated cellulose polyvinyl alcohol composite materials. *Compos. Part A Appl. Sci. Manuf.*, 39, 738, 2008.
79. Dagang, L., Tuhua, Z., Peter, R.C., Kaifu, L., Qinglin, W., Starch composites reinforced by bamboo cellulosic crystals. *Bioresour. Technol.*, 101, 2529, 2010.
80. Angles, M.N. and Dufresne, A., Plasticized starch/tunicin whiskers nanocomposite materials. 2. Mechanical behavior. *Macromolecules*, 34, 2921, 2001.
81. Mathew, A.P. and Dufresne, A., Morphological investigation of nanocomposites from sorbitol plasticized starch and tunicin whiskers. *Biomacromolecules*, 3, 609, 2002.
82. Orts, W.J., Shey, J., Imam, S.H., Glenn, G.M., Guttman, M.E., Revol, J.F., Application of cellulose microfibrils in polymer nanocomposites. *J. Polym. Environ.*, 13, 301, 2005.
83. Oksman, K., Mathew, A.P., Bondeson, D., Kvien, I., Manufacturing process of cellulose whiskers/polylactic acid nanocomposites. *Compos. Sci. Technol.*, 66, 2776, 2006.
84. Bondeson, D. and Oksman, K., Polylactic acid/cellulose whisker nanocomposites modified by polyvinyl alcohol. *Compos. Part A Appl. Sci. Manuf.*, 38, 2486, 2007.
85. Ali, A., Bano, S., Poojary, S., Chaudhary, A., Kumar, D., Negi, Y.S., Effect of cellulose nanocrystals on chitosan/PVA/nano β-TCP composite scaffold for bone tissue engineering application. *J. Biomater. Sci. Polym. Ed*, 1, 1–19, 2021.
86. Chandramohan, D. and Marimuthu, K., A review on natural fibers. *IJRRAS*, 8, 194, 2011.
87. Farzamfar, S., Naseri-Nosar, M., Vaez, A., Esmaeilpour, F., Ehterami, A., Sahrapeyma, H., Samadian, H., Hamidieh, A.-A., Ghorbani, S., Goodarzi, A., Neural tissue regeneration by a gabapentin-loaded cellulose acetate/gelatin wetelectro-spunscaffold. *Cellulose*, 25, 1229, 2018.
88. Seddiqi, H., Oliaei, E., Honarkar, H., Jin, J., Geonzon, L.C., Bacabac, R.G., Klein-Nulend, J., Cellulose and its derivatives: Towards biomedical applications. *Cellulose*, 28, 1893, 2021.
89. Xue, J., Wu, T., Dai, Y., Xia, Y., Electrospinning and electrospun nanofibers: Methods, materials, and applications. *Chem. Rev.*, 119, 5298, 2019.
90. Farzamfar, S., Salehi, M., Tavangar, S.M., Verdi, J., Mansouri, K., Ai, A., Malekshahi, Z.V., Ai, J., A novel polycaprolactone/carbon nanofiber composite as a conductive neural guidance channel: An *in vitro* and *in vivo* study. *Prog. Biomater.*, 8, 239, 2019.
91. Hasan, A., Waibhaw, G., Saxena, V., Pandey, L.M., Nanobiocomposite scaffolds of chitosan, carboxymethyl cellulose and silver nanoparticle modified cellulose nanowhiskers for bone tissue engineering applications. *Int. J. Biol. Macromol.*, 111, 923, 2018.
92. Saxena, V., Hasan, A., Pandey, L.M., Antibacterial nano-biocomposite scaffolds of chitosan, carboxymethyl cellulose and Zn & Fe integrated hydroxyapatite (Chitosan-CMC-FZO@HAp) for bone tissue engineering. *Cellulose*, 28, 9207, 2021.
93. Hurtukova, K., Fajstavrova, K., Rimpelova, S., Vokata, B., Antibacterial properties of a honeycomb-like pattern with cellulose acetate and silver nanoparticles. *Materials*, 14, 4051, 2021.

94. Sajjad, W., Khan, T., Ul-Islam, M., Khan, R., Hussain, Z., Khalid, A., Wahid, F., Development of modified montmorillonite-bacterial cellulose nanocomposites as a novel substitute for burn skin and tissue regeneration. *Carbohydr. Polym.*, 206, 548, 2019.
95. Tohamy, K.M., Mabrouk, M., Soliman, I.E., Beherei, H.H., Aboelnasr, M.A., Novel alginate/hydroxyethyl cellulose/hydroxyapatitecomposite scaffold for bone regeneration: *In vitro* cell viability and proliferation of human mesenchymal stem cells. *Int. J. Biol. Macromol.*, 112, 448, 2018.
96. Song, J., Jeong, G.J., Lee, T.J., Kim, S.W., Kim, Y.J., Bhang, S.H., Delivery of extracellular matrix-enriched stem cells encapsulated with enzyme-free pH-sensitive polymer for enhancing therapeutic angiogenesis. *J. Ind. Eng. Chem.*, 104, 381, 2021.
97. Halib, N., Ahmad, I., Grassi, M., Grassi, G., The remarkable three-dimensional network structure of bacterial cellulose for tissue engineering applications. *Int. J. Pharm.*, 566, 631, 2019.
98. Luo, H., Cha, R., Li, J., Hao, W., Zhang, Y., Zhou, F., Advances in tissue engineering of nanocellulose-based scaffolds: A review. *Carbohydr.Polym.*, 224, 115144, 2019.
99. Pal, N., Dubey, P., Gopinath, P., Pal, K., Combined effect ofcellulose nanocrystal and reduced graphene oxide intopoly-lactic acid matrix nanocomposite as a scaffold andits anti-bacterial activity. *Int. J. Biol. Macromol.*, 95, 94, 2017.
100. Abraham, E., Weber, D.E., Sharon, S., Lapidot, S., Shoseyov, O., Multifunctional cellulosic scaffolds from modifiedcellulose nanocrystals. *ACS Appl. Mater. Inter.*, 9, 2010, 2017.
101. Courtenay, J.C., Deneke, C., Lanzoni, E.M., Costa, C.A., Bae, Y., Scott, J.L., Sharma, R.I., Modulating cell response on cellulose surfaces; tunable attachment and scaffold mechanics. *Cellulose*, 25, 925, 2018.
102. Rashad, A., Mohamed-Ahmed, S., Ojansivu, M., Berstad, K., Yassin, M.A., Kivijarvi, T., Heggset, E.B., Syverud, K., Mustafa, K., Coating 3D printed polycaprolactone scaffolds with nanocellulose promotes growth and differentiation of mesenchymal stem cells. *Biomacromol.*, 19, 4307, 2018.
103. Basu, P., Saha, N., Alexandrova, R., Andonova-Lilova, B., Georgieva, M., Miloshev, G., Saha, P., Biocompatibility and biological efficiency of inorganic calcium filled bacterial cellulose based hydrogel scaffolds for bone bioengineering. *Int. J. Mol. Sci.*, 193980, 2018b.
104. Niemczyk-Soczynska, B., Gradys, A., Kolbuk, D., Krzton-Maziopa, A., Sajkiewicz, P., Crosslinking kinetics of methylcellulose aqueous solution and its potential as a scaffold for tissue engineering. *Polymers*, 11, 1772, 2019.
105. Agis, H., Beirer, B., Watzek, G., Gruber, R., Effects of carboxymethyl cellulose and hydroxypropylmethyl cellulose on the differentiation and activity of osteoclasts and osteoblasts. *J. Biomed. Mater. Res. A*, 95, 504, 2010.
106. Hokmabad, V.R., Davaran, S., Aghazadeh, M., Rahbarghazi, R., Salehi, R., Ramazani, A., Fabrication and characterization of novel ethyl cellulose-grafted-poly (εcaprolactone)/alginate nanofibrous/macroporous scaffolds incorporated with nano-hydroxyapatite for bone tissue engineering. *J. Biomater. Appl.*, 33, 1128, 2019.
107. Lukanina, K.I., Grigoriev, T.E., Krasheninnikov, S.V., Mamagulashvilli, V.G., Kamyshinsky, R.A., Chvalun, S.N., Multi-hierarchical tissue-engineering ECM-like scaffolds based on cellulose acetate with collagen and chitosan fillers. *Carbohydr. Polym.*, 191, 119, 2018.
108. Palaninathan, V., Raveendran, S., Rochani, A.K., Chauhan, N., Sakamoto, Y., Ukai, T., Maekawa, T., Kumar, D.S., Bioactive bacterial cellulose sulfate electrospun nanofibers for tissue engineering applications. *J. Tissue. Eng. Regen. Med.*, 12, 1634, 2018.
109. Henschen, J., Illergard, J., Larsson, P.A., Ek, M., Wagberg, L., Contact-active antibacterial aerogels from cellulose nanofibrils. *Colloids Surf. B*, 146, 415, 2016.
110. Wang, X., Cheng, F., Liu, J., Smatt, J.-H., Gepperth, D., Lastusaari, M., Xu, C., Hupa, L., Biocomposites of copper-containing mesoporous bioactive glass and nanofibrillated cellulose:

Biocompatibility and angiogenic promotion in chronic wound healing application. *Acta Biomater.*, 46, 286, 2016.

111. Nuutila, K., Laukkanen, A., Lindford, A., Juteau, S., Nuopponen, M., Vuola, J., Kankuri, E., Inhibition of skin wound contraction by nanofibrillar cellulose hydrogel. *Plast. Reconstr. Surg.*, 141, 357e, 2018.
112. Wutticharoenmongkol, P., Hannirojram, P., Nuthong, P., Gallic acid-loaded electrospun cellulose acetate nanofibers as potential wound dressing materials. *Polym. Adv. Technol.*, 30, 1135, 2019.
113. Kwak, M.H., Kim, J.E., Go, J., Koh, E.K., Song, S.H., Son, H.J., Kim, H.S., Yun, Y.H., Jung, Y.J., Hwang, D.Y., Bacterial cellulose membrane produced by Acetobacter sp. A10 for burnwound dressing applications. *Carbohydr. Polym.*, 122, 387, 2015.
114. Wu, C.-N., Fuh, S.-C., Lin, S.-P., Lin, Y.-Y., Chen, H.-Y., Liu, J.-M., Cheng, K.-C., Tempo-oxidized bacterial cellulosepellicle with silver nanoparticles for wound dressing. *Biomacromol.*, 19, 544, 2018.
115. Squinca, P., Berglund, L., Hanna, K., Rakar, J., Junker, J., Khalaf, H., Farinas, C.S., Oksman, K., Multifunctional ginger nanofiber hydrogels with tunable absorption: The potential for advanced wound dressing applications. *Biomacromolecules*, 22, 3202, 1–16, 2021.
116. Hakkarainen, T., Koivuniemi, R., Kosonen, M., Escobedo-Lucea, C., Sanz-Garcia, A., Vuola, J., Valtonen, J., Tammela, P., Makitie, A., Luukko, K., Nanofibrillar cellulose wound dressing in skin graft donor site treatment. *J. Control Release*, 244, 292, 2016.
117. Pandey, M., Mohamad, N., Low, W.-L., Martin, C., Amin, M.C.I.M., Microwaved bacterial cellulose-based hydrogel microparticles for the healing of partial thickness burnwounds. *Drug Delivery Transl. Res.*, 7, 89, 2017.
118. Suliwarno, A., Hydrogel based on crosslinked methylcelluloseprepared by electron beam irradiation for wound dressing application. *Indones. J. Chem.*, 14, 262, 2014.
119. Li, H., Zhang, Z., Godakanda, V.U., Chiu, Y.-J., Angkawinitwong, U., Patel, K., Stapleton, P.G., de Silva, R.M., de Silva, K.N., Zhu, L.-M., The effect of collection substrate on electrospun ciprofloxacin-loaded poly (vinylpyrrolidone) andethyl cellulose nanofibers as potential wound dressing materials. *Mater. Sci. Eng. C*, 104, 109917, 2019a.
120. Bianchi, S.E., Machado, B.E., da Silva, M.G., da Silva, M.M., DalBosco, L., Marques, M.S., Horn, A.P., Persich, L., Geller, F.C., Argenta, D., Coumestrol/hydroxypropyl-β-cyclodextrin association incorporated in hydroxypropylmethylcellulose hydrogel exhibits wound healing effect: *In vitro* and *in vivo* study. *Eur. J. Pharm. Sci.*, 119, 179, 2018.
121. Teixeira, M.A., Paiva, M.C., Amorim, M.T.P., Electrospun nanocomposites containing cellulose and its derivatives modified with specialized biomolecules for an enhanced wound healing. *Nanomaterials*, 10, 557, 2020.
122. Yan, H., Chen, X., Feng, M., Shi, Z., Zhang, W., Wang, Y., Ke, C., Lin, Q., Entrapment of bacterial cellulose nanocrystals stabilized pickering emulsions droplets in alginate beads for hydrophobic drug delivery. *Colloids Surf. B*, 177, 112, 2019.
123. Numata, K. and Kaplan, D.L., Silk-based delivery systems of bioactive molecules. *Adv. Drug Deliv. Rev.*, 62, 1497, 2010.
124. DeMuth, P.C., Min, Y., Irvine, D.J., Hammond, P.T., Implantable silk composite microneedles for programmable vaccine release kinetics and enhanced immunogenicity in transcutaneous immunization. *Adv. Healthc. Mater.*, 3, 47, 2014.
125. Hashemikia, S., Hemmatinejad, N., Ahmadi, E., Montazer, M., A novel cotton fabric with anti-bacterial and drug delivery properties using SBA-15-NH2/polysiloxane hybrid containing tetracycline. *Mater Sci. Eng. C.*, 59, 429, 2016.

126. Godakanda, V.U., Li, H., Alquezar, L., Zhao, L., Zhu, L.-M., de Silva, R., de Silva, K.N., Williams, G.R., Tunable drug release from blend poly (vinyl pyrrolidone)-ethyl cellulose nanofibers. *Int. J. Pharm.*, 562, 172, 2019.
127. Jeddi, M.K. and Mahkam, M., Magnetic nano carboxymethylcellulose-alginate/chitosan hydrogel beads as biodegradable devices for controlled drug delivery. *Int. J. Biol. Macromol.*, 135, 829, 2019.
128. Thoorens, G., Krier, F., Leclercq, B., Carlin, B., Evrard, B., Microcrystalline cellulose, a direct compression binder in a quality by design environment - A review. *Int. J. Pharm.*, 473, 64, 2014.
129. Bhandari, J., Mishra, H., Mishra, P.K., Wimmer, R., Ahmad, F.J., Talegaonkar, S., Cellulose nanofiber aerogel as apromising biomaterial for customized oral drug delivery. *Int. J. Nanomed.*, 12, 2021, 2017.
130. Mahmoudian, M. and Ganji, F., Vancomycin-loaded HPMC microparticles embedded within injectable thermosensitive chitosan hydrogels. *Prog. Biomater.*, 6, 49, 2017.
131. Guimard, N.K., Gomez, N., Schmidt, C.E., Conducting polymers in biomedical engineering. *Prog. Polym. Sci.*, 32, 876, 2007.
132. Dai, L., Liu, R., Hu, L.-Q., Wang, J.-H., Si, C.-L., Self-assembled PEG-carboxymethyl cellulose nanoparticles/α-cyclodextrin hydrogels for injectable and thermosensitive drug delivery. *RSC Adv.*, 7, 2905, 2017.
133. Wasilewska, K. and Winnicka, K., Ethylcellulose–a pharmaceutical excipient with multidirectional application in drug dosage forms development. *Materials*, 12, 1–21, 2019.
134. Li, C., Wang, Z.-H., Yu, D.-G., Williams, G.R., Tunable biphasic drug release from ethyl cellulose nanofibers fabricated using a modified coaxial electrospinning process. *Nanoscale Res. Lett.*, 9, 258, 2014.
135. Wsoo, M.A., Shahir, S., Bohari, S.P.M., Nayan, N.H.M., Abd Razak, S.I., A review on the properties of electrospun celluloseacetate and its application in drug delivery systems: A new perspective. *Carbohydr. Res.*, 491, 107978, 2020.
136. Xie, Y.-L., Wang, M.-J., Yao, S.-J., Preparation and characterizationof biocompatible microcapsules of sodium cellulose sulfate/chitosan by means of layer-by-layer self-assembly. *Langmuir*, 25, 8999, 2009.
137. Khalid, M.Y., Rashid, A.A., Arif, Z.U., Ahmed, W., Recent advances in nanocellulose-based different biomaterials: Types, properties, and emerging applications. *J. Mater. Res. Technol.*, 14, 2601, 2021.
138. Nejatian, K., Zafar, M.S., Najeeb, S. *et al.*, Dental biocomposites, in: *Biomaterials for oral and dental tissue engineering*, pp. 65–84, 2017.
139. Chandramohan, D. and Marimuthu, K., Characterization of natural fibers and their application in bone grafting substitutes. *Acta Bioeng. Biomech.*, 13, 77, 2011.
140. Nurhanisah, M.H., Saba, N., Jawaid, M., Paridah, M.T., Design of prosthetic leg socket from kenaf fibre based composites, in: *Green biocomposites: Green energy and technology*, M. Jawaid, M. Salit, O. Alothman (Eds.), Springer, Cham, 2017.
141. Abdul, K.H.P.S., Bhat, A.H., Abu, B.A., Tahir, P.M., Zaidul, I.S.M., Jawaid, M., Cellulosic nanocomposites from natural fibers for medical applications: A review, in: *Handbook of polymer nanocomposites. processing, performance and application*, pp. 475–511, Springer, Berlin Heidelberg, 2015.
142. Lindman, B.R., Clavel, M.A., Mathieu, P., Iung, B., Lancellotti, P., Otto, C.M., Pibarot, P., Calcific aortic stenosis. *Nat. Rev. Dis. Primers*, 2, 16006, 2016.
143. Azevedo, E.P., Retarekar, R., Raghavan, M.L., Kumar, V., Mechanical properties of cellulose: Chitosan blends for potential use as a coronary artery bypass graft. *J. Biomater. Sci. Polym. Ed.*, 24, 239, 2013.

144. Sheikh, Z., Sima, C., Glogauer, M., Bone replacement materials and techniques used for achieving vertical alveolar bone augmentation. *Materials*, 8, 2953, 2015.
145. Sharip, N.S., Yasim-Anuar, T.A.T., Norrrahim, M.N.F., Shazleen, S.S. *et al.*, A review on nanocellulose composites in biomedical application, in: *Composites in Biomedical Applications*, pp. 178–179, Taylor & Francis Group, LLC, 2020.
146. Abou Neel, E.A., Chrzanowski, W., Salih, V.M., Kim, H.W., Knowles, J.C., Tissue engineering in dentistry. *J. Dent.*, 42, 915, 2014.

14

Biobased Materials for Biomedical Engineering

Ioana Duceac[1], Fulga Tanasă[1], Mărioara Nechifor[1] and Carmen-Alice Teacă[2]*

[1]"Petru Poni" Institute of Macromolecular Chemistry, Polyaddition and Photochemistry Department, Iasi, Romania

[2]"Petru Poni" Institute of Macromolecular Chemistry, Center for Advanced Research in Bionanoconjugates and Biopolymers, Iasi, Romania

Abstract

Improving the human health and healthcare system by means of biomedical engineering (BME) implies the use of principles of modern engineering and intimate top knowledge of the human body in order to solve current or exceptional problems in medicine. One particular direction of research and development is the employment of biobased materials in advanced devices, materials, and therapeutic approaches, given their unique combination of properties. Thus, using a wise selection of natural polymers (*per se* and modified) and synthetic polymers in various combinations (blends and composites), new biomaterials have been obtained. Their physical and biochemical characteristics, especially their ability to trigger an identical natural response in the human organism, made them fit for applications, including tissue engineering, artificial organs, implants, and auxiliary materials. Some of the recent most relevant reports in the field will be summarized in this chapter, along with highlights concerning the prospective of these materials in BME applications.

Keywords: Biobased materials, natural polymers, synthetic polymers, biomedical engineering, tissue engineering, artificial organs, implants, auxiliary materials

14.1 Introduction

In its striving to close the gap between engineering and life sciences, with an emphasis on medicine and healthcare, the biomedical engineering (BME) brings together the advantages of engineering (from computer-assisted design to all specific skills) and knowledge in medical and biological sciences, and focus them on the advance in healthcare by solving problems in diagnosis, therapy, monitoring, prosthesis, rehabilitation, and all related issues. Being an eminently interdisciplinary field, its evolution is strongly dependent on the trends in basic sciences, while the research and development cover a wide variety of applications. Thus, highly appreciated achievements include imaging equipment, theranostic devices, biocompatible and/or bioresorbable implantables, grafts and organs, various prostheses, regenerative tissues, drugs with targeting ability and controlled discharge, etc.

**Corresponding author*: cateaca14@yahoo.com

Inamuddin and Tariq Altalhi (eds.) Handbook of Bioplastics and Biocomposites Engineering Applications: 2nd Edition, (275–298) © 2023 Scrivener Publishing LLC

Biobased materials consist of polymers derived from different renewable resources (biopolymers) [1–8], and they can be classified into three types taking into consideration their character and generation: (a) polymers of natural origin, such as polysaccharides (alginate, hyaluronic acid, cellulose, lignin, starch, chitosan, etc.), proteins (collagen, gelatin, fibroin, zein, etc.); (b) synthetic polymers obtained from biobased monomers (*e.g.*, itaconic acid, citric acid, succinic acid, etc.) by different polymerization methods; (c) bacterial biopolymers (polysaccharides, polyamides and poly(amino acid)s, polyesters—*e.g.*, polyhydroxyalkanoates, polyphosphates, DNA) obtained by enzymatic processes. For illustrative purpose, there has to be mentioned examples as the following: osteoinductive bone filling nanocomposites based on silk fibroin and strontium-doped hydroxyapatite nanoparticles [9], sericin and graphene oxide [10], collagen-chitosan, and calcium aluminate [11] or on catechol-chitosan matrix and imidazole-modified zeolite nanoparticles [12]; organs and grafts prepared by bioprinting using bioinks based on different polymers: photocurable chitosan [13], collagen-hyaluronic acid–containing gelatin nanoparticles [14] or poly(lactic acid) (PLA) combined with gelatin-alginate hydrogels [15]; injectable thermosensitive nanofibers made of chitosan and TEMPO-oxidized cellulose [16]; biomedical implants obtained from biomaterials having multiscale hierarchical architectures, such as nanocomposites based on hydroxyapatite/chitosan/collagen and boron nitride nanoparticles deposited onto biomedical alloys [17]; high-performance bioelectronic devices have been obtained from nanostructured films made of chitosan and silk nanofibrils [18, 19]; tridimensional scaffolds with regenerative properties made of calcium phosphate cements and different biopolymers (alginate, collagen, chitosan, gelatin, hyaluronic acid, cellulose derivatives, PLA, starch) [20].

The first and foremost issue to be considered when it comes to biobased materials used in BME is their biocompatibility, concretely the interaction with the living tissues and their response to the presence of these exogenous materials. The desired outcome includes not only the lack of any reaction to the inflammation process, but also an effective interoperability between the biomaterial and the host tissue/organ, a dynamic transfer of mass and energy, as well as the capacity to fully transform the biodegradable implantable devices into nontoxic, easy-to-eliminate by-products, or even the possibility to retrieve some biocompatible, but nonbiodegradable, devices used in imaging investigations, drug delivery and theranostic.

For example, tissue engineering employs scaffolds that have to host living cells and provide the nutrients and growth factors, and gas transfer required for cell multiplication and differentiation [21–23]. This chain of reactions starts with the adsorption of recognition proteic substances onto the surface of the implanted material. It is considered that this phenomenon triggers cellular adhesion, followed by proliferation, migration, and differentiation, finally yielding in a new, healthy, functional tissue identical to the host. In this regard, the literature reports indicated that biopolymers granted an enhanced control over the cellular response as compared to the synthetic ones, but biomaterials obtained combining both types of polymers benefited from their cumulated advantageous features [24].

In the case of biocompatible, but nonbiodegradable, implantable devices, organs, or grafts, the first response of the organism is the specific inflammatory reaction, which can last 3 to 15 days, when specialized cells (leukocytes, monocytes) and substances (growth factors, cytokines) act in synergy to protect the host from the intrusive material. The intensity of the reaction to any inflammation process is strongly related to the biocompatibility of the implantable material and the characteristics of the new interphase formed at the host/implantable interface [25, 26]. The next phase is the chronic reaction, which implies the presence of lymphocytes and macrophages, along with the occurrence of new blood vessels that will ensure the transfer of gases, nutrients and metabolites, growth factors, thus enabling the healing, regeneration and functionality of the implant.

Still, biocompatibility represents more than biosafety, lack of mutagenic and carcinogenic potential, low—or absent—inflammatory response. It also requires the ability of the biomaterial to imitate the natural cells microenvironment and regulate the local synthesis of the factors regulating growth processes in order to provide cells appropriate conditions for normal adhesion, migration, proliferation, differentiation and functioning [27, 28].

Last, but not least, the financial aspects must be addressed as well. BME heavily relies on advanced research in the area of materials derived from biological sources, which, in return, comes with high production costs. Nevertheless, the societal demand for biocompatible materials for tissue engineering, artificial organs, implants and auxiliaries is increasing, as indicated by market shares. A recent market report on the evolution of the medical engineered materials [29] indicated that the market size of these materials in 2020 was estimated at USD 15.86 billion, and the forecast for revenue value in 2028 was estimated at USD 44.62 billion, which means a tripling of this market. Another interesting aspect was also revealed: the projection for the development in Europe and North America indicated a steady growth, while the prognosis for the Asia-Pacific market showed an increase higher than any other. Still, dedicated investment instruments, the change in the consumers lifestyle, the increasing interest in minimally invasive procedures, government involvement in R&D and healthcare regulations may all contribute to the positive dynamics of this domain.

14.2 Biomaterials

Biomaterials may represent:

- materials derived from biological renewable resources [1, 2, 8] with fast growing development (such living systems transform the carbon dioxide emissions from the surrounding environment into biopolymers and other compounds—*e.g.*, sugars, lipids—all these being easily converted to high-value materials)
- man-made materials used in medical devices [3, 5–7] that interact with biological systems without harmful effects, while performing, augmenting, or replacing a natural function

Usually, biomaterials are employed and present versatility for many medical applications as follows:

- benign functionality (as in the case of heart valves);
- bioactivity in relation with a significant functional interaction (*e.g.* implants for hips based on hydroxyapatite coating);
- every day employment (*e.g.*, dentistry, surgery, drug delivery);
- suitability for transplantation (*e.g.*, autograft, allograft, or xenograft types)

Biomaterials, based on biopolymers or metals, are now used routinely in diverse biomedical applications—from contact lenses to artificial hearts. Biomaterials should exhibit a biocompatibility behavior under exposure to different chemical and physical environments as these are inherently present in the biomedical applications, this feature being related to their nontoxic action (*e.g.*, "smart" drug delivery systems) and suppressing the immune responses.

Generally, polymers derived from biological sources are employed to a large extent in applications, such as biomedical engineering, and drug industry. Some examples of such biopolymers and their possible potential applications are presented in Table 14.1.

Table 14.1 Examples of biopolymers and their application in biomedical engineering.

Biopolymer	Application	References
Alginates	Drug delivery, cell encapsulation and tissue engineering (scaffold)	[30]
Cellulose	Drug delivery and cell encapsulation	[31]
Chitosan	Orthopedic implants, orthodontic microimplants	[32–34]
Hyaluronates	Ophthalmologic and visco-supplementation compositions (e.g., those employed in treatment of the knee affected by osteoarthritis)	[35]
Collagen	Wound dressing and tissue recovery	[33, 36]
Peptides	Drug delivery	[37]
Polyamides	Hydrogels and blends as drug carriers [e.g. γ-PGA a homopolyamide-poly(γ-D-glutamic acid) composed of glutamic acid units, D- and/or L-type, linked by amide bonds type]	[38, 39]
Polyesters—PHAs polyhydroxyalkanoates	Regenerative medicine, engineering of implants and tissues (A significant example are sutures produced using some legally approved PHA-based products.)	[40]
Polyphosphates (PolyP)	Regenerative medicine and drug delivery	[41]
Dextran	Drug delivery, tissue engineering (scaffold), vascular applications	[42]
Xanthan	Drug delivery	[43, 44]

14.3 Biobased Materials for Implants and Tissue Engineering

Tissue engineering (TE) is an interdisciplinary domain of investigation aiming at obtainment new bioartificial systems and the development of new technologies, devices and materials capable to successfully interact with the biological, living tissues and organisms. From a classic perspective, a tissue engineering product is composed of: polymeric matrix, cells and signal molecules. The efforts invested in this area aim at regenerating a damaged tissue or organ, or substitute it from a structural and/or functional point of view. This can be achieved by growing cells (stem, from donor, from the patient, etc.) into biodegradable 3D matrices. The latter enables proper growth and development of the artificial graft tissue and maintains normal metabolic activity until implantation. Another studied strategy is that of inserting an acellular graft, which is populated with cells postimplantation from the surrounding tissue. Ideally, there is a corresponding degradation rate of the artificial matrix that occurs during the regeneration/emergence of the neoformation tissue (at the same rate with the degradation), leading to complete integration into the host organism.

On average in Europe, in each year, 41,000 patients are considered for a transplant, while new patients (around 48000) are recorded on the schedule lists. Each hour, approx. six new patients are registered for a transplant necessity. The top 3 organs that are transplanted are kidney (63%), liver (24%), and heart (6%). Thousands of patients are suffering due to the chronic shortage of biological components, namely organs, tissues, and cells. The result of the long, increasing waiting lists is the loss of patients—in Europe, approximately 21 patients died every day in 2020, in relation to transplant necessities [45]. There are several physical, chemical and biological parameters taken into consideration when designing a TE polymeric matrix. Therefore, after the contact with the wound bed, for example, the implant must adhere to the host tissue.

Biomaterials, according to the homonym journal, are defined as materials designed to be employed *per se* or as component of some multicomponent systems in order to regulate the course of healing or diagnosis actions through adjusting the interplays with cells of living bodies [46]. Biomaterials used for implants and TE can be classified as described in Figure 14.1.

For therapeutic purposes, polymers must meet multiple conditions, ensuring the intended physiological effect, good tolerance and diminished side effects. Fundamentally, the polymeric material must meet structural and purity conditions according to a specific purpose, so that it enables the intended effect, doesn't accumulate in the body, doesn't modify structurally while implanted, can resist sterilization procedures without degrading. Depending on the implantation site, new conditions emerge. These concern the capacity of

Biomaterials

Chemical nature
- Polymeric
- Metallic
- Ceramic
- Composite

Physical shape
- Injectable fluid
- Film
- Tubes
- Fibers
- Gels

Behavior in biological environment
- Resorbable
- Partially resorbable
- Non resorbable

Figure 14.1 Categories of biomaterials for implants and tissue engineering.

providing necessary mechanical and physic-chemical properties, such as elasticity/rigidity, reversible or irreversible complex formation with proteins or other biochemical components, surface properties and adhesion, etc. Last but not least, there are conditions to the polymeric materials used in TE relevant for their biological interactions: ensuring biocompatibility and hemocompatibility, lack of accumulation in tissues, absence of allergic, toxic or carcinogenic side effects [47].

Biocompatibility is an all-present aspect in TE applications. It can be approached on the surface or in bulk and the interaction between biomaterial and living beings can be addressed on 3 levels: interaction with proteins, with enzymes or with cells. It is not to be understated/minimized the effect of the biological environment on the polymeric matrix: fibrous encapsulation, biodegradation (may be enzymatic or due to the action of acids or peroxides generated by inflammatory cells, such as neutrophils, or the specific medium pH of the tissue/organ). One of the most important phenomena that take place at the surface of an implant is the protein adsorption, which determines the evolution of the biological reaction and the fate of the implant in the organism (thus being rejection, encapsulation, or integration). The contact with the blood is also of great interest since it can indicate how the artificial matrix can promote coagulation or absence of hemostasis, blood cell lesion or other phenomena. In addition, biocompatibility is determined by the type of the biopolymeric materials, the kind of tissue exposed to the implant and the duration of the exposure.

Another key factor in designing materials for TE is the architecture, which must ensure structural similarity and integration for the survival and development of cells, along with permeability or porosity for gas and metabolite exchange.

14.3.1 Skin Tissue Engineering and Wound Dressings

Skin is the body organ with the highest weight and its major role is the protection of the inner structures: muscles and organs. The major skin pathologies that are addressed by TE are burns, chronic wounds, ulcers, and traumatic injuries. Particularly much attention has been focused on treating burns due to the major impact that it has on the organism and the difficult, impaired healing that often occurs, and the high risk of infection associated with the loss of the natural protection barrier.

During the past three decades, many types of burn wound dressings have been developed. Their efficiency is much increased as a consequence of the advancement in knowledge about the regeneration process and the optimal conditions that enable high quality healing. Therefore, there have been studied and commercialized: permeable or barrier films, hydrogels, hydrocolloids, foams, silicone nonwoven, tissue sealants, alginate or collagen matrices, dressings loaded with silver ions, antibiotics or anti-inflammatory agents. For example, biobased semipermeable film wound dressings can be obtained using hyaluronic acid (Seprafilm®), regenerated oxidized cellulose (Interceed®), poly(lactic-co-caprolactone) (Topkin®, Oprafol®, Mesofol®), chitin and chitosan [48]. Hydrocolloid wound dressings are usually made of carboxymethyl cellulose or based on various types of proteins, such as collagen, gelatin, or pectin [49, 50].

Alginate is another biopolymer of interest for wound dressing applications. It can be processed as foam, sponge, hydrogel and it has the advantage of rapid cross-linking under mild conditions and the presence of calcium ions. Hence, numerous alginate-based dressings are commercially available [51]. The wound dressings can be classified according to the interaction with the biological environment:

- *Passive dressing*—represent only a protective layer over the wound;
- *Active dressing*—which are permeable to water, vapor, oxygen and impermeable to bacteria (transparent polymeric films, hydrogels, foams);
- *Bioactive dressings*—release of biologically active agents for healing (hydrocolloids, hydrogels, alginate, collagen and chitosan-based dressings).

Tissue substituents are another class of materials for skin treatment. They are used for a better regeneration of the damaged tissue, ease of pain, faster wound closure, an esthetic, functional de novo tissue, without keloid scars. In the case of extensive burns, they ensure a higher survival chance.

There are various types of skin substituents, most of which include collagen (structural support and cell adhesion), fibronectin (cell adhesion and proliferation), or growth factors to ensure successful integration with the surrounding tissue [52]. Some even enclose cells, a fact which accelerates the biological response and the healing. These are the most effective but they have specific indications of use. Bioactive tissue substituents may be of various natures as illustrated in Figure 14.2.

Temporary substituents are recommended for the use in superficial burn wounds or just for protection of the wounds until grafting. Permanent substituents replace the damaged skin, similar to natural grafts. In developing such systems there are two approaches: two-layer substituent (the interior layer integrated into the dermal tissue and the exterior layer, synthetic, is replaced with autograft or seeded with epithelial cells) or bulk matrix as a replacement for either the dermis or the epidermis. Ideally, bilayer permanent substituents are easily and safely applied, adhere to the wound bed, are readily available, are structurally similar to the skin, have fast integration, present low to no risk of infection, lead to good and cosmetic results, and have a relatively low cost. Many substituents are available commercially and in clinic:

- Apligraf and OrCel are allografts composed of two layers of collagen matrix or sponge seeded human keratinocytes and fibroblasts; they are FDA approved for the healing of injuries caused by persistent ulcer disease;
- Epicel is a one-layer autograft made of autologous keratinocytes, designed to replace the epidermis;

Bioactive tissue substituents

Composition

- Natural tissue (alograft, xenograft)
- Synthetic-biologic composite;
- Dermal equivalents based on collagen (Integra);
- Artificial tissue from *in vitro* cell culture.

Use indications

- Temporary
- Permanent

Figure 14.2 Types of bioactive tissue substituents.

- Integra is one of the first synthetic TE products, with a sandwich structure composed of an outer silicone layer and a biocompatible matrix based on collagen and glycosaminoglycans;
- GRAFTSKIN (Organogenesis Inc. Canton, USA) is a collagen network seeded with fibroblasts and covered with keratinocytes;
- TERUDERMIS (Teruneo Corporation, USA) is a sandwich-like composite made of fibrous collagen layer obtained by freeze-drying, coated with silicone;
- BIOBRANE® (Dow B.Hiekaw Inc.)—knitted nylon scaffold impregnated with porcine collagen, coated with a semipermeable silicone membrane. The material resulted is flexible, with high water vapor permeation, transparent, enabling the wound monitorization. The presence of the collagen limits the contracture of the wound.

The bioartificial skin is a product obtained by a particular technology: biopsy of fibroblasts and keratinocytes from healthy tissue followed by cell proliferation in a support matrix. There are three important aspects concerning this type of system: it must be completely biodegradable, it must ensure the development of new tissue, and it has to be stable under the considered conditions.

The polymer matrix can be of various natures:

- *Proteins*: collagen, gelatin, fibroin, fibrinogen, albumin, elastin;
- *Polysaccharides*
  - Animal-derived polysaccharides: chondroitin sulfate, hyaluronic acid, heparin chitosan;
  - Plant-derived polysaccharides: cellulose and its derivatives, starch, alginate, guar gum;
  - Bacterial-derived polysaccharides: pullulan, dextran, gellan gum, xanthan gum;
- *Synthetic polymers*: acrylic acid and its derivatives, silicone, polylactic acid PLA, polycaprolactone PCL, polyvinyl alcohol PVA, polyurethane PUR.

An interesting strategy is combination the specific features of polymer components, both natural and synthetic, aiming at biocompatible structures, but also with good physical and chemical properties, e.g., chitosan and acrylates [53, 54], PVA and cellulose or pullulan [55], leading to high water retention volumes and good *in vitro* biocompatibility.

14.3.2 Bone Tissue Engineering

Bone tissue is composed of a composite extracellular matrix (ECM) based mostly on collagen and hydroxyapatite, and a series of specific cells whose activity is that of remodeling the bone (osteocytes, osteoblasts). Bones have three major functions in the body: mechanical support for muscles, protection of organs and bone marrow, and ion source/deposit for the well-functioning of the organism. Some mechanical characteristics of native bone tissue and artificial bone substitutes are given in Table 14.2 for comparison purposes.

From a clinical perspective, the major problems that could have a TE solution include traumatic injuries, bone cancer, and congenital defects. Bone tissue is the second most

Table 14.2 Properties of native human bone tissue and artificial bone substitute.

Property / Material	Mechanical resistance (MPa)	Elastic modulus (MPa)	Young modulus (N/m^2)	Ref.
Cortical bone tissue				
Compression	180 ± 36	18.6 ± 28.8		
Elongation	151 ± 28	(* 10^3)		
Torsion	65 ± 9	7.1 ± 28.2 (* 10^3)		
Cancellous bone tissue				
Compression	1.5-12	40-400		
Composites				
Biomimetic composites made of chitosan/collagen/hyaluronic acid and calcium phosphates, SPIONs			70-275	[56]
Biomimetic composites of hydroxyapatite and star-shaped poly(2,2-dimethyl trimethylene carbonate)	1.21 (compression)	360		[57]

transplanted tissue with 643,000 annual procedures and a cost of ~$600 mil/year only in the USA. Despite the fact that the graft can be autologous, allogenic or synthetic, there are limitations regarding the graft failure, the availability at the donor site and the need of two surgical procedures for one graft (excision and implantation).

In this context, TE strategies have been investigated aiming at producing materials similar to autologous grafts, thus including cells and growth factors embedded into a collagenic mineralized matrix. The latter has to be molded into a shape similar to the area that needs to be replaced. Moreover, a material with microporous structure has more exposed surface favoring cell adhesion and development of new tissue [58].

In addition to biocompatibility and biodegradability, the polymeric support for the artificial bone should stimulate the cellular functions (modified surface with bioactive factors, drugs or genes) and must exhibit appropriate mechanical resistance and degradation rate under exposure to enzymes, cells and other factors of the biological environment.

A successful strategy has been the biomimetic production of the scaffold, based on collagen, or other biobased polymers, such as chitosan, albumin, hyaluronic acid, and gelatin, followed by *in situ* precipitation of calcium phosphates, also considering the addition of nanoparticles with magnetic properties, for guided anticancer remedial treatment [56, 59].

14.3.3 Cartilage Tissue Engineering

Cartilage is a tissue with low regeneration capacity as a consequence of the lack of vascularization. Patients with cartilage defects due to arthritis, congenital defects or trauma are treated using plastic reconstruction or implantation of prosthetics. Transplant is rarely an option and metal implants often lead to toxicity or necrosis/fibrosis/erosion of the surrounding tissues. These issues occur often on hip, knee, shoulder or elbow joint implants. Nonsurgical intervention exists that imply the injection of polymer solution aiming at increasing the articular space, lowering friction between the articular surfaces, and to decrease the associated pain. The disadvantage is that this method is suitable only for mild cartilage degradation.

From a TE perspective, chondrocytes have been seeded on biopolymers and cultured *in vitro* in Petri dishes or bioreactors for new cartilage synthesis. As polymer matrix, collagen, hyaluronic acid and PGA have been reported to ensure a 3D support and cell attachment, according to a measured and designed scaffold. Due to the favorable *in vitro* conditions, the chondrocytes grow inside the material and start to produce ECM composed of collagen and GAG. A simultaneous deterioration of the initial scaffold with the tissue regeneration at the implantation site will lead to a good biocompatibility over time.

The focus on the polymeric materials is directed towards reproducibility of the 3D porous structures with uniformly distributed pores and optimal flow/mass transfer of nutrients and metabolites during *in vitro* culture; stimulation of cell growth and production of ECM without side or toxic effects; and biodegradation with a precisely controlled rate. Artificial cartilage is an implantable ready to use tissue, which can be obtained using autologous or allograft chondrocytes or stem cells, and it offers control over the *in vitro* growth process and tissue development.

Porous scaffolds with fibrous structure have been obtained as foams and meshes from poly(glycolic acid) PGA, poly(lactic acid) PLA, and their copolymers [60], or materials from natural sources (type I, II collagen, collagen composites, hyaluronic acid [61, 62]. Despite the good interaction with the chondrocytes, these materials do not allow uniform cell distribution and thus collagen content remains low. Hydrogels resulted from polymers, both natural and synthetic ones (e.g., collagen, alginate, agarose, fibrin, PEG, hyaluronic acid, alginate, silk), are an attractive biomaterial because in hydrogel culture, the cells can be well dispersed, exhibit normal morphology and function, and the material can entrap the newly synthesized ECM [63]. In addition, much like the native tissue, hydrogels have the intrinsic hydrophilic nature as a consequence of the polymers in its structure. Interestingly, agarose is particularly of interest for cartilage TE due to its ability to maintain long-term suspension of chondrocytes, without altering their phenotype, in addition to being non-degradable and passive towards the cells. Such material has been proved to be efficient in order to mend the flaws in both articular cartilage and bone, as evidenced *in vivo* testing. Moreover, Cartipatch® is a hydrogel scaffold implant based on agarose and alginate populated with autologous chondrocytes used clinically for mending cartilage flaw and which showed an improved efficiency as revealed by a two years experimental trial [64].

Another attractive strategy is the employment of photopolymerizable hydrogels, especially for the encapsulation of cells *in vitro* and/or for *in vivo* cell delivery. The most important advantages are the ability of the material to adapt to the defect`s complex shape and the fact that the polymerization reaction can take place at ambient or body temperature.

This favors the addition and integration of the surrounding tissue, using modified natural polymers (methacrylated hyaluronic acid, gelatin, chitosan, peptides, silk fibroin, sericin and chondroitin sulfate) [65].

Current challenges regarding cartilage replacement include anatomical shape and particularities, small implant size and difficult scale-up, and unsuccessful integration with surrounding tissue due to graft bevel edge and consequent stress concentration.

14.3.4 Ligament and Tendon Implants and Tissue Engineering

Tendon and ligament wounds are one of the most common injuries that are clinically addressed on a daily basis. In the context of the increasing life expectancy and aging population, these injuries become a physical and financial burden. Both types of tissues have in common the fact that they are majorly made of collagen and they are poorly vascularized, hence slow or no regeneration is expected.

Noninvasive treatment for ligament and tendon therapy includes the use of drugs with slow action against rheumatic disorders or drugs with nonsteroidal action against inflammatory processes aiming to impede inflammation and maintain the structural integrity of the tissue. For more serious lesions, autografts, allografts or xenografts have been the solution.

However, research has been conducted toward the design of artificial ligament and tendon bioinspired and biocompatible materials that would favor regeneration or replace tissue function. For example, Migonney *et al.* have used poly(ε-caprolactone) (PCL) fibers grafted with poly(sodium styrene sulfonate) for enhanced biological properties as a promising substitute for the injured anterior cruciate ligament [66]. The employment of the second polymer in order to modify the surface was proven to be biocompatible, promoting human ligament fibroblast colonization *in vitro* [67]. Unfortunately, tendon reconstruction has also been studied using synthetic fibers, such as PLA, PET, PTFE, Dacron, nylon, PP, due to the fact that natural polymers do not exhibit optimal mechanical behavior. However, despite their good mechanical properties and their capacity to induce tissue encapsulation, they do not induce functional orientation of *de novo* tissue [68].

14.3.5 Cardiovascular Implants and Tissue Engineering

Cardiovascular diseases continue to be the leader in the top causes of illness and death in developed countries. Approximately 70% of the population is affected by them and in Europe heart conditions represent the first cause of death (41%). Among the risk factors, there are a fat-rich diet, alcohol, smoking, and sedentary lifestyle.

14.3.5.1 Valve Implants

The heart valves ensure the correct blood flow on the two distinct circuits, namely the pulmonary and the systemic circulations, directed by the tricuspid and the mitral valves, respectively.

When the cardiac valves do not control the blood flow, there are two probable causes: stenosis and regurgitation. As a consequence, the valves become more rigid and thicker or they fuse, they do not close perfectly and do not maintain the blood flow in one direction.

Two types of prosthesis are available for valve replacement: biological and mechanical. The former are basically grafts from porcine aortic grafts or bovine pericardial tissue, which are used directly for implant or pretreated using chemical methods. The latter is based on metal and synthetic polymer materials.

An ideal heart valve must be with long-term durability, physiologic hemodynamic characteristics (no protein gradient, blood flow, no turbulence), minimal complications (no thrombogenicity or hemolysis, low profile for minimal interaction with other adjacent structures), facile and rapid implantation, no anticoagulant treatment necessary, silent, low cost and easily accessible.

Unfortunately, currently there are no valves made from biobased materials due to the discrepancy in the level of mechanical properties and hemocompatibility between synthetic polymer materials and biobased materials. Most likely, R&D will take into consideration multicomponent systems with complex formulations having multiscale hierarchical architectures as valid alternatives for the materials presently in use.

14.3.5.2 Artificial Heart/Cardiac Patches

Aiming at supporting the vital function of the heart, scientists and medical personnel have designed and tested various ideas including Jarvik-7, AbioCor, or the more recent Nicholas Cohrs soft artificial heart created using 3D printed silicone [69]. However, synthetic polymers and metal parts have been exclusively used for the fabrication of these devices due to the challenging organ to be replaced—huge mechanical load, fine control of rhythm and pressure, etc.

From a biomaterial perspective, another solution has been explored for the treatment of heart attacks as cardiac patches. Such a biobased material was patented having a layered architecture with a biocompatible and biodegradable scaffold comprising synthetic and natural polymers, (polylactic acid, glycolic acid, lactone, polyglycerol sebacate, copolymer of synthetic polymers and combinations thereof, and chitosan, sodium alginate and cellulose), a hydrogel layer made of self-assembling peptides, and a polyurethane-based support. The patch proved to be totally hemocompatible and showed a behavior superior to polytetrafluoroethylene (PTFE), used for any soft tissue repair [70].

14.3.5.3 Vascular Grafts and TE

The cardiovascular atherosclerotic disease is a major medical issue in the occidental world. This pathology reduces locally the blood flow but it can evolve to the complete obstruction of a blood vessel. From the medical experience, stents have proven to be a temporary solution since the same place is often subjected to restenosis. Another solution is the grafting of the vessel with either an artificial graft or an autologous graft from the saphenous vein (for small-diameter vessels) or an artery, depending on the particular case. Cryogenic allografts have also been used for bypassing coronary arteries, but the issues related to occlusions and aneurysms have limited the use of this alternative to the cases where an autologous graft is not possible. However, the autologous graft, despite the good clinical results, is not without its disadvantages: numerous surgical interventions, the risk of damaging the graft during the transplant, the lack of donor site for the patients with amputated limbs.

In this context, there is an opportunity to design and create artificial vessel grafts, using polymeric biomaterials, such as polyurethanes (PUR). During the past decade, several PUR vascular implants have been developed, with a nonbiodegradable profile, to prevent the collapse of the graft, with reduced thrombogenic effect and good potential as TE scaffolds. For example, Chronoflex® represents a family of medical grade polymers used for such applications, precisely aliphatic or aromatic polycarbonate-based thermoplastic urethanes. These ether-free polyurethane elastomers have been designed to overcome surface degradation, such as stress-induced microfissures, are stable to biological degradation and exhibit a low modulus of elasticity, good solvent resistance and limited softening *in vivo* [71].

MyoLink™ (CardioTech International, Massachusetts) is a vascular graft resulting from poly(carbonate-urea)-urethane, which exhibits good compliance. It was clinically tested for small diameter vessel replacements, and studied with RGD or heparin-modified surface *in vitro* [72]. Currently, only synthetic polymer vascular grafts are available, such as Vectra (Thoratec Corp, Pleasanton, Calif), Gore-Tex (W.L. Gore & Associates, Inc) and Exxcel (Boston Scientific), which are made of PTFE, PUR or blend based on these polymers [73].

Currently, synthetic vascular grafts have been proved to be successfully implanted to replace vessels with large diameter, but they are not as suitable for low diameter vessels due to associated complications of thrombosis and bacterial infections. However, recent research using cells (endothelial, fibroblasts, stem), growth factors (e.g., VEGF, cytokines), synthetic (PET, PTFE, PGA, PLA, PUR) or collagenic scaffolds, or decellularized ECM, associated with mechanical stimulation of the artificial graft in pulsatile flow systems or by cyclic longitudinal deformations have proven the high potential of polymer-based materials for vascular TE [74].

14.3.6 Liver Tissue Engineering and Bioreactors

The issues related to liver transplant include: severe donor selection, and lack of clinical alternatives in hepatic deficiencies. Among the research strategies approached, there is the temporary shunt of the hepatic functions with extracorporeal bioreactors, and hepatocyte transplant for a more long-term solution.

Among the hepatic assistance devices are HepatAssist (Circuit Biomedical), Live Rx (Algenix Inc), ELAD (VitaGen), and Excorp Medical BAL. They all use hepatocytes, human or porcine, on a hollow fibers support matrix based on collagen, cellulose, gelatin, chitosan or other polymers [75]. Other configurations are also possible, such as spheroids, sandwich-like collagen systems, microcapsules and overlaid cultures.

Until nowadays, no viable artificial liver has been developed that would successfully take over the hepatic complex functions (e.g., digestive/enzymatic, metabolic, synthesis of albumin, complement proteins, and glycogen, detoxification, regulation of amino acids, lipids, and carbohydrates). However, attempts have been made to construct polymeric matrices for the transplant of a significant number of hepatocytes and sustain their viability. The polymer-cell interface is controlled using proteins (laminin, fibronectin) and factors governing growth processes in order to enable cell adhesion, viability and normal function. The challenges associated concern the fact that the artificial liver must assume 10% of the liver function, it needs a large number of hepatocytes, which are very difficult to maintain viable. In addition, there are controversies regarding the employment of porcine cell cultures due to the possible occurrence of inherent hazardously infectious processes and slightly different metabolism. Moreover, hepatocytes are susceptible to mechanical stress and have limited capacity/volume of extracting and processing nutrients and other molecules from blood.

14.3.7 Kidney Tissue Engineering and Dialysis Devices

Kidneys are susceptible to failure due to residue accumulation, acidosis, hypertension, etc. Its functions include filtration and resorption, and hormone secretion. Various dialysis systems have been developed for assisting patients with kidney problems. More recently, dialysis has been performed using an artificial kidney based on hollow cellulose fibers [76]. These systems allow a good blood flow and mechanical resistance, low resistance to flow, one-use only, low cost, no infection risk, and easy sterilization.

Implantable artificial kidney has been studied and different systems have been designed, including filter-based miniature renal assist device (RAD) with miniature MEMS silicone membrane, and stem cell artificial kidney prepared *in vitro* [77]. Finally, there are studies concerning the use of decellularized kidney ECM and their repopulation with cells [78].

14.3.8 Nervous Tissue Engineering and Implants

The TE efforts concerning the nervous system are focused on replacing a missing nervous component, regeneration of nervous tissue, stimulation of coupling elements. Delivery of neuroactive molecules requires a suitable system, which must enable the bioactive molecule stability, necessary dosage, solubility, targeted delivery, and lack of secondary effects. Apart from the classic solutions (poly(ethylene vinyl acetate), silicone elastomers, polyesters), such biopolymeric systems with controlled delivery have been studied based on poly(lactic-co-glycolic acid) [79]. The major disadvantage of these encapsulation forms is the limited volume of drug that can be loaded, as well as the release rate, which is determined by the polymer structural and performance features.

Another direction of study has been the encapsulation of cells using artificial scaffolds: cross-linked polyelectrolyte solutions, derivatives of ECM (such as Matrigel), and hydrogels. These scaffolds prevent the formation of cell aggregates, allow cell adhesion and may induce cell differentiation [80]. For example, silk-based tubular scaffolds have been filled with stem cells and a regeneration process related to the peripheral nerve was promoted *in vivo* testing [81]. In another study, porous tubular scaffolds were prepared using collagen. The biodegradable material, along with the optimal porosity obtained by freeze-drying allowed peripheral nerve regeneration [82].

However, nerve reconstruction has also been approached from another perspective—that of guiding channels. This type of implant has the role of reducing the suture tension, protecting the nerve and directing the axons. Such implants have been designed with a layered structure. The outer layer is frequently made of polymers with specific behavior and of different nature (biobased and synthetic) according to their functional purpose:

- Inert polymers: silicone elastomer, PVC, PE;
- Permeable polymers: collagen, PTFE, poly(acrylonitrile-vinyl chloride);
- Resorbable polymer: PLA, PGA, collagen;

- Polyelectrolyte polymers: PTFE, poly(vinylidene fluoride);
- Polymers releasing trophic factors: poly(ethylene-vinyl chloride).

The intermediary layer has to be designed to support the cell-seeded inner layer, meaning that it has to be highly biocompatible. This drastically limits the choice in materials and reduces it to biopolymers, such as collagen, fibrin, or glycosaminoglycans [83–85].

For this particular application it is important to take into consideration the electric properties of the implant, apart from the surface properties, degradation/resorption, and the capacity to release biologically active molecules, which are important to any of the discussed types of tissues and organs. It has been studied *in vivo* the regeneration of a sectioned nerve and it has been evidenced that the galvanic current produced in silicone channels with electrodes has improved the nerve regeneration. Recent studies have investigated the possible use of films made of blended materials based on silk fibroin and chitosan for such application [86].

14.4 Auxiliary Materials

Biomaterials have current applications in orthopedics, dentistry, ophthalmology, internal medicine, neurology, nephrology, drug delivery, cardiology, tissue replacements, as well as for monitoring and theranostic. Other applications involve wound dressings and artificial skin, dental and cochlear implants, materials for cell encapsulation, tissue screws and tacks, tissue adhesives and sealants, and tissue substitutes and sutures.

Humankind has used materials to repair the body for millennia. Natural and synthetic threads, fibers, and sutures are still used today to repair wound sites. The use of sutures at the dawn of humanity (approx. 32000 years ago) has been scarcely documented, but it has been proved that ancient Egyptians used linen threads to close wounds (approx. 4000 years ago). Early Greek literature mentioned metallic sutures for the first time. Ligatures of gold wire were used by Galen of Pergamon (c. 130–200 AD). Sushruta, an Indian surgeon, described absorbable sutures derived from animal sinews in 600 BC. In 1759, Samuel Hallowell, an English surgeon, used a wooden peg and thread to repair a brachial artery. Despite the lack of knowledge (in the sense of modern science), limitations of hygiene and sterilization protocols, and empirical procedures in surgery and postoperator care, many of the documented operations have been successful.

In the 19th century, American surgeons used lead and silver wire sutures, while other vascular sutures were performed employing ivory clamps or even silk threads. In Europe, catgut has been used as suture thread since the Middle Age and until the early 1970s only collagenous materials (such as catgut) found general acceptance as absorbable sutures. Silk has been employed as biodegradable surgical thread for centuries due to the good adhesive properties to living cells demonstrated by silk fibroin [87]. Intermolecular hydrogen network and high level of crystalline feature confer excellent fiber forming ability to nylon-type polyamides. Sutures made from single or multifilament threads and meshes made of various fabrics (woven, nonwoven or knitted) were used for implant fixation, as vessel implants, and dressings [88, 89].

Proteins and other amino acid-derived polymers are the major structural components of many tissues and, by consequence, such materials provide a high degree of biocompatibility when employed in surgery and hemostasis, tissue engineering (as scaffolds) or dug delivery (as vehicles), despite their susceptibility to undergo naturally controlled degradation processes [90]. Poly(α-ester)s have been employed in different biomedical applications, such as sutures, devices for fracture fastening, or even as supports for cell transplantation [91, 92].

The poly(glycolic acid) and nonwoven polyglycolide fabrics, although of synthetic nature, have been used as biodegradable sutures (e.g., Dexon®), in devices for the internal stabilization of bones and supports for tissue engineering due to their excellent biocompatibility and biodegradability, aside their high ability to form fibers and good prior-to-implantation mechanical properties [93]. Based on the success of these materials recorded through clinic studies, it was possible to design and develop new biodegradable polymer materials able to act as temporary orthopedic implants and related biomedical applications.

Poly(*L*-lactide) (PLLA) polymers have the ability to yield in high strength fibers, hence they have been tested in tissue engineering applications as scaffold or augmentation devices, and as suture threads with properties higher than Dexon®, with prospectives to replace nonbiodegradable fibers, such as Dacron [94, 95]. Still, the degradation of PLLA takes place very slowly (*in vivo* total resorption occurs in 2–5.6 years) due to its high molecular weight and hydrophobicity [96, 97].

Poly(lactide-*co*-glycolide) polymers (PLGA) have been used in biomedical applications (sutures, bone pins, stents, supports for drug delivery and tissue engineering) [98, 99].

Poly(lactide-*co*-glycolide)s prepared in various formulations and commercially available (PuraSorb®PLG, Vicryl®, VicrylRapide®, Panacryl®) have been successfully tested for potential applications as sutures with properties governed by their composition [100].

Polyglyconates, block copolymers of glycolides with trimethylene carbonate (TMC), are used as flexible suture materials (Maxon®) and orthopedic tacks and screws (Acufex®) [93]. Flexible fibers, used as monofilament suture threads, have been obtained by the copolymerization of ε-caprolactone and glycolide. A terpolymer prepared from glycolide, TMC, and dioxane (BioSyn®) has been used as suture materials that exhibit a reduced stiffness and moderate degradation (12–16 weeks) [101, 102]. Hydrolysis of the ester bonds is responsive for bulk erosion of PLGA. The LA/GA ratio, molecular weight, and the shape and structure of the matrix influence the rate of degradation. Good processability of these biocompatible copolymers results in materials with various architectures and degradation rates that can be managed. FDA approved these biodegradable polymers to be used in humans because their biodegradability is higher than that of the earlier suture materials [103]. Other versatile resorbable multifilament sutures are based on biodegradable polylactides and glycolides. However, multifilament sutures present a higher risk of infection ascribed to a significant friction when inserting tissues. Consequently, the fabrication of monofilament sutures has required much research into developing materials that have facilitated their obtaining. Polydioxanone (PDS) was marketed as monofilament sutures in the 1980s. When ingested, PDS is included in a degradation chain of reactions just like other polyglycolides, or is broken down to glyoxylate [104].

Accelerated tissue regeneration requires absorbable materials with porous structures. Therefore, collagen sponges have been successfully employed in such applications due to their supramolecular structure, biocompatibility and biodegradability. DuraGen® is a new commercial suture-free product intended for mending the damages of the spinal *dura mater* and its regeneration, and consists of a three-dimensional graft of collagen [105].

Symmetrical bisamide-diols have been employed to prepare poly(ester amides) by reaction with succinyl chloride. Polymers have shown good mechanical properties and were tested as potential bioresorbable suture materials. Glycolic acid and diamine alkanes (with $2 \div 12$ $-CH_2$ groups) were used to prepare different water soluble bisamide-diols [106].

Other vascular sutures and surgical meshes with good biocompatibility have been obtained from polyvinylidene fluoride (PVF) which is a rather inert polymer [107].

14.5 Conclusion and Future Trends

Biobased materials have proved to be, beyond any doubt, worthy of the interest of the R&D scientists in BME. Their relevance as materials of critical importance has been shown in this review. Ranging from tissue engineering to theranostic devices, and from artificial organs to a wide variety of implantables, most BME applications rely on biopolymers and complex materials comprising both natural and synthetic polymers.

There are still some areas where the leading role of the synthetic polymers cannot be yet challenged, such as the cardiovascular implants (valves, artificial heart, and vascular grafts), liver substitutes (extracorporeal bioreactors), nervous channels and sensory implants, or artificial kidneys. This does not mean that scientists have not already addressed the subject and initiated studies to tackle the employment of biobased materials in these applications. One promising direction of research seems to be the materials with multiscale hierarchical architectures. Basically, they consist of polymer matrices made of natural polymers (raw or appropriately modified), *per se* or in wise combinations with synthetic polymers (blends, copolymerization, crosslinking), and various fillers, single type or mixtures (*e.g.*, cellulose nanofibrils, hydroxyapatite nanoparticles, silk fibroin nanofibrils, etc.), of selected nature and properties. Such multicomponent biopolymer-based materials combine advantageous characteristics of constituents and, due to the active interphase formed at the interface with the host tissues, can be both biocompatible and high performance. On the other hand, it is possible to select suitable processing techniques that can ensure the development of new biomaterials with high purity, thus limiting the use of toxic solvents and subsequent purification stages. Recent data indicated that processing biopolymers in the presence of supercritical fluids is a promising way for the production of microstructured and nanostructured biomaterials.

At the same time, understanding the complex both *in vitro* and *in vivo* interactions of biobased materials with the living cells is of utmost importance. Thus, the in-depth studies on the main and side effects of new materials, in correlation with clinical trials, will contribute to enhance the risk management and increase the operational safety of different implants, organs and devices.

However, advances in BME come at a high cost due to the significant input of the state-of-the-art biobased materials. Despite this, the affordability of common products and devices is a requirement that needs to be met through the cooperation of different sectors of society. Thus, dedicated investment instruments, the change in the consumers lifestyle, the increasing interest in minimally invasive procedures, government involvement in the healthcare insurance system, effective support of the R&D, etc. can contribute to the positive dynamics of this domain.

References

1. Salerno, A. and Pascual, C.D., Bio-based polymers, supercritical fluids and tissue engineering. *Process Biochem.*, 50, 826, 2015.
2. Babu, R.P., O'Connor, K., Seeram, R., Current progress on bio-based polymers and their future trends. *Prog. Biomater.*, 2, 1, 2013.
3. Buwalda, S.J., Bio-based composite hydrogels for biomedical applications. *Multifunct. Mater., IOP Sci.*, 3, 022001, 2020.
4. Zare, E.N., Makvandi, P., Borzacchiello, A., Tay, F.R., Ashtari, B., Padil, V.V., Antimicrobial gum bio-based nanocomposites and their industrial and biomedical applications. *Chem. Commun.*, 55, 14871, 2019.
5. Birajdar, M.S., Joo, H., Koh, W.G., Natural bio-based monomers for biomedical applications: A review. *Biomater. Res.*, 25, 8, 2021.
6. Bedian, L., Villalba-Rodríguez, A.M., Hernández-Vargas, G., Parra-Saldivar, R., Iqbal, H.M., Bio-based materials with novel characteristics for tissue engineering applications–A review. *Int. J. Biol. Macromol.*, 98, 837, 2017.
7. Cywar, R.M., Rorrer, N.A., Hoyt, C.B., Beckham, G.T., Chen, E.Y.-X., Bio-based polymers with performance-advantaged properties. *Nat. Rev. Mater.*, 1, 83–103, 2022.
8. Moradali, M.F. and Rehm, B.H.A., Bacterial biopolymers: From pathogenesis to advanced materials. *Nat. Rev. Microbiol.*, 18, 195, 2020.
9. Wang, L., Pathak, J.L., Liang, D., Zhong, N., Guan, H., Wan, M., Miao, G., Li, Z., Ge, L., Fabrication and characterization of strontium-hydroxyapatite/silk fibroin biocomposite nanospheres for bone-tissue engineering applications. *Int. J. Biol. Macromol.*, 142, 366, 2020.
10. Qi, C., Deng, Y., Xu, L., Yang, C., Zhu, Y., Wang, G., Wang, Z., Wang, L., A sericin/graphene oxide composite scaffold as a biomimetic extracellular matrix for structural and functional repair of calvarial bone. *Theranostics*, 10, 741, 2020.
11. de Alcantara Abdala, J.M., dos Santos, P.C.O., de Vasconcellos, L.M.R., Raniero, L.J., de Castro-Raucci, L.M.S., de Oliveira, P.T., Pandolfelli, V.C., de Oliveira, I.R., Calcium aluminate cement-based blends for application to fill in bone defects. *Res. Biomed. Eng.*, 36, 429, 2020.
12. Liu, Y., Zhu, Z., Pei, X., Zhang, X., Cheng, X., Hu, S., Gao, X., Wang, J., Chen, J., Wan, Q., ZIF-8-modified multifunctional bone-adhesive hydrogels promoting angiogenesis and osteogenesis for bone regeneration. *ACS Appl. Mater. Interfaces*, 12, 36978, 2020.
13. Tonda-Turo, C., Carmagnola, I., Chiappone, A., Feng, Z., Ciardelli, G., Hakkarainen, M., Sangermano, M., Photocurable chitosan as bioink for cellularized therapies towards personalized scaffold architecture. *Bioprinting*, 18, e00082, 2020.
14. Clark, C.C., Aleman, J., Mutkus, L., Skardal, A., A mechanically robust thixotropic collagen and hyaluronic acid bioink supplemented with gelatin nanoparticles. *Bioprinting*, 16, e00058, 2019.
15. Kolan, K.C., Semon, J.A., Bindbeutel, A.T., Day, D.E., Leu, M.C., Bioprinting with bioactive glass loaded polylactic acid composite and human adipose stem cells. *Bioprinting*, 18, e00075, 2020.
16. Nguyen, T.H.M., Abueva, C., Van Ho, H., Lee, S.Y., Lee, B.T., *In vitro* and *in vivo* acute response towards injectable thermosensitive chitosan/Tempo-oxidized cellulose nanofiber hydrogel. *Carbohydr. Polym.*, 180, 246, 2018.
17. Tozar, A., Karahan, I.H., Yücel, Y., Optimization of the electrophoretic deposition parameters for biocomposite hydroxyapatite/chitosan/collagen/h-BN coatings on Ti6Al4V biomedical implants. *Metall. Mater. Trans.*, 50, 1009, 2019.
18. Huang, J., Qin, J., Zhang, P., Chen, X., You, X., Zhang, F., Yao, M., Facile preparation of a strong chitosan-silk biocomposite film. *Carbohydr. Polym.*, 229, 115515, 2020.

19. Li, L., Yang, H., Li, X., Yan, S., Xu, A., You, R., Zhang, Q., Natural silk nanofibrils as reinforcements for the preparation of chitosan-based bionanocomposites. *Carbohydr. Polym.*, 253, 117214, 2021.
20. Ruffini, A., Sandri, M., Dapporto, M., Campodoni, E., Tampieri, A., Sprio, S., Nature-inspired unconventional approaches to develop 3D bioceramic scaffolds with enhanced regenerative ability. *Biomedicines*, 9, 916, 2021.
21. O'Brien, F.J., Biomaterials & scaffolds for tissue engineering. *Mater. Today*, 14, 88, 2011.
22. Black, J., *Biological performance of materials. Fundamentals of biocompatibility*, 4th, p. 520, CRC Press, Boca Raton, 2005.
23. Ratner, B.D., The biocompatibility of implant materials, in: *Host response to biomaterials*, S. Badylak (Ed.), pp. 37–51, Academic Press, Burlington, 2015.
24. Williams, D., Biocompatibility, in: *Tissue engineering*, C. Van Blitterswijk, J. De Boer, P. Thomsen, J. Hubbell, R. Cancedda, J.D. de Bruijn, A. Lindahl, J. Sohier, D. Williams (Eds.), pp. 255–278, Academic Press, Burlington, 2008.
25. Morais, J.M., Papadimitrakopoulos, F., Burgess, D.J., Biomaterials/tissue interactions: Possible solutions to overcome foreign body response. *AAPS J.*, 12, 188, 2010.
26. Ahadian, S., Rahal, R., Ramón, A.J., Obregón, R., Hasan, A., Biomaterials in Tissue Engineering, in: *Tissue engineering for artificial organs: Regenerative medicine, smart diagnostics and personalized medicine*, A. Hasan (Ed.), pp. 35–83, Wiley-VCH Verlag GmbH & Co. KGaA, Weinheim, Germany, 2017.
27. Sands, R.W. and Mooney, D.J., Polymers to direct cell fate by controlling the microenvironment. *Curr. Opin. Biotechnol.*, 18, 448, 2007.
28. Anderson, J.M., Rodriguez, A., Chang, D.T., Foreign body reaction to biomaterials. *Semin. Immunol.*, 20, 2, 86, 2008.
29. Medical Engineered Materials Market Share, Size, Trends, Industry Analysis Report, By Type (Elastomers, Plastics, Foams, Others); By Application; By Region; Segment Forecast, Polaris Market Research, New York, USA, 111pp, 2021 - 2028, published date: Oct. 2021, report ID: PM2127, provider: POLARIS Market Research. (last accessed at http://www.polarismarketresearch.com/industry-analysis/medical-engineered-materials-market, Dec. 23, 2021).
30. Farokhi, M., Shariatzadeh, F.J., Solouk, A., Mirzadeh, H., To, Alginate based scaffolds for cartilage tissue engineering: A review. *Int. J. Polym. Mater. Polym. Biomater.*, 69, 230, 2019.
31. Gopi, S., Balakrishnan, P., Chandradhara, D., Poovathankandy, D., Thomas, S., General scenarios of cellulose and its use in the biomedical field. *Mater. Today Chem.*, 13, 59, 2019.
32. Francis, A., Yang, Y., Boccaccini, A.R., A new strategy for developing chitosan conversion coating on magnesium substrates for orthopedic implants. *Appl. Surf. Sci.*, 466, 854, 2019.
33. Dumont, V.C., Mansur, H.S., Mansur, A.A.P., Carvalho, S.M., Capanema, N.S.V., Barrioni, B.R., Glycol chitosan/nanohydroxyapatite biocomposites for potential bone tissue engineering and regenerative medicine. *Int. J. Biol. Macromol.*, 93, 1465, 2016.
34. Thikhanahly, N., Shin, H., Gupta, K.C., Kang, I.K., Yu, W., Bioactive antibacterial modification of orthodontic microimplants using chitosan biopolymer. *Macromol. Res.*, 27, 504, 2019.
35. Kim, H. *et al.*, Hyaluronate and its derivatives for customized biomedical applications. *Biomaterials*, 123, 155, 2017.
36. Tozar, A., Karahan, I.H., Yücel, Y., Optimization of the electrophoretic deposition parameters for biocomposite hydroxyapatite/chitosan/collagen/h-BN coatings on Ti6Al4V biomedical implants. *Metall. Mater. Trans.*, 50, 1009, 2019.
37. Ganguly, A., Sharma, K., Majumder, K., Peptides as biopolymers: Past, present, and future, in: *Biopolymer-based formulations*, K. Pal, I. Banerjee, P. Sarkar, D. Kim, W.-P. Deng, N.K. Dubey, K. Majumder (Eds.), pp. 87–104, Elsevier, Oxford, England, United Kingdom, 2020.

38. Hoste, K., Schacht, E., Seymour, L., New derivatives of polyglutamic acid as drug carrier systems. *J. Control. Release*, 64, 53, 2000.
39. Tolentino, A., Alla, A., Martínez de Ilarduya, A., Muñoz-Guerra, S., Complexes of polyglutamic acid and long-chain alkanoylcholines: Nanoparticle formation and drug release. *Int. J. Biol. Macromol.*, 66, 346, 2014.
40. Kai, D. and Loh, X.J., Polyhydroxyalkanoates: Chemical modifications toward biomedical applications. *ACS Sustain. Chem. Eng.*, 2, 106, 2014.
41. Wang, X., Schröder, H.C., Müller, W.E.G., Amorphous polyphosphate, a smart bioinspired nano-/bio-material for bone and cartilage regeneration: Towards a new paradigm in tissue engineering. *J. Mater. Chem. B*, 6, 2385, 2018.
42. Seker, S., Elçin, A.E., Elçin, Y.M., Macroporous elastic cryogels based on platelet lysate and oxidized dextran as tissue engineering scaffold: *In vitro* and *in vivo* evaluations. *Mater. Sci. Eng. C*, 110, 110703, 2020.
43. Singhvi, G., Hans, N., Shiva, N., Kumar Dubey, S., Xanthan gum in drug delivery applications, in: *Natural polysaccharides in drug delivery and biomedical applications*, A. Md S. Hasnain and K. Nayak (Eds.), pp. 121–144, Academic Press, Elsevier, 2019.
44. Maji, B. and Maiti, S., Chemical modification of xanthan gum through graft copolymerization: Tailored properties and potential applications in drug delivery and wastewater treatment. *Carbohydr. Polym.*, 117095, 2020.
45. Council of Europe, The Human Rights Channel is a showcase for the Council of Europe's extensive collection of video and photo content on the themes of its core activities: human rights, democracy and the rule of law. Human rights channel Organ Donation and Transplantation. https://human-rights-channel.coe.int/organ-donation-en.html.
46. https://www.sciencedirect.com/journal/biomaterials/about/aims-and-scope (last accessed: February 9, 2022).
47. Chandra, P.K., Soker, S., Atala, A., Tissue engineering: Current status and future perspectives, in: *Principles of tissue engineering*, 5th, R. Lanza, R. Langer, J.P. Vacanti, A. Atala (Eds.), pp. 1–36, Academic Press, Cambridge, Massachusetts, United States, 2020.
48. Jürgens, C., Schulz, A.P., Porté, T., Faschingbauer, M., Seide, K., Focus on biomaterials biodegradable films in trauma and orthopedic surgery. *Eur. J. Trauma*, 32, 160, 2006.
49. Abdelrahman, T. and Newton, H., Wound dressings: Principles and practice. *Surgery*, 29, 491, 2011.
50. Moura, L.I.F., Dias, A.M.A., Carvalho, E., De Sousa, H.C., Recent advances on the development of wound dressings for diabetic foot ulcer treatment - A review. *Acta Biomater.*, 9, 7093, 2013.
51. Lee, K.Y. and Mooney, D.J., Alginate: Properties and biomedical applications. *Prog. Polym. Sci.*, 37, 106, 2012.
52. Falanga, V., Bioengineered skin constructs, in: *Principles of tissue engineering*, 5th, R. Lanza, R. Langer, J.P. Vacanti, A. Atala (Eds.), pp. 1331–1352, Academic Press, Cambridge, Massachusetts, United States, 2020.
53. Duceac, I.A., Verestiuc, L., Dimitriu, C.D., Maier, V., Coseri, S., Design and preparation of new multifunctional hydrogels based on chitosan/acrylic polymers for drug delivery and wound dressing applications. *Polym. (Basel)*, 12, 1, 2020.
54. Duceac, I.A., Lobiuc, A., Coseri, S., Verestiuc, L., Tunable hydrogels based on chitosan, collagen and poly(acrylic acid) for regenerative medicine. *Proceedings of the IEEE International Conference on e-Health and Bioengineering EHB 2019*, vol. 2, pp. 3–6, 2019.
55. Baron, R.I., Culica, M.E., Biliuta, G., Bercea, M., Gherman, S., Zavastin, D., Ochiuz, L., Avadanei, M., Coseri, S., Physical hydrogels of oxidized polysaccharides and poly(vinyl alcohol) forwound dressing applications. *Mater. (Basel)*, 12, 1569, 2019.

56. Cojocaru, F.D., Balan, V., Tanase, C.E., Popa, I.M., Butnaru, M., Bredetean, O., Mares, M., Nastasa, V., Pasca, S., Verestiuc, L., Development and characterisation of microporous biomimetic scaffolds loaded with magnetic nanoparticles as bone repairing material. *Ceram. Int.*, 47, 11209, 2021.
57. Socka, M., Michalski, A., Pelin, I.M., Pawlak, A., Tanasa, F., Biela, T., Basko, M., Preparation of biomimetic composites of hydroxyapatite and star-shaped poly(2,2-dimethyl trimethylene carbonate)s terminated with carboxyl end-groups. *Polymer*, 186, 122078, 2020.
58. Duceac, I.A. and Tanasa, F., Novel chitosan-hydroxyapatite macroporous composites for biomedical applications. *Rev. Roum. Chim.*, 65, 1111, 2020.
59. Cojocaru, F.D., Balan, V., Popa, M.I., Lobiuc, A., Antoniac, A., Antoniac, I.V., Verestiuc, L., Biopolymers – calcium phosphates composites with inclusions of magnetic nanoparticles for bone tissue engineering. *Int. J. Biol. Macromol.*, 125, 612, 2019.
60. Zwingmann, J., Mehlhorn, A.T., Südkamp, N., Stark, B., Dauner, M., Schmal, H., Chondrogenic differentiation of human articular chondrocytes differs in biodegradable PGA/PLA scaffolds. *Tissue Eng.*, 13, 2335, 2007.
61. Hubka, K.M., Dahlin, R.L., Meretoja, V.V., Kasper, F.K., Mikos, A.G., Enhancing chondrogenic phenotype for cartilage tissue engineering: Monoculture and coculture of articular chondrocytes and mesenchymal stem cells. *Tissue Eng. Part B. Rev.*, 20, 641, 2014.
62. Yang, S.W., Ku, K.C., Chen, S.Y., Kuo, S.M., Chen, I.F., Wang, T.Y., Chang, S.J., Development of chondrocyte-seeded electrosprayed nanoparticles for repair of articular cartilage defects in rabbits. *J. Biomater. Appl.*, 32, 800, 2018.
63. Mauck, R.L. and Burdick, J.A., Engineering cartilage tissue, in: *Tissue engineering*, N. Pallua and C. Suscheck (Eds.), pp. 493–520, Springer, Berlin, Heidelberg, 2011.
64. Tan, A.R. and Hung, C.T., Engineering Functional Cartilage Grafts, in: *Tissue engineering in regenerative medicine*, H.S. Bernstein (Ed.), pp. 237–250, Humana Press, Springer Science+Business Media, LLC, Humana Totowa, New Jersey, USA, 2011.
65. Wei, W., Ma, Y., Yao, X., Zhou, W., Wang, X., Li, C., Lin, J., He, Q., Leptihn, S., Ouyang, H., Advanced hydrogels for the repair of cartilage defects and regeneration. *Bioact. Mater.*, 6, 998, 2021.
66. Rangel, A., Colaço, L., Nguyen, N.T., Grosset, J.F., Egles, C., Migonney, V., Adapting mechanical characterization of a biodegradable polymer to physiological approach of anterior cruciate ligament functions. *IRBM*, 43, 39, 2020.
67. Lessim, S., Migonney, V., Thoreux, P., Lutomski, D., Changotade, S., PolyNaSS bioactivation of LARS artificial ligament promotes human ligament fibroblast colonisation *in vitro*. *Biomed. Mater. Eng.*, 23, 289, 2013.
68. Zeugolis, D., II, Chan, J.C.Y., Pandit, A., Tendons: Engineering of functional tissues, in: *Tissue engineering*, N. Pallua and C. Suscheck (Eds.), pp. 537–572, Springer, Berlin, Heidelberg, 2011.
69. Cohrs, N.H., Petrou, A., Loepfe, M., Yliruka, M., Schumacher, C.M., Kohll, A.X., Starck, C.T., Schmid Daners, M., Meboldt, M., Falk, V., Stark, W.J., A soft total artificial heart—First concept evaluation on a hybrid mock circulation. *Artif. Organs*, 41, 948, 2017.
70. G. Saint-Pierre, M. Herrero Gomez, S. Martinez Creispera, A. Saiani, C. Merry, K. Meade, J.B. Guilbaud, A.F. Miller, C. Ciobanu, E. Amler, New scaffold for cardiac patch. Patent WO 2014044321 A1, assigned to University of Manchester, 2014.
71. http://www.advbiomaterials.com/products/polycarbonate/chronoflex_c.html (last accessed: February 9, 2022).
72. Tiwari, A., Salacinski, H.J., Punshon, G., Hamilton, G., Seifalian, A.M., Development of a hybrid cardiovascular graft using a tissue engineering approach. *FASEB J.*, 16, 791, 2002.
73. Kapadia, M.R., Popowich, D.A., Kibbe, M.R., Modified prosthetic vascular conduits. *Circulation*, 117, 1873, 2008.

74. Huang, A.H. and Niklason, L.E., Engineering of arteries *in vitro*. *Cell. Mol. Life Sci.*, 71, 2103, 2014.
75. Chen, A.X., Chhabra, A., Song, H.H.G., Fleming, H.E., Chen, C.S., Bhatia, S.N., Tissue engineering: Controlled apoptosis of stromal cells to engineer human microlivers. *Adv. Funct. Mater.*, 30, 2070317, 2020.
76. Yamazaki, K., Matsuda, M., Yamamoto, K., Yakushiji, T., Sakai, K., Internal and surface structure characterization of cellulose triacetate hollow-fiber dialysis membranes. *J. Memb. Sci.*, 368, 34, 2011.
77. Kim, S., Fissell, W.H., Humes, D.H., Roy, S., Current strategies and challenges in engineering a bioartificial kidney. *Front. Biosci. (Elite Ed).*, 7, 215, 2015.
78. Kim, J.H., Atala, A., Yoo, J.J., Tissue engineering of the kidney, in: *Principles of tissue engineering*, 5th, R. Lanza, R. Langer, J.P. Vacanti, A. Atala (Eds.), pp. 825–843, Academic Press, Cambridge, Massachusetts, United States, 2020.
79. O'Donnell, A., Moollan, A., Baneham, S., Ozgul, M., Pabari, R.M., Cox, D., Kirby, B.P., Ramtoola, Z., Intranasal and intravenous administration of octa-arginine modified poly(lactic-co-glycolic acid) nanoparticles facilitates central nervous system delivery of loperamide. *J. Pharm. Pharmacol.*, 67, 525, 2015.
80. Salerno, A. and Pascual, C.D., Bio-based polymers, supercritical fluids and tissue engineering. *Process Biochem.*, 50, 826, 2015.
81. Yang, Y., Yuan, X., Ding, F., Yao, D., Gu, Y., Liu, J., Gu, X., Repair of rat sciatic nerve gap by a silk fibroin-based scaffold added with bone marrow mesenchymal stem cells. *Tissue Eng. Part A*, 17, 2231, 2011.
82. Cerri, F., Salvatore, L., Memon, D., Martinelli Boneschi, F., Madaghiele, M., Brambilla, P., Del Carro, U., Taveggia, C., Riva, N., Trimarco, A., Lopez, I.D., Comi, G., Pluchino, S., Martino, G., Sannino, A., Quattrini, A., Peripheral nerve morphogenesis induced by scaffold micropatterning. *Biomaterials*, 35, 4035, 2014.
83. Dodla, M.C., Alvarado-Velez, M., Mukhatyar, V.J., Bellamkonda, R.V., Peripheral nerve regeneration, in: *Principles of regenerative medicine*, 3rd, A. Atala, R. Lanza, A.G. Mikos, R. Nerem (Eds.), pp. 1223–1236, Academic Press, Elsevier, Burlington, Massachusetts, USA, 2019.
84. Hsu, S.H. and Chen, C.W., 3D bioprinting nerve, in: *3D Bioprinting for reconstructive surgery*, D.J. Thomas, Z.M. Jessop, I.S. Whitaker (Eds.), pp. 355–366, Woodhead Publishing, Elsevier, Oxford, England, United Kingdom, 2018.
85. Enemchukwu, N.O. and García, A.J., Peptide- and Protein-Modified Surfaces (ch. 4.13), in: *Comprehensive biomaterials II (Reference module in materials science and materials engineering)*, P. Ducheyne (Ed.), pp. 200–220, Elsevier, Kidlington, Oxford, England, United Kingdom, 2017.
86. Li, L., Yang, H., Li, X., Yan, S., Xu, A., You, R., Zhang, Q., Natural silk nanofibrils as reinforcements for the preparation of chitosan-based bionanocomposites. *Carbohydr. Polym.*, 253, 117214, 2021.
87. Vepari, C. and Kaplan, D.L., Silk as a biomaterial. *Prog. Polym. Sci.*, 32, 991, 2007.
88. Nanni, F., Lamastra, F.R., Pisa, F., Gusmano, G., Synthesis and characterization of poly(ε-caprolactone) reinforced with aligned hybrid electrospun PMMA/nano-Al_2O_3 fibre mats by film stacking. *J. Mater. Sci.*, 46, 6124, 2011.
89. Kubyshkina, G., Zupančič, B., Štukelj, M., Grošelj, D., Marion, L., Emri, I., Sterilization effect on structure, thermal and time-dependent properties of polyamides, in: *Mechanics of time-dependent materials and processes in conventional and multifunctional materials*, vol. 3, T. Proulx (Ed.), pp. 11–19, Springer, New York, 2011.

90. Meinel, L., Hofmann, S., Karageorgiou, C., Kirker-Head, C., Mc Cool, J., Gronowicz, G., Zichner, L., Langer, R., Vunjak-Novakovic, G., Kaplan, D.L., The inflammatory responses to silk fibers *in vitro* and *in vivo*. *Biomaterials*, 26, 147, 2005.
91. Vert, M., Aliphatic polyesters: Great degradable polymers that cannot do everything. *Biomacromolecules*, 6, 538, 2005.
92. Albertsson, A.C. and Varma, I.K., Aliphatic polyesters, in: *Biopolymers, biology, chemistry, biotechnology, applications, polyesters III - applications and commercial products*, vol. 4, Y. Doi and A. Steinbuchel (Eds.), pp. 25–53, Wiley-VCH, Weinheim, Germany, 2002.
93. Terasaka, S., Iwasaki, Y., Shinya, N., Uchida, T., Fibrin glue and polyglycolic acid nonwoven fabric as a biocompatible dural substitute. *Neurosurgery*, 58, 134, 2006.
94. Lu, H.H., Cooper, J.A., Manuel, S., Freeman, J.W., Attawia, M.A., Ko, F.K., Laurencin, C.T., Anterior cruciate ligament regeneration using braided scaffolds: *In vitro* optimization studies. *Biomaterials*, 26, 4805, 2005.
95. Cooper, J.A., Lu, H.H., Ko, F.K., Freeman, J.W., Laurencin, C.T., Fiber-based tissue-engineered scaffold for ligament replacement: Design considerations and *in vitro* evaluation. *Biomaterials*, 26, 1523, 2005.
96. Bergsma, J.E., Rozema, F.R., Bos, R.R., Boering, G., de Bruijn, W.C., Pennings, A.J., *In vivo* degradation and biocompatibility study of *in vitro* pre-degraded as-polymerized polyactide particles. *Biomaterials*, 16, 267, 1995.
97. Middleton, J.C. and Tipton, A.J., Synthetic biodegradable polymers as orthopedic devices. *Biomaterials*, 21, 2335, 2000.
98. Lu, Y. and Chen, S.C., Micro and nano-fabrication of biodegradable polymers for drug delivery. *Adv. Drug Deliv. Rev.*, 56, 1621, 2004.
99. Katti, D.S., Robinson, K.W., Ko, F.K., Laurencin, C.T., Bioresorbable nanofiber-based systems for wound healing and drug delivery: Optimization of fabrication parameters. *J. Biomed. Mater. Res. B Appl. Biomater.*, 70, 286, 2004.
100. Nair, L.S. and Laurencin, C.T., Biodegradable polymers as biomaterials. *Prog. Polym. Sci.*, 32, 762, 2007.
101. Zhu, G., Mallery, S.R., Schwendeman, S.P., Stabilization of proteins encapsulated in injectable poly(lactide-co-glycolide). *Nat. Biotechnol.*, 18, 52, 2000.
102. Ueda, H. and Tabata, Y., Polyhydroxyalkanonate derivatives in current clinical applications and trials. *Adv. Drug Deliv. Rev.*, 55, 501, 2003.
103. Griffith, L.G., Polymeric biomaterials. *Acta Materialia*, 48, 263, 2000.
104. Maurus, P.B. and Kaeding, C.C., Bioabsorbable implant material review. *Oper. Tech. Sports Med.*, 12, 158, 2004.
105. Narotham, P.K., Jose, S., Nathoo, N., Taylon, C., Vora, Y., Collagen matrix (DuraGen) in dural repair: Analysis of a new modified technique. *Spine*, 29, 2861, 2004.
106. Lips, P.A.M., van Luyn, M.J.A., Chiellini, F., Brouwer, L.A., Velthoen, I.W., Dijkstra, P.J., Feijen, J., Biocompatibility and degradation of aliphatic segmented poly(ester amide)s: *In vitro* and *in vivo* evaluation. *J. Biomed. Mater. Res. A*, 76, 699, 2006.
107. Ahad, I.U.I., Bartnik, A., Fiedorowicz, H., Kostecki, J., Korczyc, B., Ciach, T., Brabazon, D., Surface modification of polymers for biocompatibility *via* exposure to extreme ultraviolet radiation. *J. Biomed. Mater. Res. A*, 102, 3298, 2013.

15

Applications of Bioplastics in Sports and Leisure

Radhika Malkar[1], Sneha Kagale[1], Sakshi Chavan[1], Manishkumar Tiwari[2] and Pravin Patil[3]*

[1]Department of Biotechnology Engineering, KIT's College of Engineering (Autonomous), Kolhapur, India
[2]Department of Chemical Engineering, SVKM'S NMIMS Mukesh Patel School of Technology Management & Engineering, Mumbai, Maharashtra, India
[3]Department of Basic Science and Humanities, Mukesh Patel School of Technology Management and Engineering, SVKM's NMIMS University, Mumbai, Maharashtra, India

Abstract

For decades, the global scenario of plastic pollution has been increasing and ultimately leading to an increase in greenhouse gasses emission and depletion of the ecosystem. The sporting and leisure equipment are mostly made up of plastic, contributing to environmental pollution. The sports and leisure sector impact many people, and therefore, it is the best global platform to reach and influence them to go sustainable. Therefore, bioplastics, biodegradable plastics, renewable plastics, or biopolymers are emerging solutions to reduce the ill effects of plastics. There are many types of equipments in the sport and leisure industry that can be replaced with bioplastics. This chapter summarizes various applications of bioplastics based on cornstarch, fruits, vegetables, plants, animals, etc., in leisure activities and indoor-outdoor sports equipment.

***Keywords*:** Plastic, pollution, environment, bioplastic, biodegradable plastic, sustainable, sports, leisure

15.1 Introduction

Plastic, a polymeric substance with the ability to be molded or sculpted, generally with heat and pressure. Synthetic polymers or plastics produced from nonrenewable sources like coal and petroleum have caused many environmental problems and taken decades to degrade in the environment [1]. The specific features of plastics like low electrical conductivity, transparency, low density, and toughness, provoke its production into a wide range of items. This adaptable material may be found in our appliances, computers, clothing, and many more.

Plastics have several indices of environmental impact, such as eutrophication, ecotoxicity both marine and terrestrial, global warming, acidification, etc. [2]. Due to this, environment-friendly materials were developed to replace the synthetic polymers that can be degraded by microorganisms in a limited period and also in the absence of ecotoxicity.

**Corresponding author*: dr.pravinpatil.ict@gmail.com, pravin.patil@nmims.edu

Inamuddin and Tariq Altalhi (eds.) Handbook of Bioplastics and Biocomposites Engineering Applications: 2nd Edition, (299–316) © 2023 Scrivener Publishing LLC

A recently published report by the Association of Plastics Manufacturers in Europe and the European Association of Plastics Recycling and Recovery Organizations (EPRO) states that the production of plastic has increased globally, from 359 million tonnes in 2018 to 368 million tonnes in 2019. But in Europe, the production of plastic has reduced from 61.8 million tonnes in 2018 to 57.9 million tonnes. The global production percentage of plastics is shown in Figure 15.1 which assists in understanding the need to replace plastic with Bioplastics. In the year 2019, the demand for plastic in the sport and leisure industry contributed to around 4.1% in Europe alone. Figure 15.1 highlights % of plastic production globally, [3].

As sports have immense power to affect many lives, efforts have to be focused to replace this petroleum-based plastic with biodegradable plastic. It will definitely help decrease the load of plastic recycling and greenhouse gas emissions.

15.1.1 Plastic Pollution Due to Leisure and Sports Industries

The leisure or free time of a human being is the time where a person can find enjoyment and relaxation. This is the time when a person does not think about work, business, study, or daily chores. Leisure has many health benefits and helps in reducing the stress of a person. There are many activities that can be done in this time, which can be individual or group. An individual can just sit and relax by listening to music, playing an instrument, and indoor games. There are various social and outdoor activities, such as going for a picnic or playing football, skiing, bike riding, horse riding, swimming, playing games where people can feel stress-free. All these activities contribute to the generation of a lot of plastic waste. Tourism is another major international sector, and mostly they are growing in a natural area. Therefore, people visiting such places create a huge amount of plastic waste in terms of food packets, wrappers, water bottles, ice-cream cups, clothes, and sporting goods, such as plastic balls, broken rackets, toys, etc. All these activities can also lead to noise, water, and light pollution. During the pandemic condition, it was also observed that increased use of personal protective equipment

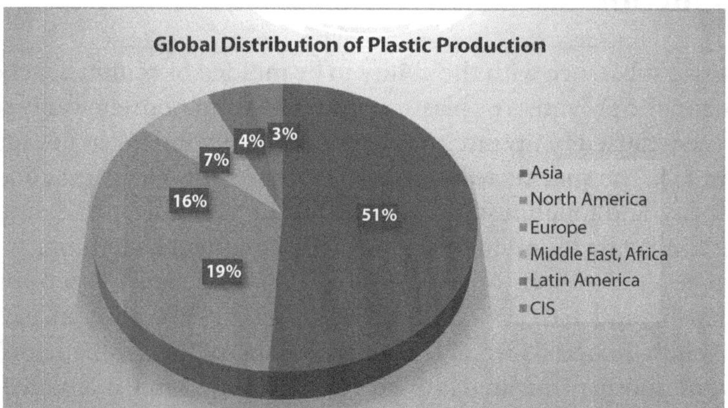

Figure 15.1 Global distribution of plastic production.

in response to coronavirus contributed to plastic pollution on beaches, parks, tourist destinations, etc. In 2016, almost 9 to 14 million tonnes of plastic had entered the ocean, and it is predicted that it will rise about 23 to 27 million tonnes by the end of 2040 [4].

More than 300 million tonnes of plastic are getting produced per year, half of which is used to produce single-use items, such as cups, straws, and shopping bags [5]. According to the 2018's UN report, around 13 million tonnes of plastic are going into the ocean [6]. Also, plastic on break down after use or incineration produces greenhouse gasses, which contributes to climate change [7]. If we look at the major sports events or venues, they are fully occupied with plastic material from athletes' clothes synthetic turf on the stadium to tickets, signage, and food stalls. It was estimated that major sports events could generate up to 750,000 plastic bottles [8]. Most of the sportswear is made from synthetic fibers of plastic. While washing these clothes, they shed tiny microfibers thinner than hair. One single polyester fleece jacket can result in the shedding of almost 1 million fibers and drains into the river. So, it can be a significant source of water pollution [9]. The global level sports events, such as Olympics, Commonwealth games, marathons, world cup matches of football and cricket, Wimbledon Tennis tournaments, are producing a lot of plastic waste, and they tend to persist in the environment or in the water body. Therefore, world sports event governing bodies are acting to reduce the contribution of sports events to plastic pollution. The International Olympic Committee and different bodies, such as world sailing, World Rugby, the International Association of Athletics Federations, and National Olympic Committees have pledged to support the UN clean sea campaign [10]. World sailing, the International Ice Hockey Federation, and the International Association of Athletics Federations encourage different waste reduction strategies. In India, a new "green protocol" is being used under which a zero-waste policy has been adopted during Indian Premier League matches. In 2018 at London's Kia Oval cricket ground, they have banned plastic straws and introduced compostable coffee cups. A similar strategy has been applied at Twickenham, which is famous for rugby [11]. The head of the UN environment Erik Solheim says that sports and the environment are interconnected and for its survival, a healthy atmosphere should be created. Therefore, besides all these practices, one more remedy a world can go for is applying biodegradable plastic to make various sports materials.

15.1.2 Bioplastics: Overview and Classification

Bioplastics have become popular as "environment friendly" compared to conventional plastics, with an estimated 25% rise in their global market till 2023 [12]. Bioplastics are the materials that assist in environmental conservation and sustainability. Bioplastics are made by different methods using different raw materials; for instance, fermentation processes/methods may be used with raw material like corn, wheat, rice, sugar, and soya, also some wastes like agricultural waste/waste from the food industry and also from the use of non-edible genetically modified plants. Bioplastics can be broadly classified (Figure 15.2) as follows:

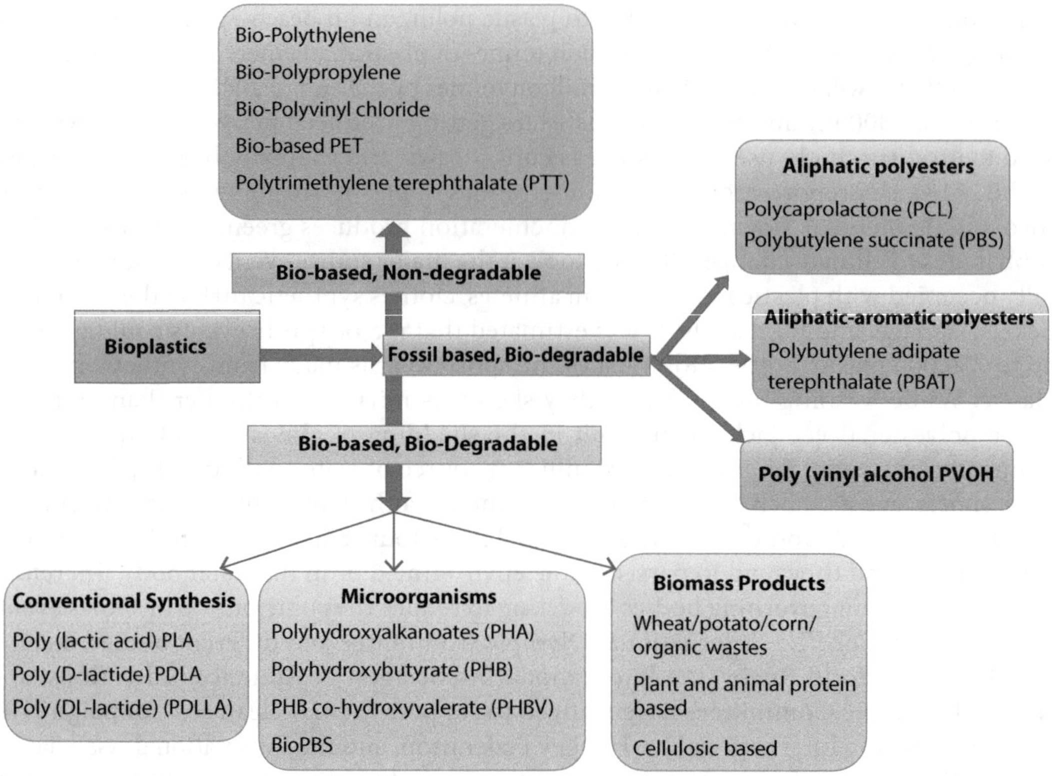

Figure 15.2 Bioplastics classification (Source, modified) [17].

15.1.2.1 Biobased Nonbiodegradable

This type of bioplastic category is also known as "drop-in bioplastics" or "bioblend" as a biodegradable component is added in smaller amounts to enhance the fragmentation of bioplasts, despite the addition of biocomponent, the impact to the environment remains alike that of conventional polymer plastics. They include plastics like Bio-Polyethylene(Bio-PE), Bio-Propylene (Bio-PP), Bio-Polyethylene terephthalate (Bio-PET), Bio-Polyvinyl chloride (Bio-PVC), and Bio-Polytrimethylene terephthalate (Bio-PTT), Bio-based polyamides (Bio-PA) whose description is given below:

15.1.2.1.1 Bio-Polyethylene (Bio-PE)

It is manufactured with ethylene polymerization using various catalysts at different temperatures and pressure. It represents 30% of the total plastics being produced [13]. Bio-PE is produced by polymerizing the bioethylene. Bioethylene can be obtained on dehydrating bioethanol with catalysts like alumina or silica-alumina, which are solid acid based at approximately the temperature of 400°C [13]. Bio-PE is being used in a similar way as plyhthene plastic, such as in the wire, cable, packaging plastic, grocery bags, bottles, toys, housewares, etc.

15.1.2.1.2 Bio-Propylene (Bio-PP)

Propylene is produced through glucose fermentation to get isobutanol, and is dehydrated to get biobutylene and biopropylene, then biopropylene can be obtained through polymerization [14]. According to Bioplastics Europe, the production of biopropylene is predicted to increase six-fold by 2024 [13]. This is a thermoplastic polymer, and it is the world's second most-produced synthetic plastic. Polypropylene is one of the most flexible thermoplastics globally, which explains its extensive use and popularity in packaging, consumer goods, and construction [13].

15.1.2.1.3 Bio-PET and Bio-PTT

Biobased PET and PTT are biopolyesters produced from biobased sources viz., terephthalic acid (TA) and ethylene glycol. Bio-PET, a non-biodegradable material but is now reported to be degraded by the enzymatic activity of a novel bacterium, its ability to utilize low-crystallinity PET as a carbon source; this indeed is an opportunity for biodegradation of PET with an enzyme called PETase. PET has applications in the sectors like packaging and textiles also others. Bio-PTT is used in consumer goods, textiles, automotive, and transport [13].

15.1.2.1.4 Bio-Poly Vinyl Chloride (Bio-PVC)

Polymerization of vinyl chloride gives poly vinyl chloride (PVC). Chlorinated polyvinyl chloride is a thermoplast and highly flexible material. Therefore it has numerous applications specifically in fluid (hot and cold) handling industries. PVC is the largest end-product consumer of chlorine. For the production of biobased PVC, bioethylene is derived from bioethanol. Like PE, biobased PVC can completely substitute conventional PVC [15]. Polyvinyl chloride (PVC, or vinyl) is a versatile plastic and cost-effective that is used in an expansive range of products in the construction and building, electronics, automobile, health care, and other industries, including siding and piping, blood bags, and tubing, windshield system components, wire, and cable insulation, and more.

15.1.2.1.5 Bio-Polyamides (Bio-PA)

Polyamides (Pas) are produced from diacid and diamine by a condensation reaction and they can also be a single repeating unit comprising of both the carboxylic and amines. Biobased polyamides are available in the market made up of undecanoic acid or sebacic acid, extracted from castor oil. For instance, polyamide 11(PA 11) can also be partially or fully bioderived, examples being PA610, PA6, PA66, PA510, PA1010, and PA12. The applications of Bio-PA are in electrics and electronics, building and construction, automotive and transportation, textiles, and consumer goods.

15.1.2.2 Biobased, Biodegradable

This type of bioplastics is produced from microbial fermentation or plant biomass and/or animal-derived polymers. Polyesters (for instance: 3-hydroxybutyric acid) are derived from microbial sources; they are produced by micro-organisms to support their growth as well as survival. Fermentation through micro-organisms for the bioplasts is limited due to the high production and recovery cost [16]. Plant-derived raw materials like starch from wheat, rice, barley, oat, soy sources from jut, hump, pineapple, banana stem, etc., vegetable oil extract thermoplastic starch lignin, cellulose, etc. are used for the production of bioplastics. Certain products that are non-vegan such as gluten, silk, chitin, casein, gelatin, wool, and fats are used in bioplastic production. Bioplastics under this category that are commercially available are Polylactic acid (PLA), Polybutylene adipate terephthalate (PBAT), Polybutylene Succinate(Bio-PBS), and Polyhydroxy butyrate(PHB) [17].

15.1.2.2.1 Polylactic Acid (PLA)

It is also known as polylactide, which is produced by the lactide polymerization. Lactide is synthesized from two lactic acid molecules with dehydration followed by condensation reaction. Glucose obtained from sugarcane, potato, tapioca, corn is used for lactic acid production via fermentation [18]. The polymerization of L-lactide and D-lactide gives semicrystalline polymers, namely poly-L-lactic acid (PLLA) and poly-D-lactic acid (PDLA) [19, 20]. Also, the meso lactide polymerization gives the amorphous poly (L,D-lactic acid) (PDLLA). PLA degradation depends on shape, size of the product, its isomer ratio, and temperature and it usually degrades in 6 to 24 months of time span. Due to its good mechanical properties, it has wide applications in various sectors including sports and leisure industry.

15.1.2.2.2 Polyhydroxyalkanoates (PHA), Polyhydroxy Butyrate (PHB), and Poly (3-Hydroxybutyrate-co-3-hydroxyvalerate) (PHBV)

PHA is an aliphatic polyester manufactured by polymerizing beta, alpha, and delta hydroxyalkanoic acids obtained by fermentation of the sugar and lipid from different feedstocks [21]. PHB has an excessive modulus-of-elasticity and acts as a good barrier for water vapor and UV seen light as compared to PLA, but it is fragile and has got inferior thermal balance within the molten state. Similar to PLA it has its application in biomedical fields and packaging. The diffused PHAs are PHB and poly (3-hydroxybutyrate-co-3-hydroxyvalerate) (PHBV). PHBV currently is not viable in the market as it leads to extensive production costs. Including starch-based and cellulose-based polymers, they have applications in packaging, mulch films, plant pots, especially PHA in the medical field.

15.1.2.3 Fossil-Based, Biodegradable

Aliphatic polyesters, such as poly-caprolactone (PCL), polyglycolic acid (PGA), and polyvinyl alcohol (PVOH), contain the polymers that possess biodegradability properties. For example, functional group stability, reactivity, hydrophilicity, molecular weight, and elasticity contribute to their biodegradability [22]. Examples of Fossil based biodegradable plastics are explained below:

15.1.2.3.1 Aliphatic Polyesters

Aliphatic, semicrystalline thermoplast, Poly(buylene succinate) (PBS) is a biodegradable and is similar to rare polymers such as poly(ethylene succinate) (PES), poly(butylene adipate) (PBA), and poly(butylene succinate-co-adipate) (PBSA). It is used in various applications such as packaging, plant containers, hygiene products, and shopping bags. PCL known as polycaprolactone is an aliphatic(linear), semicrystalline polyester that belongs to the family of poly(omega-hydroxyalkanoate). It is used in packaging, textiles, consumer goods, agriculture, and horticulture [23].

15.1.2.3.2 Aliphatic-Aromatic Polyesters

Aliphatic and aromatic copolyster called as Polybutylene adipate terephthalate (PBAT) can be obtained from a poly-condensation reaction of adipic acid, terephthalic acid, and 1,4-butanediol. PBAT is majorly applicable in packaging, agriculture, and disposable tableware organic bags and is biodegradable [24].

15.1.2.3.3 Polyvinyl Alcohol (PVA)

Another biodegradable thermoplastic viz., Polyvinyl alcohol(PVA), manufactured by hydrolyzing poly(vinyl acetate). It has vast applications in agriculture, medical, packaging, consumer goods like dyes, detergents, etc. It is also used in various spinning techniques [13].

15.2 Bioplastic in Leisure

15.2.1 Camping

During vacation time, thousands of people go camping. There is a report that almost 5,000 tents were abandoned at Glastonbury in 2015 vacation. Therefore, a Campbell (University College London) student designed a corn starch bioplastic tent, which degrades within 120 days. The graduate made it more durable and waterproof by incorporating natural fibers of bamboo or silk in between the bioplastic sheets. It is marketed at Comp-a-Tent (Corn starch-based) and KarTent (cardboard tent) [23].

15.2.2 Eyewear

The most common type of plastic is polycarbonate used for sunglasses or eyewear frames. Timberland company manufactures sunglasses and their frames, which contain 80% biobased plastic [24].

Ochis Coffee eyewear collection is made up of organic coffee and flax. These equipments decompose 100 times faster than the traditional polymer eyewear material in the soil and can become fertilizer. Besides this reshape, eco, eco eyes, blue planet, kids by safilo, awear proof, and costa are the branded eyewear made from biobased plastic or recycled plastic [25].

Kickstarter is an American platform for converting ideas into reality. They have prepared a 100% biobased, oil-free glass wear by combining cornstarch-based PLA, PHB (produced by microorganisms), and plasticizing additive. It can withstand up to 110°C. After its useful life, it degrades in less than 90 days and thus helps in reducing the emission of greenhouse gasses [26]. Their other eyewear frames are handmade from cellulose acetate of natural fibers of cotton, wood, etc. Because of the strong acetate component, these frames are impact-resistant and resilient [27]. The sports things like goggles and specs are made from biopolyamide, which also have high-quality optical lenses [28].

A top brand, Gucci made eyewear called Eyeweb from a plant source that is castor oil-based polyamides. They have also produced biodegradable sunglasses from wood or lignin and natural wax, also called liquid wood [29].

15.2.3 Toys

Children's sand toys, such as sand bucket, sand truck, windmill, shovels, set of different accessories of beach game, railway models, building blocks, puzzles, etc., are all widely available in the market and are made up of PHA or cotton cellulose or PLA. Bioplastics are widely used in the production of children's toys. For example, a company named Dantoy produces several toys varieties from the materials for the role-play to garden and kindergarten. In role-play, the bioplastic material includes BIO dinner set, green garden salad set, Dantoy coffee set.

The toys offered by a giant brand company called in Aarhus, Denmark named "Dantoy" are play cups, junior stilts, disco professionals are offered by Dantoy that are made up of bioplastics. Under vehicles, tractors with front loader and rubber wheels, fun cars, police station, tow truck, etc., are offered by Dantoy. Sit and ride toys include a Dantoy scooter, police scooter, and giant excavator. The items used for the sand and water are also eco-friendly, namely, BIO sand set, transparent bucket, sand mill, and bucket set. Toys produced for the usage in the garden of the same brand are green garden set, rocker happy dog, and wheelbarrow with two rubber wheels. Toys that can be used in the snow are bum sled, bob sleigh, snow shovel. In the kindergarten, toys offered by Dantoy are a green garden set, large building blocks, and fun cars. All the products mentioned for the toys are produced by Dantoy [30]. Other examples of bioplastic marketed toys include Binabo flexible construction strips, BiOBUDDi Arctic Whale or Seal 2-in-1- bioplastic building blocks, etc. [31].

Injection molding machines are widely used in the toy business. The petro-plastic pellets used to feed them emit a significant amount of carbon. Biobased plastic toys are created from renewable feedstock like PLA, PHA, and starch polymers, rather than fossil-based plastics [32].

Plates, silverware, and toy tractors with tilting ladders are among the seven bioplastic goods produced by the Danish toy company [33]. As children pick up and put their building blocks, figurines, and cups in their mouths, toys must be of good quality and devoid of dangerous elements. Bioplastics manufactured from renewable raw materials provide a suitable and long-term solution for plastic toys in children's toys. With FKuR bioplastics, you receive high-quality toys that satisfy high expectations and that children will enjoy for a long time [35].

JELUPLAST® can be molded three-dimensionally and has a vast range of design possibilities, but it also has the benefits of wood. As a result, the bioplastic achieves more strength and stiffness than plastic, increasing toy safety. The new material closely resembles wood in warmth, texture, and scent. JELUPLAST® is ideal for toy manufacturing since it can be entirely through-dyed [37].

The BioFactur bioplastic building blocks are pollution-free, biodegradable, and manufactured from renewable materials. Organic colors are often used to dye the fabric [38]. A bucket, a strainer, a sand mold, a rake, and a spade made of bioplastic are included in the BioFactur set of sandpit toys. BPA-free and packaged in a bag made of high-quality cotton [38].

15.2.4 Electronic Equipment and Other

The first electronic equipment made of biobased plastic included the Sony Walkman™ made up of PLA in the year 2002. In 2005, a mobile phone with external cover was made using PLA that was strengthened with Kenaf fibers or natural fibers. Nowadays, numerous electronic goods are available in the market, such as few computer parts, musical headphones, etc. which are manufactured with bioplastics [28]. There are various materials that are 3D printed with a biodegradable plastic material, such as face planter, watercolor pan, nose stud, etc. [36]. Liquid wood is a lignin-based sustainable bioplastic that has found use in various musical instruments [39].

ARBOFORM® is constructed entirely of renewable raw materials and is entirely biodegradable. This substance comprises lignin, the second most prevalent natural polymer after cellulose. When lignin is combined with natural fibers (flax, hemp, or other fiber plants) and natural additives, a fiber composite is created that can be treated at high temperatures and pressures and molded using plastic injection molding equipment. Loudspeakers and musical instruments are examples of typical applications [40].

15.3 Bioplastic in Sports

15.3.1 Shoes and Sneakers

Shoes are composed of various plastic and common leather, contributing to their significant carbon footprint. Native Shoes, a Canadian shoe firm, is attempting to reduce its use of plastic by creating a fully biodegradable shoe, as every component is manufactured from plant material. Their new Plant Shoe is crafted entirely of natural fibers, including pineapple husk for the toe and thread made from the fibrous vegetable jute soaked in olive oil. These shoes are made from locally produced braided jute that has been steeped in olive oil for softness. The Plant Shoe, like other shoes, will ultimately wear out. Instead of throwing this pair in the garbage at the end of its life, it can be placed in a compost bin, where it will begin to decompose in 45 days once exposed to bacteria and natural composting agents. And, because it is free of chemicals and synthetic additives, it will not harm the environment in the process [34].

Reebok, by Adidas (ADDDF) since 2005, announced the design for its first plant-based sports shoe, the Floatride GROW, that is durable enough to withstand the wear and tear of athletes' energetic and frequent application. The top portion of the shoe is mainly constructed of eucalyptus. Castor beans and natural rubber are used to make the soles. The majority of shoe waste comprises unrecyclable plastic, leather, and petroleum sourced rubber, which takes 25 to 80 years to disintegrate organically. According to NPD senior sports business analyst Matt Powell, "Around 35% of pollees stated they are ready to pay extra for a shoe manufactured from sustainable sources." Although it may not represent the whole market, it is unquestionably a growing area. Nike (NKE) debuted a plant-based, athletic line with the limited edition of a "100% organic" bespoke Nike Air Max 90 [34].

A socially conscious shoe brand, TOMS launched a new collection built from sustainable elements. The Earthwise line includes plant-based colors, recyclable plastic uppers, and, most importantly, the I'm green™ ethylene-vinyl acetate (EVA) developed by the Brazilian petrochemical corporation Braskem. The I'm green™ ethylene-vinyl acetate EVA rubber outsoles are a biobased substitute to the typical petroleum-based ethylene-vinyl acetate (EVA) outsoles, which do not biodegrade and are challenging to recycle, which is why many shoes end up in landfills. Ethylene-vinyl acetate (EVA) derived from sugarcane, on the other hand, is recyclable, reusable, and biodegradable, as well as carbon-negative. Sugarcane absorbs huge amount of carbon dioxide from environment while releasing oxygen. When sugarcane juice is crushed, it is converted into bioethanol, which serves as the foundation for various bioplastics [41].

PUMA InCycle comprises degradable polymers, recycled polyesters, and organic cotton. The upper portion of PUMA's biodegradable shoe "Basket" is composed of a blend of cotton and linen. The base is made up of biodegradable plastic APINAT Bio. The APINAT bioplastic line from the firm proposes hard or soft grade degradable bioplastics from synthetic raw materials (APINAT B) and biodegradable bioplastics from renewable resources (APINAT BR) with renewable content (up to roughly 60%) [29].

MoEa (Mother Earth) is a sneaker brand from France, making shoes from plants and fruits such as apple, pineapple, grapes, and corn. These new sneakers have reduced carbon emission by 89% than traditional leather sneakers. Consumers exchanges their shows with the company, and then they again recycle them into soles, and thus company saves 40% of rubber on each pair of soles. Alike MoEa some other companies such as Nike, Nothing New, and Converse Renew are using recycled plastic for making sneakers. The famous Italian brand of sneakers ID.EIGHT uses by-products of the food industry such as pineapple, grapes, seeds, and apples. The other parts of the shoe, such as the upper sole, laces, and label, are also made up of recycled material [42]. VEGEA Milan, an Italy-based company, is developing fashion, packaging automotive, and transportation materials from plant-based sources. In a similar way, they have made biker jackets and shoes of high quality and completely vegan [43].

15.3.2 Ski Boots

Arkema (France) launched its biobased Pebax® RNew 80R53 grade for ski boots. It is 50% stiffer than current Pebax® grades while retaining the vital Pebax® features and eco-friendly qualities that attract athletes. It is made using renewable raw materials having more than 90% biobased content. It is lightweight, UV resistant, and offers excellent imaginativeness for innovative designs while maintaining high quality and durability under icy situations [44]. Atomic's Renu is the world's first ski boot made entirely of renewable materials such as bioplastics, cotton, and bamboo [45]. A 100% biobased thermoplastic polyether (TPE) was introduced in 2010 for application in ski boots. The TPE material is made entirely of biobased polyamide. The top cuff of this ski boot is composed of partly biobased elastomer [28].

15.3.3 Snow Goggles

Spektrum, a Swedish sports eyewear business, introduced a new line of ski goggles built from its signature plant-based bioplastic. The frames are manufactured with a resin derived from maize and castor bean derivatives that are unsuitable for human consumption. Spektrum uses a unique method that changes the components to make polymer pellets then colored. This guarantees that all of Spektrum's frames have a rich color, and the whole process enhances durability and scratch resistance [46].

Bosky's eco-friendly sunglasses are a potential alternative to similar items that rely on plastics and foams. The vents are made of fine-weave hemp, and the frames are made of plant-based (non-petroleum) plastic. However, the lens on the test model gets scratched quickly, and after a few days of wear, one of the hemp vents starts to separate [47].

15.3.4 Surfboards and Surfskates

Stephen Mayfield, a UC San Diego professor of biology along with his team, had discovered a revolutionized algae-based bioplastic. This is used for making shoes and flip-flops. They also used it for making surfboards, skateboards, and yoga mats. The surfing industry has embraced this innovation for producing more eco-friendly products. Further, this algae-based plastic can be utilized for making polyurethane athletic shoes and even car seats and tires [48]. The Tesla surfboards use biobased epoxy resin, which is superior to traditional polyester [49]. US-based surfing equipment manufacturing company' Surfing Fins' collaborated with "Green Dot Bioplastic" for manufacturing a wide range of surfboards fins from bioplastic and biocomposites. For that purpose, they have used Green Dot's trademarked Terratek wood-plastic composite, which is a combination of wood fibers and recycled polypropylene (PP) plastic [50]. There are some other companies, such as Surftech, Glide, Starboard, Solace SUP, Infinity SUP, Invert SUP, BIC SUP, and NSP, which are trying to replace plastic with bioplastic for making surfboards eco-friendly and sustainable to reduce carbon footprint [51]. Carver has been producing surf skate since 1996. Their new Ecothane Mag wheel of a skateboard is immensely smooth and developed using non-petroleum-based material that is soyabean oil and helps in decreasing the wheel's carbon footprint [52].

15.3.5 Sportscar

Sports car racing is a road racing sport that uses enclosed sports cars. The first bioplastic car in the automobile industry was made by scholars of the Eindhoven University of Technology. The weight of this sportscar is almost half that of a regular car. This car is weather-proof and made from polylactic acid (PLA). The carriage is composed of sugars [53].

A collaboration between the Japanese government industry and academia has developed a bioplastic with improvements in exterior surface, strength, and heat resistance. This poly-lactic acid-based bioplastic is made from corn (88%) and petroleum (12%). They focused on synthesizing a new nucleating agent for crystallization and a blending compound to increase the strength and heat resistance capacity, which can be beneficial for automobile manufacturing. A nucleating agent acts as a driving force for the crystal formation process. Compared with contemporary bioplastic, it also shows three-time greater shock resistance with 25% greater heat resistance [54]. Recently another group of researchers from the University of Tokyo made a highly heat resistant bioplastic from pulp, which is usually an ingredient in paper production. It withstands up to 500°C and is lightweight in nature. It has been observed that this plastic is more heat tolerant than aluminum and magnesium, so it can be combined with these two metals to prepare different parts of the car [55].

For the first time, bioplastic is used for the development of interior parts of **"Mazda MX-5,"** which is a lightweight two-passenger roadster sports car. This car was launched internationally in 2015 by Mazda Motor Corporation, a Japanese automaker company. This biobased engineering plastic has previously been used in CX-9, Mazda3, Mazda2, and for decoration parts of MX-5 RF.

In partnership with Mitsubishi, Mazda has synthesized a plant-sourced polymer that is weather, scratch, and impact resistant to sustain harsh environmental conditions. Such plastic is suitable for the production of exterior parts of a car. The advantage of this material is that it can be painted, and thus the emission of volatile organic compounds related to the painting process gets decreased. Such pigmented material had shown a dark smooth shade with mirror-like glazing. Adoption of such plant-derived material has reduced the use of petroleum resources and thus CO_2 emissions.

For many years Mazda has been working actively to develop technologies in the biomass field. Under the name of "Mazda Biotechmaterial," the company has successfully synthesized high strength heat resistant, plant-derived bioplastic. They have applied it for the development of interior parts of an automobile. They have also synthesized the world's first plant-based biofabric for vehicle seats [56, 57].

15.3.6 Football, Baseball, Basketball, Soccer Ball, and Volleyball

Biodegradable thermoplastic polyurethane (TPU), an important polymer that amalgamates the malleability and flexibility and high strength of rubber and plastic respectively. It possesses many advantages such as recyclability, excellent mechanical and thermal strength, wear resistance, biocompatibility, etc. Therefore, TPU is environmentally friendly and allows manufacturers to use it as an alternative to plastic in various fields such as aerospace and electronics, medical, automotive, and sports. These products are non-toxic and can be broken down into primary chemical components approximately between 3 and 5 years. During the biodegradation process, it does not change pH of soil or groundwater [58]. Eco Sports is a sports equipment manufacturing company from Los Angeles, founded by Troy

Akin. They are using biodegradable TPU for the manufacturing of different sports goods such as basketballs, footballs, soccer balls, volleyballs, and baseball gloves. Even they are making sports t-shirts with 92% recycled PET [59].

One of the greatest recycling was made by Adidas in partnership with Parley for the Oceans. The company has successfully recycled 1.8 million plastic bottles for the development of the American Football Field in Miami. These bottles have been collected from beaches, seashores, islands, and coastal regions. Usually, these football turfs consist of reground rubber, which helps to bounce the football. A similar kind of impact was observed with recyclable plastic used in the form of small pellets, which also ensured the safety of the players by providing cushioning to them [60].

The official World Cup 2018 football was made from biobased EPDM rubber (ARLANXEO) which is produced by Keltan Eco, who is largest producer of rubber in the world. The EPDM rubber (ethylene-propylene-diene monomer) Keltan Eco 6950 serves as the foundation for a sponge rubber layer beneath the outer cover of the "Telstar 18" ball. It acts as a fluxuous cushion for the ball and provides excellent bounce properties through games. Materials used in this layer have to fulfill specific standards for qualities such as density, hardness, weight and be processable. However, the layer's elasticity and resilience are the essential performance characteristics. Keltan Eco an EPDM rubber is made from sugarcane (plant-based ethylene). The amount of biobased material varies between 50% and 70%, based on the ethylene concentration of every rubber grade. Adidas uses Keltan Eco 6950, distinguished by its indefinite nature and high crosslinking density. Resulting in good low-temperature characteristics and satisfies the maximum elasticity and robustness requirements [61].

Pebax® Rnew, a biosourced elastomer manufactured by Arkema in France, is designed as a durable soccer ball named "Join the Team" composed of a super-tough, bioplastic material [62].

15.3.7 Hockey

For Tokyo Olympic games 2020, the international hockey federation made a partnership with Polytan, a famous supplier of world-class hockey fields, and FIH's (International Hockey Federation) Global Partner for developing the eco-friendly and sustainable hockey turf. It was going to be prepared from 60% sugar cane Poligras Tokyo GT (Green Technology) based on Braskem's renewable I'm Green polyethylene technology [63]. This technology produces plastic from ethanol obtained from cane sugar, and this has a positive impact on carbon footprint as compared to fossil polyethylene. It saves 4 tonnes of CO_2 per tonne of a pitch, which corresponds to saving approximately 28 tonnes of CO_2 for the entire hockey pitch [64].

Apart from Tokyo 2020 Olympic games, this Poligras of Polytan was the choice for A) 7 Olympic games including Beijing, London, and Rio, B) 74% of FIH Hockey Pro League venues, C) 2018, 2022, 2026 Hockey World Cups, D) 2018 & 2022 Asian Games, E) 6 Commonwealth Games and F) 10 of the 14 Tokyo 2020 Olympic Qualifier matches. Therefore, it has become hockey's number 1 turf [65].

15.4 Conclusion

Bioplastic or biodegradable plastic is gaining tremendous value due to several ill effects of polymers on the environment. Nowadays, the bioplasts are widely used in several industries including sports and leisure, which helps reduce the greenhouse emissions. In years to come, bioplastic is the best solution for wastes generated in nature. The application of bioplastics in the manufacturing of different sporting and leisure goods, such as kid toys, sports cars, bikes, shoes, turfs, musical instruments, balls, t-shirts, googles, etc., will create awareness among people for acceptance and usage of the bioplastics. Recapitulating, with the incorporation of biopolymers in daily routine, will lead to a sustainable environment.

References

1. Emadian, S.M., Onay, T.T., Demirel, B., Biodegradation of bioplastics in natural environments. *Waste Manage.*, 59, 526, 2017.
2. Gironi, F. and Piemonte, V., Bioplastics and petroleum-based plastics: Strengths and weaknesses. *Energy Sources Part A Recovery Util. Environ. Eff.*, 33, 1949, 2011.
3. An analysis of European plastics production, demand and waste data, issue, plastics, the facts, Europe, 2020, https://issuu.com/plasticseuropeebook/docs/plastics_the_facts-web-dec2020.
4. UN Environment Programme, Reframing tourism to address plastic pollution, UNEP, UNWTO, the Ellen MacArthur Foundation and the GTPI, Kenya, 2021, https://www.unep.org/news-and-stories/story/reframing-tourism-address-plastic-pollution.
5. International Union for Conservation of Nature (IUCN), Marine plastic pollution, International Union for Conservation of Nature publications, Switzerland, 2021, https://www.iucn.org/resources/issues-briefs/marine-plastic-pollution.
6. UN environmental programme, The state of plastics: World environment day outlook 2018, United Nations Environment Programme, Kenya, 2018, https://www.unep.org/resources/report/state-plastics-world-environment-day-outlook-2018.
7. Zheng, J. and Suh, S., Strategies to reduce the global carbon footprint of plastics. *Nat. Clim. Change*, 9, 374, 2019.
8. Gov.UK, Sports industry urged to kick plastics out of sport, Environment Agency. UK, 2021, https://www.gov.uk/government/news/sports-industry-urged-to-kick-plastics-out-of-sport.
9. Pinol, P., Rodriguez, L., Brouta, M., Escamilla, M., Escudero, R., Mitigation of microplastics impact caused by textile washing processes LIFE13 ENV IT 1069. *Mermaids*, 1, 1–5, 2016.
10. The World Economic Forum, The Wimbledon tennis tournament is reducing single-use plastics, Johnny Wood, Writer, Formative Content. World Economic Forum, Switzerland, 2019, https://www.weforum.org/agenda/2019/06/wimbledon-championships-reducing-single-use-plastics/.
11. The Colosseum Sports And Fitness, 10 ways the world of sport is tackling plastic pollution, THE COLOSSEUM, SPORTS AND FITNESS, Chennai, India, 2019, https://www.thecolosseum.in/post/10-ways-the-world-of-sport-is-tackling-plastic-pollution.
12. European-bioplastics, New market data: The positive trend for the bioplastics industry remains stable, European Bioplastics. Berlin, 2018, https://www.european-bioplastics.org/new-market-data-the-positive-trend-for-the-bioplastics-industry-remains-stable/.
13. Fredi, G. and Dorigato, A., Recycling of bioplastic waste: A review. *Adv. Ind. Eng. Polym. Res.*, 4, 159, 2021.

14. Siracusa, V. and Blanco, I., Bio-polyethylene (Bio-PE), Bio-polypropylene (Bio-PP) and Bio-poly(ethylene terephthalate) (Bio-PET): Recent developments in bio-based polymers analogous to petroleum-derived ones for packaging and engineering applications. *Polym. (Basel)*, 12, 8, 2020.
15. Pascault, J.P., Höfer, R., Fuertes, P., Mono-, di-, and oligosaccharides as precursors for polymer synthesis, in: *Polymer Science: A Comprehensive Reference*, vol. 10, Elsevier B.V., Spain, 2012.
16. Luengo, J.M., García, B., Sandoval, A., Naharro, G., Olivera, E.R., Bioplastics from microorganisms. *Curr. Opin. Microbiol.*, 6, 251, 2003.
17. Bhagwat, G., Gray, K., Wilson, S.P., Muniyasamy, S., Vincent, S.G.T., Bush, R., Palanisami, T., Benchmarking bioplastics: A natural step towards a sustainable future european committee for standardization. *J. Polym. Environ.*, 28, 3055, 2020.
18. Lasprilla, A.J.R., Martinez, G.A.R., Lunelli, B.H., Jardini, A.L., Filho, R.M., Poly-lactic acid synthesis for application in biomedical devices - A review. *Biotechnol. Adv.*, 30, 321, 2012.
19. Kobayashi, J., Asahi, T., Ichiki, M., Oikawa, A., Suzuki, H., Watanabe, T., Fukada, E., Shikinami, Y., Structural and optical properties of poly lactic acids. *J. Appl. Phys.*, 77, 2957, 1995.
20. Maharana, T., Mohanty, B., Negi, Y.S., Melt-solid polycondensation of lactic acid and its biodegradability. *Prog. Polym. Sci.*, 34, 99, 2009.
21. Bhatia, S.K., Otari, S.V., Jeon, J.M., Gurav, R., Choi, Y.K., Bhatia, R.K., Pugazhendhi, A., Kumar, V., Banu, J.R., Yoon, J.J., Choi, K.Y., Biowaste-to-bioplastic (polyhydroxyalkanoates): Conversion technologies, strategies, challenges, and perspective. *Bioresour. Technol.*, 326, 124733, 2021.
22. Anderson, J.M. and Shive, M.S., Biodegradation and biocompatibility of PLA and PLGA microspheres. *Adv. Drug Delivery Rev.*, 64, 72, 2012.
23. The Guardian, Living in a box: The cardboard and bioplastic tents pitching up at festivals, The Guardian, UK, 2016, https://www.theguardian.com/sustainable-business/2016/jul/02/cardboard-bioplastic-tents-pitching-up-at-festivals-glastonbury-waste.
24. Auto news, Timberland eyewear looks great and supports the environment, Gear Patrol. NY, USA, 2021, https://autobala.com/timberland-eyewear-looks-great-and-supports-the-environment/245805/.
25. Bioplastic news, Bioplastics Eyewear, Axel Barrett, BioplasticsNews.com, Brussels, 2019, https://bioplasticsnews.com/2019/04/06/bioplastics-eyewear/.
26. Kickstarter, 100% Biobased eyewear fighting global warming, Kickstarter, NY, USA, 2017. https://www.kickstarter.com/projects/36549098/cps-100-biobased-bioplastic-designer-eyewear.
27. Kickstarter, Handmade eyewear from biodegradable and recycled materials, Kickstarter, NY, USA, 2021, https://www.kickstarter.com/projects/oceanvegansneakers/handmade-eyewear-from-biodegradable-and-recycled-materials.
28. Thielen, M., *Bioplastics*, p. 41, Fachagentur Nachwachsende Rohstoffe e. V. (FNR), Germany, 2019.
29. Green Chemical Blog, PUMA debuts biodegradable shoes, Green Chemicals Blog, UK, 2012, https://greenchemicalsblog.com/2012/10/11/puma-debuts-biodegradable-shoes/.
30. Dantoy, Dantoy's Bio life cycle, dantoy, Denmark, 2022, (Website Access date) https://dantoy.dk/en/bio/.
31. Jiminy eco toys, Bio-plastic toys, Jiminy Eco Toys, Dublin, Ireland, 2022, (Website Access date), https://jiminy.ie/collections/bio-plastic-toys.
32. Bioplastic News, Bioplastic toys and biotoys, AXEL BARRETT(Author), BioplasticsNews.com. Brussels, 2015, https://bioplasticsnews.com/bioplastics-toys/.
33. Etsy, Bioplastic, Etsy, UK/Germany, 2022, (Website Access date) https://www.etsy.com/in-en/market/bioplastic?ref=pagination&page=8.

34. Fast Company, Every single piece of these sneakers is made from plants, EILLIE ANZILOTTI, Fast Company, NY, USA, 2019, https://www.fastcompany.com/90362193/every-part-of-these-sneakers-is-made-from-a-plant.
35. FKUR, Plastic made by natural, toys from Bioplastics, FKuR, Germany/USA, 2022, https://fkur.com/en/applications/toys-from-bioplastics/.
36. Bioplastic news, Bioplastic sneakers will be a rage in 2020, AXEL BARRETT, BioplasticsNews.com. Published on edition.cnn.com, Brussels, 2019, https://bioplasticsnews.com/2019/12/17/bioplastic-sneakers/.
37. Renewable Carbon News, Bioplastic for the toy industry, RENEWABLE CARBON NEWS, Germany, 2015, https://renewable-carbon.eu/news/bioplastic-for-the-toy-industry/.
38. Eco and Upcycling Market, Greenpicks, Greenpicks, NA/INternational, 2022, (Website Access date) https://www.greenpicks.de/en/0404g-2800020-toy-blocks-made-of-bioplastics-from-biofactur.html/.
39. Alb filter, Bioplastics, Alb filter, Deutschland 2019, https://alb-filter.com/blogs/ratgeber/biokunststoffe?lang=en.
40. Bioplastics News, TECNARO, Bioplastic News, Brussels, 2022, (Website Access date) https://bioplasticsnews.com/arboform/.
41. Spring wise, Sugarcane-based bioplastic replaces rubber shoe soles, 2020, https://www.spring-wise.com/sustainability-innovation/fashion-beauty/eco-friendly-shoes-toms-eva-alternative.
42. Open Growth Community, New sustainable sneakers made from fruit waste, OpenGrowth, CA, USA, 2021, https://www.opengrowth.com/article/new-sustainable-sneakers-made-from-fruit-waste.
43. VEGEA, Biker and shoes made from VEGEA, VEGEA company. Milano, Italy 2020, https://www.vegeacompany.com/biker-and-shoes-made-from-vegea/.
44. Bioplastic news, Arkema latest solutions for ski boots, Axel Barrett, BioplasticsNews.com, Brussels, 2014, https://bioplasticsnews.com/2014/02/09/arkemas-latest-solutions-for-ski-boots/.
45. Evo, Atomic RENU 110 ski boots 2010, evo, US, Canada, 2022, (Website Access date) https://www.evo.com/outlet/alpine-ski-boots/atomic-renu110#:~:text=Atomic%20RENU%20110%20Ski%20Boots%202010%3A%20The%20world's%20first%20carbon,evo%20always%20salutes%20green%20initiatives.
46. Spring wise, High-tech ski goggles made from bioplastic, Springwise, UK, 2021, https://www.springwise.com/sustainability-innovation/sport-fitness/spektrum/ostra-ski-goggles#:~:-text=Spotted%3A%20Swedish%20sports%20eyewear%20brand,a%20number%20of%20innovative%20features.
47. Gear Institute, Bosky bioplastic snow goggle review, Gear Institute, Santa Fe, NM, 2015, https://gearinstitute.com/gear-review/bosky-bioplastic-snow-goggle/.
48. Bioplastic news, US Leads Algae Bioplastics with Flip Flops, Yoga Mats and Skateboards, Axel Barrett, BioplasticsNews.com, Brussels, 2020, https://bioplasticsnews.com/2020/07/17/us-algae-bioplastics-flip-flops-yoga-mats-skateboards/.
49. Bioplastic News, Elon musk builds bioplastics surfboards, Axel Barrett, BioplasticsNews.com. Brussels, 2018, https://bioplasticsnews.com/2018/08/07/elon-musk-bioplastics-surfboards/.
50. Composite World, Bio-composite surfboard fins, CompositesWorld, Cincinnati, USA, 2017, https://www.compositesworld.com/articles/bio-composite-surfboard-fins-.
51. Men's Journal, Seven SUP brands going seriously sustainable, Men's Journal, New York, 2022. https://www.mensjournal.com/adventure/five-eco-friendly-standup-paddleboards/.
52. Bureo, New ECOthane Wheels, Bureo, CA, USA, 2016, https://bureo.co/blogs/the-rip/ecothane.
53. Bioplastic News, First car made completely from bioplastics, Axel Barrett, Bioplastics News.com, Brussels, 2018, https://bioplasticsnews.com/2018/07/09/bioplastics-car/.

54. Japanese Research Consortium develops high-strength heat-resistant bioplastic for auto parts, Green Car Congress, Japan, 2006, https://www.greencarcongress.com/2006/05/japanese_resear.html.
55. Green Car Congress, *Japanese research consortium develops high-strength heat-resistant bioplastic for auto parts, green car congress*, 2006, https://www.greencarcongress.com/2006/05/japanese_resear.html.
56. MAZDA Newsroom, Mazda develops bio-based engineering plastic featuring high-q, Mazda Motor Corporation, Hiroshima, Japan, 2014, https://newsroom.mazda.com/en/publicity/release/2014/201412/141210a.html.
57. Mazda, Bio-based plastics, Mazda, Europe, 2022, https://www.mazda.com/en/innovation/technology/env/bioplastics/.
58. Wang, Z., Yan, J., Wang, T., Zai, Y., Qiu, L., Wang, Q., Fabrication and properties of a bio-based biodegradable thermoplastic polyurethane elastomer. *Polymers*, 11, 1121, 2019.
59. Eco Sports, Men's athletic eco-friendly shirts, Eco Sports, US, 2022. https://ecosports.com/products/athletic-shirts-eco-friendly-rpet-work-out-shirt.
60. Intelligent Living, Adidas made a football field from 1.8 million recycled plastic bottles, Intelligent Living Media Ltd, UK, 2020, https://www.intelligentliving.co/adidas-football-field-recycled-plastic-bottles/.
61. Bioplastics Magzine, Keltan Eco part of the official match football for World Cup 2018, . bioplastics MAGAZINE, Germany, 2017, https://www.bioplasticsmagazine.com/en/news/meldungen/27112017-new-football-incorparates-bio-rubber.php.
62. Bioplastics Magazine, 9, 5, 2014. 04-2010, issuu, https://issuu.com/bioplastics/docs/bioplasticsmagazine_1004/28.
63. Bioplastic News, Bioplastics hockey fields for Tokyo Olympic Games 2020, Axel Barrett, BioplasticsNews.com, Brussels, 2019, https://bioplasticsnews.com/2019/01/23/tokyo-2020-bioplastics-hockey-turf/.
64. Big Stadium Hocke, Green technology, Big Stadium Hockey, Australia, 2022, https://www.bigstadiumhockey.com/green-technology.
65. FIH (International Hockey Federation), Polytan, The International Hockey Federation (FIH), Lausanne, Switzerland, 2021, http://www.fih.ch/inside-fih/fih-quality-programme-for-hockey-turf/find-out-who-has-fih-certification/fih-preferred-suppliers/polytan/s.

16

Biocomposites in Active and Intelligent Food Packaging Applications

Ru Wei Teoh, Yin Yin Thoo and Adeline Su Yien Ting*

School of Science, Monash University Malaysia, Bandar Sunway, Selangor, Malaysia

Abstract
Biobased food packaging materials integrated with additives, which typically include antimicrobial and antioxidant agents, give rise to biocomposite materials. Biocomposite materials are considered intelligent biobased materials as they can provide real-time information on the quality of food products. Among the different additives used are plant extracts, essential oils, enzymes, and bacterocins that exhibit antimicrobial and antioxidant activities. They effectively inhibit growth of food spoilage microbes while prolonging shelf-life. Similarly, incorporation of metal particles and nanoclay into biocomposites also render excellent antimicrobial properties. These biocomposites are also capable of modifying the internal atmosphere of the packaging through scavenging of moisture, oxygen, and ethylene. Subsequently, food freshness is maintained along with microbial-free food conditions. Natural food colorants and pigments are used in the development of intelligent food packaging as they are sensitive to pH changes and gaseous production. They can detect and provide real-time information on food spoilage. This chapter aims to review the biocomposite materials used as active and intelligent food packaging materials as they have the potential to increase shelf-life and freshness of food, while providing real-time information on food quality.

***Keywords*:** Active food packaging, biocomposites, food colorant, metal particles, plant extracts

16.1 Introduction

In recent years, plastic food packaging materials have become one of the main causes of severe pollution affecting the environment and human health [1]. Ingesting microplastics leads to health implications, such as impaired metabolism, induced responses on oxidative stress pathways, and neurotoxicity [2]. It is therefore imperative that plastic materials are replaced with biobased materials to alleviate problems from plastic pollution. Biobased materials, such as polysaccharide, protein, and lipid, had gained much attention, owing to their biocompatibility, biodegradability, renewability, and nontoxic nature [3, 4]. This

**Corresponding author*: adeline.ting@monash.edu

Inamuddin and Tariq Altalhi (eds.) Handbook of Bioplastics and Biocomposites Engineering Applications: 2nd Edition, (317–360) © 2023 Scrivener Publishing LLC

resulted in the extensive development on food packaging films, coatings, and even food containers using these biobased materials [5, 6].

Although food packaging from biobased materials have great potential in replacing plastic materials, their application is relatively limited. Biobased food packaging materials are typically produced from a single source of biobased material, which often does not produce high-quality food packaging materials. Maran *et al.* [7] and Dias *et al.* [8] showed that a pure starch-based film has poor mechanical and moisture barrier properties. For pure chitosan film, high water vapor permeability and moisture content was detected [9]. This led to the development of biocomposites as the addition of other compounds into the biobased material resulted in better quality packaging materials. This was reported by Silva *et al.* [10] in which through the mixing of polysaccharide with pure whey protein isolate (WPI), the oxygen and carbon dioxide resistance of the biocomposite was significantly improved. Therefore, food packaging materials made from single source of biobased material is no longer ideal, and that biocomposites (with two or more different compounds) are now highly preferred.

Biocomposite food packaging materials yield better physical and mechanical barrier properties [11]. These improvements are inherent to the food packaging material itself and function passively, without affecting the food product. Nevertheless, biocomposites can have an indirect impact on the food product, depending on the additives incorporated. Some specific additives confer antimicrobial, antioxidant and gaseous scavenging activity, to prolong the shelf-life of the food product. Some other additives provide real-time information of food quality leading to intelligent food packaging [12].

A wide variety of additives can be incorporated into biocomposites to give rise to active and intelligent food packaging. Plant extracts and essential oils are of particular interest, as they contain bioactive compounds that have medicinal, antimicrobial and antioxidant properties [13, 14]. Incorporation of essential oils prevents microbial contamination and oxidation of food, as well as sustaining food freshness for longer period. Bacteriocins and enzymes render similar effects as essential oils, although they are used more prominently for food preservation [15, 16]. Inorganic compounds, such as metal particles and minerals, have also been explored and found to have diverse functions and benefits. These inorganic compounds have antimicrobial and antioxidant activity, as well as the ability to scavenge carbon dioxide and ethylene [17]. They control microbial growth and regulate the internal atmosphere within a package. Natural food colorants and pigments are also frequently integrated into biocomposites, especially for intelligent food packaging as they are excellent indicators of food freshness [18]. These compounds are sensitive to pH changes (through color change) and the alteration of gaseous content in the packaging atmosphere [19, 20].

The aim of this chapter is to review the recent progress and impact of biocomposites incorporated with various natural compounds, inorganic materials, natural food colorants and pigments, for beneficial use in active and intelligent food packaging applications. The current challenges of each biocomposite in food packaging applications are also addressed in this chapter. It is expected that the potential of active compounds in biocomposite food packaging is thoroughly investigated, and modifications can be done on active compounds to further enhance the development of biocomposites for food packaging.

16.2 Advances in Biocomposite Application in Active and Intelligent Food Packaging

Active food packaging is defined as the packaging approach that intentionally incorporates subsidiary constituents into the packaging material or package headspace, to enhance the performance of the packaging system [21]. On the other hand, intelligent food packaging refers to food packaging approaches where packages are integrated with a target-specific sensor, which reveals food quality attributes to the customers such as freshness and occurrence of microbial contamination [22]. Both the active and intelligent food packaging approaches are based on the same concept that is the integration of additives to create biocomposites that result in specific or multi-function of the food packaging system. Figure 16.1 shows the overall function of typical active and intelligent food packaging applications.

16.2.1 Antimicrobial and Antioxidant Properties in Active Food Packaging

Antimicrobial and antioxidant properties are essential in active food packaging applications as they prevent food spoilage caused by microorganisms. Food spoilage is a major issue in food safety as this will lead to the spread of food-borne illnesses [23]. Food oxidation is also another important factor that contributes to food spoilage, as food quality is compromised due to the alteration of organoleptic properties and nutrients in the food [24, 25]. Both factors lead to shorten shelf-life of food. Therefore, most biocomposites in active food packaging applications are equipped with antimicrobial and antioxidant properties to inhibit the growth of pathogenic microbes and to prevent oxidation of food, thereby reducing risks of contracting food-borne diseases.

To achieve this, biocomposites are integrated with constituents that have antimicrobial and antioxidant properties. However, the incorporation of a single constituent may not be ideal as the amount required to achieve desired effects may be high and exceeds safety levels.

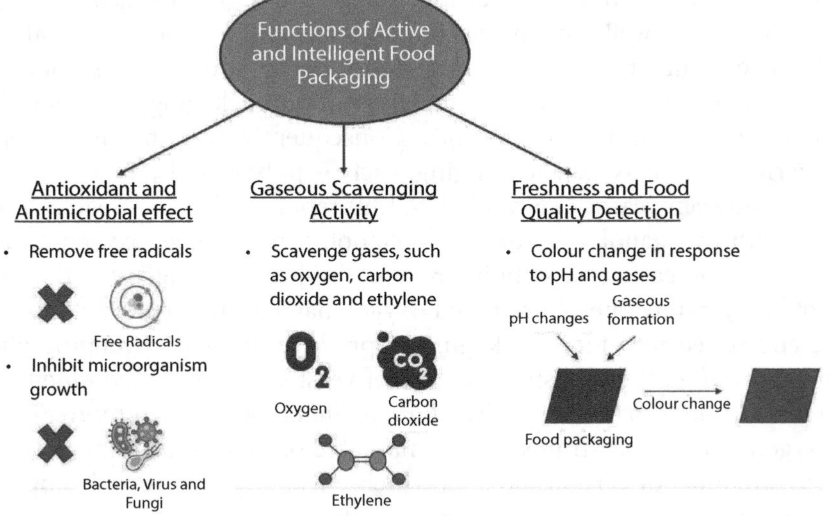

Figure 16.1 Overview of the functions of active and intelligent food packaging.

In addition, if the constituents have unique sensory profile (e.g., strong aroma, unpleasant taste), consumer acceptance may be impacted when the constituent is applied at high concentrations. To mitigate this, different combinations of constituents are often integrated into biocomposites as desired effects can be achieved with low concentrations [26]. It is also ideal if the constituents integrated into the biocomposites have sustained and controlled release mechanisms [27]. To achieve this, biobased materials are fabricated into microparticles or nanoparticles, such as cellulose nanoparticles [28], pullulan nanoparticles [29] and zein nanoparticles [30]. As these constituents gradually migrate into the headspace within the packaging or directly to the food products, the shelf-life and freshness is prolonged.

Another trending approach in fabricating biocomposites is via electrospinning and Pickering emulsion. The electrospinning technique involves the electrohydrodynamic process where biobased materials form ultrathin fibers [31] with higher total surface area, higher loading capacity and sustained release of embedded active constituents [32]. The physical and structural characteristics of biocomposites can also be tailored according to desired requirements through electrospinning [33]. For Pickering emulsion technique, this technique is suitable for constituents that are volatile and highly sensitive [34]. This technique is suitable for incorporation of essential oils, which results in food packaging that has sustained release of constituents, and improved mechanical and water barrier properties [35]. Biocomposites in active food packaging produced by the Pickering emulsion technique are considered as novel approaches, and further studies are required to further elucidate its potential and benefits in food packaging applications.

16.2.2 Gaseous Scavenging Activity in Active Food Packaging

Oxygen, carbon dioxide and ethylene are the main gaseous that affects the freshness, quality and shelf-life of food products. Oxygen causes food oxidation and encourages the growth of aerobic food pathogens. Oxygen within the food packaging must be eliminated by scavenging oxygen agents such as iron, palladium and titanium [36]. These metal compounds turn into metal oxide forms upon interaction with the humidity in food packaging, and in this metal oxide form, they can scavenge oxygen [37]. The oxygen scavenging effect by metal compounds is well recognized, but safety is still a concern. Metal compounds can cause food contamination as it can migrate to food products and is not suitable to be used in a microwave oven. In addition, metals in food packaging can potentially set off metal detectors in the production line [38]. Consequently, non-metallic compounds are sought as alternatives for oxygen scavenging, such as polyphenols, α-tocopherol, ascorbic acid, catechol, and even yeasts. Gaikwad *et al.* [39] reported that food packaging materials incorporated with pyrogallol, a natural polyphenol, can scavenge oxygen and reduce fat oxidation, thereby increasing oil stability. Bi *et al.* [40] further suggested that biocomposite films containing α-tocopherol improved oxygen barrier properties. Interestingly, yeasts can also be embedded into food packaging to produce oxygen scavenging effect as well [41]. Contarino *et al.* [42] established the role of yeast strains (*Wickerhamomyces anomalus, Metschnikowia pulcherrima, Aureobasidium pullulans,* and *Saccharomyces cerevisiae*) to scavenge oxygen. These four strains have remarkable oxygen consuming abilities and can produce large amounts of carbon dioxide, which effectively suppressed fungal growth so that food products are free from fungal contamination.

Apart from oxygen, carbon dioxide is another important parameter in maintaining food quality and freshness. The balance of carbon dioxide content in food packaging is critical as levels that are too high or too low could stimulate growth of microorganisms, induce food oxidation and promote physiological damages in perishables [43]. To manage carbon dioxide levels in packaged food, carbon dioxide scavengers such as zeolites, activated carbon, calcium hydroxide and sodium carbonate, can be incorporated [44, 45]. In addition, carbon dioxide emitters such as citric acid, sodium bicarbonate and ferrous carbonate are equally effective in regulating carbon dioxide levels [46]. To date, the progress and development of carbon dioxide scavenger or emitter in biocomposite food packaging applications is very limited, but there is tremendous potential in these applications.

Ethylene is another important gaseous that influences the quality of food, particularly fresh produce such as fruits and vegetables. This volatile compound is a phytohormone that mediates growth of fruits and vegetables, and the metabolic processes related to ripening, softening and senescence of fruits and vegetables [47]. Ethylene impacts the freshness of produce, as ethylene accelerates ripening process resulting in spoilage and reduced storage period. To mitigate this, ethylene scavengers such as potassium permanganate, is included into food packaging in the form of sachet, placed together with fruits and vegetables within a plastic film [48]. This has successfully delayed the ripening of tomatoes significantly [49]. The application of potassium permanganate, however, is limited due to toxicity. Ingestion of potassium permanganate causes gastrointestinal, respiratory and hepatic damage [50]. Additionally, potassium permanganate may stain food products and affect sensory properties. Recent studies explored the potential of minerals such as zeolite, as a new approach in scavenging ethylene. Do Nascimento Sousa *et al.* [51] revealed that impregnation of zeolite into chitosan will result in a biocomposite film that can scavenge ethylene effectively. Furthermore, Kumar *et al.* [52] reported that addition of zeolite into lignocellulosic packaging paper exhibit superior ethylene scavenging properties, with up to 62% absorbency, and with improved mechanical and water resistance properties as well. Nevertheless, usage of zeolite for human food application has not been approved [53].

16.2.3 Freshness and Food Quality Detection in Intelligent Food Packaging

Intelligent food packaging is a technology designed to monitor the quality of food products based on changes to food product, which are interpretable and readable. To some extent, real-time information can be provided as well, which will impact food supply chains and consumer demand [54] and helps in the planning for logistics [55]. There are three main components to intelligent food packaging; indicator, sensor and data carrier. Indicators detect changes in food product quality through chemical and biological indicators, which are interpreted as qualitative or semi-quantitative information of the condition of the food product [56]. Indicators provide information that are typically interpreted visually. Sensors consist of receptors that respond to specific chemical or biological compounds. This change will be detected by a transducer and this signal change will be converted into an output signal, which will then be processed by electronic devices [56]. Data carriers, such as barcode and quick response (QR) code, functions to provide information on the condition of food products. Data carriers automate traceability, which will improve food safety and boost consumer confidence [57].

Most biocomposites share similarities with indicators rather than sensors. Indicators are relatively easy to produce, do not require additional electronic device to interpret results, and are compatible with various biodegradable compounds. Natural food colorants and pigments are examples of indicators, owing to their color changing ability in response to pH changes, gaseous content, and temperature changes within the packaging environment [58]. These changes represent food freshness, quality and safety [59].

As compared to indicators, development of sensors is rather complex as signal transduction or electronic device needs to be included to interpret the data. There is still a technological gap in the application of biocomposites for intelligent food packaging using sensors and electronic devices in the market [60]. One possible approach of advancing the use of sensors in intelligent food packaging is to perhaps integrate the use of indicators (i.e. color indicators) with mobile device applications (color scanner) and Internet of Things (IoT), so that the relevant data of food conditions based on color change can be delivered to consumers [61]. Through this, biocomposites as indicators in intelligent food packaging can provide reliable information with the aid of digital technology.

16.3 Biocomposites Incorporated with Natural Compounds

Natural compounds are derived from living organisms and obtained through extraction, expression or distillation without any chemical modification [62]. Natural compounds are rich in active constituents that have excellent nutritional values. They often function as natural additives, preservatives, fragrances, flavors and are also known to have antimicrobial, antioxidant, antiaging and anticancer properties [63]. These beneficial compounds are harnessed and used in various applications such as for drug development, aromatherapy, fortification of food with enhanced nutrient content and more recently, integration into food packaging. The common beneficial compounds integrated into food packaging includes plant extracts and essential oils, and enzymes and bacteriocins. Figure 16.2

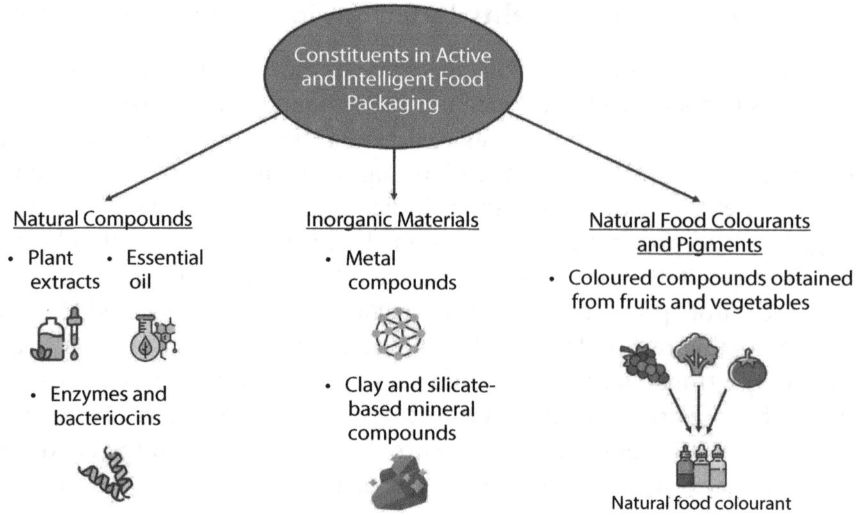

Figure 16.2 Different types of constituents incorporated into active and intelligent food packaging.

summarized the different compounds commonly added for active and intelligent food packaging applications.

16.3.1 Plant Extracts

Plant extracts are extracted from various parts of a plant, which includes leaves, flowers, fruits, twigs, bark and roots. To obtain plant extracts, the following methods are typically used: solvent extraction, distillation method or supercritical fluid extraction [64]. Different methods used will yield different amount and composition of active compounds [65]. Plant extracts impregnated into biocomposite for active food packaging applications usually have excellent antioxidant and antimicrobial activities. Both antioxidant and antimicrobial properties are critical in determining the freshness, safety, quality and shelf-life of food products. Table 16.1 summarizes the various plant extracts incorporated into biobased food packaging.

The antioxidant activities of plant extracts incorporated into biocomposites are attributed to the role of scavenging radical oxygen species (ROS) which deters oxidation of fats and lipids in food products [66]. This will ensure food products are free from toxic aldehydes and that polyunsaturated fatty acids are unaffected [67]. This also prevents lipid oxidation and the food from turning rancid or off-flavored [68]. Several studies have shown that biocomposites with several plant extracts can successfully confer antioxidant activities. Da Rosa *et al.* [69] showed that the addition of olive leaf extract into carrageenan films render the biocomposite with excellent antioxidant activity (32% increase in antioxidant activity). Kumar *et al.* [70] reported that addition of pineapple peel extract into polyvinyl alcohol/corn starch packaging film improved the 2,2-diphenyl-1-picrylhydrazyl (DPPH) radical scavenging activity by 42%. The antioxidant effect of biocomposites with plant extracts were also studied on food models. Rambabu *et al.* [71] demonstrated that mango leaf extract incorporated in chitosan film can preserve cashew nuts better than commercial plastic film. Interestingly, the antioxidant activity of the biocomposite film is 11.5 times higher than the commercial plastic films. Friedrich *et al.* [72] revealed the impact of starch/gelatin-based biodegradable coatings embedded with extracts from *Tetradenia riparia*, which successfully improved the antioxidant activities in strawberries. The authors revealed that the high antioxidant activity of strawberries was maintained within ten days of storage, with increased antioxidant effect as the concentrations of *Tetradenia riparia* extract increases.

Some plant extract has antimicrobial properties. When used in food packaging, these extracts suppress bacterial and fungal growth, which protects the food product from spoilage and microbial contamination. The antimicrobial mechanisms include disrupting cell wall or cell membrane structure, inhibiting DNA and RNA replication, and interfering with the synthesis of protein [73, 74]. Ekramian *et al.* [75] demonstrated that sago starch packaging film impregnated with extracts from *Nigella sativa* L. inhibited microbial growth. Hu *et al.* [76] also reported that extracts from the vegetable amaranth was effective in suppressing microbial growth when incorporated into the quaternary ammonium chitosan/fish gelatin biocomposite film. Amaranth extracts increased the DPPH radical scavenging activity significantly from 6.8% to 95.8%. The beneficial attributes of plant extracts have been verified in food model studies as well. Guo *et al.* [77] revealed the potential of sea buckthorn pomace extract incorporated into potato starch film, in delaying deterioration and preventing spoilage by bacteria on beef jerky. The presence of sea buckthorn pomace extracts effectively inhibited the proliferation of *Pseudomonas* spp., *Enterobacteriaceae*, lactic acid

Table 16.1 Common biocomposites incorporated with plant extracts in active food packaging applications.

Natural compound	Type of natural compound	Biobased material as main component	Effect of biocomposite film	Effects on food model systems	References
Plant extract	Olive leaf extract (OLE)	Carrageenan	High DPPH scavenging activity of 32.55% Good stretching capacity and higher water vapour permeability (WVP)	NA	[69]
	Pineapple peel extract (PPE)	Corn starch/PVA	Highest DPPH scavenging activity of approximately 42% Enhanced thermal stability and elongation at break	NA	[70]
	Mango leaf extract (MLE)	Chitosan	High FRAP, DPPH and ABTS scavenging activity, with 3.47 μg gallic acid (GA) equivalent, 8.29 μg GA equivalent and 87.16% Improved film hydrophobicity and tensile strength	Significant inhibit of cashew nuts oxidation after 28 days of storage Better reduction in fatty acid oxidation as compared to commercial plastic film	[71]
	Tetradenia riparia extract (TRE)	Starch/gelatin	Small concentration needed to reduce 50% DPPH free radicals Decrease water solubility	Reduce bacterial growth and ripening on strawberries within 10 days of storage	[72]

(*Continued*)

Table 16.1 Common biocomposites incorporated with plant extracts in active food packaging applications. (*Continued*)

Natural compound	Type of natural compound	Biobased material as main component	Effect of biocomposite film	Effects on food model systems	References
	Nigella sativa extract (NSE)	Sago starch	Effective in inhibit *E. coli* and *S. aureus*, with inhibition zone of 9.4 mm and 13.6 mm, respectively Increased elongation at break and WVP	NA	[75]
	Amaranth extract (AE)	Quaternary ammonium chitosan/fish gelatin	High antimicrobial activity against *S. aureus*, *L. monocytogenes*, *E. coli* and *S. typhimurium*, with inhibition zone 13.4 mm, 13.6 mm, 13.9 mm and 14.3 mm respectively. High DPPH scavenging activity of 95.8%	NA	[76]
	Sea buckthorn pomace extract (SBPE)	Esterified potato starch	Inhibit proliferation of *Pseudomonas* spp., *Enterobacteriaceae*, lactic acid bacteria, *L. monocytogenes*, *E. coli*, *S. aureus* and *Salmonella spp.*	Beef jerky samples have lower lipid and protein oxidation (lower TBARS and TVB-N) throughout 13 months	[77]
	Moringa oleifera seed meal extract (ME)	Chitosan	NA	Inhibit growth of *B. cereus* and *E. coli* on chicken sausage, up to nine days	[78]

(*Continued*)

Table 16.1 Common biocomposites incorporated with plant extracts in active food packaging applications. (*Continued*)

Natural compound	Type of natural compound	Biobased material as main component	Effect of biocomposite film	Effects on food model systems	References
	Pomegranate peel extract (PE) encapsulated in chitosan nanoparticles	Zein	High DPPH scavenging activity with 60.51% Sustained release rate of PE by chitosan nanoparticles Improved tensile strength	Inhibit growth of *L. monocytogenes* on pork samples up to ten storage days	[79]
	Chlorophyll from black mulberry leaf encapsulated with carboxymethyl cellulose (CMC)/silica nanoparticle	Black mulberry pectin	Improved antioxidant activity based on mathematical three-dimensional curve Enhanced inhibitory effect against *S. aureus* and *E. coli* Sustained release of chlorophyll when encapsulated with CMC	NA	[80]
	Yerba mate extract	Electrospun zein fibers	High antioxidant activity, with 83.0% DPPH free radical inhibition Sustained release of 49% YME in hydrophilic food stimulant	NA	[81]
	Resveratrol	Electrospun gelatin/zein fibers	High antibacterial efficiency against *E. coli* and *S. aureus* Sustained release was achieved in hydrophilic and acidic food stimulants	Extended pork shelf-life by three days, with lower TVB-N value. Inhibit *S. aureus*, *L. monocytogenes* and *Enterobacteriaceae* on pork samples	[82]

bacteria, *Listeria monocytogenes*, *Escherichia coli*, *Staphylococcus aureus* and *Salmonella spp.* The extract was also able to reduce incidence of lipid and protein oxidation, as lower total volatile basic nitrogen (TVB-N) and thiobarbituric acid reactive substances (TBARS) from beef jerky were detected [77]. Beef jerky incorporated with this extract was acceptable as the sensory score is satisfactory. Sharma *et al.* [78] also reported the bacteriostatic effect of chitosan-based coating impregnated with *Moringa oleifera* seed extract. Raw chicken sausages coated with this biocomposite have both Gram-negative bacteria, *E. coli* and *Bacillus cereus*, remain inactivated up to nine days.

The effectiveness of plant extracts in biocomposites is further improved by adopting the nanotechnology approach. Cui *et al.* [79] fabricated biocomposite zein film that contain chitosan nanoparticles encapsulating pomegranate peel extract and found that the biocomposite film has good thermal stability and a sustained release rate of pomegranate peel extract. The antibacterial effect was extended, which effectively suppressed the growth of *L. monocytogenes* in meat samples. In another study, chlorophyll extracts from black mulberry leaf encapsulated with carboxymethylcellulose/silica nanoparticles and embedded in black mulberry pulp pectin packaging film, resulted in enhanced thermal stability, antioxidant and antimicrobial effect against *E. coli* and *S. aureus* [80]. Aside from the nanoparticle technology in encapsulation, electrospinning technique is also an emerging approach to encapsulate plant extract for biocomposites. Biocomposites made from electrospun zein fiber blended with yerba mate extract has high antioxidant activity, up to 83% DPPH inhibition, and has sustained release of 49% yerba mate extract into hydrophilic food stimulant within 50 h [81]. Electrospun gelatin/zein fiber mat incorporated with encapsulated resveratrol from fruits and vegetables, also rendered effective antibacterial and antioxidant activities, where growth of *E. coli* and *S. aureus* were significantly suppressed [82]. In summary, the nanoparticle and electrospinning technology aids in regulating the release of plant extracts from biocomposites in the food packaging, so that a more desirable and prolonged antimicrobial and antioxidant effect can be achieved. Further work can also be done to determine the acceptable levels of plants extracts as nanoparticles in food products, in accordance with food safety and regulations [83].

16.3.2 Essential Oils

Essential oils are extracted from various parts of the plant from the extraction process. They differ from plant extracts as they are composed of mostly aromatic, volatile compounds such as terpenes, phenols, aldehydes, ketones, esters and oxides [84]. Essential oils have been used extensively in alternative natural medicine, aromatherapy and cosmetic purposes, due to their therapeutic properties [85]. The applications of essential oils in food packaging industry had increased in the recent years, due to their antimicrobial and antioxidant properties, which are beneficial in preserving food products. Table 16.2 summarizes the common essential oils incorporated into biobased food packaging.

Essential oils integrated in biocomposites are reportedly able to inhibit lipid and protein oxidation, similar to plant extracts. As a result, the quality, freshness and nutritional content of the food products are preserved [86]. Examples of essential oils incorporated include essential oil from *Artemisia campestris* into chitosan films [87], essential oil from *Origanum onites* L. into gelatin films [88], and lemon essential oil [89] into chitosan/gelatin films which has been found to have a radical scavenging capacity of 83.3%, up to 79.99% antioxidant

Table 16.2 Common biocomposites incorporated with essential oils in active food packaging applications.

Natural compound	Type of natural compound	Biobased material as main component	Effect of biocomposite film	Effects on food model systems	References
Essential oil	*Agaricus campestris* essential oil (ACEO)	Chitosan	Enhanced antioxidant effect with approximately 30% DPPH radical scavenging activity and 40% ion chelating activity Improve water resistance, thermal stability and UV-VIS light barrier	NA	[87]
	Origanum onites L. essential oil (OOEO)	Gelatin	Excellent antioxidant activity with 46.52% DPPH and 83.13% ABTS scavenging activity and 78.63% metal chelating activity Enhance UV-VIS light barrier and elongation at break	NA	[88]
	Lemon essential oil (LEO)	Gelatin/chitosan	Excellent antioxidant activity with 60.20% DPPH, 73.70% ABTS scavenging activity and 79.99% metal chelating activity Increased tensile strength and elongation at break	Significantly reduced total mesophilic aerobic bacteria count with < 10 cfu/g for seven storage days on fresh apple slices	[89]

(*Continued*)

Table 16.2 Common biocomposites incorporated with essential oils in active food packaging applications. (*Continued*)

Natural compound	Type of natural compound	Biobased material as main component	Effect of biocomposite film	Effects on food model systems	References
	Cinnamon oil (CO) + rambutan peel extract (RPE)	Cassava starch/whey protein	Exhibited good antioxidant activity with controlled release in hydrophilic food stimulant Effectively suppress growth of *B. cereus*, *E. coli* and *S. aureus*	The total viable count of salami remained low within ten days of storage	[90]
	Rockrose essential oil (REO)	Pullulan	Good antioxidant activity with 76.52% inhibition in β-carotene bleaching test Suppressed growth of *S. aureus*, *L. monocytogenes*, *E. faecalis* and *B. cereus*	NA	[93]
	Citrus limonia essential oil (CEO)	Chitosan	Significantly inhibited growth of *S. aureus*, with a large inhibition zone of 29.1 mm	NA	[94]
	Cumin essential oil (CEO)	Shahri Balangu seed mucilage	NA	Effectively reduce total viable, psychrotropic, coliforms, mold and yeast count. Inhibit *S. aureus* and *E. coli* on meat samples Shelf-life of meat samples extended to nine days with a lower peroxide and TBA value	[95]

(*Continued*)

Table 16.2 Common biocomposites incorporated with essential oils in active food packaging applications. (*Continued*)

Natural compound	Type of natural compound	Biobased material as main component	Effect of biocomposite film	Effects on food model systems	References
	Lavender essential oil (LO)	Qodume Shirazi seed mucilage	NA	Total viable, psychrotrophic, coliform, mold and yeast count Inhibit *S. aureus* and *E. coli* growth on ostrich meat Prolonged shelf-life of ostrich meat for another six days	[96]
	Garlic essential oil (GEO) nanoencapsulated with nanoliposomes	Chitosan and whey protein	NA	Reduce aerobic plate count, psychrotrophic and LAB count of sausages within 50 days of storage Reduced lipid oxidation of sausages samples with lower TBARS and PV values	[97]
	Galangal root oil (GRO) encapsulated in β-cyclodextrin	Electrospun swim bladder gelatin	Sustained release of GRO at 32.24% at 4°C and 64.87% at 25°C after seven days of storage Inhibited enterohaemorrhagic *E. coli* with an inhibition zone of 18.5 mm	Enterohaemorrhagic *E. coli* from fresh beef is effectively inhibited with a lowest microbial population of 4.04 log CFU/g	[98]

(*Continued*)

Table 16.2 Common biocomposites incorporated with essential oils in active food packaging applications. (*Continued*)

Natural compound	Type of natural compound	Biobased material as main component	Effect of biocomposite film	Effects on food model systems	References
	Clove essential oil (CLO) stabilized with zein particle in Pickering emulsion	Chitosan	Controlled release of approximately 30% of CLO within six storage days Growth of *E. coli* and *S. aureus* were significantly suppressed	NA	[100]
	Marjoram essential oil (MEO) stabilized with whey protein isolate/inulin in Pickering emulsion	Pectin	Good antioxidant activity with approximately 60% DPPH scavenging activity Sustained release rate of MEO at 4°C, 25°C and 40°C with approximately 30%, 50% and 80% Increase in tensile strength	NA	[101]

activity and 69.99% antioxidant activity, respectively. A separate study by Chollakup *et al.* [90] revealed that rambutan peel extract (RPE) with cinnamon essential oil (CEO) incorporated into cassava starch/whey protein biocomposite film, contributed to the strong antioxidant activity, as well as antibacterial effect against *B. cereus*, *E. coli* and *S. aureus*. This same biocomposite film retained freshness of salami slices for up to ten days. In short, incorporation of essential oils in biocomposites promotes antioxidant and antimicrobial properties that prevents oxidation and microbial growth, therefore prolonging shelf-life.

The antimicrobial activities of essential oils are attributed to terpenes, terpenoids, and phenylpropenes [91]. Essential oils mainly target the cell membrane of bacteria and disrupt cell structure causing leaking of intracellular compounds and lysis of bacterial cells [92]. Examples of essential oils incorporated into biocomposite for their antimicrobial activities include rockrose essential oil, *Citrus limonia* essential oil, cumin essential oil and lavender essential oil. Rockrose essential oil incorporated into pullulan-based packaging film inhibited growth of *Enterococcus faecalis*, *L. monocytogenes* and *B. cereus* [93]. Essential oil from *Citrus limonia* integrated with chitosan showed excellent antibacterial effect against *S. aureus* [94]. Impregnation of cumin essential oil in Shahri Balangu seed mucilage coating suppressed growth of *S. aureus* and *E. coli* while extending the shelf-life of beef slices up to nine days [95]. In addition, lavender essential oil blended with Qodume Shirazi seed mucilage resulted in a biocomposite coating that can impede growth of *S. aureus* and *E. coli*, as well as reducing yeast and mold count. When coated on meat samples, the shelf-life significantly increased by six days [96].

The effectiveness of biocomposite food packaging incorporated with essential oils can also be improved with nanoencapsulation and electrospinning technology. Esmaeili *et al.* [97] used nanoencapsulated garlic essential oil for incorporation into chitosan/whey protein packaging film. The resulting packaging film effectively retard lipid oxidation through sustained release of garlic essential oil. In addition, growth of lactic acid bacteria, psychrotrophic bacteria and *S. aureus* were effectively suppressed. In another study, Zhou *et al.* [98] encapsulated galangal root oil in electrospun swim bladder gelatin resulting in enhanced thermal stability and inhibition towards enterohemorrhagic *E. coli* on fresh beef for a longer period of time. Aside from electrospinning and nanoencapsulation techniques, Pickering emulsion can also be used. The benefit of this method allows hydrophobic essential oils to disperse well across the food packaging, minimizing coalescence while maximizing performance [99]. This effect was reported by Xu *et al.* [100], whereby clove essential oil incorporated via the Pickering emulsion method into chitosan-based edible film exhibit stronger antimicrobial effect against *S. aureus* and *E. coli* than pure chitosan film. Furthermore, clove essential oil in Pickering emulsion are protected from degradation, prolonging the antimicrobial effect. Another study by Almasi *et al.* [101] incorporated marjoram essential oil into pectin film in Pickering emulsion, and this biocomposite effectively inhibited the growth of *S. aureus* and *E. coli*, as compared to control film. The antimicrobial effect was also prolonged due to the slow release of marjoram essential oil. In conclusion, integration with nanoencapsulation, electrospinning, and Pickering emulsion technologies will enhance the overall performance of essential oils in biocomposite active food packaging applications.

16.3.3 Enzymes and Bacteriocins

Enzymes are used extensively in the food industry as they enhance food sensory properties, enrich nutritional values, as biocatalysts to accelerate processes, and as food preservatives [102, 103]. Bacteriocins on the other hand, are active peptides produced by microorganisms which are antimicrobial [104]. Both enzymes and bacteriocins are useful in active food packaging applications as they can confer antimicrobial properties and to a certain extent, antioxidant activities. Table 16.3 summarizes the common enzymes and bacteriocins incorporated into biobased food packaging.

Lysozyme is a hydrolytic enzyme capable of damaging the bacterial peptidoglycan layer and causes leakage of bacterial cytoplasmic components. This will lead to cell death [105]. When lysozyme is embedded into chitosan/sodium alginate film, the growth of fish spoilage bacteria, *P. fluorescens* and *Shewanella putrefaciens* are effectively suppressed [106]. Peroxidase enzymes, such as lactoperoxidase is also applied in food packaging as this enzyme can inhibit bacterial growth through oxidation of microbial proteins. This affects the metabolic processes and proliferation of bacteria [107]. Lactoperoxidase can effectively inhibit common fish spoilage bacteria, *Pseudomonas* spp. and *Shewanella* spp., when applied on chitosan packaging film [108]. Lactoperoxidase also have strong antimicrobial activity against *E. coli* and *Pseudomonas* spp. on cheese samples [109]. When this biocomposite is combined with essential oil from *Bunium persicum,* synergistic inhibitory effect towards both Gram-positive and Gram-negative bacteria was observed. The proteolytic enzyme lactoferrin is also important in food packaging. Lactoferrin that is embedded in gelatin film, binds strongly to iron resulting in antimicrobial effect against *E. coli* and *S. aureus* [110]. In short, incorporation of enzymes in biocomposite food packaging applications had shown great potential in antimicrobial activity, especially against Gram-negative bacteria.

For bacteriocins, nisin is currently the only bacteriocin considered as "Generally Regarded as Safe" (GRAS) and considered safe for consumption [111]. Nisin is produced by *Lactococcus lactis* and can exert strong antimicrobial effect against Gram-positive bacteria through disruption of cell wall synthesis and perforation of bacterial cell membrane [112]. Chitosan/cellulose bilayer film incorporated with nisin significantly suppressed the growth of Gram-positive *L. monocytogenes* on cheese for 14 days during storage [113]. Nisin blended with chitosan/carboxymethyl chitosan film is equally effective against *L. monocytogenes* [114]. The addition of nisin into chitosan film successfully inhibited *L. monocytogenes* on sliced ham samples over 24 h storage [115]. The application of nisin in biocomposite food packaging is promising, but further studies are required to explore their effectiveness towards other food-borne pathogens aside from *Listeria* spp.. There are also other promising bacteriocins to be explored such as Bac7293 from *Weissella hellenica* and bacteriocin-like substances (BLIS) from *Enterococcus faecium* ES216 [116, 117]. While both have demonstrated antimicrobial activities, their usage is limited as they are yet to receive the GRAS status.

16.3.4 Challenges in Food Packaging Applications of Biocomposites Integrated With Natural Compounds

There are several challenges in using natural compounds for biocomposites, which includes strong distinct aroma, selective antimicrobial effect, and susceptibility to degradation. Plant

Table 16.3 Common biocomposites incorporated with enzymes and bacteriocins in active food packaging applications.

Natural compound	Type of natural compound	Biobased material as main component	Effect of biocomposite packaging	Effects on food model systems	References
Enzyme	Lysozyme	Chitosan/sodium alginate	Significantly suppressed the growth of *P. fluorescens* and *S. putrefaciens* Increase tensile strength, elongation at break	NA	[106]
	Lactoperoxidase (LPS)	Sodium alginate, chitosan and gelatin	NA	Fish burger with chitosan with LPS packaging has the lowest psychrotrophic bacteria, *Pseudomonas* spp. and *Shewanella* spp. count TBARS value of fish burger is significantly low after 20 days storage	[108]
	Lactoperoxidase	Whey protein solution		Growth of *E. coli* and *Pseudomonas* spp. on Gouda cheese are suppressed effectively within 90 days of storage Lower TBARS, peroxide and free fatty acid values	[109]

(*Continued*)

Table 16.3 Common biocomposites incorporated with enzymes and bacteriocins in active food packaging applications. (*Continued*)

Natural compound	Type of natural compound	Biobased material as main component	Effect of biocomposite packaging	Effects on food model systems	References
	Lactoferrin (LTF) in chitosan nanofiber particles	Gelatin	High antioxidant activity with 67.6% DPPH scavenging activity Inhibit growth of *E. coli* and *S. aureus* with 18.0 mm and 15.0 mm inhibition zone, respectively	NA	[110]
Bacteriocin	Nisin	Chitosan/nanozinc oxide	Suppressed growth of *L. monocytogenes* within four storage weeks	Significantly suppressed growth of *L. monocytogenes* on cottage cheese samples, with reduction of 5.0 log CFU/g	[113]
	Nisin	Chitosan/carboxymethyl chitosan	Inhibited growth of *L. monocytogenes* with 15 mm of inhibition zone	NA	[114]
	Nisin	Chitosan	NA	Showed synergistic effect against *L. monocytogenes* in dry-cured ham with rice bran extract	[115]

(*Continued*)

Table 16.3 Common biocomposites incorporated with enzymes and bacteriocins in active food packaging applications. (*Continued*)

Natural compound	Type of natural compound	Biobased material as main component	Effect of biocomposite packaging	Effects on food model systems	References
	Bacteriocin 7293 (Bac7293)	Poly (lactic acid)/ sawdust	Suppressed growth of *L. monocytogenes, S. aureus, P. aeruginosa* with reduction of 4.78, 4.56 and 4.10 log CFU/cm^2 respectively Slow release of Bac7293 into different food stimulants with an average value of 3.3 mg/dm^3	*L. monocytogenes, S. aureus, P. aeruginosa, A. hydrophila, E. coli* and *S. thyphimurium* on pangasius fish fillet are reduced about 2-5 log CFU/m^2	[117]
	Bacteriocin-like substances (BLIS) from *Enterococcus faecium* ES216	Triticale flour	*L. innocua* inhibited with an inhibition zone of 6.6 mm after 45 days of incubation	*Listeria* counts on cheese samples decreased significantly by 3 log units within 15 storage days	[116]

extracts and essential oils with strong aroma and taste [118] may affect consumer acceptance of the food product [119]. Plant extracts and essential oils are also generally stronger in inhibiting Gram-positive bacteria, rather than Gram-negative bacteria [120]. This is mainly due to the difference in cell wall and membrane structure of Gram-positive and negative bacteria, whereby Gram-negative bacteria has an outer hydrophilic lipopolysaccharide layer which acts as barriers to the hydrophobic essential oils, rendering essential oils less effective [121, 122]. Thus, food products contaminated with Gram-negative bacteria may demonstrate their presence and proliferation, even in the presence of plant extracts and essential oils. Biocomposites integrated with enzymes and bacteriocins risk potential degradation. This is because enzymes are highly sensitive to a wide range of environmental factors, such as temperature, light, humidity, and oxygen [123, 124]. Variations or changes to food packaging conditions will render the enzymes less effective than desired. These factors are the main constraints to the use of natural compounds in biocomposite development and applications, and improvements are required for wider applications in future.

16.4 Biocomposites Incorporated with Inorganic Materials

Inorganic materials are generally regarded as substances that do not contain carbon compounds, such as metals and minerals. In the food industry, metals and minerals play crucial roles in food fortification and preservation. They are also known to be essential for human nutrition, as well as valued for their antibacterial and antioxidant properties [125, 126]. In food packaging, metals and minerals are integrated for the purpose of retarding food oxidation, inhibiting growth of microorganism on food, and scavenging oxygen, carbon dioxide and ethylene. Moreover, the effect of metals on food packaging can be further enhanced through modification into microparticles and nanoparticles, which improves the overall functionality of food packaging [127].

16.4.1 Metal Compounds

Metal compounds are typically incorporated into biobased materials in the form of metal nanoparticles. Examples include pure metals or metal oxide forms, such as copper, silver, gold, aluminum, copper oxide, zinc oxide and titanium dioxide [128]. Biocomposites containing these metal nanoparticles enhanced antimicrobial activities and have better gaseous scavenging activities [129]. This is attributed to metal nanoparticles having higher surface-to-volume ratio, enabling strong and sustained antimicrobial effect. Table 16.4 summaries the common metal compounds incorporated into biobased food packaging applications.

Silver nanoparticles are commonly used in biocomposites. The incorporation of silver nanoparticles into sugar palm starch-based biocomposite exhibited remarkable bacteriostatic effect, whereby growth of *Salmonella cholerasuis*, *E. coli*, and *S. aureus* was significantly inhibited [130]. In another study, silver loaded nanotitanium dioxide, blended into fish gelatin/chitosan biocomposite film showed significant antibacterial effect against *E. coli*, *S. aureus* and *Botrytis cinerea* [131]. Antimicrobial effect was also observed in silver nanoparticles impregnated into chitosan biocomposite film, which inhibited *Pseudomonas fluorescens* [132]. Similarly, silver nanoparticles incorporated into 2,2,6,6-Tetramethylpiperidin-1-yl) oxyl (TEMPO)-mediated nanocellulose packaging film also have prolonged antimicrobial

Table 16.4 Common biocomposites incorporated with metal compounds in active food packaging applications.

Inorganic materials	Type of inorganic materials	Biobased material as main component	Effect of biocomposite packaging	Effects on food model systems	References
Metal compounds	Silver nanoparticles (AgNP)	Sugar palm nanocrystalline cellulose	Significantly suppressed growth of *E. coli*, *S. aureus* and *S. cholerasuis* with inhibition zone of 7.66, 0.5 and 8.0 mm respectively	NA	[130]
	Titanium dioxide-silver (TiO_2-Ag)	Fish gelatin/chitosan	Significantly inhibit growth of *E. coli*, *S. aureus* and *B. cinerea* with inhibition zone of 20.60, 22.86 and 23.72 mm respectively	NA	[131]
	Copper oxide nanoparticles (CuONPs)	Whey protein isolate	Inhibit growth of *S. aureus* and *E. coli* with inhibition zone of 7.64 and 9.00 mm respectively	NA	[134]
	ZnO nanoparticles	Chitosan/gelatin and pectin/gelatin	Inhibition against *E. coli*, *B. subtilis* and *S. aureus* with inhibition zone of 17.17, 19.00 and 20.33 mm respectively	NA	[135]
	Biosorbed – silver nanoparticles (BSNPs)	Chitosan	Inhibit growth of *P. fluorescens* with an inhibition zone of approximately 8 mm	*P. fluorescens* counts are lower in whey protein isolate/ carrageenan hydrogel (food stimulant) when compared to control Delayed release of BSNPs into hydrogel	[132]

(*Continued*)

Table 16.4 Common biocomposites incorporated with metal compounds in active food packaging applications. (*Continued*)

Inorganic materials	Type of inorganic materials	Biobased material as main component	Effect of biocomposite packaging	Effects on food model systems	References
	Silver nanoparticles (AgNPs)	TEMPO-oxidized nano-cellulose (TNC)	AgNPs slightly increase antioxidant activity and significantly suppressed growth of *S. aureus* and *E. coli*, with inhibition zone of approximately 7 and 2.5 mm respectively Release of AgNPs is reduced with increasing pH within 14 storage days	NA	[133]
	Nanosized TiO_2	Chitosan	Effectively inhibit growth of *S. aureus*, E. *coli*, *S.* Typhimurium, *P. aeruginosa*, *Aspergillus* and *Penicillium*, with 53%, 53%, 20%, 15%, 20% and 10% growth inhibition respectively Significantly degrade ethylene by almost 12% after 200 min	NA	[140]
	TiO_2 nanotube (TNT)	Carrageenan	Lower *E. coli* and *L. monocytogenes* counts by 1.5 CFU/mL and 2.5 CFU/mL	Excellent ethylene scavenging activity in banana samples and freshness maintained for 12 storage days	[141]
	TiO_2	Zein nanofibers	Increase thermal stability and hydrophobicity	Significantly reduce the ethylene concentration of cherry tomatoes from day 3 to day 22 during storage period	[142]

effect against *S. aureus* and *E. coli* [133]. Aside from silver nanoparticles, copper oxide nanoparticles blended in whey protein isolate packaging film effectively suppressed *E. coli* and *B. cereus* [134]. Similarly, zinc oxide nanoparticles incorporated into chitosan-based coating have excellent antimicrobial effect against *E. coli*, *Bacillus subtilis*, and *S. aureus* [135].

Metal compounds in food packaging have great potential for use in gaseous scavenging activity as well, which include palladium, platinum, titanium and iron [136]. These metal compounds can scavenge oxygen, carbon dioxide and ethylene [137, 138]. Among the various metals, titanium is the most used metal compound. Titanium dioxide incorporated into biobased food packaging materials absorbs ethylene. Titanium functions as a catalyst for the photocatalytic degradation of ethylene into carbon dioxide and water, in the presence of UV radiation [139]. Nano-titanium dioxide that is blended into chitosan packaging film, effectively degrade ethylene. as well as exhibiting strong antimicrobial effect against *S. aureus* and *E. coli* [140]. Similarly, titanium dioxide, in the form of nanotube embedded into carrageenan-based packaging film, has excellent ethylene scavenging capacity to improve the freshness of banana when compared with neat carrageenan film [141]. In addition, the growth of *E. coli* and *L. monocytogenes* are also suppressed. Another study using titanium dioxide with electrospun zein nanofibers resulted in lower ethylene concentration within the packaging atmosphere [142]. The cherry tomatoes inside this packaging benefitted with a slower ripening rate and a longer shelf-life.

16.4.2 Clay and Silicate-Based Mineral Compounds

Clay minerals in food packaging consists of a few groups, such as the smectite group, kaolin group and palygorskite-sepiolite group. Biocomposites integrated with these clay mineral groups will have enhanced antioxidant and antimicrobial effects [143]. These clay mineral compounds will also interact with other constituents, such as plant extracts, essential oils or metal compounds, to improve the sustained release of these constituents into food products. Table 16.5 summarizes the common clay and silicate-based mineral compounds incorporated into biobased food packaging.

The smectite group is made up of clay and silicate compounds with a central octahedral sheet sandwiched in between two tetrahedral sheets, to give rise to a hexagonal structure. One example is the montmorillonite (MMT) clay [144]. The addition of MMT loaded-cinnamaldehyde into chitosan-based packaging film resulted in a sustained release of cinnamaldehyde active compounds and longer antimicrobial effect against *S. aureus* and *E. coli* [145]. MMT can also be incorporated with carvacol essential oil into thermoplastic starch to achieve a synergistic antimicrobial effect to inhibit *E. coli,* in contrast to the biocomposite film without MMT that has no inhibitory effect [146]. On the other hand, the kaolin group consist of a gibbsite-like dioctahedral sheet and a tetrahedral sheet arranged in layered structure, such as halloysite and kaolinite [144]. Halloysite nanotubes loaded with chlorogenic acid are blended with electrospun chitosan-based nanofibers, which successfully sustain the release of chlorogenic acid and inhibited *E. coli* and *S. aureus* [147]. When kaolinite is impregnated into chitosan packaging film, this packaging film showed excellent antimicrobial activity against *E. coli*, *S. aureus*, and *Klebsiella pneumoniae*, as well as increased DPPH radical scavenging activity [148]. For palygorskite-sepiolite, this group consists of clay minerals with unique ribbon structure formed from continuous two-dimensional tetrahedral sheets of palygorskite and sepiolite [144]. When sepiolite is incorporated

Table 16.5 Common biocomposites incorporated with clay and silicate-based mineral compounds in active food packaging applications.

Inorganic materials	Type of inorganic materials	Biobased material as main component	Effect of biocomposite packaging	Effects on food model systems	References
Clay and silicate-based mineral compounds	Sepiolite/Sepiolite hybrid with myrtle berries extract (MBE)	Alginate	Significantly enhanced antioxidant activity, with an increment from 9.2% to 62.1% DPPH scavenging activity Delay the release of MBE by 11.3% for the first hour of storage at 50% ethanol food stimulant.	NA	[149]
	montmorillonite (MMT)/ MMT loaded with cinnamaldehyde (CIN)	Chitosan	Synergistic inhibitory effect against *S. aureus* and *E. coli* when combined with CIN Cumulative release of CIN is lower by approximately 20%	NA	[145]
	Halloysite nanotubes (HNTs)/HNTs loaded with chlorogenic acid (CGA)	Electrospun Chitosan/ polycaprolactone nanofibers	Antioxidant activity and antimicrobial activity against *E. coli* and *S. aureus* increased Lower sustained release of CGA with approximately 55%	NA	[147]
	MMT/MMT hybrid with carvacrol essential oil (EO)	Thermoplastic starch	Significantly enhance the antimicrobial effect against *E. coli*	NA	[146]

(Continued)

Table 16.5 Common biocomposites incorporated with clay and silicate-based mineral compounds in active food packaging applications. (*Continued*)

Inorganic materials	Type of inorganic materials	Biobased material as main component	Effect of biocomposite packaging	Effects on food model systems	References
	Kaolinite	Chitosan	Significantly decrease the antioxidant activity, with reduction of almost 80% DPPH radical scavenging and 35% metal chelating activity Growth of *S. aureus*, *K. pneumoniae* and *E. coli* are significantly suppressed, with inhibition zone of 15.0, 12.0 and 15.0 mm	NA	[148]
	Zeolites	Cassava starch based	Increase elongation at break	Maintained colour, vitamin C content and reduce weight loss of broccolis after seven storage days	[152]

(*Continued*)

Table 16.5 Common biocomposites incorporated with clay and silicate-based mineral compounds in active food packaging applications. (*Continued*)

Inorganic materials	Type of inorganic materials	Biobased material as main component	Effect of biocomposite packaging	Effects on food model systems	References
	Zeolites	Chitosan	Adsorbed a maximum of 8.8×10^{-1} mmol g^{-1} ethylene, which is almost 30 times significantly more than control packaging	NA	[51]
	Zeolites	Chitosan/vanillin	Effectively degrade ethylene with a low ethylene concentration throughout 35 storage days	Protect mango fruits from anthracnose disease with a significantly low severity index of diseases of 56.67% up to ten storage days Reduced weight loss, maintained firmness and minimal colour changes on mango fruits	[153]

into alginate film with myrtle berries extracts (MBE), antioxidant activities were detected [149]. Therefore, clay minerals not only control the release of active compounds but also exhibit antioxidant and antimicrobial properties.

Similar to metal compounds, clay minerals in active food packaging applications can also scavenge for carbon dioxide and ethylene gaseous. Zeolite is a popular compound used in active biocomposite food packaging applications [150]. Zeolites contain open channels and cavities which can trap carbon dioxide and ethylene [151]. Mixing zeolite with cassava starch-based film reportedly extended the freshness and shelf-life of broccoli florets up to seven days [152]. The incorporation of zeolite into chitosan film resulted in high ethylene sorption capacity as compared to neat chitosan film [51]. Embedding zeolite into chitosan/vanillin coating has also been found to enhance the removal of ethylene produced from mangoes [153]. The mangoes wrapped with this biocomposite had lower ethylene concentrations for 35 days and was also protected from contracting anthracnose disease. In short, zeolite in biocomposites improves the shelf-life of fruits and vegetables and has excellent potential for further exploration as biocomposite for other commodities.

16.4.3 Challenges in Food Packaging Applications of Biocomposites Integrated with Inorganic Materials

The primary concern of using inorganic materials in food is on food safety. This is because the metal and mineral compounds risk leaching into food products over time. Studies had shown that long-term oral ingestion of metal particles can potentially induce DNA damage [154], induce hemolysis through oxidative stress [155], affect the pigmentation of skin [156], and accumulate in liver [157]. Although most studies do not reveal the limit of metal compounds that can cause *in-vitro* or *in-vivo* toxicity [158], it is generally understood that the potential risks and toxicity effects of metal compounds in food packaging exist and thorough investigations are needed prior to wide usage of these inorganics in food packaging.

16.5 Biocomposites Incorporated with Natural Food Colorants and Pigments

Intelligent food packaging applications utilize the ability of food colorants to change color in response to changes on food quality and freshness. Synthetic colored dyes are common indicators, but recent findings show that the synthetic dyes are toxic, carcinogenic, and mutagenic, causing health implications and harming the environment [159]. As an alternative, natural food colorants and pigments derived from fruits and vegetables are becoming increasingly important [160]. Examples of natural food colorants and pigments used in intelligent food packaging include anthocyanins, betacyanin, curcumin, carotenoids, and chlorophylls. These compounds are sensitive to pH changes or changes in gaseous content in the packaging atmosphere, therefore are used to represent the conditions of food quality and freshness [161]. Table 16.6 summarizes the common natural food colorants and pigments incorporated into biobased active and intelligent food packaging applications.

Some of these natural food colorants and pigments also demonstrate antioxidant and antimicrobial properties. Therefore, the use of natural food colorants and pigments present

Table 16.6 Common biocomposites incorporated with natural food colourants and pigments in intelligent and intelligent/active food packaging applications.

Indicator/ sensor	Type of indicator/ sensor	Biobased material as main component	Effect of biocomposite packaging (intelligent/ intelligent and active)	Effects on food model systems	References
Natural food colourants and pigments	Red radish extract (anthocyanins)	Gelatin	Sensitive to a large range of pH, from pH 1 - 10 More stable in refrigeration than room temperature	Detect spoilage of white prawn and chicken samples through TVB-N (ammonia), with colour changes detected throughout 14 storage days	[166]
	Blueberry residue (BR)	Cassava starch	Various colour changes, ranging from red to blue in detecting pH 5 – 11	Inconclusive spoilage test on chicken meat as colour migration occurred and colour difference is not clear within ten storage days	[167]
	Amaranthus leaf extract (ALE)	Gelatin/PVA	Significantly increase antioxidant activity with an increment of almost 40% DPPH scavenging activity *S. aureus*, *B. cereus*, *E. coli* and *P. fluorescence* are significantly inhibited with 30, 33, 36 and 25 mm inhibition zone respectively Detect pH changes from pH 1 – 9 with colour changes from pink to yellow	Total bacterial count, *S. aureus* and fecal coliforms of chicken and fish samples decreased significantly after three storage days Level of TVBN and TBARS increase slowly in chicken and fish samples for 12 storage days, which indicate prolonged freshness and delayed oxidation	[168]

(Continued)

Table 16.6 Common biocomposites incorporated with natural food colourants and pigments in intelligent and intelligent/active food packaging applications. (*Continued*)

Indicator/ sensor	Type of indicator/ sensor	Biobased material as main component	Effect of biocomposite packaging (intelligent/ intelligent and active)	Effects on food model systems	References
	Black plum peel extract (BPPE)	Chitosan/nano-TiO_2	High antioxidant activity with 72.85% DPPH scavenging activity Remarkable antimicrobial activity against *E. coli*, *Salmonella* spp., *S. aureus* and *L. monocytogenes* with inhibition zone of 6.14, 7.25, 7.26 and 7.33 mm Detect changes from pH 2 - 13 with colour changes from red to dark yellow	NA	[20]
	Curcumin (Cur)	Chitosan/oxidized chitin nanocrystal	Significantly increase antioxidant activity with DPPH scavenging activity of almost 60% Detect changes from pH 3 - 10 with different intensity of red colour	Detect freshness of fresh shrimp and silvery pomfret with the colour changes from yellow (day 0) to orange-red (day 5) and well correlated to the increasing basic TVB-N level	[174]

manifold benefits; as freshness indicator as well as for their valued antioxidant and antimicrobial effects.

16.5.1 Intelligent Food Packaging with Natural Food Colorants and Pigments

Food spoilage is commonly accompanied by a change in food pH. The change in pH is usually the result of degradation of food due to spoilage microorganisms. For example, food degradation by lactic acid bacteria (LAB), yeast or molds, results in the production of by-products such as lactic acid and acetic acid [162]. These acidic compounds have low pH and leads to sourly taste, acidic off-odor and food discoloration [163]. On the other hand, high pH also indicates food spoilage, especially for seafood and meat products. This is due to the breakdown of proteins and amino acids into ammonia and volatile amines [164]. Both acidic and alkaline conditions alter the chemical structure of natural food colorants and pigments, thereby giving rise to color change [165]. Natural food colorants and pigments can be used to represent the change in food freshness and quality due to pH change.

The incorporation of natural food colorants and pigments into biocomposites have been reported in several studies. Red radish anthocyanin incorporated into starch/gelatin film resulted in a pH-sensitive intelligent food packaging to detect freshness of white prawns [166]. Addition of blueberry residue into cassava starch-based film produced a pH-sensitive film that detects the freshness of orange juice and chicken meat. The changes of pH in both food samples are reflected by the change in color of the food packaging film [167]. Furthermore, *Amaranthus* leaf extract that is rich in betalains and impregnated into gelatin-based film, is effective in detecting the freshness of chicken and fish samples [168]. This biocomposite film changes color when the pH of food samples increases, which indicates the occurrence of food spoilage. In short, most natural food colorants and pigments are sensitive to a wide range of pH and are typically able to develop distinct colors to reflect the different pH values. Therefore, natural food colorants and pigments are suitable as freshness indicators in intelligent food packaging.

16.5.2 Potential of Natural Food Colorants and Pigments as Active and Intelligent Food Packaging

There is tremendous potential of incorporating natural food colorants and pigments in active and intelligent food packaging. Natural food colorants and pigments are not just freshness indicators, but also demonstrate desirable antioxidant and antimicrobial properties [169]. Recent studies have reported the potential of combining natural food colorants and pigments with other plant extracts or metal compounds to enhance the overall functionality of biocomposites for food packaging [170]. This includes the use of blackberry pulp, a natural food colorant, that is incorporated into plantain starch film to produce an active and intelligent biocomposite film. This packaging film is able to change color in response to different pH, has good antioxidant effect, as well as the ability to suppress the growth of *E. coli* and *S. aureus* [171]. Another example is the use of curcumin, which was added to *Lallemantia iberica* seed gum (LISG)-based film. This intelligent film provides information on pH change, as well as rendering antioxidant and antimicrobial effects against Gram-positive and Gram-negative bacteria in shrimp samples [172]. Similarly, blending betacyanin-rich red pitaya flesh extract (RPFE) with starch-based packaging film can produce a biocomposite film, which is capable of detecting shrimp freshness as well as having antioxidant activities to ensure freshness of shrimps [173].

In other examples, chitosan-based film integrated with black plum peel extract (BPPE) and nanosized titanium dioxide was equally effective as food spoilage indicator [20]. This packaging film was also effective in inhibiting *E. coli*, *Salmonella* spp., *S. aureus* and *L. monocytogenes*, as well as demonstrating ethylene scavenging attributes. Wu *et al.* [174] mixed curcumin with oxidized chitin nanocrystals into chitosan packaging film. This biocomposite film detected spoilage in fresh shrimp and pomfret samples from accumulation of alkaline amine compounds from degradation. To summarize, food packaging with natural food colorants and pigments not only detects food spoilage through pH changes, but also exhibit antimicrobial and antioxidant effects at the same time.

16.5.3 Challenges in Food Packaging Applications of Biocomposites Integrated with Natural Food Colorants and Pigments

The major limiting factor in the use of natural food colorants and pigments in food packaging is the accuracy and reliability of color change. The interpretation of food freshness is highly dependent on visual inspection of colour change, and this may be a problem as not all consumers interpret colour change equally and correctly. This may potentially mislead the consumers and create confusion [175]. Furthermore, the sensitivity of natural food colorants may vary from one food product to another, due to the different properties and characteristics of the food products. Improvements can be implemented by integrating digital platform and tools that assist in identifying the correct color profile. Examples, such as IoT and big data, can assist to provide detailed information instantly to the consumers [176, 177].

The food packaging films will also have to interact closely or directly with the food product [178]. Close contact is essential to provide an accurate assessment. However, the drawback of such close contact with the natural food colorants and pigments in the film is that the original color, taste, and appearance of the food product may be altered [179]. Consumers may reject despite the food products are still fresh.

16.6 Conclusion

Most biocomposites in food packaging are incorporated with natural and metal compounds to give rise to food packaging with remarkable antioxidant and antimicrobial effects. As a result, these biocomposites maintain food freshness and significantly control microbial growth, which prolongs the shelf-life of the food product. Integration of desired constituents into biocomposites can be achieved via nanoencapsulation and electrospinning technique, to produce films with desired protective effects, as well as a sustained release of compounds for a longer period of protection. Synergistic effects from natural and metal compounds are also achievable for some specific pairings to further improve the overall functionality of the biocomposite. Biocomposites is indeed the future in active and intelligent food packaging applications, as many benefits can be derived from their applications.

Further explorations into food safety aspects of the developed biocomposites and the constituents incorporated is needed to fully understand their potential use. Integration with digital tools and infrastructures are also highly recommended to further advance the usage and reliability of information provided by intelligent food packaging.

References

1. Walker, T.R., McGuinty, E., Charlebois, S., Music, J., Single-use plastic packaging in the Canadian food industry: Consumer behavior and perceptions. *Humanit. Soc. Sci. Commun.*, *8*, 1–11, 2021.
2. Deng, Y., Zhang, Y., Lemos, B., Ren, H., Tissue accumulation of microplastics in mice and biomarker responses suggest widespread health risks of exposure. *Sci. Rep.*, *7*, 1, 2017.
3. Adrian, G., Mihai, M., Vodnar, D.C., The use of chitosan, alginate, and pectin in the biomedical and food sector biocompatibility, bioadhesiveness, and biodegradability. *Polymers*, *11*, 1837, 2019.
4. Song, R., Murphy, M., Li, C., Ting, K., Soo, C., Zheng, Z., Current development of biodegradable polymeric materials for biomedical applications. *Drug Des. Devel. Ther.*, *12*, 3117, 2018.
5. Calva-Estrada, S.J., Jiménez-Fernández, M., Lugo-Cervantes, E., Protein-based films: Advances in the development of biomaterials applicable to food packaging. *Food Eng. Rev.*, *2019*, 78, 2019.
6. Romaric, O., Déley, S.D., Générose, V.-D., Abadjayé, F.S., Akouavi, B.F.-A., Menouwesso, H.H., Djidjoho, J.H., Paulin, A., Bio-based packaging used in food processing: A critical review. *Afr. J. Food Sci.*, *15*, 4, 131, 2021.
7. Maran, J.P., Sivakumar, V., Sridhar, R., Thirugnanasambandham, K., Development of model for barrier and optical properties of tapioca starch based edible films. *Carbohydr. Polym.*, *92*, 1335, 2013.
8. Dias, A.B., Müller, C.M.O., Larotonda, F.D.S., Laurindo, J.B., Mechanical and barrier properties of composite films based on rice flour and cellulose fibers. *LWT – Food Sci. Technol.*, *44*, 535, 2011.
9. Rui, L., Xie, M., Hu, B., Zhou, L., Yin, D., Zeng, X., A comparative study on chitosan/gelatin composite films with conjugated or incorporated gallic acid. *Carbohydr. Polym.*, *173*, 473, 2017.
10. Silva, K.S., Mauro, M.A., Gonçalves, M.P., Rocha, C.M.R., Synergistic interactions of locust bean gum with whey proteins: Effect on physicochemical and microstructural properties of whey protein-based films. *Food Hydrocolloids*, *54*, 179, 2016.
11. Youssef, A.M., El-Sayed, S.M., El-Sayed, H.S., Salama, H.H., Dufresne, A., Enhancement of Egyptian soft white cheese shelf life using a novel chitosan/carboxymethyl cellulose/zinc oxide bionanocomposite film. *Carbohydr. Polym.*, *151*, 9, 2016.
12. Soltani Firouz, M., Mohi-Alden, K., Omid, M., A critical review on intelligent and active packaging in the food industry: Research and development. *Food Res. Int.*, *141*, 1, 2021.
13. Mitropoulou, G., Sidira, M., Skitsa, M., Tsochantaridis, I., Pappa, A., Dimtsoudis, C., Proestos, C., Kourkoutas, Y., Assessment of the antimicrobial, antioxidant, and antiproliferative potential of sideritis raeseri subps. raeseri essential oil. *Foods*, *9*, 1, 2020.
14. Sritalahareuthai, V., Temviriyanukul, P., On-Nom, N., Charoenkiatkul, S., Suttisansanee, U., Phenolic profiles, antioxidant, and inhibitory activities of Kadsura heteroclita (Roxb.) craib and Kadsura coccinea (Lem.) A.C. Sm. *Foods*, *9*, 1, 2020.
15. Marvdashti, L.M., Yavarmanesh, M., Koocheki, A., Controlled release of nisin from polyvinyl alcohol - Alyssum homolocarpum seed gum composite films: Nisin kinetics. *Food Biosci.*, *28*, 133, 2019.

16. Yang, Y., Liu, H., Wu, M., Ma, J., Lu, P., Bio-based antimicrobial packaging from sugarcane bagasse nanocellulose/nisin hybrid films. *Int. J. Biol. Macromol.*, *161*, 627, 2020.
17. Byun, Y., Bae, H.J., Whiteside, S., Active warm-water fish gelatin film containing oxygen scavenging system. *Food Hydrocolloids*, *27*, 250, 2012.
18. Sani, M.A., Azizi-Lalabadi, M., Tavassoli, M., Mohammadi, K., McClements, D.J., Recent advances in the development of smart and active biodegradable packaging materials. *Nanomaterials*, *11*, 1, 2021.
19. Yong, H., Wang, X., Bai, R., Miao, Z., Zhang, X., Liu, J., Development of antioxidant and intelligent pH-sensing packaging films by incorporating purple-fleshed sweet potato extract into chitosan matrix. *Food Hydrocolloids*, *90*, 216, 2019.
20. Zhang, X., Liu, Y., Yong, H., Qin, Y., Liu, J., Liu, J., Development of multifunctional food packaging films based on chitosan, TiO_2 nanoparticles and anthocyanin-rich black plum peel extract. *Food Hydrocolloids*, *94*, 80, 2019.
21. Lim, L.T., Enzymes for food-packaging applications, in: *Improving and Tailoring Enzymes for Food Quality and Functionality*, Elsevier Ltd, Amsterdam, The Netherlands, 2015.
22. Yucel, U., Intelligent packaging, in: *Reference Module in Food Sciences*, Elsevier, Amsterdam, The Netherlands, 2016.
23. Avila-Novoa, M.G., Iñíguez-Moreno, M., Solís-Velázquez, O.A., González-Gómez, J.P., Guerrero-Medina, P.J., Gutiérrez-Lomelí, M., Biofilm formation by Staphylococcus aureus isolated from food contact surfaces in the Dairy Industry of Jalisco, Mexico. *J. Food Qual.*, *2018*, 1, 2018.
24. Samdani, G.K., McClements, D.J., Decker, E.A., Impact of phospholipids and tocopherols on the oxidative stability of soybean oil-in-water emulsions. *J. Agric. Food Chem.*, *66*, 3939, 2018.
25. Zheng, J., Han, Y., Ge, G., Zhao, M., Sun, W., Partial substitution of NaCl with chloride salt mixtures: Impact on oxidative characteristics of meat myofibrillar protein and their rheological properties. *Food Hydrocolloids*, *96*, 36, 2019.
26. Arezoo, E., Mohammadreza, E., Maryam, M., Abdorreza, M.N., The synergistic effects of cinnamon essential oil and nano TiO2 on antimicrobial and functional properties of sago starch films. *Int. J. Biol. Macromol.*, *157*, 743, 2020.
27. Peighambardoust, S.J., Peighambardoust, S.H., Mohammadzadeh Pournasir, N., Pakdel, P., Properties of active starch-based films incorporating a combination of Ag, ZnO and CuO nanoparticles for potential use in food packaging applications. *Food Packag. Shelf Life*, *22*, 1, 2019.
28. Dewangan, A.K., Mazumder, S., Perumal, Y., Chopra, K., Mazumder, S., Preparation, characterization and anti-inflammatory effects of curcumin loaded carboxymethyl cellulose acetate butyrate nanoparticles on adjuvant induced arthritis in rats. *J. Drug Deliv. Sci. Technol.*, *41*, 269, 2017.
29. Yuan, R., Zheng, F., Zhong, S., Tao, X., Zhang, Y., Gao, F., Yao, F., Chen, J., Chen, Y., Shi, G., Self-assembled nanoparticles of glycyrrhetic acid-modified pullulan as a novel carrier of curcumin. *Molecules*, *19*, 13305, 2014.
30. Li, H., Wang, D., Liu, C., Zhu, J., Fan, M., Sun, X., Wang, T., Xu, Y., Cao, Y., Fabrication of stable zein nanoparticles coated with soluble soybean polysaccharide for encapsulation of quercetin. *Food Hydrocolloids*, *87*, 342, 2019.
31. Xue, J., Wu, T., Dai, Y., Xia, Y., Electrospinning and electrospun nanofibers: Methods, materials, and applications. *Chem. Rev.*, *119*, 5298, 2019.
32. Cerqueira, M.A., Fabra, M.J., Castro-Mayorga, J.L., Bourbon, A.I., Pastrana, L.M., Vicente, A.A., Lagaron, J.M., Use of electrospinning to develop antimicrobial biodegradable multilayer systems: Encapsulation of cinnamaldehyde and their physicochemical characterization. *Food Bioprocess Technol.*, *9*, 1874, 2016.

33. Yang, Y., Li, W., Yu, D.G., Wang, G., Williams, G.R., Zhang, Z., Tunable drug release from nanofibers coated with blank cellulose acetate layers fabricated using tri-axial electrospinning. *Carbohydr. Polym.*, *203*, 228, 2019.
34. Cossu, A., Wang, M.S., Chaudhari, A., Nitin, N., Antifungal activity against Candida albicans of starch Pickering emulsion with thymol or amphotericin B in suspension and calcium alginate films. *Int. J. Pharm.*, *493*, 233, 2015.
35. Souza, A.G., Ferreira, R.R., Paula, L.C., Mitra, S.K., Rosa, D.S., Starch-based films enriched with nanocellulose-stabilized Pickering emulsions containing different essential oils for possible applications in food packaging. *Food Packag. Shelf Life*, *27*, 1, 2021.
36. Gaikwad, K.K., Singh, S., Lee, Y.S., Oxygen scavenging films in food packaging. *Environ. Chem. Lett.*, *16*, 523, 2018.
37. Foltynowicz, Z., Bardenshtein, A., Sängerlaub, S., Antvorskov, H., Kozak, W., Nanoscale, zero valent iron particles for application as oxygen scavenger in food packaging. *Food Packag. Shelf Life*, *11*, 74, 2017.
38. Golja, V., Dražić, G., Lorenzetti, M., Vidmar, J., Ščančar, J., Zalaznik, M., Kalin, M., Novak, S., Characterisation of food contact non-stick coatings containing TiO_2 nanoparticles and study of their possible release into food. *Food Addit. Contam. - Chem. Anal. Control Expo. Risk Assess.*, *34*, 421, 2017.
39. Gaikwad, K.K., Singh, S., Lee, Y.S., A new pyrogallol coated oxygen scavenging film and their effect on oxidative stability of soybean oil under different storage conditions. *Food Sci. Biotechnol.*, *26*, 1535, 2017.
40. Bi, F., Yong, H., Liu, J., Zhang, X., Shu, Y., Liu, J., Development and characterization of chitosan and D-α-tocopheryl polyethylene glycol 1000 succinate composite films containing different flavones. *Food Packag. Shelf Life*, *25*, 100531, 2020.
41. Anthierens, T., Ragaert, P., Verbrugghe, S., Ouchchen, A., de Geest, B.G., Noseda, B., Mertens, J., Beladjal, L., de Cuyper, D., Dierickx, W., du Prez, F., Devlieghere, F., Use of endospore-forming bacteria as an active oxygen scavenger in plastic packaging materials. *Innov. Food Sci. Emerg. Technol.*, *12*, 594, 2011.
42. Contarino, R., Brighina, S., Fallico, B., Cirvilleri, G., Parafati, L., Restuccia, C., Volatile organic compounds (VOCs) produced by biocontrol yeasts. *Food Microbiol.*, *82*, 70, 2019.
43. Ghidelli, C. and Pérez-Gago, M.B., Recent advances in modified atmosphere packaging and edible coatings to maintain quality of fresh-cut fruits and vegetables. *Crit. Rev. Food Sci. Nutr.*, *58*, 662, 2018.
44. Miranda, K.W.E., Bresolin, J.D., Natarelli, C.V.L., Benevides, S.D., Bastos, M., do, S.R., Mattoso, L.H.C., de Oliveira, J.E., Potential use of poly(lactic acid) nanofibers mats as Nano-sachets in postharvest of climacteric fruits and vegetables. *J. Appl. Polym. Sci.*, *138*, 1, 2021.
45. Tsironi, T., Ntzimani, A., Gogou, E., Tsevdou, M., Semenoglou, I., Dermesonlouoglou, E., Taoukis, P., Modeling the effect of active modified atmosphere packaging on the microbial stability and shelf life of gutted sea bass. *Appl. Sci.*, *9*, 1, 2019.
46. Boz, Z., Brecht, J.K., Mclamore, E., Butler, J.E., Pelletier, W., Kiker, G., Challenges and advances in development of active components to modify headspace gases in packaging of fresh produce and muscle foods. *J. Appl. Packag. Res.*, 10, 62, 2015.
47. Tucker, G., Yin, X., Zhang, A., Wang, M., Zhu, Q., Liu, X., Xie, X., Chen, K., Grierson, D., Ethylene and fruit softening. *Food Qual. Saf.*, *1*, 253, 2017.
48. Álvarez-Hernández, M.H., Martínez-Hernández, G.B., Avalos-Belmontes, F., Miranda-Molina, F.D., Artés-Hernández, F., Postharvest quality retention of apricots by using a novel sepiolite–loaded potassium permanganate ethylene scavenger. *Postharvest Biol. Technol.*, *160*, 1, 2020.
49. Warsiki, E., Application of chitosan as biomaterial for active packaging of ethylene absorber. *IOP Conf. Ser. Earth Environ. Sci.*, *141*, 1, 2018.

50. Özyiğit, F., Yıldız, E., Çetiner, M., Coşgun, S., A rare case of poisoning: Potassium permanganate toxicity. *Turk. J. Anaesthesiol. Reanim.*, *48*, 248, 2020.
51. do Nascimento Sousa, S.D., Santiago, R.G., Soares Maia, D.A., de Oliveira Silva, E., Vieira, R.S., Bastos-Neto, M., Ethylene adsorption on chitosan/zeolite composite films for packaging applications. *Food Packag. Shelf Life*, *26*, 1, 2020.
52. Kumar, A., Gupta, V., Singh, S., Saini, S., Gaikwad, K.K., Pine needles lignocellulosic ethylene scavenging paper impregnated with nanozeolite for active packaging applications. *Ind. Crops Prod.*, 170, 1, 2021.
53. Prasai, T.P., Walsh, K.B., Bhattarai, S.P., Midmore, D.J., Van, T.T.H., Moore, R.J., Stanley, D., Zeolite food supplementation reduces abundance of enterobacteria. *Microbiol. Res.*, *195*, 24, 2017.
54. Dodero, A., Escher, A., Bertucci, S., Castellano, M., Lova, P., Intelligent packaging for real-time monitoring of food-quality: Current and future developments. *Appl. Sci.*, *11*, 1, 2021.
55. Chen, L., Ao, F., Ge, X., Shen, W., Food-grade pickering emulsions: Preparation, stabilization and applications. *Molecules*, *25*, 3202, 2020.
56. Yam, K.L., Takhistov, P.T., Miltz, J., Intelligent packaging: Concepts and applications. *J. Food Sci.*, *70*, 1, 2005.
57. Dey, S., Saha, S., Singh, A.K., McDonald-Maier, K., FoodSQRBlock: Digitizing food production and the supply chain with blockchain and QR code in the cloud. *Sustainability*, *13*, 1, 2021.
58. Saliu, F. and della Pergola, R., Carbon dioxide colorimetric indicators for food packaging application: Applicability of anthocyanin and poly-lysine mixtures. *Sens. Actuators B Chem.*, *258*, 1117, 2018.
59. Price, R.E., Longtin, M., Conley-Payton, S., Osborne, J.A., Johanningsmeier, S.D., Bitzer, D., Breidt, F., Modeling buffer capacity and pH in acid and acidified foods. *J. Food Sci.*, *85*, 918, 2020.
60. Wang, L., Wu, Z., Cao, C., Technologies and fabrication of intelligent packaging for perishable products. *Appl. Sci.*, *9*, 1, 2019.
61. Patel, H. and Joy, R., IOT color based object sorting machine. *Int. J. Appl. Eng. Res.*, *13*, 7383, 2018.
62. Herman, S., Fragrance, in: *Cosmetic science and technology: Theoretical principles and applications*, p. 267, 2017.
63. Teodoro, A.J., Bioactive compounds of food: Their role in the prevention and treatment of diseases. *Oxid. Med. Cell. Longev.*, *2019*, 1, 2019.
64. Zhang, Q.W., Lin, L.G., Ye, W.C., Techniques for extraction and isolation of natural products: A comprehensive review. *Chin. Med.*, *13*, 1, 2018.
65. Ferrentino, G., Giampiccolo, S., Morozova, K., Haman, N., Spilimbergo, S., Scampicchio, M., Supercritical fluid extraction of oils from apple seeds: Process optimization, chemical characterization and comparison with a conventional solvent extraction. *Innov. Food Sci. Emerg. Technol.*, *64*, 102428, 2020.
66. Lund, M.N., Hviid, M.S., Skibsted, L.H., The combined effect of antioxidants and modified atmosphere packaging on protein and lipid oxidation in beef patties during chill storage. *Meat Sci.*, *76*, 226, 2007.
67. Pratap Singh, A., Fathordoobady, F., Guo, Y., Singh, A., Kitts, D.D., Antioxidants help favorably regulate the kinetics of lipid peroxidation, polyunsaturated fatty acids degradation and acidic cannabinoids decarboxylation in hempseed oil. *Sci. Rep.*, *10*, 1, 2020.
68. Sridhar, K. and Charles, A.L., Grape skin extracts as a sustainable source of antioxidants in an oil-in-water emulsion: An alternate natural approach to synthetic antioxidants using principal component analysis. *Int. J. Food Sci. Technol.*, *56*, 1937, 2021.

69. da Rosa, G.S., Vanga, S.K., Gariepy, Y., Raghavan, V., Development of biodegradable films with improved antioxidant properties based on the addition of carrageenan containing olive leaf extract for food packaging applications. *J. Polym. Environ.*, *28*, 123, 2020.
70. Kumar, P., Tanwar, R., Gupta, V., Upadhyay, A., Kumar, A., Gaikwad, K.K., Pineapple peel extract incorporated poly(vinyl alcohol)-corn starch film for active food packaging: Preparation, characterization and antioxidant activity. *Int. J. Biol. Macromol.*, *187*, 223, 2021.
71. Rambabu, K., Bharath, G., Banat, F., Show, P.L., Cocoletzi, H.H., Mango leaf extract incorporated chitosan antioxidant film for active food packaging. *Int. J. Biol. Macromol.*, *126*, 1234, 2019.
72. Friedrich, J.C.C., Silva, O.A., Faria, M.G.I., Colauto, N.B., Gazzin, Z.C., Colauto, G.A.L., Caetano, J., Dragunski, D.C., Improved antioxidant activity of a starch and gelatin-based biodegradable coating containing Tetradenia riparia extract. *Int. J. Biol. Macromol.*, *165*, 1038, 2020.
73. Khameneh, B., Iranshahy, M., Soheili, V., Fazly Bazzaz, B.S., Review on plant antimicrobials: A mechanistic viewpoint. *Antimicrob. Resist. Infect. Control*, *8*, 1, 2019.
74. Kumbar, V.M., Peram, M.R., Kugaji, M.S., Shah, T., Patil, S.P., Muddapur, U.M., Bhat, K.G., Effect of curcumin on growth, biofilm formation and virulence factor gene expression of Porphyromonas gingivalis. *Odontology*, *109*, 18, 2019.
75. Ekramian, S., Abbaspour, H., Roudi, B., Amjad, L., Nafchi, A.M., Influence of nigella sativa L. extract on physico-mechanical and antimicrobial properties of sago starch film. *J. Polym. Environ.*, *29*, 201, 2021.
76. Hu, H., Yao, X., Qin, Y., Yong, H., Liu, J., Development of multifunctional food packaging by incorporating betalains from vegetable amaranth (Amaranthus tricolor L.) into quaternary ammonium chitosan/fish gelatin blend films. *Int. J. Biol. Macromol.*, *159*, 675, 2020.
77. Guo, Z., Han, L., Yu, Q.l, Lin, L., Effect of a sea buckthorn pomace extract-esterified potato starch film on the quality and spoilage bacteria of beef jerky sold in supermarket. *Food Chem.*, *326*, 1, 2020.
78. Sharma, P., Wichaphon, J., Klangpetch, W., Antimicrobial and antioxidant activities of defatted Moringa oleifera seed meal extract obtained by ultrasound-assisted extraction and application as a natural antimicrobial coating for raw chicken sausages. *Int. J. Food Microbiol.*, *332*, 1, 2020.
79. Cui, H., Surendhiran, D., Li, C., Lin, L., Biodegradable zein active film containing chitosan nanoparticle encapsulated with pomegranate peel extract for food packaging. *Food Packag. Shelf Life*, *24*, 1–9, 2020.
80. Sharifi, K.A. and Pirsa, S., Biodegradable film of black mulberry pulp pectin/chlorophyll of black mulberry leaf encapsulated with carboxymethylcellulose/silica nanoparticles: Investigation of physicochemical and antimicrobial properties. *Mater. Chem. Phys.*, *267*, 1, 2021.
81. Pinheiro Bruni, G., dos Santos Acunha, T., de Oliveira, J.P., Martins Fonseca, L., Tavares da Silva, F., Martins Guimarães, V., da Rosa Zavareze, E., Electrospun protein fibers loaded with yerba mate extract for bioactive release in food packaging. *J. Sci. Food Agric.*, *100*, 3341, 2020.
82. Li, L., Wang, H., Chen, M., Jiang, S., Cheng, J., Li, X., Zhang, M., Jiang, S., Gelatin/zein fiber mats encapsulated with resveratrol: Kinetics, antibacterial activity and application for pork preservation. *Food Hydrocolloids*, *101*, 105577, 2020.
83. Istiqola, A. and Syafiuddin, A., A review of silver nanoparticles in food packaging technologies: Regulation, methods, properties, migration, and future challenges. *J. Chin. Chem. Soc.*, *67*, 1942, 2020.
84. Park, Y.L. and Tak, J.H., Chapter 6 - Essential oils for arthropod pest management in agricultural production systems, in: *Essential Oils in Food Preservation, Flavor and Safety*, V.R. Preedy (Ed.), Elsevier Inc, Amsterdam, The Netherlands, 2016.

85. Hamzeh, S., Safari-Faramani, R., Khatony, A., Effects of aromatherapy with lavender and peppermint essential oils on the sleep quality of cancer patients: A randomized controlled trial. *Evid.-Based Complement. Alternat. Med., 2020*, 1, 2020.
86. Amorati, R., Foti, M.C., Valgimigli, L., Antioxidant activity of essential oils. *J. Agric. Food Chem., 61*, 10835, 2013.
87. Moalla, S., Ammar, I., Fauconnier, M.L., Danthine, S., Blecker, C., Besbes, S., Attia, H., Development and characterization of chitosan films carrying Artemisia campestris antioxidants for potential use as active food packaging materials. *Int. J. Biol. Macromol., 183*, 254, 2021.
88. Kilinc, D., Ocak, B., Özdestan-Ocak, Ö., Preparation, characterization and antioxidant properties of gelatin films incorporated with Origanum onites L. essential oil. *J. Food Meas. Charact., 15*, 795, 2021.
89. Tügen, A., Ocak, B., Özdestan-Ocak, Ö., Development of gelatin/chitosan film incorporated with lemon essential oil with antioxidant properties. *J. Food Meas. Charact., 14*, 3010, 2020.
90. Chollakup, R., Pongburoos, S., Boonsong, W., Khanoonkon, N., Kongsin, K., Sothornvit, R., Sukyai, P., Sukatta, U., Harnkarnsujarit, N., Antioxidant and antibacterial activities of cassava starch and whey protein blend films containing rambutan peel extract and cinnamon oil for active packaging. *LWT – Food Sci. Technol., 130*, 109573, 2020.
91. Carpena, M., Nuñez-Estevez, B., Soria-Lopez, A., Garcia-Oliveira, P., Prieto, M.A., Essential oils and their application on active packaging systems: A review. *Resources, 10*, 1, 2021.
92. Yang, X.N., Khan, I., Kang, S.C., Chemical composition, mechanism of antibacterial action and antioxidant activity of leaf essential oil of Forsythia koreana deciduous shrub. *Asian Pac. J. Trop. Med., 8*, 694, 2015.
93. Luís, Â., Ramos, A., Domingues, F., Pullulan films containing rockrose essential oil for potential food packaging applications. *Antibiotics, 9*, 1, 2020.
94. Filho, J.G., de, O., de Deus, I.P.B., Valadares, A.C.F., Fernandes, C.C., Estevam, E.B.B., Egea, M.B., Chitosan film with citrus limonia essential oil: Physical and morphological properties and antibacterial activity. *Colloids Interfaces, 4*, 1, 2020.
95. Alizadeh Behbahani, B., Noshad, M., Jooyandeh, H., Improving oxidative and microbial stability of beef using Shahri Balangu seed mucilage loaded with Cumin essential oil as a bioactive edible coating. *Biocatal. Agric. Biotechnol., 24*, 1, 2020.
96. Heydari, S., Jooyandeh, H., Alizadeh Behbahani, B., Noshad, M., The impact of Qodume Shirazi seed mucilage-based edible coating containing lavender essential oil on the quality enhancement and shelf life improvement of fresh ostrich meat: An experimental and modeling study. *Food Sci. Nutr., 8*, 6497, 2020.
97. Esmaeili, H., Cheraghi, N., Khanjari, A., Rezaeigolestani, M., Basti, A.A., Kamkar, A., Aghaee, E.M., Incorporation of nanoencapsulated garlic essential oil into edible films: A novel approach for extending shelf life of vacuum-packed sausages. *Meat Sci., 166*, 1, 2020.
98. Zhou, C., Abdel-Samie, M.A., Li, C., Cui, H., Lin, L., Active packaging based on swim bladder gelatin/galangal root oil nanofibers: Preparation, properties and antibacterial application. *Food Packag. Shelf Life, 26*, 1, 2020.
99. Yang, Y., Fang, Z., Chen, X., Zhang, W., Xie, Y., Chen, Y., Liu, Z., Yuan, W., An overview of pickering emulsions: Solid-particle materials, classification, morphology, and applications. *Front. Pharmacol., 8*, 1, 2017.
100. Xu, Y., Chu, Y., Feng, X., Gao, C., Wu, D., Cheng, W., Meng, L., Zhang, Y., Tang, X., Effects of zein stabilized clove essential oil Pickering emulsion on the structure and properties of chitosan-based edible films. *Int. J. Biol. Macromol., 156*, 111, 2020.

101. Almasi, H., Azizi, S., Amjadi, S., Development and characterization of pectin films activated by nanoemulsion and Pickering emulsion stabilized marjoram (Origanum majorana L.) essential oil. *Food Hydrocolloids*, *99*, 1, 2020.
102. Fernandes, P., Enzymes in food processing: A condensed overview on strategies for better biocatalysts. *Enzyme Res.*, *2010*, 1–19, 2010.
103. Raveendran, S., Parameswaran, B., Ummalyma, S.B., Abraham, A., Mathew, A.K., Madhavan, A., Rebello, S., Pandey, A., Applications of microbial enzymes in food industry. *Food Technol. Biotechnol.*, *56*, 16, 2018.
104. Cotter, P.D., Ross, R.P., Hill, C., Bacteriocins-a viable alternative to antibiotics? *Nat. Rev. Microbiol.*, *11*, 95, 2013.
105. Pushkaran, A.C., Nataraj, N., Nair, N., Götz, F., Biswas, R., Mohan, C.G., Understanding the structure-function relationship of lysozyme resistance in Staphylococcus aureus by peptidoglycan o-acetylation using molecular docking, dynamics, and lysis assay. *J. Chem. Inf. Model.*, *55*, 760, 2015.
106. Li, Q., Xu, J., Zhang, D., Zhong, K., Sun, T., Li, X., Li, J., Preparation of a bilayer edible film incorporated with lysozyme and its effect on fish spoilage bacteria. *J. Food Saf.*, *40*, 5, 1–9, 2020.
107. Bafort, F., Parisi, O., Perraudin, J.P., Jijakli, M.H., Mode of action of lactoperoxidase as related to its antimicrobial activity: A review. *Enzyme Res.*, *2014*, 1–13, 2014.
108. Ehsani, A., Hashemi, M., Afshari, A., Aminzare, M., Raeisi, M., Zeinali, T., Effect of different types of active biodegradable films containing lactoperoxidase system or sage essential oil on the shelf life of fish burger during refrigerated storage. *LWT - Food Sci. Technol.*, *117*, 1, 2020.
109. Saravani, M., Ehsani, A., Aliakbarlu, J., Ghasempour, Z., Gouda cheese spoilage prevention: Biodegradable coating induced by Bunium persicum essential oil and lactoperoxidase system. *Food Sci. Nutr.*, *7*, 959, 2019.
110. Tavassoli, M., Sani, M.A., Khezerlou, A., Ehsani, A., McClements, D.J., Multifunctional nanocomposite active packaging materials: Immobilization of quercetin, lactoferrin, and chitosan nanofiber particles in gelatin films. *Food Hydrocolloids*, *118*, 1, 2021.
111. Müller-Auffermann, K., Grijalva, F., Jacob, F., Hutzler, M., Nisin and its usage in breweries: A review and discussion. *J. Inst. Brew.*, *121*, 309, 2015.
112. Prince, A., Sandhu, P., Kumar, P., Dash, E., Sharma, S., Arakha, M., Jha, S., Akhter, Y., Saleem, M., Lipid-II independent antimicrobial mechanism of nisin depends on its crowding and degree of oligomerization. *Sci. Rep.*, *6*, 1, 2016.
113. Divsalar, E., Tajik, H., Moradi, M., Forough, M., Lotfi, M., Kuswandi, B., Characterization of cellulosic paper coated with chitosan-zinc oxide nanocomposite containing nisin and its application in packaging of UF cheese. *Int. J. Biol. Macromol.*, *109*, 1311, 2018.
114. Zimet, P., Mombrú, Á.W., Mombrú, D., Castro, A., Villanueva, J.P., Pardo, H., Rufo, C., Physicochemical and antilisterial properties of nisin-incorporated chitosan/carboxymethyl chitosan films. *Carbohydr. Polym.*, *219*, 334, 2019.
115. Martillanes, S., Rocha-Pimienta, J., Llera-Oyola, J., Gil, M.V., Ayuso-Yuste, M.C., García-Parra, J., Delgado-Adámez, J., Control of Listeria monocytogenes in sliced dry-cured Iberian ham by high pressure processing in combination with an eco-friendly packaging based on chitosan, nisin and phytochemicals from rice bran. *Food Control*, *124*, 1, 2021.
116. Salvucci, E., Rossi, M., Colombo, A., Pérez, G., Borneo, R., Aguirre, A., Triticale flour films added with bacteriocin-like substance (BLIS) for active food packaging applications. *Food Packag. Shelf Life*, *19*, 193, 2019.
117. Woraprayote, W., Pumpuang, L., Tosukhowong, A., Zendo, T., Sonomoto, K., Benjakul, S., Visessanguan, W., Antimicrobial biodegradable food packaging impregnated with Bacteriocin 7293 for control of pathogenic bacteria in pangasius fish fillets. *LWT - Food Sci. Technol.*, *89*, 427, 2018.

118. Szymandera-Buszka, K., Waszkowiak, K., Jędrusek-Golińska, A., Hęś, M., Sensory analysis in assessing the possibility of using ethanol extracts of spices to develop new meat products. *Foods*, *9*, 1, 2020.
119. Kholibrina, C.R. and Aswandi, A., The consumer preferences for new sumatran camphor essential oil-based products using a conjoint analysis approach. *IOP Conf. Ser. Earth Environ. Sci.*, *715*, 1, 1–7, 2021.
120. Man, A., Santacroce, L., Jacob, R., Mare, A., Man, L., Antimicrobial activity of six essential oils against a group of human pathogens: A comparative study. *Pathogens*, *8*, 1, 2019.
121. Gadisa, E., Weldearegay, G., Desta, K., Tsegaye, G., Hailu, S., Jote, K., Takele, A., Combined antibacterial effect of essential oils from three most commonly used ethiopian traditional medicinal plants on selected multidrug resistant bacteria. *BMC Complement Altern. Med.*, *19*, 1, 2019.
122. Li, Z.H., Cai, M., Liu, Y.S., Sun, P.L., Luo, S.L., Antibacterial activity and mechanisms of essential oil from Citrus medica L. var. sarcodactylis. *Molecules*, *24*, 1, 2019.
123. Scanlon, M.G. and Henrich, A.W., Factors affecting enzyme activity in food processing, in: *Proteins in Food Processing*, Second Edition, R.Y. Yada (Ed.), pp. 337–365, Woodhead Publishing, Oxford, United Kingdom, 2018.
124. Wu, H., Effect of different light qualities on growth, pigment content, chlorophyll fluorescence, and antioxidant enzyme activity in the red alga Pyropia haitanensis (Bangiales, Rhodophyta). *BioMed. Res. Int.*, *2016*, 1, 2016.
125. Akhidime, I.D., Saubade, F., Benson, P.S., Butler, J.A., Olivier, S., Kelly, P., Verran, J., Whitehead, K.A., The antimicrobial effect of metal substrates on food pathogens. *Food Bioprod. Process.*, *113*, 68, 2019.
126. Olson, R., Gavin-Smith, B., Ferraboschi, C., Kraemer, K., Food fortification: The advantages, disadvantages and lessons from sight and life programs. *Nutrients*, *13*, 1, 2021.
127. Longano, D., Ditaranto, N., Cioffi, N., di Niso, F., Sibillano, T., Ancona, A., Conte, A., del Nobile, M.A., Sabbatini, L., Torsi, L., Analytical characterization of laser-generated copper nanoparticles for antibacterial composite food packaging. *Anal. Bioanal. Chem.*, *403*, 1179, 2012.
128. Hoseinnejad, M., Jafari, S.M., Katouzian, I., Inorganic and metal nanoparticles and their antimicrobial activity in food packaging applications. *Crit. Rev. Microbiol.*, *44*, 161, 2018.
129. Primožič, M., Knez, Ž., Leitgeb, M., (Bio) Nanotechnology in food science — Food packaging. *Nanomaterials*, *11*, 292, 2021.
130. Rozilah, A., Aiza Jaafar, C.N., Sapuan, S.M., Zainol, I., Ilyas, R.A., The effects of silver nanoparticles compositions on the mechanical, physiochemical, antibacterial, and morphology properties of sugar palm starch biocomposites for antibacterial coating. *Polymers*, *12*, 1, 2020.
131. Lin, D., Yang, Y., Wang, J., Yan, W., Wu, Z., Chen, H., Zhang, Q., Wu, D., Qin, W., Tu, Z., Preparation and characterization of TiO2-Ag loaded fish gelatin-chitosan antibacterial composite film for food packaging. *Int. J. Biol. Macromol.*, *154*, 123, 2020.
132. Wang, L., Periyasami, G., Aldalbahi, A., Fogliano, V., The antimicrobial activity of silver nanoparticles biocomposite films depends on the silver ions release behaviour. *Food Chem.*, *359*, 1, 2021.
133. Wu, Z., Deng, W., Luo, J., Deng, D., Multifunctional nano-cellulose composite films with grape seed extracts and immobilized silver nanoparticles. *Carbohydr. Polym.*, *205*, 447, 2019.
134. Asdagh, A., Karimi Sani, I., Pirsa, S., Amiri, S., Shariatifar, N., Eghbaljoo–Gharehgheshlaghi, H., Shabahang, Z., Taniyan, A., Production and characterization of nanocomposite film based on whey protein isolated/copper oxide nanoparticles containing coconut essential oil and paprika extract. *J. Polym. Environ.*, *29*, 335, 2021.

135. Jovanović, J., Ćirković, J., Radojković, A., Mutavdžić, D., Tanasijević, G., Joksimović, K., Bakić, G., Branković, G., Branković, Z., Chitosan and pectin-based films and coatings with active components for application in antimicrobial food packaging. *Prog. Org. Coat.*, *158*, 1, 2021.
136. Dey, A. and Neogi, S., Oxygen scavengers for food packaging applications: A review. *Trends Food Sci. Technol.*, *90*, 26, 2019.
137. Byun, Y., Bae, H.J., Whiteside, S., Active warm-water fish gelatin film containing oxygen scavenging system. *Food Hydrocolloids*, *27*, 250, 2012.
138. Faas, N., Röcker, B., Smrke, S., Yeretzian, C., Yildirim, S., Prevention of lipid oxidation in linseed oil using a palladium-based oxygen scavenging film. *Food Packag. Shelf Life*, *24*, 1, 2020.
139. Gaikwad, K.K., Singh, S., Negi, Y.S., Ethylene scavengers for active packaging of fresh food produce. *Environ. Chem. Lett.*, *18*, 269, 2020.
140. Siripatrawan, U. and Kaewklin, P., Fabrication and characterization of chitosan-titanium dioxide nanocomposite film as ethylene scavenging and antimicrobial active food packaging. *Food Hydrocolloids*, *84*, 125, 2018.
141. Ezati, P., Riahi, Z., Rhim, J.-W., Carrageenan-based functional films integrated with CuO-doped titanium nanotubes for active food-packaging applications. *ACS Sustain. Chem. Eng.*, *9*, 9300, 2021.
142. Böhmer-Maas, B.W., Fonseca, L.M., Otero, D.M., da Rosa Zavareze, E., Zambiazi, R.C., Photocatalytic zein-TiO_2 nanofibers as ethylene absorbers for storage of cherry tomatoes. *Food Packag. Shelf Life*, *24*, 100508, 2020.
143. Lee, M.H., Kim, S.Y., Park, H.J., Effect of halloysite nanoclay on the physical, mechanical, and antioxidant properties of chitosan films incorporated with clove essential oil. *Food Hydrocolloids*, *84*, 58, 2018.
144. Unalan, I.U., Cerri, G., Marcuzzo, E., Cozzolino, C.A., Farris, S., Nanocomposite films and coatings using inorganic nanobuilding blocks (NBB): Current applications and future opportunities in the food packaging sector. *RSC Adv.*, *4*, 29393, 2014.
145. Cui, R., Yan, J., Cao, J., Qin, Y., Yuan, M., Li, L., Release properties of cinnamaldehyde loaded by montmorillonite in chitosan-based antibacterial food packaging. *Int. J. Food Sci. Technol.*, *56*, 3670, 2021.
146. de Souza, A.G., dos Santos, N.M.A., da Silva Torin, R.F., dos Santos Rosa, D., Synergic antimicrobial properties of Carvacrol essential oil and montmorillonite in biodegradable starch films. *Int. J. Biol. Macromol.*, *164*, 1737, 2020.
147. Zou, Y., Zhang, C., Wang, P., Zhang, Y., Zhang, H., Electrospun chitosan/polycaprolactone nanofibers containing chlorogenic acid-loaded halloysite nanotube for active food packaging. *Carbohydr. Polym.*, *247*, 1, 2020.
148. Neji, A.B., Jridi, M., Kchaou, H., Nasri, M., Dhouib Sahnoun, R., Preparation, characterization, mechanical and barrier properties investigation of chitosan-kaolinite nanocomposite. *Polym. Test.*, *84*, 1, 2020.
149. Cheikh, D., Martín-Sampedro, R., Majdoub, H., Darder, M., Alginate bionanocomposite films containing sepiolite modified with polyphenols from myrtle berries extract. *Int. J. Biol. Macromol.*, *165*, 2079, 2020.
150. Gmoshinski, I.V., Bagryantseva, O.V., Arnautov, O.V., Khotimchenko, S.A., Nanoclays in food products: Benefits and possible risks (literature review). *Health Risk Anal.*, *2020*, 142, 2020.
151. Li, X., Shen, W., Sun, H., Meng, L., Wang, B., Zhan, C., Zhao, B., Theoretical studies on carbon dioxide adsorption in cation-exchanged molecular sieves. *RSC Adv.*, *10*, 32241, 2020.
152. Marzano-Barreda, L.A., Yamashita, F., Bilck, A.P., Effect of biodegradable active packaging with zeolites on fresh broccoli florets. *J. Food Sci. Technol.*, *58*, 197, 2021.

153. Jaimun, R. and Sangsuwan, J., Efficacy of chitosan-coated paper incorporated with vanillin and ethylene adsorbents on the control of anthracnose and the quality of Nam Dok Mai mango fruit. *Packag. Technol. Sci.*, *32*, 383, 2019.
154. de Matteis, V., Malvindi, M.A., Galeone, A., Brunetti, V., de Luca, E., Kote, S., Kshirsagar, P., Sabella, S., Bardi, G., Pompa, P.P., Negligible particle-specific toxicity mechanism of silver nanoparticles: The role of Ag+ ion release in the cytosol. *Nanomed.: Nanotechnol. Biol. Med.*, *11*, 731, 2015.
155. Ferdous, Z., Beegam, S., Tariq, S., Ali, B.H., Nemmar, A., The *in vitro* effect of polyvinylpyrrolidone and citrate coated silver nanoparticles on erythrocytic oxidative damage and eryptosis. *Cell. Physiol. Biochem.*, *49*, 1577, 2018.
156. Kim, J.J., Konkel, K., McCulley, L., Diak, I.L., Cases of argyria associated with colloidal silver use. *Ann. Pharmacother.*, *53*, 867, 2019.
157. Blanco, J., Tomás-Hernández, S., García, T., Mulero, M., Gómez, M., Domingo, J.L., Sánchez, D.J., Oral exposure to silver nanoparticles increases oxidative stress markers in the liver of male rats and deregulates the insulin signalling pathway and p53 and cleaved caspase 3 protein expression. *Food Chem. Toxicol.*, *115*, 398, 2018.
158. Dimitrijevic, M., Karabasil, N., Boskovic, M., Teodorovic, V., Vasilev, D., Djordjevic, V., Kilibarda, N., Cobanovic, N., Safety aspects of nanotechnology applications in food packaging. *Proc. Food Sci.*, *5*, 57, 2015.
159. Ardila-Leal, L.D., Poutou-Piñales, R.A., Pedroza-Rodríguez, A.M., Quevedo-Hidalgo, B.E., A brief history of colour, the environmental impact of synthetic dyes and removal by using laccases. *Molecules*, *26*, 1, 2021.
160. El-Wahab, H.M.F.A. and Moram, G.S.E.D., Toxic effects of some synthetic food colorants and/or flavor additives on male rats. *Toxicol. Ind. Health*, *29*, 224, 2013.
161. Luzardo-Ocampo, I., Ramírez-Jiménez, A.K., Yañez, J., Mojica, L., Luna-Vital, D.A., Technological applications of natural colorants in food systems: A review. *Foods*, *10*, 1, 2021.
162. Genovese, M.E., Abraham, S., Caputo, G., Nanni, G., Kumaran, S.K., Montemagno, C.D., Athanassiou, A., Fragouli, D., Photochromic paper indicators for acidic food spoilage detection. *ACS Omega*, *3*, 13484, 2018.
163. Audenaert, K., D'Haene, K., Messens, K., Ruyssen, T., Vandamme, P., Huys, G., Diversity of lactic acid bacteria from modified atmosphere packaged sliced cooked meat products at sell-by date assessed by PCR-denaturing gradient gel electrophoresis. *Food Microbiol.*, *27*, 12, 2010.
164. Ma, Q., Du, L., Wang, L., Tara gum/polyvinyl alcohol-based colorimetric NH3 indicator films incorporating curcumin for intelligent packaging. *Sens. Actuators B Chem.*, *244*, 759, 2017.
165. Castañeda-Ovando, A., Pacheco-Hernández, M., de, L., Páez-Hernández, M.E., Rodríguez, J.A., Galán-Vidal, C.A., Chemical studies of anthocyanins: A review. *Food Chem.*, *113*, 859, 2009.
166. Chayavanich, K., Thiraphibundet, P., Imyim, A., Biocompatible film sensors containing red radish extract for meat spoilage observation. *Spectrochim. Acta A Mol. Biomol. Spectrosc.*, *226*, 1, 2020.
167. Luchese, C.L., Abdalla, V.F., Spada, J.C., Tessaro, I.C., Evaluation of blueberry residue incorporated cassava starch film as pH indicator in different simulants and foodstuffs. *Food Hydrocolloids*, *82*, 209, 2018.
168. Kanatt, S.R., Development of active/intelligent food packaging film containing Amaranthus leaf extract for shelf life extension of chicken/fish during chilled storage. *Food Packag. Shelf Life*, *24*, 1, 2020.
169. Dobrucka, R. and Przekop, R., New perspectives in active and intelligent food packaging. *J. Food Process. Preserv.*, *43*, 1, 2019.

170. Priyadarshi, R., Ezati, P., Rhim, J.-W., Recent advances in intelligent food packaging applications using natural food colorants. *ACS Food Sci. Technol.*, *1*, 124, 2021.
171. Gutiérrez, T.J., Active and intelligent films made from starchy sources/blackberry pulp. *J. Polym. Environ.*, *26*, 2374, 2018.
172. Taghinia, P., Abdolshahi, A., Sedaghati, S., Shokrollahi, B., Smart edible films based on mucilage of lallemantia iberica seed incorporated with curcumin for freshness monitoring. *Food Sci. Nutr.*, *9*, 1222, 2021.
173. Yao, X., Qin, Y., Zhang, M., Zhang, J., Qian, C., Liu, J., Development of active and smart packaging films based on starch, polyvinyl alcohol and betacyanins from different plant sources. *Int. J. Biol. Macromol.*, *183*, 358, 2021.
174. Wu, C., Sun, J., Chen, M., Ge, Y., Ma, J., Hu, Y., Pang, J., Yan, Z., Effect of oxidized chitin nanocrystals and curcumin into chitosan films for seafood freshness monitoring. *Food Hydrocolloids*, *95*, 308, 2019.
175. Tiekstra, S., Dopico-Parada, A., Koivula, H., Lahti, J., Buntinx, M., Holistic approach to a successful market implementation of active and intelligent food packaging. *Foods*, *10*, 465, 2021.
176. Jin, C., Bouzembrak, Y., Zhou, J., Liang, Q., van den Bulk, L.M., Gavai, A., Liu, N., van den Heuvel, L.J., Hoenderdaal, W., Marvin, H.J.P., Big data in food safety - A review. *Curr. Opin. Food Sci.*, *36*, 24, 2020.
177. Mallik, A., Karim, A.B., Md Z.H., Md M.A., Monitoring food storage humidity and temperature data using IoT. *MOJ Food Process. Technol.*, *6*, 399, 2018.
178. Park, Y.W., Kim, S.M., Lee, J.Y., Jang, W., Application of biosensors in smart packaging. *Mol. Cell. Toxicol.*, *11*, 277, 2015.
179. Landim Neves, M.I., Silva, E.K., Meireles, M.A.A., Natural blue food colorants: Consumer acceptance, current alternatives, trends, challenges, and future strategies. *Trends Food Sci. Technol.*, *112*, 163, 2021.

17

Biofoams for Packaging Applications

Vinod V.T. Padil

Institute for Nanomaterials, Advanced Technologies, and Innovation (CXI), Technical University of Liberec (TUL), Studentská, Liberec, Czech Republic

Abstract

This chapter extensively describes an overview of biobased polymers' composite foams for packaging appliances. The essential features of the biofoams sourced from botanicals, trees, animals, seaweed, bacteria, and others and their blending with natural/synthetic biodegradable polymers are expanded. The merits and demerits of biofoams with petroleum-based plastics in their preparation, properties, biodegradability, and uses are compared. Numerous important biofoams are generated from starch, cellulose, tree gums (Arabic, Karaya, and Kondagogu), carrageenan, xanthan, chitosan, and polylactic acid are broadly discussed. Notably, the chemical structure, blending, preparation method, use of various additives or plasticizers, and foams in multiple types, such as fibers, films, and sponges' types and their physical, chemical, morphological, mechanical, thermal, antibacterial efficiency and biodegradability toward the various packaging appliances, including food packaging of biofoams have been displayed in this chapter.

***Keywords*:** Biofoams, biodegradable, additives, packaging, starch, cellulose, gums

17.1 Introduction

Biobased packaging materials from various natural and synthetic origins expand the growing demand for potential appliances in various packaging industries. Due to their "green" and sustainable nature, many newer materials and composites are added to these biobased materials. The biobased material should be completed from renewable resources and have enough strength to get processed into packaging products. Various parameters like melt strength, tensile strength, elongation, processability, flexibility, thermal stability, biodegradability, and composability decide the market acceptability and make any biobased material commercially viable [1–3]. Packaging industries comprise the packaging of food, consumer good, materials for building, interior components, and automotive parts. The development of green and sustainable goals achieve by addressing the global issues by utilizing waste materials into valuable products, reduction of CO_2 emission, and keeping unique natural resources.

Email: vinod.padil@tul.cz

Inamuddin and Tariq Altalhi (eds.) Handbook of Bioplastics and Biocomposites Engineering Applications: 2nd Edition, (361–382) © 2023 Scrivener Publishing LLC

Among the current packaging scenario, petrochemical plastics such as high and low-density polyethylene, polypropylene, polystyrene, polyurethane, etc. have prominent places. The central focus area in the packaging is the replacement of oil-based plastics and developing biobased plastic and their industrial-scale manufacturing, recyclability, and compostability [4, 5]. The depletion of oil resources, increasing the prizes for these materials, nonbiodegradability, and improper waste disposal have influenced the minimum uses of petrochemical plastics in the packaging sector [6, 7]. The concept of sustainable packaging demands the minimum use of packaging material, recyclability, and its most minuscule carbon footprint [8]. The use of plastic packaging waste contributed to 46% of plastic debris. It has been estimated that by 2024, the demand for the cost of packaging will go USD1.05 trillion [9]. The roadmap to curtail the extensive use of packing wastes in public and the environment, many government agencies of countries like the UK China have adopted policies. Among them are fines to the people who discard the rubbish into the environment, charging the packaging company to insist their product quality, written information about handling and recycling, and fining to the public who mismanage with the packaging waste, etc. [8].

It is well accepted by all sorts of the public and government. Further, there are many issues regarding developing biobased plastic for food, and other applications require high mechanical, thermal, barrier properties. Additionally, enzymes are added to improve the compostability of the bioplastic and use natural additives for enhancing stability, barrier coating, crosslinking, and biodegradability [10].

The new packaging solutions made from renewable raw materials and upcycling applications will further contribute to the circular economy of packaging, providing optimum protection for food and drinks and our environment and its limited resources. The majority of the bioplastic formulations are categorized into three types, such as biofoams, biodegradable and compostable, and biofoams and nonbiodegradable. The present chapter highlighted biofoams of natural materials from plant, botanical, bacterial, and animal sources and are all biodegradable categories. The various designs of biofoams such as films, sponges, and fibers based on these natural sources are exemplified their structural, physicochemical properties, and potential applications in packaging industries.

Further, this chapter demonstrates the origin of various biosources, their chemical structures, and processes of the packaging materials for developing in packaging technology. In this context, the overall physicochemical, mechanical, thermal, and barrier properties of the biofoams with different sources are broadly discussed in this chapter. This chapter reflects the environmental manifestation of biobased reinforced materials and delivers their critical challenges in packaging fields. Furthermore, an overview of innovative uses of bioform packaging in the food industry field and provide information on their biodegradability and other regulations. Moreover, the heat sealability, biopolymer blends, focusing on miscibility, compatibility, property relationships, and printability of biobased materials as par with the plastic materials used for packaging applications are critically evaluated.

17.2 Biofoams from Botanical and Plant Sources

In this section, the starch and its blends, cellulose and tree gums, and their application in mainly food packaging areas are described.

17.3 Starch and Their Blends

According to the plastic industry, starch and its blends were ranked among the biofoams and biodegradable packaging materials. According to the plastic industry, starch and its combinations placed the essential commodity to reach their use up to 1.3 Mt (metric tons) by 2020 [11]. The various modifiers/stabilizers, such as sorbitol or glycerol, foaming agents, nanoparticles, inorganic/organic materials, antibacterial agents, etc., are added into starch blend constituted biobased foams for food packaging applications [12]. The edible foams of the starch consist of waxy corn, cassava, potato, wheat, and corn starch [13].

The development of thermoplastic starch relies on the additives, plasticizers (usually glycerol or water), their critical melting temperature, and other parameters [14]. Starch is an abundant plant-derived polysaccharide used as food and in many industrial activities. The major components present in starch are amylose and amylopectin. Both amorphous and crystalline regions are observed in the starch structure. Amylopectin, a double helix structure, contributes to the crystallinity of starch, whereas amylose is present in the amorphous form [15, 16]. Structurally, starch is amorphous and cannot form films alone due to low mechanical and physicochemical properties. Starch with other materials could improve the plasticization via blending, an additive, such as inorganic, organic, chemical modification, or nanoparticle incorporation [12]. Extrusion, freeze-drying, microwave, solvent exchange, hot pressing, etc., are the general methods for preparing starch foams [17]. Figure 17.1 represents the various primary techniques employed to prepare foams and their appliances.

The following approaches have been demonstrated to improve the structural, mechanical, and barrier properties of foams: (1) incorporation of salts such as $CaCO_3$, talc, NaCl, and $CaCl_2$) have shown to reduce processing time and enhanced the foam cell structure and mechanical attributes [18–21]; (2) blending with natural/synthetic polymers; (3) the additives and water content which influenced the various properties (melting temperature, crystallinity, foaming process, cell structure, as well as mechanical) of starch-based material [22].

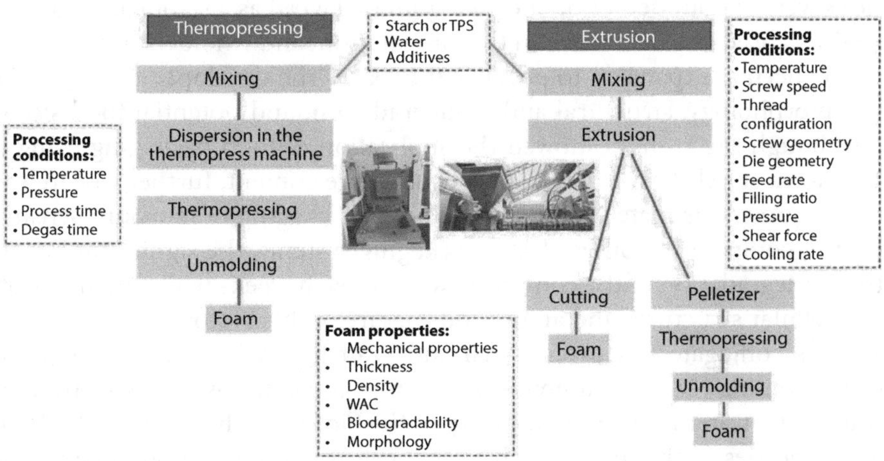

Figure 17.1 Foams processing of starch and their conditions and properties. (Reprinted with permission from Ref. [12].)

A biodegradable foam based on starch (based on wet by-product from processing steamed potatoes with xanthan gum as a plasticizer and oat fibers used as a natural reinforcing agent was prepared via thermal pressing [23]. Starch-based foam developed with 20% of each plasticizer and oat fibers showed a high water absorption and solubility index. The addition of additives entirely controlled the mechanical property of the foam, and the complete biodegradation of the samples was observed after 20 days.

Starch with PLA was combined to produce lightweight foam [24]. The physicochemical mechanical parameters such as expansion, unit and bulk density, water solubility index, and compressibility were optimized for the foam preparation. The result showed that the physicochemical and mechanical attributes are directly proportional to the amount of PLA in the composite. However, the higher amount of moisture content undesirably affected foam properties.

An alternative sustainable and green foam based on the extrusion method of potato, corn, and wheat starch was demonstrated [25]. This material could be replaced by expanded polystyrene loose-fill foam products (EPS), which are not environmentally friendly and possess environmental pollution, poor composability, and nonbiodegradable. The corn thermoplastic starch (TPS)-based foams containing 3% plasticizer (glycerol) and 1% polyvinyl alcohol shows excellent foams. The produced foam had a density in the range of 18.7 to 30.5 kg/m^3 and was 2.5 times denser than EPS-based foams. The disadvantage of the starch-based packaging materials compared to EPS is their open cellular structure with higher moisture absorption. However, their mechanical properties did not affect moisture absorption.

Biofoam properties and development entirely depend on the starch type, foaming process conditions, additive concentration, economic viability, antibacterial efficiency [12]. Recently, due to the spread of COVID-19 pandemic conditions, the food packaging materials are also undergoing stringent rules and regulations and high demand for developing such materials that could address the safety and antimicrobial defensive packaging system for the benefit of society [26].

For the development of food packaging foams, based on agrochemical agroindustrial sources that are renewable, abundant, and low cost, cassava starch incorporated grape stalks formulations were expounded [27]. The guar gum (0.4 wt%) as a stabilizing agent and glycerol (5 wt%) was used as a plasticizer in this preparation, and whole the mixed components were done by thermal expansion to produce the foam. The developed packaging biofoams tested their morphology, structural and biodegradation, and potential food storage capability. Even though the authors showed the application of food packaging using the thermoplastic cassava starch-based foams for low moisture content, further research would be expected to improve this material for suitability for packaging of moisture-sensitive materials. However, compared to other plastic packaging systems, the significant benefit of this foam is the biodegradation of the composite within seven weeks which might be attributed to the open cellular structures and amorphous nature of this biofoam.

Biopolymers conjugate based on starch and kraft lignin bio foams was prepared via compression molding [28]. The amount of lignin in the composite was restricted to 20%, which reduced the water absorption capacity of the biofoam. This research highlighted the significant properties of the foams, such as reduced water absorption enhanced mechanical features. As a hydrophobic material, lignin has been established as a reinforcing agent with starch. The physicochemical and water barrier properties of kraft lignin with urea-crosslinked tapioca starch foam were also demonstrated via the solvent casting method [29]. It has been shown that incorporating lignin in starch matrices enhances most of its properties, including mechanical and flexibility [30–32].

Starch-cellulose foam was prepared by compression molding using environmentally friendly materials, such as citric acid as a crosslinking agent cellulose fiber as a reinforcing medium [33]. The crosslinking agent (citric acid) has influenced the mechanical properties of the conjugate foam. With the addition of 5%(w/w) citric acid, the tensile strength increased to 2.25 MPa. Still, the flexural modulus and flexural modulus were 601.1 MPa and 7.61 MPa of the starch/cellulose composite foam, respectively. Furthermore, improved moisture barrier performance of the crosslinked foam with enhanced thermal stability is suitable for the packaging of fruits and vegetables.

17.4 Cellulose-Based Biofoams for Packaging Application

Cellulose nanofibers (CNF) are used as reinforcement fillers to improve the mechanical and physicochemical properties of the biofoams. The excellent properties of CNF, such as high aspect ratio, biodegradability, flexibility, and crystallinity, are suitable for foam formation [34]. However, compression and injection molding, extrusion, solvent exchange, and spin coating are not ideal for CNF due to their aggregation [35]. In this context, the ice-templating method is mainly preferred for the preparation of CNF based foams for improving their dispersion in a suitable solvent [36].

CNF (2 wt%—a byproduct from paper mills sludge) and PVA (2 wt%) mixtures were prepared with sodium tetraborate decahydrate (0.5 wt%), a crosslinking agent [37]. The ensued foam had a porosity (97.9 ± 0.3%), pore size (3.2 ± 0.9 μm), density (0.03 g cm^{-3}), and compressive strength of 116 kPa at 20% strain suitable for single-use packaging application. However, water vapor permeability (WVP) needs to be studied elaborately to improve the water barrier properties of the biofoam.

In another study, a biofoam consisting of cellulose and surface-modified montmorillonite (SM-MMT) was fabricated via the high shear homogenizer [38]. The cellulose foam porosity was controlled by adding Tween-80 (0.1%, 0.6%, and 1.1%) as a forming agent to stabilize SM-MMT dispersion and glycerin solution (8.0 wt%) as a plasticizer. The resultant foam had enhanced thermal insulating, with improved mechanical, thermal, and barrier properties.

Cellulose biofoam was fabricated using citric acid and chitosan to produce a low density (33–66 kg/m^3) hydrophobic foam with potential antimicrobial and antifungal properties [39]. Sodium dodecyl sulfate (SDS) was used as a foam-forming agent in this preparation. The chitosan or polyvinylamine contributes to the foam's water stability and antibacterial efficiency. This research highlighted the development of environmentally sustainable low-density packaging material, replacing non-biodegradable, toxic plastic packaging material.

17.5 Packaging Foams from Animal-Based Polysaccharides

Biofoams based on chitin and chitosan has shown immense potential in food packaging due to their safe, economical, highly available, biodegradable, antibacterial, and biocompatible polymer [40–43]. Chitosan is an abundantly available biopolymer after cellulose. Its film-forming property is confirmed by its appliances in the food packaging sector via direct casting, coating, dipping, layer-by-layer assembly, and extrusion process [44]. Chitosan could be easily converted into films, sponges, and fibers. The chitosan composite with various additives such as nanoparticles, clays, or inorganic/organic materials exhibits excellent antioxidant and antimicrobial properties for multiple products [45, 46]. A biofoam combination of chitosan and pine needle extract (PNE) was developed for food packaging with enhanced antioxidant and barrier properties [47].

Chitosan foams have produced the blending with other biopolymer such as starch, cellulose, gums, and other forms, such as cyclodextrin, konjac glucomannan, glucose, xylan, kefiran, and including proteins, for improvement of physical, chemical, structural. Morphological and barrier properties [43, 48–50].

Various additives, such as nanoparticles, clay, and graphene oxide (GO), have been used to improve the properties of chitosan food packing foams [51]. Chitosan and GO biocomposite film with electrical conductivity, antioxidant activity, improved water solubility, and mechanical features are herein.

17.6 Seaweed-Based Biofoams

A composite biofoam based on alginate, a seaweed polysaccharide, with potato starch as copolymer support and sepiolite (a clay mineral) as a reinforcing agent, was prepared by freeze-drying method [52]. The interaction between the polymeric functional groups (-OH) and clay minerals (silanol groups) is effectively demonstrated using the formation of biofoam. The mechanical attribute of the biofoam showed after the addition of clay. It was ~29 MPa in foams containing 10% starch, 40% sepiolite, and 50% alginate composition. The resultant product will eventually apply to both insulating materials and packaging.

In another work, antimicrobial food packaging foams comprised of starch, modified starch (cationic starch), and sodium alginate polyelectrolyte was developed [53]. The ensued biofoam did not present any other potential antimicrobial agent. The composite biofoam had a high thermal stability water contact angle apart from improved char yield and glass transition temperature as determined by DSC analysis. Furthermore, the ensued foam showed a potential antimicrobial effect with enhanced structural physical properties projected as a food packaging material.

Cellulose nanocrystals (CNS) extracted from sugarcane bagasse were reinforced with k-carrageenan composite, and the resultant foam was illustrated [54]. Various films with CNC contents (1, 3, 5, and 8 wt%) loaded with k-carrageenan were fabricated. The addition of CNS has reduced the optical transmittance of the films. However, the mechanical properties (Young's modulus, tensile strength, and toughness parameters) tremendously increased (~ 101% and 74% for Young's modulus and tensile strength, respectively) by the addition of CNS (8 wt.%). These investigations progressed to expose the use of this composite foam in the field of food packaging.

Carrageenan is obtained from the seaweed Rhodophyceae family. It applies in various formulations as thickeners, emulsifiers, and stabilizers and is profoundly used as an edible coating agent [55]. Carrageenan foams have been developed for films and coating for food packaging, blending with other biopolymers or nanoparticles, improving shelf life, barrier, and mechanical attributes [56].

A composite biofoam consisting of carrageenan, agar, and TiO_2 nanoparticles was fabricated as food packaging. Incorporating TiO_2 enhances the antibacterial, food brightener, and color additive and protects food from light-induced oxidation [57].

Carrageenan polymer, refined and semirefined forms with konjac glucomannan composite biofoams, was prepared for sesame oil packaging [58]. The additives' various physical, chemical, mechanical, and barrier properties in the composite formation and their application for oil storage were demonstrated. Adding konjac glucomannan into the carrageenan polymer blends improved the foam's overall performance, such as water vapor transmission rate, mechanical (tensile strength and elongation at break), and thermal properties.

17.7 Polylactic Acid

It is one of the most successful and commercially viable biobased and biodegradable polymers for the sustainable packaging industry. It has been marketed commercially since

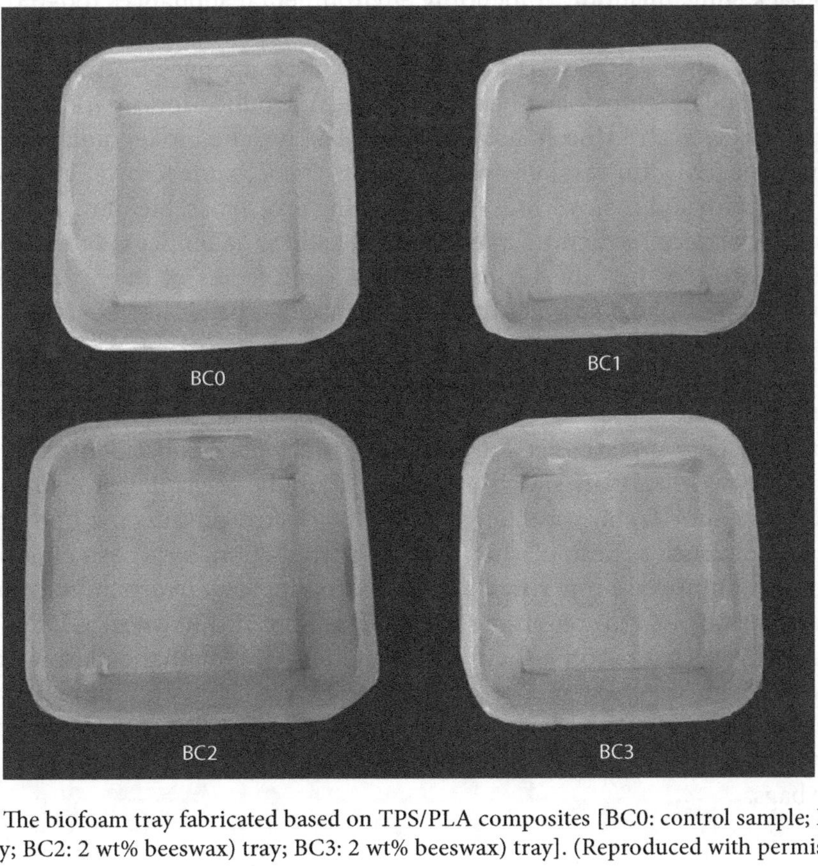

Figure 17.2 The biofoam tray fabricated based on TPS/PLA composites [BC0: control sample; BC1: 1 wt% beeswax) tray; BC2: 2 wt% beeswax) tray; BC3: 2 wt% beeswax) tray]. (Reproduced with permission from Ref. [59].)

2012 by Natureworks. Currently, there are commercial plants to manufacture polylactic acid (PLA) worldwide, and they are being used in the sustainable packaging industry significantly. It can be seen in food packaging, carry bags, medical textile/product, 3D printing, etc., and requires an industrial composting facility in an excellent end-of-life scenario. Industry and research institutes claim the technological development for PLA recycling, but not yet seen in the market. Commercial technology to convert sugar into PLA is also a strategic solution by industry experts.

Biodegradable foam based on PLA and thermoplastic starch was prepared by flat extrusion, calendering, and thermopressing [59]. The emulsion was prepared using various concentrations of coating agent (1 to 3 wt. % of beeswax in the mixture of ethanol and Tween 80) were applied to enhance the barrier and mechanical properties of the composite biofoam. The resultant biofoam with a 1-wt.% coating agent (beeswax) had a higher tensile strength (11.5 ± 1.0 MPa) and a lower water solubility (22.8 ± 0.8%). The ensured biofoam tray (Figure 17.2) could be an excellent product for packaging fresh fruits and vegetables.

17.8 Tree Gum-Based Foams

Tree gum-based lightweight foams have garnered attention in food, biomedical, and pharmaceutical packaging, including impending environmental appliances [60–63]. Tree gums are biodegradable and economical materials available in plenty for the sustainable process for developing bioforms of various kinds such as films, sponges, and fibers [64, 65]. So far, many tree exudate gums have been achieved to produce food packaging films with enhanced properties such as mechanical, thermal, barrier, and biodegradable. In this chapter, the dominant tree gum exudates such as gum Arabic, gum Karaya, gum Kondagogu, and gum Tragacanth and their various foam-forming properties, blending nature, physical, chemical, morphological, structural, mechanical, barrier, and biodegradable properties, are critically examined. Further, the varieties of packaging forms of these gums in different packing roles such as films, sponges, coatings, wrapping, bags, and fibers are extensively illustrated.

The packaging industry is looking for lightweight, recyclable, and biodegradable materials. Therefore, tree gums are options to replace the traditional non-biodegradable petroleum-based materials in packaging applications. Further, they contribute a substantial share with other biopolymers, such as PLA, PHB, starch, and cellulose, in the packaging industry.

The bioform of gum Arabic was fabricated via the electrospinning method. In this work, various treatments, such as heat, plasma, and γ-ray irradiation, have been employed to produce fibers with improved properties such as hydrophobicity, hydrophilicity, mechanical, thermal, barrier, etc. [66]. The bioplastic fibers of gum Arabic (20 wt.%) and PVA (10 wt.%) were prepared by electrospinning their blend solution. The methane plasma-treated fiber foam showed a high-water contact angle than γ-ray irradiated and untreated fibers. The antibacterial, antioxidant, high barrier, and mechanical properties of the treated fiber foams suggested that they are potentially helpful for breathing food-packaging material for biodegradable bags.

The modified gum Arabic with PLA composite foam was prepared by casting and leaching process in another work [67]. The structure-property relationship of the ensued product, such as thermal, crystallization, thermomechanical, and wettability was investigated. The plasticizing effect of modified gum Arabic and the foaming effect of sucrose improved the property of the composite PLA foam with lower density, decreased crystallinity, enhanced surface, and water contact angle, and reduced pore size.

17.9 Karaya Gum-Based Foams

Gum-based three-dimensional hierarchical complex forms, such as sponges or fibers, have potential benefits in many more applications due to their structural and physicochemical properties. Varieties of functional groups present in the gum structures interact with other polymeric entities to form complexes and entirely depend on polymer concentration, ionic

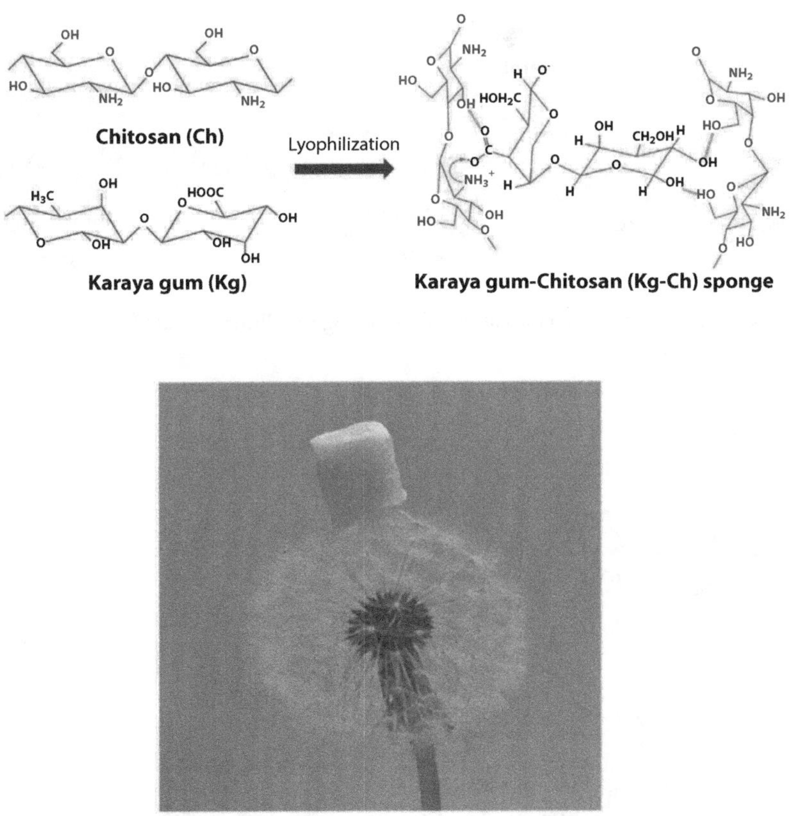

Figure 17.3 Karaya gum and chitosan interaction produce Karaya-chitosan foam (top) and the resultant lightweight foam (down). (Reprinted with permission from Ref. [73].)

strength, pH of the reaction medium, nature, and position of ionic groups on the polymeric chain, etc. [68].

Karaya gum and chitosan mixed in the ratio of I:4 wt.% to fabricate composite sponge and physical, chemical, mechanical thermal, and biodegradation properties were assessed [69]. The foam was prepared via the freeze-drying method using the respective aqueous solutions of polymers. The electrostatic interaction of Karaya and chitosan is the principal mechanism of forming a lightweight sponge, as illustrated in Figure 17.3. The ensued foam was used for environmental remediation of toxic dyes from water.

17.10 Kondagogu Gum-Based Foams

Gum kondagogu and sodium alginate generated green and sustainable foam in a facile synthesis [70] (Figure 17.4). The blending of biopolymeric aqueous solution using glycerol as a plasticizer was investigated to produce films or foams with enhanced mechanical, thermal, and barrier properties. The various functional groups present in kondagogu and

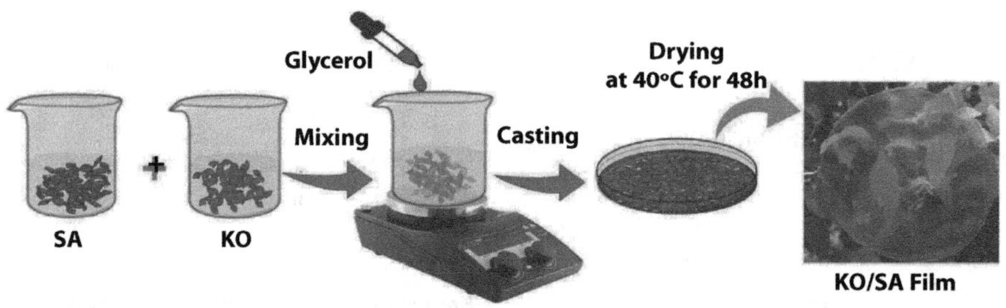

Figure 17.4 The steps involved in the fabrication of kondagogu-sodium alginate foam. (Reprinted with permission from Ref. [74].)

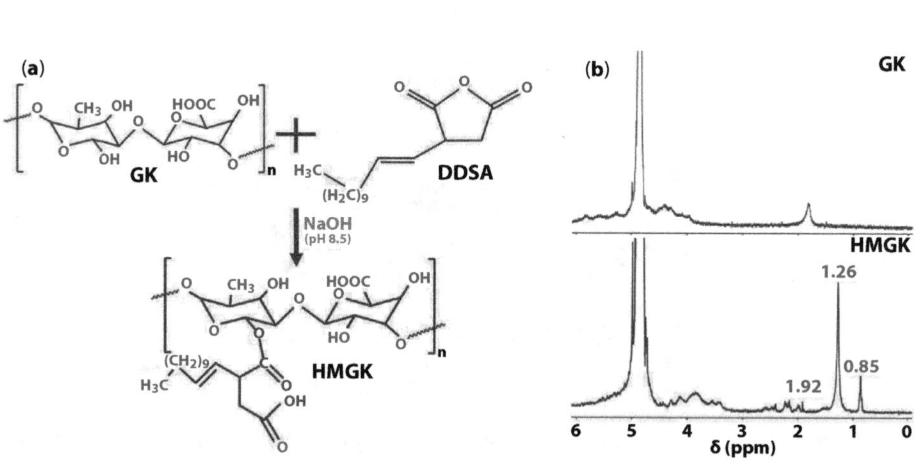

Figure 17.5 (a) The schematic presentation of the interaction of GK with DDSA to form DDS A-modified gum karaya (HMGK); (b) 1H proton NMR spectra of GK (top) and HMGK (bottom), respectively. (Reprinted with permission from Ref. [66].)

alginate and their interaction to form hydrogen bonding are the mechanism of formation that improved the conjugated foam's structural integrity. The incorporation of sodium alginate into the blended film with gum kondagogu had a higher water contact angle, tensile strength (~24 MPa), and thermal characteristics. Furthermore, the biodegradation of the blended foam within one month time was recorded to be more than 90%. The resulting karaya-alginate foam can be used for food packaging applications.

In another research, the chemical modification of karaya gum with dodecenyl succinic anhydride (DDSA) produced a flexible, structurally stable, mechanically robust foam [62]. The esterification between gum kondagogu with DDSA was confirmed via spectroscopic methods (FTIR and NMR). The schematic preparation of DDSA-modified gum kondagogu (HMGK) is presented in Figure 17.5. Further, the foam has improved its mechanical, barrier (water vapor and oxygen), optical, antibacterial, and thermal characteristics by incorporating nanocellulose into the DDSA-modified foam [71]. Furthermore, the DDSA-Kondagogu with 15% nano-cellulose foam showed potential antibacterial efficacy toward *S. aureus* and *E. coli* type of bacteria. The additional benefits favored by this foam are biodegradable under natural environmental conditions.

17.11 Microbial Gum-Based Packaging Foams

The food packaging sector has successfully utilized bionanocomposites and nanomaterials in packaging applications for enhancing their properties. such as antioxidant and antibacterial. Combining three biopolymeric blends, such as k-Carrageenan, xanthan gum, and gellan gum, contributed composite film via solvent casting [72]. The advantages of mixing the various biobased polymers ensured soft texture, excellent elasticity, and mechanically robustness. Integration of TiO2 nanoparticles into the composite foam significantly altered the mechanical (tensile strength, tensile modulus) and thermal properties, including glass transition temperature of the k-Carrageenan/xanthan/gellan gum biofoam. Furthermore, the hydrophobicity and water stability of the films have increased upon the addition of TiO_2 into the composite foam. The other essential features of this composite foam exhibited antibacterial properties and shielding of UV light, which contributed to developing food packaging areas.

Carrageenan/SiO_2 nanofoam was designed, and their potential appliances in food additive, anti-foaming agents, etc., were extensively studied. According to the United States Food and Drug Administration, silica particles up to 2% are considered safe for food applications [73]. The nanofillers of SiO_2 enhance the barrier, mechanical and optical properties. The incorporation of SiO_2 into the Kappa-Carrageenan increased the tensile strength of the composite foam from the range of 53.9 MPa compared to pristine polymer (46.8 MPa). Further, the antibacterial efficacy of the composite film with SiO_2 nanoparticles against *Staphylococcus aureus* and *Escherichia coli*. This investigation exposes Kappa-Carrageenan/ SiO_2 foam to the shelf-life of foods in biodegradable packaging.

Xanthan gum is used for packaging applications in nano-coating material to enhance food safety and quality [72]. Emerging biofoams based on biodegradable, "green," and sustainable materials for developing practical packaging applications are illustrated in Table 17.1.

Table 17.1 Biofoams based on biobased materials and their preparation, properties, and appliances.

Biofoam	**Preparation method**	**Additives/reinforcing agent**	**Specific application/ properties**	**Ref.**
Citrus pectin/ cellulose nanofibers	Pickering emulsion gel template method/ freeze casting/ chemical cross-linking.	Thymol, CNF	- antibacterial - sustained release of thymol - packaging of edible fungi	[74]
Starch/LCNF nanocomposite	Microfluidization/ freeze-drying	Lignin-containing cellulosic nanofibrils (LCNF)	Lightweight packaging material.	[75]
Wheat-gluten	*in situ* polymerization	silica	cell size: 2–400 μm; density: 60–163 kg/ m3, and compression modulus: 1–9 MPa - storage/release of natural oil or surfactants - antimicrobial	[76]
Wheat flour	freeze-drying	carbon nanotubes (CNTs), glycerol	- EMI Shielding - pressure sensor.	[77]
Mycelium based biofoam	incubation	agar, and cellulose	Biodegradable packaging materials	[78]
Mycelia of the macrofungi	Incubation	Sawdust, wheat bran, and $CaCO_3$	packaging	[79]
Cassava starch	Mixing and casting	Carboxymethyl cellulose (CMC)/ *Lactobacillus plantarum* and *Pedocococcus pentosaceus*/ glycerol	Food packaging	[80]

(*Continued*)

Table 17.1 Biofoams based on biobased materials and their preparation, properties, and appliances. (*Continued*)

Biofoam	Preparation method	Additives/reinforcing agent	Specific application/ properties	Ref.
Starch-based arracacha	thermopressing	Glycerol (plasticizer), and magnesium stearate (foaming agent)	Dry food commercial packaging.	[81]
Cassava starch	Thermal expansion process	sesame cake, glycerol, guar gum, and magnesium stearate	Dry foods and foods with low moisture packaging.	[82]
Carrageenan	Green synthesis	Laponite and Ag nanoparticles	- improved barrier and mechanical properties. - antibacterial efficiency. - antibacterial food packaging. - enhanced antimicrobial properties.	[83]
Kappa-carrageenan	Homogenization/ casting	Bentonite nanoclay/ glycerol	Fresh fruits and vegetables packaging.	[84]
Carrageenan	Solution casting	ZnO nanoparticles/ melanin	Antibacterial food packaging	[85]
k-carrageenan	Sonication/ solution casting	Gelatin, nano-SiO2 and sorbitol/glycerol	packaging	[86]
Karaya gum	Sonication/ solution casting	*Schisandra chinensis* oil/ glycerol	Biodegradable film	[87]
Karaya gum	Homogenization, sonication and casting	Cloisite Na+, glycerol and cinnamaldehyde	Biodegradable and antibacterial film	[88]
Gum Arabic	Casting	Cellulose nanocrystals and peel extracts (grape skin, pomegranate and red pitaya)	Food packaging	[89]

(*Continued*)

Table 17.1 Biofoams based on biobased materials and their preparation, properties, and appliances. (*Continued*)

Biofoam	Preparation method	Additives/reinforcing agent	Specific application/ properties	Ref.
Gum Arabic	Casting method	Konjac glucomannan/ agar/virgin coconut oil	polysaccharide-based edible film in cucumber packaging.	[90]
Octenyl succinic anhydride (OSA) modified gum Arabic	Mixing, dialyzed and free-drying	Cinnamon essential oil, chitosan	Antibacterial poly-electrolytes film	[91]
Gum Arabic	Homogenization, casting and drying	cellulose nanocrystals/ glycerol	"Green" packaging system	[92]
Tragacanth	Ultrasonication and casting	Gelatin, zinc oxide nanoparticles	antimicrobial food packaging	[93]
Tragacanth	Solution casting and solvent evaporation	gallic acid, Poly (vinyl alcohol)	Antioxidant packaging	[94]
Xanthan	Homogenization, ultrasonication and casting	hydroxypropyl methylcellulose, glycerin, and sorbitol	Food preservation and packaging.	[95]
Agar	Solution casting	Carrageenan, ZnS nanoparticles, nanocellulose and Pickering emulsion of tea tree essential oil	antibacterial and antioxidant film for food packaging	[96]
PLA	Extrusion and Compression molding	Soybean oil	Disposable packaging material.	[97]
PLA	Blending	Ethylene vinyl acetate	Environmental applications	[98]
Cassava starch	Thermopressing	Peanut skin, glycerol, guar gum, magnesium stearate	Multiple applications	[99]

(*Continued*)

Table 17.1 Biofoams based on biobased materials and their preparation, properties, and appliances. (*Continued*)

Biofoam	Preparation method	Additives/reinforcing agent	Specific application/ properties	Ref.
Arracacha starches	Thermopressing process	Sweet potato, oca	Dry food packaging	[81]
Sweet potato starch	Thermopressing	Oregano/thyme essential oil	Antibacterial food packaging	[100]
Corn	Thermopressing	Cellulose, xylan, lignin fraction of sorghum, PVA, corn starch	Biodegradable food trays	[101]
Cellulose	Compression molding	Starch, citric acid, carnauba wax	Packaging of fruits and vegetables	[102]

17.12 Conclusion and Outlooks

The packaging industry encourages to use and looking for biodegradable, lightweight, economical, reusable, ecofriendly, and sustainable foams to replace petroleum-based nonbiodegradable polymers. In this chapter, biodegradable natural biobased polymers of the types such as starch, cellulose, tree gums, chitosan, carrageenan, and xanthan their processing, characterization, application are described. Furthermore, the various process techniques, blending ratio, different plasticizers, and additives to improve these foams' structural, mechanical, barrier and biodegradation, and antibacterial properties are extensively deliberated. Additionally, nanotechnology and interdisciplinary sciences that can be applied in foamed products and their potential applications in packaging fields are to be investigated. Apart from this, many challenges underwent to use biobased polymers to replace the existing expanded polystyrene in foam packaging. The significant problems, such as low mechanical, barrier, water resistance, porosity, cell size, cell density, biodegradation, and flexibility, are much to be studied and optimized to improve the overall performance of the biobased foams. Importantly, biowastes materials and other food waste are to be explored to develop as a packaging foam to reduce the overall cost of packaging materials in addition to their biodegradable and sustainable nature.

References

1. Arias, A., González-García, S., González-Rodríguez, S., Feijoo, G., Moreira, M.T., Cradle-to-gate life cycle assessment of bio-adhesives for the wood panel industry. A comparison with petrochemical alternatives. *Sci. Total Environ.*, 738, 140357, 2020.
2. Jem, K.J. and Tan, B., The development and challenges of poly (lactic acid) and poly (glycolic acid). *Adv. Ind. Eng. Polym. Res.*, 3, 60, 2020.
3. Ates, B., Koytepe, S., Ulu, A., Gurses, C., Kumar Thakur, V., Chemistry, structures, and advanced applications of nanocomposites from biorenewable resources. *Chem. Rev.*, 120, 9304, 2020.

4. Sid, S., Mor, R.S., Kishore, A., Sharanagat, V.S., Bio-sourced polymers as alternatives to conventional food packaging materials: A review. *Trends Food Sci. Technol.*, 115, 87, 2021.
5. Stoyanova, N., Paneva, D., Mincheva, R., Toncheva, A., Manolova, N., Dubois, P., Rashkov, I., Poly(l-lactide) and poly(butylene succinate) immiscible blends: From electrospinning to biologically active materials. *Mater. Sci. Eng. C*, 41, 119, 2014.
6. Dubey, S.P., Thakur, V.K., Krishnaswamy, S., Abhyankar, H.A., Marchante, V., Brighton, J.L., Progress in environmental-friendly polymer nanocomposite material from PLA: Synthesis, processing and applications. *Vacuum.*, 146, 655, 2017.
7. Brockhaus, S., Petersen, M., Kersten, W., A crossroads for bioplastics: Exploring product developers' challenges to move beyond petroleum-based plastics. *J. Cleaner Prod.*, 127, 84, 2016.
8. You, S., Li, W., Yan, W.-C., Sonne, C., Green strategies for sustainable packaging. *Science*, 372, 802, 2021.
9. Kang, P., Song, G., Xu, M., Miller, T.R., Wang, H., Zhang, H., Liu, G., Zhou, Y., Ren, J., Zhong, R., Duan, H., Low-carbon pathways for the booming express delivery sector in China. *Nat. Commun.*, 12, 1, 2021.
10. Hu, B., Lightweight materials from biopolymers and biofibers, in: *ACS Symposium Series*, vol. 1175, pp. 239–255, 2014.
11. Imre, B. and Pukánszky, B., Compatibilization in bio-based and biodegradable polymer blends. *Eur. Polym. J.*, 49, 1215, 2013.
12. Tapia-Blácido, D.R., Aguilar, G.J., de Andrade, M.T., Rodrigues-Júnior, M.F., Guareschi-Martins, F.C., Trends and challenges of starch-based foams for use as food packaging and food container. *Trends Food Sci. Technol.*, 119, 257, 2022.
13. Acosta, S., Chiralt, A., Santamarina, P., Rosello, J., González-Martínez, C., Cháfer, M., Antifungal films based on starch-gelatin blend, containing essential oils. *Food Hydrocolloids*, 61, 233, 2016.
14. Svagan, A.J., Azizi Samir, M.A.S., Berglund, L.A., Biomimetic polysaccharide nanocomposites of high cellulose content and high toughness. *Biomacromolecules*, 8, 2556, 2007.
15. Pérez, S., Baldwin, P.M., Gallant, D.J., Structural features of starch granules I, in: *Starch*, p. 149, 2009.
16. Madrigal, L., Sandoval, A.J., Müller, A.J., Effects of corn oil on glass transition temperatures of cassava starch. *Carbohydr. Polym.*, 85, 875, 2011.
17. Soykeabkaew, N., Thanomsilp, C., Suwantong, O., A review: Starch-based composite foams. *Compos. Part A Appl. Sci. Manuf.*, 78, 246, 2015.
18. Zhou, J., Song, J., Parker, R., Microwave-assisted moulding using expandable extruded pellets from wheat flours and starch. *Carbohydr. Polym.*, 69, 445, 2007.
19. Heartwin Amaladhas Pushpadass, B., Suresh Babu, G., Weber, R.W., Hanna, M.A., Extrusion of starch-based Loose-fill packaging foams #: Effects of temperature, moisture and talc on physical properties. *Packag. Technol. Sci. Packag. Technol. Sci.*, 21, 171, 2008.
20. Combrzyński, M., Combrzyński, C., Matwijczuk, A., Wójtowicz, A., Oniszczuk, T., Karcz, D., Szponar, J., Niemczynowicz, A., Bober, D., Mitrus, M., Kupryaniuk, K., Stasiak, M., Dobrzański, B., Oniszczuk, A., Wyszyński, S., Wyszyński, W., Potato starch utilization in ecological loose-fill packaging materials-sustainability and characterization. *Mater. (Basel)*, 13, 1390, 2020.
21. Zhou, J., Song, J., Parker, R., Structure and properties of starch-based foams prepared by microwave heating from extruded pellets. *Carbohydr. Polym.*, 63, 466, 2006.
22. Meng, L., Liu, H., Yu, L., Duan, Q., Chen, L., Liu, F., Shao, Z., Shi, K., Lin, X., How water acting as both blowing agent and plasticizer affect on starch-based foam. *Ind. Crops Prod.*, 134, 43, 2019.
23. Hamine, N., Rodrigues, P., Taveira De Souza, J., Rodrigues, R.L., Canteri, M.H.G., Mara, S., Tramontin, K., Carlos De Francisco, A., Starch-based foam packaging developed from a by-product of potato industrialization (Solanum tuberosum L.), *Appl. Sci.*, 10, 2235, 2020.

24. Fang, Q. and Hanna, M.A., Functional properties of polylactic acid starch-based loose-fill packaging foams. *Cereal Chem.*, 77, 779–789, 2000. https://doi.org/10.1094/CCHEM.2000.77.6.779
25. Mitrus, M. and Moscicki, L., Extrusion-cooking of starch protective loose-fill foams. *Chem. Eng. Res. Des.*, 92, 778, 2014.
26. Chin, A.W.H., Chu, J.T.S., Perera, M.R.A., Hui, K.P.Y., Yen, H.-L., Chan, M.C.W., Peiris, M., Poon, L.L.M., Stability of SARS-CoV-2 in different environmental conditions. *Lancet Microbe*, 1, e10, 2020.
27. Engel, J.B., Ambrosi, A., Tessaro, I.C., Development of biodegradable starch-based foams incorporated with grape stalks for food packaging. *Carbohydr. Polym.*, 225, 115234, 2019.
28. Yan, Q., Arango, R., Li, J., Cai, Z., Fabrication and characterization of carbon foams using 100% Kraft lignin. *Mater. Des.*, 201, 109460, 2021.
29. Mariana, M., Alfatah, T., Abdul Khalil, H.P.S., Yahya, E.B., Olaiya, N.G., Nuryawan, A., Mistar, E.M., Abdullah, C.K., Abdulmadjid, S.N., Ismail, H., A current advancement on the role of lignin as sustainable reinforcement material in biopolymeric blends. *J. Mater. Res. Technol.*, 15, 2287, 2021.
30. Lo, H., Kadiyala, S., Guggino, S.E., Leong, K.W., Poly(L-lactic acid) foams with cell seeding and controlled-release capacity. *J. Biomed. Mater. Res.*, 30, 475, 1996.
31. Kang, S., Li, X., Fan, J., Chang, J., Hydrothermal conversion of lignin: A review. *Renewable Sustainable Energy Rev.*, 27, 546, 2013.
32. Kaewtatip, K. and Thongmee, J., The effects of cross-linked starch on the properties of thermoplastic starch. *Mater. Des.*, 45, 586, 2013.
33. Hassan, M.M., Tucker, N., Le Guen, M.J., Thermal, mechanical and viscoelastic properties of citric acid-crosslinked starch/cellulose composite foams. *Carbohydr. Polym.*, 230, 115675, 2020.
34. Moon, R.J., Martini, A., Nairn, J., Simonsen, J., Youngblood, J., Cellulose nanomaterials review: Structure, properties and nanocomposites. *Chem. Soc. Rev.*, 40, 3941, 2011.
35. Dlouhá, J., Suryanegara, L., Yano, H., Cellulose nanofibre–poly(lactic acid) microcellular foams exhibiting high tensile toughness. *React. Funct. Polym.*, 85, 201, 2014.
36. Deville, S., Ice-templating, freeze casting: Beyond materials processing. *J. Mater. Res. 2013 2817*, 28, 2202, 2013.
37. Adu, C., Rahatekar, S., Filby, J., Ayre, D., Jolly, M., Structural packaging foams prepared by uni-directional freezing of paper sludge cellulose nanofibres and poly (vinyl alcohol). *Mater. Lett.*, 253, 242, 2019.
38. Ahmadzadeh, S., Nasirpour, A., Keramat, J., Hamdami, N., Behzad, T., Desobry, S., Nanoporous cellulose nanocomposite foams as high insulated food packaging materials. *Colloids Surf. A Physicochem. Eng. Asp.*, 468, 201, 2015.
39. Ottenhall, A., Seppänen, T., Ek, M., Water-stable cellulose fiber foam with antimicrobial properties for bio based low-density materials. *Cellulose*, 25, 2599, 2018.
40. Singh, G., Singh, S., Kumar, B., Gaikwad, K.K., Active barrier chitosan films containing gallic acid based oxygen scavenger. *Food Measure*, 15, 585, 2021.
41. Tian, B. and Liu, Y., Chitosan-based biomaterials: From discovery to food application. *Polym. Adv. Technol.*, 31, 2408, 2020.
42. Siripatrawan, U. and Harte, B.R., Physical properties and antioxidant activity of an active film from chitosan incorporated with green tea extract. *Food Hydrocolloids*, 24, 770, 2010.
43. Priyadarshi, R. and Rhim, J.W., Chitosan-based biodegradable functional films for food packaging applications. *Innov. Food Sci. Emerg. Technol.*, 62, 102346, 2020.
44. Wang, H., Qian, J., Ding, F., Emerging chitosan-based films for food packaging applications. *J. Agric. Food Chem.*, 66, 395, 2018.

45. Tien, N.D., Lyngstadaas, S.P., Mano, J.F., Blaker, J.J., Haugen, H.J., Mano, S.P., Blaker, J.F., Haugen, J.J., Recent developments in chitosan-based micro/nanofibers for sustainable food packaging, smart textiles, cosmeceuticals, and biomedical applications. *Molecules*, 26, 2683, 2021.
46. Merzendorfer, H., Cohen, E. Chitin/Chitosan: Versatile ecological, industrial, and biomedical applications. in: *Extracellular Sugar-Based Biopolymers Matrices. Biologically-Inspired Systems*. Cohen, E., Merzendorfer, H. (eds.), vol 12, Springer, Cham, 2019. https://doi.org/10.1007/978-3-030-12919-4_14
47. Kadam, A.A., Singh, S., Gaikwad, K.K., Chitosan based antioxidant films incorporated with pine needles (Cedrus deodara) extract for active food packaging applications. *Food Control*, 124, 107877, 2021.
48. Souza, M.P., Vaz, A.F.M., Cerqueira, M.A., Texeira, J.A., Vicente, A.A., Carneiro-da-Cunha, M.G., Effect of an edible nanomultilayer coating by electrostatic self-assembly on the shelf life of fresh-cut mangoes. *Food Bioprocess Technol.*, 8, 647–654, 2015.
49. Kamdem, D.P., Shen, Z., Nabinejad, O., Development of biodegradable composite chitosan-based films incorporated with xylan and carvacrol for food packaging application. *Food Packag. Shelf Life*, 21, 100344, 2019.
50. Gravel, A. and Doyen, A., The use of edible insect proteins in food: Challenges and issues related to their functional properties. *Innov. Food Sci. Emerg. Technol.*, 59, 102272, 2020.
51. Barra, A., Ferreira, N.M., Martins, M.A., Lazar, O., Pantazi, A., Jderu, A.A., Neumayer, S.M., Rodriguez, B.J., Enăchescu, M., Ferreira, P., Nunes, C., Eco-friendly preparation of electrically conductive chitosan - reduced graphene oxide flexible bionanocomposites for food packaging and biological applications. *Compos. Sci. Technol.*, 173, 53, 2019.
52. Darder, M., Matos, C.R.S., Aranda, P., Gouveia, R.F., Ruiz-Hitzky, E., Bionanocomposite foams based on the assembly of starch and alginate with sepiolite fibrous clay. *Carbohydr. Polym.*, 157, 1933, 2017.
53. Şen, F., Uzunsoy, İ., Baştürk, E., Kahraman, M.V., Antimicrobial agent-free hybrid cationic starch/sodium alginate polyelectrolyte films for food packaging materials. *Carbohydr. Polym.*, 170, 264, 2017.
54. Kassab, Z., Aziz, F., Hannache, H., Ben Youcef, H., El Achaby, M., Improved mechanical properties of k-carrageenan-based nanocomposite films reinforced with cellulose nanocrystals. *Int. J. Biol. Macromol.*, 123, 1248, 2019.
55. Bertolino, V., Cavallaro, G., Lazzara, G., Milioto, S., Parisi, F., Halloysite nanotubes sandwiched between chitosan layers: Novel bionanocomposites with multilayer structures. *New J. Chem.*, 42, 8384, 2018.
56. Aga, M.B., Dar, A.H., Nayik, G.A., Panesar, P.S., Allai, F., Khan, S.A., Shams, R., Kennedy, J.F., Altaf, A., Recent insights into carrageenan-based bio-nanocomposite polymers in food applications: A review. *Int. J. Biol. Macromol.*, 192, 197, 2021.
57. Wang, S.X., Liu, Y., Zhang, G.Q., Zhao, S., Xu, F., Geng, X.L., Wang, H.X., Cordysobin, a novel alkaline serine protease with HIV-1 reverse transcriptase inhibitory activity from the medicinal mushroom Cordyceps sobolifera. *J. Biosci. Bioeng.*, 113, 42–47, 2012.
58. Ganesan, A.R., Shanmugam, M., Ilansuriyan, P., Anandhakumar, R., Balasubramanian, B., Composite film for edible oil packaging from carrageenan derivative and konjac glucomannan: Application and quality evaluation. *Polym. Test.*, 78, 105936, 2019.
59. Reis, M.O., Olivato, J.B., Bilck, A.P., Zanela, J., Grossmann, M.V.E., Yamashita, F., Biodegradable trays of thermoplastic starch/poly (lactic acid) coated with beeswax. *Ind. Crops Prod.*, 112, 481, 2018.

60. Padil, V.V.T., Wacławek, S., Černík, M., Varma, R.S., Tree gum-based renewable materials: Sustainable applications in nanotechnology, biomedical and environmental fields. *Biotechnol. Adv.*, 36, 1984–2016, 2018.
61. Padinjareveetil, A.K.K., Padil, V.V.T., Černík, M., Graphene oxide-based nanocomposite and their biomedical applications. in: *Nanoengineering of Biomaterieals*. S. Jana and S. Jana (eds.), pp. 427–456S, 2022. https://doi.org/10.1002/9783527832095.ch31
62. Venkateshaiah, A., Havlíček, K., Timmins, R.L., Röhrl, M., Wacławek, S., Nguyen, N.H.A., Černík, M., Padil, V.V.T., Agarwal, S., Alkenyl succinic anhydride modified tree-gum kondagogu: A bio-based material with potential for food packaging. *Carbohydr. Polym.*, 266, 118126, 2021.
63. Raj, V., Lee, J.H., Shim, J.J., Lee, J., Recent findings and future directions of grafted gum karaya polysaccharides and their various applications: A review. *Carbohydr. Polym.*, 258, 117687, 2021.
64. Akshay Kumar, K.P., Zare, E.N., Torres-Mendieta, R., Wacławek, S., Makvandi, P., Černík, M., Padil, V.V.T., Varma, R.S., Electrospun fibers based on botanical, seaweed, microbial, and animal sourced biomacromolecules and their multidimensional applications. *Int. J. Biol. Macromol.*, 171, 130, 2021.
65. Padil, V.V.T., Cheong, J.Y., KP, A.K., Makvandi, P., Zare, E.N., Torres-Mendieta, R., Wacławek, S., Černík, M., Kim, I.D., Varma, R.S., Electrospun fibers based on carbohydrate gum polymers and their multifaceted applications. *Carbohydr. Polym.*, 247, 116705, 2020.
66. Padil, V.V.T., Senan, C., Wacławek, S., Černík, M., Agarwal, S., Varma, R.S., Bioplastic fibers from gum arabic for greener food wrapping applications. *ACS Sustainable Chem. Eng.*, 7, 5900, 2019.
67. Borkotoky, S.S., Ghosh, T., Bhagabati, P., Katiyar, V., Poly (lactic acid)/modified gum arabic (MG) based microcellular composite foam: Effect of MG on foam properties, thermal and crystallization behavior. *Int. J. Biol. Macromol.*, 125, 159, 2019.
68. Costa, M.P.M., Prates, L.M., Baptista, L., Cruz, M.T.M., Ferreira, I.L.M., Interaction of polyelectrolyte complex between sodium alginate and chitosan dimers with a single glyphosate molecule: A DFT and NBO study. *Carbohydr. Polym.*, 198, 51, 2018.
69. Ramakrishnan, R.K., Padil, V.V.T., Wacławek, S., Černík, M., Varma, R.S., Eco-Friendly and economic, adsorptive removal of cationic and anionic dyes by bio-based Karaya gum—chitosan sponge. *Polym.*, 13, 251, 2021.
70. Ramakrishnan, R.K., Wacławek, S., Černík, M., Padil, V.V.T., Biomacromolecule assembly based on gum kondagogu-sodium alginate composites and their expediency in flexible packaging films. *Int. J. Biol. Macromol.*, 177, 526, 2021.
71. Wang, J., Gardner, D.J., Stark, N.M., Bousfield, D.W., Tajvidi, M., Cai, Z., Moisture and oxygen barrier properties of cellulose nanomaterial-based films. *ACS Sustainable Chem. Eng.*, 6, 49, 2017.
72. Balasubramanian, R., Kim, S.S., Lee, J., Lee, J., Effect of TiO2 on highly elastic, stretchable UV protective nanocomposite films formed by using a combination of k-Carrageenan, xanthan gum and gellan gum. *Int. J. Biol. Macromol.*, 123, 1020, 2019.
73. Venkatesan, R., Rajeswari, N., Thiyagu, T.T., Preparation, characterization and mechanical properties of K-carrageenan/SiO2 nanocomposite films for antimicrobial food packaging. *Bull. Mater. Sci.*, 40, 609, 2017.
74. Wu, W., Wu, Y., Lin, Y., Shao, P., Facile fabrication of multifunctional citrus pectin aerogel fortified with cellulose nanofiber as controlled packaging of edible fungi. *Food Chem.*, 374, 131763, 2022.
75. Ago, M., Ferrer, A., J. Rojas, O., Starch-based biofoams reinforced with lignocellulose nanofibrils from residual palm empty fruit bunches: Water sorption and mechanical strength. *ACS Sustainable Chem. Eng.*, 4, 5546, 2016.

76. Wu, Q., Yu, S., Kollert, M., Mtimet, M., Roth, V.S., Gedde, U.W., Johansson, E., Olsson, R.T., Hedenqvist, M.S., Highly absorbing antimicrobial biofoams based on wheat gluten and its biohybrids. *ACS Sustainable Chem. Eng.*, 4, 2395, 2016.
77. Chen, Y., Liu, Y., Li, Y., Qi, H., Highly sensitive, flexible, stable, and hydrophobic biofoam based on wheat flour for multifunctional sensor and adjustable EMI shielding applications. *ACS Appl. Mater. Interfaces*, 13, 30020, 2021.
78. Joshi, K., Meher, M.K., Poluri, K.M., Fabrication and characterization of bioblocks from agricultural waste using fungal mycelium for renewable and sustainable applications. *ACS Appl. Biol. Mater.*, 3, 1884, 2020.
79. Bruscato, C., Malvessi, E., Brandalise, R.N., Camassola, M., High performance of macrofungi in the production of mycelium-based biofoams using sawdust — Sustainable technology for waste reduction. *J. Cleaner Prod.*, 234, 225, 2019.
80. Li, S., Ma, Y., Ji, T., Sameen, D.E., Ahmed, S., Qin, W., Dai, J., Li, S., Liu, Y., Cassava starch/carboxymethylcellulose edible films embedded with lactic acid bacteria to extend the shelf life of banana. *Carbohydr. Polym.*, 248, 116805, 2020.
81. Cruz-Tirado, J.P., Vejarano, R., Tapia-Blácido, D.R., Barraza-Jáuregui, G., Siche, R., Biodegradable foam tray based on starches isolated from different Peruvian species. *Int. J. Biol. Macromol.*, 125, 800, 2019.
82. Machado, C.M., Benelli, P., Tessaro, I.C., Sesame cake incorporation on cassava starch foams for packaging use. *Ind. Crops Prod.*, 102, 115, 2017.
83. Vishnuvarthanan, M. and Rajeswari, N., Preparation and characterization of carrageenan/silver nanoparticles/Laponite nanocomposite coating on oxygen plasma surface modified polypropylene for food packaging. *J. Food Sci. Technol.*, 56, 2545, 2019.
84. Dogaru, B.I., Simionescu, B., Popescu, M.C., Synthesis and characterization of κ-carrageenan bio-nanocomposite films reinforced with bentonite nanoclay. *Int. J. Biol. Macromol.*, 154, 9, 2020.
85. Roy, S. and Rhim, J.W., Carrageenan-based antimicrobial bionanocomposite films incorporated with ZnO nanoparticles stabilized by melanin. *Food Hydrocolloids*, 90, 500, 2019.
86. Hashemi Tabatabaei, R., Jafari, S.M., Mirzaei, H., Mohammadi Nafchi, A., Dehnad, D., Preparation and characterization of nano-SiO2 reinforced gelatin-κ-carrageenan biocomposites. *Int. J. Biol. Macromol.*, 111, 1091, 2018.
87. Yousuf, B., Wu, S., Gao, Y., Characteristics of karaya gum based films: Amelioration by inclusion of Schisandra chinensis oil and its oleogel in the film formulation. *Food Chem.*, 345, 128859, 2021.
88. Cao, T.L. and Song, K.B., Active gum karaya/Cloisite Na+ nanocomposite films containing cinnamaldehyde. *Food Hydrocolloids*, 89, 453, 2019.
89. Kang, S., Xiao, Y., Wang, K., Cui, M., Chen, D., Guan, D., Luan, G., Xu, H., Development and evaluation of gum arabic-based antioxidant nanocomposite films incorporated with cellulose nanocrystals and fruit peel extracts. *Food Packag. Shelf Life*, 30, 100768, 2021.
90. Xiao, M., Luo, L., Tang, B., Qin, J., Wu, K., Jiang, F., Physical, structural, and water barrier properties of emulsified blend film based on konjac glucomannan/agar/gum arabic incorporating virgin coconut oil. *LWT*, 154, 112683, 2022.
91. Xu, T., Gao, C.C., Feng, X., Wu, D., Meng, L., Cheng, W., Zhang, Y., Tang, X., Characterization of chitosan based polyelectrolyte films incorporated with OSA-modified gum arabic-stabilized cinnamon essential oil emulsions. *Int. J. Biol. Macromol.*, 150, 362, 2020.
92. Kang, S., Xiao, Y., Guo, X., Huang, A., Xu, H., Development of gum arabic-based nanocomposite films reinforced with cellulose nanocrystals for strawberry preservation. *Food Chem.*, 350, 129199, 2021.

93. Shahvalizadeh, R., Ahmadi, R., Davandeh, I., Pezeshki, A., Seyed Moslemi, S.A., Karimi, S., Rahimi, M., Hamishehkar, H., Mohammadi, M., Antimicrobial bio-nanocomposite films based on gelatin, tragacanth, and zinc oxide nanoparticles – Microstructural, mechanical, thermo-physical, and barrier properties. *Food Chem.*, 354, 129492, 2021.
94. Goudar, N., Vanjeri, V.N., Dixit, S., Hiremani, V., Sataraddi, S., Gasti, T., Vootla, S.K., Masti, S.P., Chougale, R.B., Evaluation of multifunctional properties of gallic acid crosslinked Poly (vinyl alcohol)/Tragacanth Gum blend films for food packaging applications. *Int. J. Biol. Macromol.*, 158, 139, 2020.
95. Zheng, M., Chen, J., Tan, K.B., Chen, M., Zhu, Y., Development of hydroxypropyl methylcellulose film with xanthan gum and its application as an excellent food packaging bio-material in enhancing the shelf life of banana. *Food Chem.*, 374, 131794, 2022.
96. Vinod, V.T.P., Saravanan, P., Sreedhar, B., Devi, D.K., Sashidhar, R.B., A facile synthesis and characterization of Ag, Au and Pt nanoparticles using a natural hydrocolloid gum kondagogu (Cochlospermum gossypium). *Colloids Surf. B Biointerfaces*, 83, 291, 2011.
97. Li, P., Zhu, X., Kong, M., Lv, Y., Huang, Y., Yang, Q., Li, G., Fully biodegradable polylactide foams with ultrahigh expansion ratio and heat resistance for green packaging. *Int. J. Biol. Macromol.*, 183, 222, 2021.
98. Han, D.H., Choi, M.C., Nagappan, S., Kim, Y.M., Kim, H.S., Ethylene vinyl acetate (EVA)/poly(lactic acid) (PLA) blends and their foams. *Mol. Cryst. Liq. Cryst.*, 707, 38, 2020.
99. Machado, C.M., Benelli, P., Tessaro, I.C., Study of interactions between cassava starch and peanut skin on biodegradable foams. *Int. J. Biol. Macromol.*, 147, 1343, 2020.
100. Cruz-Tirado, J.P., Barros Ferreira, R.S., Lizárraga, E., Tapia-Blácido, D.R., Silva, N.C.C., Angelats-Silva, L., Siche, R., Bioactive Andean sweet potato starch-based foam incorporated with oregano or thyme essential oil. *Food Packag. Shelf Life*, 23, 100457, 2020.
101. Luna, P., Darniadi, S., Chatzifragkou, A., Charalampopoulos, D., Biodegradable foams based on extracted fractions from sorghum by-products. *IOP Conf. Ser. Earth Environ. Sci.*, 749, 012057, 2021.
102. Díaz-Montes, E., Castro-Muñoz, R., Edible films and coatings as food-quality preservers: An overview. *Foods*, 10, 249, 2021. https://doi.org/10.3390/foods10020249.

18

Biobased and Biodegradable Packaging Plastics for Food Preservation

Carolina Caicedo[1], Alma Berenice Jasso-Salcedo[2,5], Lluvia de Abril Alexandra Soriano-Melgar[2,5], Claudio Alonso Díaz-Cruz[3], Enrique Javier Jiménez-Regalado[4] and Rocio Yaneli Aguirre-Loredo[4,5]*

[1]*Grupo de Investigación en Química y Biotecnología (QUIBIO), Facultad de Ciencias Básicas, Universidad Santiago de Cali, Pampalinda, Santiago de Cali, Colombia*
[2]*Departamento de Biociencias y Agrotecnología, Centro de Investigación en Química Aplicada (CIQA). Blvd. Enrique Reyna Hermosillo 140, San José de los Cerritos, Saltillo, Coahuila de Zaragoza, México*
[3]*Departamento de Ingeniería Química, Facultad de Ciencias Químicas, Universidad Autónoma de Coahuila, Blvd. Venustiano Carranza SN, Saltillo, Coahuila, México*
[4]*Departamento de Procesos de Polimerización, Centro de Investigación en Química Aplicada (CIQA) Blvd. Enrique Reyna Hermosillo 140, San José de los Cerritos, Saltillo, Coahuila de Zaragoza, México*
[5]*Investigadoras por México-CONACyT, Ciudad de México, México*

Abstract
The biobased, biodegradable, and compostable advanced materials are one of the emerging technologies in the area of packaging, mainly due to the current need to use alternative materials to those based on fossil fuels, such as petroleum, both in the food industry and in all those industries in which plastic materials are used. Therefore, this chapter addresses the different polymers and their natural sources, as well as their properties, additives, and formulations that present an improvement and advantages for the design of new packaging that can be a viable and natural alternative for use in food preservation, including those considered as active or intelligent packaging. Additionally, the functional properties of packaging, the current state of the market for biobased materials and packaging, and additional perspectives to this field of new biomaterials generation are addressed.

***Keywords*:** Biopolymers, food packaging, shelf life, natural fibers, barrier properties, mechanical performance, biodegradable packaging, sustainability

18.1 Introduction

Preserving food has been a difficulty that humanity has faced since ancient times. Over time, this process has been improved and become more demanding, reaching the food

**Corresponding author*: yaneli.aguirre@ciqa.edu.mx

Inamuddin and Tariq Altalhi (eds.) Handbook of Bioplastics and Biocomposites Engineering Applications: 2nd Edition, (383–424) © 2023 Scrivener Publishing LLC

packaging. Packaging performs an essential role in consumer purchasing decisions, so the food industry is committed to meeting these complex demands. Today the consumer demands greater freshness in the products, and a high nutritional value, in addition to paying more attention to the type and functionality of the packaging.

A package seeks to provide efficient solutions in terms of cost, to increase the useful life of food, without neglecting the convenience of the consumer. Likewise, it seeks to use fewer natural resources, reduce food waste, as well as greater recyclability, all from the raw material to the actual performance of the packaging. Special emphasis is placed on this since the packaging matrix can be made up of renewable sources or conventional synthetic materials. Taking into account these current needs to reduce the use of fossil resources, significant efforts have been made to develop packaging products from renewable sources that are more environmentally friendly. Among these new, more sustainable alternatives, one can find packaging made from biobased polymers, from natural or biodegradable sources, and synthetic sources.

Natural Biobased Polymers: these polymers are synthesized by living organisms, such as polysaccharides, cellulose, vegetable oils, proteins, among others, which after extraction and purification can be used directly in various industrial applications. These packages, after being discarded, can be completely biodegraded by microorganisms without damaging the environment.

Synthetic Biobased Polymers: monomers are derived from renewable sources of energy, but for their conversion into polymers, a chemical transformation is required. This is the case of starch, which can be hydrolyzed and converted into lactic acid from which polylactic acid (PLA) can be used in the manufacture of packaging.

This chapter will address the different materials and polymers used for the production of biobased and biodegradable packaging for use in the food industry, as well as their functionality and evaluation.

18.2 Sources for Obtaining Polymers

There are two aspects in the type of food packaging that currently exist, synthetic and biobased, which can also be classified according to their capacity and type of biodegradation, as well as the type of polymer of origin. Polymers of biological or biobased origin can be extracted from natural sources, synthesized from the fermentation of microorganisms or chemically (Figure 18.1).

18.2.1 Polymers Extracted from Natural Sources

Natural polymers are biomacromolecules that are widely found in our environment, either in plants or animals. Among them are proteins, cellulose, natural rubber, silk, wool, and starch [1]. These polymers turn out to be biodegradable as they do not present structural modifications or combinations with other compounds [2]. Biodegradable polymeric materials have physicochemical properties that deteriorate due to chain breakage when exposed to the enzymatic process of certain microorganisms, which under an aerobic process result in carbon dioxide, while under an anaerobic process they produce methane. For both cases, the following are by-products: water, inorganic compounds and biomass [3]. It should be

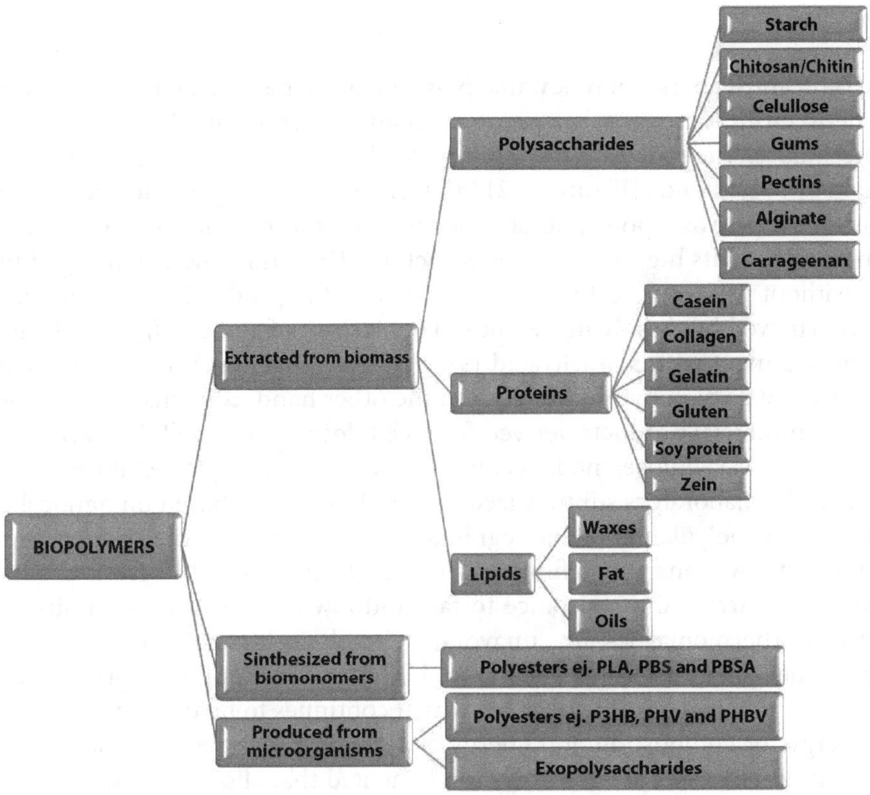

Figure 18.1 Classification of biodegradable polymeric materials according to their source and production method.

noted that not only biobased polymers are biodegradable but also various synthetic polymers can be transformed into simple substances through partial or complete biodegradation [4]. Microorganisms secrete enzymes that can form complexes with specific substrates, which are determined by their active site. Figure 18.1 presents the classification of biopolymers grouped according to their source [5].

The European Bioplastics Organization [6] has defined as bioplastic all those plastic materials that derive from biomass as a renewable resource, this allows the following classification: biopolymers extracted from natural sources, polymers produced by chemical synthesis from biomonomers, and polymeric materials produced by microorganisms. It is estimated that 170 billion metric tons of biomass are produced annually, within which the consumption of this material for the food and nonfood sector is less than 5%. In this case, carbohydrates derived from biomass represent approximately 75% of this source [7], hence for the future of green chemistry, and specifically for the packaging sector, it has been projected as a "first-generation raw material" for bioplastics. In general, the structure of these materials has high crystallinity and polar functional groups, which has an effect on low mechanical properties, a barrier to water vapor, thermal stability, and processing properties. Therefore, the polymers described below require processing aids to compete with traditional petroleum-derived plastics.

a) Polysaccharides

Cellulose

Cellulose is the most abundant renewable polymer, it can be isolated from plants and wood, contains between 40% and 50%, this being the main source, normally associated with lignin and hemicellulose. The structure of cellulose is a linear polymer of repeating β-D-glucose units linked by β-1,4 bonds (Figure 18.2) [8]. Cellulose is arranged into ordered fibers with a crystalline structure and poorly soluble due to strong intramolecular and intermolecular hydrogen bonds and its highly organized structure. This limits the obtaining of filmogenic solutions without previous treatments. In particular, the production of films, such as cellophane, is achieved by dissolving cellulose in mixtures of sodium hydroxide and carbon disulfide in treatment with sulfuric acid [9]. However, this material requires coating with waxes to decrease its high hydrophilicity. On the other hand, condensation reactions allow a variety of commercial products derived from cellulose, such as cellulose acetate, ethylcellulose, hydroxyethyl cellulose, methylcellulose, and hydroxypropyl cellulose, among others [10]. Farris and collaborators summarized at least 10 coatings based on nanocellulose that either attached to polyolefins or biodegradable PLA controlled the oxygen permeability [11]. These partially transparent films are applied to dry food and fresh products. They have a good gas barrier and resistance to fats and low resistance to humidity. As for the manufacture of thermoplastics, it is unavoidable to obtain the derivatives in powder form and then extrude them using plasticizers and additives. Among the agricultural residues that stand out today is sugarcane bagasse, which continues to be used in the production of thermal energy for combustion and energy generation processes in a controlled manner. Biotechnological processes are gradually implemented that allow the extraction of components, such as cellulose. Currently, for the extraction of cellulose methods of vapor explosion, solvent extraction and alkaline treatment are used.

Starch

It is the reserve carbohydrate in plants and represents an important source of energy for humans, it is found in various cereals, rhizomes, roots, and tubers. From the nutritional point of view, starch is a carbohydrate with great importance after glucose, and other sugars, and over time, it has become viable in a wide range of possibilities, opening a gap in scientific research in the world [12]. On an industrial level, starch can be converted into

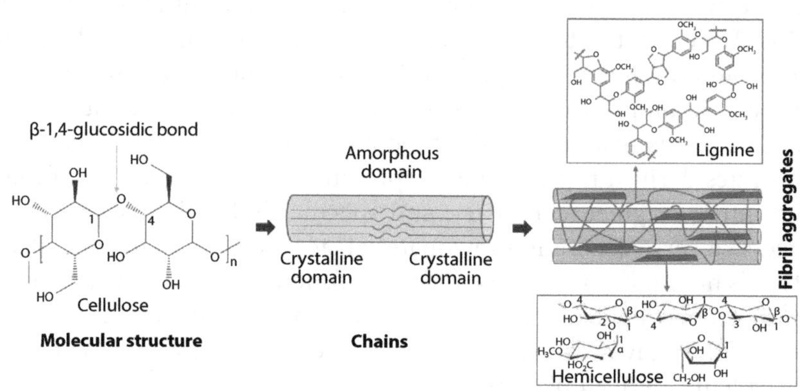

Figure 18.2 Structure of cellulose in different dimensions.

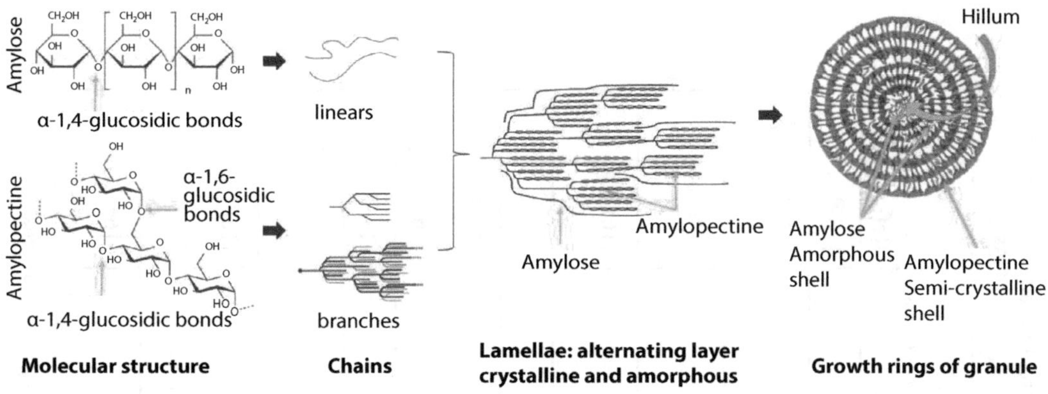

Figure 18.3 Structure of starch in different dimensions [156, 157].

chemicals, such as ethanol, acetone, and organic acids through fermentation processes. It is used in its natural form as a filler material in thermoplastic matrix composites and can also be modified with a variety of plasticizers and reagents to produce new polymeric materials [13]. Starch is not a uniform product, so it has required an arduous physico-chemical characterization in each species and botanical subspecies to achieve better use [14]. In each type of seed, there are different proportions of amylose and amylopectin (Figure 18.3).

Amylose is a polysaccharide characterized by glucose units, which present α-(1,4) and branched α-(1,6) glycosidic bonds. It is a molecule that tends to form micelles because it is insoluble in water, whose micelles are generated by the ability of these molecules to generate hydrogen bonds and helical structures for the formation of different complexes. Amylopectin is a polysaccharide that differs from amylose in that it presents glucose branches, formed in the same way by glucose units, mainly through α-(1,4) bonds (between 94% and 96%) and to a lesser degree (4–6%) α-(1,6) bonds [15, 16]. The branches of amylopectin correspond to 5% of the molecule, resulting in a complex structure that allows partial solubility in water. In general, starch in legumes usually contains ~30% amylose, while in cereals, such as wheat, corn, barley and long grain rice, its content is of ~20% [17]. In starch granules from conventional sources, amylopectin is the main constituent. However, there are some exceptions in genetically modified starches that can contain high concentrations of amylose (>50%) [18]. Waxy starches have a low proportion of amylose (<10%) and have a waxy appearance in the endosperm in cereals [19]. Table 18.1 lists the amylose content, thermal, and morphological properties of the starch granule obtained from different botanical sources.

Chitin/Chitosan

Chitin is the second most abundant biopolymer in nature. It is obtained from by-products of the fishing industry, mainly crab, shrimp, and lobster shells. Likewise, it can also be extracted from the exoskeleton of some insects or by fungal fermentation. Chitin is found at the molecular level and at the structural level tissues made up of fibers wrapped by proteins (Figure 18.4). The structure of this polysaccharide is 2-acetamido-2-deoxy-D-glucose

Table 18.1 Amylose content, thermal properties and morphological data of the starch granule from different botanical sources. Adapted from Srichuwong, Sathaporn, Titi Candra Sunarti, Takashi Mishima, Naoto Isono and Makoto Hisamatsu [158].

Starch source	Type*	Amylose content (%)	Gelatinization temperature range (°C)	ΔH_g (J/g)	Granule shape	Granule average size (nm)	Ref.
Barley	A	22–27	56.3–59.5	10.0 ± 0.3	Round, oval	17.2	[159]
Cassava	A	15–26	64.3–68.3	14.7 ± 0.7	Round, truncated	47.2	[158]
Corn	A	20–28	64.1–69.4	12.3 ± 0.0	Round, polygonal	15.0	[14, 160]
Malanga or Cocoyam	A	13–24	61.8–69.4	14.2 ± 0.5	Spherical, Polygonal	20.7	[161]
Oat	A	19–23	60.4–67.0	3.0 ± 0.2	Polygonal	5.1	[162]
Potato	B	25–31	58.2–62.6	15.8 ± 1.2	Oval, Spherical	42.3	[14, 160, 163]
Rice	A	17–23	70.3–76.2	13.2 ± 0.6	Polygonal	4.7	[162, 164]
Sago	C	24–31	67.0–73.9	15.7 ± 0.7	Oval, spherical	34.3	[165]
Sweet potato	A	19–29	55.2 – 61.3	9.2 ± 0.1	Round, polygonal	22.7	[166, 167]
Taro coconut or Chinese taro	A	13–28	67.3–72.9	14.9 ± 0.8	Polygonal	5.3	[168]
Wheat	A	17–34	57.1–61.6	10.7 ± 0.2	Round, lenticular	22.5	[166, 169, 170]

*According to its source, the type of X-ray diffraction pattern of starch is A from cereals, B from tubers, and C from legumes and seeds.

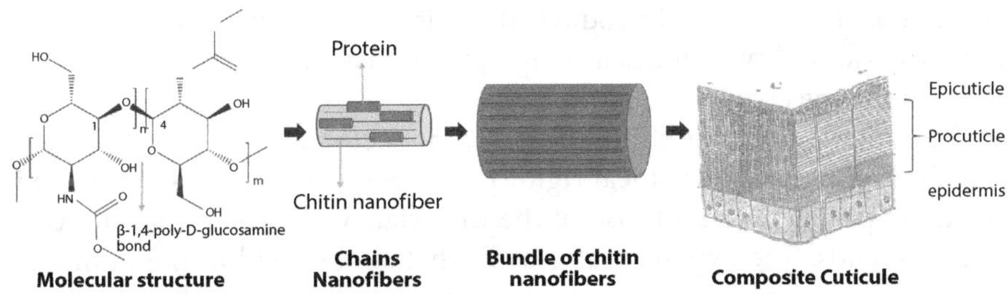

Figure 18.4 Structure of chitin in different dimensions.

via a β (1-4) linkage and presents three available forms α-, β-, and γ-chitin, the alpha form being the most studied [20]. Acetyl groups are responsible for the limitations in achieving wide applications of chitin largely due to their extremely hydrophobic nature. Therefore, deacetylation reactions are oriented, in order to transform chitin (poly (β-N-acetylglucosamine) into chitosan at ~50% partially and 100% completely. The amino groups (-NH_2) in chitosan once protonated by interaction in dilute acid or base medium allow the solubility of the biopolymer to be improved. They have been explained from the mechanism of destruction of networks with hydrogen bonds within chitosan molecules by electrostatic repulsion between positive charges [21]. Acetic acid is what is commonly used to dissolve chitosan. The parameters that influence the dissolution process of chitosan are: deacetylation degree, pH, concentration of the polymer, polymer molecular weight, and temperature [22, 23].

In the packaging sector, chitosan is widely applied, and this is related to the fluidity of the filmogenic solution in the processes. Therefore, it seeks to improve the morphology of the films which compromises an optimization in the dynamic rheology of the solution. Some previously mentioned parameters have a direct influence. The properties that shown dependence on the deacetylation degree are solubility and antimicrobial activity. Chitosan has antimicrobial capacity due to charged amino groups, which link with the negative charges of the cell membrane of microorganisms, generating electrostatic interactions and causing death [24, 25]. On the other hand, the free amino group allows a wide mode of interaction among which is the formation of complexes with ions or nanoparticles, these provide different functionalities as active ingredients [26]. Another property is the gelling or crosslinking that can occur between partially deacetylated polymer chains or the addition of crosslinking agent. Once the polymer gels under drying conditions it can form films or coatings, these show moisture barrier properties that are desirable for food packaging [27]. Biodegradation is a property that is currently being promoted in all-new food packaging materials. Single-use packaging, when discarded, can be degraded by enzymes or bacteria. The use of chitin residues to obtain pure chitin or chitosan and the respective oligosaccharide derivatives can involve two different processes: chemical and biotechnological. Due to its high moisture content and chemical composition, industrially usable waste or raw material to produce chitin and chitosan constitute highly perishable biomass. From a microbiological point of view, it is also very susceptible to enzymatic degradation, especially under the climatic conditions prevailing most of the year in many of the regions where they are generated. Because it is a natural product, there is no constant relationship between the components. The variability includes its length, the percentage of amino-acetyl groups, and their position along the chain.

Among the limitations for the chitosan packaging sector are high sensitivity to water, low thermal stability, and high mechanical rigidity [20]. Some strategies have made it possible to improve these properties and those of the materials with which they are combined by compatibilized blends. The exponential growth that is observed in the number of studies published in the last decade is due to the multiple functionalities that they derive. These show the obtaining, characterization, design, and applications of chitosan mixed with other biodegradable and synthetic polymers [28]. Compatibility in mixture with other natural polymers, mainly polysaccharides and proteins (such as pectin, alginate, starch, carrageenan, xanthan gum, gelatin, collagen, cellulose, and their derivatives) [29] depends on the interaction through the hydrogen bonds of chitosan.

b) Proteins

Another type of films that have become of increasing research interest in recent years are those based on proteins. The ability of protein films to form and cohere, as well as their lower cost and biodegradability, are some of the factors that increase their interest. These compounds have good characteristics as barrier agents against gases, such as oxygen and flavors. The main problem with protein-based films is their high permeability to water vapor due to the hydrophilic nature of proteins [30]. The main protein-based films applied today are based on gelatin, gluten, soy proteins, corn proteins, casein, and some others, such as whey proteins. Some of these are discussed below:

Gelatin

Gelatin is obtained from the processing of collagen with some dilute acids or alkaline groups. This collagen is obtained from animal sources, as it is present in skin, connective tissues, and bones (Figure 18.5) [31]. The predominant hydrogen bridges by the carboxyl and amino groups in the structure allow the stabilization of triple helixes. Its chemical

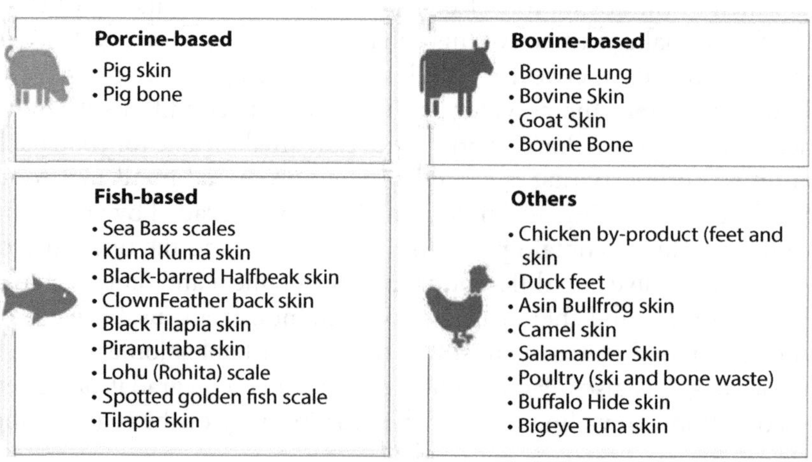

Figure 18.5 Collagen sources for gelatin extraction.

structure is based on polypeptide chains of α (one polymer/single chain, ~90,000 g/mol), β (two α-crosslinked chains, ~180,000 g/mol), and γ (three chains α-crosslinked, ~300,000 g/mol) chains [32]. Gelatin has a wide range of applications, including acting as a gelling agent that forms transparent elastic thermoreversible gels. Furthermore, gelatin's gelling properties, the main characteristics that can support the consideration of this biopolymer as a good film-forming agent are its ability to form and stabilize emulsions, its adhesion characteristics, and its dissolution behavior. The sensitivity of these molecules to interact with humidity causes deficiencies in the mechanical properties that limit their application as packaging material. Some novel methodologies for strengthening gelatin-based materials, such as vapor crosslinking, and the use of nanomaterial fillers, including hydroxyapatite nanoparticles and tricalcium phosphate [33, 34]. Likewise, it has been sought to improve the functional properties of biodegradable gelatin films by mixing with biopolymers, such as chitosan, crosslinking them with the help of the microbial transglutaminase enzyme [35].

Gluten

This protein complex comprises glutenin and gliadin. Gliadins are primarily monomeric polypeptides with molecular weights between 15,000 and 80,000. This confers greater viscosity and extensibility to the products. It is soluble in alcoholic solutions. While glutenins are polymeric, they have disulfide-linked polymer chains with molecular weights between 80,000 and several million. This component is responsible for elasticity and toughness and is soluble in dilute acids (insoluble in most solvents due to its high molecular weight). Gluten has superior mechanical properties compared to other vegetable proteins. Gluten proteins show reactivity to different chemicals and temperatures that allow the generation of different materials from two different processes: melting (solvent-based) and thermal processes. Therefore, it is considered partially thermoplastic. The poor nutritional value of gluten, its characteristic of being a renewable raw material, highly available, biodegradable, biocompatible, and economica, is what generates its attractiveness for the use and elaboration of different packaging with potential application in the food area and to be considered as sustainable. Gluten-based bioplastics allow the use of conventional processing methods, such as extrusion [36, 37], lamination, roller milling, or thermoforming [38, 39]. Strategies to improve the oxygen barrier that continues to be a limitation for packaging applications will be presented in the following sections.

18.2.2 Biopolymers Synthesized by Microorganisms

Polyhydroxyalkanoates

Polyhydroxyalkanoates (PHA) are biodegradable polyesters synthesized by microorganisms intracellularly as a reserve for energy, and carbon storage. PHA can be found in the form of granules dispersed within bacterial cells [40]. They can be synthesized by certain bacteria, archaea, and microalgae. For the synthesis of PHA, different options of substrates and bacterial strains have been evaluated to optimize their production and reduce costs. Using a bacterial strain of marine origin as *Halomonas hydrothermalis* increases PHA production, making the process more economically profitable, and exploring the use of more water instead of freshwater or more expensive commercial media [41]. PHA can be classified according to the size of the monomer. Short-chain (scl-PHA) is composed of monomers of three to five carbon atoms, medium-chain (mcl-PHA) of 6 to 14 carbons, and long-chain (lcl-PHA) for those composed of monomers of 15 or more carbon atoms [42]. The resulting type of PHA depends on the microorganism used for its synthesis. PHAs are mainly obtained from raw materials such as cane sucrose, sugar beet sucrose, or glucose based on corn starch, as well as vegetable oils from soybean, palm, and corn.

Materials made from PHA have various properties; crystalline and highly brittle thermoplasts, such as poly (3-(R)-hydroxybutyrate) homopolyester (PHB), which significantly limits their applicability [43]. Despite this, PHB-based materials have similar thermal and mechanical properties to conventional synthetic plastics [43]. However, more elastomeric materials can also be found, such as the copolyesters of 3-hydroxybutyrate (3HB) and 3-hydroxy valerate (3HV) or 4-hydroxybutyrate (4HB), which are scl-PHA [44].

Xanthan gum

Xanthan gum is an anionic extracellular polysaccharide, composed of glucose, mannose, and glucuronic acid, of high molecular weight (2-20 × 10^3 kg mol^{-1} M_w), and water-soluble [45, 46]. This gum is produced by the aerobic fermentation of certain species of *Xanthomonas campestris* as a secondary metabolite [47]. It is widely used in the food industry to control the rheology of the final product due to its high viscosity at low concentrations [46]. Films composed of different proportions of xanthan gum and hydroxypropylmethylcellulose (HPMC) biopolymers showed good mechanical and water vapor transmission behavior. This formulation was successfully applied in the postharvest preservation of bananas, preserving the original appearance of the fruits for 18 days [48].

Bacterial cellulose

Cellulose is a polymer that can be obtained from bacterial fermentation processes (*Acetobacter*, *Achromobacter, Dickeya*, *Alcaligenes*, *Aerobacter, Rhizobium*, *Agrobacterium*, *Rhodobacter,* and *Sarcina*) [49]. Although its chemical structure is the same as that of plant origin, it has the different conformation and physicochemical properties [49]. The main advantage of bacterial cellulose is that it can be obtained more cleanly, with more efficient processes than those carried out for the extraction of vegetable cellulose. While cellulose from plant sources requires the removal of lignin and other compounds, bacterial cellulose is obtained almost cleanly, and directly. This reduction in the steps of the extraction process means that bacterial cellulose can be obtained more purely, in good quantity, and a viable candidate for getting biodegradable materials for use in industries such as food, medicine, and catalysis. A bacterial cellulose-based smart film was developed for the preservation of tilapia fillets in combination with pelargonidin dye [50]. This smart packaging had the ability to change color, being an indicator of the loss of freshness of the food, efficiently and quickly alerting the consumer. Bacterial cellulose can also be used to obtain nanosized compounds, such as nanofibrils and nanocrystals, these nanomaterials can be used as structural matrices and as reinforcement packaging materials, respectively [51].

Alginates

This linear polysaccharide is made up of two types of monosaccharides, mannuronic acid (M), and guluronic acid (G). Alginate is extracted mainly from algae and bacteria mainly from *Azotobacter*, and they make up the cell wall. The polysaccharide structures fraction, molecular weight, and order of M and G units determine its physicochemical properties (inert nature, low toxicity, solubility, viscosity, optical rotation, gel formation, among others). In particular, the gelation potential exhibits a strong dependence on the length and distribution of the G blocks, due to the ability to complex with cations (e.g., Ca^{2+}, Zn^{2+}, Ba^{2+}, Fe^{3+}, Al^{3+}) [52]. Available carboxylates can coordinate with ions in the internal cavities of polymer networks. This crosslinking makes it possible to increase the functional properties of the casting films [53]. In the packaging sector, there is evidence of the participation of alginates as active packaging, encapsulated for controlled drug release and mainly as food coatings. Techniques for conventional polymer processing, such as 3D printing and extrusion, are used in plasticized, binartial, and tertiary blends or composites [54, 55]. The main properties they provide are hydrophilicity, biocompatibility, biodegradability, mucoadhesiveness, and pH-responsive, characteristics that could be used for the development of smart packaging. Alginates have been studied extensively due to their multiple characteristics, as well as low cost, which make them a good choice for use in the food packaging industry, as seen in the recently developed materials, shown in Table 18.2. Biodegradable films made of alginate exhibit good flexibility, transparency, gloss, and low oil permeability. Alginate films with thymol oil proved to be effective for the preservation of cut apples. The materials presented an improved mechanical behavior, with an excellent capacity to block UV light, antioxidant capacity, avoiding the color change of the cut fruit, in addition to inhibiting the growth of pathogenic bacteria [56].

Table 18.2 Alginate-based packaging materials and coatings used for the preservation of various foods in the last five years (2018–2022).

Film/coating formulation and food application	Elaboration method	Year	Ref.
Chemical, physical and antibacterial properties of edible films from alginate and different proportions of castor oil.	Edible films/ casting	2018	[171]
Morphological and optical properties, barrier to water vapor, and antiviral and antioxidant capacity of films made from alginate and extract of grape seed and green tea.	Edible films/ emulsion-casting	2018	[172]
Physical and antioxidant properties of films from alginates and black chokeberry extract.	Edible films/ emulsion-casting	2018	[173]
Physical, morphological and structural properties of gelatin-sodium alginate with polyphenols from tea.	Edible films/ emulsion-casting	2018	[174]
Thermal, mechanical and permeability and antibacterial, optical properties of films from carboxymethyl cellulose/alginate/chitosan. This coating was applied to tomato to evaluate preservation properties.	Edible film and coating/casting	2019	[175]

(*Continued*)

Table 18.2 Alginate-based packaging materials and coatings used for the preservation of various foods in the last five years (2018–2022). (*Continued*)

Film/coating formulation and food application	Elaboration method	Year	Ref.
Preservation properties of figs (*Ficus carica*) using a coating of alginate and chitosan blends.	Edible films/ casting	2019	[176]
Preservation properties of peach coated with alginate and rhubarb extract.	Edible films/ casting	2019	[177]
Preservation properties of the *Pholiota nameko* mushroom used coating based on sodium alginate added with essential oil of thyme, *L*-cysteine and nisin.	Edible films/ casting	2019	[178]
Physical, antimicrobial capacity of alginate/*aloe vera* films reinforced with TiO_2. Applied to tomatoes and the preservation of this fruit was studied.	Edible films/ casting	2020	[179]
Mechanical, physicochemical properties and biological activities of edible films based on alginate and medicinal essential oil.	Edible films/ casting	2020	[180]
UV barrier ability, improves gas barrier, antifungal, and antimicrobial properties of a film/coating of alginate/ *aloe vera* with frankincense oil. Coating applied on green peppers.	Edible coatings/ casting	2021	[181]
Structural, UV barrier, mechanical and thermal properties of edible films based on alginate/*aloe vera* and garlic oil. Preservation properties were evaluated on tomatoes.	Edible coatings/ casting	2021	[182]
Improvement of the mechanical performance and barrier properties of alginate/$CaCl_2$ edible films. Chilling injury delay, reduction in respiratory rate and weight loss of coated apple.	Edible coatings/ casting	2022	[183]

18.2.3 Biopolymers Obtained by Chemical Synthesis

Polylactic Acid (PLA)

PLA is an aliphatic linear polymer [poly(α-ester)], also known as α-hydroxyalkanoic polyester derivative that is produced by the polymerization of lactide in an ionic manner. Lactide is a cyclic lactic acid compound obtained from dehydration-condensation processes [40]. Another process to obtain lactide is from starch by fermentation processes. PLA is a very versatile biopolymer; currently, several products made with it can be found on the market, such as foams, bottles, apparel, textiles, 3D printing, and food packaging. The low melting viscosity of PLA makes it excellent for molding plastic pieces [40]. One of the disadvantages that it presents, and its use is not yet widely extended, is the poor crystallization rate when subjected to long molding cycles. Likewise, the materials have high gas permeation and poor mechanical performance [40]. Because of this, a way has been sought to improve the functional behavior of PLA-based materials, mixing with other polymers, both synthetic and natural, as well as with various natural fibers [57].

Polyethylene Brassylate (PEB)
Polyethylene brassylate (PEB) is a novel biodegradable polymer source that produces materials with good mechanical performance [58], it is obtained from the monomer ethylene brassylate, a commercially available lactone, synthesized from the castor bean, being cheaper than other macrolactones [59, 60].

18.3 Additives in Packaging Materials

The use of various additives or fillers (such as fibers, silicates, metal oxides, nanocomposites, among others) applied to the packaging sector emerges as an alternative for the generation of a new class of biocomposite polymeric materials try to face the environmental impact caused by plastic from synthetic origin.

18.3.1 Natural Origin

The use of natural additives in plastic products, both for packaging, automotive, construction, and other services, has been widely explored because they can act as a structural reinforcement material. Natural fillers in packaging materials: animal as silk, wool, hair; mineral as asbestos and clays, vegetal as wood, seed, leaf, fruit, stalk, grass, essential oils,

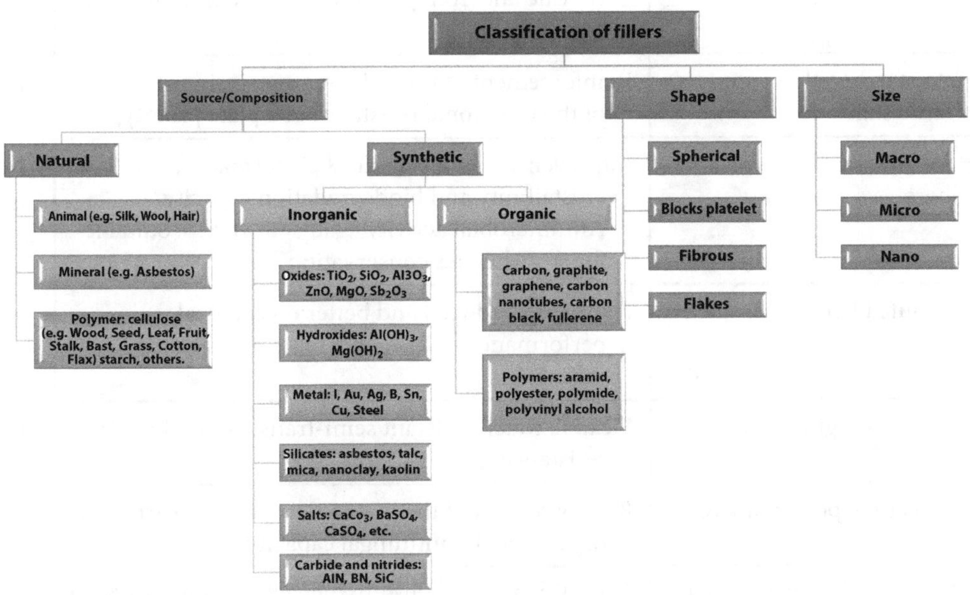

Figure 18.6 Classification of fillers or additives in polymeric composite materials.

plants extracts (Figure 18.6). Among the various existing reinforcement materials of natural origin, these can be classified into organic and inorganic.

a) Vegetable Fibers

Polymer compounds reinforced with plant derivatives have gained relevance due to their high biodegradability, low cost, and specific mechanical performance. Vegetable fibers are hydrophilic in nature, while synthetic polymers are hydrophobic, their use in this type of plastics provides an improved disintegration capacity. While integration of vegetable fibers into polymer from renewable sources provide an improvement in their tensile and elongation properties, which are usually very weak. The vegetable fibers are obtained mainly from

Table 18.3 Composition and functional properties of biobased food packaging reinforced with vegetable fibers, developed in the last 3 years (2020–2022).

Polymeric-fiber matrix	Material properties	Ref.
Palm leaf	Biodegradable tableware made from 100% palm leaf. Areca palm leaves pressed and molded into plates and bowls. Washable and reusable.	[184]
Cocoa husk cellulose—sugarcane bagasse	Material with low susceptibility to moisture damage, low water absorption, and water vapor permeability.	[185]
Corn starch—Agave fibers	Modification of the structural matrix of starch films, material with a good barrier to water vapor.	[186]
Dioscorea hispida starch—*D. hispida* fibers	Higher thermal resistance, mechanical performance (tensile and Young's modulus), and gas barrier of starch films.	[187]
Potato starch/gellan gum—Garlic skin	Reinforcement of the polymer matrix, improvement of the functional resistance of a plate prototype.	[188]
PHBV—*Ceiba pentandra* fibers	Improvement of mechanical performance, crystallinity and biodegradation capacity. Antimicrobial activity against *S. aureus*. Suitable for strawberries conservation.	[102]
PLA—Jute fibers	Hydrolysis reduced and better mechanical performance of PLA-short jute fibers based materials.	[57]
PVA—Cotton gin trash	Flexible water-resistant semi-transparent film, UV resistance.	[189]
PVA/pectin—pomegranate fibers	Reinforcement of the material, suitable as carrier of organic acids, antifungal capacity.	[190]

LLDPE, linear low density polyethylene; PHBV, poly(3-hydroxybutyrate-co-3-hydroxyvalerate); PLA, polylactic acid; PVA, poly (vinyl alcohol).

agricultural waste; therefore, they do not compete with food, and an added value is given to a material considered initially waste. The vegetable fibers most used as reinforcement material in polymers include bamboo, palm, flax, banana, agave, cotton, and sisal [61, 62]. An ideal vegetable fiber selection based on a systematic approach relies on available data of cost, availability, mechanical and barrier properties, and a multicriteria decision analysis method. An improvement of the multicriteria decision analysis method may be the inclusion of the geographical distribution data, extraction time and chemicals, energy consumption, and fiber compatibility [63]. Each criterion is weighted by experts from the industrial sector and the vegetable fibers ranked either using the Analytical Hierarchy Process (AHP) analysis or the Technique for Order Preference by Similarity to Ideal Solution (TOPSIS) method [63, 64]. Remarkably, fruit-based fibers like coir, pineapple, and banana ranked first on such studies. Table 18.3 shows some biobased materials developed with vegetable fibers and the functional properties obtained. The fiber that is extracted from bananas has important properties of mechanical resistance, which is an excellent choice as a polymer reinforcing material [65, 66]. It is extracted from industrial waste, thus reducing the environmental impact even more. Composites developed with these fibers can be processed using conventional compression, injection, and hot-pressing molding techniques [66]. The main parameters that influence the performance of polymeric compounds reinforced with natural fibers are the length of the fiber, surface morphology, orientation, configuration, moisture content, and treatment temperature [65].

b) Plant Extracts and Essential Oils

Today's consumers are looking for healthier alternatives to food additives, which is why new natural antimicrobial additives, mainly organic acids, and essential oils, have been sought in the last two decades. There are many plants from which organic acids can be extracted, which are beneficial in providing functional properties (antibacterial and antioxidant) to food packaging materials. These organic acids come from edible plant extracts such as cloves [67], coffee [68], cinnamon [69, 70], thyme [71], roselle [72, 73], oregano [53], pomegranate [74, 75], rosemary [76], garlic [77], as well as orange [78], black pepper [79], and ginger essential oils [79, 80]. The use of biobased packaging composed of derivatives extracted from plants has been shown to extend the lifetime of meats, fish, cheeses, fruits, and vegetables, inhibiting microbial growth, reducing the rate of respiration, maturation, and oxidation, maintaining the texture and nutritional quality of the products for a longer time.

c) Inorganic Materials

Several studies have reported that by reducing the particle size of the reinforcement, it is possible to induce a uniform dispersion, increasing the specific surface area of the nano additive, such as nanocellulose [81, 82] and nanoclays [83, 84]. In addition, some nanocomposites materials can exhibit improved thermal, mechanical, and barrier properties with lower reinforcement loads (1 to 5% by volume) [60]. Nanoclays have a structural arrangement in the form of superimposed layers; this structure can be modified, through the separation of layers, a process that helps alter the functional properties of the packaging materials. The chemical structure of these minerals consists of tetrahedral sheets of Si-O with intercalations of octahedral Al-O sheets [83, 85]. The multilayer arrangement of nanoclays that difficult for gas molecules to diffuse through the material in which they are incorporated, improving moisture resistance and reducing permeability.

The most widely used natural clay is montmorillonite (MMT), which has a moderate negative surface charge. The use of low concentrations of MMT has improved the gas barrier capabilities and mechanical performance of biodegradable starch-based materials [83, 86]. Another nanoclay is the halloysite, an aluminosilicate clay, in addition to the typical structure of these clays, halloysite can also be found in spherical, flat, and other forms, which can be used to confer different properties on polymeric materials [85]. The incorporation of these clays is more effective when combined with processes such as sonication, which help to separate the layers and distribute them more homogeneously in the material matrix [83].

18.3.2 Synthetic Origin

Some metallic nanoparticles (NP's) as silver, zinc, copper, and gold [83, 87–89], can be added into the biobased packaging matrix, some of which can also provide antimicrobial capacity to the composite material [90, 91]. NP's can be classified according to their shape, as shown in Figure 18.6. The effect of copper sulfide nanoparticles on starch has been investigated, obtaining a significant contribution to the UV barrier, in addition to the mechanical and vapor barrier properties. The results derived from *in vitro* analysis showed excellent nanocomposite biocompatibility in cutaneous fibroblast cell lines, and antimicrobial activity [92]. Similarly, improvement of these physical properties has been found in edible starch and pectin-based biofilms with inclusions of nanotitania [93]. On the other hand, the degree of exfoliation of bentonite nanoclays modified with surfactants, which have been incorporated into plasticized starch matrices, has been evaluated [94]. Films with a higher degree of exfoliation improved their barrier property, reducing water vapor permeability (WVP). This also generated a delay in the temperature for starch degradation and an improvement in the performance of the material due to the nano reinforcements. Additionally, films of gelatin-based, and bionanocomposites/chitosan nanofiber/ZnO NP's/betanin nanoliposomes showed synergy between the components by providing high antibacterial activity and inhibitory activity of 2,2-diphenyl-1-picryl-hydrazyl-hydrate (DPPH) [95]. These materials are designed as containers for the preservation of food, especially meat.

18.4 Active Packaging

Among the materials destined for packaging to which additives have been added for their improvement, it is possible to find the so-called active packaging (AP) and intelligent or smart packaging (SP). This type of packaging, due to its importance in the food industry, is essential to highlight as a separate section. In addition to the growing interest in the search for biodegradable, biocompatible, and edible packaging, the study of active and intelligent packaging has increased in recent years to improve and extend the shelf life of food, reduce the use of chemical preservatives and provide more information on traceability and food safety to the consumer [96]. AP is one that has improved characteristics over conventional packaging by continuously releasing substances that present a beneficial action to the food contained in the package or by absorbing undesirable compounds. These advantages help extend or improve food's shelf life and commercialization by reducing spoilage caused by different atmospheric, chemical, or biological conditions [97]. In case polymeric or biobased materials cannot fulfill the barrier property to gases and aromas, desiccant and absorbent materials are added to the packaging to capture oxygen and other gases (scavengers), such as ethylene hormone, which causes fruit ripening. Intelligent or smart packages are those that feature intelligent functions, such as sensing and recording information important for food safety and quality preservation [96, 98].

Food deteriorates rapidly due to oxidation processes, which affect the appearance, color, taste, smell, and the loss of nutraceutical compounds. To counteract these effects, different treatments are applied, the main active compounds added to biobased food packaging consist of substances with antimicrobial capacity and antioxidant agents since they are the leading causes of food spoilage. However, many of these compounds can be lost quickly during food treatment processes or during storage, which is why one of the most widely used alternatives is the use of different functionalized containers with antioxidant compounds, also known as active packaging. AP is known to be a novel method of increasing the shelf life of foods, especially highly perishable foods while maintaining their quality and safety through controlled interaction with the contents. Additionally, AP of consumable products can reduce the addition of preservatives or chemical additives to foods. The active substances placed in the materials have a certain release rate, which can be regulated, meaning that the concentrations of the active substances prevent oxidation and spoilage of the food, thus prolonging its shelf life [99]. Therefore, the following sections review some of the latest research reports on different packaging with antioxidants and antimicrobial compounds, as well as the control mechanisms of pathogenic microorganisms, the regulation of active compounds, the different biobased matrices, and the main considerations necessary for the generation and design of the packaging.

18.4.1 Antioxidants in Biobased Active Packaging

The antioxidant compounds are the most used active additives to prevent oxidation or rancidity of food. Antioxidants are chemical compounds capable of reducing the oxidation of a substrate by intercepting and reacting with free radicals at a higher rate than the oxidizable substrate. In both cases, the antioxidants act through the exact mechanisms of action, conditioned by the chemical structure of the selected antioxidant. The use of antioxidants has a particular interest in polymeric materials intended for food packaging and a double objective since many polymer formulations include the addition of antioxidants to avoid thermo-oxidative degradation during the process and extend the shelf life of the food during the service of the material. To develop an antioxidant container, it is necessary to add an antioxidant substance that resists the process temperature, which can later be released from the polymeric matrix.

The antioxidant action is generated by the elimination of free radicals, which is another method of preventing oxidation, in addition to eliminating the presence of oxygen inside the package. The antioxidant compounds used in packaging, generally of natural origin, act by eliminating radicals, trapping them in such a way that they do not cause oxidation processes. This means that packaging materials do not have to be as effective in their gas or vacuum barrier properties, thus improving the different types of packaging or increasing the benefits of those materials of biological origin with low barrier properties. Active packages with

Table 18.4 Natural antioxidant additives incorporated in biobased packaging materials for the preservation of different foods.

Type of food	Biobased-antioxidant matrix	Benefits	Ref.
Apple (fresh-cut)	Sodium alginate—thymol	Mechanical resistance, UV-vis light block, reduced microbial growth, reduce weight loss	[56]
Blueberries fruits	Corn starch/chitosan—lemon essential oil	Control of the respiration rate of the fruit and reduction of weight loss	[191]
Cheese and ham slices	Agar—Stevia	Stability for three months' storage. Retards growth of microorganisms	[192]
Chicken nuggets	Carrageenan—anise, caraway, and nutmeg essential oils	Positive antioxidant and antimicrobial effects on the shelf life	[193]
Fresh fish burgers	Chitosan—seaweed extract	Controlled water activity changes, reduced microbial growth and lipid oxidation	[194]
Lamb meat	Carrageenan—Olive leaf extract	Reducing the count of psychrophiles and increase the shelf life of lamb meat	[195]
Agaricus bisporus Mushrooms	Chitosan—callic acid	Postharvest quality maintained	[196]
Agaricus bisporus Mushrooms	Cellulose nanocrystals loaded in gellan gum—gallic acid	Mechanical resistance, prevent ripening	[82]
Palm oil	Cassava starch—coffee extract and cocoa powder	Decreased water vapor permeability and dark pigmentation	[68]
Tilapia fish fillet	Gelatin/Carrageenan—pomegranate peel extract	Quality maintenance and prolongation of storage	[75]

antioxidants are based on releasing these from the polymeric matrix into the food. However, some antioxidants can act as free radical scavengers in the package's headspace [99].

Antioxidant compounds can be classified according to their nature as either natural or synthetic, and there are a large number of compounds in both classifications, both being used in different materials. The current trend is toward the use of mainly natural antioxidants, however, the advantages of the properties of synthetic antioxidants mean that they are still used today. The most commonly used synthetic antioxidants are butylated hydroxyanisole (BHA) and butylated hydroxytoluene (BHT), which give stability to materials and increase the shelf life of containers [99].

Inside natural antioxidants, there is a huge list of compounds, highlighting, in a general way, three major groups: polyphenols, vitamins (α-tocopherol and ascorbic acid), and pigments (carotenoids and xanthophylls) [97, 100, 101]. All of them can come from different sources, such as plant extracts (including tea), agroindustrial residues (fruits, leaves, and stems), and even antioxidants obtained from herbs and spices [102]. The incorporation of extracts obtained from natural sources in biobased materials is part of current trends [101]. Their natural origin is more attractive to consumers, but also the possibilities of their application, as they can be used in packaging for food preservation of both plant and animal origin. Other compounds with diverse properties that allow the development of formulations of biobased materials that may have antioxidant capacity are polysaccharides, including chitosan [103]. Recently, cannabidiol, a nonpsychotropic compound derived from the cannabis plant, has been studied [104]. This compound has shown an important antioxidant capacity in food packaging materials made from biopolymers such as PLA.

The mobility of antioxidants will depend on the composition and concentration of these components, giving a greater diffusion of antioxidants within the polymeric materials with less branching of their chains and with larger free spaces, likewise, the smaller antioxidant compounds will be able to permeate more easily through the polymeric structures, making their diffusion faster [99]. In turn, environmental factors may be acting on the release of different antioxidants both synthetic and natural [97]. Therefore, changes in pH, temperature, relative humidity, and other factors external to the packaging will act as stimulants for the liberation of antioxidant agents. Different active packaging containing antioxidant compounds of biological origin for the preservation of different foods are presented in Table 18.4.

18.4.2 Active Packaging Biobased with Antimicrobial Agents

Food safety is of concern to consumers and the entire food industry, since the development of microorganisms not only contaminates food and can cause illness, but they are also the main cause of food losses and, therefore, economic losses. Hence, the importance of using antimicrobial agents in different food packaging [105]. Due to this, AP seeks the incorporation of antimicrobial and antioxidant agents. Sometimes, the same compounds can have both properties (function as antioxidant or as antimicrobial compounds). Therefore, the incorporation of different antimicrobials into packaging materials continues to be considered among the emerging packaging technologies [106]. The mode of action of these compounds can be considered as inactivating agents, which retard the development and growth of microorganisms [105, 107]. Like bioactive packaging with antioxidant additives, bioactive packaging with antimicrobial agents can act by releasing directly into the food or headspace [102].

The efficacy of AP with antimicrobial additives will depend on the ability of the agent to inactivate or inhibit particular microorganisms, its diffusion to food, and the characteristics of the food itself that allow it to act or enhance the antimicrobial effect [108, 109]. However, despite the importance of the diffusion or permeability of antimicrobial compounds through different polymers and materials, this parameter is little studied [110]. The diffusivity of antimicrobial agents in complex colloidal systems is a poorly explored area. At the same time, those processes can be reasonably hypothesized that directly influences the antimicrobial efficacy of the AP.

The food industry increasingly requests metallic nanoparticles that provide antimicrobial capacity to the packaging materials because their addition in low concentration and low cost makes them an excellent alternative as an active additive for AP. Antimicrobial biomaterial based on chitosan/PVA/SiO_2 NPs doped with zinc oxide NPs (ZnO) [111], positive results were obtained. The increase in the shelf life of the packaged food (bread), due to the protection against Gram-positive and Gram-negative bacteria. Similar results were achieved with alginate cross-linked with $CaSO_4$ and ZnO [112]. The protection time has been evaluated in cheeses with films based on chitosan/guar gum/ZnO based on Roselle calyx extract [73]. On the other hand, biomaterials from chitosan/polyethylene oxide and hybrid NPs (Ag/TiO_2) have shown antibacterial activity [113]. Likewise, dapsone-coated TiO_2 (DAP-TiO_2) NPs were prepared as a decorated biological agent to improve the UV blocking properties of biodegradable PVA nanocomposites [114]. This coating made it possible to better disperse the nanoparticles and connect with the polymer to avoid the migration of NPs to food [112]. PLA/clay bionanomaterials exhibited excellent antimicrobial properties, and in the disintegration process under composting conditions, the active compounds caused acceleration of this process [115]. Regarding biodegradability, an accelerated disintegration has been evidenced in the Ecovio biopolymer (a mixture of PLA and polyesters) with the incorporation of microparticles and nanoparticles to ZnO [116].

Some of the most commonly used antimicrobial compounds include organic acids (lactic, acetic, and citric), phenolic compounds, metal oxide compounds (copper, zinc, titanium, silver, calcium oxides, etc.), natural extracts and essential oils (from seeds, spices, herbs, and aromatic plants), sulfites, terpenes, lipids, peptides, chitosan, nanoparticles and nanoclays, chlorine dioxide, to name a few [96, 97, 105, 107, 117–120]. Whether active packaging with antimicrobial compounds of natural or synthetic origin, these packaging are widely employed as they allow increasing the shelf life of food [121]. The antimicrobial effect of essential oils and phenolic compounds is given by the destabilization and increased permeability of the cell membrane of the microorganism, generating the release of cell contents and, with it, death [122, 123].

With nanotechnology and its increasing application in food preservation, different compounds with nanometer size, such as silver nanoparticles, are part of the new antimicrobial compounds used in the packaging industry [124]. However, chitosan and its derivatives (including chitosan nanoparticles) are the materials with antimicrobial properties and other characteristics that make it the ideal material for the formulation and design of packaging materials with potential application in food preservation [119, 125].

To potentiate the effect of antimicrobial compounds or to regulate their action and permeability towards food, five application methods have been proposed 1) antimicrobial compounds incorporated in the polymeric matrix, 2) by encapsulation or antimicrobial sachets inserted inside the package, 3) by volatile compounds added in the headspace or generate its volatilization to the headspace, 4) coatings on the surface of the matrix packaging, and 5) in direct contact with the product [106, 123]. Therefore, the diffusion of the antimicrobial compound, as well as the polymeric matrix, could alter the efficacy of antimicrobial agents [110], as well as their availability and efficacy over time, since antimicrobial compounds may not be effective during the entire storage period [118].

In the particular case of essential oils, they present a greater resistance to the processing of containers, having good chemical and thermal stability, however, being highly hydrophobic presents low solubility in polar agents, such as water, which has led them to be added in containers through encapsulation, even generating a slower and more constant release [126]. Table 18.5 describes some antimicrobial compounds and the biobased packaging matrices in which they were added for food preservation.

Table 18.5 Antimicrobial additives incorporated in biobased packaging for the preservation of different foods.

Type of food	Biobased-antioxidant matrix	Benefits	Ref.
Apple	ß-Cyclodextrin—palmarosa and star anise essential oils	Extend fruit shelf life	[197]
Beef meat	Chitosan—*Zataria multiflora* Boiss oil	Good sensorial analysis and extend shelf life	[198]
Beef patties	Carboxymethyl cellulose—Apple peel powder and tartaric acid	Good sensorial analysis and extend shelf life	[199]
Black grape fruits	Chitosan and cellulose acetate phthalate—ZnO NPs	Extend shelf life	[200]
Cheese	Alginate—ZnO NPs and citronella essential oil	Extend shelf life	[201]
Chicken breast meat	Chitosan—propolis extract	Good sensorial analysis and decrease deterioration of chicken meat	[202]
Cucumber	Corn starch/gelatin—citric acid	Solubility, swelling index, water vapor permeability, and mechanical properties	[103]
Poultry meat	Chitosan—ZnO NPs	Reduction of microbial growth and oxidation of meat fats	[203]

(*Continued*)

Table 18.5 Antimicrobial additives incorporated in biobased packaging for the preservation of different foods. (*Continued*)

Type of food	Biobased-antioxidant matrix	Benefits	Ref.
Green grape fruits	Agar—ZnO NPs	Extend shelf life	[204]
Lamb meat	Chitosan-*Satureja khuzestanica* essential oil	Improved sensory attributes and extend shelf life	[205]
Marinated salmon	Corn starch/gelatin—Ethyl lauroyl arginate	Antilisterial activity	[206]
Minced meat	Fish gelatin/chickpea protein—*Nigella sativa* essential oil and copper sulfide NPs	Antimicrobial functionality against *E. coli* and *S. aureus*. Extend shelf life	[207]
Mini kiwi fruits	Furcellaran/gelatin—Selenium and silver NPs	Antibacterial activity against *S. aureus* and *E. coli*. Extend shelf life	[208]
Pork loin	Gelatin-Propolis extract	Extend shelf life	[209]
Sliced cooked ham.	Starch/PLA – Ag NPs	Mechanical properties and water vapor permeability were not affected	[210]
Strawberry fruits	Carboxymethyl cellulose/guar gum—Ag NPs	Freshness, extend shelf life and decreased weight loss	[211]
Strawberry fruits	Starch – chitosan NPs loaded with cinnamon essential oil	Film limit the moisture transfer, water permeability improved, antimicrobial effect against *E. coli* and *Bacillus subtillis*. Extend shelf life.	[212]
Strawberry fruits	Wood cellulose nanocrystals and nanofibers, sugarcane cellulose, chitin and chitosan nanofibers	Preserving fruit freshness	[213]

18.5 Smart Packaging

The trend of the digital era and the use of artificial intelligence to facilitate some daily activities has generated that even the area of materials and packaging is oriented in this direction. Thanks to these advances, smart or intelligent packaging (SP) emerged, a new technology focused on quality control, safety, and improvement in food traceability, whose indicators provide important information in a visual and simple way for all those involved in the marketing chain and the final consumer [127].

The information provided by the active packaging is of great relevance in the food field, as it can detect changes in the physiology of the food due to an increase in the microbiological load and chemical changes that occur during the decomposition of the food [124, 127]. Some of the components of smart packaging include immobilized enzymes, antimicrobial agents, and immobilized bioactive components, which will allow the monitoring of product characteristics, such as spoilage, growth of microorganisms, color changes or physicochemical characteristics of food, etc. [97, 98, 124]. Trends in this area include food tracking, from material and product sourcing, processing, distribution, storage, and sale, where quality and safety can always be monitored remotely. To this end, smart packaging is based on the application of indicators and sensors [127].

18.5.1 Indicators

Indicators are mechanisms that allow the qualitative measurement of important parameters in food, which can be placed on the external or internal part of the package, considered as indirect or direct indicators that provide relevant information [127, 128]. According to the information they provide or measure, they can be classified into 1) time and temperature indicators, 2) mechanical damage indicators, 3) pathogenic microorganism presence indicators, 4) color change indicators, 5) gas leakage indicators, 6) freshness indicators, among others. Thus, the indicators can be used to know the product's current state or estimate the remaining useful life of the food contained in these packages [96, 98, 129]. Most indicators work on the principle of a significant color change so that visually it is easy to determine the status of the quality and shelf life of food, whose devices consist of a label, laminate, or coating printed on the package [96, 128].

18.5.2 Biosensors

In the context of biosensors, these are tools that detect and transmit the information, which consists mainly of a receptor and its translator, whose origin is biological. The main materials are enzymes, hormones, and other metabolites, whose action is to recognize the analyte of interest so that the translator converts the chemical signal into an electrical one. That is why these biosensors are particularly sensitive to pH changes and some physicochemical changes; being more effective packaging components compared to indicators [96].

Changes in pH originate during food degradation and the release of nitrogen in gaseous form, so unlike indicators, these sensors are usually placed far away from the food, but with enough space to achieve detection of the changes occurring in the products. Due to the knowledge of biosensors and the properties of some natural pigments such as anthocyanins, which are sensitive to pH change and therefore change color, these compounds have been developed as biosensors [127]. The use of compounds, such as anthocyanins, would have advantages, such as their natural origin and their extraction, can be performed through agroindustrial waste, in addition to taking advantage of their antioxidant and antimicrobial properties. Therefore, the study of biosensors and other natural compounds that have stability and can be implemented in packaging requires further studies.

18.6 Functional Properties of Biobased Packaging and Their Effect on Food Preservation

For the preservation of food, whether of vegetable or animal origin, the packaging is required to protect it from environmental conditions, microbiological attacks, physical damage, chemical changes, among others [130]. In addition, the packaging is of great importance during the transportation and marketing of products. Thence, the packaging industry is very relevant, and faces constant challenges along with the needs that are generated for food preservation; one of them is the search for more environmentally friendly, biodegradable, and biocompostable materials, among others [96]. Thus, the materials that will dominate the market will be biobased, but this origin should not modify or reduce their properties. Some studies have shown strengths and great benefits from processability, improvement in the performance of physical properties, biodegradability, antimicrobial activity to economic profitability. Among the most relevant properties and characteristics of the materials used for food preservation included: physical and mechanical properties, barrier properties to gas (water vapor, oxygen, ethylene, CO_2). However, the use for food packaging represents insecurity for the consumer.

Now the monomaterial packaging is a trend aiming for a reduction of residues through improved recollection and recycling. The right material choice is challenging to fulfill since one material may not possess the appropriate barrier properties apart from the mechanical and thermal properties. A continuous search for greener and/or biodegradable packaging options is published in several food packaging journals and discussed in conventions and seminars dedicated to scientists, developers, and industrials, in the search for materials that can compete with those from petrochemical sources. Biodegradable packaging materials have not yet matched the functional properties of conventional synthetic plastics such as polyethylene (PE), polyethylene terephthalate (PET), polypropylene (PP), among others, even when this come from natural sources. Besides the current processing technologies using lamination, coextrusion, rolling are difficult to replace or substitute by for example electrodeposition, foaming, 3D printing to processing such high temperature sensitive biomaterials. Despite this, it is necessary to continue researching in this regard, to reach the goal of having materials that are completely friendly to the environment from renewable and sustainable sources.

18.6.1 Physical and Mechanical Properties

The main objectives of packaging are to facilitate the collection and packaging of food, protecting it from physical and environmental risks during storage and transport. The materials used in the different types of packaging must have specific characteristics [105], among them, material appearance and mechanical resistance [131]. These properties are most relevant during the processing of the package, the packaging of food and during the handling of the package throughout its shelf life. Therefore, tensile and elongation strength, puncture or tear resistance, as well as its resistance to different impacts and bending are relevant in packaging [99]. The physical and mechanical properties are mainly due to the materials used in the packaging. First, the transparency of the package was highly valued by the consumer, then biodegradability and/or compostability has been asked by a mindful buyer. In the particular case of petroleum-derived plastics, their mechanical, thermal, and gas permeability properties have allowed them to be widely used and distributed in different applications since, in the case of biopolymers or biobased packaging, they have reduced physical and mechanical properties [131]. In the search for alternatives to the use of conventional plastics, one of the most promising materials is PLA. PLA exhibits mechanical properties similar to other conventional plastics (such as PET, and PP) [132]. Thermoplastic starch has poor mechanical and barrier properties [14, 131]. Another material widely studied in recent years, nanocellulose, shows strength properties very similar to PET, with low density, and biocompatible [131]. One option to modify and increase the properties of bio-based materials is to use natural fibers from waste by-products, which reduces density, increases tensile strength, and produces biodegradable materials [133]. Polysaccharides provide good mechanical properties to materials and are a good source for blending with other additives. However, the addition of essential oils in different biomaterials has increased in recent years; oil is a plasticizer that reduces stiffness and tensile strength, increased extensibility or elasticity [107]. For all these reasons, biobased materials are still widely studied, whose commercial applications are limited, and should be oriented according to the desired application [131], as these qualities can directly influence the quality of the packaged food [101].

18.6.2 Susceptibility to Moisture

Water vapor and oxygen are two important environmental components to control during storage, as they can modify product characteristics or reduce shelf life. They can easily penetrate through different polymers and biobased materials, so it is important to take into account the susceptibility to moisture or water vapor permeability. Thus, water vapor diffusion is generated in the hydrophilic parts of the materials [107], so this property (moisture susceptibility) will also depend on the packaging components [101, 134]. PLA presents a low susceptibility to moisture which, depending on the application given to PLA, these characteristics might not be an advantage [132]. Natural fibers increase water absorption due to fiber agglomeration, being another problem to be solved to employ natural fibers as reinforcing agents in the packaging [107]. Polysaccharides present an effective barrier against moisture, whereas protein components are hydrophilic and therefore increase permeability and susceptibility to water. For this reason, hydrophobic additives, such as essential oils, are used to increase the water barrier property. However, this property will be modified depending on the ratio between materials and essential oils, the porosity, and the resistance generated by these ratios [107]. Natural fibers are also hydrophilic, so these materials can absorb moisture [133]. In the particular case of nanocellulose, it is one of the poorest biopolymers as a barrier to water vapor, although its mechanical and gas barrier properties are superior [131]. Certain additives incorporated into starch open up new opportunities to improve its performance. It has been shown that the incorporation of a triblock copolymer, such as pluronic F127, improves the functional properties of starch-chitosan composite films, mainly in the water solubility resistance and barrier capacity to water vapor, mechanical and thermal [134]. The main mechanisms of water absorption by different polymers can occur from pores or cracks in the materials, by capillary transport, or by diffusion of water between the polymer chains; therefore, the composition of hydrophobic and hydrophilic agents will mainly modify this property in different packaging [133, 134]. In this context, plant-based foods are the products that require the most moisture control to prevent the development of microorganisms. Thus, they will require packaging made of materials with lower susceptibility to moisture [101]. Likewise, these foods require packaging material with good gas barrier properties.

18.6.3 Gas Barrier

The barrier property to gases, mainly oxygen, water vapor, nitrogen, ethylene, and carbon dioxide, are the most relevant for food preservation, besides some other gaseous molecules like volatile odors, aromas, and flavors are also controlled by the packaging materials. Since the presence of oxygen increases food oxidation and, with it, organoleptic and nutraceutical properties decrease, which can lead to bad odors and rancidity of the lipid compounds present [99, 105]. Therefore, retaining the aroma and organoleptic properties of foods will result in flavor preservation and improved consumer acceptance. Therefore, retaining aromas or being a barrier to aroma is also an inherent property of packaging, and will be a key factor when considering packaging design and manufacturing [97].

On the other hand, films and coatings for plant-based foods function as a barrier against gaseous respiration, but the conditions should not generate a completely anoxic environment, as this would lead to fermentation and the development of bad odors in the food [101]. Gas permeability will occur in conjunction with material thickness, crystallinity, and particle size [99, 131]. The size or type of gas is also relevant, as it can permeate the material by dissolving it or penetrating through pores or holes in the polymer matrix. Therefore, in addition to the water vapor barrier property, the permeability to different gases must be studied in-depth and taken into account for the modeling and design of packaging materials. For example, in terms of vegetables, it is required 1% to 5% of oxygen to continue with their respiration postharvest, while dry products like coffee and high humid ready to eat food require low oxygen to prevent oxidation and spoilage by aerobic microorganisms.

Considering the biobased materials, polysaccharides used for packaging formulation show excellent barrier properties against O_2 and CO_2, while proteins have limited barrier action against them. Cellulose polymer exhibits different characteristics and properties depending on its application [110], although particularly, cellulose and starch exhibit limited barrier properties [135]. The natural biodegradable polyesters (PHA and PHB), they also exhibit inferior barrier properties compared to petroleum-derived materials [135]. Therefore, recent research focuses on the improvement of the different properties of biomaterials, which consists of the addition of different components that confer new properties without modifying the particularity of remaining biodegradable and/or biocompostable. Poor mechanical and barrier properties are part of the limitations for the commercial application of different biobased materials, as well as, the still higher cost than petroleum-derived materials [96, 110], so there is a need for further improvement of materials through different additives [136].

In particular, with the increasing use of nanotechnology, new materials of nanometer sizes have started to be used for the modification of pre-existing materials [135]. Nanocellulose shows low oxygen permeability [131]. Nanocellulose had demonstrated great oxygen barrier properties for dry as well as humid conditions up to the limit of 70% relative humidity (RH) [137]. Furthermore, nanocellulose barrier properties can be enhanced by modifying the surface functional groups. For instance, aminosilanated nanocellulose films show better oxygen transmission rate (OTR) than bare nanocellulose films at 97% RH. The authors attributed this performance to the efficient condensation of siloxanes and formation of a cross-linked layer over nanocellulose that inhibits the interactions with gas and vapor [138]. The obtained OTR values (0.17 to 1 mL mm m^{-2} day^{-1} atm^{-1}) were comparable to that of wax coated nanocellulose films and much lower than the permeability of PET, PE and PP (2, 38 and 80 mL mm m^{-2} day^{-1} atm^{-1} of the oriented polyolefin, respectively) [138].

Another essential gas in the food industry, especially for beverage packaging, is CO_2. The packaging materials are intended to be an excellent barrier to this gas to prevent carbonated beverages from being damaged by their migration. PHB materials with CO_2 permeability values similar to those of PET and polyvinyl chloride (PVC) have been reported [139]. The different determinations methods of the gas barrier properties through permeation testing under international standards are listed in Table 18.6. In such determinations is vital to simulate temperature and humidity conditions that the package actually will experience in the supply chain, for instance, dry and cold storage or high humidity shelf life in warm climates. Preferable to the test in the shaped and finished package instead of the films.

Table 18.6 Standards for determination of barrier properties in biomaterials intended for food package.

Gas of interest	Standard	Country	Type of material to be tested
O_2	ASTM D1434	International	Gas permeability of plastic film and sheeting
	ASTM D3985	International	Oxygen permeability through plastic films
	ASTM F1307	International	Oxygen transmission rate through dry packages
	ASTM F1927	International	Oxygen transmission rate, permeability controlled RH on barrier materials
	DIN 53380-3	German	Oxygen transmission rate of plastic films and plastics moldings
	ISO 15105-2	International	Gas transmission rate of any plastic material: film, sheeting, laminate, co-extruded, and flexible plastic-coated material
	JIS K7126	Japanese	Oxygen transmission rate in plastic film

(*Continued*)

Table 18.6 Standards for determination of barrier properties in biomaterials intended for food package. (*Continued*)

Gas of interest	Standard	Country	Type of material to be tested
CO_2	ASTM F2476	International	CO_2 transmission rate through all barrier materials: film, sheeting, laminates, coextrusions, or plastic-coated papers or fabrics
Water vapor	ASTM E96	International	Plastic films, paper, sheet materials, fiberboards, and wood products. Materials with thickness < 32mm
	ASTM F1249	International	Flexible barrier materials, film and sheet < 3 mm in thickness
	ASTM F372	International	Single-layer or multilayer synthetic or natural polymers and metal foils including coated materials. Materials with thickness < 3 mm
	ISO 2528	International	Transmission rate of sheet materials, thickness < 3 mm or > 1 g/m^2 per day
	ISO 9932	International	Transmission rate of sheet materials, maximum thickness of 38 mm and 0.05 to 65 g/m^2 per day
	ASTM DI653	International	Transmission through organic coating films
	DIN 53122	German	Sheet materials: plastic films, elastomer films, paper
	TAPPI T 448	USA	Paper and paperboard < 3 mm at 23°C and 50% RH
	TAPPI T 464	USA	Paper and paperboard < 3 mm at high temperature and humidity

RH, relative humidity.

18.7 Current State of the Biobased Packaging Market

Among the bioplastic sources there is a compostable option brand named Mater-Bi® from an Italian company. The product contains up to 60% of starch of at least two different crops harvest in Italy. This material has been widely used by other companies to transform it into usable packages for dry and humid items. Here we will mention few cases that reached the market place this year. The first was a thermoformed tray was processed for meat preservation by Fileni that resist low temperatures. A coextrusion process of Mater-Bi for lamination or coating into a series of food packages has been tested by Sacchital. In another case, Mater-Bi was added with corn and cellulose resulting in a clear package ClearCompost® intended for dry food storage like pasta of Agnesi and Misura brand [140].

A commercial gum Glyate® obtained from tamarind seed with low viscosity was chemically modified to be used as a natural modifier of a pre-existing lacquer mixture. The bioORMOCER® barrier contains 10%wt. of the aminated hemicellulose and was used as lacquer of polyolefins, PLA, and PLA/PBAT films [141]. The resulting cured coating showed a thickness of 2-4 μm and preserve the film flexibility. The OTR of a package made of bioORMOCER® and polyolefins was 25 $cm^3\ m^{-2}\ d^{-1}\ bar^{-1}$ at 23 °C and 50% RH that is about four times higher than the food film polyvinylidene chloride of Saran® trade name.

A Skalax® XH11 based on xylan obtained from agricultural residues of flax, barley, spelt, and oat commercialized by a Swedish company has been tested as a coating for parchment paper sealed packages showing an OTR of 10 $cm^3\ m^{-2}\ d^{-1}\ bar^{-1}$ [142]. Also, the product composed of arabinoxylan and water as a plasticizer can form a free standing Skalax® film of 30 to 40 μm of 0.19 $cm^3\ m^{-2}\ d^{-1}\ bar^{-1}$ OTR displaying a promising out layer film in food packaging [143, 144]. A coating of 350 μm composed of commercially available starch ECOFILM® was tested as free standing film and also in a multiple layer package composed of HPDE showing OTR of 0.5 and 0.21 $cm^3\ m^{-2}\ d^{-1}\ bar^{-1}$, respectively [145]. The food packages were used to preserve whole chicken at 4 °C for more than 13 days. The Australian company has a variety of patented packaging materials with the brand name PLANTIC™ with excellent oxygen barrier properties to packaging chocolate and meat [146]. Other potential biopolymers for food packages like PHBs whose brand name Kaneka PHBH® shows a reduced oxygen and carbon dioxide permeability compared with PLA and PBAT films. For example, 5 $cm^3\ mm\ m^{-2}\ d^{-1}$ atm versus 40 and 20 $cm^3\ mm\ m^{-2}\ d^{-1}$ atm of PBAT and PLA, respectively [147].

A three-layer package of 30 μm in thickness based on κ-carrageenan from FUJIFILM Wako Pure Chemical Corporation as the inner layer and out layer and a polysaccharide pullulan from Hayashibara Co., Ltd. as in between gas barrier layer showed a great oxygen barrier property of 11.5 $cm^3\ m^{-2}\ d^{-1}\ bar^{-1}$, compared to their individuals 88.8 and 40.5 $cm^3\ m^{-2}\ d^{-1}\ bar^{-1}$, respectively. The japanese package composed of corn preserved food and dissolve in warm and hot water, furthermore, the inventors claim their edible properties, for example, as miso containers for instant soups [148].

Biopolymers of animal sources are rare but exist in the market, that is the case of a whey protein based packaging has been developed by a German company. The inventors use a formulation containing whey protein, at least two denaturing agents like sodium dodecyl sulfate and urea as well as glycerol as a plasticizer. The thermoformed PET package coated with a 15-μm-thick protein layer shows an OTR of 13 $cm^3\ m^{-2}\ d^{-1}\ bar^{-1}$, while on PE the OTR is reduced up to 40 times [149].

PLA recycling has been recently carried out at industrial scale. Companies like Futerro® and Looplife Polymers locate in Belgium and Total Corbion PLA of Netherlands offer their technology for recycling of bioplastics. For instance, Futerro uses the technology named LOOPLA to scale up the production of PLA using recycled PLA to 180 thousand ton per year in collaboration with a BBCA Biochemical for the Asian market [150]. The technology is based on polymer chain breakage by hydrolysis using ethyl acetate and water at 137 °C and 2.4 bar. The recovery is about 97% of lactic acid. The PLA may contain traces of PET and PP (up to 2%) and yet be able to recycle [151, 152].

Luminy® PLA is a product obtained from a similar chemically recycling process launched in October 2021. The technology behind require of collection and cleaning of PLA return by the consumer and from industrial residues [150]. The company offers in exchange a heat-resistant bioplastic with at least 20% content of recycled PLA, a successful proof of concept has been reported using water bottles recovered by Sansu Korea.

Like the previous example, there are a series of biomaterials in the global market to start with to produce new composites with the requirements and specifications for food contact. Either by combining two or more bioplastics or incorporating vegetal and animal fibers, their use will explode in the next decade thanks to the new processing and recycling technologies. The German Association of Organic Food Manufacturers show a list of European providers and manufactures of cellulose, PLA and starch based materials for food packaging [153]. Other two biobased materials like bio-PE and bio-PET are also enlisted. Among the biopolymers we highlight those comprehend into PLA (NatureWorks®, Futerro®, and Lacea), PHAs (BioMer®, Nodax™, CJ PHA®, TerraBio®, and Veolia) and PHBs (PHBH™, Enmat™, Biopol®, Biogreen®, and Biocycle®) [40, 138, 154, 155].

18.8 Prospects for Food Packaging and the Use of Biobased Materials

Due to the great need for more sustainable food packaging materials, worldwide scientific research aimed at generating new biobased and biodegradable materials has increased. It is a fact that in the coming years and even decades, we will continue to see an increase in this area of study. The design of new formulations, the search for new biodegradable polymers, their structural and functional evaluation until reaching their use in food, are fundamental aspects that need to be evaluated. These biobased and biodegradable materials must present superior characteristics, similar to those of current synthetic plastics to be able to extend their uses even to other areas, such as medical, automotive, and construction. Active and biobased materials will continue to be emerging technologies due to their benefits and excellent applications, which provide advantages over conventional packaging. However, studies on the different polymeric matrices and the permeability of active agents (antioxidants and antimicrobials) will be necessary to determine the optimal design of the materials, as well as their applied study in real challenges, such as different types of food of animal and vegetable origin. The investigations must be developed by researchers from different areas of knowledge to achieve the goal. The development of sustainable materials can increase the life span of food, guaranteeing society's access to adequate food without damaging the environment.

References

1. Thomas, S., Durand, D., Chassenieux, C., Jyotishkumar, P., *Handbook of biopolymer-based materials: From blends and composites to gels and complex networks*, John Wiley & Sons, Weinheim, Germany, 2013.
2. Green, S., Chapter 3—Compounds and composite materials, in: *PEEK Biomaterials Handbook*, S.M. Kurtz (Ed.), pp. 23–48, William Andrew Publishing, Oxford, 2012.

3. Mohan, K., Microbial deterioration and degradation of polymeric materials. *J. Biochem. Technol.*, 2, 210, 2011.
4. Lambert, S. and Wagner, M., Environmental performance of bio-based and biodegradable plastics: The road ahead. *Chem. Soc*, 46, 6855, 2017.
5. Kaplan, D.L., Introduction to biopolymers from renewable resources, in: *Biopolymers from Renewable Resources*, D.L. Kaplan (Ed.), pp. 1–29, Springer Berlin Heidelberg, Berlin, Heidelberg, 1998.
6. European-Bioplastics, *Bioplastics market data 2018*, Berlin, European Bioplastics, Marienstraße 19–20, 2018.
7. Fisheries, F.A.O. Aquaculture Department, FishStatJ–Software for fishery and aquaculture statistical time series, 2019, https://www.fao.org/fishery/en/statistics/software/fishstatj/en.
8. Habibi, Y., Lucia, L.A., Rojas, O.J., Cellulose nanocrystals: Chemistry, self-assembly, and applications. *Chem. Rev.*, 110, 3479, 2010.
9. Petersen, K., Væggemose Nielsen, P., Bertelsen, G., Lawther, M., Olsen, M.B., Nilsson, N.H., Mortensen, G., Potential of biobased materials for food packaging. *Trends Food Sci. Technol.*, 10, 52, 1999.
10. vom Stein, T., Grande, P., Sibilla, F., Commandeur, U., Fischer, R., Leitner, W., Domínguez de María, P., Salt-assisted organic-acid-catalyzed depolymerization of cellulose. *Green Chem.*, 12, 1844, 2010.
11. Rovera, C., Ghaani, M., Farris, S., Nano-inspired oxygen barrier coatings for food packaging applications: An overview. *Trends Food Sci. Technol.*, 97, 210, 2020.
12. Fan, Y. and Picchioni, F., Modification of starch: A review on the application of "green" solvents and controlled functionalization. *Carbohydr. Polym.*, 241, 116350, 2020.
13. Kargarzadeh, H., Huang, J., Lin, N., Ahmad, I., Mariano, M., Dufresne, A., Thomas, S., Gałęski, A., Recent developments in nanocellulose-based biodegradable polymers, thermoplastic polymers, and porous nanocomposites. *Prog. Polym. Sci.*, 87, 197, 2018.
14. Gómez-Aldapa, C.A., Velazquez, G., Gutierrez, M.C., Castro-Rosas, J., Jiménez-Regalado, E.J., Aguirre-Loredo, R.Y., Characterization of functional properties of biodegradable films based on starches from different botanical sources. *Starch - Stärke*, 72, 1900282, 2020.
15. Bertoft, E., Piyachomkwan, K., Chatakanonda, P., Sriroth, K., Internal unit chain composition in amylopectins. *Carbohydr. Polym.*, 74, 527, 2008.
16. Hanashiro, I., Abe, J.-i., Hizukuri, S., A periodic distribution of the chain length of amylopectin as revealed by high-performance anion-exchange chromatography. *Carbohydr. Res.*, 283, 151, 1996.
17. Madhusudhan, B. and Tharanathan, R.N., Legume and cereal starches—Why differrences in digestibility? Part 1: isolation and composition of legume (greengram and bengalgram) starches. *Starch-Stärke*, 47, 165, 1995.
18. Li, H., Gidley, M.J., Dhital, S., High-amylose starches to bridge the "fiber gap": development, structure, and nutritional functionality. *Compr. Rev. Food Sci. Food Saf.*, 18, 362, 2019.
19. Noda, T., Nishiba, Y., Sato, T., Suda, I., Properties of Starches from several low-amylose rice cultivars. *Cereal Chem.*, 80, 193, 2003.
20. Aguirre-Loredo, R.Y., Rodríguez-Hernández, A.I., Morales-Sánchez, E., Gómez-Aldapa, C.A., Velazquez, G., Effect of equilibrium moisture content on barrier, mechanical and thermal properties of chitosan films. *Food Chem.*, 196, 560, 2016.
21. Pillai, C.K.S., Paul, W., Sharma, C.P., Chitin and chitosan polymers: Chemistry, solubility and fiber formation. *Prog. Polym. Sci.*, 34, 641, 2009.
22. Rinaudo, M., Chitin and chitosan: Properties and applications. *Prog. Polym. Sci.*, 31, 603, 2006.

23. Cho, J., Grant, J., Piquette-Miller, M., Allen, C., Synthesis and physicochemical and dynamic mechanical properties of a water-soluble chitosan derivative as a biomaterial. *Biomacromolecules*, 7, 2845, 2006.
24. Latou, E., Mexis, S.F., Badeka, A.V., Kontakos, S., Kontominas, M.G., Combined effect of chitosan and modified atmosphere packaging for shelf life extension of chicken breast fillets. *LWT*, 55, 263, 2014.
25. Liu, H., Du, Y., Wang, X., Sun, L., Chitosan kills bacteria through cell membrane damage. *Int. J. Food Microbiol.*, 95, 147, 2004.
26. Valencia, A.M., Valencia, C.H., Zuluaga, F., Grande-Tovar, C.D., Synthesis and fabrication of films including graphene oxide functionalized with chitosan for regenerative medicine applications. *Heliyon.*, 7, e07058, 2021.
27. Priyadarshi, R., Sauraj, Kumar, B., Deeba, F., Kulshreshtha, A., Negi, Y.S., Chitosan films incorporated with Apricot (Prunus armeniaca) kernel essential oil as active food packaging material. *Food Hydrocolloids*, 85, 158, 2018.
28. Haghighi, H., Licciardello, F., Fava, P., Siesler, H.W., Pulvirenti, A., Recent advances on chitosan-based films for sustainable food packaging applications. *Food Packag. Shelf Life*, 26, 100551, 2020.
29. Wang, H., Qian, J., Ding, F., Emerging chitosan-based films for food packaging applications. *J. Agric. Food Chem.*, 66, 395, 2018.
30. Ruggeri, E., Farè, S., De Nardo, L., Marelli, B., Edible biopolymers for food preservation, in: *Sustainable Food Packaging Technology*, A. Athanassiou (Ed.), pp. 57–105, Wiley-VCH GmbH, Weinheim, Germany, 2021.
31. Luo, Q., Hossen, M.A., Zeng, Y., Dai, J., Li, S., Qin, W., Liu, Y., Gelatin-based composite films and their application in food packaging: A review. *J. Food Eng.*, 313, 110762, 2022.
32. Alipal, J., Mohd Pu'ad, N.A.S., Lee, T.C., Nayan, N.H.M., Sahari, N., Basri, H., Idris, M.I., Abdullah, H.Z., A review of gelatin: Properties, sources, process, applications, and commercialisation. *Mater. Today: Proc.*, 42, 240, 2021.
33. Salem, A., Jridi, M., Abdelhedi, O., Fakhfakh, N., Nasri, M., Debeaufort, F., Zouari, N., Development and characterization of fish gelatin-based biodegradable film enriched with Lepidium sativum extract as active packaging for cheese preservation. *Heliyon*, 7, e08099, 2021.
34. Rangaraj, V.M., Devaraju, S., Rambabu, K., Banat, F., Mittal, V., Silver-sepiolite (Ag-Sep) hybrid reinforced active gelatin/date waste extract (DSWE) blend composite films for food packaging application. *Food Chem.*, 369, 130983, 2022.
35. Alvarado, S., Sandoval, G., Palos, I., Tellez, S., Aguirre-Loredo, Y., Velazquez, G., The effect of relative humidity on tensile strength and water vapor permeability in chitosan, fish gelatin and transglutaminase edible films. *Food Sci. Technol.*, 35, 690, 2015.
36. Hochstetter, A., Talja, R.A., Helén, H.J., Hyvönen, L., Jouppila, K., Properties of gluten-based sheet produced by twin-screw extruder. *LWT*, 39, 893, 2006.
37. Redl, A., Morel, M.H., Bonicel, J., Vergnes, B., Guilbert, S., Extrusion of wheat gluten plasticized with glycerol: Influence of process conditions on flow behavior, rheological properties, and molecular size distribution. *Cereal Chem.*, 76, 361, 1999.
38. Angellier-Coussy, H., Torres-Giner, S., Morel, M.H., Gontard, N., Gastaldi, E., Functional properties of thermoformed wheat gluten/montmorillonite materials with respect to formulation and processing conditions. *J. Appl. Polym. Sci.*, 107, 487, 2008.
39. Gällstedt, M., Mattozzi, A., Johansson, E., Hedenqvist, M.S., Transport and tensile properties of compression-molded wheat gluten films. *Biomacromolecules*, 5, 2020, 2004.
40. Naser, A.Z., Deiab, I., Defersha, F., Yang, S., Expanding poly(lactic acid) (PLA) and polyhydroxyalkanoates (PHAs) applications: A review on modifications and effects. *Polymers*, 13, 4271, 2021.

41. Dubey, S. and Mishra, S., Natural sea salt based polyhydroxyalkanoate production by wild Halomonas hydrothermalis strain. *Fuel*, 311, 1, 122593, 2021.
42. Wang, S., Chen, W., Xiang, H., Yang, J., Zhou, Z., Zhu, M., Modification and potential application of short-chain-length polyhydroxyalkanoate (SCL-PHA). *Polymers*, 8, 273, 2016.
43. Olejnik, O., Masek, A., Zawadziłło, J., Processability and mechanical properties of thermoplastic polylactide/polyhydroxybutyrate (PLA/PHB) bioblends. *Materials*, 14, 898, 2021.
44. Koller, M., Poly(hydroxyalkanoates) for food packaging: Application and attempts towards implementation. *Appl. Food Biotechnol.*, 1, 3, 2014.
45. Ahmad, M., Ritzoulis, C., Chen, J., Meigui, H., Bushra, R., Jin, Y., Xiao, H., Xanthan gum – mucin complexation: Molecular interactions, thermodynamics, and rheological analysis. *Food Hydrocolloids*, 114, 106579, 2021.
46. Meydanju, N., Pirsa, S., Farzi, J., Biodegradable film based on lemon peel powder containing xanthan gum and TiO2–Ag nanoparticles: Investigation of physicochemical and antibacterial properties. *Polym. Test.*, 106, 107445, 2022.
47. Ramos de Souza, E., Rodrigues, P.D., Sampaio, I.C.F., Bacic, E., Crugeira, P.J.L., Vasconcelos, A.C., dos Santos Silva, M., dos Santos, J.N., Quintella, C.M., Pinheiro, A.L.B., Almeida, P.F.d., Xanthan gum produced by Xanthomonas campestris using produced water and crude glycerin as an environmentally friendlier agent to enhance oil recovery. *Fuel*, 310, 122421, 2022.
48. Zheng, M., Chen, J., Tan, K.B., Chen, M., Zhu, Y., Development of hydroxypropyl methylcellulose film with xanthan gum and its application as an excellent food packaging bio-material in enhancing the shelf life of banana. *Food Chem.*, 374, 131794, 2022.
49. Carreño Pineda, L.D., Caicedo Mesa, L.A., Martínez Riascos, C.A., Técnicas de fermentación y aplicaciones de la celulosa bacteriana: Una revisión. *Ing. Cienc.*, 8, 307, 2012.
50. Liu, H., Shi, C., Sun, X., Zhang, J., Ji, Z., Intelligent colorimetric indicator film based on bacterial cellulose and pelargonidin dye to indicate the freshness of tilapia fillets. *Food Packag. Shelf Life*, 29, 100712, 2021.
51. Nascimento, E.S., Barros, M.O., Cerqueira, M.A., Lima, H.L., Borges, M.d.F., Pastrana, L.M., Gama, F.M., Rosa, M.F., Azeredo, H.M.C., Gonçalves, C., All-cellulose nanocomposite films based on bacterial cellulose nanofibrils and nanocrystals. *Food Packag. Shelf Life*, 29, 100715, 2021.
52. Sinha, P., Ubaidulla, U., Hasnain, M.S., Nayak, A.K., Rama, B., Alginate-okra gum blend beads of diclofenac sodium from aqueous template using ZnSO4 as a cross-linker. *Int. J. Biol. Macromol.*, 79, 555, 2015.
53. Benavides, S., Villalobos-Carvajal, R., Reyes, J.E., Physical, mechanical and antibacterial properties of alginate film: Effect of the crosslinking degree and oregano essential oil concentration. *J. Food Eng.*, 110, 232, 2012.
54. Falcone, G., Saviano, M., Aquino, R.P., Del Gaudio, P., Russo, P., Coaxial semi-solid extrusion and ionotropic alginate gelation: A successful duo for personalized floating formulations via 3D printing. *Carbohydr. Polym.*, 260, 117791, 2021.
55. Zhang, J., Liu, L., Jiang, Y., Shah, F., Xu, Y., Wang, Q., High-moisture extrusion of peanut protein-/carrageenan/sodium alginate/wheat starch mixtures: Effect of different exogenous polysaccharides on the process forming a fibrous structure. *Food Hydrocolloids*, 99, 105311, 2020.
56. Chen, J., Wu, A., Yang, M., Ge, Y., Pristijono, P., Li, J., Xu, B., Mi, H., Characterization of sodium alginate-based films incorporated with thymol for fresh-cut apple packaging. *Food Control*, 126, 108063, 2021.
57. Jiang, N., Li, Y., Li, Y., Yu, T., Li, Y., Li, D., Xu, J., Wang, C., Shi, Y., Effect of short jute fibers on the hydrolytic degradation behavior of poly(lactic acid). *Polym. Degrad. Stab.*, 178, 109214, 2020.

58. Wei, Z., Jin, C., Xu, Q., Leng, X., Wang, Y., Li, Y., Synthesis, microstructure and mechanical properties of partially biobased biodegradable poly(ethylene brassylate-co-ε-caprolactone) copolyesters. *J. Mech. Behav. Biomed. Mater.*, 91, 255, 2019.
59. Pascual, A., Sardon, H., Veloso, A., Ruipérez, F., Mecerreyes, D., Organocatalyzed synthesis of aliphatic polyesters from ethylene brassylate: A cheap and renewable macrolactone. *ACS Macro Lett.*, 3, 849, 2014.
60. Butron, A., Llorente, O., Fernandez, J., Meaurio, E., Sarasua, J.-R., Morphology and mechanical properties of poly(ethylene brassylate)/cellulose nanocrystal composites. *Carbohydr. Polym.*, 221, 137, 2019.
61. Qian, S., Sheng, K., Yu, K., Xu, L., Lopez, C.A.F., Improved properties of PLA biocomposites toughened with bamboo cellulose nanowhiskers through silane modification. *J. Mater. Sci.*, 5, 10920, 2018.
62. Zhang, Z., Li, Y., Fu, K., Li, Q., Determination of interfacial properties of cellulose nanocrystal-modified sisal fibre in epoxy by cyclic single-fibre pull-out. *Compos. Sci. Technol.*, 193, 108142, 2020.
63. Salwa, H.N., Sapuan, S.M., Mastura, M.T., Zuhri, M.Y.M., Analytic Hierarchy Process (AHP)-Based materials selection system for natural fiber as reinforcement in biopolymer composites for food packaging. *BioResources*, 14, 33, 2019.
64. Ali, H.T., Alghtani, A.H., Felemban, B.F., Abd El-Aziz, K., Saber, D., Ahmed, E.M., Megahed, M., Fotouhi, M., Multivariable analysis for selection of natural fibers as fillers for a sustainable food packaging industry. *Mater. Res. Express*, 8, 095504, 2021.
65. Laxshaman Rao, B., Makode, Y., Tiwari, A., Dubey, O., Sharma, S., Mishra, V., Review on properties of banana fiber reinforced polymer composites. *Mater. Today: Proc.*, 47, 2825, 2021.
66. Jamil, M. and Hasan, M., Effect of chemical treatment on the properties of banana fiber reinforced polymer composites, in: *Reference Module in Materials Science and Materials Engineering*, Elsevier, Amsterdam, Netherlands, 2021.
67. Kumar, A. and Saini, C.S., Edible composite bi-layer coating based on whey protein isolate, xanthan gum and clove oil for prolonging shelf life of tomatoes. *Measurement: Food*, 2, 100005, 2021.
68. Veiga-Santos, P., Silva, L.T., de Souza, C.O., da Silva, J.R., Albuquerque, E.C., Druzian, J.I., Coffee-cocoa additives for bio-based antioxidant packaging. *Food Packag. Shelf Life*, 18, 37, 2018.
69. Solís-Contreras, G.A., Rodríguez-Guillermo, M.C., Reyes-Vega, M.d.l.L., Aguilar, C.N., Rebolloso-Padilla, O.N., Corona-Flores, J., Soriano-Melgar, L.d.A.A., Ruelas-Chacon, X., Extending shelf-life and quality of minimally processed golden delicious apples with three bioactive coatings combined with cinnamon essential oil. *Foods*, 10, 597, 2021.
70. Choo, K.W., Lin, M., Mustapha, A., Chitosan/acetylated starch composite films incorporated with essential oils: Physiochemical and antimicrobial properties. *Food Biosci.*, 43, 101287, 2021.
71. Talón, E., Trifkovic, K.T., Nedovic, V.A., Bugarski, B.M., Vargas, M., Chiralt, A., González-Martínez, C., Antioxidant edible films based on chitosan and starch containing polyphenols from thyme extracts. *Carbohydr. Polym.*, 157, 1153, 2017.
72. Gómez-Aldapa, C.A., Díaz-Cruz, C.A., Castro-Rosas, J., Jiménez-Regalado, E.J., Velazquez, G., Gutierrez, M.C., Aguirre-Loredo, R.Y., Development of antimicrobial biodegradable films based on corn starch with aqueous extract of hibiscus sabdariffa L. *Starch - Stärke*, 73, 2000096, 2021.
73. El-Sayed, S.M., El-Sayed, H.S., Ibrahim, O.A., Youssef, A.M., Rational design of chitosan/guar gum/zinc oxide bionanocomposites based on Roselle calyx extract for Ras cheese coating. *Carbohydr. Polym.*, 239, 116234, 2020.

74. Gull, A., Bhat, N., Wani, S.M., Masoodi, F.A., Amin, T., Ganai, S.A., Shelf life extension of apricot fruit by application of nanochitosan emulsion coatings containing pomegranate peel extract. *Food Chem.*, 349, 129149, 2021.
75. Khojah, S.M., Bio-based coating from fish gelatin, K-Carrageenan and extract of pomegranate peels for maintaining the overall qualities of fish fillet. *J. Aquat. Food Prod. Technol.*, 29, 810, 2020.
76. Yeddes, W., Djebali, K., Aidi Wannes, W., Horchani-Naifer, K., Hammami, M., Younes, I., Saidani Tounsi, M., Gelatin-chitosan-pectin films incorporated with rosemary essential oil: Optimized formulation using mixture design and response surface methodology. *Int. J. Biol. Macromol.*, 154, 92, 2020.
77. Chaudhary, B.U., Lingayat, S., Banerjee, A.N., Kale, R.D., Development of multifunctional food packaging films based on waste Garlic peel extract and Chitosan. *Int. J. Biol. Macromol.*, 192, 479, 2021.
78. do Evangelho, J.A., da Silva Dannenberg, G., Biduski, B., el Halal, S.L.M., Kringel, D.H., Gularte, M.A., Fiorentini, A.M., da Rosa Zavareze, E., Antibacterial activity, optical, mechanical, and barrier properties of corn starch films containing orange essential oil. *Carbohydr. Polym.*, 222, 114981, 2019.
79. Amalraj, A., Haponiuk, J.T., Thomas, S., Gopi, S., Preparation, characterization and antimicrobial activity of polyvinyl alcohol/gum arabic/chitosan composite films incorporated with black pepper essential oil and ginger essential oil. *Int. J. Biol. Macromol.*, 151, 366, 2020.
80. Awe, F.B., Fagbemi, T.N., Ifesan, B.O.T., Badejo, A.A., Antioxidant properties of cold and hot water extracts of cocoa, Hibiscus flower extract, and ginger beverage blends. *Food Res. Int.*, 52, 490, 2013.
81. Alanis, A., Valdés, J.H., María Guadalupe, N.-V., Lopez, R., Mendoza, R., Mathew, A.P., Díaz de León, R., Valencia, L., Plasma surface-modification of cellulose nanocrystals: A green alternative towards mechanical reinforcement of ABS. *RSC Adv.*, 9, 17417, 2019.
82. Criado, P., Fraschini, C., Becher, D., Mendonca Pereira, F.G., Salmieri, S., Lacroix, M., Modified cellulose nanocrystals (CNCs) loaded in gellan gum matrix enhance the preservation of Agaricus bisporus mushrooms. *J. Food Process. Preserv.*, 44, e14846, 2020.
83. Calambas, H.L., Fonseca, A., Adames, D., Aguirre-Loredo, Y., Caicedo, C., Physical-mechanical behavior and water-barrier properties of biopolymers-clay nanocomposites. *Molecules*, 26, 6734, 2021.
84. Dehnad, D., Emam-Djomeh, Z., Mirzaei, H., Jafari, S.-M., Dadashi, S., Optimization of physical and mechanical properties for chitosan–nanocellulose biocomposites. *Carbohydr. Polym.*, 105, 222, 2014.
85. Danyliuk, N., Tomaszewska, J., Tatarchuk, T., Halloysite nanotubes and halloysite-based composites for environmental and biomedical applications. *J. Mol. Liq.*, 309, 113077, 2020.
86. Romero-Bastida, C.A., Bello-Pérez, L.A., Velazquez, G., Alvarez-Ramirez, J., Effect of the addition order and amylose content on mechanical, barrier and structural properties of films made with starch and montmorillonite. *Carbohydr. Polym.*, 127, 195, 2015.
87. Chen, Q., Shi, Y., Chen, G., Cai, M., Enhanced mechanical and hydrophobic properties of composite cassava starch films with stearic acid modified MCC (microcrystalline cellulose)/NCC (nanocellulose) as strength agent. *Int. J. Biol. Macromol.*, 142, 846, 2020.
88. Montero, B., Rico, M., Rodríguez-Llamazares, S., Barral, L., Bouza, R., Effect of nanocellulose as a filler on biodegradable thermoplastic starch films from tuber, cereal and legume. *Carbohydr. Polym.*, 157, 1094, 2017.
89. Ortega, F., Arce, V.B., Garcia, M.A., Nanocomposite starch-based films containing silver nanoparticles synthesized with lemon juice as reducing and stabilizing agent. *Carbohydr. Polym.*, 252, 117208, 2021.

90. Lazić, V., Vivod, V., Peršin, Z., Stoiljković, M., Ratnayake, I.S., Ahrenkiel, P.S., Nedeljković, J.M., Kokol, V., Dextran-coated silver nanoparticles for improved barrier and controlled antimicrobial properties of nanocellulose films used in food packaging. *Food Packag. Shelf Life*, 26, 100575, 2020.
91. Reshmy, R., Madhavan, A., Philip, E., Paul, S.A., Sindhu, R., Binod, P., Pugazhendhi, A., Sirohi, R., Pandey, A., Sugarcane bagasse derived nanocellulose reinforced with frankincense (Boswellia serrata): Physicochemical properties, biodegradability and antimicrobial effect for controlling microbial growth for food packaging application. *Environ. Technol. Innov.*, 21, 101335, 2021.
92. Roy, S., Rhim, J.-W., Jaiswal, L., Bioactive agar-based functional composite film incorporated with copper sulfide nanoparticles. *Food Hydrocolloids*, 93, 156, 2019.
93. Farhoodi, M., Mousavi, S.M., Sotudeh-Gharebagh, R., Emam-Djomeh, Z., Oromiehie, A., Migration of aluminum and silicon from PET/Clay nanocomposite bottles into acidic food simulant. *Packag. Technol. Sci.*, 27, 161, 2014.
94. Monteiro, M.K.S., de Oliveira, V.R.L., dos Santos, F.K.G., de Barros Neto, E.L., de Lima Leite, R.H., Aroucha, E.M.M., de Oliveira Silva, K.N., Synergistic effect of the sequential intercalation of three types of surfactants in the exfoliation degree of bentonite clay in films of cassava. *J. Mol. Liq.*, 266, 770, 2018.
95. Amjadi, S., Nazari, M., Alizadeh, S.A., Hamishehkar, H., Multifunctional betanin nanoliposomes-incorporated gelatin/chitosan nanofiber/ZnO nanoparticles nanocomposite film for fresh beef preservation. *Meat Sci.*, 167, 108161, 2020.
96. Sharma, R. and Ghoshal, G., Emerging trends in food packaging. *Nutr. Food Sci.*, 48, 764, 2018.
97. Ghoshal, G., Recent trends in active, smart, and intelligent packaging for food products, in: *Handbook of Food Bioengineering, Food Packaging and Preservation*, A.M. Grumezescu and A.M. Holban, (Eds.), pp. 343–374, Academic Press, Cambridge, Massachusetts, United States, 2018.
98. Roya, A.Q. and Elham, M., Intelligent food packaging: Concepts and innovations. *Int. J. Chemtech Res.*, 9, 669, 2016.
99. Kuai, L., Liu, F., Chiou, B.-S., Avena-Bustillos, R.J., McHugh, T.H., Zhong, F., Controlled release of antioxidants from active food packaging: A review. *Food Hydrocolloids*, 120, 106992, 2021.
100. Rangaraj, V.M., Rambabu, K., Banat, F., Mittal, V., Natural antioxidants-based edible active food packaging: An overview of current advancements. *Food Biosci.*, 43, 101251, 2021.
101. Ribeiro, A.M., Estevinho, B.N., Rocha, F., Preparation and incorporation of functional ingredients in edible films and coatings. *Food Bioprocess Technol.*, 14, 209, 2021.
102. Varghese, S.A., Pulikkalparambil, H., Rangappa, S.M., Siengchin, S., Parameswaranpillai, J., Novel biodegradable polymer films based on poly(3-hydroxybutyrate-co-3-hydroxyvalerate) and Ceiba pentandra natural fibers for packaging applications. *Food Packag. Shelf Life*, 25, 100538, 2020.
103. Kumar, R., Ghoshal, G., Goyal, M., Biodegradable composite films/coatings of modified corn starch/gelatin for shelf life improvement of cucumber. *J. Food Sci. Technol.*, 1, 58, 1227–1237, 2020.
104. Plota, A. and Masek, A., Plant-origin stabilizer as an alternative of natural additive to polymers used in packaging materials. *Int. J. Mol. Sci.*, 22, 4012, 2021.
105. Hong, L.G., Yuhana, N.Y., Zawawi, E.Z.E., Review of bioplastics as food packaging materials. *AIMS Mater. Sci.*, 8, 166, 2021.
106. Chawla, R., Sivakumar, S., Kaur, H., Antimicrobial edible films in food packaging: Current scenario and recent nanotechnological advancements- a review. *Carbohydr. Polym. Technol. Appl.*, 2, 100024, 2021.

107. Varghese, S.A., Siengchin, S., Parameswaranpillai, J., Essential oils as antimicrobial agents in biopolymer-based food packaging - A comprehensive review. *Food Biosci.*, 38, 100785, 2020.
108. Fernández-Pan, I., Maté, J.I., Gardrat, C., Coma, V., Effect of chitosan molecular weight on the antimicrobial activity and release rate of carvacrol-enriched films. *Food Hydrocolloids*, 51, 60, 2015.
109. Singh, S., Lee, M.H., Shin, Y., Lee, Y.S., Antimicrobial seafood packaging: A review. *J. Food Sci. Technol.*, 53, 2505, 2016.
110. Wang, J., Euring, M., Ostendorf, K., Zhang, K., Biobased materials for food packaging. *J. Bioresour. Bioprod.*, 7, 1, 1–13, 2022.
111. Al-Tayyar, N.A., Youssef, A.M., Al-Hindi, R.R., Antimicrobial packaging efficiency of ZnO-SiO2 nanocomposites infused into PVA/CS film for enhancing the shelf life of food products. *Food Packag. Shelf Life*, 25, 100523, 2020.
112. Satriaji, K.P., Garcia, C.V., Kim, G.H., Shin, G.H., Kim, J.T., Antibacterial bionanocomposite films based on CaSO4-crosslinked alginate and zinc oxide nanoparticles. *Food Packag. Shelf Life*, 24, 100510, 2020.
113. Abutalib, M. and Rajeh, A., Enhanced structural, electrical, mechanical properties and antibacterial activity of Cs/PEO doped mixed nanoparticles (Ag/TiO2) for food packaging applications. *Polym. Test.*, 93, 107013, 2021.
114. Moustafa, H., Karmalawi, A.M., Youssef, A.M., Development of dapsone-capped TiO2 hybrid nanocomposites and their effects on the UV radiation, mechanical, thermal properties and antibacterial activity of PVA bionanocomposites. *Environ. Nanotechnol. Monit. Manage.*, 16, 100482, 2021.
115. Villegas, C., Arrieta, M.P., Rojas, A., Torres, A., Faba, S., Toledo, M., Gutierrez, M., Zavalla, E., Romero, J., Galotto, M., PLA/organoclay bionanocomposites impregnated with thymol and cinnamaldehyde by supercritical impregnation for active and sustainable food packaging. *Compos. B. Eng.*, 176, 107336, 2019.
116. del Campo, A., de Lucas-Gil, E., Rubio-Marcos, F., Arrieta, M.P., Fernández-García, M., Fernández, J.F., Muñoz-Bonilla, A., Accelerated disintegration of compostable Ecovio polymer by using ZnO particles as filler. *Polym. Degrad. Stab.*, 185, 109501, 2021.
117. Sharma, R., Jafari, S.M., Sharma, S., Antimicrobial bio-nanocomposites and their potential applications in food packaging. *Food Control*, 112, 107086, 2020.
118. El-Saber Batiha, G., Hussein, D.E., Algammal, A.M., George, T.T., Jeandet, P., Al-Snafi, A.E., Tiwari, A., Pagnossa, J.P., Lima, C.M., Thorat, N.D., Zahoor, M., El-Esawi, M., Dey, A., Alghamdi, S., Hetta, H.F., Cruz-Martins, N., Application of natural antimicrobials in food preservation: Recent views. *Food Control*, 126, 108066, 2021.
119. Kongkaoroptham, P., Piroonpan, T., Pasanphan, W., Chitosan nanoparticles based on their derivatives as antioxidant and antibacterial additives for active bioplastic packaging. *Carbohydr. Polym.*, 257, 117610, 2021.
120. Pateiro, M., Munekata, P.E.S., Sant'Ana, A.S., Domínguez, R., Rodríguez-Lázaro, D., Lorenzo, J.M., Application of essential oils as antimicrobial agents against spoilage and pathogenic microorganisms in meat products. *Int. J. Food Microbiol.*, 337, 108966, 2021.
121. Jideani, V.A. and Vogt, K., Antimicrobial packaging for extending the shelf life of bread—A review. *Crit. Rev. Food Sci. Nutr.*, 56, 1313, Washington, DC, United States, 2016.
122. Fan, X., Ngo, H., Wu, C., Natural and bio-based antimicrobials: A review, in: *Natural and Bio-Based Antimicrobials for Food Applications*, X. Fan, H. Ngo, C. Wu, (Eds.), pp. 1–24, American Chemical Society, Washington, DC, United States, 2018.
123. Ahmed, S., Sameen, D.E., Lu, R., Li, R., Dai, J., Qin, W., Liu, Y., Research progress on antimicrobial materials for food packaging. *Crit. Rev. Food Sci. Nutr.*, 1, 3088–3102, 2020.

124. Adeyeye, S.A.O., Food packaging and nanotechnology: Safeguarding consumer health and safety. *Nutr. Food Sci.*, 49, 1164, 2019.
125. Kumar, S., Mukherjee, A., Dutta, J., Chitosan based nanocomposite films and coatings: Emerging antimicrobial food packaging alternatives. *Trends Food Sci. Technol.*, 97, 196, 2020.
126. Zanetti, M., Carniel, T.K., Dalcanton, F., dos Anjos, R.S., Gracher Riella, H., de Araújo, P.H.H., de Oliveira, D., Antônio Fiori, M., Use of encapsulated natural compounds as antimicrobial additives in food packaging: A brief review. *Trends Food Sci. Technol.*, 81, 51, 2018.
127. Rodrigues, C., Souza, V.G.L., Coelhoso, I., Fernando, A.L., Bio-based sensors for smart food packaging—Current applications and future trends. *Sensors*, 21, 2148, 2021.
128. Stark, N.M. and Matuana, L.M., Trends in sustainable biobased packaging materials: A mini review. *Mater. Today Sustainability*, 15, 100084, 2021.
129. Pavelková, A., Time temperature indicators as devices intelligent packaging. *Acta Univ. Agric. Silvic. Mendelianae Brun.*, 61, 245, 2013.
130. Atta, O.M., Manan, S., Shahzad, A., Ul-Islam, M., Ullah, M.W., Yang, G., Biobased materials for active food packaging: A review. *Food Hydrocolloids*, 125, 107419, 2021.
131. Silva, F.A., Dourado, F., Gama, M., Poças, F., Nanocellulose bio-based composites for food packaging. *Nanomaterials*, 10, 2041, 2020.
132. Reichert, C.L., Bugnicourt, E., Coltelli, M.-B., Cinelli, P., Lazzeri, A., Canesi, I., Braca, F., Martínez, B.M., Alonso, R., Agostinis, L., Verstichel, S., Six, L., Mets, S.D., Gómez, E.C., Ißbrücker, C., Geerinck, R., Nettleton, D.F., Campos, I., Sauter, E., Pieczyk, P., Schmid, M., Bio-based packaging: Materials, modifications, industrial applications and sustainability. *Polymers*, 12, 1558, 2020.
133. Väisänen, T., Das, O., Tomppo, L., A review on new bio-based constituents for natural fiber-polymer composites. *J. Cleaner Prod.*, 149, 582, 2017.
134. Fonseca-García, A., Jiménez-Regalado, E.J., Aguirre-Loredo, R.Y., Preparation of a novel biodegradable packaging film based on corn starch-chitosan and poloxamers. *Carbohydr. Polym.*, 251, 117009, 2021.
135. Jabeen, N., Majid, I., Nayik, G.A., Bioplastics and food packaging: A review. *Cogent Food Agric.*, 1, 1117749, 2015.
136. Asgher, M., Qamar, S.A., Bilal, M., Iqbal, H.M., Bio-based active food packaging materials: Sustainable alternative to conventional petrochemical-based packaging materials. *Food Res. Int.*, 137, 109625, 2020.
137. Wang, J., Gardner, D.J., Stark, N.M., Bousfield, D.W., Tajvidi, M., Cai, Z., Moisture and oxygen barrier properties of cellulose nanomaterial-based films. *ACS Sustain. Chem. Eng.*, 6, 49, 2018.
138. Jammernegg, K., Stelzer, F., Spirk, S., Polyhydroxyalkanoates–A prospective food packaging material: Overview of the state of the art, recent developments and potentials, in: *Biopackaging*, p. 317, 2017.
139. Mensitieri, G., Di Maio, E., Buonocore, G.G., Nedi, I., Oliviero, M., Sansone, L., Iannace, S., Processing and shelf life issues of selected food packaging materials and structures from renewable resources. *Trends Food Sci. Technol.*, 22, 72, 2011.
140. ITSCOMPOSTABLE, They are compostable too, 2020, https://www.itscompostable.com/.
141. Emmert, K., Amberg-Schwab, S., Braca, F., Bazzichi, A., Cecchi, A., Somorowsky, F., bioORMOCER®—Compostable functional barrier coatings for food packaging. *Polymers*, 13, 8, 1257, 2021.
142. C.-H. Fernandi, Sealed package comprising parchment paper and a polysaccharide-based coating. WO2020209781A1, 2020.
143. Seelution, Renewable barriers, 2020, https://www.seelution.se/products/renewable-barriers/.
144. Gröndahl, M., Bindgard, L., Gatenholm, P., Hjertberg, T., Polymeric film or coating comprising hemicellulose, US8557033B2, assigned to Seelution AB, United States Patent and Trademark Office, Alexandria, Virginia, USA, 2007.

145. Morris, B.L. and Jacka, D.C., Food package, AU2013202416B2, assigned to Plantic Technologies Ltd, Australian Patent Office, Canberra, Australia, 2014.
146. Kuraray, PLANTIC. Product information, 2003, https://www.kuraray.com/products/plantic.
147. Network, C.T.C. KANEKA Biodegradable Polymer PHBH, 2020, https://www.kanekabiopolymers.com.
148. Otsuki, A., Ishihara, T., Tashiro, Y., Mori, K., Soluble package having oxygen barrier properties, WO/2018/003738, assigned to TOYO SEIKAN GROUP HOLDINGS, LTD, Japan Patent Office, Tokyo, Japan, 2018.
149. A. Stäbler, and M. Schmid, Thermoformable multi-layer composite with oxygen barrier and protein-containing formulation for producing the multi-layer composite, 3241677A1, 2017.
150. TotalCorbion-PLA, About PLA, 2020, https://www.total-corbion.com/.
151. Sherwood, J., Closed-loop recycling of polymers using solvents. *Johnson Matthey Technol. Rev.*, 64, 1, 4–15, 2020.
152. Coszach, P., Bogaert, J.-C., Willocq, J., Chemical recycling of PLA by hydrolysis, US20120142958A1, assigned to Futerro SA, United States Patent and Trademark Office, Alexandria, Virginia, USA, 2012.
153. Biokunststofftool, Materials – Biokunststofftool, 2021, https://www.biokunststofftool.de/materials/?lang=en.
154. Green-Serendipity, Supply chain - Green serendipity, 2021, https://www.greenserendipity.nl/projects/bioplastics/supply-chain/.
155. K. Kamoun, Manufacture of packages, profiles and other biodegradable products, EP3415569A1, European Patent Office, Munich, Germany, 2017.
156. Myers, A.M., Morell, M.K., James, M.G., Ball, S.G., Recent progress toward understanding biosynthesis of the amylopectin crystal. *Plant Physiol.*, 122, 989, 2000.
157. Zhong, Y., Liu, L., Qu, J., Blennow, A., Hansen, A.R., Wu, Y., Guo, D., Liu, X., Amylose content and specific fine structures affect lamellar structure and digestibility of maize starches. *Food Hydrocolloids*, 108, 105994, 2020.
158. Srichuwong, S., Sunarti, T.C., Mishima, T., Isono, N., Hisamatsu, M., Starches from different botanical sources I: Contribution of amylopectin fine structure to thermal properties and enzyme digestibility. *Carbohydr. Polym.*, 60, 529, 2005.
159. Punia, S., Barley starch modifications: Physical, chemical and enzymatic - A review. *Int. J. Biol. Macromol.*, 144, 578, 2020.
160. Torres, F.G., Troncoso, O.P., Díaz, D.A., Amaya, E., Morphological and thermal characterization of native starches from Andean crops. *Starch - Stärke*, 63, 381, 2011.
161. Hernández-Medina, M., Torruco-Uco, J.G., Chel-Guerrero, L., Betancur-Ancona, D., Caracterización fisicoquímica de almidones de tubérculos cultivados en Yucatán, México. *Food Sci. Technol.*, 28, 718, 2008.
162. Tester, R.F., Karkalas, J., Qi, X., Starch—Composition, fine structure and architecture. *J. Cereal Sci.*, 39, 151, 2004.
163. Wang, L. and Wang, Y.-J., Structures and physicochemical properties of acid-thinned corn, potato and rice starches. *Starch—Stärke*, 53, 570, 2001.
164. Arns, B., Bartz, J., Radunz, M., Evangelho, J.A.d., Pinto, V.Z., Zavareze, E.d.R., Dias, A.R.G., Impact of heat-moisture treatment on rice starch, applied directly in grain paddy rice or in isolated starch. *LWT*, 60, 708, 2015.
165. Zhu, F., Recent advances in modifications and applications of sago starch. *Food Hydrocolloids*, 96, 412, 2019.
166. Takeda, Y., Takeda, C., Suzuki, A., Hizukuri, S., Structures and properties of Sago Starches with low and high viscosities on amylography. *J. Food Sci.*, 54, 177, 1989.

167. Abdorreza, M., Robal, M., Cheng, L., Tajul, A., Karim, A., Physicochemical, thermal, and rheological properties of acid-hydrolyzed sago (Metroxylon sagu) starch. *LWT*, 46, 135, 2012.
168. Ríos-Ríos, K., González-Mondragón, E., Segura-Campos, M., Ramírez-Jiménez, M., López-Luna, J., López-Martínez, I., Betancur-Ancona, D., Physicochemical and nutritional characterization of starch isolated from Colocasia antiquorum cultivated in Oaxaca, Mexico. *J. Chem.*, 2016, 6721418, 2016.
169. Collado, L.S., Mabesa, R.C., Corke, H., Genetic variation in the physical properties of sweet potato starch. *J. Agric. Food Chem.*, 47, 4195, 1999.
170. Zhang, T. and Oates, C.G., Relationship between α-amylase degradation and physico-chemical properties of sweet potato starches. *Food Chem.*, 65, 157, 1999.
171. Abdel Aziz, M.S., Salama, H.E., Sabaa, M.W., Biobased alginate/castor oil edible films for active food packaging. *LWT*, 96, 455, 2018.
172. Fabra, M.J., Falcó, I., Randazzo, W., Sánchez, G., López-Rubio, A., Antiviral and antioxidant properties of active alginate edible films containing phenolic extracts. *Food Hydrocolloids*, 81, 96, 2018.
173. Kim, S., Baek, S.-K., Song, K.B., Physical and antioxidant properties of alginate films prepared from Sargassum fulvellum with black chokeberry extract. *Food Packag. Shelf Life*, 18, 157, 2018.
174. Dou, L., Li, B., Zhang, K., Chu, X., Hou, H., Physical properties and antioxidant activity of gelatin-sodium alginate edible films with tea polyphenols. *Int. J. Biol. Macromol.*, 118, 1377, 2018.
175. Salama, H.E., Abdel Aziz, M.S., Alsehli, M., Carboxymethyl cellulose/sodium alginate/chitosan biguanidine hydrochloride ternary system for edible coatings. *Int. J. Biol. Macromol.*, 139, 614, 2019.
176. Reyes-Avalos, M.C., Minjares-Fuentes, R., Femenia, A., Contreras-Esquivel, J.C., Quintero-Ramos, A., Esparza-Rivera, J.R., Meza-Velázquez, J.A., Application of an alginate–Chitosan edible film on figs (ficus carica): Effect on bioactive compounds and antioxidant capacity. *Food Bioprocess Tech.*, 12, 499, 2019.
177. Li, X.-y., Du, X.-l., Liu, Y., Tong, L.-j., Wang, Q., Li, J.-l., Rhubarb extract incorporated into an alginate-based edible coating for peach preservation. *Sci. Hortic.*, 257, 108685, 2019.
178. Zhu, D., Guo, R., Li, W., Song, J., Cheng, F., Improved postharvest preservation effects of Pholiota nameko mushroom by sodium alginate–Based edible composite coating. *Food Bioprocess Tech.*, 12, 587, 2019.
179. Salama, H.E. and Abdel Aziz, M.S., Optimized alginate and Aloe vera gel edible coating reinforced with nTiO2 for the shelf-life extension of tomatoes. *Int. J. Biol. Macromol.*, 165, 2693, 2020.
180. Mahcene, Z., Khelil, A., Hasni, S., Akman, P.K., Bozkurt, F., Birech, K., Goudjil, M.B., Tornuk, F., Development and characterization of sodium alginate based active edible films incorporated with essential oils of some medicinal plants. *Int. J. Biol. Macromol.*, 145, 124, 2020.
181. Salama, H.E. and Aziz, M.S.A., Development of active edible coating of alginate and aloe vera enriched with frankincense oil for retarding the senescence of green capsicums. *LWT*, 145, 111341, 2021.
182. Abdel Aziz, M.S. and Salama, H.E., Developing multifunctional edible coatings based on alginate for active food packaging. *Int. J. Biol. Macromol.*, 190, 837, 2021.
183. Duong, N.T.C., Uthairatanakij, A., Laohakunjit, N., Jitareerat, P., Kaisangsri, N., An innovative single step of cross-linked alginate-based edible coating for maintaining postharvest quality and reducing chilling injury in rose apple cv. 'Tabtimchan' (Syzygium samarangenese). *Sci. Hortic.*, 292, 110648, 2022.
184. Evergreen-Eco-Concepts, Process, 2021, http://www.arecaleafplates.com/about.html.

185. Azmin, S.N.H.M., Hayat, N.A.b.M., Nor, M.S.M., Development and characterization of food packaging bioplastic film from cocoa pod husk cellulose incorporated with sugarcane bagasse fibre. *J. Bioresour. Bioprod.*, 5, 248, 2020.
186. Sifuentes-Nieves, I., Yáñez Macías, R., Neira Velázquez, G., Velázquez, G., Garcia Hernandez, Z., Gonzalez Morones, P., Hernández-Hernández, E., Biobased sustainable materials made from starch and plasma/ultrasound modified Agave fibers: Structural and water barrier performance. *Int. J. Biol. Macromol.*, 193, 2374, 2021.
187. Hazrati, K.Z., Sapuan, S.M., Zuhri, M.Y.M., Jumaidin, R., Preparation and characterization of starch-based biocomposite films reinforced by Dioscorea hispida fibers. *J. Mater. Res. Technol.*, 15, 1342, 2021.
188. Hernández-Varela, J.D., Chanona-Pérez, J.J., Resendis-Hernández, P., Gonzalez Victoriano, L., Méndez-Méndez, J.V., Cárdenas-Pérez, S., Calderón Benavides, H.A., Development and characterization of biopolymers films mechanically reinforced with garlic skin waste for fabrication of compostable dishes. *Food Hydrocolloids*, 124, 107252, 2022.
189. Haque, A.N.M.A. and Naebe, M., Flexible water-resistant semi-transparent cotton gin trash/ poly (vinyl alcohol) bio-plastic for packaging application: Effect of plasticisers on physicochemical properties. *J. Cleaner Prod.*, 303, 126983, 2021.
190. Nandhavathy, G., Dharini, V., Anand Babu, P., Nambiar, R.B., Periyar Selvam, S., Sadiku, E.R., Mahesh Kumar, M., Determination of antifungal activities of essential oils incorporated-pomegranate peel fibers reinforced-polyvinyl alcohol biocomposite film against mango postharvest pathogens. *Mater. Today: Proc.*, 38, 923, 2021.
191. Bof, M.J., Laurent, F.E., Massolo, F., Locaso, D.E., Versino, F., García, M.A., Bio-packaging material impact on blueberries quality attributes under transport and marketing conditions. *Polymers*, 13, 481, 2021.
192. Amariei, S., Norocel, L., Gutt, G., New edible packaging material with function in shelf life extension: Applications for the meat and cheese industries. *Foods*, 9, 5, 562, 2020.
193. Bharti, S.K., Pathak, V., Alam, T., Arya, A., Singh, V.K., Verma, A.K., Rajkumar, V., Materialization of novel composite bio-based active edible film functionalized with essential oils on antimicrobial and antioxidative aspect of chicken nuggets during extended storage. *J. Food Sci.*, 85, 2857, 2020.
194. Albertos, I., Martin-Diana, A., Burón, M., Rico, D., Development of functional bio-based seaweed (Himanthalia elongata and Palmaria palmata) edible films for extending the shelflife of fresh fish burgers. *Food Packag. Shelf Life*, 22, 100382, 2019.
195. Martiny, T.R., Raghavan, V., Moraes, C.C.d., Rosa, G.S.d., Dotto, G.L., Bio-based active packaging: Carrageenan film with olive leaf extract for lamb meat preservation. *Foods*, 9, 1759, 2020.
196. Liu, J., Liu, S., Zhang, X., Kan, J., Jin, C., Effect of gallic acid grafted chitosan film packaging on the postharvest quality of white button mushroom (Agaricus bisporus). *Postharvest Biol. Technol.*, 147, 39, 2019.
197. da Rocha Neto, A.C., Beaudry, R., Maraschin, M., Di Piero, R.M., Almenar, E., Double-bottom antimicrobial packaging for apple shelf-life extension. *Food Chem.*, 279, 379, 2019.
198. Mojaddar Langroodi, A., Tajik, H., Mehdizadeh, T., Preservative effects of sumac hydroalcoholic extract and chitosan coating enriched along with Zataria multiflora Boiss essential oil on the quality of beef during storage. *Vet. Res. Forum*, 9, 153, 2018.
199. Shin, S.-H., Chang, Y., Lacroix, M., Han, J., Control of microbial growth and lipid oxidation on beef product using an apple peel-based edible coating treatment. *LWT*, 84, 183, 2017.
200. Indumathi, M., Sarojini, K.S., Rajarajeswari, G., Antimicrobial and biodegradable chitosan/ cellulose acetate phthalate/ZnO nano composite films with optimal oxygen permeability and hydrophobicity for extending the shelf life of black grape fruits. *Int. J. Biol. Macromol.*, 132, 1112, 2019.

201. Motelica, L., Ficai, D., Oprea, O., Ficai, A., Trusca, R.-D., Andronescu, E., Holban, A.M., Biodegradable alginate films with ZnO nanoparticles and citronella essential oil—A novel antimicrobial structure. *Pharmaceutics*, 13, 1020, 2021.
202. Mehdizadeh, T. and Langroodi, A.M., Chitosan coatings incorporated with propolis extract and Zataria multiflora Boiss oil for active packaging of chicken breast meat. *Int. J. Biol. Macromol.*, 141, 401, 2019.
203. Souza, V.G.L., Rodrigues, C., Valente, S., Pimenta, C., Pires, J.R.A., Alves, M.M., Santos, C.F., Coelhoso, I.M., Fernando, A.L., Eco-friendly ZnO/Chitosan bionanocomposites films for packaging of fresh poultry meat. *Coatings*, 10, 110, 2020.
204. Kumar, S., Boro, J.C., Ray, D., Mukherjee, A., Dutta, J., Bionanocomposite films of agar incorporated with ZnO nanoparticles as an active packaging material for shelf life extension of green grape. *Heliyon*, 5, e01867, 2019.
205. Pabast, M., Shariatifar, N., Beikzadeh, S., Jahed, G., Effects of chitosan coatings incorporating with free or nano-encapsulated Satureja plant essential oil on quality characteristics of lamb meat. *Food Control*, 91, 185, 2018.
206. Moreno, O., Gil, À., Atarés, L., Chiralt, A., Active starch-gelatin films for shelf-life extension of marinated salmon. *LWT*, 84, 189, 2017.
207. Rasul, N.H., Asdagh, A., Pirsa, S., Ghazanfarirad, N., Sani, I.K., Development of antimicrobial/antioxidant nanocomposite film based on fish skin gelatin and chickpea protein isolated containing Microencapsulated Nigella sativa essential oil and copper sulfide nanoparticles for extending minced meat shelf life. *Mater. Res. Express*, 9, 2, 2022.
208. Jamróz, E., Kopel, P., Juszczak, L., Kawecka, A., Bytesnikova, Z., Milosavljevic, V., Makarewicz, M., Development of furcellaran-gelatin films with Se-AgNPs as an active packaging system for extension of mini kiwi shelf life. *Food Packag. Shelf Life*, 21, 100339, 2019.
209. Marcinkowska-Lesiak, M., Wojtasik-Kalinowska, I., Onopiuk, A., Zalewska, M., Poltorak, A., Application of Propolis Extract in Gelatin Coatings as Environmentally Friendly Method for Extending the Shelf Life of Pork Loin. *Coatings*, 11, 979, 2021.
210. da Silva Rodrigues, R.M., de Souza, S.J., Balan, G.C., Droval, A.A., Yamashita, F., Gonçalves, O.H., Leimann, F.V., Shirai, M.A., Production of antimicrobial biobased packaging and application in sliced cooked ham. *Acta Sci. Technol.*, 43, e51399, 2021.
211. Kanikireddy, V., Varaprasad, K., Rani, M.S., Venkataswamy, P., Reddy, B.J.M., Vithal, M., Biosynthesis of CMC-Guar gum-Ag0 nanocomposites for inactivation of food pathogenic microbes and its effect on the shelf life of strawberries. *Carbohydr. Polym.*, 236, 116053, 2020.
212. Ferreira, R.R., Souza, A.G., Quispe, Y.M., Rosa, D.S., Essential oils loaded-chitosan nanocapsules incorporation in biodegradable starch films: A strategy to improve fruits shelf life. *Int. J. Biol. Macromol.*, 188, 628, 2021.
213. Sun, X., Wu, Q., Picha, D.H., Ferguson, M.H., Ndukwe, I.E., Azadi, P., Comparative performance of bio-based coatings formulated with cellulose, chitin, and chitosan nanomaterials suitable for fruit preservation. *Carbohydr. Polym.*, 259, 117764, 2021.

19

Bioplastics-Based Nanocomposites for Packaging Applications

Xiaoying Zhao[1]* and Yael Vodovotz[2]†

[1]*College of Chemistry and Materials Engineering, Beijing Technology and Business University, Beijing, China*

[2]*Ohio State University, Department of Food Science and Technology, Columbus, OH, USA*

Abstract

Bioplastic-based nanocomposites (bionanocomposites) have attracted great attention as they are renewable, biodegradable, and have improved performance namely strength, heat resistance, and barrier properties compared with other bioplastics. Bionanocomposites can also offer functionalities, such as antimicrobial, antifouling, and sensing properties, making them suitable to develop active and intelligent food packaging that can prolong food shelf life, detect food spoilage, monitor food storage conditions, and alert consumers to the deteriorated conditions of the food. The development and applications of bionanocomposites for packaging are reviewed in this chapter, as well as public concerns and regulations of the nanostructured materials, especially for food packaging use.

***Keywords*:** Bioplastics, nanocomposites, bionanocomposites, sustainable packaging, food packaging

19.1 Introduction

Packaging accounts for the largest plastic consumption and is the main contribution to municipal solid waste [1]. Currently, more than 90% of the plastics used for packaging are petroleum-based and nonbiodegradable [2], highlighting the lack of sustainability of the plastic industry [2]. Therefore, the packaging industry is searching for bioalternatives to conventional plastics. According to European Bioplastics, the definition of bioplastics are those polymers that are either fully/partially biobased, or biodegradable, or both [1]. The bioplastic packaging market is growing rapidly driven by the increasing consumer awareness of the impact of traditional plastics on the environment and growing concern of fossil consumption [1]. However, currently, bioplastic only accounts for less than 1% of the packaging market, due to hurdles such as inadequate material properties, poor processability,

**Corresponding author*: zhaoxy@btbu.edu.cn

†*Corresponding author*: vodovotz.1@osu.edu

Inamuddin and Tariq Altalhi (eds.) Handbook of Bioplastics and Biocomposites Engineering Applications: 2nd Edition, (425–444) © 2023 Scrivener Publishing LLC

and high costs [3]. Particularly, the applications of bioplastics for packaging is restricted by their brittleness, low water vapor and oxygen barrier, and inadequate thermal and rheological properties for thermal processing [3]. Therefore, advancement within the bioplastic industry requires improvement in bioplastic properties and functionalities to make them comparable with their petroleum-based counterparts.

Nanotechnology, a concept brought up by Richard Feynman in 1959 [4, 5], has been proved promising in improving plastic performance, including thermal, mechanical, and physicochemical properties [3]. It involves creating and utilizing structures with at least one dimension on the scale of nanometers (1–100 nm), i.e., nanostructures [4–6]. The nanostructures can create new properties not observed in either bulk materials or isolated molecules [5]. Specifically, the use of nanotechnology in polymers refers to a multidisciplinary field of developing, manufacturing, processing, and applying polymer materials filled with components which have nanostructures, the resulting product is known as nanocomposite [5, 7]. In the late 1980s, Toyota researchers reported that incorporation of nanoclay (5 wt.%) significantly improved the barrier, thermal, and mechanical performance of the nanocomposites [5]. Subsequently, research on nanocomposites for packaging of food, pharmaceutical, cosmetic, electronic, and other products has expanded dramatically [8–10]. Particularly, there is a growing interest in applying polymer nanocomposites for food packaging (Figure 19.1), which is the largest (more than one third) packaging segment [11] and constantly seeks better packages with new functionalities to improve food safety and extend food shelf-life, driven by the changes in production practices and consumers' lifestyles [12].

Nanoparticles can enhance food packaging materials' barrier properties by creating tortuous diffusion paths for gas and moisture molecules [4]. Nanoparticles can also carry active substances like antimicrobial agents to manufacture active food packaging [13]. The most used nanoparticles for active packaging are metal oxide and carbon nanotubes [13]. Additionally, nanosensors have been developed and incorporated into polymer matrices to monitor the changes in food and its environment and provide real-time information reflecting food quality [13]. Examples of these nanosensors include indicators and sensors of

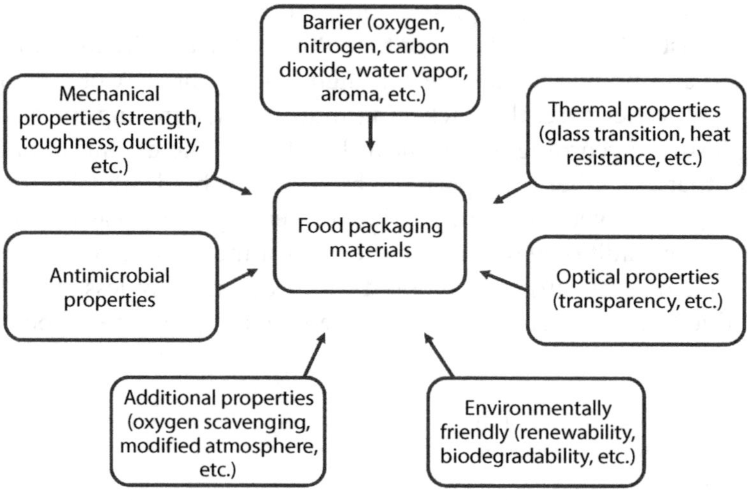

Figure 19.1 Properties generally required for food packaging materials.

Table 19.1 Some of the bioplastics and nanoparticles used to manufacture bionanocomposites.

Type of bioplastics and nanofillers			**Packaging applications**
Bioplastics [1]	**Biobased nonbiodegradable** [1]	Biobased PE, biobased PET, etc. [1]	Bioalternatives for petroleum-based PE and PET for food packaging, such as bottles for drinks, wrapping films, pouches, etc. [1, 17]
	Biobased biodegradable [1]	Polyhydroxyalkanoates (PHA), Polylactide (PLA), Thermoplastic starch (TPS) blends, Polybutylene succinate (PBS), etc. [1]	Films, bags, bottles, cups, bricks, bowls, jars, trays [1]
	Petroleum-based biodegradable [1]	Poly (butylene adipate-co terephthalate) (PBAT), etc. [1]	Bags, films [1]
Nanofillers	**Organic**	**Clay**: MMT, Cloisite 30B, Cloisite 20A, Cloisite Na+, Cloisite 10A [18]	Enhance mechanical, thermal, and barrier properties for packaging materials, especially for food products with long shelf life [18]
		Carbon: carbon nanowall (CNW), carbon nanotube (CNT), etc. [19]	Enhance mechanical , thermal conductivity, electronic conductivity, antimicrobial properties for food packaging and electronic packaging [19]
		Natural biopolymer derived: cellulose, chitosan, etc.	Enhance mechanical such as rigidity and tensile strength, barrier and biodegradation properties of packaging materials [18, 19]
	Inorganic	**Metal**: gold, copper, silver	Antimicrobial agents for active and intelligent food packaging [18]
		Metal oxide: MgO, ZnO, TiO_2, Ag_2O	

freshness, oxygen, and pathogen [13]. Currently, the most used polymers for the manufacture of the packaging nanocomposites are mainly petroleum based, such as poly (ethylene terephthalate) (PET), polystyrene (PS), polypropylene (PP), polyethylene (PE), etc. [14].

Bionanocomposite, in which bioplastic serves as a polymer matrix for nanoparticles, can more readily meet the demand for more sustainable and ecofriendly packaging materials [15, 16]. Example bioplastic matrices include polylactic acid (PLA), polyhydroxyalkanoates (PHA), starch and chitosan, polybutylene succinate (PBS), and poly-caprolactone (PCL) [16]. Nanofillers for bionanocomposites include MMT (montmorillonite), AgO, TiO_2, SiO_2, ZnO, and others [16]. This chapter describes the most recent research and development in bioplastic-based nanocomposites, including the properties and functions of the different bioplastic matrices and nanofillers, their applications in food packaging and their safety regulations (Table 19.1).

19.2 Bioplastic-Based Nanocomposites

Bioplastics are categorized into three groups according to their resource and biodegradability: i. biobased and nonbiodegradable, i.e., "drop-in" replacements of conventional plastics, such as biobased PE and biobased PET; ii. biobased biodegradable, such as polyhydroxyalkanoates (PHA), polylactide (PLA), thermoplastic starch (TPS) blends, and polybutylene succinate (PBS); iii. petroleum-based biodegradable, such as poly ε-caprolactone (PCL) and poly (butylene adipate-co terephthalate) (PBAT). It is worth mentioning that bioplastics and biopolymers are exclusive terms although in some cases they are used interchangeably [20]. Biopolymers are produced from natural resources and are either naturally located in the living organisms or synthesized by using the living organisms [20]. Some examples of biopolymers include starch, pectin, cellulose, gelatin, chitin, keratin, and etc. [21] Bioplastics can be fabricated from biopolymers via polymerization [20], such as in the case of polyhydroxyalkanoates (PHAs) and thermoplastic starch (TPS) [20]. Another type of biodegradable plastic is oxo-degradable bioplastic, which will not be covered in this chapter as its applications in packaging is debatable because of the potential formation of microplastics [1]. This chapter will mainly discuss the bionanocomposites made from the most used bioplastic matrix including PLA, PHA, starch, and PBS.

19.2.1 PLA Bionanocomposites

PLA is biobased, industrially compostable and has polystyrene-like strength and modulus [1]. However, PLA has limited packaging applications as it is brittle, thermally unstable, and has low gas and moisture vapor barrier, and low heat resistance [22]. Thus, the use of nanofillers, such as clay, metal/metallic hydroxides, carbonaceous, cellulose, and chitin/chitosan nanofillers can improve PLA performance to expand its food packaging applications [22].

Clay-based nanofillers, also known as aluminosilicate-based nanofillers, are cost-effective and have been employed to prepare PLA bionanocomposites [23, 24]. The most investigated clay nanofillers include layered silicates like montmorillonite and fibrillar silicates, like halloysite and sepiolite [24, 25]. Layered silicate nanofillers can significantly improve the thermal stability, barrier, and mechanical properties of PLA [25], where the improvements depend on the level of dispersion of the nanofillers within the PLA matrix [25], which is mainly determined by two factors, i.e., the polymer-nanofiller affinity and the methods of preparing the nanocomposites[25], including *in situ* intercalation, melt intercalation and solvent intercalation [25]. Clay surfaces are usually chemically modified with organic compounds to be hydrophobic to enhance the intercalation/exfoliation processes in a polymer matrix and polymer-clay compatibility [26]. For example, Cloisite® 30B, a layered silicate modified with a quaternary ammonium salt [26], has good affinity with the PLA matrix as its diols can favorably interact with the carbonyl moieties on the PLA backbone [25]. PLA/Cloisite 30B (15 wt.%) nanocomposites had a 61% increase in storage modulus compared with pristine PLA [25]. Ray *et al.* reported that PLA nanocomposites filled with montmorillonite modified with trimethyl octadecylammonium had improved mechanical and barrier properties [27], where the chemically modified montmorillonite had a layered distribution in nanocomposites [27], due to the formation of mesoscopic structures between the filler and the matrix [27].

Addition of montmorillonite can also accelerate PLA biodegradation under composting conditions such as industrial composting conditions (high relative humidity at above 50°C) [25]. The PLA biodegradation was correlated to the homogeneous dispersion of the montmorillonite layers, i.e., higher dispersion lead to higher biodegradation rate [25]. PLA photo-oxidation can also be intensified by the addition of layered silicates [25]. Addition of Fe(III)-montmorillonite nanofillers promoted the cleavage of PLA backbone by catalytic decomposition of the hydroperoxide and formation of anhydride [25]. Layered silicates and cellulose nanowhiskers have been used together as an efficient way to enhance PLA barrier properties [27]. For example, multilayered montmorillonite and chitosan were deposited on PLA films via electrostatic forces utilizing a layer-by-layer procedure [27, 28]. The PLA films were covered with 70 montmorillonite/chitosan bilayers and had significant decrease in oxygen permeability [27, 28]. The PLA films with good oxygen barrier and good optical clarity are promising for food packaging applications. Nanofillers with sheet-like morphology are mostly efficient in improving PLA barrier properties as they have larger length-to-width ratio to create a more tortuous pathway [27, 28].

Metallic-based nanoparticles, such as zinc oxide (ZnO), titanium dioxide (TiO_2) and silver (Ag) have been used to impart antibacterial functionality to PLA for active food packaging applications. These nanoparticles were reported to inhibit microbial growth through generating reactive oxygen and/or destroying the membrane integrity and permeability of the microbes [29, 30]. For example, PLA films incorporated with multilayered ZnO nanoparticles inhibited the growth of microbes including *Candida albicans, Bacillus atrophaeus, Escherichia coli, Bacillus cereus,* and *Staphylococcus* [31]. PLA/TiO_2 nanocomposites limited the growth of DH5α *Escherichia coli,* with the inhibition distances being doubled when 1% TiO_2 nanoparticles were added [29]. PLA/silver nanocomposite with 8 to 32 wt.% Ag nanoparticles (Ag-NPs) had a strong antibacterial activity which increased with the Ag-NPs loading [32]. Chitosan has been used to mediate silver nanoparticles as a reducing agent of silver nitrate [33]. The PLA/chitosan-silver nanocomposites can significantly inhibit the growth of Gram-negative bacteria and Gram-positive *S. aureus* compared to the pristine PLA. The enhanced antimicrobial properties of the nanocomposites were attributed to the synergic effect of the silver and chitosan fillers. In addition, the nanocomposites had increased biodegradation rate, improved ductility and toughness, increased crystallinity, and enhanced thermal stability. These improved properties suggested potential for their food packaging applications [33].

Other commonly used nanofillers for PLA nanocomposites include cellulose and carbon nanofillers. Nanocellulose is generated from cellulose macromolecule containing both crystalline and amorphous regions connected by β-1, 4 linked glucopyranose units [34]. Nanocellulose has gained great research interest as it is biocompatible, nontoxic, can self-assemble, has low density, high surface area, and hydroxyl groups that can be easily modified [34]. It has been reported that PLA/nanocellulose nanocomposites made via melt blending with maleated-PLA as a compatibilizer had a 131%, 138%, and 40% increase in impact strength, tensile strength, and Young's modulus, respectively, compared with pristine PLA [35].

Carbon nanofillers, such as graphene (Ga)/graphene oxide (GO) and carbon nanotubes (CNTs), have also been increasingly used for PLA nanocomposite manufacturing. Carbon nanofillers have good thermal, mechanical, electrical, and flame-retardant properties [36]. Ga usually has large surface area that can enhance its interfacial adhesion with PLA matrix while CNTs usually possess high aspect ratio which can increase their bonding with PLA [36]. It has been reported that PLA/CNTs/AgNPs nanocomposites prepared via solvent casting showed improved thermal stability and tensile strength compared with neat PLA [37]. Addition of GO via solution casting also improved PLA thermal stability, as well as hydrophobicity, and slightly decreased PLA tensile strength [38].

19.2.2 PHA Bionanocomposites

Polyhydroxyalkanoates (PHAs) is a type of renewable and biodegradable aliphatic polyesters made by microbial fermentation [1]. Polyhydroxybutyrate (PHB) is one of the most studied PHAs, which has limited industrial use due to its brittleness and poor processability [1]. PHB properties were improved by incorporating 3-hydroxyvalerate (HV) monomers to form a copolymer poly(3-hydroxybutyrate-3-hydroxyvalerate) (PHBV) [1]. Packaging is the fastest growing sector for PHAs [1] yet research on improving PHAs flexibility and processability and reducing its cost is needed to expand its packaging use [1]. Nanofillers have been used to reinforce the PHAs properties. The commonly used fillers for PHA bionanocomposites include inorganic nanofillers like silica, clay, graphene, and carbon nanotubes which can increase PHAs crystallization rate, thermal stability, and stiffness, and polysaccharide nanocrystals, such as cellulose, starch and chitin [39, 40].

Halloysite nanotubes (HNT) with a length of 100 to 2000 nm and a diameter of 10 to 50 nm were used to manufacture PHBV bionanocomposites. HNT has high modulus and high length to diameter ratio, which make it a promising reinforcing agent. HNT can make hydrogen bonds with biopolymers as it has siloxane in its structure [41]. It has been reported that HNT at low concentration increased the crystallization rate of PHBV while HNT at high concentration restricted PHBV crystallization [42, 43]. When PHBV grafted with maleic acid, i.e., PHBV-g-MA, was added as a compatibilizing agent, good interaction between HNT and the matrix was observed [41, 44]. Another popular nanoclay filler for PHBV biocomposites is montmorillonite (Mt) from smectite group. To have a better interaction with the biopolymer matrix, Mt has been chemically modified with alkyl ammonium cations to form organophilic montmorillonite (OMMT). It has been reported that OMMT can improve PHBV thermal stability and biodegradation rate at both industrial composting temperature (60°C) and room temperature [41, 45]. PHBV/OMMT nanocomposite films were reported to have lower biodegradation rate compared with pristine PHBV in composting conditions (at 58°C for 70 days), due to the barrier influence of the OMMT fillers [45]. Interestingly, different types of nanoclays affected PHBV biodegradation differently, e.g., PHBV/Cloisite Na+© bionanocomposite had a higher biodegradation rate than PHBV/OMMT bionanocomposite [45]. Hybrid nanoparticles, OMMT-Cloisite 30B was incorporated into PHBV/PLA bioplastic blend with PHBV-g-MA as a compatibilizer [45]. The thermal and tensile properties such as Young's modulus of the bioplastic were improved. The incorporation of the Cloisite 30B improved the miscibility of the composite components [45]. Cloisite 30B also improved the mechanical properties and thermal stability, as well as suppressed the crystallization of PHBV/PBAT bioplastic blends [45]. Cloisite 30B was hybridized with HNT nanoparticles to obtain good filler dispersion and enhance PHBV thermal stability [45]. PHBV/hydroxyapatite (HA) bionanocomposite with high porosity and a core-shell structure was synthesized. The successful bond between the HA nanoparticles and PHBV matrix endowed the composites with good strength [46, 47]. Kaolinite/silica core-shell nanoparticles modified with alkylepoxysilane were used to enhance mechanical properties of PHB-PHBHHx (3-hydroxyhexanoate) co-polyesters [48]. The core-shell structure resulted in fine nanoparticles' dispersion in the polymer matrix while the chemical modification lead to covalent bonding between the polymer chains and the alkylepoxysilane, with both contributing to improved performance of the composites [48]. Another common used clay nanofiller is halloysite, which has good thermal and rheological properties and is a good reinforcing agent for PHBV [45].

Polysaccharide nanofillers such as cellulose and chitin nanocrystals are gaining interest due to their abundance, renewability, biodegradability, and high specific strength [49]. Cellulose nanocrystals (CNC) were used to produce PHBV bionanocomposites through extrusion. The composite had a unique orthorhombic α-type crystalline structure and higher crystallization rate than PHBV. Composites with CNC at a lower loading had a good filler dispersion while higher CNC loading caused agglomerations [49]. The bionanocomposites were considered suitable for plastic packaging applications [49]. Pristine chitin nanocrystals cannot bond to the polymer matrix very well and tend to aggregate due to their cationic structure [50], lessening their efficiency as reinforcing agents, thus chemical modification is usually used to improve their hydrophobicity and dispersion [50].

Metallic-based nanoparticles, including Ag, TiO_2, and ZnO, have also been used to produce PHA bionanocomposites with antimicrobial properties for active food packaging applications. In addition, PHAs is an effective biobased template for synthesizing Ag nanoparticles with reducing agents $NaBH_4$ and $Ag(NO_3)$ [51]. The PHA matrix was able to stabilize Ag nanoparticles for up to 30 days [51]. PHA was also used to synthesize Au nanoparticles [52]. Au nanoparticles were first coated with enzyme to synthesize PHA on the Au surface. Monomer 3-HB-CoA was then added to produce Au nanoparticles encapsulated by PHB (Figure 19.2) [52]. The obtained Au-PHB nanocomposite can be used as theranostic agents in both diagnosis and therapeutic cases [53] as it can enhance imaging contrast. In addition, TiO_2 nanofillers were used to enhance PHBV/PCL blend by increasing the blend thermal stability and mechanical performance [66].

Another type of nanofillers for PHA bioplastic is carbon nanofillers such as carbon nanowall (CNW) and carbon nanotube (CNT). Surface modification and proper nanocomposite synthesis procedure are usually needed to improve the enhancing efficiency of the carbon nanofillers [54]. Depending on the concentration, the carbon nanofillers have shown either nucleating or restricting effects on PHBV crystallization [54].

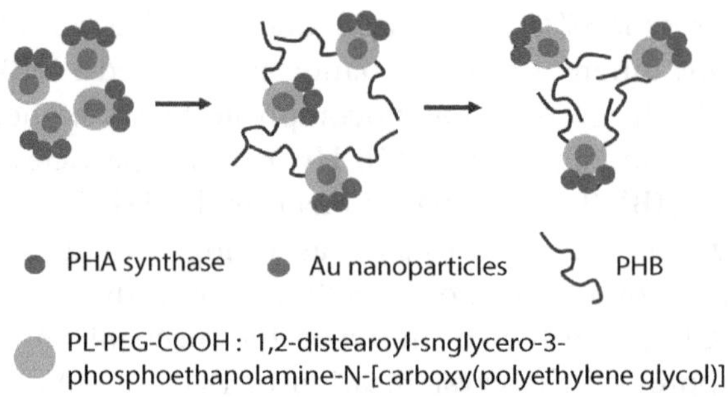

Figure 19.2 Formation of PHB granule coated with AuNP [52]. PHA synthase first adheres to Au nanoparticles (AuNP). Then PHA synthase dimers are formed AuNP surface at the presence of 3HBCoA. Finally, PHB is polymerized and the nanoparticles coalesce via hydrophobic interactions [52].

19.2.3 Starch Bionanocomposites

Starch-based materials is one of the most popular bioplastics as they are abundant, inexpensive, cheap, renewable, biodegradable, and biocompatible [1]. Starch itself cannot be used as a thermoplastic material and requires the use of plasticizers for processing, leading to the formation of thermoplastic starch (TPS) [1]. To advance the industrial applications of the TPS, nanofillers such as polysaccharides, phyllosilicates, metals and carbonaceous fillers have been used to improve TPS performance, including thermal, mechanical, barrier, optical, antimicrobial, electrical, magnetic, and biocompatible properties, to target different applications [55]. The addition of nanofillers usually increases TPS modulus, strength, and biodegradability but decreases ductility and transparency [56]. The key factors for good performance of the nanofiller-reinforced TPS nanocomposites include good filler/matrix compatibility and proper nanofillers' distribution and dispersion in the TPS [56].

Polysaccharide nanofillers used to enhance TPS mainly include cellulose fillers, such as cellulose nanofibers (CNFs), cellulose nanocrystals (CNCs), and cellulose microfibrils (MFC), and starch nanofillers, which are made by precipitation of the amorphous starch followed by enzymatic hydrolysis and microfluidization [55]. Corn-derived TPS was enhanced by cellulose nanofibers made from acetylated cellulose nanofibers and Kenaf bast fibers [55]. The two cellulose nanofillers both increased TPS mechanical properties and moisture barrier, with the acetylated cellulose nanofibers having more significant effect [55]. Similar results were observed in CNC-reinforced potato starch films [55]. The enhancing effect of the cellulose nanofillers in different forms, including nanocrystals, nanofibers and microfibrils, were compared as well [55]. Microfibrillated cellulose showed the highest efficiency in improving maize starch thermal, mechanical, and transparency properties, as it has high aspect ratio and network structure of aggregated cellulose microfibrils [55]. Waxy starch nanocrystals were incorporated into starch to improve mechanical and crystallization properties as well as moisture barrier [57, 58]. It has been reported that the starch nanoparticles are more efficient than cellulose nanofillers in improving the mechanical properties of the starch-based materials, as the starch/starch bionanocomposties have better filler/filler and filler/matrix interactions than the starch/cellulose bionanocomposites [58]. In a starch/cellulose bionanocomposite, the plasticizers can migrate to the filler-matrix interface, making the starch chains crystallize on the surfaces of the cellulose fillers, i.e., transcrystallization, preventing the percolation of the nanofillers in the starch matrix and hindering the stress transfer between the fillers and the matrix [58].

Phyllosilicate nanofillers, such as montmorillonite clays, hectorite, kaolinite, bentonite, etc., have been used to manufacture starch bionanocomposites as they are abundant, inexpensive, and have versatile structures, which can impart different properties to the composites for targeted applications [59]. Among them, MMT nanofillers, which has a structure of two tetrahedral silica layers and an octahedral alumina layer between them, are the most used for starch bionanocomposites. It has been reported that addition of MMT (5%) increased TPS Young's modulus and tensile strength by 198% and 87%, respectively [60], with this significant improvement attributing to the good exfoliation and dispersion of the MMT nanofillers [55]. However, MMT addition usually decreases TPS tensile elongation due to their high rigidity and effect of restricting starch chain mobility [61]. MMT addition can improve the water resistance of the TPS due to the strong hydrogen bonds between TPS and MMT, with a higher MMT loading resulting in a lower TPS water uptake [62]. Additionally, incorporation of MMT decreased TPS transparency as the nanofillers increase the diffuse reflectance and dispersion of the light [56]. Addition of MMT also accelerated TPS biodegradation under the soil-burial testing condition [63]. To improve the starch/nanofiller compatibility, organically modified MMT, i.e., O-MMT, which has exfoliated/intercalated structures of proper interlayer distances has been used [63]. O-MMT nanofillers are more efficient in restricting starch crystallization than MMT fillers [63]. It is generally believed that MMT nanofillers at low concentrations (up to 3%) can improve TPS modulus and strength while at high concentrations have contradicting effects due to filler agglomeration in the matrix [63]. Compared with CNC nanofillers, MMT nanofillers are more efficient in improving TPS modulus and strength but decreasing the tensile elongation since MMT fillers have greater exfoliation effects [63].

Other nanofillers for TPS composites include metallic and carbon nanoparticles. Zinc oxide nanoparticles were used to manufacture TPS composites with antimicrobial functionality [64]. Paper coated with starch/ZnO bionanocomposites had high brightness, good printability, good oil absorption, UV resistance, and antifungal properties. Addition of ZnO nanofillers also increased the electrical conductivity of the starch materials [65]. Silver nanoparticles (AgNPs) were used to produce starch nanocomposites with improved tensile strength, tensile elongation, and antimicrobial properties [66]. Carbon nanoparticles (CNPs), such as carbon black (CB), graphite and carbon nanotubes (CNT), have been applied to manufacture starch bionanocomposites. These fillers can improve mechanical and barrier properties as well as the electrical conductivity of the starch [67] due to the presence of CNT-OH bonds in the composites [68]. Surface modification has been used to improve dispersion and interaction of the carbon nanofillers with the starch matrix. For example, glucose-modified multi-walled carbon nanotubes (MWCNT) nanofillers were used to enhance corn starch [69]. Additionally, carbonaceous nanofillers were combined with cellulose nanofibers to enhance starch bioplastics. Corn starch bionanocomposites with graphene oxide nanoplatelets (3 wt.%) and cellulose nanofibers (15 wt.%) had a 300% increase in tensile strength, a 1260% increase in elastic modulus, and a 78% decrease in oxygen permeability compared with neat corn starch [70]. The good gas barrier makes the starch film promising for food packaging applications [70].

19.2.4 PBS Bionanocomposites

Poly (butylene succinate) (PBS) is biodegradable and mostly petroleum-based as its building blocks 1–4 butanediol (BDO) and succinic acid are mainly petroleum-derived [1]. PBS has advantages including high heat deflection temperature, good processability and thermal stability, mechanical performance like PP and PE, as well as higher biodegradation rate than other aliphatic polyesters like PLA [1]. PBS has high degree of crystallinity, and its products become brittle during use due to its cold crystallization [1]. Nanofillers have thus been incorporated into PBS to improve its crystallization and mechanical performance. Silica nanofillers, such as organomontmorillonite (OMMT) and organically modified layered silicate (OMLS), were used to manufacture PBS bionanocomposites to improve PBS strength, modulus, gas barrier, and flame-retardant properties [71]. Conflicting reports of the effect of OMMT on PBS biodegradability have been reported: either accelerating or slowing down biodegradability in natural organic compost soil for 180 days [72]. Addition of compatibilizing agents, such as PBS-g-MA can accelerate biodegradation of the PBS/OMMT nanocomposites [72]. Organoclay nanofillers (6.9 vol %) were used to improve PBS water vapor and gas barriers (by 45%), thermal stability, melt viscosity, and mechanical properties while retain PBS biocompatibility and biodegradability [73]. Addition of silicon carbide nanofillers (SiC, 3.5 wt.%) increased PBS tensile elongation and by strength 436 and 45%, respectively, improved PBS thermal conductivity by 160%, and significantly improved PBS storage and Young's modulus [74]. SiC nanoparticles also decreased PBS crystallinity and significantly accelerated its biodegradation [74].

Cellulose nanofillers like cellulose nanocrystals (CNCs), cellulose nanofibers (CNFs), nanofibrillated cellulose (NFC), microfibrillated cellulose (MFC), nanocrystalline cellulose (NNC), have been used to manufacture PBS nanocomposites. Incorporation of CNCs (0.8 wt.%) increased PBS tensile strength by 17% while higher CNCs loading (1.5%) decreased the reinforcing efficiency and had minimal effect on PBS strength [75]. The change in the reinforcing efficiency was associated with the change in the composite rheological properties, where two rheological percolation thresholds at the CNC loading of 0.8 and 1.5 wt.% were observed [75]. The addition of CNC (> 1.5 wt.%) also accelerated PBS biodegradation [75]. Similarly, the addition of CNCs changed the crystalline structures of PBS and altered its strength [74]. Acetylchloride (AC)-treated MFC nanofillers were used to manufacture PBS nanocomposites [76]. The AC treatment weakened the hydrogen bonding interactions of the MFC, leading to a good nanofiller dispersion in the matrix [76] and improved composite performance [76]. Compared with pristine PBS, PBS/AC-MFC nanocomposites had a 233% and 123% increase in modulus and tensile strength, respectively [76]. Without AC treatment, the composite tensile strength and modulus were increased by 7 and 46%, respectively, over the pristine PBS [76]. Research on PBS/NCC bionanocomposites reported that NCC nanoparticles were well dispersed in the PBS matrix due to the strong interfacial bonding [76]. NCC addition slightly increased the glass transition and crystallization temperatures and crystallization rate of PBS as the NCC nanofillers acted as nucleating agents, while NCC addition had no significant effect on the crystal structure and crystallization mechanism [76]. PBS/NFC nanocomposites were prepared through melt blending with a premixed PBS/NFC solution as a masterbatch [77]. The masterbatch manufacturing method allowed the incorporation of a very high NFC loading (5 - 50 wt.%) with good dispersion in PBS. The PBS/NFC nanocomposite had an 180 and 250% improvement in elastic and storage modulus, respectively, compared with pristine PBS [77]. NFC addition accelerated PBS crystallization rate, decreased the degree of crystallinity, and significantly improved PBS biodegradation [77]. Films from the PBS/NFC nanocomposites were proposed promising for packaging applications [77]. Similarly, addition of CNFs (2 wt.%) improved PBS tensile and flexural modulus, with a higher filler loading resulting in a higher composite modulus, due to the formation of nanofiber network at higher filler loadings [74]. However, the toughness and ductility of the PBS were decreased with the addition of CNFs [74].

Chitin nanofillers are another type of renewable and biodegradable nanofillers which are attracting academic and industrial interests. Compared with cellulose nanofillers, chitosan nanowhiskers (CWs) have similar effect in improving PBS strength but are more efficient in improving PBS ductility and toughness [74]. For example, an addition of 0.2 wt.% CWs improved PBS tensile elongation and toughness by 530 and 320%, respectively [74]. This is probably because CWs have highly polar surfaces which can strongly bind to PBS polymer chains, promoting the formation of fibrillar and micro-voided structures of the composites, improving filler-matrix stress transfer [78].

Another type of nanofillers for PBS is carbon nanofiller, such as carbon nanotubes (CNTs), multi-walled carbon nanotubes (MWCNTs), graphene oxide (GO), graphene nanoplatelets (GNs), etc. Research on PBS/MWCNTs nanocomposites made from melt blending demonstrated that the incorporation of MWCNTs increased the storage and elastic modulus of the PBS by 100 and 88%, respectively [71]. Additionally, the electrical conductivity of the PBS was significantly improved [71]. CNTs modified with N,N0-dicyclohexylcarbodiimide (DCC) dehydrating agent were incorporated into PBS via melt blending [79]. The nanofillers had a fine dispersion in PBS and the obtained nanocomposites had enhanced mechanical and thermal performance compared with pristine PBS, i.e., addition of 3 wt.% CNTs increased PBS thermal degradation temperature by 12°C and storage and modulus of the PBS by 120 and 55%, respectively [79]. The nanocomposites also had improved conductivity and anti-static efficiency, making them potential bioalernatives for electronic packaging materials [79]. Graphene oxide (GO) nanofillers modified with octadecylamine were used to manufacture PBS nanocomposites with structures of mono- to few-layers [80]. The addition of the GO nanofillers (0.1 wt.%) improved PBS modulus and tensile strength by 50 and 58.9%, respectively while had no significant effect on PBS crystallinity [80]. The improvement in the mechanical performance was attributed to the good filler-matrix interfacial bonding, due to the reaction of nucleophilic substitution between the carboxylic group on the PBS and the amine group on the GO [80].

Currently, some of the nanocomposite materials with improved gas barrier and other properties are commercially available and have been used for food packaging applications such as bottles for drinks, and thermoformed multilayer-nanocomposites food containers [81]. For example, Aegis® OX is a commercial nanocomposite with clay nanoparticles as oxygen scavengers and active plastic layer to enhance the gas and aroma barrier performance [81]. However, the widespread applications of the nanocomposites for packaging applications have yet to be realized, requiring multidisciplinary research and collaborations to overcome the hurdles such as safety, technology, regulation and standardization.

19.3 Packaging Applications

Food packaging provides functions to protect the food from physical, chemical, biological and other pollutions during its storage, handling, and transportation during the food chain. Additionally, the properties of the food packaging materials including strength and stiffness, barrier to oxygen and carbon dioxide.

Moisture barrier and resistance to food component need to be tailored to meet the requirement of different food products and to prolong food shelf life and emphasize food preservation [82]. As discussed above, the incorporation of nanofillers can enhance bioplastic materials by improving the thermal, mechanical, barrier and biodegradation properties and endowing bioplastics with special functionalities such as antimicrobial and detecting chemical and biological alterations [83]. The bionanocomposites with improved performance are promising bioalternatives for packaging applications. Particularly, bionanocomposites with good barrier and smart (e.g. antibacterial) properties are promising for food packaging applications while bionanocomposites with improved conducting and fire retardant properties are promising for electronic packaging applications. High barrier to gases and vapors apart from good mechanical properties are the key attribute in food packaging materials. Bionanocomposites can provide good balance in mechanical, barrier, thermal and properties. As discussed above, the incorporation of nanofillers can enhance barrier properties of the bioplastic materials by generating a more tortuous diffusive path of vapor and gas molecules, addressing the inherent disadvantages of the bioplastics. Bionanocomposites' barrier performance is also dependent on the level of dispersion and orientation of the nanofillers in the matrix [83, 84]. Bionanocomposites with improved barrier properties against water vapor, gas and flavor compounds are particularly suitable to package food products sensitive to moisture and oxygen such as powders and high-fat products [83, 84]. Additionally, bionanocomposites with improved properties are also expected to tolerate higher temperatures during thermal processing and disinfection. Currently, the most favorable nanofillers used to manufacture bionanocomposites for food packaging use include layered silicate nanoclays such as kaolinite and MMT, metallic nanoparticles such as zinc oxide (ZnO-NPs), titanium dioxide (TiO2-NPs) and silver nanoparticles (Ag-NPs). The most attractive bioplastic matrix for bionanocomposties include PLA, PHA and starch-based materials, e.g., PLA/clay bionanocomposites were employed to produce cutlery and food containers [83].

Polymer bionanocomposites with nanofillers that can offer active properties such as oxygen scavenging capability and antimicrobial activity have been demonstrated suitable for food packaging applications. Such packaging is generally referred as active packaging. Bionanocomposites with antimicrobial activity are efficient in enhancing food safety and elongating food shelf life suitable by restricting the growth of post-processing pathogenic microorganisms. The most promising applications thus far of the bionanocomposites with antimicrobial and bioactive properties is for shelf-life extension of meat products [83, 84]. Some nanofillers such as TiO_2 also possess oxygen scavenging properties, which could decrease and maintain the oxygen level in packaging retarding food oxidation or deterioration. This property is especially useful for packaging of fruits and high-fat foods which become brown and have rancid odors and flavors when oxidized [85]. Bionanocomposites containing chitosan, carboxymethyl cellulose, and ZnO nanoparticles were used to package Egyptian soft white cheese. The moisture, pH, titratable acidity, rheology and color of the packaged cheese had minimal change after 4 weeks at 7 °C while the current shelf life of soft cheese is 2 to 3 weeks at 5°C to 7°C [86]. The bionanocomposites also had effective antimicrobial performance against yeast, mold, and bacterial in the packaged cheese products [84].

Bionanocomposites with reactive nanofillers can also be used as microbiological and biochemical sensors and indicators, a requirement of intelligent packaging. These biosensors can detect and react to the alterations in the packaging environments like oxygen/carbon dioxide level, humidity, temperature, and contamination/deterioration of the food products [83, 84]. The monitoring and indicator functions are important to ensure the freshness and safety of the food during storage. For example, in cases where food packaging has sealing defects that go undetected during manufacturing, smart packaging that include nano-sensors can alert consumers of impact on food quality due to the exposure to undesirable conditions such as high levels of oxygen or improper temperatures [83, 84]. As intelligent packaging allows the real-time monitoring of food quality throughout the supply chain, it can also help reduce food waste. For example, most consumers mistakenly believe date labels on food packaging are safety indicators rather than quality indicators. Around 90% of the consumers discard food safe to eat once the "expiration" or "best before" dates arrive. Intelligent packaging can communicate to consumers accurate food quality information and helps them make rational decisions about discarding food [83, 84]. Green, active, and intelligent packaging technologies can work together to yield a multipurpose food-packaging system, to reduce food loss and create a more sustainable food system, an attainable goal for food packaging technology.

19.4 Safety Issue and Regulations

Although the nanotechnology is attracting attention from the packaging industry, the public concern over the safety of the nanocomposite materials for packaging especially food packaging is limiting their commercial use. It is generally believed that the effects of nanomaterials on human health depend on the toxicity, migration rate, morphology, size and the consumption dose of the nanoparticles, areas that are continually debated [87]. The migration of nanoparticles into food matrix also depends on their compatibility with foods. Particles at nanoscale normally display activity different from those at macrolevel, as they can more easily move throughout the human body [88]. The first test of evaluating the safety of the nanocomposites is assessing migration into the foods and the overall risk is low if none are detected in the foods [89]. However, if the nanoparticles can migrate into the food matrix, their solubility in both the food and gastrointestinal passage needs to be determined to evaluate the health effect of the nanomaterials [89]. For the nanoparticles that can be fully dissolved in the food matrices, their toxicity upon exposure is like their bulk form[89]. In addition, the toxicity of nanoparticles and their dissolved ions are different, e.g., silver nanoparticles are less toxic than silver ions [89].

Currently, there are only a handful of methods with enough selectivity and sensitivity to detect and quantify the nanomaterial migrants from the package [88, 89]. The most used technique of analyzing nanomaterials within a polymer matrix include high-resolution imaging such as transmission electron microscopy (TEM), X-ray diffraction (XRD) and X-ray fluorescence (XRF) [88]. Techniques used to detect nanomaterials in food simulants is mainly the asymmetric flow field-flow fractionation (AF4) equipped with element-specific techniques or light scattering detectors which allow the separation and precise measurement of nanomaterials based on their size [88]. The three primary routes of human exposure to nanoparticles include dermal oral ingestion, inhalation, and contact [88]. *In vitro*, such as rodent and human cell lines from skin, lung, liver, intestine, and *in vivo* model systems are applied to evaluate nanoparticle toxicity [88]. Nanoparticles' absorption, metabolism, distribution, and excretion (ADME) are usually determined to understand their kinetics and accumulation in the body [88]. Currently, most of the tests performed in rodents have not clearly demonstrate systemic toxicity of nanoparticles at doses as high as of 5000 mg/kg [90]. Although the scientific findings currently suggest that the toxicity of nanomaterials cannot be definitively predicted [87], some food products and pesticides containing invisible and unlabeled nanoscale additives are already commercially available [91].

Currently, there is no specific legislation for the nanoparticles [92] while world widely, their use is usually initiated by the private sectors, academia, and governments [89]. The terms "nanoscale," "nanotechnology," "nanomaterial," and other related terms are not yet clearly defined or regulated by the U.S. Food and Drug Administration (FDA) or any labeling requirements [89]. The European Commission (EC) claimed that risk assessment of the nanomaterials should be undertaken as for other ingredients which might be toxic in certain cases [89]. The European Food Safety Authority (EFSA) has published guidelines on assessing the hazard of utilizing nanotechnologies and nanoscience in feed and food chains regarding their effect on human and animal health. According to the Plastic Food Contact Materials Regulation (EU) 10/2011, for food contact use of the nanomaterials, an application containing information about the toxicology, migration and exposure of the nanoparticles need to be submitted to European Food Safety Authority (EFSA) for their authorization [89]. The Australian National Industrial Chemicals Notification and Assessment Scheme (NICNAS) has defined industrial nanomaterials and published a guidance on assessing safety of nanomaterials [89].

In the future, more conclusive evidence about the migration, identification and chronic exposure of the nanomaterials is needed for a better understanding of the effects of nanoparticles on human health and environment and an adequate regulation of their use for packaging (especially food packaging) applications [87, 91]. Additionally, as the regulatory approval of the use of novel nanomaterials for food packaging is a long and complex process, it would be prudent for companies to collaborate with government to make sure they can meet the regulatory requirements at both research and commercialization stages [88].

19.5 Conclusions

Increasing environmental concerns have driven the research on bioalternatives for packaging applications. Nanotechnology is a promising technology to benefit collectively society and the environment by the fabrication of bionanocomposites with enhanced thermal, mechanical, barrier, biodegradation, fire retardant, and other performance parameters at a relatively small amount of nanoparticle loading. The industrial use of bionanocomposites is still limited by challenges of meeting end use requirements and relatively high cost. The incorporation of nanofillers into bioplastics for the development of active/intelligent food packaging generally increases the cost of the package. For example, intelligent passive tags cost (5–10¢ each) more than the conventional barcode label (less than 1¢ each) [93]. In the food packaging sector, this is especially critical since the profit margins are relatively low compared with other products and packaging should account for no more than 10% of the total cost [88]. The safety concerns of the consumers are another factor affecting adoptability of bionanocomposite packaging with manufacturers bearing an ethical responsibility to assure minimal safety risks.

Currently, most of the research is about studying the effect of the nanofillers on the bioplastic properties while research on overcoming the challenge of improving the compatibility between nanofillers and polymer matrix for nanocomposite performance and meeting food contact regulations is needed. There is also a lack of governmental guidance and policies in place for the production and applications of nanocomposites nor a clear definition for nanoparticle size. Therefore, a combination of functionality, cost and safety challenges and a lack of clear guidance and policies has impeded adoption of this potential groundbreaking technology.

References

1. Zhao, X., Cornish, K., Vodovotz, Y., Narrowing the gap for bioplastic use in food packaging: An update. *Environ. Sci. Technol.*, 54, 4712, 2020.
2. Zhao, X., Venoor, V., Koelling, K., Cornish, K., Vodovotz, Y., Bio-based blends from poly(3-hydroxybutyrate-co-3-hydroxyvalerate) and natural rubber for packaging applications. *J. Appl. Polym. Sci.*, 136, 47334, 2019.
3. Wróblewska-Krepsztul, J., Rydzkowski, T., Borowski, G., Szczypiński, M., Klepka, T., Thakur, V.K., Recent progress in biodegradable polymers and nanocomposite-based packaging materials for sustainable environment. *Int. J. Polym. Anal. Charact.*, 23, 383, 2018.
4. Enescu, D., Cerqueira, M.A., Fucinos, P., Pastrana, L.M., Recent advances and challenges on applications of nanotechnology in food packaging. A literature review. *Food Chem. Toxicol.*, 134, 110814, 2019.
5. Lagaron, J.M. and Lopez-Rubio, A., Nanotechnology for bioplastics: Opportunities, challenges and strategies. *Trends Food Sci. Technol.*, 22, 611, 2011.
6. Jeevanandam, J., Barhoum, A., Chan, Y.S., Dufresne, A., Danquah, M.K., Review on nanoparticles and nanostructured materials: History, sources, toxicity and regulations. *Beilstein J. Nanotechnol.*, 9, 1050, 2018.
7. González, A., Contreras, C.B., Igarzabal, C.I.A., Strumia, M.C., *Food packaging*, p. 265, Elsevier, The Netherlands, Amsterdam, The Netherlands, 2017.
8. Connolly, M., Zhang, Y., Brown, D.M., Ortuno, N., Jorda-Beneyto, M., Stone, V., Fernandes, T.F., Johnston, H.J., Novel polylactic acid (PLA)-organoclay nanocomposite bio-packaging for the cosmetic industry; migration studies and *in vitro* assessment of the dermal toxicity of migration extracts. *Polym. Degrad. Stab.*, 168, 108938, 2019.
9. Das, R.N., Lin, H., Lauffer, J.M., Rowlands, M., Card, N., Markovich, V.R., Printable nanocomposites for electronic packaging, p. 1093, Endicott Interconnect Technologies, Inc, 2008.
10. Vasile, C., Polymeric nanocomposites and nanocoatings for food packaging: A review. *Materials*, 11, 1834, 2018.
11. Plastemart, Food packaging accounts for almost 35% of the global packaging market, 2016, http://www.plastemart.com/plastic-technical-articles/food-packaging-accounts-for-almost-35-of-the-global-packaging-market/2337.

12. Salgado, P.R., Di Giorgio, L., Musso, Y.S., Mauri, A.N., Recent developments in smart food packaging focused on biobased and biodegradable polymers. *Front. Sustainable Food Syst.*, 5, 125, 2021.
13. Youssef, A.M., Polymer nanocomposites as a new trend for packaging applications. *Polym. Plast. Technol. Eng.*, 52, 635, 2013.
14. Xia, Y., Rubino, M., Auras, R., Interaction of nanoclay-reinforced packaging nanocomposites with food simulants and compost environments. *Adv. Food Nutr.*, 88, 275, 2019.
15. Salvetat, J.P., Briggs, G.A.D., Bonard, J.M., Bacsa, R.R., Kulik, A.J., Stöckli, T., Burnham, N.A., Forró, L., Review article polymer-matrix nanocomposites, processing, manufacturing, and application an overview. *Phys. Rev. Lett.*, 82, 944, 1999.
16. Sharma, R., Jafari, S.M., Sharma, S., Antimicrobial bio-nanocomposites and their potential applications in food packaging. *Food Control*, 112, 107086, 2020.
17. Dimitrie, S., Alina-Mihaela, C., Stela, I.A., Maricica, S., Antimicrobial nanostructures for food packaging design. *WayScience*, 477 &scy, 176, 2021.
18. Priyadarshi, R., Roy, S., Ghosh, T., Biswas, D., Rhim, J.-W., Antimicrobial nanofillers reinforced biopolymer composite films for active food packaging applications-A review. *Sustain. Mater. Technol.*, e00353, 2021.
19. Díez-Pascual, A.M. and Diez-Vicente, A.L., ZnO-reinforced poly (3-hydroxybutyrate-co-3-hydroxyvalerate) bionanocomposites with antimicrobial function for food packaging. *ACS Appl. Mater. Interfaces*, 6, 9822, 2014.
20. Mohapatra, S., Vishwakarma, K., Joshi, N.C., Maity, S., Kumar, R., Ramchander, M., Pattnaik, S., Samantaray, D.P., A review on PHAs: The future biopolymer, in: *Environ. and Agri. Microbiol.: Applications for Sustainability*, p. 83, 2021.
21. Muneer, F., Nadeem, H., Arif, A., Zaheer, W., Bioplastics from biopolymers: An eco-friendly and sustainable solution of plastic pollution. *Polym. Sci. Ser. C*, 63, 47, 2021.
22. Lagarón, J.-M., *Multifunctional and nanoreinforced polymers for food packaging*, pp. 485–497, Elsevier, Amsterdam, The Netherlands, 2011.
23. Ahmed, J. and Varshney, S.K., Polylactides—chemistry, properties and green packaging technology: A review. *Int. J. Food Prop.*, 14, 37, 2011.
24. Armentano, I., Bitinis, N., Fortunati, E., Mattioli, S., Rescignano, N., Verdejo, R., López-Manchado, M.A., Kenny, J.M., Multifunctional nanostructured PLA materials for packaging and tissue engineering. *Prog. Polym. Sci.*, 38, 1720, 2013.
25. Raquez, J.-M., Habibi, Y., Murariu, M., Dubois, P., Polylactide (PLA)-based nanocomposites. *Prog. Polym. Sci.*, 38, 1504, 2013.
26. Sharma, A.K., Mortensen, A., Schmidt, B., Frandsen, H., Hadrup, N., Larsen, E.H., Binderup, M.L., *In-vivo* study of genotoxic and inflammatory effects of the organo-modified Montmorillonite Cloisite® 30B. *Mutat. Res. Genet. Toxicol. Environ. Mutagen.*, 770, 66, 2014.
27. Rhim, J.-W., Park, H.-M., Ha, C.-S., Bio-nanocomposites for food packaging applications. *Prog. Polym. Sci.*, 38, 1629, 2013.
28. Svagan, A.J., Åkesson, A., Cárdenas, M., Bulut, S., Knudsen, J.C., Risbo, J., Plackett, D., Transparent films based on PLA and montmorillonite with tunable oxygen barrier properties. *Biomacromolecules*, 13, 397, 2012.
29. Segura González, E.A., Olmos, D., Lorente, M.Á., Vélaz, I., González-Benito, J., Preparation and characterization of polymer composite materials based on PLA/TiO2 for antibacterial packaging. *Polymers*, 10, 1365, 2018.
30. Dasgupta, N. and Ramalingam, C., Silver nanoparticle antimicrobial activity explained by membrane rupture and reactive oxygen generation. *Environ. Chem. Lett.*, 14, 477, 2016.

31. Mizielińska, M., Kowalska, U., Jarosz, M., Sumińska, P., Landercy, N., Duquesne, E., The effect of UV aging on antimicrobial and mechanical properties of PLA films with incorporated zinc oxide nanoparticles. *Int. J. Environ. Res.*, 15, 794, 2018.
32. Shameli, K., Ahmad, M.B., Yunus, W.M.Z.W., Ibrahim, N.A., Rahman, R.A., Jokar, M., Darroudi, M., Silver/poly (lactic acid) nanocomposites: Preparation, characterization, and antibacterial activity. *Int. J. Nanomed.*, 5, 573, 2010.
33. Sonseca, A. *et al.*, Multifunctional PLA blends containing chitosan mediated silver nanoparticles: Thermal, mechanical, antibacterial, and degradation properties. *Nanomaterials*, 10, 22, 2020.
34. Kian, L., Saba, N., Jawaid, M., Sultan, M., A review on processing techniques of bast fibers nanocellulose and its polylactic acid (PLA) nanocomposites. *Int. J. Biol. Macromol.*, 121, 1314, 2019.
35. Ghasemi, S., Behrooz, R., Ghasemi, I., Yassar, R.S., Long, F., Development of nanocellulose-reinforced PLA nanocomposite by using maleated PLA (PLA-g-MA). *J. Thermoplast. Compos. Mater.*, 31, 1090, 2018.
36. Singh, N.P., Gupta, V., Singh, A.P., Graphene and carbon nanotube reinforced epoxy nanocomposites: A review. *Polymer*, 180, 121724, 2019.
37. de Carvalho, A.P.A. and Junior, C.A.C., Green strategies for active food packagings: A systematic review on active properties of graphene-based nanomaterials and biodegradable polymers. *Trends Food Sci. Technol.*, 103, 130–143, 2020.
38. Gong, M., Zhao, Q., Dai, L., Li, Y., Jiang, T., Fabrication of polylactic acid/hydroxyapatite/graphene oxide composite and their thermal stability, hydrophobic and mechanical properties. *J. Asian Ceram. Soc.*, 5, 160, 2017.
39. Buzarovska, A., Bogoeva-Gaceva, G., Grozdanov, A., Avella, M., Crystallization behavior of polyhydroxybutyrate in model composites with kenaf fibers. *J. Appl. Polym. Sci.*, 102, 804, 2006.
40. Ten, E., Turtle, J., Bahr, D., Jiang, L., Wolcott, M., Thermal and mechanical properties of poly (3-hydroxybutyrate-co-3-hydroxyvalerate)/cellulose nanowhiskers composites. *Polymer*, 51, 2652, 2010.
41. Carli, L.N., Crespo, J.S., Mauler, R.S., PHBV nanocomposites based on organomodified montmorillonite and halloysite: The effect of clay type on the morphology and thermal and mechanical properties. *Compos. Part A Appl. Sci.*, 42, 1601, 2011.
42. Bondeson, D., Mathew, A., Oksman, K., Optimization of the isolation of nanocrystals from microcrystalline cellulose by acid hydrolysis. *Cellulose*, 13, 171, 2006.
43. Rivera-Briso, A.L. and Serrano-Aroca, Á., Poly (3-Hydroxybutyrate-co-3-Hydroxyvalerate): Enhancement strategies for advanced applications. *Polymers*, 10, 732, 2018.
44. Kennouche, S., Le Moigne, N., Kaci, M., Quantin, J.C., Caro-Bretelle, A.S., Delaite, C., Lopez-Cuesta, J.M., Morphological characterization and thermal properties of compatibilized poly (3-hydroxybutyrate-co-3-hydroxyvalerate)(PHBV)/poly (butylene succinate)(PBS)/halloysite ternary nanocomposites. *Eur. Polym. J.*, 75, 142, 2016.
45. Iggui, K., Le Moigne, N., Kaci, M., Cambe, S., Degorce-Dumas, J.R., Bergeret, A., A biodegradation study of poly (3-hydroxybutyrate-co-3-hydroxyvalerate)/organoclay nanocomposites in various environmental conditions. *Polym. Degrad. Stab.*, 119, 77, 2015.
46. Doyle, C., Tanner, E., Bonfield, W., *In vitro* and *in vivo* evaluation of polyhydroxybutyrate and of polyhydroxybutyrate reinforced with hydroxyapatite. *Biomaterials*, 12, 841, 1991.
47. Jack, K.S., Velayudhan, S., Luckman, P., Trau, M., Grøndahl, L., Cooper-White, J., The fabrication and characterization of biodegradable HA/PHBV nanoparticle–polymer composite scaffolds. *Acta Biomater.*, 5, 2657, 2009.

48. Zhang, Q., Liu, Q., Mark, J.E., Noda, I., A novel biodegradable nanocomposite based on poly (3-hydroxybutyrate-co-3-hydroxyhexanoate) and silylated kaolinite/silica core–shell nanoparticles. *Appl. Clay Sci.*, 46, 51, 2009.
49. Malmir, S., Montero, B., Rico, M., Barral, L., Bouza, R., Farrag, Y., PHBV/CNC bionanocomposites processed by extrusion: Structural characterization and properties. *Polym. Compos.*, 40, E275, 2019.
50. Wang, J., Wang, Z., Li, J., Wang, B., Liu, J., Chen, P., Miao, M., Gu, Q., Chitin nanocrystals grafted with poly (3-hydroxybutyrate-co-3-hydroxyvalerate) and their effects on thermal behavior of PHBV. *Carbohydr. Polym.*, 87, 784, 2012.
51. Phukon, P., Saikia, J.P., Konwar, B.K., Enhancing the stability of colloidal silver nanoparticles using polyhydroxyalkanoates (PHA) from Bacillus circulans (MTCC 8167) isolated from crude oil contaminated soil. *Colloids Surf. B*, 86, 314, 2011.
52. Rey, D.A., Strickland, A.D., Kirui, D., Niamsiri, N., Batt, C.A., *In vitro* self-assembly of gold nanoparticle-coated poly (3-hydroxybutyrate) granules exhibiting plasmon-induced thermo-optical enhancements. *ACS Appl. Mater. Interfaces*, 2, 1804, 2010.
53. Luo, S., Yang, X., Shi, C., Newly emerging theranostic agents for simultaneous cancertargeted imaging and therapy. *Curr. Med. Chem.*, 23, 483, 2016.
54. Capadona, J.R., Shanmuganathan, K., Trittschuh, S., Seidel, S., Rowan, S.J., Weder, C., Polymer nanocomposites with nanowhiskers isolated from microcrystalline cellulose. *Biomacromolecules*, 10, 712, 2009.
55. de Freitas, A.D.S.M., da Silva, A.P.B., Montagna, L.S., Nogueira, I.A., Carvalho, N.K., de Faria, V.S., Dos Santos, N.B., Lemes, A.P., Thermoplastic starch nanocomposites: Sources, production and applications-A review. *J. Biomater. Sci. Polym. Ed.*, 33, 7, 900–945, 2021..
56. Rivadeneira-Velasco, K.E., Utreras-Silva, C.A., Díaz-Barrios, A., Sommer-Márquez, A.E., Tafur, J.P., Michell, R.M., Green nanocomposites based on thermoplastic starch: A review. *Polymers*, 13, 3227, 2021.
57. Garcia, N.L., Ribba, L., Dufresne, A., Aranguren, M.I., Goyanes, S., Physico-mechanical properties of biodegradable starch nanocomposites. *Macromol. Mater. Eng.*, 294, 169, 2009.
58. Angellier, H., Molina-Boisseau, S., Dole, P., Dufresne, A., Thermoplastic starch– waxy maize starch nanocrystals nanocomposites. *Biomacromolecules*, 7, 531, 2006.
59. Abdullah, Z.W. and Dong, Y., Recent advances and perspectives on starch nanocomposites for packaging applications. *J. Mater. Sci.*, 53, 15319, 2018.
60. Aouada, F.A., Mattoso, L.H., Longo, E., New strategies in the preparation of exfoliated thermoplastic starch–montmorillonite nanocomposites. *Ind. Crops Prod.*, 34, 1502, 2011.
61. Müller, C.M., Laurindo, J.B., Yamashita, F., Composites of thermoplastic starch and nanoclays produced by extrusion and thermopressing. *Carbohydr. Polym.*, 89, 504, 2012.
62. Huang, M.-F., Yu, J.-G., Ma, X.-F., Studies on the properties of montmorillonite-reinforced thermoplastic starch composites. *Polymer*, 45, 7017, 2004.
63. Behera, A.K., *IOP Conference Series: Materials Science and Engineering*, vol. 410, IOP Publishing, United Kingdom, Bristol, United Kingdom, p. 012001, 2018.
64. Rahman, M.A.A., Mahmud, S., Alias, A.K., Nor, A.F.M., Effect of nanorod zinc oxide on electrical and optical properties of starch-based polymer nanocomposites. *J. Phys. Sci.*, 24, 17, 2013.
65. Prasad, V., Shaikh, A.J., Kathe, A.A., Bisoyi, D.K., Verma, A.K., Vigneshwaran, N., Functional behaviour of paper coated with zinc oxide–soluble starch nanocomposites. *J. Mater. Process. Technol.*, 210, 1962, 2010.
66. Ortega, F., García, M.A., Arce, V.B., Nanocomposite films with silver nanoparticles synthesized *in situ*: Effect of corn starch content. *Food Hydrocolloids*, 97, 105200, 2019.

67. Shahbazi, M., Rajabzadeh, G., Sotoodeh, S., Functional characteristics, wettability properties and cytotoxic effect of starch film incorporated with multi-walled and hydroxylated multi-walled carbon nanotubes. *Int. J. Biol. Macromol.*, 104, 597, 2017.
68. Mallakpour, S., Ultrasonic-assisted fabrication of starch/MWCNT-glucose nanocomposites for drug delivery. *Ultrason. Sonochem.*, 40, 402, 2018.
69. Mallakpour, S. and Rashidimoghadam, S., Starch/MWCNT-vitamin C nanocomposites: Electrical, thermal properties and their utilization for removal of methyl orange. *Carbohydr. Polym.*, 169, 23, 2017.
70. Ramezani, H., Behzad, T., Bagheri, R., Synergistic effect of graphene oxide nanoplatelets and cellulose nanofibers on mechanical, thermal, and barrier properties of thermoplastic starch. *Polym. Adv. Technol.*, 31, 553, 2020.
71. Rafiqah, S., Khalina, A., Harmaen, A.S., Tawakkal, I.A., Zaman, K., Asim, M., Nurrazi, M.N., Lee, C.H., A review on properties and application of bio-based poly (butylene succinate). *Polymers*, 13, 1436, 2021.
72. Phua, Y., Lau, N., Sudesh, K., Chow, W., Ishak, Z.M., Biodegradability studies of poly (butylene succinate)/organo-montmorillonite nanocomposites under controlled compost soil conditions: Effects of clay loading and compatibiliser. *Polym. Degrad. Stab.*, 97, 1345, 2012.
73. Ilsouk, M., Raihane, M., Rhouta, B., Meri, R.M., Zicans, J., Vecstaudža, J., Lahcini, M., The relationship of structure, thermal and water vapor permeability barrier properties of poly (butylene succinate)/organomodified beidellite clay bionanocomposites prepared by *in situ* polycondensation. *RSC Adv.*, 10, 37314, 2020.
74. Platnieks, O., Gaidukovs, S., Thakur, V.K., Barkane, A., Beluns, S., Bio-based poly (butylene succinate): Recent progress, challenges and future opportunities. *Eur. Polym. J.*, 161, 110855, 2021.
75. Kim, H.J., Choi, Y.H., Jeong, J.H., Kim, H., Yang, H.S., Hwang, S.Y., Koo, J.M., Eom, Y., Rheological percolation of cellulose nanocrystals in biodegradable poly (butylene succinate) nanocomposites: A novel approach for tailoring the mechanical and hydrolytic properties. *Macromol. Res.*, 29, 720, 2021.
76. Nazrin, A., Sapuan, S.M., Zuhri, M.Y.M., Ilyas, R.A., Syafiq, R.S.F.K.S., Sherwani, S.F.K., Nanocellulose reinforced thermoplastic starch (TPS), polylactic acid (PLA), and polybutylene succinate (PBS) for food packaging applications. *Front. Chem.*, 8, 213, 2020.
77. Platnieks, O., Sereda, A., Gaidukovs, S., Thakur, V.K., Barkane, A., Gaidukova, G., Filipova, I., Ogurcovs, A., Fridrihsone, V., Adding value to poly (butylene succinate) and nanofibrillated cellulose-based sustainable nanocomposites by applying masterbatch process. *Ind. Crops Prod.*, 169, 113669, 2021.
78. Kim, H., Shin, M.S., Jeon, H., Koo, J.M., Eom, Y., Choi, S., Shin, G., Oh, D.X., Hwang, S.Y., Park, J., Highly reinforced poly (butylene succinate) nanocomposites prepared from chitosan nanowhiskers by *in-situ* polymerization. *Int. J. Biol. Macromol.*, 173, 128, 2021.
79. Shih, Y.-F., Chen, L., Jeng, R., Preparation and properties of biodegradable PBS/multi-walled carbon nanotube nanocomposites. *Polymer*, 49, 4602, 2008.
80. Abidin, A.S.Z., Yusoh, K., Jamari, S.S., Abdullah, A.H., Ismail, Z., Surface functionalization of graphene oxide with octadecylamine for improved thermal and mechanical properties in polybutylene succinate nanocomposite. *Polym. Bull. (Berlin)*, 75, 3499, 2018.
81. Shankar, S. and Rhim, J.W., Polymer nanocomposites for food packaging applications, in: *Func. Phy. Pro. Polym. Compos.*, p. 29, 2016.
82. Silvestre, C., Duraccio, D., Cimmino, S., Food packaging based on polymer nanomaterials. *Prog. Polym. Sci.*, 36, 1766, 2011.
83. Mousa, M.H., Dong, Y., Davies, I.J., Recent advances in bionanocomposites: Preparation, properties, and applications. *Int. J. Polym. Mater.*, 65, 225, 2016.

84. Youssef, A.M. and El-Sayed, S.M., Bionanocomposites materials for food packaging applications: Concepts and future outlook. *Carbohydr. Polym.*, 193, 19, 2018.
85. Othman, S.H., Bio-nanocomposite materials for food packaging applications: Types of biopolymer and nano-sized filler. *Agric. Agric. Sci. Proc.*, 2, 296, 2014.
86. Hamad, M.N.F., *Evaluation of methods used for making soft cheese*, Mansoura University, Dakahlia Governorate, Egypt, Doctor Thesis, 2002.
87. Sothornvit, R., Nanostructured materials for food packaging systems: New functional properties. *Curr. Opin. Food Sci.*, 25, 82, 2019.
88. Mihindukulasuriya, S. and Lim, L.-T., Nanotechnology development in food packaging: A review. *Trends Food Sci. Technol.*, 40, 149, 2014.
89. Sarfraz, J., Gulin-Sarfraz, T., Nilsen-Nygaard, J., Pettersen, M.K., Nanocomposites for food packaging applications: An overview. *Nanomaterials*, 11, 10, 2021.
90. Maisanaba, S., Pichardo, S., Puerto, M., Gutierrez-Praena, D., Camean, A.M., Jos, A., Toxicological evaluation of clay minerals and derived nanocomposites: A review. *Environ. Res.*, 138, 233, 2015.
91. Sorrentino, A., Gorrasi, G., Vittoria, V., Potential perspectives of bio-nanocomposites for food packaging applications. *Trends Food Sci. Technol.*, 18, 84, 2007.
92. Garcia, C.V., Shin, G.H., Kim, J.T., Metal oxide-based nanocomposites in food packaging: Applications, migration, and regulations. *Trends Food Sci. Technol.*, 82, 21, 2018.
93. Lee, S.J. and Rahman, A.M., *Innovations in food packaging*, pp. 171–209, Elsevier, The Netherlands, Amsterdam, The Netherlands, 2014.

20

Applications of Bioplastics in Disposable Products

Mahrukh Aslam[1], Habibullah Nadeem[1], Farrukh Azeem[1], Muhammad Zubair[1], Ijaz Rasul[1], Saima Muzammil[2], Muhammad Afzal[1] and Muhammad Hussnain Siddique[1]*

[1]Department of Bioinformatics and Biotechnology, Government College University, Faisalabad, Pakistan
[2]Department of Microbiology, Government College University, Faisalabad, Pakistan

Abstract

Environmental pollution is a big concern for the world, and it is caused by a variety of sources, one of which is plastic. The demand of plastic is increasing as a result of urbanization and industrialization. About 30 families of the plastics are known that are used on a commercial scale. The problem of using plastic is that they are not degradable, cost-effective, and environment friendly. The production of waste plastics is projected to be around 300 million tons per year, which is a massive amount. Plastics that are used nowadays are made up of nonrenewable and thermochemical resources, for which huge amount of energy is required for its synthesis, which make them uneconomical and also hazardous for the living world. An ideal substitute for these problems is the formation of biobased polymers, also known as the bioplastics. Bioplast is made up of renewable and biodegradable biomass, such as corn, starch, vegetable fat, and food waste. The total carbon emission from it is about 69% after industrial composting, which is not harmful for the environment. A growing amount of attention of bioplast, its types, how it is better from conventional plastics and its applications in various fields and products are being discussed in this chapter.

Keywords: Bioplast, biopolymers, bioplastics, thermochemical, biodegradable, PHAs, PHB, renewable, lignocelluloses

20.1 Introduction

As the world's population is increasing, so does the demand for plastic, which is now used in a wide range of applications. There is remarkable increase in demand of plastics since last decade because of its multifaceted properties [1]. The manufacturing of conventional plastics by petrochemicals exceeds 300 million tons per year, and these plastics are not environmentally friendly. As a result, the bioplastic industry has become more sensitive.

Bioplastics are the plastics that are obtained from the biological source instead of petrochemicals and are more ecofriendly as they can minimize the atmospheric carbon to 30% to 70% in contrast to synthetic plastics and are also cost effective [2]. The components that are

**Corresponding author*: mhs1049@gmail.com

Inamuddin and Tariq Altalhi (eds.) Handbook of Bioplastics and Biocomposites Engineering Applications: 2nd Edition, (445–456) © 2023 Scrivener Publishing LLC

used in the formation of the bioplastics are vegetable fats, agricultural waste, straw, cornstarch rather than the conventional components, like wood, stone, horn, glass, leather, etc. Bioplastics are far better than the conventional plastic because of following reasons: 1) The materials that are used for the synthesis of bioplastics are green and renewable. 2) It utilizes less energy for its formation. 3) They are more aversion to the heat, organic solvents, oxidation, and microorganisms. 4) The bioplastic products have long life duration as compared to the conventional plastics.

There are two types of the bioplastics on the basis of degradability: 1) biodegradable 2) nonbiodegradable. Biobased biodegradable plastics include PHB, PHBV, PLA, CA, CAB, CAP, CN. Biobased nonbiodegradable plastic contain PE, PA 11, PA 12, PET, PTT. Nonbiodegradable bioplastics are slightly biodegradable but they are not as much efficient as biodegradable bioplastics.

There are many applications of bioplastics in different industries where they are playing crucial role for the welfare of mankind. Bioplastics are being used in the food industry for packaging to enhance the shelf life of a food product and to avoid the growth of unwanted microorganisms [3]. They are also used in packaging industries, in cosmetic industry as microbeads. Automotive industries are also utilizing the bioplastics to make the vehicles stronger and more sustainable. There are some other applications like in textile, toys, nursery products, agricultural and horticultural products etc.

20.2 Plastics vs Bioplastics

Inadequate active solid waste management identifies that the plastic garbage is causing serious environmental problems and are also affecting the health of the human beings and animals. As a result, there is lot of curiosity about it. These are some points that make bioplastics more important from the ordinary plastics, which are also represent in Figure 20.1 [4].

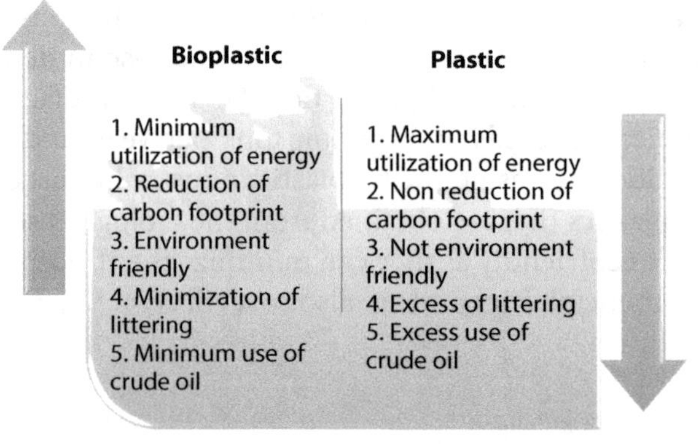

Figure 20.1 Comparison of bioplastic and plastic.

20.2.1 Minimum Utilization of Energy

Bioplastics is extremely significant from the conventional plastic because it requires less amount of energy for its production as the cost of the petroleum is skyrocketing and its supply will run out in the near future.

20.2.2 Reduction of Carbon Footprint

The other benefit of using bioplastic is that the carbon footprint can be significantly reduced. The carbon dioxide absorbed by the plant in the photosynthesis process is isolated by plastic manufactured from biological source [5]. This isolation is reversible if the generated bioplastic dissolves back into the carbon dioxide and water.

20.2.3 Environment Friendly

Scientists are trying to make plastics from the green materials, such as plants and animals, as a main component for its production that have biodegradability property.

20.2.4 Littering Minimization

Shopping bags are the biggest concern of environmental pollution. Most of the shopping bags that are excessively being used in our daily life are not disposable and are polluting the oceans. As a result, marine life is endangered, but bioplastic, because to its biodegradability, has overcome this problem.

20.2.5 Not Usage of Crude Oil

The manufacture of conventional plastic using crude oil necessitates a 20-kW hour of energy; however, the production of bioplastic does not necessitate, such a large amount of energy.

20.3 Types of Bioplastics

Bioplastics are a type of biobased polymer that has a diverse set of properties and applications. On a regular basis, new materials are discovered and tested. Bioplast is divided into five categories, which is presented in Figure 20.2.

20.3.1 Starch-Based

Corn, wheat, rice, potato and other botanical resources are the great source of starch and it is present in most of the plant parts like the roots, seeds and the tuber [6]. Starch is a polysaccharide and about 75% of the plants that are on the earth have starch in it and it is the major component of energy for the plants [7]. The structure of starch contains amylose and the amylopectin and its quantity is mainly depended on the source from which the starch is extracted. Plastic that are made from the starch are the multiplex fuse of degradable plastic, such as PLA, PBAT, PBS, PCL, and PHAs. Starch-based bioplastics have very phenomenal properties like mechanical properties, water resistance, and processing properties [8].

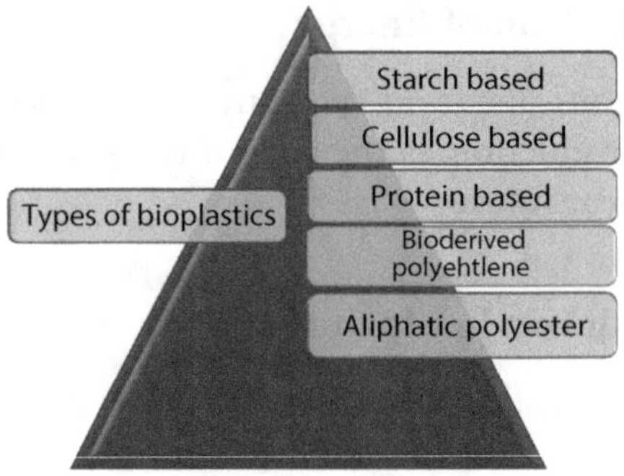

Figure 20.2 Types of bioplastics.

Starch-based bioplastics are widely used in the package industries and in the formation of the nanocomposite films.

20.3.2 Cellulose-Based

Cellulose is the component that is normally found in the most varieties of plants and nowadays plastics are being produced by the fusion of the cellulose with it. Cellulose-based bioplastics have many advantages over the conventional plastic like it is biodegradable by either aerobic and anaerobic means. This type of plastics have applications in packaging, biomedical, automotive, and in aeronautics industries [9].

20.3.3 Protein-Based

Protein-based bioplastics are in high demand these days in a variety of industries, and these proteins are primarily derived from animals (such as milk proteins, collagen, and gelatin), as well as plants (soybean, wheat gluten, and maize) [10, 11]. Protein-based bioplastic has high efficiency of degradation as comparison to the other polymers [12].

- Wheat gluten proteins are playing vital role in the formation of the film. These films are being produced by arranging the fine layer of the preotic solution and dry it with the help of the sulfite. The alternate ways of producing biofilm from the protein is boiling of the protein solution as it results the skin, which can be used as biofilm or by using the turbulent agent [13].
- By boiling nonfat milk, milk proteins can be used as a bioplastic in the form of a film for packaging [14].
- Collagen and gelatin are obtained from the frozen meat and both have wide range of applications like in medical, pharmaceutical industry and in the pipe industry [15].

20.3.4 Bioderived Polyethylene

- Rather than fossil fuels, polyethylene is made from the fermentation of basic agricultural products like sugarcane and corn.

- Bioderived polyethylene has application in many fields like in the formation of crates, tray, bottles of milk, toys, and in the making of industrial bulk containers.

20.3.5 Aliphatic Polyesters

Aliphatic biopolymer is a biodegradable polymer and there are various factors that affect its physical properties like availability of polar groups, molecular mass, crystallinity, orientation and the flexibility of chains [16]. The examples of aliphatic polymers are PLA, PHB and PGA. These biopolymers are being widely used in the pharmaceutical technology and in the medicine.

20.4 Applications of Bioplast

Bioplastic is the game changer for this modern world and does wonders in different areas of life due to its unlimited benefits over the conventional plastics as it is highly sustainable, biodegradable, biocompatibility, renewable, environment friendly, and have long shelf life in contrast to normal plastics. There are many applications of the bioplastics, such as in medical field, pharmaceuticals, drug delivery, in agriculture, packaging industry, in construction, bioremediation, in the production of the biofuels, and many others.

20.4.1 Medical Applications

Bioplast, especially PHAs, has significant role in the medical field in comparison to others. These bioplastics are involved in the engineering of the cardiovascular tissues, treatment of the fractures, micro encapsulation, radio potentiator, neural tissue engineering, urological stunts, for the diagnosis of narcolepsy, and also, it has anticancer ability [17–19]. PHAs and other bioplastics are being used for the treatment of cancer as this disease requires lots of money for its treatment. The benefit of using PHAs for the treatment of the cancer is that it is biocompatible; formed by the biomaterials due to this, it coordinates with the biological system really well and shows the great output of host [20]. It is expecting that the PHAs will do miracles in the biomedical area in the near future according to the FDA [21, 22]. FDA has also revealed that PHB is very helpful in the discovery of cancerous cells because the normal cells have less bounding capacity to the PHB as compared to the cancerous cells. Contact angle technique enables us to know about this property of PHB. In contrast to biopsy, the diagnosis of cancerous cells by using PHB is a more time-efficient and nonabrasive technique.

20.4.2 Wound Dressing Application

Wound dressing is really important for the treatment of the skin wound either it is big or small. The dressing that we use for the treatment of the wound should have properties like microbial resistance, less harmful to the wound, sustainability of the moisture content, water resistance, absorbs discharge coming out from the wound, etc. To get all these qualities, biobased polymers incorporated with the electro spun nanofibers are best for the wound dressing, and it is the emerging field for research in the future [23].

20.4.3 Drug Delivery Application

Bioplast are being used for the targeted delivery of the drugs. PHB is one of the bioplastic that has a great ability for the transmission of the drug to its actual location because of its good elasticity and the biocompatibility but in contrast to this it has some drawbacks like it is chromophore, and the time for deterioration is less, which makes it less suitable [24]. This problem is sorted out by mixing chitosan with the PHB, and this mixture is used in the tissue engineering of cartilage and in the dressing that are used after surgeries [25, 26]. Bioplastic in the form of the ligand is the new journey in the drug delivery, for example, the ligand of the polypeptide combined with the PhaP and with the nanoparticles of the PHA. This PhaPPHA ligand is involved in the treatment for cancer [27]. The other example is PHA synthase, which is combined with the nanoparticles of PHB that is inserted in the antibody consisting of single chain for the diagnosis of the cancer of colon. Nanoparticles of PHBHHx are also used in the proper transfer of insulin, and this has a long-term effect [28].

20.4.4 Agricultural Applications

The application of bioplast is much less in agriculture as compared to the industrial and the medical field. Only the net of plastic and the plastic substances, such as greenhouse films, are used in the agriculture sector, and these films and net plastics are important to maintain the standard of the crop, to enhance its production and to shield the crops against the pathogens [29, 30]. Generally, in agriculture, conventional polymers are being used, which are made from the petroleum source, such as high-density polyethylene (HDPE). It has great strength of elasticity. As HDPE is petroleum-derived plastic, it has many concerns like it is not easy to degrade and it is the serious threat for the animals as well as for environment. To overcome this problem, bioplastics and their blend are used for this purpose. PHA blends are used to make plastic nets more precisely, such as a blend of PLA/PHA and 4 hydroxy butyrate, and its elastic strength is higher than HDPE (25–1000%) [31].

In agriculture, plastic bags, also known as seedling bags, are often used in nurseries for the early growth of plants.

These bags are not biodegradable, so they are being replaced by biobased polymers that are less hazardous to the environment and easier to breakdown. The example of such bioplast is low density polyethylene (LDPE) because it maintains the water and the moisture level of soil and also maintains the temperature of soil for the better growth of the plants [32].

Mulch films are of two types: traditional and plastic-based, and both serve an important role in agriculture. Conventional films have some drawbacks like it needs high number of components for its production and they are not good to stop the growth of unwanted plants. The basic use of these mulch films are to sustain the soil content, keep the water and the moisture of the soil at equilibrium, and to stop the growth of unwanted plants [33].

20.4.5 3D Printing

The cost of PHA production is very expensive so fillers are added during the manufacturing process to help alleviate this difficulty. The most common filler that is added in the PHA is lignin, which is used in the formation of clay minerals and wooden floors [34]. Lignin that is used for the fillers are derived from the industries and it has replaced the petroleum-based additives. Lignin-based fillers do not require any kind of premodifications, for example, residues obtained from the corn, bagasse and *Arundo donax* [35]. The mechanism of blending lignin does not require complex modifications. It is used for the 3D printing after the trimming of PHB-lignin blend.

20.4.6 Applications in Packaging Industry

Packaging industry is one of the biggest industries in which the plastics are being used in a high amount. Normally, most of the packaging industries are using plastics that are produced from petrochemicals, and these plastics are nonbiodegradable, which cause environmental pollution.

Food industries need plastics for the packaging of the food to sustain the shelf life of the food and to conserve it without any contamination. There are many petrochemical-based plastics that are extensively being used for food packaging like polypropylene, polyethylene and poly-amide. These three plastics are being used widely in the formation of milk, oil, water and tea containers. The problem of using conventional plastics is that the harmful toxins from these plastics enter into the food products and deteriorate the quality of the food. Therefore, scientists are interested to develop the biobased plastics that have nontoxic effect and are also environment friendly.

Among all the biobased plastic, PHB, which is the polymer of PHA, is widely used in the food packaging. The existence of 4HB linear chain, poly3-hydoxybutyrate-co-4-hydroxybutyrate P(3HB-co-4HB), has exceptional toughness and processing capabilities among PHBs. The advantages of these polymers are that the linear chain lowers down the transparency, melting temperature and they have great elasticity. Nondegradable plastics are now converted into biodegradable plastics by using PHB films, and these films are reported in the packaging of salmon to keep it frozen. PHB film in the packaging of food has a great importance because it has great elasticity strength, high oxygen permeability (OP), high radical scavenging rate, high water vapor transmission rate (WVTR), and antioxidant and antibacterial characteristics [36].

Table 20.1 Bioplastics in food packaging.

Packaging application	Bioplastic	Advantages
Frozen salmon	PHB films	Great elasticity strength, high oxygen permeability, antioxidant and antibacterial properties
Fresh cut fruits and vegetables	Nano-coatings, nano-clays, nano-laminates based biopolymer	Long shelf life, ecofriendly, biodegradable, ameliorate the growth of pathogenic microorganisms, toxin detection
Milk chocolates	Corn-based biopolymer	Enhance the physical characteristics of packaging material, maintain the quality of the food, maintain the texture of the food
Potato chips	Metalized cellulose film	Maintain the moisture content, long shelf life, exceptional toughness

Nanoparticles are used in many fields due to their remarkable properties, such as large surface area, magnetic property, and ecofriendly. In the food industry, these nanoparticles are added into the bioplastics or biobased films for maintaining the quality of food. Nanoparticles are added into the dispersed polymer molecules to ameliorate barrier properties [37]. Some of the bioplastics used in the food packaging and their advantages are listed in Table 20.1.

20.4.7 Bioremediation Applications

Bioplastics are excessively used for the treatment of the wastewater. Moving bed biofilm carrier (MBBC) is one of the prominent examples of bioplastics that are being used for the bioremediation of the wastewater. It has ability to treat the heavy metals and the toxic pollutants that are present in the industrial and domestic wastewater. These carrier molecules are inoculated in the microorganisms that have ability to remove or to degrade the toxic and harmful substances from the wastewater. Accinelli *et al.* [38] reported the confiscation of the monomer bisphenol that is used in the plastic industry in a huge quantity and atrazine that is present in the agriculture wastewater.

20.4.8 Biofuel Applications

Biopolymer is gaining attention in the production of the biofuels, which are more ecofriendly as compared to the other polymers. 2G biofuel, which is generally obtained from the lignocellulosic biopolymer, is an example of such biopolymer. Industrial and municipal waste is the major source of plastic, which is not easy to degrade and cause serious damage to the environment. This problem is solved by pyrolysis.

Feedstock like biopolymers (cellulose, lignocellulose) and nonbiopolymers (polyethylene, polypropylene and polystyrene etc.) are used in the pyrolysis for the production of biofuel. It is the thermochemical process in which waste plastic is changed into the hydrocarbon fuel like oils and gas without any production of harmful by product. There are two ways of converting the waste plastic into the biofuels, which include SMUDA process and thermofuel process. The biofuels obtained from these processes are really good for the transportation and electricity purposes [39] (Figure 20.3).

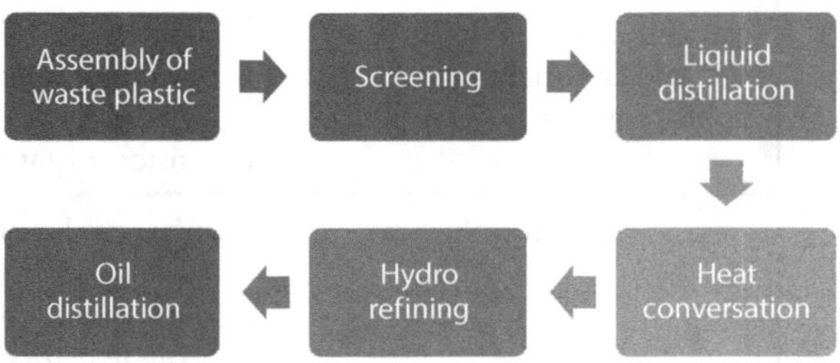

Figure 20.3 Production of biofuel from bioplast.

20.5 Conclusion

This chapter includes how plastics are being used worldwide in a large quantity and its comparison with the conventional plastics as these conventional plastics are not good to use because they are not ecofriendly and are less economical. Bioplast is gaining the attention of the researchers due to their biodegradable properties, low cost, and their production from renewable resources, like agricultural waste, food waste, cellulose, starch, etc. Major applications of bioplast are discussed in this chapter, but further studies are required to find its applications in many other fields.

References

1. Lagaron, J.M. and Lopez-Rubio, A., Nanotechnology for bioplastics: Opportunities, challenges and strategies. *Trends Food Sci. Technol.*, 22, 611, 2011.
2. Rahmatiah Al Faruqy, M.S. and Liew, K.C., Properties of bioplastic sheets made from different types of starch incorporated with recycled newspaper pulp. *J. Trans. Sci. Technol.*, 3, 257, 2016.
3. Thakur, S., Chaudhary, J., Sharma, B., Verma, A., Tamulevicius, S., Thakur, V.K., Sustainability of bioplastics: Opportunities and challenges. *Curr. Opin. Green Sustain. Chem.*, 13, 68, 2018.
4. Pilla, S., *Handbook of bioplastics and biocomposites engineering applications*, vol. 81, John Wiley & Sons, United Kingdom (UK), 2011.
5. Gill, M., Bioplastic: A better alternative to plastics. *Int. J. Res. Appl. Nat. Soc. Sci.*, 2, 115, 2014.
6. Ali, A., Yu, L., Liu, H., Khalid, S., Meng, L., Chen, L., Preparation and characterization of starch-based composite films reinforced by corn and wheat hulls. *J. Appl. Polym. Sci.*, 134, 45159, 2017.
7. Asaf Kleopas, S., *Synthesis and properties of starch based bio-materials*, Univ. Groningen, Groningen, 2008.
8. Ali, A. *et al.*, Preparation and characterization of starch-based composite films reinfoced by polysaccharide-based crystals. *Compos. Part B Eng.*, 133, 122, 2018.
9. Yaradoddi, J.S. *et al.*, Biodegradable carboxymethyl cellulose based material for sustainable packaging application. *Sci. Rep.*, 10, 1, 2020.
10. Pommet, M., Redl, A., Morel, M.-H., Guilbert, S., Study of wheat gluten plasticization with fatty acids. *Polym. (Guildf)*, 44, 115, 2003.
11. Cuq, B., Gontard, N., Guilbert, S., Proteins as agricultural polymers for packaging production. *Cereal Chem.*, 75, 1, 1998.
12. Jerez, A., Partal, P., Martínez, I., Gallegos, C., Guerrero, A., Protein-based bioplastics: Effect of thermo-mechanical processing. *Rheol. Acta*, 46, 711, 2007.
13. Gontard, N. and Guilbert, S., Bio-packaging: Technology and properties of edible and/or biodegradable material of agricultural origin, in: *Food Packaging and Preservation*, p. 159, Springer, United States, 1994.
14. Mabesa, R.C., Marshall, R.T., Anderson, M.E., Factors influencing tenacity of dried milk films exposed to high humidity. *J. Food Prot.*, 42, 631, 1979.
15. Cavallaro, J.F., Kemp, P.D., Kraus, K.H., Collagen fabrics as biomaterials. *Biotechnol. Bioeng.*, 43, 781, 1994.
16. Albertsson, A.-C. and Varma, I.K., Aliphatic polyesters: Synthesis, properties and applications, in: *Degrad. Aliphatic Polyesters*, p. 1, 2002.
17. Steinbüchel, A. and Füchtenbusch, B., Bacterial and other biological systems for polyester production. *Trends Biotechnol.*, 16, 419, 1998.

18. Angelova, N. and Hunkeler, D., Rationalizing the design of polymeric biomaterials. *Trends Biotechnol.*, 17, 409, 1999.
19. Ha, C.-S. and Cho, W.-J., Miscibility, properties, and biodegradability of microbial polyester containing blends. *Prog. Polym. Sci.*, 27, 759, 2002.
20. Williams, D.F., On the nature of biomaterials. *Biomaterials*, 30, 5897–5909, 2009.
21. Chen, G.-Q., A microbial polyhydroxyalkanoates (PHA) based bio- and materials industry. *Chem. Soc. Rev.*, 38, 2434, 2009.
22. Sabarinathan, D., Chandrika, S., Venkatraman, P.V., Easwaran, M., Sureka, C.S., Preethi, K., Production of polyhydroxybutyrate (PHB) from Pseudomonas plecoglossicida and its application towards cancer detection. *Inform. Med. Unlocked*, 11, 61, 2018.
23. Chen, S., Liu, B., Carlson, M.A., Gombart, A.F., Reilly, D.A., Xie, J., Recent advances in electrospun nanofibers for wound healing. *Nanomedicine*, 12, 1335, 2017.
24. Wang, A. *et al.*, Piezoelectric nanofibrous scaffolds as *in vivo* energy harvesters for modifying fibroblast alignment and proliferation in wound healing. *Nano Energy*, 43, 63, 2018.
25. Sadeghi, D., Karbasi, S., Razavi, S., Mohammadi, S., Shokrgozar, M.A., Bonakdar, S., Electrospun poly (hydroxybutyrate)/chitosan blend fibrous scaffolds for cartilage tissue engineering. *J. Appl. Polym. Sci.*, 133, 44171, 2016.
26. Amini, F., Semnani, D., Karbasi, S., Banitaba, S.N., A novel bilayer drug-loaded wound dressing of PVDF and PHB/chitosan nanofibers applicable for post-surgical ulcers. *Int. J. Polym. Mater. Polym. Biomater.*, 68, 772, 2019.
27. Yao, Y.-C. *et al.*, A specific drug targeting system based on polyhydroxyalkanoate granule binding protein PhaP fused with targeted cell ligands. *Biomaterials*, 29, 4823, 2008.
28. Peng, Q., Zhang, Z.-R., Gong, T., Chen, G.-Q., Sun, X., A rapid-acting, long-acting insulin formulation based on a phospholipid complex loaded PHBHHx nanoparticles. *Biomaterials*, 33, 1583, 2012.
29. Niaounakis, M., *Biopolymers: Applications and trends*, William Andrew, Norwich, New York, 2015.
30. Guerrini, S., Borreani, G., Voojis, H., Biodegradable materials in agriculture: Case histories and perspectives, in: *Soil Degradable Bioplastics for a Sustainable Modern Agriculture*, pp. 35–65, Springer, New York, United States, 2017.
31. Williams, S.F., Rizk, S., Martin, D.P., Poly-4-hydroxybutyrate (P4HB): A new generation of resorbable medical devices for tissue repair and regeneration. *Biomed. Tech. Eng.*, 58, 439, 2013.
32. Adane, L. and Muleta, D., Survey on the usage of plastic bags, their disposal and adverse impacts on environment: A case study in Jimma City, Southwestern Ethiopia. *J. Toxicol. Environ. Health Sci.*, 3, 234, 2011.
33. Rydz, J., Sikorska, W., Kyulavska, M., Christova, D., Polyester-based (bio) degradable polymers as environmentally friendly materials for sustainable development. *Int. J. Mol. Sci.*, 16, 564, 2015.
34. Bittmann, B., Bouza, R., Barral, L., Diez, J., Ramirez, C., Poly (3-hydroxybutyrate-co-3-hydroxyvalerate)/clay nanocomposites for replacement of mineral oil based materials. *Polym. Compos.*, 34, 1033, 2013.
35. Angelini, S., Cerruti, P., Scarinzi, G., Malinconico, M., Extraction and fractionation of a lignocellulosic biomass and its use as a bio-filler in poly (3-hydroxybutyrate). *Cellul. Chem. Technol.*, 50, 429, 2016.
36. Ma, Y., Li, L., Wang, Y., Development of PLA-PHB-based biodegradable active packaging and its application to salmon. *Packag. Technol. Sci.*, 31, 739, 2018.
37. Silvestre, C., Duraccio, D., Cimmino, S., Food packaging based on polymer nanomaterials. *Prog. Polym. Sci.*, 36, 1766, 2011.

38. Accinelli, C., Saccà, M.L., Mencarelli, M., Vicari, A., Application of bioplastic moving bed biofilm carriers for the removal of synthetic pollutants from wastewater. *Bioresour. Technol.*, 120, 180, 2012.
39. Saha, S., Sharma, A., Purkayastha, S., Pandey, K., Dhingra, S., Bio-plastics and biofuel: Is it the way in future development for end users?, in: *Plastics to Energy*, pp. 365–376, Elsevier, Amsterdam, Netherlands, 2019.

21

Bioplastic-Based Nanocomposites for Smart Materials

Marya Raji[1], Abdellah Halloub[2], Abou el Kacem Qaiss[1] and Rachid Bouhfid[1]*

[1]Composites and Nanocomposites Center (CNC), Moroccan Foundation for Advanced Science, Innovation and Research (MAScIR), Rabat Design Center, Rue Mohamed El Jazouli, Madinat El Irfane, Rabat, Morocco

[2]Université Mohammed V de Rabat, Faculté des Sciences, Laboratoire de matériaux Nanotechnologies et Environnement (LMNE), Rabat, Morocco

Abstract

Bioplastics have unique qualities and may be produced from biosubstances, such as cellulose, proteins, and starch, as well as among crop waste and industrial waste. Because of environmental issues, replacing conventional plastics with bioplastics is a top goal, and it is even better if they are made directly from agricultural waste. Moreover, the use of bioplastics in smart materials start attracting researchers' attention, where smart materials technology permits us to overcome environmental issues by applying its functions. This chapter aims to a brief definition of the different kinds of bioplastics, fields application, main manufacturing technics, and finally, the application of bioplastics as smart materials.

Keywords: Bioplastics, biosubstances, manufacturing technics, smart materials

21.1 Introduction

Bioplastics are strong materials that are being studied to replace petroleum-based traditional plastics to make them more sustainable. They are produced from renewable resources and maybe recycled organically through biological processes, saving limited natural resources and lowering CO_2 gas emissions. Bioplastics are being used in a variety of industries, including food packaging, composting bags, agriculture, and hygiene [1]. Three types of biobased materials are often used: Polymers directly removed from biomass, including chitosan, starch, lignin, hemicellulose, and cellulose, as well as proteins like zein, soy, gluten, and lipids. A second family of biodegradable and/or renewable polymers, such as thermoplastics and thermosets made from vegetable oils, combines biomass-derived monomers with traditional chemical synthesis processes to get the final biodegradable and/or regenerative polymers. The last category is polymers formed via natural or genetically

**Corresponding author*: r.bouhfid@mascir.ma

Inamuddin and Tariq Altalhi (eds.) Handbook of Bioplastics and Biocomposites Engineering Applications: 2nd Edition, (457–470) © 2023 Scrivener Publishing LLC

altered micro-organisms for example polyhydroxyalkanoates (PHA) and polypeptides like the elastin-like polymers.

Even though bioplastics have a lot of promise to replace petroleum-based materials and reduce pollution, they still have a lot of property and processing issues that prevent them from being used in a lot of applications, especially in the packaging industry. The causes for that are usually attached to their weaker gas and vapor barrier qualities, high water sensitivity, lesser heat resistance, shorter shelf-life stability owing to aging, migration, as well as a diversity of processability concerns still related to the usage of bioplastics. Polymer blends and composites are often explored to enhance the characteristics of bioplastics and substitute a huge amount of unsuited versatile polymers, where inorganic reinforcement (e.g., calcium carbonate, montmorillonite, kaolinite, sepiolite, talc), natural fibers (both wood and plant fibers), and other kinds of fillers, such as graphene nanosheets, carbon nanotubes, have all been examined for polymer composites or biocomposites.

The development of novel biocomposite materials that are smart or intelligent, versatile, more survivable, adaptable, and environmentally friendly has garnered significant attention in the last decades and has become a vital research topic for all scientific disciplines ranging from technology to engineering [2]. The concept of smart and adaptive structures was created in the mid-1980s as a new age of the greatest intriguing category of materials with large-scale systems. These intelligent products have a wide range of properties and behaviors, including self-alterable dynamic properties, shape memory, self-healing, self-sensing, pH-sensitive, thermosensitive, and self-adaptability, and are sensitive to any exterior stimuli in their environments, such as moisture, light, pressure, pH, electric, temperature, stress, or magnetic field [3–5]. Smart products, which will have the capacity to be active, clever, sharp, trendy, and sophisticated, may now be used in many aspects of human life and technology. Furthermore, a lot of researchers have looked at the usage of their potential in many technical applications that might be advantageous in the whole of the world.

21.2 Biopolymer

21.2.1 Natural Polymers

a. Cellulose

Cellulose is a nonbranched polymer made up of repeated units of D-glucose linked together by glycosidic bonds. In vegetables, green plants, algae, and some bacteria, cellulose is the main structural component of cell walls [6]. water-insoluble, resistant to enzymes digesting, and it may be subjected to fermentation in the presence of colonic bacteria [7]. Cellulose as biocomposites shows a low WVB (water vapor barrier) property due to their hydrophilic nature, which limits their application in food packaging. Due to its hydrophilic nature, cellulose as biocomposites has low WVB qualities, which restricts its use in the field of food packaging [8].

b. Chitosan

Chitosan is one of the notable biopolymers with interesting characteristics, including biocompatibility and biodegradability, and is designated as a "generally regarded as safe" material from the Food and Drug Administration in the United States. Chitosan is a positively charged linear polysaccharide that contains a random combination of -(1–4)-linked

D-glucosamine and N-acetyl-D-glucosamine. Chitosan is a chitin derivative that is formed when chitin is deaminated. Furthermore, the amino groups found on the chitosan surface allow it to interact chemically and physically with nanoparticles, medicines, polymers, and cells [9]. This material is being studied for a variety of biological applications since it is biocompatible, biodegradable, non-toxic, and has wound healing and antibacterial capabilities [10]. Chitosan has a natural antibacterial property that prevents bacteria and fungus from growing. Chitosan treatment, for example, produces structural modifications in the named membrane–wall complex into Staphylococcus aureus cultures, resulting in bacterial destruction and damage of surface cell structures [11]. As result, the biocomposites based on chitosan had antibacterial properties, low water barrier, low oxygen permeability, and strong mechanical qualities.

c. Starch

Green plants manufacture starch in two organelles, which is the primary storage carbohydrate [12]. Starch is a homopolysaccharide with glucose as its fundamental unit. It is classified as a biopolymer since it comes from nature and is made of two primary substances: amylopectin and amylose [13]. Amylose is an important lineal polymer that conforms to the amorphous areas of starch granules by joining glucose units with -(1–4) bonds and having few branching sites [14]. Amylopectin, however, is a branched component in which glucose units are connected by -(1–4) bonds in lineal sections and -(1–6) bonds at branching sites [14]. Understanding the physicochemical and functional properties of starch is necessary for finding appropriate uses for it in diverse applications. To determine the optimum uses for native and modified starches, it is required to have a fundamental understanding of some of their qualities, such as pasting profile, gelatinization, retrogradation, and rheological characteristics.

d. Hemicellulose

Hemicellulose is a side-chained heteropolymer made up of xylans, alternating mannose and glucose units (mannans or glucomannans), or galactose units (galactans) [15]. For different feedstock sources, the number of sugar monomers might vary substantially. Hemicelluloses are classed as xylans, mannans, or glucans based on the major sugar that remains in the backbone [16]. Several classes of hemicellulose can be detected depending on the plant type, developmental phase, and tissue character [11]. According to the fibers hydration, these various subtypes may be divided into two groups. Hemicelluloses are components that may be precipitated by ethanol after being extracted from the cell wall with a dilute alkali solution. They are extracted using an aqueous solution and calcium chelator when the pectin present in the walls has been depleted [17].

e. Lignin

Lignin, the second most prevalent organic component in plants, accounts for 30% of all non-fossil organic carbon in nature [18]. It is also the natural raw material with the highest aromatic moieties concentration. Guaiacyl lignins and guaiacyl-syringyl lignins are the two major types of lignin. Softwoods have guaiacyl lignin, whereas hardwoods contain guaiacyl-syringyl lignin. Guaiacyl lignin is largely made up of units of coniferyl alcohol, but guaiacyl-syringe lignin has monomeric units made up of both sinapyl and coniferyl alcohols [19]. Despite the massive volumes of lignin generated each year, the lignin applied as a renewable material for the production of high-value chemicals and consumer goods is still in its infancy. Basic and applied lignin research will be a major issue for academics and companies in the next years, especially with a future without oil on the horizon [18].

The development of effective and selective catalysts for lignin transformation into aromatic compounds with realistic yields for the petrochemical sector and improved utilization of lignin will most likely be the most significant task for the next few years.

21.2.2 Synthetic Polymers

a. Polylactic acid

PLA (polylactic acid) is a biodegradable and recyclable polyester derived from renewable feedstock. LA (lactic acid) is produced from sucrose or glucose fermentation and purified to a higher purity [20]. Polylactic acid has been utilized in food packaging development, textiles, and more recently engineering polymers. The basics of lactic acid manufacturing, PLA characteristics, production method, producers, conversion procedures, and market potential are all covered. PLA is currently a niche product with significant development potential [21, 20]. Transparency, high mechanical strength, strong barrier qualities, low toxicity, biocompatibility, and processability are among PLA's distinctive characteristics, making it a viable contender for food packaging applications. PLA blends make use of PLA's consistent miscibility with other polyesters. Most PLA mixes are utilized for film materials with a PLA percentage that is somewhat lower than the polyester component. By modifying the blend components or the quantities of other blend components or other conceivable materials components, the qualities of a PLA polyester mix may be greatly altered. On top of that, there are the PLA's changeable qualities.

b. Polycaprolactone (PCL)

PCL is a biodegradable polyester with a Tg (glass transition temperature) of approximately –60°C and a low melting temperature of around 60°C [22]. PCL is made via ring-opening polymerization of -caprolactone with stannous octanoate as a catalyst. In physiological environments (such as those found in the human body), PCL is destroyed by hydrolysis of its ester bonds, and as a result, it has attracted a lot of interest as a biomaterial for implantation [23]. It is particularly interesting for long-term implanted device preparation. For controlled release and targeted medication delivery, several medicines have been enclosed within PCL beads [24]. To get an excellent biodegradable material at a reasonable cost, PCL is frequently combined with starch [25].

c. Poly(propylene carbonate)

The result of propylene oxide and CO_2 alternating copolymerization is poly(propylene carbonate) (PPC). For the reason that the PPC ester structure on its backbone, PPC is a biodegradable amorphous polymer. Various ways for synthesizing PPC with propylene or oxide phosgene were known, which was inconvenient due to safety concerns [26]. Zinc carboxylate catalysts have been used to copolymerize propylene oxide with CO_2 to produce high molecular weight (HMW) PPC [26]. Transition metal complexes have been produced for the copolymerization of carbon dioxide with epoxides; however, they have yet to be extensively used and/or optimized in the development of better PPC materials. It has been the subject of much research, and several companies have looked at using it as a commodity thermoplastic. PPC has only been marketed as a sacrificial polymer in applications that benefit from PPC's clean thermal degradation. PPC has high compatibility with impact resistance, and it may be melt treated in the same way as commercial polyethylene. However, poor thermal and processing qualities have hampered its development for thermoplastic applications.

21.3 Biopolymer-Based Nanocomposites

Several reviews on bioplastic composites have been published in recent years, addressing various aspects of production, processing, characteristics (physical, mechanical, thermomechanical, functional, etc.), and applications. The science of filler reinforced composites has revealed that the filler size has an important effect on the composite's properties [27]. This is because the filler size affects the surface interactions, adhesion, filler motion, dispersion, bonding, and other properties. Because some of these effects become more pronounced as filler size decreases, they have a greater impact on nanoscale properties [28]. Nanoscale-filled bioplastic composites with filler less than 100 nm and an aspect ratio of at least 100 have been developed to overcome the constraints of typical micrometer-scale bioplastic composites. It is well known that there is a small interphase layer around the fillers in bioplastic matrix composites in which the plastic chains have a different conformation than their bulk counterparts [29]. In addition, there is a small interphase layer around the fillers in matrix bioplastics in which the polymer chains have a different conformation than their bulk counterparts [30, 31].

There are many methods for fabricating nanoscale filled reinforced bioplastics depending on the final applications including solvent casting, melt compounding, thermoforming… etc. The excellent dispersion/distribution of nanofiller into the bioplastic matrix, the type of the bioplastics, and the production procedures utilized all influence the attributes of developed nanofiller bioplastic. The development of those bioplastics has been the subject of several investigations. The solvent casting method (see Figure 21.1) was invented over a century ago and is still the oldest plastic technology [32]. After that, alternative bioplastic composite-making techniques took over, reducing the necessity of casting-evaporation. However, because of the exceptionally high-quality requirements of the generated biocomposites, this technology is gaining popularity. Furthermore, it is a widely utilized biocomposite-making technique in the literature [33, 34]. The biocomposites produced using this method are notable for their transparency, uniformity, and simplicity of demolding. Furthermore, the volatile nature of the solvents utilized in this approach is a distinguishing feature. The evaporation phase of the solvent should be slow in order to create a nice surface free of damage and air bubbles [35]. The intense connection between the nanofiller might settle during this phase, promoting the creation of a stiff percolating network.

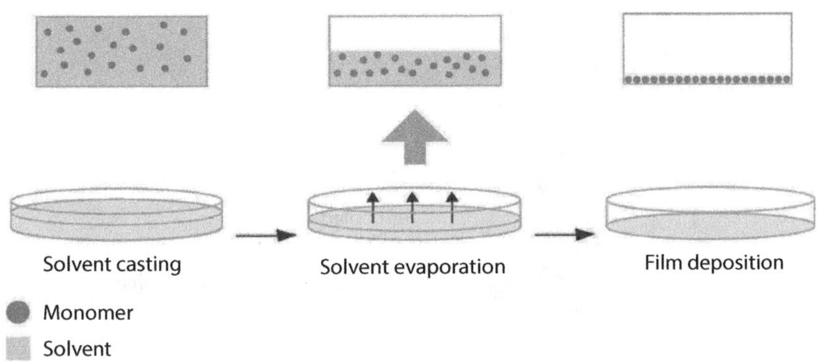

Figure 21.1 Preparation of films via solvent casting method.

Moreover, the biocomposites packed with nanoparticles are multiphase solid materials that come in a variety of morphologies (cast, flat film, and blown film) and are made up of distributed nanoparticles in a bioplastic matrix to improve the overall characteristics of plastic materials [36–38]. The properties of bioplastic materials can be shaped in sheets or films, and they are greatly influenced by the manufacturing process as well as the processing parameters. Extrusion, calendering, injection molding, compression molding, and other methods are used to process bioplastics reinforced with nanoparticles on an industrial scale.

Otherwise, calendaring (Figure 21.2) is a high-precision, continuous technique for producing large quantities of rolled sheets or films with the desired thickness and appearance. This method is used to make bioplastics, as well as some modified thermoplastics, flooring, and applying a plastic coating on paper, textiles, and other supporting products [39]. In this process, either the Banbury mixer which is an enclosed mixing chamber, and two-roll mills are employed to discharge the bioplastic materials, this step is called pre-blending and in which the mixing of bioplastic materials with other ingredients as like plasticizers, stabilizers, lubricants, pigment, and fillers is typically performed either in a continuous strip or batches, then a large melt compounding extruder is utilized for fluxing and the extrudate is generally fed directly to the calendar chamber. Calendars are made up of two or more counter-rotating cylinders or horizontal rolls, also known as bowls, in which a continuous film of hot plastic materials is squeezed between heated rollers of various sizes rotating at slightly different speeds before being stripped off the last calendar roll with a small, faster-stripping roll. The heated film is air-cooled as it passes through a succession of cooling drums before being sliced into individual sheets or twisted into a continuous roll. The characteristics of the bioplastic films are influenced by the temperature and speed of the rollers. V. L. D. [40].

In addition, Injection molding is one of the main methods for producing bioplastic components of various sizes, complexity, and applications. This procedure was simple and consisted of three phases [41]. The bioplastic pallets are first placed in a hopper, which then feeds them to a heated injection device, where the bioplastic ingredients are softened and made fluid. A reciprocating screw is applied to force bioplastic materials through a ram extruder or a screw extruder. Finally, the molten bioplastic is driven through a nozzle at

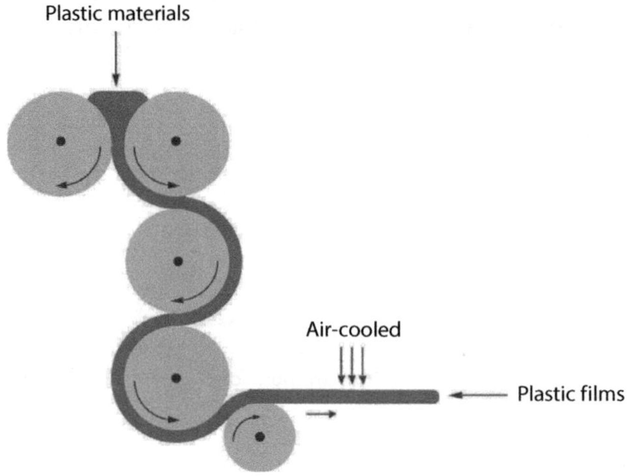

Figure 21.2 Preparation of films via calendaring method.

the chamber's end into a closed mold, which is usually kept shut by a clamping mechanism. The mold walls are already cold at this point, allowing the plastic portion to solidify and cool. When the mold is opened, the ejection system automatically removes the required biocomposites, and the procedure is repeated [42]. Further, because of its cheap cost and high volume manufacturing, the compression molding method is extensively utilized to create big biocomposites films [41]. Pressure and heat are used to compress the bioplastic granules into the desired form in this procedure. The bioplastic materials are initially put between two hold mold faces, and the mold is subsequently closed using a hydraulic press.

21.4 Bioplastics-Based Nanocomposites for Smart Materials

The use of bioplastics nanocomposites in smart materials was improved in several material types, such as pH-sensitive bioplastics, thermosensitive bioplastics, antimicrobial bioplastics, and thermochromic bioplastics, those materials are paired with many applications (see Table 21.1) to improve quality of life like packaging materials, biomedical applications, textile, 3D printing... Furthermore, they had unavoidable good health, environmental, and socioeconomic benefits, that might be magnified even further by implementing intelligent functionalities through renewable biomaterials and reliable technology [2]. Intelligent polymers, responsive polymers, sensitive polymers, stimuli-sensitive polymers, stimuli-responsive polymers, environmentally sensitive polymers, and environmentally responsive plastics are all names for these one-of-a-kind macromolecules [43]. In detail, the stimuli are classified into three groups; physical comprise temperature, electric/magnetic fields, and light; and chemical stimuli contain pH and ionic strength, and biochemical stimuli include enzymes and antigens. Depending on the use, these parameters have been manufactured in a variety of shapes. Polymer chains dispersed in fluids, chemically crosslinked

Table 21.1 Examples of the stimuli-responsive applications.

Types	Applications	Ref.
Chemical Stimuli-Responsive	Ionic-Strength Responsive	[45]
	Redox-responsive	[46]
	Reactive Oxygen Species (ROS) Responsive	[42, 47]
	pH-Responsive	[48]
Physical Stimuli-Responsive	Light stimuli-responsive	[49]
	Temperature-Responsive	[50]
	Magnetic Field-Responsive	[51]
Biological Stimuli-Responsive	Glucose Responsiveness	[52]
	Photomodulation	[52]
	Hydrolytic/pH-Triggered Degradation	[53]

hydrogels, physical gels, micelles, and chains fixed or grafted onto solid surfaces are all examples. Furthermore, regardless of their physical structure, intelligent biopolymers can be conjugated with biological or synthetic substances, and the activity of the conjugates will be determined by polymer-conjugate interactions and the polymer's reaction to the applied stimulus. Several smart polymers have been created to integrate two or more mechanisms of responsiveness in a single material in order to improve and expand their usefulness [44]. A popular method for creating smart polymers requires mimicking their properties in synthetic systems to get a modification in one property over an external trigger.

21.5 Physical Stimuli-Responsive Biopolymer

Temperature and light are two interesting examples of physical stimuli. Temperature-sensitive materials are amongst the most researched stimuli-responsive materials over several decades as a result of their capacity to adapt properties, such as solubility, conformation, and hydrophobic/hydrophilic balancing, thus offering the ability for a varied choice of application fields [54]. Thermoresponsive polymers may react to outside temperature gradients by changing certain of their macroscopic characteristics significantly: when the biopolymers' transition temperature is close to the usage temperature, they can be employed as smart systems. This sort of system has a specific solution temperature (usually in water) at that the phase of the polymer and solution changes depending on their composition [55, 56]. Systems with one phase overhead a precise temperature and phase separation under that temperature have an upper critical solution temperature (UCST). Polymer solutions that seem monophasic below a certain temperature and biphasic above it, on the other hand, often display the so-called lower critical solution temperature (LCST). Thermoresponsive polymers have gotten a lot of interest recently since their characteristics can be controlled by changing molecular properties and their enhanced conversion processes. A thermo-sensitive biopolymer, physically cross-linked injectable hydrogel was established for the effective and sustained administration of disulfiram to cancer cells, as there is currently no hydrogel formulation available. Because of the unique qualities of hydrogels, which are three-dimensional structures made of diverse polymers, they offer an advantage over other drug delivery methods. Bioplastics were proposed to develop an injectable hydrogel system for the sustained problem of cancer cells. Because of their thermoresponsive characteristics, cellulose derivatives are being researched. Lee *et al.* explored the thermosensitive gelation of aqueous hydrophobically modified MC (HM-MC) solutions for its potential applications in the pharmaceutical and biological areas [43]. Aqueous solutions of ethyl(hydroxyethyl) cellulose (EHEC) display thermoresponsive activity, and in situ–forming structures may be generated by adding ionic surfactants (e.g., sodium dodecyl sulfate, SDS). EHEC/surfactant systems have been suggested for the administration of local anesthetic agents to the periodontal pocket, as well as for the delivery of nasal and ocular drugs.

21.6 Chemical Stimuli-Responsive Biopolymers

Among chemically sensitive drug-delivering methods, pH and oxidation/reduction potential shifts are two frequent stimuli used to cause abrupt changes in a material's characteristics.

The variation in local pH levels between distinct tissues and cellular compartments has repeatedly catalyzed building pH-responsive delivery vehicles for extracellular or intracellular medication administration [57, 58]. The reaction of natural polymers to pH variations is a widely researched smart behavior that they may display; pH-responsive materials are of particular interest because of the variances in pH that generally exist or that may arise in the body as a result of physiological or pathological processes. Consumer preferences and the need for safe and high-quality food goods have resulted in new inventions and alterations to many types of packaging materials. Packaging is significant in the food sector because it protects the product from environmental impacts, interacts with the customer as a marketing tool, gives increased ease of use and timesaving convenience to the consumer, and includes items of varied sizes and forms [59, 60]. A food product's quality and safety might be compromised at numerous points throughout distribution and transportation. As a result, any mismanagement of the food product can have a substantial influence on its safety and overall quality, particularly if the package's integrity is not preserved. The sensors based on bioplastics have at least one biobased component, which might be the substrate or the element that had the sensing properties. The majority of bioplastics sensors applied in the food packaging industry target colorimetric analyte detection. Similar, the anthocyanins and curcumin are familiar as pH indicators, the benefit of those molecules is their large presence in wide flowers, vegetables, and fruits. Choi *et al.* (2017), for example, created a pH sensor consisting of starch and agar extracted from sweet potato, where the revealed color fluctuations were at pH of 2–10 after the ad of anthocyanin extracts from potato purple. Moreover, Zhai *et al.* [61] employed the matrix of gelatin-gellan including red radish anthocyanin with a pH range of 2-12. Although colorimetric pH-sensitive sensors are inconvenient for selective analysis, the change in pH is frequently enough to assess food quality, as degrading proteins create alkaline volatile nitrogen molecules. Redox-sensitive nanoparticles are especially intriguing since cancer cells have greater amounts of glutathione (GSH) than healthy ones. The use of bioplastics films as a support for different active substances (antibacterials) for example natural extracts which may include throughout the package production procedure is known as active food packaging. Antimicrobial inclusion might result in a material having antibacterial action, preventing germs from growing on the surface. The antimicrobial substance present in the bioplastic composites may partially travel or completely via a slow diffusion into the food surface or drugs, the thing that supports the protection from the microbes. In some cases, they prefer to not migrate and only act when they are in direct contact with the food surface [62]. Suryanegara *et al.* [63] studied the antimicrobial propriety into the bioplastic based on PLA-chitosan-TiO2 and PLA-chitosan- ZnO. They found that PLA-chitosan-ZnO bioplastic offered good stability in the environment present bacteria, such as Gram-positive, Gram-negative, *Aspergillus niger*, and *Candida albicans*.

21.7 Biological Stimuli-Responsive Biopolymers

Macromolecules which react to biological stimuli may be useful for a variety of applications, such as drug and gene delivery, imaging, sensing, and biomimicry, as synthetic membranes. Physical and chemical reactions in macromolecules can be induced by biochemical stimuli, such as sol-gel and gel-sol transitions, triggered assembly or disassembly, structural rearrangement, and surface functionalization [64]. The benefit of adding biochemical stimuli-induced reactions into a molecular design is that they have the potential to be biomedically specific, especially when paired with a second cooperative biochemical stimulus or complementing physical or chemical stimuli. The biological stimuli bioplastics are also useful for tissue engineering (TE) as a smart application. Tissue engineering is based on the use of chemical, biological, and engineering concepts in order to repair, regenerate, or restore live tissue by employing cells, biomaterials, and components separately or in combination [65]. The biomaterial scaffold supports tissue regeneration in tissue engineering (TE) by providing mechanical and porous network structural support, form, and hierarchy architecture with surface chemistry enabling cell attachment, cell-cell communication through the porous network, proliferation, and differentiation. To date, the majority of synthetic biomaterials used in TE are generated based on lactic acid (LA), caprolactone monomers to make poly(L-Lactide) (PLL), poly(L-glycolic acid), and/or their combination to introduce co-polymers or physical mixing of these synthetic polymers. Chitosan, alginate, hyaluronic acid, starch, collagen, cellulose, and derivatives of these polymers are employed [65]. To manage the forms, sizes, porosities, and topologies of biomaterial scaffolds for TE applications, a range of methodologies and procedures have been developed. Gene delivery, drug delivery, aimed at treating numerous genetic illnesses, and tissue engineering (TE) to renew or substitute biologically injured or unhealthy tissue or produce substituted organs for a varied variety of medical conditions, such as heart diseases, diabetes, cirrhosis, osteoarthritis, and disfigurement, are just a few of the biomedical applications of smart biopolymers that can respond to different external stimuli [66]. The capacity to produce very precise multifaceted structures that may distort in reply to varied conditions is one of the benefits of combining 3D printing and shape-memory polymers. As a result, diverse methodologies may be applied to manufacture more biomimetic and multi-functional biomedical goods, which will have a wide range of possible uses in the biomedical industry in the future.

21.8 Conclusion

The potential of using biopolymers as smart materials is based on the properties used in polymers and the manufacturing process. For this reason, by integrating contemporary materials science design methodologies into biomaterials science, new materials with specialized and crucial features have been generated. This chapter summarized several advantages that may be offered by biopolymers concerning availability, variety, adaptability, and compatibility. Furthermore, it is the most known technique used for producing bionano composites at nanoscales. And finally, a brief presentation of some of the smart applications that have made advantage of a wide range of diverse structures and materials which were developed to easily interact with one another via unique mechanisms.

References

1. S. Pilla, *Handbook of Bioplastics and Biocomposites Engineering Applications*. Salem, Wiley-Scrivener, 2011.
2. Bouhfid, R. and Jawaid, M., *Polymer Nanocomposite-Based Smart Materials*. Oxford, Woodhead Publishing, 2020.
3. Roy, S. and Rhim, J.W., Anthocyanin food colorant and its application in pH-responsive color change indicator films. *Crit. Rev. Food Sci. Nutr.*, 61, 1, 2297–2325, 2020.
4. Tang, J., Quinlan, P.J., Tam, K.C., Stimuli-responsive pickering emulsions: Recent advances and potential applications. *Soft Matter*, 11, 3512, 2015.
5. Pucci, A., Bizzarri, R., Ruggeri, G., Polymer composites with smart optical properties. *Soft Matter*, 7, 3689, 2011.
6. Butardo Jr., V.M. and Sreenivasulu, N., Tailoring grain storage reserves for a healthier rice diet and its comparative status with other cereals, vol. 323. Philippines, Elsevier Inc, 2016.
7. Kalász, H., Báthori, M., Valkó, K.L., *Basis and pharmaceutical applications of thin-layer chromatography*. Budapest: Elsevier B.V., 2020.
8. De Moura, M.R., Mattoso, L.H.C., Zucolotto, V., Development of cellulose-based bactericidal nanocomposites containing silver nanoparticles and their use as active food packaging. *J. Food Eng.*, 109, 520, 2012.
9. Agama-acevedo, E. and Perez, L.A.B., *Cereal starch production for food applications*. Yautepec, Elsevier Inc., 2018.
10. Revuri, V., Huh, K.M., Lee, Y., *Polysaccharide based nano/microformulation: An effective and versatile oral drug delivery system*. Daejeon, Elsevier Inc., 2017.
11. El Bourakadi, K. *et al.*, Chitosan/polyvinyl alcohol/thiabendazoluim-montmorillonite bio-nanocomposite films: Mechanical, morphological and antimicrobial properties. *Compos. B Eng.*, 172, 103–110, 2019.
12. Bukit, N., Frida, E., Harahap, M.H., Preparation natural bentonite in nano particle material as filler nanocomposite high density poliethylene (Hdpe), *IISTE Int. Journals*, 3, 13, 2013.
13. Jawaid, M. and Kumar, S., *Bionanocomposites for Packaging Applications*. Gewerbestrasse, Springer International Publishing, 2018.
14. Pfister, T.B. and Zeeman, S.C., Formation of starch in plant cells, *Cell. Mol. Life Sci.*, 73, 2781–2807, 2016.
15. Brunner, G., *Processing of Biomass with Hydrothermal and Supercritical Water*, 5. Hamburg, Elsevier B.V., 2014.
16. Frassoldati, A. and Ranzi, E., *Modeling of Thermochemical Conversion of Biomasses*. Milano, Elsevier Inc., 2019.
17. Gregory, A. and Bolwell, G.P., *Hemicelluloses*. Egham, Elsevier Ltd., 1999.
18. Calvo-Flores, F.G., Group, M.D., Sciences, F., Lignin as renewable raw material. *ChemSusChem* 3, 1227–1235, 2010.
19. Ghalia, M.A. and Dahman, Y., *Synthesis and utilization of natural fiber-reinforced poly (lactic acid) bionanocomposites*. Toronto, Elsevier Ltd., 2017.
20. L. T. Sin, *Polylactic Acid*, vol. 10. Malaysia, William Andrew, 2012.
21. Ibarra, V.G., Sendón, R., De Quirós, A.R., *Antimicrobial Food Packaging Based on Biodegradable Materials*. Lugo, Elsevier Inc., 2016.
22. Tang, X., Thankappan, S.K., Lee, P., Fard, S.E., Harmon, M.D., *Polymeric Biomaterials in Tissue Engineering and Regenerative Medicine*. Farmington, Elsevier Inc., 2014.
23. Arakawa, C.K. and Deforest, C.A., *Polymer Design and Development*. Seattle, Elsevier Inc., 2017.

24. Miller, K., Hsu, J.E., Soslowsky, L.J., *Soslowsky, Materials in Tendon and Ligament Repair.* Philadelphia, Elsevier Ltd., 2011.
25. Endres, H. and Ifbb, B., *Performance Profile of Biopolymers Compared to Conventional Plastics,* vol. 10. Hannover, Elsevier B.V., 2012.
26. Niaounakis, M., *Definitions of Terms and Types of Biopolymers.* Elsevier Inc., 2015.
27. Raji, M., Essabir, H., Rodrigue, D., Bouhfid, R., Qaiss, A.E.K., Influence of graphene oxide- and graphene nanosheet on the properties of polyvinylidene fluoride nanocomposites. *Polym. Polym. Compos.*, 39, 2932–2941, 2017.
28. Raji, M., Essabir, H., Essassi, E.M., Rodrigue, D., Bouhfid, R., Qaiss, A.E.K., Morphological, thermal, mechanical, and rheological properties of high density polyethylene reinforced with Illite clay. *Polym. Compos.*, 39, 1522, 2016.
29. Raji, M., Mekhzoum, M.E.M., Rodrigue, D., El, A., Qaiss, K., Bouhfid, R., Effect of silane functionalization on properties of polypropylene/clay nanocomposites, *Compos. Part B Eng.*, 146, 106–115, 2018.
30. Raji, M., Qaiss, A.E.K., Bouhfid, R., Effects of bleaching and functionalization of kaolinite on the mechanical and thermal properties of polyamide 6 nanocomposites. *RSC Adv.*, 10, 4916, 2020.
31. Mekhzoum, M.E.M., Raji, M., Rodrigue, D., Qaiss, A.E.K., Bouhfid, R., The effect of benzothiazolium surfactant modified montmorillonite content on the properties of polyamide 6 nanocomposites. *Appl. Clay Sci.*, 185, 105417, 2020.
32. Siemann, U., Solvent cast technology - A versatile tool for thin film production, in: *Progress in Colloid and Polymer Science*, vol. 130, p. 1, 2005.
33. Govindasamy, K., Ramli, M.H., Pasbakhsh, P., Pushpamalar, V., Salamatinia, B., Chitosan/cellulose/halloysite membranes produced using solvent casting method. *Polym. Polym. Compos.*, 23, 325, 2014.
34. Sanchez-Garcia, M.D., Gimenez, E., Lagaron, J.M., Morphology and barrier properties of solvent cast composites of thermoplastic biopolymers and purified cellulose fibers. *Carbohydr. Polym.*, 71, 235, 2008.
35. Oksman, K., Aitomäki, Y., Mathew, A.P., Siqueira, G., Zhou, Q., Butylina, S., Tanpichai, S., Zhou, X., Hooshmand, S., Review of the recent developments in cellulose nanocomposite processing. *Compos. Part A Appl. Sci. Manuf.*, 83, 2, 2016.
36. Essabir, H., Raji, M., Laaziz, S.A., Rodrigue, D., Bouhfid, R., Thermo-mechanical performances of polypropylene biocomposites based on untreated, treated and compatibilized spent coffee grounds, *Compos. Part B*, 149, 1–11, 2018.
37. Sana, A., Laaziz, A., Raji, M., Hilali, E., Essabir, H., Rodrigue, D., Bouhfid, R., Bio-composites based on polylactic acid and argan nut shell: Production and properties. *Int. J. Biol. Macromol.*, 104, 30, 2017.
38. Essabir, H., Raji, M., Bouhfid, R., el Kacem Qaiss, A., *Rheological Properties of Functionalized Graphene and Polymeric Matrices-Based Nanocomposites.* Rabat, Elsevier Inc., 2018.
39. S. H. Goodman, *Handbook of Thermoset Plastics.* California, Noyes Publications, 2000.
40. Costa, V.L.D., Costa, A.P., Amaral, M.E., Oliveira, C., Gama, M., Dourado, F., Simo, R.M., Effect of hot calendering on physical properties and water vapor transfer resistance of bacterial cellulose films. *J. Mater. Sci.*, 51, 9562, 2016.
41. Denn, M.M., *Polymer melt processing: Foundations in fluid mechanics and heat transfer.* New York, Cambridge University Press, 2008.
42. Pia, W.-K., Hannes, O., Jarmo, R., Injection molding of thermoplastic cellulose esters and their compatibility with poly (lactic acid) and polyethylene. *Materials (Basel).* 11, 2358, 2018.

43. Altomare, L., Bonetti, L., Campiglio, C.E., De Nardo, L., Draghi, L., Tana, F., Farè, S., Biopolymer-based strategies in the design of smart medical devices and artificial organs. *Int. J. Artif. Organs*, 41, 337, 2018.
44. Halonen, N., Pálvölgyi, P.S., Bassani, A., Fiorentini, C., Nair, R., Spigno, G., Kordas, K., Bio-based smart materials for food packaging and sensors – A review. *Front. Mater.*, 7, 1, 2020.
45. Tao, X., Ting, L., Wei-Feng, Z., Cheng-Sheng, Z., Ionic-strength responsive Zwitterionic copolymer hydrogels with tunable swelling and adsorption behaviors. *Langmuir, 35*, 1146–1155, 2018.
46. Guo, X., Cheng, Y., Zhao, X., Luo, Y., Chen, J., Yuan, W.E., Advances in redox - responsive drug delivery systems of tumor microenvironment. *J. Nanobiotechnology*, 16, 74, 2018.
47. Yu, L.-Y., Su, G.-M., Chen, C.-K., Chiang, Y.-T., Lo, C.-L., Specific cancer cytosolic drug delivery triggered by reactive oxygen species-responsive micelles, *Biomacromolecules*, 17, 3040–3047, 2016.
48. Kocak, G., Tuncer, C., Bütün, V., pH-responsive polymers. *Polym. Chem.*, 8, 144, 2017.
49. Wang, M., Wang, T., Wang, D., Jiang, W., Fu, J., Acid and light stimuli-responsive mesoporous silica nanoparticles for controlled release. *J. Mater. Sci.*, 54, 6199–6211, 2019.
50. Zheng, Y. and Zheng, S., Reactive & functional polymers poly (ethylene oxide)-grafted poly (N-isopropylacrylamide) networks : Preparation, characterization and rapid deswelling and reswelling behavior of hydrogels. *React. Funct. Polym.*, 72, 176, 2012.
51. Filipcsei, G., Csetneki, I., Szilágyi, A., Zrínyi, M., Magnetic field-responsive smart polymer composites. *Oligomers - Polym. Compos. - Mol. Imprinting*, 206, 137–189, 2007.
52. Kim, H., Kang, Y.J., Jeong, E.S., Kang, S., Kim, K.T., Glucose-responsive disassembly of polymersomes of sequence-specific boroxole-containing block copolymers under physiologically relevant conditions. *ACS Macro Lett.*, 1, 1194–1198, 2012.
53. Jun, Z., Song, S., Zong, M., Li, C., Dual-modal tumor imaging via long-circulating biodegradable core-cross-linked polymeric micelles. *NIH Public Access*, 1, 150–153, 2013.
54. Seeboth, A., Lötzsch, D., Ruhmann, R., Muehling, O., Thermochromic polymers-function by design. *Chem. Rev.*, 114, 3037, 2014.
55. White, M.A. and Leblanc, M., Thermochromism in commercial products. *J. Chem. Educ.*, 76, 1201, 1999.
56. Seeboth, A., Kriwanek, J., Vetter, R., The first example of thermochromism of dyes embedded in transparent polymer gel networks. *J. Mater. Chem.*, 9, 2277, 1999.
57. Li, G., Liu, G., Kang, E.T., Neoh, K.G., Yang, X., pH-responsive hollow polymeric microspheres and concentric hollow silica microspheres from silica-polymer core-shell microspheres. *Langmuir*, 24, 9050, 2008.
58. Chemin, M., Beaumal, B., Cathala, B., Villares, A., pH-responsive properties of asymmetric nanopapers of nanofibrillated cellulose. *Nanomaterials*, 10, 1, 2020.
59. Krämer, M., Stumbé, J.F., Türk, H., Krause, S., Komp, A., Delineau, L., Prokhorova, S., Kautz, H., Haag, R., pH-responsive molecular nanocarriers based on dendritic core-shell architectures. *Angew. Chemie Int. Ed.*, 41, 4252, 2002.
60. Mei, L., Xie, R., Yang, C., Ju, X., Wang, W., Wang, J., Chu, L., pH-responsive Ca-alginate-based capsule membranes with grafted poly (methacrylic acid) brushes for controllable enzyme reaction. *Chem. Eng. J.*, 232, 573, 2013.
61. Zhai, X., Li, Z., Zhang, J., Shi, J., Zou, X., Huang, X., Zhang, D., Sun, Y., Yang, Z., Holmes, M., Gong, Y., Povey, M., Natural biomaterial-based edible and pH-sensitive films combined with electrochemical writing for intelligent food packaging. *J. Agric. Food Chem.*, 66, 12836, 2018.
62. Brockgreitens, J. and Abbas, A., Responsive food packaging: Recent progress and technological prospects. *Compr. Rev. Food Sci. Food Saf.*, 15, 3–15, 2016.

63. Suryanegara, L. *et al.,* Novel antimicrobial bioplastic based on PLA-chitosan by addition of TiO2 and ZnO. *J. Environ. Heal. Sci. Eng.,* 19, 415–425, 2021.
64. Lim, H.L., Hwang, Y., Kar, M., Varghese, S., Smart hydrogels as functional biomimetic systems, *Biomater. Sci.,* 2, 603–618, 2014.
65. Bártolo, P.J., Chua, C.K., Almeida, H.A., Chou, S.M., Lim, A.S.C., Biomanufacturing for tissue engineering: Present and future trends. *Virtual Phys. Prototyp.*, 4, 203, 2009.
66. Hollinger, J.O., Einhorn, T.A., Doll, B.A., Sfeir, C., *Bone Tissue Engineering,* vol. 18, p. 1, Raton, CRC Press, 2004.

Part III

INDUSTRIAL APPLICATION, SUSTAINABLITY AND RECYCLING OF BIOPLASTICS

22

Applications of Biobased Composites in Optical Devices

Reshmy R.[1,2], Vaisakh P.H.[2], Eapen Philip[2], Parameswaran Binod[3], Aravind Madavan[4], Mukesh Kumar Awasthi[5], Ashok Pandey[6,7] and Raveendran Sindhu[3,8]*

[1]*Department of Science and Humanities, Providence College of Engineering, Chengannur, Kerala, India*
[2]*Post Graduate and Research Department of Chemistry, Bishop Moore College, Mavelikara, Kerala, India*
[3]*Microbial Processes and Technology Division, CSIR-National Institute for Interdisciplinary Science and Technology (CSIR-NIIST), Trivandrum, India*
[4]*School of Biotechnology, Amrita Vishwa Vidyapeetham, Amritapuri, Kollam, Kerala, India*
[5]*College of Natural Resources and Environment, Northwest A & F University, Yangling, Shaanxi, China*
[6]*Centre for Innovation and Translational Research, CSIR- Indian Institute for Toxicology Research (CSIR-IITR), Lucknow, India*
[7]*Centre for Energy and Environmental Sustainability, Lucknow, Uttar Pradesh, India*
[8]*Department of Food Technology, T K M Institute of Technology, Kollam, Kerala, India*

Abstract
Biocomposites can effectively be contributed to the development of optical devices to design sustainability, biocompatibility, flexibility, and sensing ability. Many biocomposite-derived optical biosensors have been technologically advanced in the recognition of biomarkers in malignancies, such as lung cancer, breast cancer, lymphoma cancer, and various other disorders including celiac disease and Alzheimer's disease. This chapter discusses in detail about the possibilities and challenges of three different types of biopolymers, namely polysaccharides, protein/polypeptides, and polynucleotides in the development of optical devices for versatile applications.

Keywords: Polysaccharides, polypeptides, polynucleotides, biocomposite, biosensor

22.1 Introduction

Nanoparticles (NPs) with enhanced mechanical, electrical, and optical capabilities, as well as potential durability, bioactivity, and survivability, can be used to pave the way for future contemporary technology gadget development. This multifunctional device is not only robust, flexible, and compressible but also chemically detecting, actuation, in addition to smart shielding of gas, humidity, and radiation [1–4]. These nanomaterials' improved biocompatibility make them great candidates for a wide range of demanding nanotechnology

**Corresponding author*: sindhurgcb@gmail.com; sindhur@tkmit.ac.in

Inamuddin and Tariq Altalhi (eds.) Handbook of Bioplastics and Biocomposites Engineering Applications: 2nd Edition, (473–490) © 2023 Scrivener Publishing LLC

interaction and biomedical applications, as of wearable biosensors to well-designed implants in addition to body security [5].

Environment is an imaginary resource for creating the next generation of useful bionanomaterials, since it provides novel designs and durable building blocks. Spiders produce silk thread having flexibility that surpasses the hardest commercial carbon and other fibers. These networks are similarly quite sticky, which serves as a secondary purpose of catching prey. Natural wood or bamboo, one on either hand, provide an outstanding mix of hardness and strength that is seldom found in synthetic materials and constructions [6–9]. Understanding how natural material's processes and design elements result in remarkable qualities can contribute to the development of increased bioderived chemically synthesized nanocomposites with innovative functionality and performance.

The monomeric unit composition of relevant biopolymers covered in this study may be split into three classes. They are polysaccharide-, protein-, and polynucleotide-based biocomposites. Three frequent polysaccharides bound by glycosidic links are cellulose, chitin/chitosan (CS), and alginate (ALG), while silk and collagen remain characteristic proteins made up of extended chains of amino acids. The biopolymers are biodegradable and come from natural origin, such as skin, plants, silkworm cocoons, hair, and spider webs [10]. Deoxyribonucleic acid (DNA) holds genetic codes are polynucleotides that direct the development, activity, and reproduction of living organisms, as well as numerous diseases.

Biophotonic application enabling chemical characterization and disease detection includes optical biosensing. Chemical or biological factors create optical reactions in optical biosensors, which signify a functional state or concern. The functionalization of biological materials to an optical device has significant impacts on the selectivity and sensitivity of optical signals. Optical biomaterials were used to identify biomarkers in different cancers, as well as illnesses, such as Alzheimer's disease and celiac disease. Moreover, biophotonics stimulates the growth of light-activated therapies, which has shown promise as a usage for a range of medical complications. Photodynamic therapy (PDT) is a broad-spectrum treatment method designed for conditions including endodontics and malignancies, like skin cancer, breast cancer, leukaemia cancer, and head and neck cancer that employs photosensitizing chemicals in conjunction through light to destroy tumor tissues, germs, viruses, and fungi [11]. Biophotonics, on the other hand, has been addressed as one of the best promising biomedical tools due to its ability to attain its objectives. Because optical qualities, mechanical features, structural properties, and biological capacities all have a significant influence on system performance, the accessibility of appropriate optical resources is important for realizing the full potential of biophotonic technology. A critical need for designing optical devices is to deliver effective transmission of light with minimum side effects during application.

Silicate-based glass is still the most often used optical materials framework today. Silica-based tools offer admirable optical properties, such as high transmittance across a whole wide wavelength, including visible to near infrared, permitting materials to also be employed in a variety of optical sensors [12]. The material's mechanical fragility and softness, on the other hand, provide a risk of injury to biological tissues, making it unsuitable for many biological applications. Biological applications, on the other hand, demand constituents with higher biocompatibility plus biological activities that are stimulating to produce through standard silica materials. Because implanted medical devices should ideally be dissolved and removed after usage, nondegradability is another key hurdle to *in vivo* use of optical systems based on silica. This involves the implementation of biocompatible optical materials to substitute silica. Innovative biocompatible polymers through flexible method, good optical productivity, adjustable dreadful conditions, and plan adaptability are mandatory to realize the full possibility of biophotonics, as well as generate the next age band of optical detecting, imaging, and therapeutic devices [13]. Because of recent advances in biopolymers, several biophotonic devices by required features in addition capabilities for a variety of medical claims have been produced. Biological components, such as bacteria, have been proven to be used in nature-inspired optical systems [14]; however, these systems have limited supplies, are difficult to build, and are inherently variable. Due to these restrictions, natural or synthetic polymer-based optical materials have been developed with more design freedom, allowing for a wider range of physical, chemical, and biological characteristics and functions. This chapter aims to facilitate an overview of the recent advancements in the development of optical devices from different types of bioderived materials.

22.2 Characteristics and Advantages of Biobased Composites in Optical Devices

Optical materials are required for the fabrication of optical elements, such as waveguides and lenses for biophotonics applications, including optical imaging, light-activated treatment, and optical sensing. Optical materials are those that have the capacity to alter or transform electromagnetic radiation in the visible, ultraviolet (UV), and infrared (IR) spectrum ranges. Biophotonic materials must possess several optical, biological, chemical and mechanical characteristics [15–17]. The specific characteristics of biocomposites for optical device applications are depicted in Figure 22.1.

The refractive index and the level of transparency, as well as their spectrum dependence, are frequently the most crucial parameters in the selection of an optical material. High transparency materials feature low light reflection, absorption, and scattering, all of which contribute to reduced optical loss. Light scattering can occur in silica fibers due to atomic

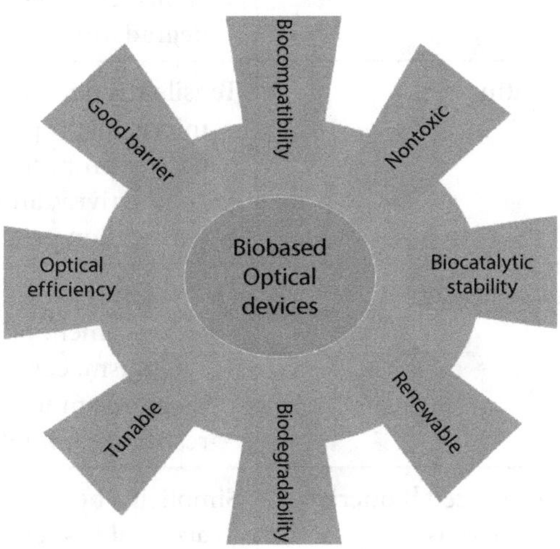

Figure 22.1 Specific features of biomaterials for optical device applications.

structural defects, which diminishes with increasing wavelength. Silica fibers with low optical loss are extensively used for long-haul fiber-optic communications due to technological breakthroughs in glass fabrication with long wavelength light around 1550 nm. The disordered network of polymeric chains and macromolecules generates in homogeneity in polymeric biomaterials [18]. Moreover, linked polymers are typically not considered for optical device applications because organic compounds display significant absorption at shorter wavelengths around 350 to 400 nm, and certain linked structures ought to absorption peaks in the visible spectrum. The optical loss of frequently manufactured transparent biopolymers ranges from 0.3 to 1 dB/cm, which is considerably greater than that of silica glass. Optical natural polymers, regardless of their optical loss, may be able to meet the distance requirements in some biomedical devices. The refractive index of a materials, which would be described as the fraction of the velocity of light happening vacuity to the phase velocity of light in a substance or medium, has an impact on optical loss. In single-material light waves,

Table 22.1 Advantages of various types of biocomposites for optical device applications.

Biocomposites	Types	Advantages	References
Polysaccharide	Cellulose	High strength, stiffness, renewability	[22, 23]
	Chitin	Renewable, nontoxic, antibacterial polysaccharide	[35, 46]
	Alginates	Biocompatible, nontoxic, cost-effective properties	[36, 47]
Protein	Silk	Remarkable mechanical properties, biocompatibility	[13, 37]
	Collagen	Hydrophilicity, swelling ability, controlled degradation rate	[48, 49]
	Gelatin	Tensile strength, anti-microbial properties, barrier properties conductivity, anti-water permeation properties	[48, 50]
Polynucleotides and carbonized	DNA origami	Drug delivery for enhancement of synergism, cancer diagnosis multidrug resistance inhibition	[40, 41]
	Carbonized bioderived materials	Simplicity, better adhesion, acceptable specific strength properties	[43, 44]

the refractive index of the substance must stay larger than that of the neighboring tissue to accomplish total internal reflection. Other parameters, such as physical surface quality and winding degrees, may impact the optical loss of light directorial materials in addition to optical transparency and refractive index. In addition to their optical characteristics, biopolymers' physical-chemical and biological properties stay similarly vital f applications in biophotonics. Optically transparent natural polymers, as related to inorganic silica optical materials, are gaining traction as key optical materials due to their ability designate invented and synthesized with soft mechanical properties, biocompatibility, biodegradability, and required chemical functionalities. Polymeric optical materials may also be easily produced into complex forms that are highly efficient optically. As a consequence of the advantages stated above, researchers are increasingly focused on the creation of complex polymeric optical biomaterials. The advantages of different categories of biobased materials for optical device applications are summarized in Table 22.1.

22.3 Polysaccharide-Based Biocomposite

Morphological and storage polysaccharides are the two types of polysaccharides. Plant cellulose and crustacean chitin are examples of structural polysaccharides, whereas starch and glycogen are examples of storage polysaccharides. They are found in almost all living things, and it is roughly 70% of the dry weight of the total biomass. Although polysaccharide composites are more ecologically friendly than petropolymers, they still have substantial drawbacks that must be overcome in order to make them a viable option. The mechanical characteristics of polysaccharide are inferior to those of ordinary polymers. Some polysaccharides have a high degree of hydrophilicity, which can lead to premature rupture. As a result, polysaccharide composites have been thoroughly investigated in order to address this issue and get extra features for certain applications. Polysaccharide-based composites

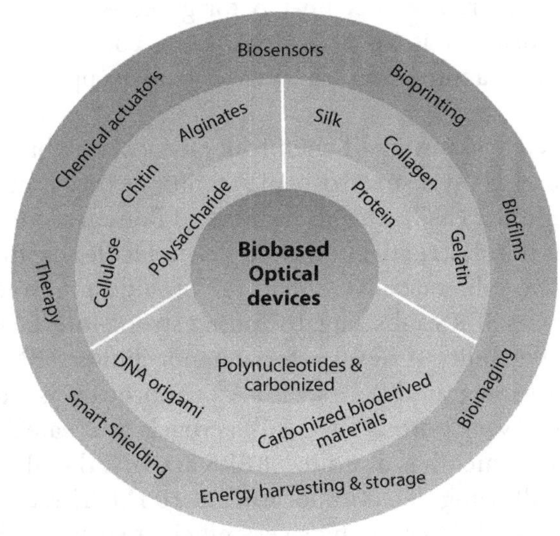

Figure 22.2 Pictorial representation of major biobased optical devices and its field of applications.

have become a current research area in the areas of materials and engineering due to its biocompatibility, renewability, and sustainability. The detailed pictorial representation of types of biomaterials utilized in the design of optical devices and its application in a number of divisions are obtainable in Figure 22.2. The authors aim to examine and analyze recent work in the development of polysaccharide-based composites for optical applications in this section.

22.3.1 Cellulose

Cellulose is a vital and plentiful renewable source for the production of novel products since it is a key constituent of plant cell walls and the most widely used natural polymers [19–21]. It is in the forms of microcrystalline cellulose, paper fibers, regenerated fibers, Kraft pulp fibers, hydrogels, transparent sheets, nanocrystals, and nanofibrillated cellulose (NFC), due to its large quantity and chemical and structural capabilities, has been widely used as a biopolymeric template for the fabrication of new luminescence biomaterials.

It has been employed in the creation of fiber optics and sensors due to its visible region transmittance and favorable permeability of several ions. The hydroxyl linkages of cellulose can be partly or completely treated with a variety of compounds to produce cellulose derivatives with beneficial qualities, like as cellulose acetate (CA), cellulose triacetate, and cellulose butyrate esters. To make a novel microstructured polymeric optical fiber from two categories of the degradable cellulose with varied glass transition temperatures, a spongy double-cladding arrangement with two cellulose butyrate tubes divided by hydroxypropyl cellulose units was used. The internal structure is a cellulose tube with a gap in the centre that may be deflated for laser transmission or kept open for medication administration. Optical, drug release, and microfluidic functions may all be combined with the nanostructured cellulose fibre [22]. In order to build optical probes, researchers used cellulose as a surface coating layer for microstructured polymeric optical fibre (MPOF). For example, the rhodamine 6G (Rh 6G)doped CA was applied to the sidewalls of the arrays pores in an MPOF to create a new nitrite (NO_2) detection probe. A liquid-phase coating was used to create another MOPF is altered with eosin-CA for gaseous ammonia fluorescence detection. The biosensor's sensitivity is linear across a large concentration range of 50 to 400 ppm, with a rapid 500-ms reaction time. This research opens up new avenues for optical gas sensing with MPOFs.

Optical and colorimetric scanning of chemical, physical and biological variations using fluorescence, Raman, and UV-vis spectroscopic techniques are among the optical systems constructed on nanocellulose (NC). Surface-enhanced Raman scattering (SERS), for example, may detect analytes nondestructively and single-molecule sensitively using noble metal nanomaterials as effective substrates. NC substrates provide a number of benefits over typical glass/plastic thin films SERS substrate, including strong interfacial contact with NPs, a large specific area, and flexibility. A new bacterial nanocellulose (BNC) thin films-based on SERS nanomaterials that integrated BNCs with Au nanorods was studied. Metal nanorods were also combined with very porous BNC/silk aerogels to create 3D plasmonic aerogels. Unlike the 2D substrate, which has a modest SERS activity, the 3D aerogel has a substantially better sensitivity, allowing detection of R6G at 10 fM. These bioplasmonic materials also showed additional capabilities, such as very efficient solar steam production via photothermal heating.

By reducing noble metal salts *in situ* inside NC materials, plasmonic nanomaterials for SERS may be produced. For example, AuNPs with a size of 60 nm can be created consistently all across the BNC matrix, by gently heating $HAuCl_4$ loaded BNC hydrogel in sodium citrate mixtures. The resultant composite nanogels is tightly filled with AuNPs and exhibits the two expanded adjustable localized surface plasmon resonance spectrum, indicating substantial interparticle interaction [23]. The highly spongy morphology is critical for permitting ensemble of plasmonic NPs that are currently impossible to achieve with conventional suspension-based SERS devices. Under such circumstances, conventional NP suspensions are likely to flocculate uncontrollably, and typical paper-based interfaces are not likely to retain their initial 3D structure [13, 14]. The porous structure of the swollen hydrogel ensures maximum sample solution adsorption on the AuNPs surface, while subsequent evaporation causes strong interactions between the plasmonic NPs to create active SERS hotspots [15].

Enhanced chiral optical nanostructures may be built using NC components as substrates. These exceptional chemical moieties could be used to make plasmonic NPs that coincide along the twisted NPs. At the single nanocrystal region, the resulting nanostructures have chiral plasmonic characteristics. Hence, the real implementation of this theory is more complicated. Cationic spherical AuNPs may be built on sulphate ester surface groups in practice. According to recent electron tomography and AFM studies, single cellulose nanocrystal (CNC) rods are twisting in the right-hand direction along their axial. The size of the NP has a strong influence on the chiral optical characteristics of the formed structures [26].

Quantum dots (QDs) are very fluorescent nanocrystals that have generated a lot of interest because of their potential applications like as bioimaging, optical biomaterials, biosensors, bioimaging, anticounterfeiting and biomarker development [27, 28]. To create translucent, smooth, and bright nanopaper materials, several types of QDs were integrated into NC surfaces. Carbon QDs, for example, can be covalently linked to NC and are nontoxic, nanoscale, luminous, and easy to modify the properties [17]. The water-soluble ZnSe QDs coated with positive charged polymers stayed applied to the cellulose nanopaper surfaces using vacuum aided filtration. The nanopaper has a good optical transmittance and high photoluminescent intensity as prepared. Due to longer chiral nematic ordering, CNC films exhibit iridescent structural color, which has significant promise for biooptical equipment like as displays and biosensors. The pitch of helical structures changes when they are exposed to diverse external conditions, resulting in structural color variations in CNC films. It has been claimed that by assembling CNCs with different hydrogel materials, a responsive nanocomposite hydrogel may be created [29]. These novel chiral nanogels have adjustable hues and may react to a variability of stimuli, for instance solvents, pH, as well as temperature, causing pitches fluctuation

Furthermore, when CNC elements are coupled with other NPs, chiral nematic frameworks generated by CNCs has attracted interest as either an LC pattern or a host for producing dynamic optical biomaterials. The orientation of Au nanorods co-dispersed in nematic phase changes as the CNC concentration rises. With the vertical and horizontal modes of Au nanorods individually stimulated, the resultant film revealed a different polarization-dependent extinction profile [30]. Composite films like this can have a variety of intriguing characteristics depending on the polymer component used. For example, because polyethylene glycol (PEG) seems to have a strong affinity for water molecules, CNC/PEG nanocomposites would absorb moisture differently [31]. More crucially, water penetration causes the PEG surface to inflate, changing to pitches size and, as a result, the color of the cholesteric structure, resulting in stimuli-responsive color shifts.

The polymers may also be used to create mesoporous photonic cellulose (MPC) films as a co-template. CNC involves assembling urea–formaldehyde (UF) precursors into a chiral nematic form, then curing the layering UF to form a resin. The polymeric matrix is then eliminated using a KOH etching technique, leaving a porous CNC structure with maintained nematic structure. Because the color change is totally reversible, it might be used for simple color-based pressure detection. The structural color of MPC film appears to be particularly sensitive to the solvent's chemical composition [32].

22.3.2 Chitin

Chitin is found in shrimps, crabs, and marine crustaceans, and stays the next most common biopolymer generated in nature later cellulose. Chitins constitute of repeating 1,4-N-acetyl glycosaminoglycan composition [33]. It can be found in the exoskeletons of arthropods as well as the yeasts and fungi cell walls, contains highly structured crystalline microfibrils that are used to create these structural components, which provide functional protection and stability [34]. The natural chitin source is processed with acid and alkaline to eliminate calcium carbonate and soluble protein constituents, respectively, to obtain pure colorless chitin. A decolorization process is used to remove any remaining pigments.

CS is a natural hydrophilic polysaccharide that is biocompatible and biodegradable. It enables reversible adsorption of water and excellent bonding to the both anionic and cationic versions of brilliant and transitions metal ions, modifying optical characteristics, because of the abundance of hydroxyl and amino functional groups [35]. CS is a viable choice for optoelectronic devices and sensors because of its exceptional film forming capabilities. A minimal Fabry–Perot biosensor for humidity was created using CS as a responsive layer on the surface of a materials. This biosensor's detecting the relative humidity (RH) and with good accuracy and consistency. Due to the CS film's small size, this has a rapid response time of 380ms and a sensitive of 0.13 nm/% of RH. It was used to create a waveguide that served as a sensing material in a RH sensor [24]. This integrated-optical sensors, which are constructed on waveguide films of the both salty and neutrality CS, have an excellent sensing and a rapid response time. The doping polymeric surfaces with the Ag^+ ions, accompanied "*in situ*" reductions and generation of AgNPs, could be used to modify the optical properties of the CS thin films. The doping level can be precisely controlled by changing the Ag^+ concentration of ions and doping duration. There was no trace of opacity in the thin films. The CS/Ag thin films with a higher index of refraction could be utilized for the design of planar optical waveguides (POW) and the optical sensing substrate. Moreover, $[AuCl_4]$ and Ag^+ ions was absorbed and reduced on 150nm thickness to spin-coated the CS thin films to produce CS/Au and CS/Ag nanocomposite films. Utilization of CS/Au and CS/Ag nanocomposites films as sensitivity coatings on POW is able to detect H_2S gas at rates. The detecting signal of the H_2S gas sensors was extremely sensitive, selective, and repeatable [25]. The biomedical implications of CS-based optoelectronic devices, because of its biocompatibility and biodegradability, CS has a great potentiality in the biophotonic applications.

22.3.3 Alginate

Alginate is a biopolymer with an anionic, which is obtained by chemical treatments of brown seaweed that is widely utilized in the biomedical field due to its high biocompatibility, inexpensive, and processability. It is a linear copolymer made up of blocks of (1-4)-linked β-D-mannuronate and α-L-guluronate molecules that are covalently attached in diverse patterns, and the chemical composition is extremely dependent on natural resources. Due to its abundance of carboxyl groups there in molecular chains, ALG is also water-soluble natural polymer that could be cross linking by the divalent cations like Ca^{2+} for development of the hydrogels [26].

ALGs have been investigated to be used in the fabrication of luminescent nanocomposites including lanthanide ions, as well as magnetic/luminescent components depending on the CdS and CdTe. The photoluminescent ALG aerogels containing Eu_{31},Tb_{31}, Tb_{31}/Eu_{31}, and an ALG Eu_{31}xerogel were fabricated. The dropping sodium ALG in trivalent solutions, steady gel particles were formed; alcogel then was acquired by the solvent exchange process by immersing hydrogel in the ethanol with significant reduction in volume; and at last, solvent was evaporated by air drying, xerogel was formed, with the significant volume reduction, fabricating a compact material, or by the supercritical CO_2 fabrication method [36]. The 5D quantum efficiency of the Eu_{31}–ALG aerogel and the alcogel has been the maximum, although the hydrogel and xerogel obtained lower values. The Tb_{31}/Eu_{31} composites are multi-wavelength emitters that can be controlled over the chromaticity spectrum from red to yellowish green spectral range. In a slightly different manner, a unique procedure integrating emulsification and self-assembly methods was successfully used to create highly magnetic luminous ALG-template composite NPs [28]. Finally, sodium ALG grafted copolymer with the poly(2-dimethylaminoethyl methacrylate) (PDMA) was employed in passivity of the layer of the colloidal CdS–thyoglycerol QDs. For thepolydentate ligand features of the CdS QDs were attributed to its surface interaction with ALG-g-PDMA. The polycrystalline QDs that emerged possessed improved emission properties [29]. Furthermore, the ALG nanocomposite are likely to be excellent materials for biomedical field implications like as bioimaging, immunoassay, and biosensing, as well as cancer detection and treatment.

22.4 Protein-Based Biocomposite

Proteins determine the texture and suppleness of human hair and skin. When the primary structures, like as protein chains, folded and intertwining, in the secondary structures with organized materials like as helices beta sheets, and turns, as well as amorphous coils are generated. Fibrous, globular, membrane, and disordered structures are the four most prevalent tertiary structures. Proteins that are disordered or globular likely to act as the catalysts /enzymes. Membrane proteins promote in the passage of substances between both the extracellular matrix and the carriers. Fibrous proteins provide tissue and bone with support and suppleness. Collagen fibres provide human skin its resilience and flexibility.

22.4.1 Silk

Silks are the natural fibrous proteins made by larvae of silkworms, spiders, moths, and butterflies. Their hydrophobic B-sheet arrangements gain a lot of hardness and mechanical properties. They could be created with biochemical functions owing to their structural designability and minimal aqueous process; they can remain paired by way of integrated dyes, growth regulators, peptides/cells [30]. It has remarkable manufacturing flexibility has contributed toward the growth of silk-based gels, thin films, and 3D skeletons for the biomedical applications [31]. Due to their low superficial roughness in addition to good transparent through the visual spectrum, silk thin films with a width of 20 to 100 m were shown to be suitable for optical components as well as biophotonic applications. It has been evolved addicted to waveguides diffraction gratings, and lenses, designed for light transmission also *in vivo* implanting uses during the last several decades. The nanopatterned silk optical devices, like as lenses, pattern generators diffraction gratings, and microlens arrays, and were produced by inducing crystallization of thin films on suitable substrates [32]. The surfaces structure of this optics could be adjusted with accuracy down to 125 nm over considerable length scales. The optical properties of films with thickness varying from 30 to 50 m were found to be compatible at 1.55. Silk thin films with thickness of 100 m were measured to have near-100 % transmission over the visible spectrum. Bioactive molecules, such as hemoglobin and peroxidase, were included during manufacturing, resulting in silk optical fibers with biocompatible, good optical efficiency and biological functionality. Silk fibroin-based nano/micro-patterned films were also created using a simple altered soft lithography method [33]. This technology enables the manufacture of optically clear and mechanically resilient films with a transversal design resolution of less than 40 nm, clearing the opportunities for sustainable development of extremely efficient biomaterial with the complex 2D and 3D nanopatterns/micropatterns. Furthermore, a low-cost, high-throughput nanoprinting technology for printing nanopatterned silk thin films in short durations with reduced manufacturing stress was designed. The autosensing optofluidic devices was developed using nanoprinted lysed red blood cell impregnated silk solution. Studies have recognized a method to create a unique implantable, durable, and biocompatible optical device integrating increased bioimaging of malignancies treatment and quantified based on silk [34] using advanced understanding and unique fabrication techniques [37]. This high-quality, versatile device exemplifies customized modern medical future generation idea and serves as a foundation for new medical equipment designs. The pure as well asrhodamine 6G-loaded micro silk optical waveguides through regulated shape and contents were constructed by direct-write fabrication of a determined form silk aqueous environment. The linear and wavy uncontaminated silk waveguides enclosed by air had optical losses [35]. The use of direct ink printing allows for the creation of complex biophotonic instruments with a variety of morphologies and bioactivities. Although silk elements have little optical losses in the air, the dealings of light accompanied by these light waves through varying as well as complex encircling tissues *in vivo* causes absorption and scattering optical losses, limiting the capacity to accomplish *in vivo* profound treatment of tissue, therapies, post-surgical checking employing single-material waveguides. The biocompatible step factor optical fibre with a silk inner and silk hydrogel coating was designed to solve this issue, through a transmission loss of roughly 2 dB/cm. The developed optical fiber's inner/coating construction minimized sensitivities to the local environment, and this method might be employed to transmit light at the organ dimension.

22.4.2 Collagen

The fibrous proteins contain different components, which implies various process conditions and results in various functionalities. The collagen is other prominent types of fibrous proteins. Despite containing various ingredients, they have analogous origins, extraction processes, and a diverse variety of the applications. It is the key structural protein in living things connective tissues and accounts for the tensile properties in cartilage, bones, blood vessels and other tissues, as well as fibrils and membrane proteins *in vivo*. Various kinds of collagen occur in the animals because the alpha chains can be the different/same. Collagen molecules generated keen on the extracellular medium arrange top to bottom in a staggered arrangement, placing particle bonding spots closer together. The age is linked to the site by the oxidative amination. The solubility is determined by its age, or cross linking amount. The freshly prepared collagen is soluble in salt solution because it is not cross linked and cross linked collagen is largely acid soluble and has a high degree of cross linking, it should be proteolysis by pepsin.

Collagen is widely utilized in tissue scaffolds as matrices. These are used as plasticizer to incorporate with gelatin and CS to generate flexible, biodegradable composite films. Because of the abundance of aromatic groups within gelatin, it can act as a UV light barrier [36]. The collagen and gelatin have limited applications, and their use outside of scaffolding and capsules is unique in comparison to silk fibroin. Furthermore, with much more time, its usability should broaden to fulfil the commercial and research needs of silk fibroin. They are biocompatible and lightweight, just like silk. Their proline residue serves as binding or loading site for optically, biologically, or electrically active components. They were largely employed in tissue regeneration scaffolds. Moreno *et al.* reported the collagen's applicability in the flexible electronics as an e-beam to deposit electrodes straight on the collagen, showing the durability of collagen over silk, which has electrodes made differently and then placed onto the film [37].

22.4.3 Gelatin

The solvent casting was used to make gelatine-based nanocomposite thin films with Ag_2O NPs. After that, Ag NPs were created by reducing Ag_2O NPs with a femtosecond laser. The light emitting rates of the laser-activated area far surpassed the luminescence intensity of the similarly triggered simple Ag_2O NPs films. The dimensions of the Ag NPs were found to be dependent on the laser period of exposure, resulting in the creation of metallic Ag regions in the nanocomposite material. The nanocomposites materials had superior mechanical characteristics and the luminescent nanostructures in the matrix remained stable over time, indicating that they could be used by means of light-emitting components in entirely optical logic systems and data storage devices. Moreover, the improved strength of the luminous clusters as a result of an interaction with the gelatin matrix implies that gelatine encapsulated Ag clusters could be useful biocompatible luminescent biosensors for the biomedical uses. A recent study was reported the fabrication of gelatin-derived nanocomposite films containing uniformly dispersed SWCNTs that were compatible with sodium docecylsulfate (SDS). The film's mechanical elongation caused significant uniaxial placement also separation of particular SWCNTs, resulting in a substantial through very much polarized absorbance as well as photoluminescence. This finding endorses up the concept that unique SWCNTs are uniformly distributed in the gelatin composite, allowing for novel applicability in optoelectronic materials [38]. In alternative study, established nanocomposite gelatin films using CdSe QDs and revealed the effect of QDs to the disruption of top Cd atoms links with gelatin functional groups [38]. The miniaturized light-emitting biosensors could benefit from these materials. Nanocomposite fibers enclosing up conversion constant and monodispersed NPs of luminescent $NaYF_4$:Yb_{31} Er_{31} summarized in mesoporous silica shells using a core/shell method were fabricated by electrospinning employing poly(e-caprolactone) plus gelatin. A drug, doxorubicin, was inserted on the encapsulated NPs, while another drug, indomethacin, was inserted on the polymeric material. Both drugs discharged from the electrospinning fiber composites with unique characteristics, implying that the hybrid nanocomposite materials might be used as a drug delivery mechanism. Indomethacin had a rapid-release profile, whereas doxorubicin had a slow-release one. It has furthermore been involved in the manufacturing and/or coating of CdTe and CdSe QDs, as well as CdSe/Ag_2Se and CdSe/HgSe core/shell, and CdSe and Au NPs for optoelectronic applications. Now the hydrothermal method of apple-like Eu31 also Tb31codoped LaCO3OH hierarchical nanoparticles with controllable emission of white light, gelatin was also employed as a reactant and structure-directing agent. It has been utilized to begin the hydrothermal preparation of extremely luminous carbon QDs. Gelatin serves as an effective matrix in nanocomposite material for improving the biocompatibility while keeping powerful optical capabilities in bioimaging applications.

22.5 Polynucleotides and Carbonized-Based Biocomposite

Polynucleotides were extensively used as materials in nanocomposites due to their strong helix structure and quantity of accessible functional groups, which allow for interactions with the other composite materials. They are a type of nucleic acid that is made up of many different nucleotides. The human body practices DNA and RNA to make biomolecules, decide their activities, and cause destroy cells. It also influence a variety of genetic dispositions associated with diabetics and protection of the environment in the production of melanin. RNA possesses a single polynucleotidic structure; while DNA is a helically coiled combined of polynucleotides. Each nucleotide has at minimum one phosphate group, as well as a nitrogenous base and a 5-carbon sugar. Protecting groups are modified *in vitro* to generate nucleotides. In addition, high-throughput manufacturing makes use of microbial gene expression.

22.5.1 DNA Origami

The polynucleotides, namely DNA ought to been widely investigated for their structure into tailored geometries using guided self-assembly methods. Lithographic techniques are now employed to create more complex hierarchical nanostructures; nevertheless, the difficulties in inducing exactly self-assembly have hindered bottom-up scaling. DNA-mediated regulated assemblies into such well-defined designs that are naturally formed will be accurately duplicates into composites 3D structures in vast quantities as one strategy to overcome this difficulty.

Advanced DNA nanotechnology revealed 2D and 3D crystals, nanoforms, and nanoscale gadgets, among other things. DNA origami is a technique for constructing desired 2D and 3D nanostructures by allowing pre-designed blends of polynucleotide sequences to self-assemble also folding in a self-controlled method below regulated temperature and chemical conditions [39]. This DNA origami method takes advantage of extremely precise hybridization among matching polynucleotide patterns, increasingly availability of polynucleotide sequences, and extremely powerful processing suites like caDNAnoto to create more complex polynucleotide frameworks. Engineered DNA can be used as a framework to build complex 1D, 2D, and 3D structures.

Utilizing complementary interactions, as well as the spatial frames of templated DNA constructs, directed assembly may be utilized to assemble silver and gold particles, as well as certain additional optical materials, mad about complex super lattices to produce metamaterials. Young *et al.* created bimetallic silver and gold super lattices using DNA with comparable self-assembly capabilities. Initially, the scientists used electrodynamic modeling to identify suitable constructions with potentially intriguing optical activity, predicting that the chosen structures will display epsilon-near-zero behavior and several optical metallic reactions. Young and coworkers used the helical DNA structure to develop chirality in noble metal nanostructures in liquid crystal biomaterials, which were subsequently dried to produce dense thin films [40]. Electrodynamic calculations projected that the final coating will have a metallic glossy surface and dielectric characteristics. Another advantage of DNA-mediated arrangement as optical biomaterials would be the durability of DNA assembly for precise nailing down of definite compositions and organization with exceptional functions.

Some have utilized DNA-templated nanostructures to modify fluorescence in the same way that they have been used to alter visible light. In most cases, biomolecule identification is restricted to labels with different colors in the visible region. DNA, on the other hand, allows the creation of metafluorophores that are angstrom-sized to adjust the brightness and color. The DNA constructed metafluorphores may produce dynamic intensity-dependent barcodes to change the color and brightness by particular biomolecular triggers that enables biomolecule identification of geometric barcoding.

DNA assembly may be utilized to manufacture complicated nanodevices by designing 2D and 3D nanoelectrical circuits, in addition to assisting in the synthesis of optically active nanomaterials. DNA was used as a framework for loading conductive metal NPs in the early investigations. Recent investigations had begun to open the path for the use of DNA backbone network as nanostructures without the need for further modification or biometallization. The polyelectrolytic behavior of DNA molecules made it difficult to explain its conductivity at first [41]. Moreover, it has already been established that helical structure of DNA can performance as a conduit for extended range charge transporter over extended distances, having helical DNA nanowires demonstrating both vacancy and electron transmission.

Magro *et al.* built on this to develop a technique for triggering DNA electrical properties. They used surface active maghemite NPs (SAMN) to immobilize DNA and generate nanobioconjugates that self-assembled into one dimensional, two dimensional, and three dimensional structures. When DNA was in solution, it showed that it could carry electricity between an electroactive solute and SAMN. When free DNA was used, the reverse result was obtained, demonstrating that NP triggering was required for electrical conduction via DNA. At the mesoscale, 3D metamaterials revealed electrical conductance equivalent to carbon paste as well as durability, with metamaterials withstanding voltammetry cycles without losing efficiency [42]. Though still a long way from becoming a gadget, a research like this makes it possible to use DNA in bioelectronics and produce better sensitive biosensors. These findings contribute to a better knowledge of DNA backbone transfer that is important aimed at DNA nanowires & DNA-templated biometallized three dimensional nanocircuits.

22.5.2 Carbon Nanomaterials

Carbon nanocomponents embedded in biocomposite matrices give amazing mechanical reinforcing and functional capabilities, such as regulated thermal and electrical conductivities, gas barrier qualities, sensing abilities, as well as regulated molecular porosity [43]. The performance of biopolymer matrices can be greatly increased even at minimal loading. For example, revealed a new low electrical percolation threshold on behalf of GO-gelatin nanocomposites, owing to the uniform thinning out of GO nanosheets inside the gelatin matrix. Li *et al.* described a unique CNT/protein core-shell nanostructure with a condensed spherical structure and a hydrophobic core manufactured by sonication of CNTs and -lactoglobulin protein. The sonication should expose the protein's core hydrophobic domain to the hydrophobic CNTs, allowing hydrophobic-hydrophobic interactions to generate stable CNT dispersions. Furthermore, partial denaturation of protein can increase hydrophobic-hydrophobic interactions between CNT as well as protein components, resulting in a larger amount of hydrophobic regions. Similarly, the hydrophobic-hydrophobic interactions are most important interactions between polysaccharides and carbon nanostructures [44].

Due to high dispersibility in aqueous conditions, GO remains a graphene derivative frequently employed as a ancestor for graphene. The thermal oxidation process, initially proposed by Hummers and improved by successors, is commonly used to make it from graphite flakes. The resulting high aspect ratio single monolayer carbon materials generate flexible sheet with such a great density of epoxy in addition to hydroxyl linkages on both sides of the horizontal plane as well as carboxyl linkages all over the place its plane borders. These functionalities can moreover be partly eliminated by using well-known GO reduction methods as hydrazine, hydroiodic acid, elevated temperatures, or microwaves to create conducting polymer. Because GO includes rich polar functionalities capable of hydrogic bonding, ionic, covalent and polar-polar interactions with biopolymers, it is simpler to integrate into biopolymers than graphene and CNT elements. A heterogeneous amphiphilic surface is also created by the coexistence of hydrophobic graphitic patches and hydrophilic oxidized regions [45]. This intriguing trait has two key implications: (1) GO sheets may form strong bonds with one or the other hydrophobic/hydrophilic biopolymers (2) the strong suit of the edges can be boosted further if a biopolymer matrix that matches the GO sheets is chosen. Hydrogen bonding as well as hydrophobic interactions of structured micellar structures maintain the combination of GO and amphiphilic biopolymers. Wan *et al.*, for example, revealed that the significant interfacial strength between GO and CS increased the tensile strength and toughness of GO films by a factor of ten.

22.6 Future Trends and Perspective

There have been significant developments in the uses of naturally generated nanocomposites over the last few decades are reported, still a number of important concerns that must be solved. NC, spider silk, as well as DNA components currently lack low-cost, scalable, repeatable extraction, synthesis, purification, and modification procedures that can provide highly pure components with minimal dispersity in sizes and content. Traditional NC material synthesis processes, such as acid hydrolysis and mechanical exfoliation, are ineffective and absence of dimensional control, as well as involving significant energy consumption, toxic ingredients, and capital-intensive industrial machinery.

Current bionanocomposites primarily exhibit the intrinsic optical features of every component, such as NC transparency, chirality of proteins with biomolecules, also photonics, absorbance, as well as light emission of nanomaterials. Many plants and animals, on the other hand, may produce vibrant structural colors by controlling light reflection and scattering by using extremely periodic photonic nanostructured materials in a laborious manner on a vast spatial scale. The incorporation of conducting bionanocomponents within bioelectronic devices necessitates a considerable amount of conductive components, including carbon nanotubes and metal NPs, which often limits mechanical compliance, flexibility, and biocompatibility. Several membranes through selective besides effective stimulating ion channels remain densely packed in their specialized electric organs, allowing electricity to be generated from the ion flow caused by ionic gradient. Overall, when considering far-reaching hard work in future ventures related to the matter of this overview for solving issues to the world's most pressing issues that will facilitate healthcare, food preservation, and dynamic/protective optical devices.

22.7 Conclusion

Biopolymers may generate complimentary interfaces with diverse synthetic components with specific functionality due to their availability of functional groups, multileveled structured secondary structure, and readily reconfigurable secondary structure. Consequently, the use of biopolymers with manmade components frequently results in physical improvements and elevated multi-functionality. Controlling the molecular in addition to supramolecular structure of biopolymeric materials on grapheme faces, for example, provides for effective load transmission at the synthetic and biological interface, making it easier to build bendable wearable electrical biobased devices with improved strong suit and resilience. Inclusion with additional synthesized well-designed elements, for instance, metallic NPs, results in the addition of new physical as well as chemical properties of concern, for example, barrier transport, optical activity, catalytic ability, and high electrical conductivity, which could be useful in sensing, power conversion and storage, biomedicine, and nanostructured materials.

References

1. Hu, K., Xiong, R., Guo, H., Ma, R., Zhang, S., Wang, Z.L., Tsukruk, V.V., Self-powered electronic skin with biotactile selectivity. *Adv. Mater.*, 28, 3549, 2016.
2. Ranjan, C., Bhattacharya, R., Mukhopadhyay, D., Mukherjee, P., Fabrication of gold nanoparticles for targeted therapy in pancreatic cancer. *Adv. Drug Deliv. Rev.*, 62, 346, 2010.
3. Tang, L., Zhao, L. *et al.*, Graphene/polymer composite materials: Processing, properties and applications, in: *Advanced Composite Materials: Properties and Applications*, E. Bafekrpour (Ed.), pp. 349–419, De Gruyter Open Poland, Warsaw, Poland, 2017.
4. Lee, J.H., Kim, H., Kim, J.H., Lee, S.H., Soft implantable microelectrodes for future medicine: Prosthetics, neural signal recording and neuromodulation. *Lab Chip.*, 16, 959, 2016.
5. Yang, W., Zhao, Z., Wu, K., Huang, R., Liu, T., Jiang, H., Chen, F., Fu, Q., Ultrathin flexible reduced graphene oxide/cellulose nanofiber composite films with strongly anisotropic thermal conductivity and efficient electromagnetic interference shielding. *J. Mater. Chem. C*, 5, 3748, 2017.
6. Gosline, J.M., Guerette, P.A., Ortlepp., C.S., Savage., K.N., The mechanical design of spider silks: From fibroin sequence to mechanical function. *J. Exp. Biol.*, 202, 3295, 1999.
7. Traykova, T., Aparicio, C., Ginebra, M.P., Planell, J.A., Bioceramics as nanomaterials. *Nanomed. J.*, 1, 91, 2006.
8. Choudhury, R.R., Sahoo, S.K., Gohil, J.M., Potential of bioinspired cellulose nanomaterials and nanocomposite membranes thereof for water treatment and fuel cell applications. *Cellulose*, 27, 6719, 2020.
9. Savolainen, K., Pylkkänen, L., Norppa, H., Falck, G., Lindberg, H., Tuomi, T., Vippola, M., Alenius, H., Hämeri, K., Koivisto, J., Brouwer, D., Mark, D., Bard, D., Berges, M., Jankowska, E., Posniak, M., Farmer, P., Singh, R., Krombach, F., Bihari, P., Kasper, G., Seipenbusch, M., Nanotechnologies, engineered nanomaterials and occupational health and safety – A review. *Saf. Sci.*, 48, 957, 2010.
10. Abdul Khalil, H.P.S., Bhat, A.H., Ireana, Y.A.F., Green composites from sustainable cellulose nanofibrils: A review. *Carbohydr. Polym.*, 87, 963, 2012.
11. Ribaut, C., Loyez, M., Larrieu, J.C., Chevineau, S., Lambert, P., Remmelink, M., Wattiez, R., Caucheteur, C., Cancer biomarker sensing using packaged plasmonic optical fiber gratings: Towards *in vivo* diagnosis. *Biosens. Bioelectron.*, 92, 449, 2017.
12. Balaji, A. and Zhang, J., Electrochemical and optical biosensors for early-stage cancer diagnosis by using graphene and graphene oxide. *Cancer Nanotechnol.*, 8, 1, 1–12, 2017.
13. Tao, H., Kainerstorfer, J.M., Siebert, S.M., Pritchard, E.M., Sassaroli, A., Panilaitis, B.J.B., Brenckle, M.A., Amsden, J.J., Levitt, J., Fantini, S., Kaplan, D.L., Omenetto, F.G., Implantable, multifunctional, bioresorbable optics. *Proc. Natl. Acad. Sci.*, 109, 19584, 2012.
14. Xin, H., Li, Y., Liu, X., Li, B., Escherichia coli-based biophotonic waveguides. *Nano Lett.*, 13, 3408, 2013.
15. Bao, H.T.L., Minh, T., Nguyen, D., Minh, D., Naturally derived biomaterials: Preparation and application, in: *Regenerative Medicine and Tissue Engineering*, p. 314, 2013.
16. Reshmy, R., Philip, E., Paul, S.A., Madhavan, A., Sindhu, R., Binod, P., Pandey, A., Sirohi, R., Nanocellulose-based products for sustainable applications-recent trends and possibilities. *Rev. Environ. Sci. Biotechnol.*, 19, 779, 2020.
17. Reshmy, R., Philip, E., Madhavan, A., Sindhu, R., Pugazhendhi, A., Binod, P., Sirohi, R., Awasthi, M.K., Tarafdar, A., Pandey, A., Advanced biomaterials for sustainable applications in the food industry: Updates and challenges. *Environ. Pollut.*, 283, 117071, 2021.

18. Bassan, P., Byrne, H.J., Bonnier, F., Lee, J., Dumas, P., Gardner, P., Resonant Mie scattering in infrared spectroscopy of biological materials-Understanding the dispersion artefact. *Analyst*, 134, 1586, 2009.
19. Cazón, P., Vázquez, M., Velazquez, G., Composite films of regenerate cellulose with chitosan and polyvinyl alcohol: Evaluation of water adsorption, mechanical and optical properties. *Int. J. Biol. Macromol.*, 117, 235–246, 2018.
20. Schlesinger, M., Hamad, W.Y., Maclachlan, M.J., Optically tunable chiral nematic mesoporous cellulose films. *Soft Matter.*, 15, 9, 4686–4694, 2015.
21. Sharma, A., Thakur, M., Bhattacharya, M., Mandal, T., Goswami, S., Commercial application of cellulose nano-composites—A review. *Biotechnol. Rep.*, 21, 316, 2019.
22. Luo, S.C., Sivashanmugan, K., Liao, J.D., Yao, C.K., Peng, H.C., Nanofabricated SERS-active substrates for single-molecule to virus detection *in vitro*: A review. *Biosens. Bioelectron.*, 61, 232, 2014.
23. Alvarez-Puebla, R.A., Arceo, E., Goulet, P.J.G., Garrido, J.J., Aroca, R.F., Role of nanoparticle surface charge in surface-enhanced raman scattering. *J. Phys. Chem. B*, 109, 3787, 2005.
24. Hassinen, J., Liljeström, V., Kostiainen, M.A., Ras, R.H.A., Rapid cationization of gold nanoparticles by two-step phase transfer. *Angew. Chem.*, 127, 8101, 2015.
25. Park, M., Chang, H., Jeong, D.H., Hyun, J., Spatial deformation of nanocellulose hydrogel enhances SERS. *Biochip J.*, 7, 234, 2013.
26. Elazzouzi, H.S., Nishiyama, Y., Putaux, J.L., Heux, L., Dubreuil, F., Rochas, C., The shape and size distribution of crystalline nanoparticles prepared by acid hydrolysis of native cellulose. *Biomacromolecules*, 9, 57, 2008.
27. Harish, K., Venkatesh, N., Bhowmik, H., Kuila, A., Metallic nanoparticle: A review. *Biomed. J. Sci. Tech. Res.*, 4, 3765, 2018.
28. Thakur, V., Guleria, A., Kumar, S., Sharma, S., Singh, K., Recent advances in nanocellulose processing, functionalization and applications: A review. *Mater. Adv.*, 2, 1872, 2021.
29. Kelly, J.A., Shukaliak, A.M., Cheung, C.C.Y., Shopsowitz, K.E., Hamad, W.Y., MacLachlan, M.J., Responsive photonic hydrogels based on nanocrystalline cellulose. *Angew. Chemie Int. Ed.*, 52, 8912, 2013.
30. Liu, Q., Campbell, M.G., Evans, J.S., Smalyukh, I.I., Orientationally ordered colloidal co-dispersions of gold nanorods and cellulose nanocrystals. *Adv. Mater.*, 26, 7178, 2014.
31. Yao, K., Meng, Q., Bulone, V., Zhou, Q., Flexible and responsive chiral nematic cellulose nanocrystal/poly(ethylene glycol) composite films with uniform and tunable structural color. *Adv. Mater.*, 29, 1, 2017.
32. Giese, M., Blusch, L.K., Khan, M.K., Hamad, W.Y., MacLachlan, M.J., Responsive mesoporous photonic cellulose films by supramolecular cotemplating. *Angew. Chem.*, 126, 9026, 2014.
33. Deepthi, S., Venkatesan, J., Kim, S.K., Bumgardner, J.D., Jayakumar, B.R., An overview of chitin or chitosan/nano ceramic composite scaffolds for bone tissue engineering. *Int. J. Biol. Macromol.*, 93, 1338–1353, 2016.
34. Ifuku, S. and Saimoto, H., Chitin nanofibers: Preparations, modifications, and applications. *Nanoscale*, 4, 3308, 2012.
35. Sergeev, A.A., Voznesenskiy, S.S., Bratskaya, S.Y., Mironenko, A.Y., Lagutkin, R.V., Investigation of humidity influence upon waveguide features of chitosan thin films. *Phys. Proc.*, 23, 115, 2012.
36. Liu, F., Carlos, L.D., Ferreira, R.A.S., Rocha, J., Gaudino, M.C., Robitzer, M., Quignard, F., Photoluminescent porous alginate hybrid materials containing lanthanide ions. *Biomacromolecules*, 9, 1945, 2008.
37. Parker, S.T., Domachuk, P., Amsden, J., Bressner, J., Lewis, J.A., Kaplan, D.L., Omenetto, F.C., Biocompatible silk printed optical waveguides. *Adv. Mater.*, 21, 2411, 2009.

38. Borkovska, L.V., Korsunska, N.O., Stara, T.R., Dzhagan, V.M., Stroyuk, O.L., Raevska, O.Y., Kryshtab, T.G., Modification by thermal annealing of the luminescent characteristics of cdse quantum dots in gelatin films. *Phys. Status Solidi Curr. Top. Solid State Phys.*, 9, 1779, 2012.
39. Tørring, T., Voigt, N.V., Nangreave, J., Yan, H., Gothelf, K.V., Themed issue: Advances in DNA-based nanotechnology DNA origami: A quantum leap for self-assembly of complex structures. *Chem. Soc. Rev.*, 40, 12, 306, 2011.
40. Woehrstein, J.B., Strauss, M.T., Ong, L.L., Wei, B., Zhang, D.Y., Jungmann, R., Yin, P., Sub–100-nm metafluorophores with digitally tunable optical properties self-assembled from DNA. *Sci. Adv.*, 3, 18, 2017.
41. Tan, B., Hodak, M., Lu, W., Bernholc, J., Charge transport in DNA nanowires connected to carbon nanotubes. *Phys. Rev. B*, 92, 1, 2015.
42. Magro, M., Baratella, D., Jakubec, P., Zoppellaro, G., Tucek, J., Aparicio, C., Venerando, R., Sartori, G., Francescato, F., Mion, F., Gabellini, N., Zboril, R., Vianello, F., Triggering mechanism for DNA electrical conductivity: Reversible electron transfer between DNA and iron oxide nanoparticles. *Adv. Funct. Mater.*, 25, 1822, 2015.
43. Chen, L.F., Huang, Z.H., Liang, H.W., Gao, H.L., Yu, S.H., Three-dimensional heteroatom-doped carbon nanofiber networks derived from bacterial cellulose for supercapacitors. *Adv. Funct. Mater.*, 24, 5104, 2014.
44. Jin, H., Wang, X., Gu, Z., Fan, Q., Luo, B., A facile method for preparing nitrogen-doped graphene and its application in supercapacitors. *J. Power Sources*, 273, 1156, 2015.
45. Xu, N., Hu, X., Xu, W., Li, X., Zhou, L., Zhu, S., Zhu, J., Mushrooms as efficient solar steam-generation devices. *Adv. Mater.*, 29, 1, 2017.
46. Voznesenskiy, S.S., Sergeev, A.A., Mironenko, A.Y., Bratskaya, S.Y., Kulchin, Y.N., Integrated-optical sensors based on chitosan waveguide films for relative humidity measurements. *Sens. Actuators B Chem.*, 188, 482, 2013.
47. Higham, A.K., Bonino, C.A., Raghavan, S.R., Khan, S.A., Photo-activated ionic gelation of alginate hydrogel: Real-time rheological monitoring of the two-step crosslinking mechanism. *Soft Matter*, 10, 4990, 2014.
48. Broderick, E.P., O'Halloran, D.M., Rochev, Y.A., Griffin, M., Collighan, R.J., Pandit, A.S., Enzymatic stabilization of gelatin-based scaffolds. *J. Biomed. Mater. Res. Part B Appl. Biomater.*, 72, 37, 2005.
49. Ahmed, S. and Ikram, S., Chitosan and gelatin based biodegradable packaging films with UV-light protection. *J. Photochem. Photobiol. B Biol.*, 163, 115, 2016.
50. Kim, Y., Minami, N., Kazaoui, S., Highly polarized absorption and photoluminescence of stretch-aligned single-wall carbon nanotubes dispersed in gelatin films. *Appl. Phys. Lett.*, 86, 1, 2005.

23

Biocomposites and Bioplastics in Electrochemical Applications

Sema Aslan[1]* and Derya Bal Altuntaş[2]

[1]*Department of Chemistry, Faculty of Science, Mugla Sitki Kocman University, Mugla, Turkey*
[2]*Department of Bioengineering, Faculty of Architecture and Engineering, Recep Tayyip Erdogan University, Rize, Turkey*

Abstract
Electrochemical studies, including sensors, biosensors, fuel cells, supercapacitors, batteries, corrosion require very much catalytic modifications upon the high costs of the noble metals or nanomaterials. Besides, because the world's energy source crisis reached very high levels, people tend to find new routes for green synthesis and application of biosourced materials. Chitosan, cellulose, wool, silk, pectin, and many of the biocomposite and bioplastic produced by nature are less hazardous, renewable, sustainable, innovative, low cost, and carbon-neutral, preferred as immobilization matrix and/or as a source as biomass for nanomaterial productions or used as the main component of an electrochemical system, such as a membrane in a fuel cell. It is known that materials originating from biological materials are less harmful to nature in different ways compared to fossil fuels. In addition, it has been reported in many studies that biomass-based biocomposites can be replaced by many materials produced with petroleum-based resources. In this sense, the chapter includes electrode modifications or the electrochemical applications of biocomposites and bioplastics.

Keywords: Biocomposite, bioplastic, electrochemistry, nanomaterial, biosensor, energy, supercapacitor

23.1 Introduction

The development of green materials has gained huge attention after the rapidly growing environmental crisis. Especially scientists focused on finding new routes to solve the problems in more biocompatible ways. Composite materials that are less harmful to the environment are being developed in all industrial areas of life, from health to energy, from food to space technologies.

The use of composite materials in electrochemical studies is a very useful option. These composites, which are mostly formed with nanomaterials, can be designed to be used sometimes as a supporting material to improve the electrochemical response, and sometimes to be used as a

**Corresponding author*: sozcan87@gmail.com

Inamuddin and Tariq Altalhi (eds.) Handbook of Bioplastics and Biocomposites Engineering Applications: 2nd Edition, (491–512) © 2023 Scrivener Publishing LLC

part of the system themselves. At this point, the use of biocomposites and bioplastics is preferred because they are less harmful to the environment and are economical. It is known that materials originating from biological materials are less harmful to nature in different ways compared to fossil fuels. In addition, it has been reported in many studies that biomass-based biocomposites can be used instead of many materials produced with petroleum-based resources. Similarly, the production of bioplastics with biological agents or obtaining them from biomass is very convenient and easy. One of the important points in electrochemical applications is to increase electrocatalytic activity. For this purpose, the total electrical efficiency can be increased by making modifications to many components of the measurement system. In this sense, fiber-structured bioplastics or biocomposite materials containing nanomaterials and biological materials, such as enzymes or microorganisms with nature-friendly properties are very useful.

In this section, information about biocomposites and bioplastics used in electrochemical studies will be mentioned. Different examples of these materials of different biological origins, which are applied in various fields of electrochemistry, will be given.

23.2 Electrochemistry

23.2.1 General Aspects

Electrochemistry can be broadly defined as electron-matter interaction [1]. Electrochemical events cause chemical changes in the structure of a substance and create electron flow [2]. In electroanalytical chemistry, chemical changes are studied by the analysis of electrons in a wide range of applications, such as sensors, biosensors, energy, biochemistry, environment, biomedical, and so on [3, 4]. The fundamentals of electroanalytical chemistry were constructed in the early 1980s with the combination of ultramicroelectrodes with biological materials. Due to the analysis type and the precision needed, several techniques were developed, such as linear voltammetry, differential pulse voltammetry, square wave voltammetry, stripping voltammetry, potentiometry, impedimetry. The merge of the nanomaterials, such as carbon nanotubes into the field made a significant impact [5, 6].

Depending on the analysis type, measurements can be taken in systems, such as stationary, mixed, or flow injection. Electrode types are divided into general classes as working, reference, and counter electrodes. Working electrodes from these electrodes are the electrodes used to measure the current difference. The performance of this electrode directly affects the sensitivity and the precision of the measurement. Generally, this goal is achieved by the addition of nanomaterials into the electrode or modification of the electroanalytical system. The main electroanalytical applications, such as biosensors, sensors, fuel cell, supercapacitor, corrosion, etc... improve their analytical performances in such a solution instead of using expensive catalysts or electrodes.

23.3 Nanomaterials in Biocomposite Applications

Nanomaterials are defined as substances, which have a size of 100 nm or less. Due to their unique surface-to-volume ratio, they can exhibit improved features, such as excellent conductivity, catalytic effect, electrochemical activity, inertness, antibacterial properties, and so on.

The evolution of the nanoparticle production ways into green synthesis is another emerging field. Beyond the biocomposite and bioplastic usage as the immobilization matrix, novel routes that including any or very little amount of hazardous chemicals are the subject of interest in nanomaterial productions. Thanks to the three-dimensional networks of these natural biocomposites, such as gelatin hydrogels, cellulose fibers, chitosan, they provide an excellent buffering condition and serve as successful stabilizing agents [7], for nanomaterial production. Also, they provide excellent synergistic effects when composed with any nanomaterials. There have been reported many of the example applications of divergent nanomaterials composed with biocomposites or bioplastics in different fields given in the following chapters.

23.4 Electrochemical Applications

23.4.1 Biosensors

Biosensors are defined as the sensors, which create a signal due to the biological recognition interaction (Figure 23.1) [8]. A biomaterial, such as an enzyme, antibody, bacteria, nucleic acid, etc… can be used as a recognition element on the electrode or the secondary metabolites can be realized by the transducer electrode. The type of the recognition method can vary according to the suitable method for that analytical determination. Sensors do not need a biological element but detect specific analytes in a matrix with the help of electroactive reactions. Here and in the following section, electrochemical biosensors and sensors will be discussed in the manner of biocomposite and bioplastic using studies.

Biocomposites are utilized as useful enzyme immobilization tools. Since the biocomposite medium is very rich in electrons, the precision of the developed biosensors' responses is enhanced.

Shieh *et al.* (2015) reported high activity in hydrogen peroxide degradation even in the presence of protease proteinase K by embedding the catalase molecules into uniformly sized ZIF-90 crystals with biocomposite formation [9]. They named this strategy the *de novo* approach. In the study reported by Hsu *et al.* (2021) catalase and α-chymotrypsin

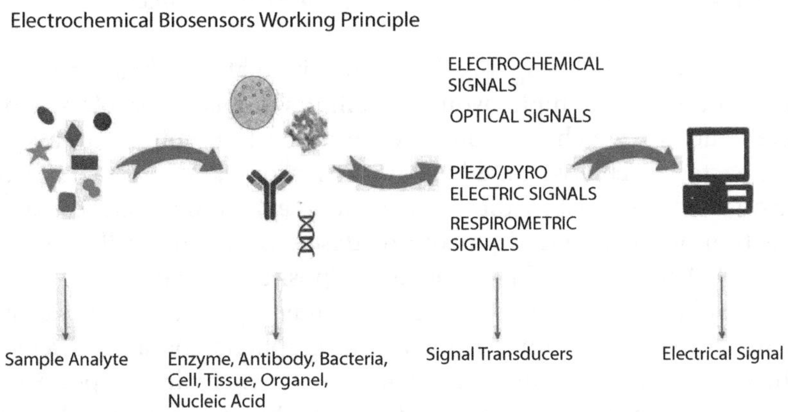

Figure 23.1 General schematic illustration of a biosensor [8].

were used together with Zn-MOF-74 nanostructures by applying the de novo method. A successful peroxidase and peptide digestion activity retention was achieved with the help of Zn-MOF-74 biocomposites [10]. Also regarding this approach, better biocatalytic activities were achieved and the enzyme loading capacity was increased.

The choice of the biocomposite and bioplastic varies according to the nature of the biocomponent, which is used in the construction of the sensing system. In general gold-based materials and sulfur-modified surfaces are utilized in immunosensors due to the strong binding affinity between sulfur and gold [11]. Aptamer and DNA-based biosensors need higher precisions so that graphene and their derivatives are largely applied as modifiers, in live-cell studies or enzymatic biosensors, the priority of the modification comes forward as biocompatible and antifouling construction [6, 12, 13]. Some electrochemical biosensor studies that have been developed with biocomposites and bioplastics are given in Table 23.1. Wipawakarn *et al.* reported a label-free voltammetric genosensor study in which chitosan, carbon nanofiber, and glutaraldehyde were used as electrode modifiers. Here chitosan serves as the main matrix for other facilitating materials. Although polysaccharides, triglycerides, or amino acids are all natural hydrogel nanocomposite supports, chitosan stands out due to its rich cross-links, which allows a large amount of water uptake. Therefore, chitosan possesses an antimicrobial and biocompatible environment for the ingredients, which are variable materials from enzymes to quantum dots. Additionally, a highly stable and elastic cross-link network of chitosan hydrogel allows the electrochemical processes to be easily carried out. Thus, chitosan serves well due to its mild and electrochemically generous conditions among various applications [13]. Another interesting peptide-based food-borne pathogen detecting voltammetric biosensor was reported by Eissa *et al.* (2020) [14]. They took the advantage of the streptavidin-biotin interaction and measured the activity of proteolytic enzymes originating from *Listeria monocytogenes* and *Staphylococcus aureus* pathogens. A catalase-based biosensor was reported for hydrogen peroxide detection in which gelatin was utilized as an immobilized and biocomposite matrix. Multiwalled carbon nanotubes and enzymes were mixed in the gelatin and applied as a successful biosensor [15]. Similarly, Farooq *et al.* (2020) reported an *S. aureus* biosensor that achieved a reasonable detection of live/dead cells in the medium [16]. They used a bacterial cellulose matrix to maintain the highly porous and fibrous surface for bacteriophage immobilization. Also carboxylated multiwalled carbon nanotubes are used to facilitate electrochemical response. They also examined the effect of polyethyleneimine successfully in that biocomposite medium.

Bacterial cellulose is a widely applied biocomposite material that is preferred due to its biocompatible nature, for example, wound healing systems, wearable sensors, etc. They have a flexible structure, which promotes their modification options. In another study, an amperometric bacterial cellulose-based lactate biosensor was developed. It exhibited great performance on artificial sweat mediums [17]. Gong *et al.* (2021) suggested an unfamiliar affinity interaction between tagged glucose oxidase enzyme and cellulose modified electrode surface [18]. Another useful form of biocomposites is nanofibers [19].

Cellulose nanofibers are one of a type. In the study reported by Velusamy *et al.* (2017) hemoglobin molecules were immobilized within graphene containing cellulose microfibers [20]. The developed biosensor is based on monitoring hydrogen peroxide production according to the electrochemical behavior of hemoglobin. Here they highlighted the practical and stable structure of the cellulose microfibers as biocomposites.

Table 23.1 Biosensor and sensor examples using biocomposites and bioplastics.

Biosensors					
Transducer	**Biocomposite/ bioplastic**	**Analyte/method**	**Range**	**LOD**	**Reference**
GCE/Chitosan-carbon nanofibers	Glutaraldehyde	DNA/Voltammetry	0.5 and 40 nM	88 pM	[13]
SPE/Gold nanoparticle-magnetic nanoparticles-peptide-protease	Peptide	*L. monocytogenes* and *S. aureus*/Amperometry Flow biosensor	10 to 10^8 CFU mL^{-1}	9 CFU mL^{-1} for *L. monocytogenes* and 3 CFU mL^{-1} for *S. aureus*	[14]
GCE/gelatin multiwalled carbon nanotubes. Catalase	Gelatin-Glutaraldehyde	H_2O_2/Cyclic voltammetry-Electrochemical impedance spectroscopy	0.2 to 5.0 mM	0.001 mM	[15]
Bacterial cellulose/ carboxylated multiwalled carbon nanotubes (c-MWCNTs)bacteriophage	Bacterial cellulose	*S. aureus*/Differential pulse voltammetry	0-7 CFU mL^{-1}	3 and 5 CFU·mL^{-1} of *S. aureus* in PBS and milk	[16]
SPE/bacterial cellulose-lactate oxidase	Bacterial cellulose	Hydrogen peroxide/ Amperometry	1-24 mmol L^{-1}	1.31 mmol L^{-1}	[17]
GCE/graphene-cellulose microfiber hemoglobin	Cellulose microfiber	Hydrogen peroxide/ Amperometry	0.05 to 926 M	0.01 M	[20]
GCPE/SBA 15 or MCM 41 silica powder -Uricase	Glutaraldehyde activated $-NH_2$ functionalized MCM–41	Uric acid/Amperometry Flow injection	50–800 µmol L^{-1}	18.5 µmol L^{-1}	[33]

(Continued)

Table 23.1 Biosensor and sensor examples using biocomposites and bioplastics. (*Continued*)

Biosensors					
Transducer	**Biocomposite/ bioplastic**	**Analyte/method**	**Range**	**LOD**	**Reference**
Carbon Rod/microporous carbon fibers (CFs)-Laccase	Cellulose	Phenolic compounds/ Amperometry	-	-	[34]
Sensors					
Transducer	**Biocomposite/ bioplastic**	**Analyte/method**	**Range**	**LOD**	**Reference**
GCE/CuCP–AuNPs	Pectin	H_2O_2/Cyclic Voltammetry, Amperometry	1–2110 μM	0.22 μM	[24]
CPE/CS-MIP	Chitosan	Lactic acid/Cyclic-Differential Voltammetry	0.01–10.0 μM and 10.0–500.0 μM	0.005 μM	[25]
GCE/CT nano Cu/N, P, S@PC	Chitosan	H_2O_2/Cyclic Voltammetry, Amperometry	3 μM to 4 mM,	1.5 μM	[30]

(*Continued*)

Table 23.1 Biosensor and sensor examples using biocomposites and bioplastics. (*Continued*)

Biosensors					
Transducer	**Biocomposite/ bioplastic**	**Analyte/method**	**Range**	**LOD**	**Reference**
GCE/ZrO_2-CS-rGOA	Chitosan	Luteolin/Cyclic-Differential Voltammetry	5 nM to 1000 nM	1 nM	[31]
SPAuE/rGO/ DWCNTs/Oct-Fe_3O_4/Cs	Chitosan	Glyphosate/Square wave Voltammetry	0.1–1000 ppb.	0.08 ppb	[32]
* Gold nanoparticle-decorated copper cross-linked pectin (CuCP), chitosan-based magnetic molecularly imprinted polymer (CS-MIP), N; P, S codoped porous carbon (N, P, S@PC), chitosan/reduced graphene oxide aerogel with dispersed ZrO_2 nanoparticles modified glassy carbon electrode (GCE/ZrO_2-CS-rGOA), octahedral-Fe_3O_4 (Oct-Fe_3O_4), chitosan (Cs).					
Biosensors					
Transducer	**Biocomposite/ bioplastic**	**Analyte/method**	**Range**	**LOD**	**Reference**
GCE/Chitosan-carbon nanofibers	Glutaraldehyde	DNA/Voltammetry	0.5 and 40 nM	88 pM	[13]
SPE/Gold nanoparticle-magnetic nanoparticles-peptide-protease	Peptide	*L. monocytogenes* and *S. aureus*/Amperometry Flow biosensor	10 to 10^8 CFU mL^{-1}	9 CFU mL^{-1} for *L. monocytogenes* and 3 CFU mL^{-1} for *S. aureus*	[14]

(*Continued*)

Table 23.1 Biosensor and sensor examples using biocomposites and bioplastics. (*Continued*)

Biosensors					
Transducer	**Biocomposite/ bioplastic**	**Analyte/method**	**Range**	**LOD**	**Reference**
GCE/gelatin multiwalled carbon nanotubes. Catalase	Gelatin-Glutaraldehyde	H_2O_2/Cyclic voltammetry-Electrochemical impedance spectroscopy	0.2 to 5.0 mM	0.001 mM	[15]
bacterial cellulose/ carboxylated multiwalled carbon nanotubes (c-MWCNTs)bacteriophage	Bacterial cellulose	*S. aureus*/Differential pulse voltammetry	0-7 CFU mL^{-1}	3 and 5 CFU·mL^{-1} of *S. aureus* in PBS and milk	[16]
SPE/bacterial cellulose-lactate oxidase	Bacterial cellulose	Hydrogen peroxide/ Amperometry	1-24 mmol L^{-1}	1.31 mmol L^{-1}	[17]
GCE/graphene-cellulose microfiber hemoglobin	Cellulose microfiber	Hydrogen peroxide/ Amperometry	0.05 to 926 M	0.01 M	[20]

(*Continued*)

Table 23.1 Biosensor and sensor examples using biocomposites and bioplastics. (*Continued*)

Biosensors					
Transducer	**Biocomposite/ bioplastic**	**Analyte/method**	**Range**	**LOD**	**Reference**
GCPE/ SBA 15 or MCM 41 silica powder -Uricase	Glutaraldehyde activated $-NH_2$ functionalized MCM–41	Uric acid/Amperometry Flow injection	50–800 μmol L^{-1}	18.5 μmol L^{-1}	[33]
Carbon Rod/microporous carbon fibers (CFs)-Laccase	cellulose	Phenolic compounds/ Amperometry	-	-	[34]
Sensors					
Transducer	**Biocomposite/ bioplastic**	**Analyte/method**	**Range**	**LOD**	**Reference**
GCE/CuCP–AuNPs	Pectin	H_2O_2/Cyclic Voltammetry, Amperometry	1–2110 μM	0.22 μM	[24]
CPE/CS-MIP	Chitosan	Lactic acid/Cyclic-Differential Voltammetry	0.01–10.0 μM and 10.0–500.0 μM	0.005 μM	[25]

(*Continued*)

Table 23.1 Biosensor and sensor examples using biocomposites and bioplastics. (*Continued*)

Biosensors					
Transducer	**Biocomposite/ bioplastic**	**Analyte/method**	**Range**	**LOD**	**Reference**
GCE/CT nano Cu/N, P, S@PC	Chitosan	H_2O_2/Cyclic Voltammetry, Amperometry	3 µM to 4 mM,	1.5 µM	[30]
GCE/ZrO_2-CS-rGOA	Chitosan	Luteolin/Cyclic-Differential Voltammetry	5 nM to 1000 nM	1 nM	[31]
SPAuE/rGO/ DWCNTs/Oct-Fe_3O_4/Cs	Chitosan	Glyphosate/Square wave Voltammetry	0.1–1000 ppb.	0.08 ppb	[32]

* Gold nanoparticle-decorated copper cross-linked pectin (CuCP), chitosan-based magnetic molecularly imprinted polymer (CS-MIP), N; P, S co-doped porous carbon (N, P, S@PC), chitosan/reduced graphene oxide aerogel with dispersed ZrO_2 nanoparticles modified glassy carbon electrode (GCE/ZrO_2-CS-rGOA), octahedral-Fe_3O_4 (Oct-Fe_3O_4), chitosan (Cs).

23.4.2 Sensors

The usage of the natural sourced hybrid materials in sensors mainly provides high stability, non-toxicity, and biodegradable material usage opportunities (Table 23.1). Especially metal-organic compositions for nanocomposites, precursors, immobilizers are originated natural polymers of biodegradable carbohydrates, amino acids, hydrogels, and so on. Pectin is one of those materials that belongs to the polysaccharide family. It is mainly composed of galacturonic acids. Generally, this carboxylic acid-rich structure of pectin provides a very convenient recognizing agent for electrochemically active and ionic analytes [21]. Consequently, the analytical performances, such as detection range, the limit of detection, the limit of quantification, relative standard deviation, etc... of the analytes by the pectin modified electrodes are improved. One of the impressive examples is reported by Alves *et al.* (2021) in a citrus originated pectin based metal sensor [22]. In another glycerol sensor developed by the modification of electrode with nanoparticle including pectin composite [23].

Pectin is used as a combinatory agent for copper and gold nanoparticles. Here it has served as a stable scaffold for uniform Au NP production. Then the produced hybrid biocomposite nanostructure was utilized as an H_2O_2 sensor [24]. Chitosan is also widely used for sensor applications as a positively charged immobilization agent. Mahmoudi *et al.* (2021) recently reported a chitosan-based magnetic molecularly imprinted polymer for the determination of lactic acid was produced due to the spoilage of milk samples [25].

The beginning of the chitosan application in the electrochemical sensor field is counted to the year 1994. In that study adrenaline and dopa were analyzed in a metal ion containing chitosan matrix [26]. After 3 years, a composite membrane of chitosan was created and applied for the detection of some halogenides and iron ions as an ionophore material. Because of this anionic chelate-making capability of the chitosan [27]. Another milestone of the chitosan using sensors reported in 2005, about the cell and protein immobilization [28]. And the rest of the studies on chitinous proteins and chitosan hydrogels especially enzymatic biosensors came like a sock ripping and every discovery made a huge contribution to the electrochemical sensor field [29]. Chitosan is often applied as biocomposite support for nanostructures. In one of those studies reported by Zhai *et al.* (2021) copper nanoparticles were deposited on the chitosan and applied to the N, P, S doped carbon electrode. Then, the obtained electrode was used as an electrochemical H_2O_2 sensor. Obtained Cu NPs were reported to be 70 nm-sized [30]. An efficient electron transfer was achieved with the reduced graphene oxide and ZrO_2/Chitosan aerogel in another study for the detection of luteolin. They proposed a very effective and practical luteolin sensor [31].

Sometimes, the synergetic effect of the two or more combinations is reported. Another reduced graphene oxide and the chitosan-containing medium were enriched with the addition of highly redox-active octahedral-Fe_3O_4 and electroactive double-walled carbon nanotubes. Here the electroactive surface area was enriched up to 1.7 folds due to the double-walled carbon nanotube addition to the biocomposite. The main property of the chitosan comes from the enhancement of the surface loading capacity in terms of the divergent composites. Additionally, the proposed glyphosate sensor achieved a reliable LOD value in a wide linear range [32].

23.4.3 Corrosion

Determination and prevention of corrosion in metals is an important study subject. A great contribution has been made to this field with electrochemical corrosion measurements. A general corrosion scheme is given in Figure 23.2 [35].

It has become a priority for corrosion inhibitors to be economical and harmless to the environment. As in every field, the use of green corrosion inhibitors has been focused on in this field as well. Thus, methods that are eco-friendly, safe, and easily biodegradable, inexpensive, stable, and less toxic than synthetic organic inhibitors are recommended. Biocomposite and bioplastic contents used in studies related to this field have also been reported. Guar gum is a member of the polysaccharide family, which contains a large moiety of pyrene groups. These hetero oxygen-bearing groups facilitate the electron transfer and adsorption on steel. Thus, a stable and useful inhibition effect was achieved on the Helmholtz interface of the electrode/solution inner layer. Abdallah (2004) [35] reported the successful inhibitor effect of guar gum on carbon steel. Also, Buchweishaija and Mhinzi reported that gum exudates from *Acacia seyal* var. showed a good corrosion inhibition effect for drinking water by the temperature up to 30°C [36].

As it has been denoted before in this chapter, active carboxylic acid and hydroxyl groups of pectin possess plenty of lone pairs and also π-bond conjugations in its natural form. Therefore, it can compose a backbonding, which prevents the acidic corrosion easily by providing an inhibition effect to the hydrolysis [37]. Additionally, the main composition of the pectin structure varies up to its main origin. It may have a larger amount of saccharides or long chains, or less amount of polyphenols or different unique organic moieties. It can be isolated from a wide variety of plant species, especially from fruit peels [38]. The corrosion applications were reported for different metals in different conditions for divergent acids (Table 23.2). Abou-Elseoud *et al.*, reported an inhibition study of the pectin obtained from different ways [39], whereas Khairou and Sayed have reported a series of composite inhibitors including pectin on the Cadmium corrosion in an acidic

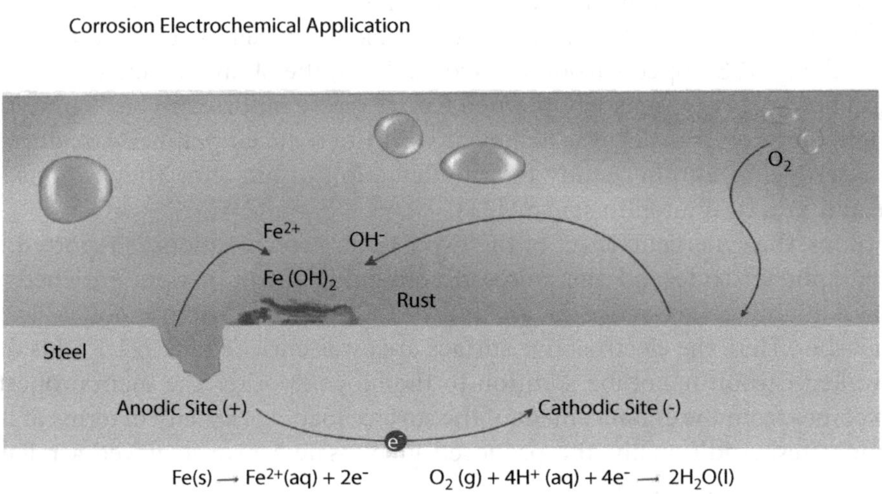

Figure 23.2 Electrochemical changes during a corrosion measurement [28].

Table 23.2 Some of the biocomposites and bioplastics utilized in corrosion inhibition.

Substrate	Inhibitor	Medium	Reference
Carbon steel	Guar gum	-	[35]
Mild steel	Gum exudate	Drinking water	[36]
Mild steel	Chitosan	0.1 M HCl solution	[44]
Aluminum alloy	Chitosan	3.65% NaCl	[45]
Copper	Chitosan	HCl solution	[46]
Carbon steel	Modified Chitosans	1 M HCl solution	[47]
Mild steel	polyethylene glycol (PEG) cross-linked chitosan	1 M sulphamic acid solution	[48]
Copper	Plant extracts	Saline water (3.4% NaCl)	[50]

medium [40, 41]. Most of the reported modifications with pectin aims to enhance the adhesion properties onto metal [42].

Chitosan is another useful natural polymer that bears amidogen and hydroxyl functional groups. These groups are opening new routes for modifications with super-anticorrosive properties since they have active hydrogens and very reactive lone electron pairs. Additionally, the hydrogel network of the chitosan exceeds the adsorption limits, which is an important parameter for a good candidate for corrosion inhibitor. So that, different derivatives of chitosan are used a corrosion inhibitor [43]. Also, it was applied as a corrosion inhibitor for steel in HCl medium [44] A6063 alloy [45], copper in the acidic medium [46]. Hussein *et al.* (2013) reported very successful types of modified chitosan for corrosion inhibition [47]. The PEG crosslinked form of the chitosan was also reported by Chauhan *et al.* (2018) as an effective inhibitor tool in the sugar industry [48].

Silk is also examined for corrosion inhibition as a natural fiber coating [49]. Hence, the required surface adhesion strength on the metal surface could not be achieved so that, silk could not late last long in the field of corrosion inhibition. The other related natural composite using studies are given in Table 23.2.

23.4.4 Energy Applications

The usage of the biomaterial varies upon the desired feature for the energy application device. In terms of fuel cells, biocomposites can be utilized as electrode-modifying material, membrane, medium, or in the whole process depending on the desired examination subject. A general scheme of a microbial fuel cell is given in Figure 23.3 [51]. The priority of the fuel cell studies is to reach cost-effective and environmentally friendly, easily late-lasting materials with the desired amount of energy productive materials. There have been many attempts to include natural materials in the energy field from various areas (Table 23.3). Natural fibers, biopolymers including chitosan [52] cellulose, and carrageenan [53], biofilms [54], biocomposites, cotton, silk, wool, etc… [55–58].

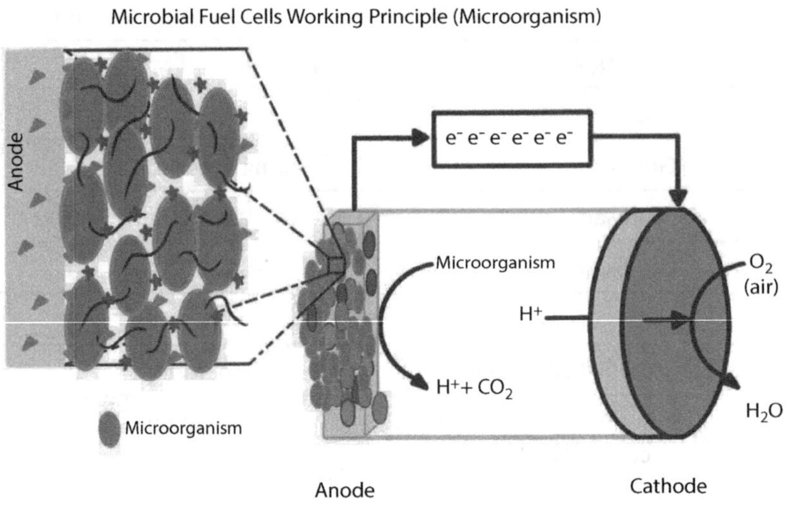

Figure 23.3 A general scheme of a microbial fuel cell [38].

Table 23.3 Some of the selected biomaterial using energy studies.

Biomaterial	Application type	Yield	Reference
Pectin with ammonium thiocyanate	PEM for fuel cell	1.5×10^{-3} S cm^{-1}	[63]
Chitosan –gelatin	Chitosan-gelatin derived hybrid porous carbon electrode	331 F g^{-1}	[85]
$LiClO_4$-doped plasticized chitosan and poly(ethylene glycol)	$LiClO_4$-doped plasticized chitosan and poly(ethylene glycol) blend as electrolyte	47 F g^{-1}	[86]
Chitosan-ZnO/PANI	CS-ZnO/PANI electrode	587.15 F g^{-1}	[87]
Chitosan	N,O-Chitosan porous carbon electrode	287.94 F g^{-1}	[88]
Chitosan	Glucose oxidase-based Enzymatic Fuel Cell Membrane	3 mA cm^{-2} Current density	[89]
Chitosan-sulfonated polyvinylidene fluoride	Direct Methanol Fuel Cell Membrane	2.85×10^{-2} S cm^{-1} at 90 °C.	[90]
Diethanolamine modified pectin	PEM for fuel cell	2.536×10^{-2} S cm^{-1}	[91]

One of the main problems of the fuel cells or batteries is the increase of the temperature, therefore the spoilage of the materials or prevention of the charge transfer efficacy occur. Chitosan is obtained from the chitin, which has too many acetylated or deacetylated functional groups [59].

The maintenance of these groups makes chitosan resistant to temperature changes up to 200 °C. Another important advantage of the chitosan for energy-producing devices is its high tensile strength and significant water uptake capability [60, 61].

So, the chitosan is reported in various fields, such as a membrane or electrode modifier up to intermediate temperatures in microbial, enzymatic, alkaline, polymer electrolyte membrane (PEM), or direct methanol fuel cells [52]. Replacing petroleum-originated polymers, such as Nafion with eco-friendly biopolymer-based membranes [53, 54] is extremely important [55]. Adiera *et al.* (2020) reported a chitosan membrane as a better conductive polymer to Perfluorosulphonic acid-based Nafion membrane [56].

Other methods including chemical modifications of chitosan with functional polymeric substances applied in PEM fabrication are another point of interest among fuel cell applications [53, 62]. These modifications are also performed for pectin, for example, biopolymer pectin with ammonium thiocyanate was synthesized to serve as polymer electrolyte [63]. Kiruthika *et al.* (2020) reported a series of such biopolymers modified with $Mg(NO_3)_2$ and examined their electrolyte performances [64]. Perumal and Selvin also reported a sustained pectin electrolyte [65].

In general, pectin has a very large amount of carboxylic acid groups with other α-galacturonic acid and methyl ester groups, which have plenty of donor electrons that are very convenient to produce metal complexes, carbonate derivatives, or metal oxides [66]. So, pectin is generally used as a highly porous material precursor for energy applications. Fan *et al.* [67] and Zhou *et al.* [68] reported the use of pectin in the same manner for supercapacitor applications.

Nanomaterial addition is another type of biocomposite production method. Especially in the self-assembled monolayer formation on natural fibers, such as cotton, silk, or wool, they gain hyper-properties like thermal, chemical, and mechanical resistance, superior antibacterial electrochemical, and flexibility properties [69]. These nanoequipped composites are successfully applied in PEMs, sensors, energy harvesters, or flexible supercapacitors. They are utilized as wearable electronic devices and provide very practical use, such as being machine-washable [70]. Stretchable supercapacitors have gained huge attention in the field of wearable electronic health devices due to their high energy supply in a very short time, low cost, safety, and ease of use [71]. Figure 23.4 shows the basic supercapacitor surface components [18].

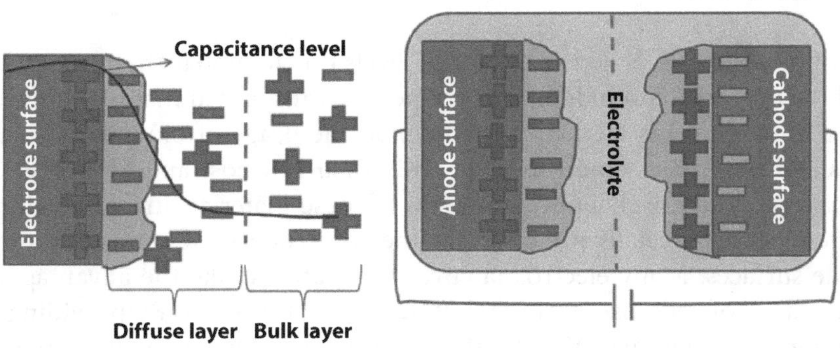

Figure 23.4 General illustration of an electronic double layer supercapacitor [18].

Silk is produced in ß-folded fibrinous protein by Bombyx mori worms including high N content [72]. It has nonwoven, random 3D fibrils fixated by sericin and not crosslinked as chitosan has [73]. Due to this ß-folded structural feature, silk provides a very high stability that long-lasting degradation. So that it is widely adressed for especially biocompatibility and long-life needed studies, such as wound healing, enzyme or live cell including fuel cells, membrane technologies. In one of those studies Pasternak *et al.*, developed a silk-based separator in a fuel cell [72]. In another bioanode production flexible silk cocoons were used [73], and Mukherjee *et al.* used the silk as a precursor for porous bioanode development [74].

In a study reported by Hwang *et al.* (2020) Ag nanowires combined with PEDOT: PPSS composite coating. Here silk yarn was used as the substrate for Ag nanowire production [71]. The conductivity of these yarns achieved attainable results that are promising for energy applications.

Biomass-derived carbon-rich materials are also used as electrode materials after the calcination and ball-milling process [75]. In this type of study Rubber seed shells (RSS) were used for this purpose and activated by KOH in addition to achieving high energy yields [76].

One of the most useful water-soluble linear polysaccharide widely used in energy applications is carrageenan. The main sources of carrageenans are seaweeds and algae. They have three types as iota, kappa, and lambda, according to the ester sulfate groups' location but overall the carrageenan structure includes very reactive hydroxyl and oxygen groups especially with cations. The film-forming ability of this polysaccharide is very appreciated however, the hydrophilic structure of the film has to be modified to improve the tensile strength. So that they are mostly preferred at the electrode surface modifications [77, 78] then blended with different polymers like polypyrrole or ionic liquids to obtain promising electrolytes or more stable membranes [78–80]. Cellulose is another promising biopolymer that can be synthesized in microcrystalline or nanocrystalline or fiber forms according to the need of the application. This water-insoluble biopolymer is preferred in electrochemical devices due to its controllable size and ready to modify structure in a cost-effective way instead of using expensive nanomaterials [81]. Generally, water-insoluble gelatin structures composed with the water-soluble carrageenan for example in the fabrication of promising membranes [74, 82], enzyme immobilization [83] or composed with ionic liquids [84].

23.5 Conclusion

The usage of biosources reached very crucial levels due to global energy source depletion. Because of the apparent and hidden costs of petroleum sourced material usage, the World has reached an end and is alert. In this manner, the usage of biomaterials, such as biocomposites and bioplastics emerged. Especially their low cost and recycling ability leads the researchers to find novel application areas of these materials including electrochemistry. Biomaterials are especially preferred for the construction of nanocomposite networks on electrode surfaces. Many electroanalytical applications take the advantage of the biocompatible nature of cellulose, carbohydrates, fibers, and so on. By providing this highly porous support, obtained detection limits are improved. Also, in energy applications, the energy yields of the devices are increased upon the featured surface properties of electrodes.

Overall, biocomposites and bioplastics are very valuable sources and matrixes for the synthesis or immobilization of the next generation, low-cost and green electrode materials for electrochemical studies.

References

1. Fung, K.W. and Mamantov, G., Electroanalytical chemistry in molten salts—A review of recent developments, in: *Advances in Molten Salt Chemistry*, J. Braunstein, G. Mamantov, G.P. Smith (Eds.), pp. 199–254, Springer, Boston, MA, 1973.
2. Scholz, F., Books on fundamental electrochemistry and electroanalytical techniques, in: *Electroanalytical Methods*, F. Scholz (Ed.), pp. 343–345, Springer, Berlin, Heidelberg, 2010.
3. Maciel, J.V., Fava, E.L., Silva, T.A., Dias, D., Fatibello-Filho, O., A combination of voltammetry of immobilized microparticles and carbon black-based crosslinked chitosan films deposited on glassy carbon electrodefor the quantification of hydroquinone in dermatologic cream samples. *J. Solid State Electrochem.*, 21, 2859, 2017.
4. Mandler, D. and Kraus-Ophir, S., Self-assembled monolayers (SAMs) for electrochemical sensing. *J. Solid State Electrochem.*, 15, 1535, 2011.
5. Uslu, B., Aboul-Enein, H.Y., Ozkan, S.A., Modern analytical electrochemistry: Fundamentals, experimental techniques, and applications. *Int. J. Electrochem.*, 2011, 2090, 2011.
6. Moretto, L.M. and Kalcher, K. (Eds.), *Environmental Analysis by Electrochemical Sensors and Biosensors*, Springer, New York, 2014.
7. Chekin, F. and Ghasemi, S., Silver nanoparticles prepared in presence of ascorbic acid and gelatin, and their electrocatalytic application. *Bull. Mater. Sci.*, 37, 1433, 2014.
8. Parkhey, P. and Mohan, S.V., Biosensing applications of microbial fuel cell: Approach toward miniaturization, in: *Biomass, Biofuels and Biochemicals, Microbial Electrochemical Technology*, S.V. Mohan, S. Varjani, A. Pandey (Eds.), pp. 977–997, Elsevier, Holland, 2019.
9. Shieh, F.K., Wang, S.C., Yen, C.I., Wu, C.C., Dutta, S., Chou, L.Y., Morabito, J.V., Hu, P., Hsu, M.-H., Wu, K.C.W., Tsung, C.K., Imparting functionality to biocatalysts via embedding enzymes into nanoporous materials by a de novo approach: Size-selective sheltering of catalase in metal–organic framework microcrystals. *J. Am. Chem. Soc.*, 137, 4276, 2015.
10. Hsu, P.H., Chang, C.C., Wang, T.H., Lam, P.K., Wei, M.Y., Chen, C.T., Chen, C.Y., Chou, L.Y., Shieh, F.K., Rapid fabrication of biocomposites by encapsulating enzymes into Zn-MOF-74 via a mild water-based approach. *ACS Appl. Mater. Interfaces*, 13, 44, 52014, 2021.
11. Borisova, B., Villalonga, M.L., Arévalo-Villena, M., Boujakhrout, A., Sánchez, A., Parrado, C., Pingarrón, J.M., Briones-Pére, A., Villalonga, R., Disposable electrochemical immunosensor for Brettanomyces bruxellensis based on nanogold-reduced graphene oxide hybrid nanomaterial. *Anal. Bioanal. Chem.*, 409, 5667, 2017.
12. Briones, C. and Moreno, M., Applications of peptide nucleic acids (PNAs) and locked nucleic acids (LNAs) in biosensor development. *Anal. Bioanal. Chem.*, 402, 3071, 2012.
13. Wipawakarn, P., Ju, H., Wong, D.K., A label-free electrochemical DNA biosensor based on a Zr (IV)-coordinated DNA duplex immobilised on a carbon nanofibre| chitosan layer. *Anal. Bioanal. Chem.*, 402, 2817, 2012.
14. Eissa, S. and Zourob, M., Ultrasensitive peptide-based multiplexed electrochemical biosensor for the simultaneous detection of Listeria monocytogenes and Staphylococcus aureus. *Microchim. Acta*, 187, 1, 2020.
15. Wang, Y., Li, T., Zhang, W., Huang, Y., A hydrogen peroxide biosensor with high stability based on gelatin-multiwalled carbon nanotubes modified glassy carbon electrode. *J. Solid State Electrochem.*, 18, 1981, 2014.

16. Farooq, U., Ullah, M.W., Yang, Q., Aziz, A., Xu, J., Zhou, L., Wang, S., High-density phage particles immobilization in surface-modified bacterial cellulose for ultra-sensitive and selective electrochemical detection of Staphylococcus aureus. *Biosens. Bioelectron.*, 157, 112163, 2020.
17. Gomes, N.O., Carrilho, E., Machado, S.A.S., Sgobbi, L.F., Bacterial cellulose-based electrochemical sensing platform: A smart material for miniaturized biosensors. *Electrochim. Acta*, 349, 136341, 2020.
18. Gong, W., Han, Q., Chen, Y., Wang, B., Shi, J., Wang, L., Cai, L., Meng, Q., Zhang, Z., Liu, Q., Yang, Y., Yang, J., Zheng, L., Li, Y., Ma, Y., A glucose biosensor based on glucose oxidase fused to a carbohydrate binding module family 2 tag that specifically binds to the cellulose-modified electrode. *Enzyme Microb. Technol.*, 150, 109869, 2021.
19. Aslan, S., Bal Altuntaş, D., Koçak, C., Subaşat, H.K., Electrochemical evaluation of titanium (IV) oxide/polyacrylonitrile electrospun discharged battery coals as supercapacitor electrodes. *Electroanalysis*, 33, 120, 2021.
20. Velusamy, V., Palanisamy, S., Chen, S.M., Chen, T.W., Selvam, S., Ramaraj, S.K., Lou, B.S., Graphene dispersed cellulose microfibers composite for efficient immobilization of hemoglobin and selective biosensor for detection of hydrogen peroxide. *Sens. Actuators B Chem.*, 252, 175, 2017.
21. Sartori, R., Layering of a film of carboxymethyl-botryosphaeran onto carbon black as a novel sensitive electrochemical platform on glassy carbon electrodes for the improvement in the simultaneous determination of phenolic compounds. *Sens. Actuators B Chem.*, 287, 18, 2019.
22. Alves, G.M., da Silva, J.L., Stradiotto, N.R., A novel citrus pectin-modified carbon paste electrochemical sensor used for copper determination in biofuel. *Measurement*, 169, 108356, 2021.
23. Arévalo, F.J., Osuna, Y., Sandoval-Cortez, J., Di Tocco, A., Granero, A.M., Robledo, S.N., Zon, M.A., Vettorazzi, N.R., Martinez, J.L., Segura, E.P., Ilina, A., Fernandez, H., Development of an electrochemical sensor for the determination of glycerol based on glassy carbon electrodes modified with a copper oxide nanoparticles/multiwalled carbon nanotubes/pectin composite. *Sens. Actuators B Chem.*, 244, 949, 2017.
24. Rani, K.K., Liu, Y.X., Devasenathipathy, R., Yang, C., Wang, S.F., Simple preparation of gold nanoparticle-decorated copper cross-linked pectin for the sensitive determination of hydrogen peroxide. *Ionics*, 25, 309, 2019.
25. Mahmoudi, Z., Tashkhourian, J., Hemmateenejad, B., Voltammetric determination of lactic acid in milk samples using carbon paste electrode modified with chitosan-based magnetic molecularly imprinted polymer. *J. Appl. Electrochem.*, 52, 35, 2022.
26. Chiessi, E. and Pispisa, B., Polymer-supported catalysis: Oxidation of catecholamines by Fe (III) and Cu (II) complexes immobilized to chitosan. *J. Mol. Catal.*, 87, 177, 1994.
27. Zhao, C.Z., Egashira, N., Kurauchi, Y., Ohga, K., Substrate selectivity of an electrochemiluminescence Pt electrode coated with a $Ru(bpy)_3^{2+}$-modified chitosan/silica gel membrane. *Anal. Sci.*, 13, 333, 1997.
28. Zhang, M., Smith, A., Gorski, W., Carbon nanotube-chitosan system for electrochemical sensing based on dehydrogenase enzymes. *Anal. Chem.*, 76, 5045, 2005.
29. Pandey, A. and Raja, A.N., Recent development in chitosan-based electrochemical sensors and its sensing application. *Int. J. Biol. Macromol.*, 164, 4231, 2020.
30. Zhai, X., Cao, Y., Liu, H., Determination of hydrogen peroxide using electrochemical sensor modified with N, P, S co-doped porous carbon/chitosan-nano copper. *J. Anal. Chem.*, 76, 891, 2021.
31. Hou, X., Wu, W., Zhao, F., Xie, W., Yang, Q., Construction of an electrochemical sensor with graphene aerogel doped with ZrO_2 nanoparticles and chitosan for the selective detection of luteolin. *Microchim. Acta*, 86, 188, 1, 2021.

32. Thanh, C.T., Binh, N.H., Duoc, P.N.D., Thu, V.T., Van Trinh, P., Anh, N.N., Van Tu, N., Tuyen, N.V., Van Quynh, N., Tu, V.C., Thao, B.T.P., Thang, P.D., Abe, H., Van Chuc, N., Electrochemical sensor based on reduced graphene oxide/double-walled carbon nanotubes/octahedral Fe_3O_4/chitosan composite for glyphosate detection. *Bull. Environ. Contam. Toxicol.*, 106, 1017, 2021.
33. Tvorynska, S., Barek, J., Josypčuk, B., Flow amperometric uric acid biosensors based on different enzymatic mini-reactors: A comparative study of uricase immobilization. *Sens. Actuators B Chem.*, 344, 130252, 2021.
34. Kavetskyy, T., Smutok, O., Demkiv, O., Maťko, I., Švajdlenková, H., Šauša, O., Novake, I., Bereke, D., Čechovad, K., Peczf, M., Nykolaishyn-Dytsoa, O., Wojnarowska-Nowakg, R., Brodah, D., Gonchar, M., Zgardzińska, B., Microporous carbon fibers as electroconductive immobilization matrixes: Effect of their structure on operational parameters of laccase-based amperometric biosensor. *Mater. Sci. Eng. C*, 109, 110570, 2020.
35. Abdallah, M., Guar gum as corrosion inhibitor for carbon steel in sulphuric acid solutions. *Port. Electrochim. Acta*, 22, 161, 2004.
36. Buchweishaija, I. and Mhinzi, G.S., Natural products as a source of environmentally friendly corrosion inhibitors: The case of gum exudates from Acacia seyal var. seyal. *Port. Electrochim. Acta*, 26, 257, 2008.
37. Lim, J., Yoo, J., Ko, S., Lee, S., Extraction and characterization of pectin from Yuza (Citrus junos) pomace: A comparison of conventional-chemical and combined physical–enzymatic extractions. *Food Hydrocoll.*, 29, 160, 2012.
38. Min, B., Lim, J., Ko, S., Lee, K.G., Lee, S.H., Lee, S., Environmentally friendly preparation of pectins from agricultural byproducts and their structural/rheological characterization. *Bioresour. Technol.*, 102, 3855, 2011.
39. Abou-Elseoud, W.S., Abdel-karim, A.M., Hassan, E.A., Hassan, M.L., Enzyme- and acid-extracted sugar beet pectin as green corrosion inhibitors for mild steel in hydrochloric acid solution. *Carbohydr. Polym. Technol. Appl.*, 2, 100072, 2021.
40. Khairou, K.S. and El Sayed, A., Inhibition effect of some polymers on the corrosion of Cd in a hydrochloric acid solutions. *J. Appl. Polym. Sci.*, 88, 866, 2003.
41. Arthur, D.E., Jonathan, A., Ameh, P.O., Anya, C., A review on the assessment of polymeric materials used as corrosion inhibitor of metals and alloys. *Int. J. Ind. Chem.*, 4, 1, 2013.
42. Geethanjali, R., Sabirneeza, A., Subhashini, S., Water-soluble and biodegradable pectin-grafted polyacrylamide and pectin-grafted polyacrylic acid: Electrochemical investigation of corrosion-inhibition behaviour on mild steel in 3.5% NaCl media. *Ind. J. Mater. Sci.*, 2014, 356075, 2014.
43. Wang, D., Li, Y., Chang, T., Luo, A., Experimental and theoretical studies of chitosan derivatives as green corrosion inhibitor for oil and gas well acid acidizing. *Colloids Surf. A*, 628, 127308, 2021.
44. Umoren, S.A., Banera, M.J., Alonso-Garcia, T., Gervasi, C.A., Mirífico, M.V., Inhibition of mild steel corrosion in HCl solution using chitosan. *Cellulose*, 20, 2529, 2013.
45. Fayomi, O.S.I., Akande, I.G., Popoola, A.P.I., Corrosion protection effect of chitosan on the performance characteristics of A6063 alloy. *J. Bio-Tribo-Corros.*, 4, 73, 2018.
46. Jmiai, A., El Ibrahimi, B., Tara, A., Oukhrib, R., El Issami, S., Jbara, O., Bazzi, L., Hilali, M., Chitosan as an eco-friendly inhibitor for copper corrosion in acidic medium: Protocol and characterization. *Cellulose*, 24, 3843, 2017.
47. Hussein, M.H., El-Hady, M.F., Shehata, H.A., Hegazy, M.A., Hefni, H.H., Preparation of some eco-friendly corrosion inhibitors having antibacterial activity from seafood waste. *J. Surfactants Deterg.*, 16, 233, 2013.
48. Chauhan, D.S., Srivastava, V., Joshi, P.G., Quraishi, M.A., PEG cross-linked chitosan: A biomacromolecule as corrosion inhibitor for sugar industry. *Int. J. Ind. Chem.*, 9, 363, 2018.

49. Fang, H., Wang, C., Zhou, S., Li, G., Tian, Y., Suga, T., Exploration of the enhanced performances for silk fibroin/sodium alginate composite coatings on biodegradable Mg–Zn–Ca alloy. *J. Magnes. Alloy.*, 9, 1578, 2021.
50. El-Dahan, H.A., Natural products as corrosion inhibitors for copper in saline water in presence of sulphide. *Egypt. J. Chem.*, 49, 589, 2006.
51. Anık, U., Tutum, M., Aslan, S., Fabrication of biofuel cell based on nanomaterial modified composite glassy carbon paste electrode. *Curr. Anal. Chem.*, 12, 54, 2016.
52. Ma, J. and Sahai, Y., Chitosan biopolymer for fuel cell applications. *Carbohydr. Polym.*, 92, 955, 2013.
53. Rosli, N.A.H., Loh, K.S., Wong, W.Y., Yunus, R.M., Lee, T.K., Ahmad, A., Chong, S.T., Review of chitosan-based polymers as proton exchange membranes and roles of chitosan-supported ionic liquids. *Int. J. Mol. Sci.*, 21, 632, 2020.
54. Şengül, E.E., Göktürk, T., Topkaya, C.G., Gup, R., Synthesis, characterization and dna interaction of Cu (II) complexes with hydrazone-Schiff base ligands bearing alkyl quaternary ammonium salts. *J. Chil. Chem. Soc.*, 65, 4754, 2020.
55. Vinod, A., Sanjay, M.R., Suchart, S., Jyotishkumar, P., Renewable and sustainable biobased materials: An assessment on biofibers, biofilms, biopolymers and biocomposites. *J. Clean. Prod.*, 258, 120978, 2020.
56. Adiera, N.R.H., Loh, K.S., Wong, W.Y., Yunus, R.M., Lee, T.K., Ahmad, A., Chong, S.T., Review of chitosan-based polymers as proton exchange membranes and roles of chitosan-supported ionic liquids. *Int. J. Mol. Sci.*, 21, 632, 2020.
57. Mehra, R., Bhushan, S., Gill, B.S., Rehman, W.U., Bast, F., Algae-based composites and their applications, in: *Biocomposites*, S. Ahmed, S. Ikram, S. Kanchi, K. Bisetty (Eds.), pp. 163–179, Jenny Stanford Publishing, Boca Raton, 2018.
58. Işık, C., Saraç, N., Teke, M., Uğur, A., A new bioremediation method for removal of wastewater containing oils with high oleic acid composition: Acinetobacter haemolyticus lipase immobilized on eggshell membrane with improved stabilities. *New J. Chem.*, 45, 1984, 2021.
59. Chai, L., Qu, Q., Zhang, L., Shen, M., , Zhang, Li Chitosan, a new and environmental benign electrode binder for use with graphite anode in lithium-ion batteries. *Electrochim. Acta*, 105, 378, 2013.
60. Salleh, N.A., Kheawhom, S., Mohamad, A.A., Chitosan as biopolymer binder for graphene in supercapacitor electrode. *Results Phys.*, 25, 104244, 2021.
61. Ng, L. and Mohamad, A., Effect of temperature on the performance of proton batteries based on chitosan–NH_4NO_3–EC membrane. *J. Membr. Sci.*, 325, 653, 2008.
62. Ikram, S., Ahmed, S. *et al.*, Chitosan-based polymer electrolyte membranes for fuel cell applications, in: *Organic-Inorganic Composite Polymer Electrolyte Membranes*, D. Inamuddin, A. Mohammad, A. Asiri (Eds.), pp. 381–398, Springer, Cham, 2017.
63. Muthukrishnan, M., Shanthi, C., Selvasekarapandian, S., Manjuladevi, R., Perumal, P., Selvin, P.C., Synthesis and characterization of pectin-based biopolymer electrolyte for electrochemical applications. *Ionics*, 25, 203, 2019.
64. Kiruthika, S., Malathi, M., Selvasekarapandian, S., Tamilarasan, K., Moniha, V., Manjuladevi, R., Eco-friendly biopolymer electrolyte, pectin with magnesium nitrate salt, for application in electrochemical devices. *J. Solid State Electrochem.*, 23, 2181, 2019.
65. Perumal, P. and Selvin, P.C., Boosting the performance of electric double layer capacitor via engaging pectin macromolecular electrolyte with elevated ionic conductivity and potential window stability. *Chem. Eng. J. Adv.*, 8, 100178, 2021.
66. Naqash, F., Masoodi, F.A., Rather, S.A., Wani, S.M., Gani, A., Emerging concepts in the nutraceutical and functional properties of pectin a review. *Carbohydr. Polym.*, 168, 227, 2017.

67. Fan, Y., Liu, P.F., Yang, Z.J., Jiang, T.W., Yao, K.L., Han, R., Huo, X.X., Xiong, Y.Y., Bi-functional porous carbon spheres derived from pectin as electrode material for supercapacitors and support material for Pt nanowires towards electrocatalytic methanol and ethanol oxidation. *Electrochim. Acta*, 163, 140, 2015.
68. Zhou, Y., Ren, X., Du, Y., Jiang, Y., Wan, J., Ma, F., In-situ template cooperated with urea to construct pectin-derived hierarchical porous carbon with optimized pore structure for supercapacitor. *Electrochim. Acta*, 355, 136801, 2020.
69. Gao, L., Feng, J., Xu, S., Shi, M., Yao, L., Wang, L., Yang, Z., General strategy to prepare single-layered Ag–Au–Pt nanocrystal ternary-coated biomass textiles through polymer-driven self-assembly. *Nanomaterials*, 10, 495, 2020.
70. Shao, G., Yu, R., Chen, N., Ye, M., Liu, X.Y., Stretchable supercapacitors: From materials and structures to devices. *Small Methods*, 5, 2000853, 2021.
71. Hwang, B., Lund, A., Tian, Y., Darabi, S., Muller, C., Machine-washable conductive silk yarns with a composite coating of Ag nanowires and PEDOT: PSS. *ACS Appl. Mater. Interfaces*, 12, 27537, 2020.
72. Pasternak, G., Yang, Y., Santos, B.B., Brunello, F., Hanczyc, M.M., Motta, A., Regenerated silk fibroin membranes as separators for transparent microbial fuel cells. *Bioelectrochemistry*, 126, 146, 2019.
73. Lu, M., Qian, Y., Yang, C., Huang, X., Li, H., Xie, X., Huang, L., Huang, W., Nitrogen-enriched pseudographitic anode derived from silk cocoon with tunable flexibility for microbial fuel cells. *Nano Energy*, 32, 382, 2017.
74. Mukherjee, P., Mishra, N.S., Saravanan, P., Polydopamine modified silk fibroin 3-D anode for enhanced microbial fuel cell operation. *Sustain. Energy Technol. Assess.*, 49, 101696, 2022.
75. Bal Altuntaş, D., Aslan, S., Akyol, Y., Nevruzoglu, V., Synthesis of new carbon material produced from human hair and its evaluation as electrochemical supercapacitor. *Energ. Source Part A*, 42, 2346, 2020.
76. Farma, R., Husni, H., Apriyani, I., Awitdrus, A., Taer, E., Biomass waste-derived rubber seed shell functionalized porous carbon as an inexpensive and sustainable energy material for supercapacitors. *J. Electron. Mater.*, 50, 6910, 2021.
77. Christwardana, M., Handayani, A.S., Yudianti, R., Cellulose–carrageenan coated carbon felt as potential anode structure for yeast microbial fuel cell. *Int. J. Hydrogen Energy*, 46, 6076, 2021.
78. Sedayu, B.B., Cran, M.J., Bigger, S.W., A review of property enhancement techniques for carrageenan-based films and coatings. *Carbohydr. Polym.*, 216, 287, 2019.
79. Esmaeili, C., Ghasemi, M., Heng, L.Y., Hassan, S.H., Abdi, M.M., Daud, W.R.W., Ilbeygi, H., Ismail, A.F., Synthesis and application of polypyrrole/carrageenan nano-bio composite as a cathode catalyst in microbial fuel cells. *Carbohydr. Polym.*, 114, 253, 2014.
80. Moniha, V., Alagar, M., Selvasekarapandian, S., Sundaresan, B., Boopathi, G., Conductive bio-polymer electrolyte iota-carrageenan with ammonium nitrate for application in electrochemical devices. *J. Non-Cryst. Solids*, 481, 424, 2018.
81. Azizi Samir, M.A., Alloin, F., Dufresne, A., Review of recent research into cellulosic whiskers, their properties and their application in nanocomposite field. *Biomacromolecules*, 6, 612, 2005.
82. Kassab, Z., Aziz, F., Hannache, H., Youcef, H.B., El Achaby, M., Improved mechanical properties of k-carrageenan-based nanocomposite films reinforced with cellulose nanocrystals. *Int. J. Biol. Macromol.*, 123, 1248, 2019.
83. van de Velde, F., Lourenco, N.D., Pinheiro, H.M., Bakker, M., Carrageenan: A food-grade and biocompatible support for immobilisation techniques. *Adv. Synth. Catal.*, 344, 815, 2002.
84. Prasad, K., Kaneko, Y., Kadokawa, J.I., Novel gelling systems of k-, i- and l-carrageenans and their composite gels with cellulose using ionic liquid. *Macromol. Biosci.*, 9, 376, 2009.

85. Yuan, M., Zhang, Y., Niu, B., Jiang, F., Yang, X., Li, M., Chitosan-derived hybrid porous carbon with the novel tangerine pith-like surface as supercapacitor electrode. *J. Mater. Sci.*, 54, 14456, 2019.
86. Sudhakar, Y.N., Selvakumar, M., Bhat, D.K., $LiClO_4$-doped plasticized chitosan and poly (ethylene glycol) blend as biodegradable polymer electrolyte for supercapacitors. *Ionics*, 19, 277, 2013.
87. Pandiselvi, K. and Thambidurai, S., Chitosan-ZnO/polyaniline ternary nanocomposite for high-performance supercapacitor. *Ionics*, 20, 551, 2014.
88. Zhang, X.L., Feng, C.N., Li, H.P., Zheng, X., C, N, O self-codoped hierarchical porous carbon from chitosan for supercapacitor electrode active materials. *Cellulose*, 28, 437, 2021.
89. Bahar, T., Development of reasonably stable chitosan based proton exchange membranes for a glucose oxidase based enzymatic biofuel cell. *Electroanalysis*, 32, 536, 2020.
90. Vijayalekshmi, V. and Khastgir, D., Chitosan/partially sulfonated poly (vinylidene fluoride) blends as polymer electrolyte membranes for direct methanol fuel cell applications. *Cellulose*, 25, 661, 2018.
91. Mishra, R.K., Datt, M., Banthia, A.K., Majeed, A.B.A., Development of novel pectin based membranes as proton conducting material. *Int. J. Plast. Technol.*, 16, 80, 2012.

24

Biofibers and Their Composites for Industrial Applications

Meshude Akbulut Söylemez[1], Kemal Özer[2] and Demet Ozer[1]*

[1]Hacettepe University, Department of Chemistry, Ankara, Turkey
[2]YKC Plastic Packaging and Caps, Sincan OSB, Ankara, Turkey

Abstract
Biofibers have a superior potential with their unique properties in industrial applications and the use of biofibers is an inevitable assignment for researchers due to the depletion of fossil fuels, climate change, CO_2 emissions, and the carbon footprint of plastics. In this chapter, the properties of seed, leaf, bast, and stalk fibers were summarized. The main drawbacks of these natural fibers as reinforcing agents are fiber quality, fire resistance, moisture adsorption, fiber/matrix adhesion, processing, and manufacturing conditions. To overcome these limitations, various chemical and physical methods can be applied. The chemical (alkalization, silanization, benzoylation, acetylation) and the physical modification methods (plasma, ultrasound, ultraviolet treatments) were explained in detail. The main goals of these methods are the cleaning or/and modifying of the fiber surface to improve hydrophobicity, compatibility, sustainability, and durability. All these modification methods enlarge the mechanical durability and thermal features of the biocomposites through the progress of the interfacial adhesion forces among the modified fiber and polymers. Therefore, after pretreatment, the biofibers and their composites with better properties can be used as a crucial alternative instead of the petroleum-based composite in industrial applications.

***Keywords*:** Biofiber, biocomposite, natural fiber, types of biofiber, modification of biofiber, industrial applications

24.1 Introduction

In addition to the high rate of the consumption of petroleum sources, the awareness in ecological, social, and economical sites causes the search for green and sustainable materials. By 2020, the US Department of Agriculture and Energy aims to produce leastways 10% of overall fundamental chemicals from biobased resources, with a target of reaching 50% by 2050 [1]. Insufficient waste disposal and restrictive policies of the European about the cleaner and safer environment push the researchers to develop new materials, which are biocompatible and biodegradable. In general, it is possible to describe these materials as biopolymers or biocomposites due to their compositions. Biopolymers are polymers that

**Corresponding author*: demetbaykan@hacettepe.edu.tr

Inamuddin and Tariq Altalhi (eds.) Handbook of Bioplastics and Biocomposites Engineering Applications: 2nd Edition, (513–538) © 2023 Scrivener Publishing LLC

are prepared from natural resources. Some of these biopolymers are biocompatible polymers that can be used in or with biological systems.

Natural fibers are a crucial component of biocomposites due to their simple availability and abundance, as well as their multiple benefits, such as low cost, biodegradability, and low density [2]. The main advantages of biofibers are their mechanical properties and lower densities than synthetic fibers and metals. The mechanical and thermal features of composites made from synthetic or natural polymers are improved by these lighter and stronger materials [3]. Biocomposites can be prepared by reinforcing synthetic or natural polymers with various natural fibers like cotton, kenaf, sisal, jute, and hemp [3]. However, the natural fibers need to be modified before usage in the production of composites. Natural fibers' main problem is their hydrophilic characteristic, which makes them inconsonant with polymeric matrixes. To get over this constraint, physical, chemical, and coupling agents can be used to increase the hydrophobicity of the fiber surfaces, resulting in greater interfacial adhesion between the natural fiber and polymer. The modification of the natural fibers can be characterized by employing several spectroscopic and microscopic methods. Alkali treatment provides the effective removal of the noncellulosic constituents [4]. The removal of the amorphous component increases the crystallinity of the natural fibers, which improves the biocomposites' mechanical and thermal stability [3]. The silanization provides the surface coating of the fibers by a hydrophobic film which provides the reduction of the water adsorption while a raising in the fiber adhesion to the polymer matrix [5]. The acetylation and benzoylation produce the hydrophobic species on the surface of the fiber that improve the compatibility between fiber and polymer [6]. Physical treatments are employed to clean fiber surfaces. Plasma treatment provides an increase in the roughness and introduces active species, which provides the functionalization of the fiber surface with various functional groups, such as COOH, OH, NH_2, C-F species [7]. All the modifications improve the mechanical stability, thermal stability, and water resistance of the biocomposites prepared by natural fiber reinforcing. All these improvements offer a wide variety of the applications like automotive, household, cosmetics, construction, aerospace, etc. [8].

In this chapter, the properties of biofibers were presented in detail. The impacts of the chemical and physical methods on the thermal, mechanical, and water uptake capacity of the fibers were also discussed. The examined chemical treatments are alkalization, silanization, benzoylation, and acetylation while plasma, ultrasound, and ultraviolet modifications are the physical methods. The changes in the structure and morphology of the biocomposites which can be investigated by FTIR and SEM were mentioned. The increment of the thermal and mechanical stability and the reduction of the water uptake capacity of the biocomposites which are prepared by employing various polymers as matrices were also discussed. In addition, the industrial uses of biocomposites, processing challenges, and future perspectives of biocomposites were assessed.

24.2 Types of Biofibers

In recent years, natural fibers and their composites have gained great attention and seemed like a great alternative for synthetic fibers due to their superior biodegradability, recyclability, renewability, rigidity, nontoxicity, eco-friendly, low-energy consumption, and low cost. The biofibers are classified into four main categories as seed fibers, leaf fibers, bast fibers, and stalk fibers as given in Figure 24.1 [9, 10]. Plant-based fibers are made up of cellulose, hemicellulose, lignin, pectin, wax, ash, and moisture. The percentages of the components differ from each other (Table 24.1) and some properties of various biofibers [11] were given in Table 24.2. According to their chemical composition, their composites show different structural, mechanical, surface, and thermal properties, and gain a great potential for various industrial applications.

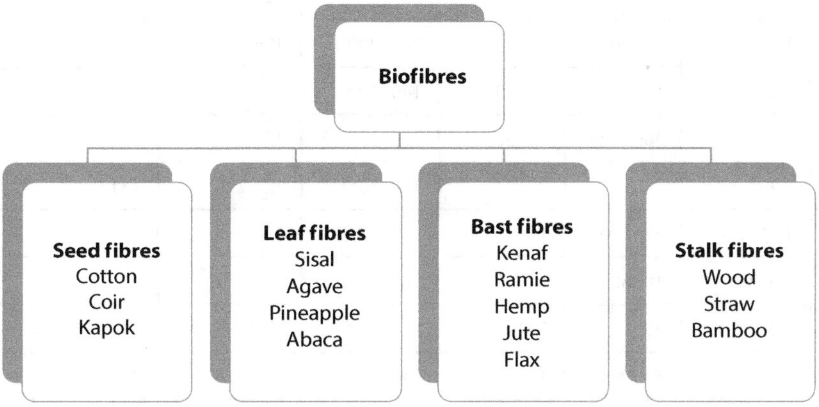

Figure 24.1 Types of biofibers.

Table 24.1 The chemical compositions of biofibers [9, 10].

Fibres	Cellulose %	Hemi-cellulose %	Lignin %	Pectin %	Waxes %	Ash	Moisture
Cotton	89	4	0.75	6	0.6	–	–
Coir	32–43	0.15–0.25	40–45	3–4	–	–	8
Kapok	38.1	–	14.1	–	2.3	–	–
Sisal	47–77	10–24	7–11	10	–	1.0	11
Pineapple	70–82	–	5–12	–	–	1.1	–
Abaca	78	–	8.8	–	–	1.1	–
Kenaf	53.5	21	17	2	–	2–5	–
Ramie	68.6–91	5–16.7	0.6–0.7	1.9	–	–	9
Hemp	57–77	14–22.4	3.7–13	0.9	–	0.8	9
Jute	45–71.5	13.6–21	12–26	0.2	–	1.5	12
Flax	71	18.6–20.6	2.2	2.3	1.7	5–10	8–12
Bamboo	73.83	12.49	10.15	0.37	–	–	3.16–8.9

Table 24.2 Physical and mechanical properties of biofibers (Copyright [11] permission from Elsevier).

Fibre type	Density (g/cm³)	Length (mm)	Diameter (μm)	Tensile strength (MPa)	Tensile modulus (GPa)	Specific modulos (approx.)	Elongation (%)
E-glass	2.5–2.59	–	<17	2000–3500	70–76	29	1..8–4.8
Abaca	1.5	–	–	400–980	6.2–20	9	1.0–10
Alfa	0.89	–	–	35	22	25	5.8
Bagasse	1.25	10–300	10–34	222–290	17–27.1	18	1.1
Bamboo	0.6–1.1	1.5–4	25–40	140–800	11–32	25	2.5–3.7
Banana	1.35	300–900	12–30	500	12	9	1.5–9
Betelnut	0.2–0.4	–	–	120–166	1.3–2.6	1–1.9	22–24
Coir	1.15–1.46	20–150	10–460	95–230	2.8–6	4	15–51.14
Cotton	1.5–1.6	10–60	10–45	287–800	5.5–12.6	6	3–10
Curaua	1.4	35	7–10	87–1150	11.8–96	39	1.3–4.9
Flax	1.4–1.5	5–900	12–600	343–2000	27.6–103	45	1.2–3.3
Hemp	1.4–1.5	5–55	25–500	270–900	23.5–90	40	1–3.5
Henequen	1.2	–	–	430–570	10.1–16.3	11	3.7–5.9
Isora	1.2–1.3	–	–	500–600	–	–	5–6
Jute	1.3–1.49	1.5–120	20–200	320–800	8–78	30	1–1.8
Kenaf	1.4	–	–	223–930	14.5–53	24	1.5–2.7
Nettle	–	–	–	650	38	–	1.7
Oil palm	0.7–1.55	–	150–500	80–248	0.5–3.2	2	17–25
Piassava	1.4	–	–	134–143	1.07–4.59	2	7.8–21.9
Pineapple	1.56	–	–	150–1627	11–82	7.8	2.4
PALF	0.8–1.6	900–1500	20–80	180–1627	1.44–82.5	35	1.6–14.5
Ramie	1–1.55	900–1200	20–80	400–1000	24.5–128	60	1.2–4.0
Sisal	1.33–1.5	900	8–200	363–700	9.0–38	17	2.0–7.0

24.2.1 Seed Fibers

Cotton is a natural fiber obtained from the cotton plant (Figure 24.2). Cotton, which has been used since ancient times, is used primarily as a textile product today. Cotton likes alluvial and strong soils. It is planted in deeply plowed and well-fertilized soils. Cotton is planted in February in warm regions and March to April in cold regions. It is harvested in August and September. The biggest danger for cotton is the rains. Rains cause a decrease in yield and quality. The structure of cotton consists almost entirely of pure cellulose. Cotton fiber has

Figure 24.2 The photographs of seed fibers.

the appearance of a twisted strip, scanned under a microscope. Commercial cotton can be collected in three basic groups. 1) Fibers with a staple length between 30 and 60 mm: Kinds of cotton in this group are of the highest quality (Corn and Sea-Island cotton); 2) Fibers with the staple length between 20 and 30 mm: Cotton in this group are bulky and medium-length (American-Upland cotton); 3) Fibers with a staple length of less than 20 mm: Cotton in this group are coarse and of low quality (Asian and Indian cotton). Apart from its classical use in spinning, cotton has various uses in the chemical industry, oil industry, and manufacture of artificial textiles. It is also used for treatment as combed hydrophilic cotton.

Coir fiber, also known as coconut fiber, is obtained from coconut shells (Figure 24.2). Coir fibers are good temperature insulation, noncombustible, unaltered by moisture, resistant to microbial/fungus deterioration, robust and durable, nontoxic, and easy to clean [12]. Although coconut trees are found across the world's tropical areas, Indonesia, the Philippines, India, Brazil, Thailand, Vietnam, and Malaysia generate the great majority of commercially manufactured coir fibers [13]. Up to now, the coir fiber is generally used for making floor furnishing and insulation materials, wires, carpets, cushions, brushes, sacking, textiles, garden articles, automotive, etc.

Kapok fiber is a single-celled, shiny, silky-like, light fiber obtained from the seeds of the kapok tree (Kapok, Java kapo, Ceiba pentandra) such as cotton (Figure 24.2). The kapok tree is a native plant of Tropical America, East Africa, West Africa, India, Ceylon, Java Island, Sumatra, Philippines, and Indonesia. It also grows in regions with tropical climates. In general, the regions 15 degrees north and south of the equator are the regions where good kapok products are obtained. Those grown at an altitude of up to 450 m above sea level give the most yield and best quality products. Kapok fibers are exceptionally lustrous, cream-yellow in color, and smooth in texture. It is a perishable, soft, and elastic fiber. Kapok fibers are single-celled, visible under the microscope as thin longitudinal ribbons. Its cross-section is oval or round. It is used in making bedding, pillows, and as a filler because of lightness. It is also used in the manufacture of pilot suits. Good air and heat insulation are provided due to the pores in the fiber structure. It is also the best-known sound holder. These characteristics are used in airplanes because of their lightweight and ability to act as sound insulation. It does not soak in water for a long time, it dries quickly when wet. It does not get wet quickly because the surface of the fiber is covered with wax. Kapok fiber is a good heat retainer. It is used in mattresses as it does not get wet easily and reduces the discomfort caused by sweating and humidity in the mattress with its moisture repellent structure. It is used in life jackets and lifebuoys, as it has very little moisture and water absorption. 1 kg of kapok comfortably holds a weight of 35 kg on water.

24.2.2 Leaf Fibers

Sisal is a plant species of the Agavaceae family that resembles hemp and is commonly used in the textile thanks to the strong fiber (Figure 24.3). It grows in hot and humid climates, it has good moisture adsorption and stretching properties, so it is used in ship rope construction, agriculture, and maritime. The sisal plant is a plant that grows in the Americas in the cold wind, hard unproductive soil, and hot sun. Its leaves are very fibrous and are used in weaving.

Agave leaf fiber is abundant in cellulose, relatively inexpensive, and plentifully available (Figure 24.3) [14]. Because they are found in a wide range of geographical places, Agave Americana is most likely the most weather tolerant of all Agave species. After 5–8 years, Agave Americana produces blooms [15].

Pineapple is obtained by pooling from the leaves of the Anunnus cosmos plant, which belongs to the Bromeliaceae family (Figure 24.3). Pineapple leaf fibers (PALF) are multicellular and lignocellulosic fibers like jute fiber. An edible member of the Bromeliaceae family, pineapple has about 2,000 varieties. It is the only source of bromelain, a complex proteolytic enzyme, and this enzyme is used in the pharmaceutical industry. The stem and leaves of the pineapple are a source of white, creamy, and silky fiber. Pineapple fiber has also been used in paper production due to its fineness, softness, and flexibility. Although PALF is finer than jute, it is 10 times coarser than cotton. The use of PALF, which is seen as an important fiber for the construction of agriculture-based industries, has not yet become fully widespread due to the lack of appropriate physical and chemical property information. It also contains oil, wax, uronic anhydride, pentosan, colored substances, and inorganic substances [10]. PALF is in a strip-like structure and is adhered to each other by substances that contribute to the strength of the fiber, such as lignin and pentosan. Due to its 70% cellulose content, this fiber is remarkably comparable to cotton cellulosic fibers. By limiting soil erosion, it aids in climate restoration and soil quality [16].

Abaca fibers can be made of the abaca leaf, which is a member of the Vulgaris family, used in the production of staff, such as hats and ropes, with creamy white to brown colors, durable, coarse, it is a hard fiber type (Figure 24.3). Abaka is a tropical plant from the Musaceae family, which resembles a date palm and has lived for 8 to 20 years. Adult plant length is between 4 and 6.7 m. Although unrelated to the cannabis plant, it is also known as Cebu hemp or Davao hemp. 94% of the world's abaca fiber manufacture is in the Philippines, it is also grown in Ecuador and Costa Rica. Abaca fibers have been used in the manufacture of sail cloths, ship ropes, nautical and sailor goods due to their strength and lack

Figure 24.3 The photographs of leaf fibers.

of moisture-adsorbing properties. In addition, coarse woven fabrics and pillow fillers are made.

24.2.3 Bast Fibers

Kenaf is a mallow-like plant that is grown in South Asia. Kenaf is the annual herbaceous plant with a woody base (Figure 24.4). The stems of the kenaf plant, which can grow 1.5 to 3.5 m, are 1 to 2 cm in diameter. The leaves are 10 to 15 cm lengthy and come in a variety of shapes, with three to seven strongly lobed leaves near the stem's base and shallowly lobed or lanceolate leaves near the stem's top. The flowers of the Kenaf plant are 8 to 15 cm in diameter. The fruit is a 2-cm diameter capsule including several seeds. Kenaf fiber is a soft bast fiber obtained from the kenaf plant. The fibers are long, varied in color from light yellow to grey, and are harder and brighter than jute. It is used for twine, rope, pulp production, and other industrial uses, just as jute. Kenaf is primarily grown in India, Bangladesh, Central Asia, Africa, and a few Central American countries, as well as a tiny amount in southeastern Europe. Kenaf has been grown in Africa for 4000 years but did not receive significant attention until World War II. The shortage of jute that occurred during and after the war increased the cultivation of kenaf and fiber production in other countries, such as the USA, Cuba, Mexico. Kenaf has similar properties to jute and can be used instead of jute as a substitute fiber. Since kenaf contains less lignin in its structure compared to similar raw materials, it is used in paper production as it becomes pulp faster and easier with fewer chemicals and can be bleached. Kenaf is an efficient renewable timber source as it grows rapidly annually.

Ramie is a shrubby perennial herb (Figure 24.4). Its fibers are obtained from the stems by cutting them every year. After harvesting for fiber, the plant can come back into shape for fiber in 50 to 60 days. Ramie can be harvested for fiber five to six times a year. Plant lifespan varies between 6 and 20 years. Ramie is used in fine and coarse weaving. It is used in feeding silkworms because its leaves are large and fleshy. Fabrics woven with fibers mixed with wool or cotton are in demand, especially in upholstery, as they are hard and durable. Ramie fibers are five times stronger than cotton fibers, two times stronger than hemp fibers, and four times stronger than flax fibers. Ramie fibers are mixed with cotton and linen fibers and used in textiles. Fabrics made of 55% ramie and 45% cotton fibers are highly resistant to machine washing. Due to the 20% moisture retention of ramie fibers, it is used in the production of seawater-resistant ropes and adsorbents. The fibers obtained from the ramie plant are resistant to bacteria, mildew, and insects, adsorb sweat, dry easily, and become stronger when wet. However, these fibers are not flexible, easily wrinkled, and wear out more quickly than cotton. Ramie fibers are widely used in fishing nets, making sports nets and bath towels.

Figure 24.4 The photographs of bast fibers.

Hemp is one of the most popular textile fiber due to its advanced thermomechanical properties. Besides this, hemp has been widely used as a building material and plastic part in industrial applications. It provides biodegradability, renewability, and sustainability to products. Hemp is called the fibers used in the production of materials, such as fabric, packaging yarn, composite obtained from the stem of an annual woody plant belonging to the family of cannabinaceae, close to nettles. Cannabis (Cannabis) is an annual plant genus belonging to the Cannabaceae family. It is a type of textile plant, which is the oldest source of vegetable raw material in human history, also called hemp, whose fibers in its stems are used in making yarn, weaving, and fabric, and the pulp part is used in paper-making. Its homeland is Central Asia. It is grown and cultivated in temperate and tropical regions. Type: It is hard, bushy, hollow stem, palmate-leaved, dioecious, and annual. It is harvested 120 to 140 days after sowing. As in flax, the fiber cells are in bundles in the bark part. Fiber production is carried out by rotting, forging, and carding processes as in flax. Fiber length is 40 to 45 mm. It is bright yellow or brown (Figure 24.4). Its fibers are durable and quite long. When lignin material accumulates in the fibers, the elasticity property decreases. The fibers of the genus are used in coarse weaving (such as warp threads of carpets, sacks, ropes, bags, net making, tent cloth). Its leaves have been used in medicine and cosmetics. Its seed is used as a fuel in terms of being very oily and as a food because it is very nutritious. Its seeds are also used in soap making and dye making. Its seeds are one of the favorite foods of birds. Today, most of the pesticides in the world are used in cotton cultivation, which is used in fiber production. Cannabis is a very powerful herb and hardly needs any medication. The vested interests leading to banning cannabis are also thought to be linked to the oil and pharmaceutical industries. Tetra-hydro-cannabinol (THC), which is concentrated in the bud tops of the unfertilized plant, which is shown as the reason for the ban of female cannabis, has been used as "marijuana" since the existence of humanity, and there are many studies published in scientific journals showing its medicinal uses. The active substance does not occur in the female plant, which is in the same environment as the male plant. Today, the use and breeding of male and female species have been liberalized in many countries.

Jute is a strong and coarse fiber derived from the plant stem in the Corchorus family that is primarily grown in India and Bangladesh (Figure 24.4). The jute plant, Corchorus capsularis, and Corchorus olitorius of the Linden (Tiliaceae) family are annual fiber plants. The homeland of the jute plant in East India. It is grown to obtain textile fiber. The jute plant is often referred to as the "Golden Fiber" because it retains its natural shine even after processing. Jute is an easy plant to work with its antistatic quality feature. In terms of weather conditions, the banks of the Ganges River are the most ideal jute growing region. Although it is generally used for decorative purposes, it is also used for protective purposes in the wood and construction industry. It is used as a decorative ornament fabric in many areas, such as home decoration, wedding candy, baby candy, bag, pouch, runner, door decoration, gift wrap, burlap, placemat, package decorations. Most of the jute fiber is used to make rope, sacks, rope, twine, and cover fabrics. Carpets, which are produced with fibers obtained from the jute plant and attract attention with their recyclable environmental features, have been produced intensively in recent years.

Flax fibers have been obtained from the stem. They are stronger than cotton. The mechanical properties are higher than glass fibers while their density is lower than glass fibers.

Figure 24.5 The photographs of stalk fibers.

Canada, France, Belgium, Netherlands are the leaders in the production of flax fibers. Flax fibers have a short life cycle (Figure 24.4). They are growing 100 days between March and July [17]. Fine and long flax fibers are generally used for furnishing linens textile while short flax is applied in automotive furniture.

24.2.4 Stalk Fibers

In the aging wood, the group of cells that harden and cannot conduct nutrient conduction forms the **wood fiber** (Figure 24.5). Cellulosic components, such as wood fibers, are taken from plants and used to manufacture products, such as paper.

Straw is a new alternative for eco-building materials and systems (Figure 24.5). Straw is an agricultural byproduct made up of the dry stalks of cereal plants that have been stripped of their grain and chaff. It may be used for a variety of things, including firewood, animal bedding and feed, thatching, and basket manufacturing.

Bamboo is a plant (Figure 24.5) that grows in tropical climates, reaches maturity in a short period of 3 to 4 years, and is used in many fields, such as paper, furniture, construction, food, chemical industry. The bamboo fiber used in textile is of Asian origin in terms of development and production and is obtained from the bamboo species called Moso bamboo (Phyllostachys heterocycle pubescent). Bamboo fiber finds a wide area of use in textiles and creates an alternative to other cellulosic fibers, thanks to its natural antibacterial properties, biodegradability, high moisture-holding capacity, softness and brightness, and UV rays' refraction.

24.3 Chemical and Physical Modification of Biofibers as Reinforcing Materials for Biocomposites

There is a certain need to treat natural fibers before using them as reinforcing materials to produce biocomposites. The low compatibility of the fibers with the polymer necessitates their treatment or modification. Furthermore, one of the most significant disadvantages of using fibers as reinforcing materials is their high moisture adsorption capability. Surface modification methods advance the interaction between the fabric and polymer. By the way, hemicellulose, which is mainly responsible for water adsorption, can be removed successfully by physical or chemical treatments. In this section, the physical and chemical treatment methods will be summarized.

24.3.1 Chemical Treatment Processes

24.3.1.1 Alkalization

The alkali treatment or mercerization of the natural fibers is the incubation of the fibers in an aqueous solution of NaOH. This treatment can be applied at high or ambient temperatures [4]. The alkali treatment provides to remove the noncellulosic parts of the fibers like hemicellulose, lignin, pectin, wax, etc. (Figure 24.6) [18]. Also, the alkali treatment provides the ionization of the surface hydroxyl groups to the alkoxides (Figure 24.7). The parameters like concentration of sodium hydroxide, immersion time, and temperature should be optimized [6]. Otherwise, some undesired changes of the structure of the fibers can occur like the formation of pores, defibrillation, and embrittlement of the fibers [4].

Since cellulose is the more crystalline component of the fibers, the alkaline treatment causes a significant increase in density. The effective surface area of the fibers also shows a certain increase which enhances the adhesion of the fibers to the matrices due to the increase in the roughness of the fibers because of the separation of the fibers from each other. By improving the crystallinity of natural fibers, the alkaline treatment improves their mechanical properties [19].

Cotton fibers were treated by dilute alkali solutions at high temperatures by Faisal and coworkers [20]. The tensile strength of modified fibers has been examined to rise, while the roundness of the fibers has improved. In another work, cotton fibers were treated by heat and concentrated alkaline solution [21]. The optimum condition was determined as 10% NaOH and 60 °C or 90 °C for removal of noncellulosic parts of the fibers like hemicellulose, lignin, and oxidized cellulose which is the reason for the wettability of cotton. This condition improves thermal stability and crystallinity while lowering the contact angle significantly (from 123.6° to 107.0°).

The main effects of alkali treatment on jute fibers are the disordering of hydrogen bonds in the structure and the enhancement of hydroxyl group ionization to alkoxide [22]. The treatment improves surface roughness by eliminating a substantial amount of lignin, wax,

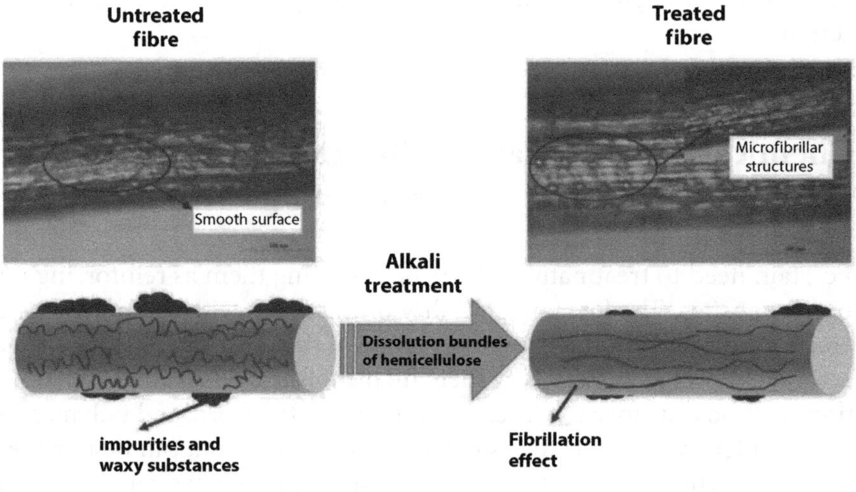

Figure 24.6 Alkali treatment for natural fibers is depicted schematically. Copyright from [18] Creative Common CC BY license.

$$\text{Fibre—OH} + \text{NaOH} \longrightarrow \text{Fibre—O}^{-}\text{Na}^{+} + H_2O$$

Figure 24.7 The mercerization mechanism of the natural fibers.

and oil from the fibers' external surface. The result of the treatment which is carried out by using 4% NaOH solution at ambient temperature is a certain increase in the tensile strength of the jute fibers.

The mercerization of the sisal fiber improves the adhesion properties of the fibers by boosting surface tension and roughness [23]. The improved adhesion of the fibers results in the formation of more dense composites with lower porosity.

The alkalization process has been shown to impact the structural, morphological, physical, and mechanical assets of bamboo [24, 25]. The transformation from cellulose I to II was determined by the IR index [24]. The thermal stability, crystallinity, and wettability of the individual bamboo fibers were improved depending on the sodium hydroxide solution concentration [26].

The alkali treatment affects a decrease in the crystallinity of the wood which can be determined by XRD measurements [27]. The reason for the decrease in crystallinity can be explained by the penetration of the sodium hydroxide into the wall of the wood cells results in swelling. The mercerization of the wood causes the removal of the hemicellulose and lignin without a change in structure [28].

The main aim of the alkali treatment of hemp fibers is the improvement of the interfacial adhesion of the fibers to the polymers employed as matrices in biocomposites [29]. There has been numerous research on the impacts of alkali modification on hemp's physical, mechanical, and thermal assets [29, 30]. Because noncellulosic elements of the fiber, such as hemicellulose, lignin, pectin, wax, and oils, can be effectively removed through mercerization, the crystallinity of the hemp increases significantly. As a result of the increase in crystallinity of the hemp due to the increase in cellulose which is the strongest and the most ordered component of the fiber, a significant increase in thermal stability and modulus is observed [31].

The biocomposite was prepared by employing PLA and alkali-treated hemp fibers as polymer matrix and reinforcing agents, respectively [29]. The alkali treatment causes the improvement of the bonding between hemp fiber and PLA. The crystallinity of the alkali-treated hemp reinforced PLA composites was determined as higher than the crystallinity of the composites reinforced with untreated hemp. It was supposed that the reason for the increase in crystallinity of the biocomposite is the nucleating capability of the treated hemp. Interfacial shear strength measurements prove that the interfacial bonding properties are enhanced through the alkalization process that results in the enhancement of the mechanical assets of the biocomposites.

24.3.1.2 Silanization

The hydrophilicity of natural fibers is the fundamental cause for their incompatibility with polymer matrices. To reduce the hydrophilicity of natural fibers, silanization coupling agents were used [32]. The bifunctional composition of the organosilanes provides the interaction between fibers and polymer matrix by forming a bridge between them (Figure 24.8). This interaction is continuous through the interface of the fiber and polymer [4].

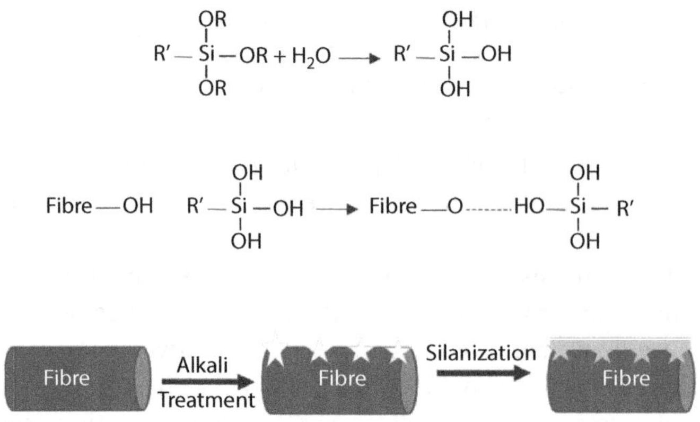

Figure 24.8 Silanization mechanism of the natural fibers.

Silane behaves like a surface coating and its hydrocarbon chain inhibits the swelling of the fiber in the matrices [4].

Silanization of the cotton fibers was carried out by employing different silanes with various substituents [33]. The water contact angle measurements show that the super-hydrophobicity for the cotton fibers was obtained because of the surface modification by employing chlorosilanes with different substituents. Another study employed a low-concentration binary solution of methyl trichlorosilane (MTS) and octadecyl trichlorosilane (OTS) in hexane to increase the hydrophobicity of cotton fibers. The water contact angle of the silver-coated silanized cotton fibers was measured as 153±2° [34]. As can be noticed silanization of the cotton fibers in a waterless medium causes a significant increase in hydrophobicity of the cotton fibers.

By blocking the hydroxyl groups of fiber with hydrophobic functional moieties of silanes, the adhesion of jute fibers to the polymeric matrix can be enhanced. It is possible to follow the treatment by FTIR spectroscopy due to the Si-O-C bond at 1205 cm^{-1} [35]. The hydroxyl group is blocked, which decreases the jute's water intake and hydrophilicity. The water contact angle of the jute was reported as 111° and the drop remains on the jute for 60 seconds while the water contact angle of the untreated fiber is 106° and the drop remains on the jute only for a few seconds [35]. Because the silane-coupling agent was uniformly coated on the surface of the jute fibers, the surface was smoother following treatment. In addition, the tensile strength of the jute shows a significant enhancement but not too much. In another example, the green composite was prepared by using modified jute fibers by prealkali treatment and then silanization and polyamide 11 [36]. The findings show that modifying the jute improves the biocomposites' mechanical and thermal stability significantly. When compared to untreated jute reinforced biocomposite, the silanized jute fibers reinforced biocomposite increased its tensile strength, impact strength, and Young's modulus by 44.1%, 26.5%, and 55.6%, respectively. Additionally, the silanization of the jute fibers improves the water adsorption capacity and weathering resistance of the biocomposites.

The bamboo fibers were treated by employing three silanes with different end groups. The condensation reaction between three different siloxanes and bamboo fibers can be proved by FTIR by the formation of new Si-O-C peaks [32]. The grafting of the siloxanes on the bamboo produces a smoother surface by forming a thin film which composed of the deposition of the silane coupling agents. This film covers the surface impurities which are rich in lignin [37]. The decrease in the hydrophilicity of the bamboo fibers enhances the adhesion property to the polymer matrices in their composites. The silanized bamboo fibers were used to prepare composites in which PP was employed as a polymeric matrix [32]. The silane modification enhances bamboo fiber adherence to the polymer, improving the composites' thermal stability, tensile and flexural strength significantly. Furthermore, due to the silane alteration, the crystallinity degree of the bamboo-PP biocomposites has decreased. It can be explained by the inhibition of the diffusion and migration of polymer chains to the nucleation center [32].

One of the ways for improving the hydrophobicity and compatibility of wood with polymer matrices is to silanized the wood fibers. Before the silanization, there must be a pretreatment of the wood with alkali solution and water to increase the accessibility of the hydroxyl groups which can be reacted with siloxanes [38]. The silanization of the wood increases the tensile and impact strength because of the improvement in the adhesion of the fibers [39]. The thermal stability of the wood fibers also increases due to the silane coating of the wood surface which causes a delay in degradation. The hydrophobicity of wood, as evidenced by a greater contact angle, improves the dispersion qualities of wood fibers in the polymeric phase [40]. With silanized wood content, the fiber-polymer adhesion increased and reduced the fracture and void development in composites [40]. It causes an enhancement in a variety of properties like tensile and impact strength of the composite prepared by modified wood as a reinforcing agent. The poor adhesion property of the unmodified wood and polymer phase can be seen by SEM images. The difference between the hydrophobicity of the wood and polymer matrix causes the formation of the aggregates [40]. The spaces between the two phases are readily visible. The silanization of wood fibers improves the interfacial adhesion of the wood and polymer, preventing cracks and gaps from forming [40].

The hemp fiber was modified first alkali solution and then three different organosilanes [41]. Hemp fibers were tested for mechanical and thermal conductivity. The results show that the binary modification provides effective removal of the hemicellulose content from the cell wall of the hemp which results in a certain improvement into tensile strength and Young modulus. Furthermore, the change boosts the thermal conductivity of the fibers [41]. The silanization of the hemp fiber can be proved by the appearance of the new peaks between 800 and 1800 cm^{-1} [42]. The specific peak at 1200 cm^{-1} can be attributed to Si O-cellulose which is evidence of the interaction between the silanols and the hydroxyl groups [43]. The formation of the new groups on the hemp surface due to the silanization inhibits the segmental mobility and improves the thermal stability fibers. The silanization of the hemp fibers provides significant protection of the surface against moisture because of the surface coating via chemical bonding [44]. The silanization of the hemp causes a decrease in the diameter of the fibers; however, the thin silane films cannot be seen by SEM [45].

24.3.1.3 Acetylation

The acetylation of the fibers is the esterification reaction via hydroxyl groups of the fibers. Esterification can be carried out with or without acid catalysis (Figure 24.9). The treatment results in the production of acetyl groups, which are less polar than hydroxyl groups, and

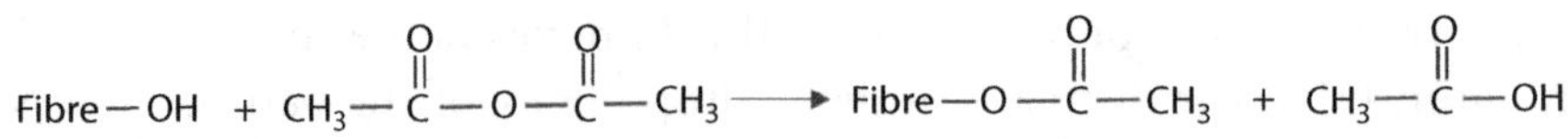

Figure 24.9 Acetylation mechanism of the natural fibers.

increase the hydrophobicity of the fibers [46]. The main effects of acetylation on the natural fibers are the reduction of moisture adsorption, increase in roughness of the surface, and improvement in mechanical properties. Acetic acid and acetic anhydride are generally used together to provide sufficient acetyl (CH_3CO-) grafting onto the natural fibers. Because the hemicellulose and lignin constituents of natural fibers are removed, pretreatment with alkali solutions can sometimes improve the efficacy of the acetyl treatment [4].

Acetylation of the jute fibers was carried out by employing acetic anhydride and N-bromosuccinimide as a catalyst. The process was optimized and the impact of the acetylation on the oil absorptivity of the jute was investigated. The results show that the oil adsorption capacity of the acetylated jute was higher than the unmodified fiber due to the increase in the hydrophobicity of the fiber surface. Because acetyl treatments block hydroxyl groups, which are essential for intermolecular interactions between polymer chains, the heat stability of jute is reduced [47]. Acetylation also provides a smoother surface by covering the gaps between the fibers which causes a remarkable decrease in water adsorption [48].

The acetyl treatment alters the sisal fibers surface chemistry and shapes without affecting their chemical structure. These changes can be investigated by spectroscopic and microscopic methods. Acetylated sisal fibers show a significant decrease in water adsorption capacity due to the increase in dimensional stability of the fibers [49]. The replacing of the hydroxyl groups with acetyl groups decreases the affinity of fibers against the water because of the increase in macromolecular stiffness. Interfibrillar failure of the sisal fibers and the formation of the voids on the surface also increases with acetyl treatment. As a result of the constraint of segmental mobility affected by the presence of large acetyl groups, modified sisal fibers have higher thermal stability than unmodified fibers [50].

The acetylation of the bamboo fibers was carried out by employing various catalysts [51]. To evaluate the dimensional stability of the bamboo fibers water adsorption rate and volume swelling efficiency experiments were carried out. The results show that the acetylation of the bamboo independent from the type of the catalyst reduces the water adsorption and volume swelling efficiency [51]. Acetylated bamboo shows a change in color due to the decrease in the holocellulose and α-cellulose and a rise in lignin content [52]. The contact angle value of the bamboo increases because of the reduction of the number of hydroxyl groups after esterification [51].

The acetylation of wood provides the esterification of 90% of lignin and 25% of holocellulose while the readily accessible hydroxyl of the hemicellulose was completely acetylated [53]. Acetyl treatment of the wood improves the bacterial resistance of the fiber [54]. The acetylation causes a significant reduction in swelling and shrinkage of the wood [55]. It can be described by the rise in the number of acetyl units on the cell wall.

The acetylation of the hemp fiber can be easily followed by FTIR spectroscopy. The intensity of the unconjugated carbonyl stretching peak of the hemicellulose shows a decrease (1730 cm^{-1}). The decrease in the adsorbed water can be seen in the reduction of the peak intensity at 1630 to 1650 cm^{-1} [44]. The acetylation peak that can be seen at 1228 to 1235 and 1740 cm^{-1} can be attributed to the C-O and C=O stretching of the ligninacetyl and uronic acid groups on the hemicellulose [56]. Due to cell wall damage, the thermal stability of acetylated hemp fibers was found lower than pristine hemp [44]. The humidity of the hemp fibers was also decreased significantly by acetylation. The surface area of the hemp fibers increases while the diameter of the fibers decreases after acetylation [6]. In addition, the surface of the modified fibers becomes smoother than the unmodified hemp [44].

24.3.1.4 Benzoylation

To activate the hydroxyl groups on the surface of natural fibers, alkali pretreatment is required. The hydroxyl groups are more accessible once the noncellulosic component of the natural fibers is removed, which improves the reaction with benzoyl chloride (Figure 24.10). The treatment increases the hydrophobicity of the fibers and improves thermal stability [57]. The treatment makes the surface of the natural fibers more reactive with benzoyl groups [58].

The benzoylation of the sisal after alkali treatment triggers a decrease in the diameter of the fibers and an increase in the roughness of the fiber surface due to the removal of the hemicellulose, lignin, and wax content of the natural fibers [59]. In addition, the formation of the small voids on the sisal surface because of the benzoyl substitution causes the enhancement of the roughness of the fiber surface [58]. The thermal stability and the crystallinity degree of sisal fibers also improve due to benzoylation [59].

The benzoylation process of the bamboo was optimized and the treated bamboo fibers were characterized by Li and coworkers [60] and showed that the thermal degradation temperature increased with substitution. The benzoyl treatment of the wood was carried out in an ionic liquid without an alkali pretreatment [61]. The treatment decreases the water content of the wood due to the increase in the hydrophobic character of the benzoyl-modified wood while the photostability and thermal stability increase.

24.3.2 Physical Treatment Processes

Physical methods, such as plasma, ultrasound, and ultraviolet light treatments, have been used to modify and improve the compatibility of biofibers for composite applications. These treatments also allow the fiber bundles to be separated into individual filaments.

24.3.2.1 Plasma Treatment

By removing contaminants from the surface, plasma treatment improves the surface attributes of natural fibers, such as hydrophobicity, roughness, flammability, and so on. Furthermore, the creation of reactive sides on the fiber surface to produce a chemical

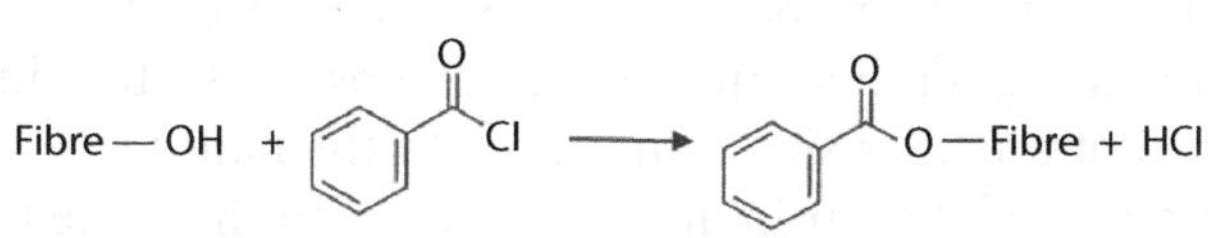

Figure 24.10 The chemical reaction between natural fibers and benzoyl chloride.

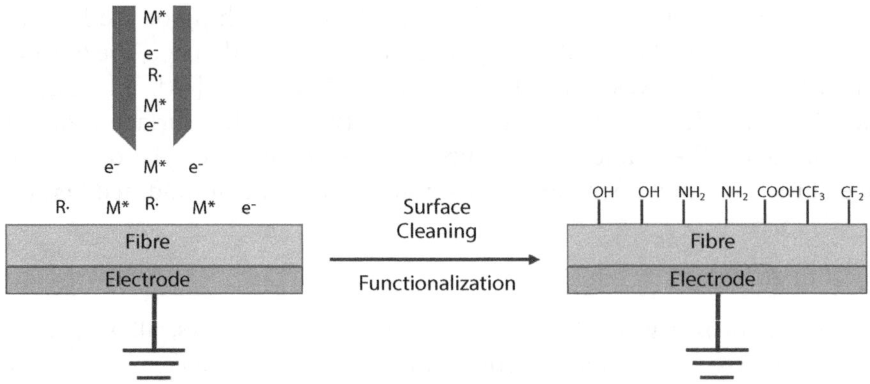

Figure 24.11 Cleaning and functionalization of the natural fiber surface via plasma treatment.

interaction with oxygen or other gases improves the fiber's hydrophilicity or hydrophobicity (Figure 24.11) [62].

The used gases can affect the surface properties of the natural fibers. Kamlangkla and co-workers used the SF6 plasma to improve the surface and mechanical properties of cotton [63]. Because of the creation of C-F bonds, the treatment significantly improves hydrophobicity. In addition, the treatment improves the cotton fabric's mechanical characteristics.

The surface of the jute fibers was treated by oxygen and argon plasma [64]. XRD confirmed the increase in crystallite size besides the degree of crystallinity of the fibers. Oxygen plasma treatment of the jute fibers causes the rupture tendency due to the higher strain compared with argon plasma. It was reported that the smoother and more compact surface due to the reduction of the width was obtained by argon plasma.

Oxygen plasma was applied to the vegetable fibers (sisal, coconut, and abaca) to modify the surface [65]. The oxygen plasma produces functional groups with oxygen. By treatment, the wax and lignin content can be removed from the fiber surface. Besides, the degradation of lignin occurs. The regularity of the segments of natural fibers increases as the surface layers of the fibers are removed. The average of the tensile strength increases while the elongation at break decreases.

24.3.2.2 Ultrasound Treatment

The ultrasound treatments of the natural fibers are not used as commonly as plasma. However, effective removal of the surface impurities of the fibers is possible with these treatments. The success of the treatment depends on the formation of microbubbles and collapse in an ultrasonic field which is called acoustic cavitation [66]. During the acoustic cavitation process, various physical effects are generated, such as shock waves, turbulence, microstreaming, etc. [67]. These physical effects can be useful to remove the impurities from the surface of the natural fiber [68, 69]. The ultrasonic treatment significantly changes the crystallinity and melting temperature of the cellulose [68]. In addition, the treatment provides the reduction of the fiber size from micro to the nanoscale. Ultrasonic treatment can help the extraction of the lignin in acidic, basic medium or the combination of these while it provides the degradation of the hemicellulose in distilled water [70]. Rosales and co-workers analyzed the results of the alkali and ultrasound treatment on the structural,

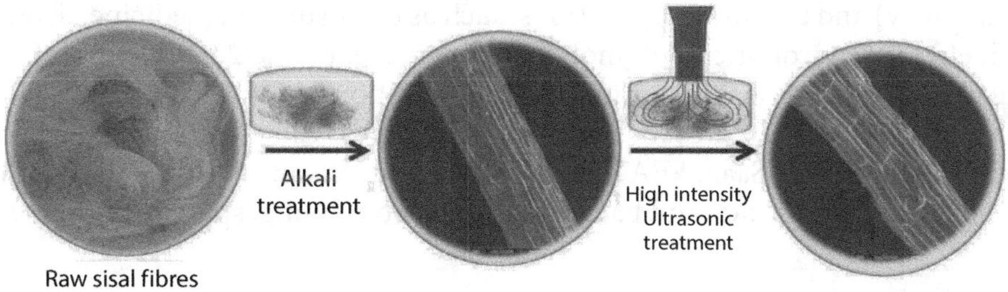

Figure 24.12 The effects of alkali and high-intensity ultrasonic treatments on sisal morphology. Copyright [72] with permission from Elsevier.

thermal, and mechanical properties of the agave tequilana [71]. The results show that the crystallinity of the fibers increases without change in the crystalline pattern and the roughness of the fibers shows a significant increase after binary treatment. The treatment also improves the thermal stability of the fibers which allows them to be used in applications with high temperatures like the manufacturing of the components of automobiles [71].

Sisal fibers were modified by alkali treatment and high-intensity ultrasound by Krishnaiah *et al.* [72]. The results prove that the thermal and mechanical stability improve remarkably by alkali and high-intensity ultrasound. The treatments were successful in removing the amorphous sections of the fiber. After chemical treatment, the high-intensity ultrasonic treatment allows the production of well-separated fiber strands (Figure 24.12). It improves the tensile strength and hydrophobicity of natural fiber reinforced composites by increasing interfacial adhesion and interaction between the fiber and polymer matrix [73].

24.3.2.3 Ultraviolet Treatment

The use of ultraviolet light to improve the physicomechanical characteristics of natural fibers is a cost-effective and environmentally friendly technology. UV radiation causes chemical reactions on the surface of biofibers, altering their appearance. Chemical pretreatments can be utilized to enhance the activity of natural fibers in some circumstances [74]. It was reported that the tensile strength, elongation at break, and Young's modulus increase with an increase in irradiation dose of UV until a certain point due to the crosslinking of cellulose [75]. In the case of the higher doses, all the mechanical properties show a decrease because of the degradation of the cellulose [74].

24.4 Biofiber Composites for Industrial Applications

For the formation of composites, biofibers have gained environmental, economic, thermal, and mechanical advantages compared with traditional reinforcements and are used especially in important industrial applications. Because of remarkable developments in science and technology, manufacturing companies have been transitioning to more sustainable, environmentally friendly economic materials. In this section, biofibers and their polymer matrices' (thermoplastic, thermoset, and biodegradable polymers) composite were briefly explained with changes in properties (mechanical, thermal, water resistance, flame retardancy) and potential applications, such as construction, medicine, dentistry, textile, food, electronics, robotics, automotive industry, etc. (Table 24.3) [76–82]. Despite the potential of biofiber composites, for industrial applications, biofibers generally apply in the automobile. AUDİ, BMW, Citroen, Ford, Lotus, Daimler/Chrysler, Mercedes/Benz, Opel, Peugeot, Renault, Rover, Saab, SEAT, Toyota, Volkswagen, and Volvo were used biofibers for trunk liner, seatbacks, side and back door panel, boot lining, spare tire lining, headliner

Table 24.3 Various applications of biopolymer-polymer composites.

Biopolymer	Polymer matrices	Application	Reference
Kenaf	Epoxy	Car bumper beam	[76]
Hemp	Polypropylene	Nonwoven mats	[86]
Coir fibre	Polypropylene	Automotove interior	[12]
Abaca	Polypropylene	Floor protection	[77]
Cotton	Polyetilenimin	Fabric	[78]
Kenaf	PLA	Cushioning	[79]
Hemp	Polyethylene	Grocery bag	[88]
Sisal	Polyethylene	Drywall and insulated panels	[90]
Bast fibre (Hemp, Flax, Jute, Sisal etc.)	Polyethylene Polyester	Carrier for door panel, covered insert, bactseat panel, door bolsters, side and back walls, rear deck tray, centre console, trunk trim, pillars, load floors, etc.	[80]
Coconut	Natural rubber	Seat bottom, back cushions, interior trim, seat cushioning, seat surfaces/backrests, etc.	[80]
Cotton	PP/PET	Sound proofing, trunk panels, insulation	[81]
Flax/hemp	Epoxy resin	Carrier for covered door panels	[81]
Flax/sisal	Thermoset resins	Door linings and panels	[82]
Kenaf	Polypropylene	Door inner panel	[82]
Wood	Acrylic resin or synthetic fibres	Carrier for door panels, covered door panels, instrument panels, coverd inserts and components, covered backseat panels, etc.	[80]
Wood flour	Leather	Upholstery, seat cover	[82]
Woven kenaf Kevlar	Epoxy	Amine hardener	[18]

panel, windshield, dashboard, business table, pillar cover panel, interior carpets, internal engine cover and interior insulation, bumper, roof shelter, etc. [83, 84].

For industrial applications, the extensively used thermoplastic is polypropylene (PP) which has been used due to low density, low price, high toughness, excellent processing, design flexibility, good resistance, and mechanical properties [17]. Ayrilmis and coworkers prepared coir fiber-polypropylene panels with maleic anhydride grafted PP (3 wt.%) in the role of a coupling agent in a rotary drum blender through the dry blending method for automotive interior applications. With increasing coir fiber amount to 60%, flexural strength (26%), tensile strength (30%), hardness, and flame retardancy of composite (7%) were increased on the contrary of water resistance and internal bonding strength [12]. Munde and coworkers prepared coir fiber-polypropylene composite using injection process with various fiber weight fraction and with increasing weight ratio, the damping ratio decreased. The composite is a potential candidate for automotive structural components [85]. Hemp-polypropylene composite prepared by a hot-pressing method as nonwoven mats. The increase in fiber content (from 30% to 70%) increased water uptake capacity (from 7% to 53%), and double-carded mats had better water resistance and bending qualities than single-carded mats [86]. Yadav and coworkers prepared polyethylene-polypropylene-hemp composite by solvent casting method as a packaging film using central composite design response surface methodology. When hemp properties were compared with composite, contact angle (30%), impact strength (24%), and transparency (66%) were increased and water vapor permeability (57%) was decreased [87]. They showed that the alkali treatment of hemp enhanced the binding efficiency between hemp and polymer matrix and the thermal, packaging, and hydrophobic properties of the biocomposite films.

For an environmentally friendly grocery bag, a polyethylene-hemp composite was formed in the form of films or sheets, like packaging and shrinkage film [88]. When the high-density polyethylene-hemp composite was prepared alkali-treated hemp in various compositions (20%–40%) through an extrusion and compression molding method, up to 40% hemp, the tensile, and flexural strength were enhanced. The produced composite has a lot of potential in civil engineering for light-load applications like short-span bridges and hurricane-proof panels [89]. The introduction of woven sisal fibers in high-density polyethylene improved tensile strength and ultimate strain by 33.30%. The sisal fiber-reinforced green HDPE composites showed a higher modulus of elasticity and has potential in structural applications, like drywall and insulated panels [90].

The biolaminates which are low density, lighter mass, and biodegradable materials can be prepared by natural fiber reinforcing of polymers. Torres-Arellano and coworkers prepared the biolaminates by employing two different biobased epoxy resins (Sicomin Resin SR GreenPoxySicomin, GP and the epoxidized vegetable oil, EVO) and three different natural fibers (ixtle, henequen, and jute) as reinforcing agents by VARI process [91]. The biolaminate's mechanical qualities were shown to be superior to laminates made with commercial glass fiber as a reinforcing material. Because of the higher cellulose content of the natural fiber and the volume fraction of the fiber, EVO/jute composites have the best mechanical characteristics. In another research, the biocomposites of epoxy resin by reinforcing alkali-treated flax, linen, and bamboo fibers by employing a vacuum bagging technique were prepared by Yan and coworkers [92]. The results show that the treatment improves the tensile and flexural properties of the biocomposites. The physical interlocking and bonding enhance the interfacial adhesion of the fiber to the epoxy resin.

24.5 Challenges and Perspectives for Future Research

For the formation of biofiber composites, several challenges have been found. Quality variations, decreased strength, increased moisture adsorption, limited processing temperatures, decreased durability, and poor fire resistance is only a few of them. The fiber quality changed to four fiber fabrication processes. Plant growth (plant species, crop culture, location, and local climate), harvesting period (maturity), fiber extraction step (decortication process, retting method), and supply stage (transportation and storage situations, and age of fiber) [11]. Differences in the production period of the plant, the combination rates that change according to the region of the plant, and the parameters that change during the production processes create problems in the sustainability and durability of the prepared products. As to the variety of compositions of fibers, the fire resistance can also convert. When the amount of cellulose is higher than hemicellulose, the flammability of biofiber increases. This also affects the processing temperature. High reaction temperatures and long processing times decrease the mechanical properties of biofibers, and high temperatures also aggravate the color and odor of the material [93]. The hydrophilic character of the biofibers binds weakly with the hydrophobic polymer during the composite formation, preventing the fiber/matrix adhesion and creating agglomeration problems. Besides this, the hydrophilicity of biofiber and swelling behavior cause higher moisture adsorption and form poor moisture resistance [94]. Adsorbed water was binding with the hydroxyl group of fiber and decreased the mechanical properties of the composite. As a result of these challenges, the manufacturing difficulties are the inevitable end.

To overcome these limitations physical [65] and chemical modification methods [4] to the natural fibers have been applied before the composite preparations. The treatments improve the natural fibers' thermal and mechanical stability, as well as their surface adherence to the polymer matrix. Furthermore, as the hydrophobicity increases due to the formation of hydrophobic species, such as benzoyl, acetyl, and etc., on the fiber surface, the water (moisture) adsorption of the fibers decreases [4]. These improvements provide the production of composites with better physical, thermal, and mechanical properties. The most used treatments were summarized in the previous sections. The effect of these treatments on the composites in which modified natural fibers were employed as reinforcing agents have been studied for various polymers, such as PP, PE, PLA, polyamides, epoxy resins, etc.

Although these modifications are still insufficient for industry applications, the economic acquirements, environmental impacts, developing technology, and increasing needs necessitate a return to nature. Natural fibers are a promising candidate material for biocomposites since they are biodegradable, light in weight, cost-effective, and environmentally benign. In terms of sustainability, biocomposites are promising materials because they can replace renewable resources with petrochemicals, resulting in lower greenhouse gas emissions, biodegradations, lower costs, and new industrial applications. Future research would be aimed to produce solutions for highlighted challenge areas.

24.6 Conclusion

In the last few decades, the research about the modification of natural fibers for employing them as reinforcing agents to produce biocomposites has been increased. These eco-friendly and low-cost materials have various advantages when compared to synthetic counterparts. The main reason for the rise in the tendency of the usage of natural fibers is the rapidity of the consumption of petroleum-based sources. Natural fibers are an appealing alternative to using in the manufacturing of composites that may be employed in a variety of industries due to their availability and sustainability. Furthermore, one of the benefits of using a reinforcing agent to improve composite qualities is the ability to use chemical and physical approaches to improve natural fiber mechanical and thermal properties. The reduction of the weight, density, and water adsorption capacity of the composites are useful in some industrial applications, such as automotive. Because the decrease in weight provides significant benefits in terms of energy consumption. Natural fibers are biodegradable, recyclable, and cheap materials. However, there are certain limitations in the application of them as reinforcing agents. The hydrophilicity and high-water adsorption capacity of the fibers cause incompatibility and low interfacial adhesion between the fiber and the polymer matrix. Chemical and physical modification approaches are routinely employed to improve these qualities. The main effect of treatments is the cleaning of the amorphous hemicellulose part of the natural fiber, which is responsible for water uptake. The structural and compositional differences in the natural fibers necessitate the optimization of the experimental conditions like concentration of the chemicals, such as NaOH, benzoyl chlorides, silanization coupling agents, time, and temperature of the process. Several studies have been carried out to optimize chemical and physical modification methods for natural fibers for use as reinforcing agents in biocomposites made with natural and synthetic polymers. All investigations found that reinforcing modified biofibers improved the mechanical and thermal assets of biocomposites. Furthermore, the water uptake and hydrophobicity of the biocomposites decrease dramatically, resulting in a significant improvement in natural fiber interfacial adhesion to polymeric matrices. The reduction of the hydrophilicity of the fiber undergoes different mechanisms due to the modification methods. Alkalization successfully eliminates the noncellulosic amorphous section of the fiber, while silanization produces a thin hydrophobic coating on the fiber's surface. By employing the acetylation and benzoylation formation of the hydrophobic moieties on the fiber surface is provided. However, an alkali pretreatment is typically employed in conjunction with these treatments to activate and clean the surface of the fiber, resulting in increased yield and effectiveness. The treatments also result in the creation of rougher surfaces, which improve the mechanical attachment of the fiber to the matrix.

To summarize, biofibers are promising materials that can be used to make biocomposites and can be substituted for petroleum-based alternatives. There are still several processing problems to overcome, which can be mitigated by using natural fiber modification methods. These challenges cause the restriction of the area of the application. Fortunately, the biocomposites with high mechanical and thermal stability manufactured by reinforcing polymers with cellulose-based natural fibers show promise in a large spectrum of applications.

References

1. Mohanty, A.K., Misra, M., Drzal, L.T., *Natural Fibers, Biopolymers, and Biocomposites*, CRC Press, Taylor&Francis Group, Boca Raton, FL, 2005.
2. Zafar, M.T., Maiti, S.N., Ghosh, A.K., Effect of surface treatments of jute fibers on the microstructural and mechanical responses of poly (lactic acid)/jute fiber biocomposites. *RSC Adv.*, 6, 73373, 2016.
3. Faruk, O., Bledzki, A.K., Fink, H.-P., Sain, M., Biocomposites reinforced with natural fibers: 2000–2010. *Prog. Polym. Sci.*, 37, 1552, 2012.
4. Kabir, M., Wang, H., Lau, K., Cardona, F., Chemical treatments on plant-based natural fibre reinforced polymer composites: An overview. *Compos. B Eng.*, 43, 2883, 2012.
5. Liu, Z. and Tisserat, B.H., Coating applications to natural fiber composites to improve their physical, surface and water absorption characters. *Ind. Crops Prod.*, 112, 196, 2018.
6. Li, X., Tabil, L.G., Panigrahi, S., Chemical treatments of natural fiber for use in natural fiber-reinforced composites: A review. *J. Polym. Environ.*, 15, 25, 2007.
7. Yuan, X., Jayaraman, K., Bhattacharyya, D., Effects of plasma treatment in enhancing the performance of woodfibre-polypropylene composites. *Compos. Part A Appl. Sci. Manuf.*, 35, 1363, 2004.
8. Pickering, K.L., Efendy, M.A., Le, T.M., A review of recent developments in natural fibre composites and their mechanical performance. *Compos. Part A Appl. Sci. Manuf.*, 83, 98, 2016.
9. Hasan, K., Horváth, P.G., Alpár, T., Potential natural fiber polymeric nanobiocomposites: A review. *Polymers*, 12, 1072, 2020.
10. Bozaci, E., Öktem, T., Seventekİn, N., Ananas yaprak lifi. *Tekst. ve Konfeksiyon*, 17, 167, 2007.
11. Dittenber, D.B. and Gangarao, H.V., Critical review of recent publications on use of natural composites in infrastructure. *Compos. Part A Appl. Sci. Manuf.*, 43, 1419, 2012.
12. Ayrilmis, N., Jarusombuti, S., Fueangvivat, V., Bauchongkol, P., White, R.H., Coir fiber reinforced polypropylene composite panel for automotive interior applications. *Fiber. Polym.*, 12, 919, 2011.
13. Arancon, R., *Natural Fiber Production and Food Securtiy: Coir in Asia and the Pacific*, p. 63, Asian and Pacific Coconut Community, Jakarta, 2007.
14. Mylsamy, K. and Rajendran, I., Investigation on physio-chemical and mechanical properties of raw and alkali-treated Agave Americana fiber. *J. Reinf. Plast. Comp.*, 29, 2925, 2010.
15. Escobar-Guzman, R.E., Hernandez, F.Z., Vega, K.G., Simpson, J., Seed production and gametophyte formation in Agave Tequilana and Agave Americana. *Botany*, 86, 1343, 2008.
16. Pandit, P., Pandey, R., Singha, K., Shrivastava, S., Gupta, V., Jose, S., Pineapple leaf fibre: Cultivation and production, in: *Pineapple Leaf Fibers*, p. 1, Springer, Singapore, 2020.
17. Yan, L., Chouw, N., Jayaraman, K., Flax fibre and its composites–A review. *Compos. B Eng.*, 56, 296, 2014.
18. Nurazzi, N., Asyraf, M., Khalina, A., Abdullah, N., Aisyah, H., Rafiqah, S., Sabaruddin, F., Kamarudin, S., Norrrahim, M., Ilyas, R., A review on natural fiber reinforced polymer composite for bullet proof and ballistic applications. *Polymers*, 13, 646, 2021.
19. Valášek, P., Müller, M., Sleger, V., Kolář, V., Hromasová, M., D'amato, R., Ruggiero, A., Influence of alkali treatment on the microstructure and mechanical properties of coir and abaca fibers. *Materials*, 14, 2636, 2021.
20. Azeem, F., Mao, N., Faisal, S., Effect of alkali treatment of lower concentrations on the structure and tensile properties of Pakistan's coarse cotton fibre. *J. Text. Inst.*, 110, 1499, 2019.
21. Macedo, J., Carmona, V., Dos Santos Rosa, D., Differences of alkali and thermal treatment on cotton residues applicable to green composites. *Mater. Sci. Eng. Adv. Res.*, 1, 13, 2016.

22. Rawal, A. and Sayeed, M., Tailoring the structure and properties of jute blended nonwoven geotextiles via alkali treatment of jute fibers. *Mater. Des.*, 53, 701, 2014.
23. Bisanda, E., The effect of alkali treatment on the adhesion characteristics of sisal fibres. *Appl. Compos. Mater.*, 7, 331, 2000.
24. Das, M. and Chakraborty, D., Influence of alkali treatment on the fine structure and morphology of bamboo fibers. *J. Appl. Polym. Sci.*, 102, 5050, 2006.
25. Das, M. and Chakraborty, D., Evaluation of improvement of physical and mechanical properties of bamboo fibers due to alkali treatment. *J. Appl. Polym. Sci.*, 107, 522, 2008.
26. Chen, H., Zhang, W., Wang, X., Wang, H., Wu, Y., Zhong, T., Fei, B., Effect of alkali treatment on wettability and thermal stability of individual bamboo fibers. *J. Wood Sci.*, 64, 398, 2018.
27. Rojo, E., Alonso, M.V., Dominguez, J.C., Saz-Orozco, B.D., Oliet, M., Rodriguez, F., Alkali treatment of viscose cellulosic fibers from eucalyptus wood: structural, morphological, and thermal analysis. *J. Appl. Polym. Sci.*, 130, 2198, 2013.
28. Chen, C., Wang, Y., Wu, Q., Wan, Z., Li, D., Jin, Y., Highly strong and flexible composite hydrogel reinforced by aligned wood cellulose skeleton via alkali treatment for muscle-like sensors. *Chem. Eng. J.*, 400, 125876, 2020.
29. Islam, M., Pickering, K., Foreman, N., Influence of alkali treatment on the interfacial and physico-mechanical properties of industrial hemp fibre reinforced polylactic acid composites. *Compos. Part A Appl. Sci. Manuf.*, 41, 596, 2010.
30. Pickering, K.L., Beckermann, G., Alam, S., Foreman, N.J., Optimising industrial hemp fibre for composites. *Compos. Part A Appl. Sci. Manuf.*, 38, 461, 2007.
31. Pickering, K.L., Li, Y., Farrell, R.L., Lay, M., Interfacial modification of hemp fiber reinforced composites using fungal and alkali treatment. *J. Biobased Mater. Bioenergy*, 1, 109, 2007.
32. Wang, Q., Zhang, Y., Liang, W., Wang, J., Chen, Y., Effect of silane treatment on mechanical properties and thermal behavior of bamboo fibers reinforced polypropylene composites. *J. Eng. Fiber. Fabr.*, 15, 1, 2020.
33. Makowski, T., Hydrophobization of cotton fabric with silanes with different substituents. *Cellulose*, 27, 1, 2020.
34. Shen, W., Zhang, L., Li, X., Yu, H.-Z., Binary silanization and silver nanoparticle encapsulation to create superhydrophobic cotton fabrics with antimicrobial capability. *Sci. Rep.*, 9, 1, 2019.
35. Gogoi, R. and Tyagi, A.K., Surface modification of jute fabric by treating with silane coupling agent for reducing its moisture regain characteristics. *J. Nat. Fibers*, 18, 803, 2021.
36. Giang, N.V., Trung, T.H., Hoang, T., Lam, T.D., Tham, D.Q., Huong, N.T., Biocomposites from polyamide 11 reinforced by organic silane modified jute fibers: Fabrication and characterization. *J. Appl. Polym. Sci.*, 139, 51795, 2021.
37. Asim, M., Jawaid, M., Abdan, K., Ishak, M.R., Effect of alkali and silane treatments on mechanical and fibre-matrix bond strength of kenaf and pineapple leaf fibres. *J. Bionic Eng.*, 13, 426, 2016.
38. Zhang, H.H., Cui, Y., Zhang, Z., Chemical treatment of wood fiber and its reinforced unsaturated polyester composites. *J. Vinyl Addit. Technol.*, 19, 18, 2013.
39. Chauhan, V., Kärki, T., Varis, J., Effect of fiber content and silane treatment on the mechanical properties of recycled acrylonitrile-butadiene-styrene fiber composites. *Chemistry*, 3, 1258, 2021.
40. Bijaisoradat, O., Yue, L., Manas-Zloczower, I., Manuspiya, H., Wood flour-high density polyethylene composites: Influence of silanization and esterification on mechanical properties. *J. Appl. Polym. Sci.*, 138, 50197, 2021.
41. Sair, S., Oushabi, A., Kammouni, A., Tanane, O., Abboud, Y., Hassani, F.O., Laachachi, A., El Bouari, A., Effect of surface modification on morphological, mechanical and thermal

conductivity of hemp fiber: Characterization of the interface of hemp–Polyurethane composite. *Case Stud. Therm. Eng.*, 10, 550, 2017.

42. Abdelmouleh, M., Boufi, S., Belgacem, M., Duarte, A., Salah, A.B., Gandini, A., Modification of cellulosic fibres with functionalised silanes: Development of surface properties. *Int. J. Adhes. Adhes.*, 24, 43, 2004.
43. Valadez-Gonzalez, A., Cervantes-Uc, J., Olayo, R., Herrera-Franco, P., Chemical modification of henequen fibers with an organosilane coupling agent. *Compos. B Eng.*, 30, 321, 1999.
44. Matykiewicz, D., Barczewski, M., Mysiukiewicz, O., Skórczewska, K., Comparison of various chemical treatments efficiency in relation to the properties of flax, hemp fibers and cotton trichomes. *J. Nat. Fibers*, 18, 735, 2021.
45. Sawpan, M.A., Pickering, K.L., Fernyhough, A., Flexural properties of hemp fibre reinforced polylactide and unsaturated polyester composites. *Compos. Part A Appl. Sci. Manuf.*, 43, 519, 2012.
46. El Boustani, M., Lebrun, G., Brouillette, F., Belfkira, A., Effect of a solvent-free acetylation treatment on reinforcements permeability and tensile behaviour of flax/epoxy and flax/wood fibre/epoxy composites. *Can. J. Chem. Eng.*, 95, 1082, 2017.
47. Teli, M. and Valia, S.P., Acetylation of jute fiber to improve oil absorbency. *Fiber. Polym.*, 14, 915, 2013.
48. Rana, A., Basak, R., Mitra, B., Lawther, M., Banerjee, A., Studies of acetylation of jute using simplified procedure and its characterization. *J. Appl. Polym. Sci.*, 64, 1517, 1997.
49. Alvarez, V.A. and Vázquez, A., Influence of fiber chemical modification procedure on the mechanical properties and water absorption of MaterBi-Y/sisal fiber composites. *Compos. Part A Appl. Sci. Manuf.*, 37, 1672, 2006.
50. Albano, C., Gonzalez, J., Ichazo, M., Kaiser, D., Thermal stability of blends of polyolefins and sisal fiber. *Polym. Degrad. Stab.*, 66, 179, 1999.
51. Huang, S., Ma, Z., Nie, Y., Lu, F., Ma, L., Comparative study of the performance of acetylated bamboo with different catalysts. *BioRes.*, 14, 44, 2019.
52. Hung, K.-C., Chen, Y.-L., Wu, J.-H., Natural weathering properties of acetylated bamboo plastic composites. *Polym. Degrad. Stab*, 97, 1680, 2012.
53. Rowell, R.M., Simonson, R., Hess, S., Plackett, D.V., Cronshaw, D., Dunningham, E., Acetyl distribution in acetylated whole wood and reactivity of isolated wood cell-wall components to acetic anhydride. *Wood Fiber Sci.*, 26, 11, 1994.
54. Rowell, R., Chemical modification of wood: A short review. *Wood Mater. Sci. Eng.*, 1, 1, 29–33, 2006.
55. Digaitis, R., Thybring, E.E., Thygesen, L.G., Fredriksson, M., Targeted acetylation of wood: A tool for tuning wood-water interactions. *Cellulose*, 28, 8009, 2021.
56. Hajiha, H., Sain, M., Mei, L.H., Modification and characterization of hemp and sisal fibers. *J. Nat. Fibers*, 11, 144, 2014.
57. Zakaria, S. and Poh, L., Polystyrene-benzoylated EFB reinforced composites. *Polym. Plast. Technol. Eng.*, 41, 951, 2002.
58. Sreekumar, P., Thomas, S.P., Marc Saiter, J., Joseph, K., Unnikrishnan, G., Thomas, S., Effect of fiber surface modification on the mechanical and water absorption characteristics of sisal/polyester composites fabricated by resin transfer molding. *Compos. Part A Appl. Sci. Manuf.*, 40, 1777, 2009.
59. Kalia, S., Kaushik, V.K., Sharma, R.K., Effect of benzoylation and graft copolymerization on morphology, thermal stability, and crystallinity of sisal fibers. *J. Nat. Fibers*, 8, 27, 2011.
60. Li, M.-F., Sun, S.-N., Xu, F., Sun, R., Benzylation and characterization of cold NaOH/urea preswelled bamboo. *BioRes.*, 7, 1876, 2012.

61. Yuan, T.-Q., Zhang, L.-M., Xu, F., Sun, R.-C., Enhanced photostability and thermal stability of wood by benzoylation in an ionic liquid system. *Ind.Crops Prod.*, 45, 36, 2013.
62. Skundric, P., Kostic, M., Medovic, A., Pejic, B., Kuraica, M., Vuckovic, A., Obradovic, B., Mitrakovic, D., Puric, J., Wetting properties of hemp fibres modified by plasma treatment. *J. Nat. Fibers*, 4, 25, 2007.
63. Kamlangkla, K., Paosawatyanyong, B., Pavarajarn, V., Hodak, J.H., Hodak, S.K., Mechanical strength and hydrophobicity of cotton fabric after SF6 plasma treatment. *Appl. Surf. Sci.*, 256, 5888, 2010.
64. Morshed, M., Alam, M., Daniels, S., Plasma treatment of natural jute fibre by RIE 80 plus plasma tool. *Plasma Sci. Technol.*, 12, 325, 2010.
65. Valášek, P., Müller, M., Sleger, V., Influence of plasma treatment on mechanical properties of cellulose-based fibres and their interfacial interaction in composite systems. *BioRes.*, 12, 5449, 2017.
66. Ashokkumar, M., The characterization of acoustic cavitation bubbles–An overview. *Ultrason. Sonochem.*, 18, 864, 2011.
67. Yusof, N.S.M., Babgi, B., Alghamdi, Y., Aksu, M., Madhavan, J., Ashokkumar, M., Physical and chemical effects of acoustic cavitation in selected ultrasonic cleaning applications. *Ultrason. Sonochem.*, 29, 568, 2016.
68. Pinjari, D.V. and Pandit, A.B., Cavitation milling of natural cellulose to nanofibrils. *Ultrason. Sonochem.*, 17, 845, 2010.
69. Pamidi, T.R.K., Johansson, Ö., Löfqvist, T., Shankar, V., Comparison of two different ultrasound reactors for the treatment of cellulose fibers. *Ultrason. Sonochem.*, 62, 104841, 2020.
70. Renouard, S., Hano, C., Doussot, J., Blondeau, J.-P., Lainé, E., Characterization of ultrasonic impact on coir, flax and hemp fibers. *Mater. Lett.*, 129, 137, 2014.
71. Solís Rosales, S.G., Naranjo Naranjo, L., Fonseca-Florido, H.A., González-Morones, P., Hernández, Z.G., Macías, R.Y., Romero, G.Y., Gallardo-Vega, C.A., Sifuentes-Nieves, I., Hernández-Hernández, E., Alkali/ultrasound treatment as alternative to modify structural and thermal properties of Agave Tequilana fibers. *J. Nat. Fibers*, 1, 2021.
72. Krishnaiah, P., Ratnam, C.T., Manickam, S., Enhancements in crystallinity, thermal stability, tensile modulus and strength of sisal fibres and their PP composites induced by the synergistic effects of alkali and high intensity ultrasound (HIU) treatments. *Ultrason. Sonochem.*, 34, 729, 2017.
73. Ghosh, R., Ramakrishna, A., Reena, G., Ravindra, A., Verma, A., Water absorption kinetics and mechanical properties of ultrasonic treated banana fiber reinforced-vinyl ester composites. *Proc. Mater. Sci.*, 5, 311, 2014.
74. Rahman, M.M. and Khan, M.A., Surface treatment of coir (Cocos nucifera) fibers and its influence on the fibers' physico-mechanical properties. *Compos. Sci. Technol.*, 67, 2369, 2007.
75. Komuraiah, A., Kumar, N.S., Prasad, B.D., Chemical composition of natural fibers and its influence on their mechanical properties. *Mech. Compos. Mater.*, 50, 359, 2014.
76. Davoodi, M., Sapuan, S., Ahmad, D., Ali, A., Khalina, A., Jonoobi, M., Mechanical properties of hybrid kenaf/glass reinforced epoxy composite for passenger car bumper beam. *Mater. Des.*, 31, 4927, 2010.
77. Bledzki, A.K., Mamun, A.A., Jaszkiewicz, A., Erdmann, K., Polypropylene composites with enzyme modified abaca fibre. *Compos. Sci. Technol.*, 70, 854, 2010.
78. Hassabo, A.G., Mohamed, A.L., Shaarawy, S., Hebeish, A., Novel micro-composites based on phosphorylated biopolymer/polyethyleneimine/clay mixture for cotton multi-functionalities performance. *Biosci. Res.*, 15, 2568, 2018.
79. Moyo, M., Kanny, K., Velmurugan, R., Performance of kenaf bon-woven mat/PLA biocomposites under medium velocity impact. *Fiber. Polym.*, 21, 2642, 2020.

80. Huda, M., Drzal, L., Ray, D., Mohanty, A., Mishra, M., Natural-fiber composites in the automotive sector, in: *Properties and Performance of Natural-Fibre Composites*, p. 221, Elsevier, 2008.
81. Abbas, M. and Jeon, H.-Y., *Generation, Development and Modifications of Natural Fibers*, Amsterdam, The Netherlands, 2020.
82. Witayakran, S., Smitthipong, W., Wangpradid, R., Chollakup, R., Clouston, P., Natural fiber composites: Review of recent automotive trends, in: *Materials Science and Materials Engineering*, Elsevier Publishing, Amherst, MA, USA, 2017.
83. Hill, K., Swiecki, B., Cregger, J., The Bio-Based Materials Automotive Value Chain, p. 112, Center for Automotive Research, Business Research, Canada, USA, 2012.
84. Ahmad, F., Choi, H.S., Park, M.K., A review: Natural fiber composites selection in view of mechanical, light weight, and economic properties. *Macromol. Mater. Eng.*, 300, 10, 2015.
85. Munde, Y.S., Ingle, R.B., Siva, I., Investigation to appraise the vibration and damping characteristics of coir fibre reinforced polypropylene composites. *Adv. Mater. Process. Technol.*, 4, 639, 2018.
86. Hargitai, H., Rácz, I., Anandjiwala, R.D., Development of hemp fiber reinforced polypropylene composites. *J. Thermoplast. Compos. Mater.*, 21, 165, 2008.
87. Dixit, S., Mishra, G., Yadav, V.L., Optimization of novel bio-composite packaging film based on alkali-treated Hemp fiber/polyethylene/polypropylene using response surface methodology approach. *Polym. Bull.*, 79, 2559, 2022.
88. Angulo, C., Brahma, S., Espinosa-Dzib, A., Peters, R., Stewart, K.M., Pillay, S., Ning, H., Development of hemp fiber composites with recycled high density polyethylene grocery bags. *Environ. Prog. Sustain. Energy*, 40, e13617, 2021.
89. Lu, N., Swan Jr., R.H., Ferguson, I., Composition, structure, and mechanical properties of hemp fiber reinforced composite with recycled high-density polyethylene matrix. *J. Compos. Mater.*, 46, 1915, 2012.
90. De Castro, B.D., De Faria, P.E., Vieira, L.M.G., Rubio, C.V.C., Maziero, R., Rodrigues, P.C.D.M., Rubio, J.C.C., Recycled green PE composites reinforced with woven and randomly arranged sisal fibres processed by hot compression moulding. *Acta Technol. Agric.*, 23, 81, 2020.
91. Torres-Arellano, M., Renteria-Rodríguez, V., Franco-Urquiza, E., Mechanical properties of natural-fiber-reinforced biobased epoxy resins manufactured by resin infusion process. *Polymers*, 12, 2841, 2020.
92. Yan, L., Chouw, N., Yuan, X., Improving the mechanical properties of natural fibre fabric reinforced epoxy composites by alkali treatment. *J. Reinf. Plast. Comp.*, 31, 425, 2012.
93. Kumar, R., Ul Haq, M.I., Raina, A., Anand, A., Industrial applications of natural fibre-reinforced polymer composites–challenges and opportunities. *Int. J. Sustain. Eng.*, 12, 212, 2019.
94. Mohanty, A., Misra, M., Drzal, L.T., Surface modifications of natural fibers and performance of the resulting biocomposites: An overview. *Compos. Interface.*, 8, 313, 2001.

25

Bioplastics and Biocomposites in Flame-Retardant Applications

L. Magunga, M. Mohapi, A. Kaleni, S. Magagula*, M.J. Mochane† and M.T. Motloung

Department of Life Sciences, Central University of Technology, Free State, Private Bag, Bloemfontein, South Africa

Abstract

Excellent flame retardant properties are a requirement for materials used in public spaces. These materials should not be susceptible to burning when exposed to an open flame. Since biocomposites and bioplastics have poor flame retardant properties, various methods have been employed to enhance their flame retardancy. However, the most common method is the incorporation of flame retardant additives. Now, with the aim of achieving optimum properties, the flame retardant additives must be homogenously mixed with the biocomposites and bioplastics. Furthermore, with the aim of maintaining the ecologically benign nature of the biocomposites and bioplastics, recent novel research has moved from the usage of halogen-based flame retardant agents to Nanosized environmentally friendly flame retardant agents. The chapter is aimed at covering the latest literature on flame retardant bioplastics and biocomposite's fabrication and application.

Keywords: Biocomposites, bioplastics, flame retardants, fire resistance, applications

25.1 Introduction

Biodegradability and eco-friendliness are two of the most sought after properties of in many industrial circles nowadays. Due to their environmentally benign nature and biodegradability, bioplastics and biocomposites are constantly receiving a lot of attention for many industrial applications. Today, biocomposites are commonly used to manufacture construction materials used in the aviation, automotive and civil engineering industries as well as in the electrical and electronics engineering, astronautics and pharmaceutical industries [1]. Contrastingly, bioplastics are commonly used in disposable materials such as in packaging materials, straws, bottles, containers and bags as well as in materials that are nondisposable materials such as in carpets, medical implants, car insulation, 3D printing, phone casings and plastic piping. Bioplastics have several advantages over conventional plastics. Some of

*Corresponding author: sifisom61@gmail.com

†Corresponding author: mmochane@cut.ac.za; Mochane.jonas@gmail.com

Inamuddin and Tariq Altalhi (eds.) Handbook of Bioplastics and Biocomposites Engineering Applications: 2nd Edition, (539–574) © 2023 Scrivener Publishing LLC

the advantages include, lower carbon dioxide emissions [2], lower carbon footprint [3] and recyclability [4]. In spite of their superior properties over conventional or petroleum-based plastics, bioplastics still occupy a small percentage of the global total plastic production. For example, in 2020, the global bioplastic production was about 2.1 million tons, which was 1 % less than the worldwide production of plastics [5]. The competitiveness of bioplastics with conventional plastics is currently hindered by several factors of which high production costs is one of them. This further hinders the replacement of conventional plastics with bioplastics [5]. Generally, bioplastics can be described as plastic materials that can be biodegradable, bio-based or have a combination of both properties (see Figure 25.1) [6]. The prefix "bio" before the word "plastic" does not necessarily mean that all bioplastics are biodegradable or compostable. It only means that bioplastics are carbon source materials [5]. Originally, bioplastics are manufactured from several food wastes which are mostly fruit and vegetable wastes via simple or complex processes. These manufactured bioplastics can be reinforced with natural fillers in order to obtain biocomposites [7]. Generally, a biocomposite can be described as a material that consists of more than one component. In these materials, one of the components is at least bio-based or biodegradable. However, most of the time, the components in biocomposites are both bio-based and biodegradable [1].

Now, it is important to note that there is a distinction between the terms bio-based and biodegradable. The term bio-based speaks to the material's origin while the term biodegradable is associated with the material's end-of-life properties [6]. Even though bioplastics and biocomposite materials have a promising future, they have one major drawback. They are susceptible to burning when exposed to a flame [6, 8]. Considering the stringent specifications of the majority of materials used in public spaces, bioplastics and biocomposites need to be developed in such a way that they are resistant to fire [6]. Various approaches are utilized to improve fire retardant characteristics of polymers and polymer composites. Such approaches include: fire retardant additives, less flammable polymers and engineering approach [9, 10]. The engineering approach method is cheap. However, it lacks reliability.

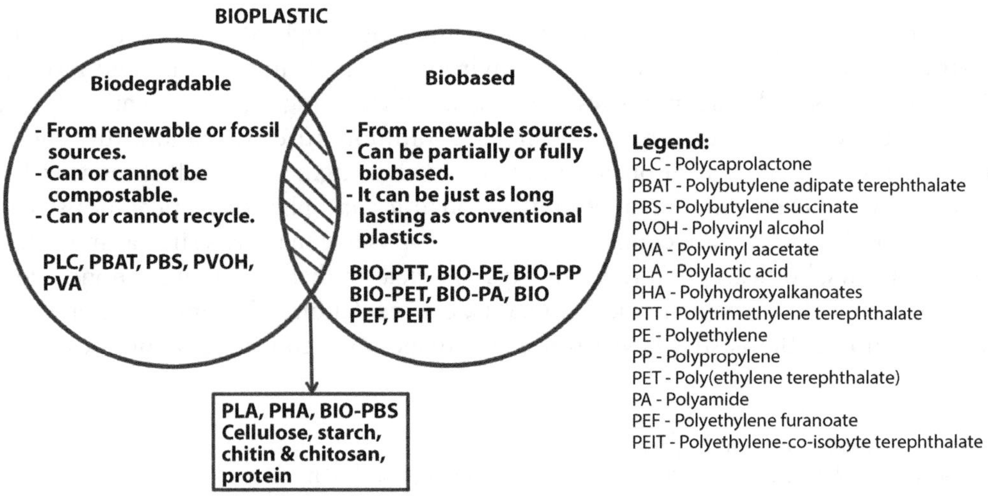

Figure 25.1 Types of bioplastics that are biodegradable, biobased or feature both properties [6]. Adapted from MDPI open access.

Polymers that are less combustible are long-lasting yet very costly. The inclusion of flame retardant additives in composite materials is a more familiar and well investigated approach [9]. The flame retardant additives can be described as micro- or Nanosized powdered fillers, which use various mechanisms of action to stop fire propagation [9, 11–14]. The addition of flame retardant additives into polymer composites needs a special and careful preparation procedure. The most significant factor in the fabrication procedure is achieving a homogenous mix between the additive particles and other components of the composites. Agglomeration occurs when flame retardant chemicals are mixed in an inhomogeneous manner, compromising the composite's mechanical and water absorption capabilities [9, 15–17]. Some approaches of mitigating the unwanted effects from flame retardant additives have been reported in literature [9, 15, 16]. The main function of a flame retardant is to negate harmful effects of radiant and contact heat, molten metal dripping, flames, sparks as well as vapours and hot gases [6, 18]. For the longest of time, the most popularly used flame retardants were halogen-based, but nowadays more strict legislations have enforced the evolution of a new generation of Nanosized flame retardants that are ecologically friendly [6]. The latest novel research works place more emphasis on Nanosized FR agents due to their environmental friendliness and improved flame resistance properties [9]. More research work is still required to further optimize the properties of Flame retardant agents. The idea is to develop flame retardants with a multifunctionality of enhancing the thermal and mechanical features of bioplastics and biocomposites while simultaneously enhancing their flame resistance properties as well. The scope of this chapter is to cover the latest literature on the use of multifunctional flame retardant agents in bioplastics and biocomposites. The chapter will also cover the applications of these flame-retardant bioplastics and biocomposites.

25.2 A Brief Introduction to Bioplastics and Biocomposites

A biocomposite can be described as a composite that is made up of at least one natural resource. Biocomposites made of natural reinforcements, such as vegetable fibers, can be either partially environmentally friendly with conventional polymer matrix that is not biodegradable, such as epoxy, polyester, polyethylene, polypropylene and so on, or completely environmentally friendly with a polymer matrix that is biodegradable, such as soy, cellulose, starch, polybutylene succinate (PBS), polylactic acid (PLA), polycaprolactone (PLC) and so on (Figure 25.2). Biocomposites derived from natural fibers and biopolymers are more ecologically friendly and are referred to as "green" composites (see Figure 25.2) [6, 19]. Table 25.1 shows the features of some common biopolymers. Furthermore, bioplastics can be characterized as a class of plastics that is either biodegradable, bio-based or a combination of both [6]. It is important to distinguish between the terms biodegradable and bio-based first. Bio-based properties are associated with the origin of the product, whereas biodegradable properties address product end-of-life issues. Notwithstanding the existence of bioplastics, plastics are still in existence. Because of its unique property, malleability or plasticity, plastic has gradually become one of the most ubiquitous products in the human world since its inception. It is easily shaped into (e.g. fibres, bottles, plates, films, boxes, etc.). Plastics are widely used in meeting our day-to-day needs nowadays, beginning with our clothes, household utilities, home electronics, medical equipment, cosmetic

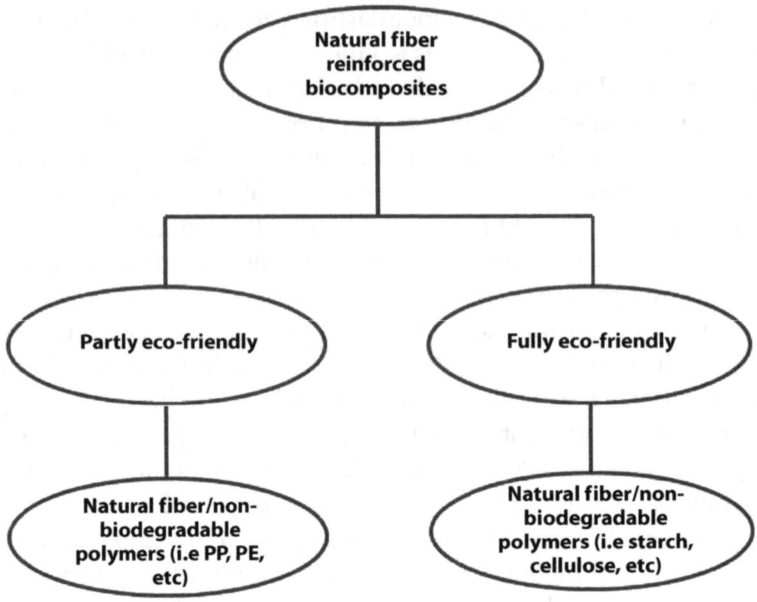

Figure 25.2 Classification of natural fiber reinforced biocomposites.

Table 25.1 Properties of commonly used biopolymers.

Biopolymer	Properties	Refs
PLA (polylactide)	- Renewable, biocompatible and biodegradable. - Synthesized via ring opening polymerization or condensation polymerization of lactic acid. - Impact properties and thermal resistance are less than those of commonly used polymers. - It is used in packaging, biomedical and textiles applications.	[28–36]
PHA (polyhydroxyalkanoates)	- Renewable, biocompatible and biodegradable - Manufactured by micro-organisms using carbon source materials. - Shear and temperature sensitive.	
Starch	- It is made up of amylopectin and amylose polysaccharides. - Starch can be used to make thermoplastic starch (TPS) by using plasticizers to disturb molecular interactions or by using complex methods such as melt mixing, morphology control and devolatization. - Starch is extremely delicate to moisture, therefore it has limited practical applications. In order to overcome this drawback, starch is blended with other less sensitive polymers and additives.	

(*Continued*)

Table 25.1 Properties of commonly used biopolymers. (*Continued*)

Biopolymer	Properties	Refs
Cellulose	- Biodegradable, biocompatible and renewable. - Extracted from cotton, wood or agricultural by-products such as crop straws, stalks and bagasse. - Regenerated cellulose is used to make fibers and films in industry while cellulose esters are used in packaging, coatings and biomedical applications.	
Chitosan and Chitin	- Biodegradable, renewable and biocompatible. - They have excellent absorption features. - Chitin is a natural polysaccharide and it is found in the skeletal structures of most invertebrates such as crustaceans and insects. - Chitosan is made from the deacetylation of chitin. - Chitosan is used in films and fibers as well as in biomedical applications.	
Proteins	- Renewable, biodegradable and biocompatible. - Can be obtained from animal and plant proteins. - Protein-based bio-products are usually obtained from wet and dry processing methods. - Protein-based bio-products are mostly utilized in pharmaceutical and food packaging.	
Bio-PBS (Bio-polybutylene succinate)	- Renewable, compostable and biodegradable. - Polymerization of 1,4- butanediol/succinic acid creates Bio-PBS. - Biobased-PBS is food contact approved and so it is mainly used in packaging and product containers.	
PBAT/PLA (polybutylene adipate-co-terephthalate)	- Biodegradable and compostable. - It has properties that are similar to those of low density polyethylene (LDPE). - Polybutylene Adipate-co-Terephthalate is synthesized from fossil materials. - Its components (i.e., PLA, starch) have a 30% biobased carbon content. - They are used for making flexible films and packaging materials e.g. compostable shopping bags.	

containers, cars, toys, etc. Plastics' overwhelming popularity aids in the development of large industries. Though it was initially thought to be a very beneficial innovation to us, with the passage of time and the passage of years, it has become more venomous to the environment [20]. In 2019, the global production of plastics was predicted to be 368 million tonnes [21], after more than 50 years of continuous growth. As a result, economic and environmental sustainability are being called into question, as fossil fuels are still widely used in the manufacture of plastic. The primary sources of those plastics are petrochemical-based substances, which cause severe environmental pollution and even the deaths of millions of animals each year due to improper disposal of its organic substrates and wastes. It does not only endanger marine mammals [22, 23], but it also causes terrestrial eco-toxicity, such as reducing soil fertility and contributing to eutrophication. Improper disposal of petrochemical plastic wastes causes major problems in a variety of metropolitan cities around the world [20, 24]. Plastics such as polyvinyl chloride (PVC), polystyrene (PS), polyethylene (PE) and others are derived from petrochemical sources (see Figure 25.3). It contributes significantly to global warming because it is derived from fossil fuels and emits greenhouse gases [20, 25]. Burning plastics is also harmful to the environment because it emits hazardous dioxin materials [20, 26], which cause acute respiratory problems in humans; thus, sustainable waste management has become a critical issue in overcoming the environmental crisis. Because of improper disposal and handling, conventional plastics frequently degrade into extremely small pieces known as microplastics. When such particles mix with water, they severely harm aquatic organisms, and humans may also suffer because of micro plastic infiltration. As a result, we require an eco-friendly and worthy substitute for petrochemical plastics, something that is simple to manage, prepare and is also economically preferable. Therefore, the introduction of bioplastic by scientists is playing a huge positive impact in the environment and helping to overcome the drawbacks of non-biobased plastics hence it has gained a widespread attraction in industrial applications. Some of the uses of bioplastics and biocomposites include, construction, building, civil, packaging, automotive, biomedical, etc. [27]. Bioplastics are environmentally friendly, biocompatible, and biodegradable. They contribute to less reliance on petroleum-based polymers, less reliance on fossil fuels, less environmental pollution by plastic waste, and minimal carbon dioxide emissions in the environment (CO_2 neutral). Scientists are looking into ways to enhance the performance properties of bioplastics, by compounding or blending bioplastics with other polymers and fibers. Blending bioplastics with natural fibers "green composites" provides a sustainable alternative with 100% degradability; scientists are also experimenting with blending bioplastics with synthetic fillers or natural fibers "Biocomposites." These "Biocomposites" materials not only have excellent properties in a variety of applications, but they also have

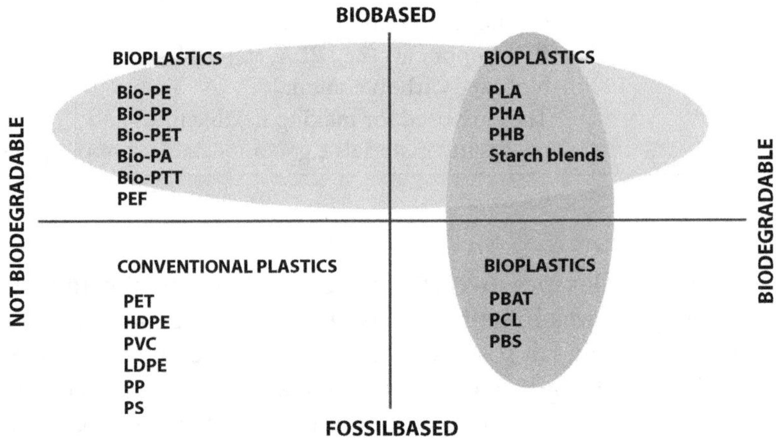

Figure 25.3 Classification of conventional plastics and the three types of bioplastics [5]. MDPI open access.

a lower carbon footprint on the environment. They also offer unique features that are naturally nonexistent [27]. Figure 25.3 shows a classification of the three types of bioplastics and common plastics.

Regardless of their origin (naturally or synthetic), polymers have poor flame retardant properties. This is because they consist of carbon as their main constituent [9]. However, in composites, the polymer matrix helps the reinforcement to fight against fire propagation and thermal loads [9, 37]. According to literature [38–48], the fire resistant features of composites and polymers have substantially researched. The flammability properties of polymers are usually explained in terms of parameters such as limiting oxygen index (LOI) and heat release rate (HRR) [9]. The limiting oxygen index (LOI) is the least amount of oxygen required for the burning of a polymer [9]. It is determined by passing an oxygen/nitrogen mixture over a burning specimen and then gradually lowering the oxygen concentration until a critical level is reached. Heat release rate (HRR) can simply defined as the rate at which fire releases energy. It is synonymous to power and is measured in Watts (W).

25.3 Flame Retardants Used in Polymer Materials

The utilization of flame retardants in polymer composites has been a hot topic since early historic times [49]. In the 1970s thermoset resins such as epoxy and polyester were regarded the major flame retardant polymers due to the use of alumina hydrates and halogen compounds as additives [49]. Traditional flame retardant chemicals (FR) used in polymer systems are seen in Figure 25.4 below.

The main aim of flame retardants materials and/or chemicals is to interfere with the combustion process either during (i) flame spread, (ii) ignition, (iii) decomposition or (iv) heating. Regulated by the nature of the flame retardant, this can either occur in gas, liquid or solid phase [49]. Conventional flame retardant chemicals such as halogens and organophosphates are accompanied by drawbacks linked to their toxicity when combusted. Available evidence suggests that organophosphates may lead to the following neurotoxic activities: (1) Organophosphate-Induced Chronic Neurotoxicity (OPICN), (2) Cholinergic Neurotoxicity and (3) Organophosphate-Delayed Chronic Neurotoxicity (OPICN) [50]. Moreover, this mentioned flame retardants are nonbiodegradable and do not align with the relevant sustainable development goals as they accumulate in the environment, breeding pollution. In as much as mineral-hydrated fillers such as magnesium hydroxide and aluminum tri-hydroxide offer an effective deviation from halogenated flame retardants, they require high filler loading of up to 50% in order to provide sufficient flame resistance

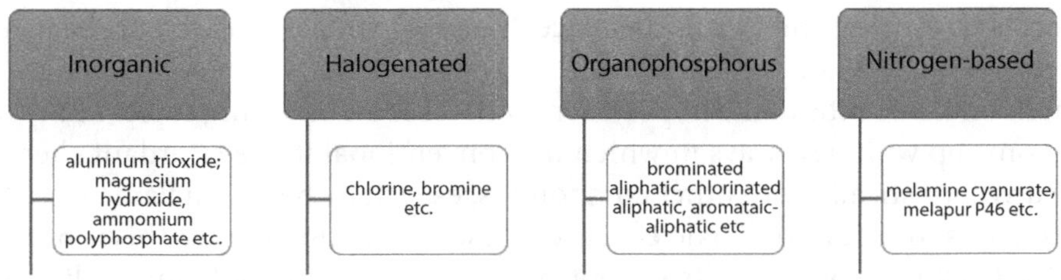

Figure 25.4 Conventional flame retardant chemicals used in polymer systems.

in a polymer system [51]. This created an opportunity for other eco-friendly flame retardant materials which required lower filler loadings to be explored. Which brought us to the introduction of bio-based and nano-based FR agents. Bio-based FR agents involve the use of biodegradable materials like modified cellulose, lignin, alginates, chitosan, starch and deoxyribonucleic acid (DNA) as flame retardant fillers in composite materials. These bio-based agents influence the flame resistance of a composite through the concept of intumescent formulation, the swelling up of a material/substance so that the underlying material is protected from heat when subjected to fire, or by producing a protective char layer that prevents the polymer from easily catching fire [51]. Other bio-based flame retardant include phytic acid, tannic acid, isosorbide, β-cyclodextrin and diphenolic acid used in epoxy, polylactic acid (PLA) and ethylene-vinyl acetate (EMA) polymer matrices [52]. During combustion, char-forming FRs undergo a process of dehydration in the under heat, causing it to form a double-bonded layer of thermal insulation protecting the base material [53].

Flame retardant fillers can be classified into (i) additive and (ii) reactive flame retardants. Additive FRs are physically incorporated during the manufacturing phase so they do not interact chemically with the polymer. Most common additive FRs are halogenated chlorine and bromine. Despite their controversies regarding toxicity, halogenated FRs are still in widespread use due to cost efficiency and high performance [54]. Halogen acids operate through the concept of engulfing active radical groups and organic chains thereby interfering with combustion causing flame retardancy [54]. Reactive FRs operate by chemically binding a FR agent to the end of a polymer chain in the form of polymer precursors or monomers. They can be regarded as superior to additive flame retardants as they form covalent bonds between the FR substance and the polymer and therefore do not disintegrate. Reactive FRs are inclusive of silicones, halogens, phospho-nitrogen, boron, nitrogen etc [54]. In addition to biobased FR researchers have begun to explore nanofillers in polymer reinforcement. The utilization of natural nanoclays such as montmorillonite, kaoline as well as other nanoparticles such as carbon nanotubes and grapheme in polymer systems has increased rapidly over the years. Nanofillers are seen to increase the thermal resistance of polymer materials due to its charring effect such as is recorded with bio-based FRs. Zhou *et al.* explored the synthesis of MoS_2@TiO_2 nanohybrids as a nano-additive for enhancing high performance bismaleimide/diallyl bisphenol A (BD), a bismaleimide resin modified with allyl. The process of MoS_2@TiO_2 nanohybrid formation involves a two-step process (Figure 25.5). The initial step includes exfoliation of MoS_2, followed by hydrolysis of tetrabutyl titanate by means of solvothermal nucleation leading to *in situ* growth of TiO_2 nanosheets [55].

The author analyzed the thermal performance of MoS_2@TiO_2/BD nanocomposite and discovered that the peak heat release rate (pHRR) of 314.25 kW/m^2 and total heat release (THR) of 53,58 MJ/m^2 were reduced by 32,5% and 15.0% respectively when compared to neat BD [55]. Moreover, the MoS_2@TiO_2 nanosheets resulted in a char formation layer during the pyrolysis process leading to less smoke and heat release as illustrated in Figure 25.6.

Research on flame retardants in polymer matrices is an on-going topic. Many researchers have come up with new ways in which the conventional flame retardant chemicals can be utilized synergistically with other compounds, especially biobased flame retardants and nano-based FRs to reduce the toxic effects of these conventional FRs. This is because most biobased FRs and nanobased FRs do not offer high amounts of flammability resistance as compared to the conventional FRs. Not only does the inclusion of biobased FRs and

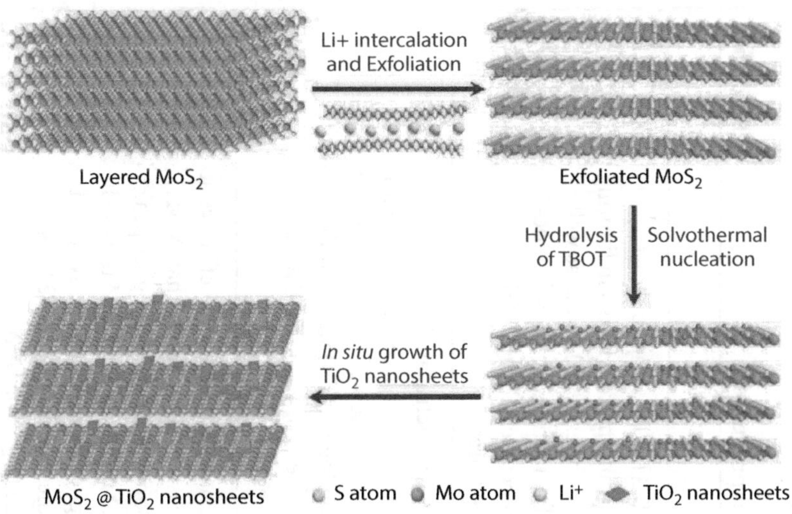

Figure 25.5 Synthesis of MoS_2@TiO_2/BD nanosheets [55]. Copyrights Elsevier.

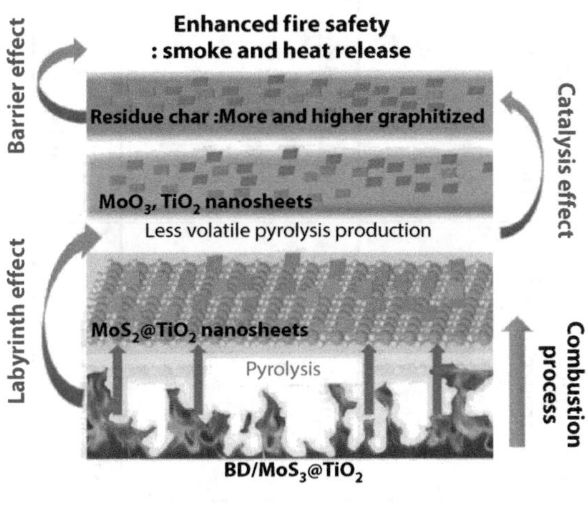

Figure 25.6 Illustration of proposed flame retardant process [55]. Copyrights Elsevier, 2019.

nanobased FRs lead us toward more sustainable and eco-friendly development, it also discovered to significantly affect the ability of the FR to increase heat stability by decreasing total heat release (THR), peak heat release rate (pHRR) etc. while lowering limiting oxygen index (LOI) at relatively low filler loading. Wang *et al.* [56] investigated the flame retardancy of poly-lactic acid (PLA) and starch biocomposites. The authors prepared samples based on PLA resin, potato starch, melamine (MA) and ammonium polyphosphate microencapsulated (MCAPP). The outcome of starch addition was an increase in LOI values. The PLA with 10 wt% starch content exhibited a LOI of as much as 41,0%. The samples containing 15 wt% MCAPP, 7,5 wt% MA, 5 wt% and 10 wt% starch reached a V-0 UL-94 rating, rendering them flame retardant. The mentioned results can cause to interpret that starch is an efficient filler for PLA with regard to protective char formation so that flame retardancy is enhanced [56]. Shi *et al.* [57] studied the flammability

Table 25.2 Flame retardants used in polymer composites.

Polymer system	Flame retardant	Type of flame retardant	Preparation method	Thermal properties	Reference
Flame retardants used in nonbiodegradable polymers					
Hydroxyl silicone oil-melamine formaldehyde/ polypropylene (HSO-MF/PP)	Ammonium polyphosphate (APP), melamine, pentaerythritol	Inorganic, nitrogen-based	Melt mixing, Hot pressing	HSO-MFAPP/PP samples showed significant improvements in limiting oxygen index (LOI) values as well as UL-94 ratings of 32 and grade-0 when pentaerythritol (PER) was incorporated.	[58]
Sisal fibre/ polypropylene	Zinc borate, Ammonium phosphate, magnesium hydroxide,	Inorganic	Rotation mixing, Injection molding	Thermogravemetric analysis reveal increased char residue at 600°C with the introduction of FRs. The limiting oxygen index (LOI) was recorded the highest for PP/30sisal/40APP samples at 31%.	[59]
Epoxy resin	Carbon nanotubes, montmorillonite (MMT)	Nano-based	Incident radiation	Carbon nanotubes are more effective at decreasing flammability than MMT. Carbon nanotubes had higher peak loss of mass rate than MMTs as well as higher thermal conductivity and thickness of the protective layer during the pyrolysis process.	[60]

(Continued)

Table 25.2 Flame retardants used in polymer composites. (*Continued*)

Polymer system	Flame retardant	Type of flame retardant	Preparation method	Thermal properties	Reference
Glass fibre/ Polyamide 6	Cloisite 30B, Clariant Exolit OP1312	Nano-based, organophosphates	Melt compounding, Injection molding	Flame retarded/glass fibre reinforced nanocomposites showed great improvements in LOI values and low filler loading. Introduction of nanoclay into the flame retarded composite significantly supressed peak heat release rate.	[61]
PHYPI/ polypropylene	Phytic acid salt	Bio-based flame retardant	Melt mixing	The PP/PH YPI sample passed the UL 94 V0 rating with a LOI of 25% at a molar ratio of 1/3. These results showed that PHYPI, a bio-based phytic acid salt, is an excellent PP flame retardant.	[62]
Unsaturated polyester resin (UPR)	Intumescent flame retardants(IFR) [Ammonium-polyphosphate (APP), pentaerythritol (PER), melamine (MEL)], montmorillonite-modified phytic acid (PA-oMMT)	Inorganic, nitrogen-based, nano-based, bio-.	Mechanical mixing	The combustion analyses shows that incorporation of IFRs and PA-oMMT into UPR reduced the total heat release value from 101.6 to 81.8 kW/m^2. The LOI raised in value from 19.1% to 29.2% when IFRs/1.5 PA-MMT was introduced.	[63]

(*Continued*)

Table 25.2 Flame retardants used in polymer composites. (*Continued*)

Polymer system	Flame retardant	Type of flame retardant	Preparation method	Thermal properties	Reference
Epoxy/polyester/ fiber glass	Aluminum hydroxide, magnesium hydroxide	Inorganic	Compounding, Compression molding	Samples containing epoxy/glass fibre and magnesium hydroxide did not have a flammability index of zero meaning they were not self-extinguishing according to ISO based 3795. Both aluminium hydroxide and magnesium played a role in reducing the flammability index of the composites with aluminium hydroxide showing to be a better flame retardant than magnesium hydroxide across all samples.	[64]
Flame retardants in biodegradable polymers					
Polylactic acid (PLA)	Tannic acid, phosphorus based compounds, organically modified montmorillonite	Bio-based, inorganic, nano-based	Internal mixing , compression molding	Tannic acid (TA) alone does is not a flame retardant agent. However with the incorporation on nanoclay (MMT) was able to reduce the peak of heat release rate (pHRR) by up to 50%. Phosphorylated TA using reactive pathways lead to 58% reduction in pHRR without affecting time-to-ignition. Phosphorylated TA also induced thermal degradation of PLA.	[65]

(*Continued*)

Table 25.2 Flame retardants used in polymer composites. (*Continued*)

Polymer system	Flame retardant	Type of flame retardant	Preparation method	Thermal properties	Reference
Polylactic acid (PLA)/ nonwoven hemp fibre mat	Carbon nanotubes, sepiolite nanoclay	Nano-based	Compounding with twin-screw extruder, hot pressing	Reduction in pHRR observed with the introduction if hemp fibre mat. Sepioliite/MWNTs/PLA nanocomoposite displayed the best performance resulting in a decrease of 58% in pHRR. Nevertheless, this value decreased with the addition of hemp fibre to 45%.	[66]
Polylactic acid (PLA)	Ammonium polyphosphate (APP), chitosan (CS), modified phytic acid (PA)	Inorganic, nano-based, bio-based	Mechanical stirring, Melt blending	At 10% of APP@CS@PA-Na addition into PLA, increased flame retardancy was realized with UL94 V-0 rating and a LOI of 30,5%. The APP@CS@PA-Na reveals its flame retardant nature in mostly in the condensation phase reducing the pHRR and THR by 33% and 19% respectively.	[67]
Polybutylene succinate	[Ammonium polyphosphate (APP) + melamine (MA)] IFR, montmorillonite (OMMT)	Inorganic, nitrogen-based, nano-based	Co-rotation with a twin-screw extruder, Injection molding and Hot press.	The optimum recorded content for OMMT in the PBS/IFR20 was 1.5 wt% reaching an LOI of 40.1% as well as a UL-94 V-0 rating. However, excessive amount of OMMT led to a fragile carbonaceous char surface.	[68]

(*Continued*)

Table 25.2 Flame retardants used in polymer composites. (*Continued*)

Polymer system	Flame retardant	Type of flame retardant	Preparation method	Thermal properties	Reference
Poly (vinyl alcohol)	(Chitosan, pentoxide, methanesulfonic acid, urea) UPCS	Nano-based, inorganic	Solution mixing	Onset temperatures reduced with the introduction of UPCS from 265°C to 246°C. This is ascribed to the removal of acetate groups and waster when UPCS degrades. The UPCS/PVA nanocompsite showed greater thermal stability at 370°C. Char residual also increased from 0.7% to 14.3%.	[69]
Polylactic acid	Ammonium polyphosphate (APP), lignin	inorganic, nano-based	Two-roll mill mixing, compression molding. LSH was prepared by sol-gel method, incorporating hydrochloric acid, ethanol and lignin	Samples that had App/lignin had lower LOI values than those containing APP/LSH at the same ratios. PLA/APP/LSH composites reached a UL-94 rating of V-0. TGA curve analyses revealed that PLA/APP/LSH samples had char formations of 13.1 wt% at 800°C as compared to PLA/APP/lignin samples with only 8.9 wt% char formation.	[70]
Polylactic acid	Modified sepiolite (SEP-DOPO)	Nano-based	Mechanical stirring, micro-extrusion.	The addition of 10 wt% SEP-DOPO resulted in a LOI increase from 20.2 % to 31.5 % making the UL-94 test to reach V-0.	[71]

(*Continued*)

Table 25.2 Flame retardants used in polymer composites. (*Continued*)

Polymer system	Flame retardant	Type of flame retardant	Preparation method	Thermal properties	Reference
Polybutylene succinate	CP-lignin (Lignin, lipase N435, Cyanuric chloride, trioxa-1-phophabicyclo-[2,2,2] octane-4-methanol-1-oxide)	Bio-based, inorganic	Solution mixing	The addition of CP-lignin to PBS at 30 wt% resulted in pHRR and THR improvements of up to 27% and 31% respectively. The ration of weight loss decreased with increasing CP-lignin content.	[72]
Poly(vinyl alcohol)	Glycerol-plasticized thermoplastic starch (TPS), MCAPP (APP, MA, formaldehyde)	Bio-based, inorganic	Solution mixing, hot press	MCAPP/PVA composites reached a UL-94 rating of V-0 at only 2 wt%. Pristine TPS did not reach the flame retardant value. The THR had a significant drop with increasing MCAPP content. However the pHRR increased slightly. A protective char was formed composed of polyphosphate and/or pyrophosphate.	[73]

properties of functionalized graphene oxide (GO) as a filler in poly-lactic acid (PLA). A natural flake graphene having an average size of 44μm was utilized for the experiment. A novel 10-(2,5-dihydroxyphenyl)-9, 10-dihydro-9-oxa-10-phosphaphenanthrene-10-oxide (DOPO-HQ) was used to functionalize the graphene oxide. When unmodified GO is incorporated to PLA at 6 wt% there is an increase of 2.48 wt% char yield. The mass loss, however, is more than that of neat PLA. Inclusion of 10 wt% functionalized graphene oxide (FGO-HQ) resulted in a 2.15 wt% char yield but also decreased the maximum mass loss rate by 19.0%. Neat PLA does not have a UL-94 rating of V-0, however, FGO-HQ contents of 2.0 wt%-10 wt% resulted in a V-0 UL-94 rating. Furthermore, LOI reached values of 20.0 % to 27.5% with the addition of 10 wt% FGO-HQ. Thus, it can be said that FGO-HQ is an effective FR reinforcement to PLA [57].

Table 25.2 shows some studies conducted on FRs used in nonbiodegradable polymer composites as well as biodegradable polymer composites and the thermal properties obtained for each.

25.4 Action Mechanisms of Flame Retardants

Flame retardants can interfere with the combustion process at multiple stages, and their mechanism is largely determined by which step of combustion they interfere with [9]. Various researchers have investigated the various mechanisms of biocomposites and bioplastics. Tawiah *et al.* [74]. investigated the mechanism and properties of flame resistant PLA composites bonded with recycled wool fibers. Forced Flame combustion of the composites was used in their study as one of the other components on flammability, and the mechanisms used were the condensed mechanism and the extreme gas phase mechanism. As per their findings, the presence of most flame retardants (FRs) in polymers resulted in increased carbon monoxide (CO), carbon dioxide (CO2), dioxins, and hydrogen cyanide (HCN) generation during combustion [75]. CO2 generation followed a similar trajectory to the peak heat release rate (PHRR) n due to the considerable transformation of incomplete oxidization product (CO) to complete oxidization product (CO2) [76]. PLA/phenylphosphonic 3(2-aminobenzothiazole) (P-TAB) and PLA/P-TAB/waste wool fibers (WF) composites had the lowest carbon monoxide production (PCOP) and produced the most CO in the first 25 to 50 seconds. However, CO production for PLA/P-TAB and PLA/P-TAB/WF composites increased marginally towards the end of the combustion process (around 180–250 s) compared to neat PLA as a result of flaming caused by incomplete combustion because of the concentrated phase mechanism of P-TAB and, conversely, the extreme gas phase mechanism of WF. The enhanced char residue caused by WF was insufficiently compact to reinforce the phosphorus condensed phase chemistry, resulting in the evolution of a higher CO. Tawiah *et al.* [77], conducted another work in which he investigated the combustion behavior of PLA/HNETP composites. The PHRR was reduced by 51.3%, the total heat released (THR) by 43.1, and carbon monoxide (CO) production was reduced by 46.5% with 3wt% HNETP loading. Higher limiting oxygen index (LOI) values of 26.5%, 28.0%, and 32.5% were achieved with the 1–3 wt percent HNETP loading, according to the results. The influence of both the condensed and gas phases coupled HNET mechanisms was attributed to the biocomposites' increased FR performance. Another study was done by Xue *et al.* [78]. On the manufacture of a quality flame retardant PLA with submicron-grained material using one-pot scalable oligomeric phosphoramide. To act as a flame retardant, 3% phenyl phosphonic piperazine (PPP) was added. PLA/3 percent PPP's effective heats of combustion (EHC) value was reduced from 2.00 MJ/m2 g to 1.96 MJ/m2 g., according to the findings. The decrease in EHC indicated that the major gaseous flame inhibition mechanism was engaged in the flame retardant process. However, due to the high char residue of PPP, which is about 10.8 wt% at 650°C, the residual mass of PLA composites after cone tests showed that adding PPP can significantly raise the final char residue, which was also consistent with the thermogravimetric analysis (TGA) results. To some extent, the higher char generated during combustion can add to the lower combustion. In the gaseous phase, PPP's good flame retardancy was attributed to radical inhibition and gas dilution effects, as well as char residue protection in the condensed phase. Wu *et al.* [79] investigated the use of a highly efficient flame retardant poly (phenylphosphoryl phenylenediamine) (PPDA) in biodegradable poly(lactic acid) to improve its fire safety. In the gas phase, PPDA's flame retardant action was based on a combination of radical inhibition and inert gases, with the char residue functioning as a barrier in the condensed phase. PPDA was developed by condensation of phenylphosphoryl dichloride (PPDCl) with p-phenylenediamine to improve the fire safety of biodegradable PLA material. When 3 wt% and 5 wt% PPDA were added to clean PLA, the limited oxygen index (LOI) values increased from 20.0% to 25.5% and 30.5%, respectively. Furthermore, the increase in total smoke release rate (TSR) and decrease in effective heats of combustion (EHC) revealed that PPDA largely hindered gaseous flames. The presence of a radical inhibition mechanism in the gaseous phase and the protection of char residues in the condensed phase were thus linked to the favorable effects of PPDA. Saba *et al.*, 2019 [80], presented the possible flame retardant mechanism of Montmorilllonite (MMT) clay and Organo-Montmorillonite (OMMT) during combustion in another study (Figure 25.7). Because of the inclusion of hydrophobic functional groups, OMMT are organophilic, have a stronger polymer attraction, have a longer distance between MMT layers, and exfoliate [81]. The epoxy glue in the nanocomposites easily intercalates between the MMT layers, shielding them from the heat radiation. External energy is stored and distributed more effectively and reliably throughout the OMMT layers. As soon as composites are burnt at temperatures bet ween 380 and 500 degrees Celsius, OMMT forms a structure that is more complicated, compact, stable, dense, and carbonaceous known as char, which becomes more homogeneous, carbonated, and thicker as the temperature rises [82]. On a surface, the generated silicate char yield acts as an excellent heat barrier, keeping the burning substrate out of the flame zone and preventing further thermal oxidative reactions [83]. However, nanoclay migration must be completed before the charring effect may begin. During the formation of gas bubbles, organoclay sheets convect upward, as they are triggered through deterioration of the polymer chains and quaternary ammonium organo modifier [82]. MMT, on the other hand, has a high aspect ratio, which results in injected nanocomposites, though not as high as OMMT [84]. Nanoclay particles migrated towards the flame surface when exposed to heat or during combustion, covering and shielding the underlying polymer [85]. The dense char layer formation in MMT nanocomposites serves as a deterrent to energy transfer, reducing the volatilization of flammable gases, inhibiting the diffusion of oxygen, and reradiating heat. To combust the epoxy/Kenaf/MMT hybrid system, the critical oxygen concentration for composite combustion has been raised, resulting in a decrease in polymer breakdown rates. However, because of the more convoluted channel given by exfoliated OMMT clay

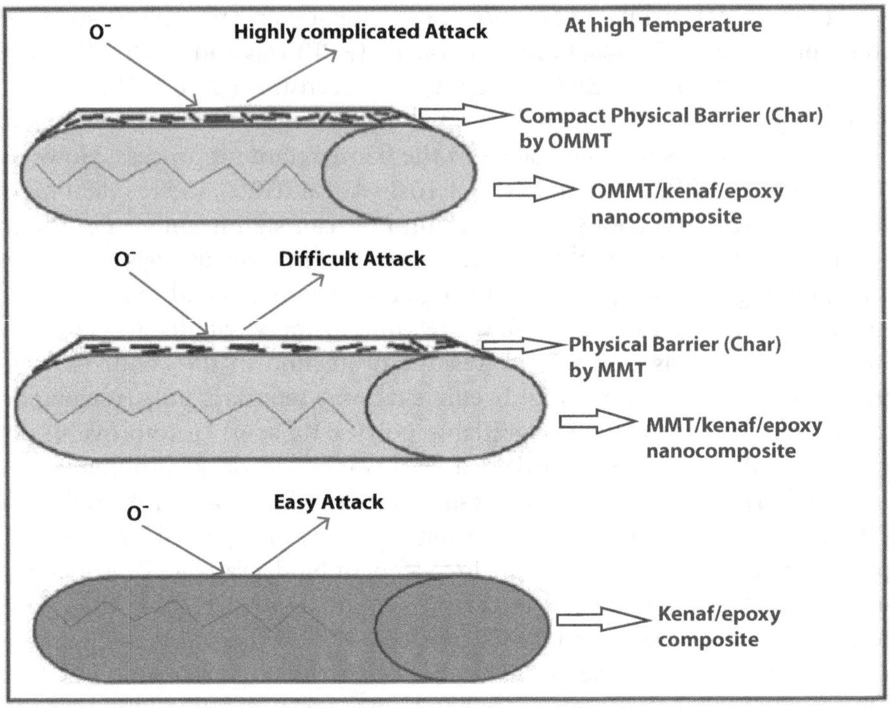

Figure 25.7 Flame retardant mechanism of clays (MMT and OMMT) during combustion [80]. Obtained with permission from Elsevier.

layers, volatile diffusion and heat transport is more restricted in OMMT hybrids than in MMT hybrids at higher temperatures. Below are descriptions of Various flame retardants and their different modes of action.

25.4.1 Char-Formation

In this method, FR additives provide a carbon - a coating of char on the polymer chain, which can prevent carbon atoms in the polymer from oxidizing and thereby protecting it from additional flame action. The carbon - containing char coating also enhances thermal insulation, minimizes mass volatiles, and prevents flammable gases from escaping. This method is mainly used by halogen-free FR additives like phosphorus and nitrogen [9]. Figure 25.6 shows the char generating mechanism in polymers [55].

25.4.2 Inert Gas

FR additives evaporate and release large volumes of nonflammable/inert gases (such as nitrogen, ammonia, and carbon dioxide) into the gas phase or by condensation in this process. These inert gases contaminate oxygen closer to the site of combustion, reducing the amount of combustible fuel available for burning. This is how metal hydroxides, metal carbonates, and some nitrogen-producing compounds act [9].

25.4.3 Contact of Chemicals

This is a reactive process in which free radicals produced during polymer ignition interact with and are held by FR additives or their by-products. The mechanism cools and resists/ reduces flammability by neutralizing free radicals. Most halogenated (brominated) FR additives function in this manner. When free radicals formed by polymers, such as hydroxide and hydrogen, come into contact with halogen free radicals in the gas stage, fire suppression occurs [9].

25.4.4 Restriction of Vapor Phase Burning

In order for vapour phase combustion to take place, substances that are volatile and are generated by a polymer's oxidative thermal degradation are necessary. To avoid vapour phase combustion, flame retardants created particularly to disintegrate at temperatures similar to those at which the polymer begins to deteriorate and produce less reactive radicals can be used (also known as free radicals). The combustion process is hindered because these radicals do not react with the volatile products in the vapour phase [51, 86]. Due to their low reactivity, these free radicals are unable to complete the oxidation process. When the oxidation process is stopped, the heat transmission back to the polymer is decreased, which slows the combustion process and finally extinguishes the flame. Flame retardants based on organo-bromine and organo-chlorine are a good illustration of this [51, 87].

25.5 Compatibility of Flame Retardants With Polymer Matrices

Polymer blending is generally accepted in the industry for producing polymeric materials with specific uses in a cost-effective manner that would otherwise be impossible to achieve with a single polymer [88]. The degree of compatibility of the polymers combined is an essential regulating parameter in this scenario. The degree of compatibility, on the other hand, is mostly determined by the interaction between the polymeric phases of the poly-blend as a result, by altering the interaction forces among the system's elements, one can obtain a material with the desired mechanical properties [89]. One of the reasons why miscible polymer blends are becoming increasingly popular in industrial applications is because of this; however, the most difficult aspect of combining polymers is determining the degree of compatibility. Various strategies for producing polymer blends and/or determining their compatibility are currently being developed [90]. Many studies have been conducted on the compatibility of flame retardant biocomposites and one of the leading factors determining good compatibility in biocomposites is interfacial surface energy. Interfacial energy is when the surface tension or friction is proportional to the energy at an interface or the measure of intermolecular bonds disrupted at a surface level [91]. Another factor responsible for compatibility between elements in polymer blending is the use of coupling agents in order to modify the interfacial region and form a connection between the elements in a polymer blend. The type of coupling agent used will also determine the properties that the biocomposite will have. The most widely used coupling agent is silane and the use of silane in flame retardant biocomposites has shown to increase the interfacial adhesion [92]. Matko *et al.* [93] used silane treatment to improve the compatibility of different biocomposites namely polypropylene, polyurethane and plasticized thermoplastic starch reinforced with ammonium polyphosphate flame retardant. 140°C heat was used to create silylated natural cellulose fibre with 20% Tetra Ethoxy Silane and a catalytic quantity of dibutyl tin dilaurate in the presence of dibutyl tin dilaurate. The results showed that surface treatment with 20% TES enhanced the interaction and bonding between polypropylene and the natural fibre. The results further showed that the use of 10% ammonium polyphosphate was enough to increase the heat properties of the plasticized thermoplastic starch. Ding *et al.* [94] modified ammonium polyphosphate with silane. Silane was subsequently added into high impact polystyrene together with carboxylic-functionalized Multi Walled Carbon Nanotubes to improve its compatibility with the high impact polystyrene matrix. High Impact Polystyrene and COOH–Multi Walled Carbon Nanotubes are hydrophobic, whereas Ammonium polyphosphate is an inorganic compound that is hydrophilic. The production process and attributes of composites would be influenced by poor dispersibility and compatibility of hydrophilic components with hydrophobic materials. Surface modification of APP with a silane coupling agent has been shown to improve this situation (see Figure 25.8). The results found that to be true and it was also discovered that silane modified ammonium polyphosphate and COOH–MWCNTs collaborated to minimize the peak and total heat release rates of high-impact polystyrene nanocomposites, with the synergistic effect becoming stronger as the COOH–MWCNTs concentration increased. Interestingly, the amount of silane modified ammonium polyphosphate added in nanocomposites was revealed to be associated to the stability of the char layer generated by COOH–MWCNTs.

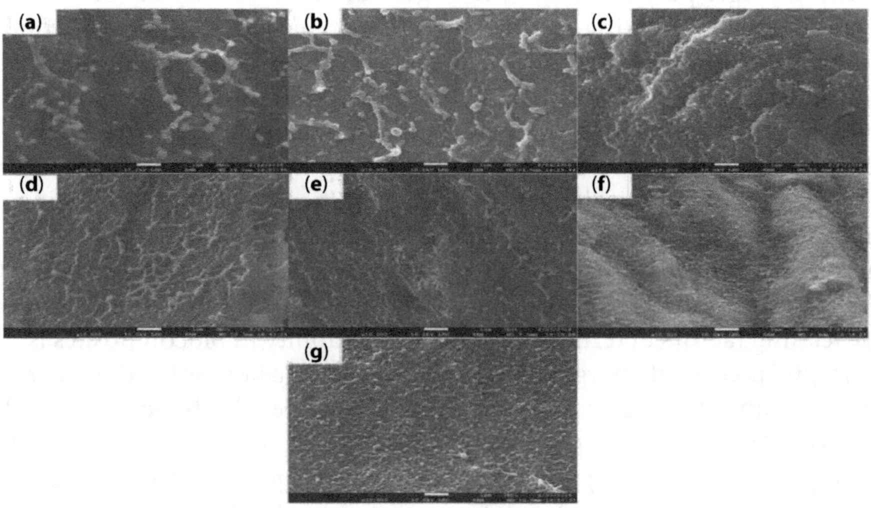

Figure 25.8 SEM images of high impact polystyrene composites [94]. MDPI Open Access.

The introduction of a flame retardant has also been shown to disturb the compatibility between components of polymer blend, e.g., biocomposite and natural fiber therefore special considerations need to be taken when flame retarding a biocomposite like the ability of the flame retardant to be integrated in the matrix without compromising the biocomposite's mechanical properties therefore The type of flame retardant used has a great influence on its compatibility with the biocomposite used [94], e.g., most organophosphorus flame retardants create phosphoric acid at high temperatures, which inhibits the substrate from burning gas. Ammonium polyphosphate is one of the most widely used intumescing flame retardants. As a result of condensed-phase mechanisms Polyphosphate of ammonium (APP), which is frequently seen in the presence of char-forming chemicals and nitrogen-containing components [6]. Shumao *et al.* [95] explored the role of ammonium polyphosphate on the characteristics of Polylactic acid reinforced ramie fiber biocomposites. In order to achieve total compatibility of the flame retardant with the natural fiber and biocomposite, ammonium polyphosphate was loaded in low quantities because during impact testing, high APP loading induces fracture propagation, which substantially affects the strength, which influences the impact strength, and hence the compact strength for PLA/ramie fiber.

25.6 Preparation of Flame-Retardant Biocomposites and Bioplastics

The need for flame retardant bio-composite products has risen significantly in the household and industrial sectors during the last years [96]. For successful production, researchers have devised a variety of fabrication methods of bio-composite materials However, in order to get fault free products, fabrication methods must be greatly improved [96, 97]. Various preparations methods are employed in order to create flame retardant biocomposite materials and the type of preparation method used plays a role in the properties that a biocomposite will have. The different preparation techniques also determine the type of application that the flame retardant biocomposite will have, e.g., injection molding method has a quick turnaround time which make its ideal for large scale production of automobile body panel, bumpers etc. [98]. Sun *et al.* [99] studied the flame retardancy of polylactic acid incorporated with polyethylene glycol and ammonia polyphosphate as a flame retardant, the biocomposite was prepared using the melt blending process PLA's flame retardancy was improved by adding 15% ammonium polyphosphate and polyethylene, resulting in a V-0 grade, as well as good mechanical results with a tensile toughness of 45 mj/m^3 and an impact strength of 14.9 kj/m^2. The melt blending method is gaining a lot of attention from researchers because of its ability to be compatible with other industrial processes such as injection molding and extrusion technique etc. and because melt blending is environmentally friendly due to its lack of organic solvents [100]. Zhang *et al.* [101] used a novel multifunctional bio derived polyphosphate to enhance the properties of PLA using the melt blending method and found that addition of a mere 6wt% polyphosphate unlocked a UL-94V-0 grading for PLA and Polyphosphate has a low melting point of 155°C, which allows it to melt with PLA during melt processing, and it also greatly increased PLA's ultra violet resistance at its lowest loading, resulting in a fine dispersion with a stage volume fraction of less than 400 nm (Figure 25.9). Jing *et al.* [102] used a simple but uncommon method called layer by layer which is a simple, versatile, and low-cost approach of covering both morphologically flat and quasi (e.g., colloidal core-shell particles) areas with nano-components from diverse groups (see Figure 25.10). The researchers used this method to fabricate a core shell flame resistant material and graphene oxide to enhance the characteristics of PLA using melt blending method as well. The results showed introduction of the grafted graphene oxide drastically improved the flame retardancy and toughness of PLA with an 86.7% improvement in impact strength at 10% graphene oxide and UL-94V-0 grade at 10 and 15 wt%.

Other common methods used to prepare flame retardant biocomposites include injection moulding, compression moulding, hot pressing etc. (see Table 25.3) [103]. Young and Tsao [104] investigated the properties and flame retardancy of PLA reinforced with bamboo fiber, the natural fiber was chemically treated with alkali solution to improve its properties and an injection molding method was employed to prepare the composite. The study's findings revealed that Tensile strength, modulus of tensile strength, flexural strength, modulus of flexural strength, and impact strength all increased.

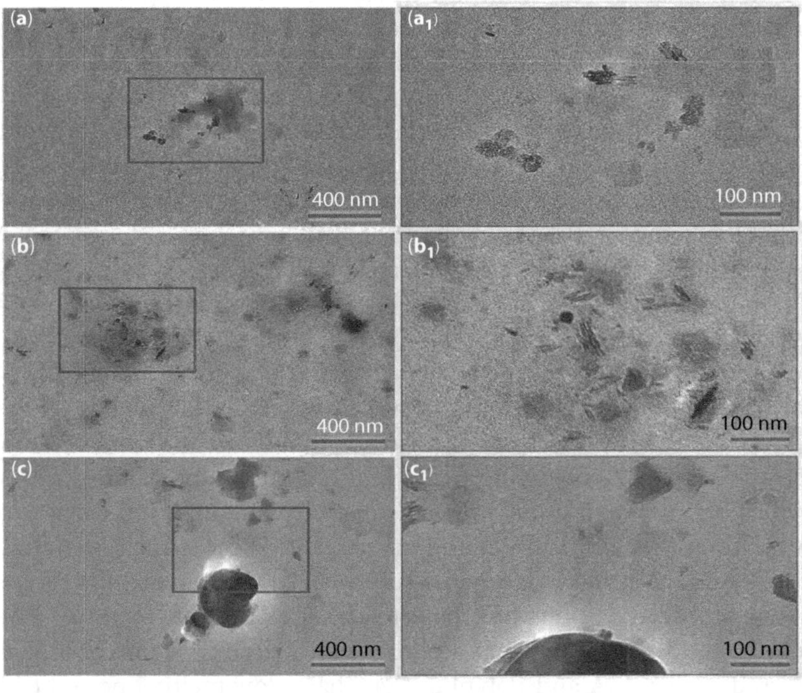

Figure 25.9 TEM images of interaction between PLA and (a, a_1) 3% polyphosphate (b, b_1) 6% polyphosphate (c, c_1) 9% polyphosphate [101]. Obtained with permission from Elsevier.

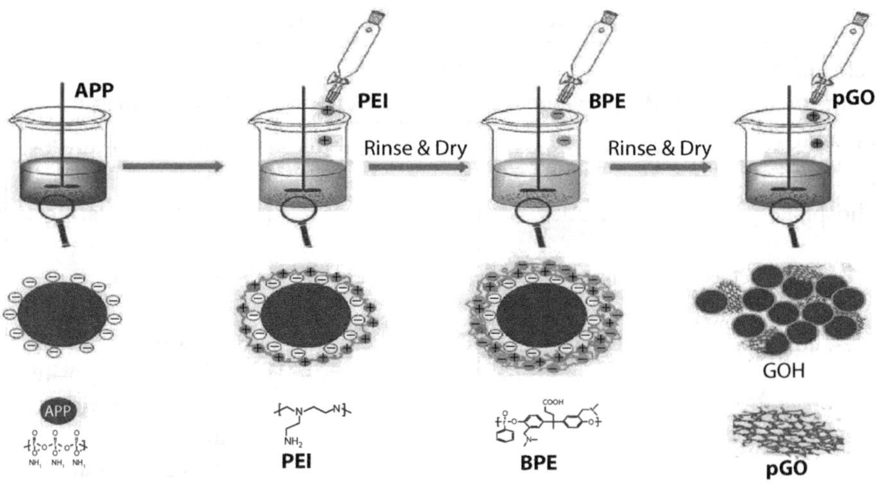

Figure 25.10 Layer by layer fabrication process [102]. Obtained with permission from Elsevier.

25.7 Applications of Flame-Retardant Bioplastics and Biocomposites

The use of petroleum based plastics has vastly aided human life but it has also taken a toll on the environment. In 1950, the world's annual production was only 2 million tonnes. Annual production has nearly doubled since then, hitting 381 million tonnes in 2015 [111]. with the worlds demand for plastics growing and with the annual production of plastics increasing each year, the accumulation of plastic waste also increases meaning numerous methods to reinforce plastics with better mechanical properties and more biodegradable and eco-friendly plastics are needed [112].

The production of flame retardant bioplastics fills the gap that traditional plastics were lacking with one of the most notable improvements being flame retardancy against fire hazards. Fire-retardancy is a property that protects against the spread of fire. Certain known characteristics such as time to ignition, heat release rate, fire propagation rate, smoke volume, and carbon dioxide emission are utilized to analyse the fire dangers of a certain polymer [51]. The use of flame retardant bioplastics does not only offer different industries a polymer that is biodegradable but one that is also effective in resistance against heat.

The application of flame retardant biocomposites is mainly dependent on the type of biocomposites and their properties, preparation technique used, e.g., polylactic acid (PLA) can be utilized for medical implants since it degrades into lactic acid. Depending on the type of material utilized, it might take anywhere from 6 months to 2 years to degrade. This means that once a PLA support structure heals, these products might progressively transfer a load to the body and PLA prepared through injection molding process can be utilized in packaging material [113]. Table 25.4 shows the intended applications of different types of flame retardant biocomposites.

Table 25.3 Summary preparation of various types of flame retardant biocomposites.

Flame retardant biocomposites	Preparation	Surface modification	Obtained properties	Ref.
Ramie fiber/PLA/ ammonium polyphosphate	Co-rotating twin screw extruder	Silane coupling	• The biocomposite obtained the UL-94V-0 rating at 10.5wt% ammonium polyphosphate loading. • The biocomposite reinforced with ammonium polyphosphate also has improved tensile strength and impact strength.	[95]
PLA/Azo-boron-BPA/ Polydopamine	Hot compression molding	Silane coupling	The synergistic AZ-A/PDA obtained a V-0 rating with a LOI value of 23.7 % at 10% load in a 1:1 ratio, while the separate components received only a V-2. The AZ-A/ PDA combination resulted in a considerable drop in PHRR with increasing residue (up to 19%). (up to 17.7%)	[105]
Polypropylene/tri-(triethoxysilylpropyl) phosphamide	Sol gel reaction	None	In comparison to pure Polypropylene, the integral procedural decomposition temperature of the flame retardant biocomposite was enhanced by 35%.	[106]

(*Continued*)

Table 25.3 Summary preparation of various types of flame retardant biocomposites. (*Continued*)

Flame retardant biocomposites	Preparation	Surface modification	Obtained properties	Ref.
Polybutylene/flax/ phosphorus compounds	Compression moulding	Ethanol	• Given the proximity of flax and phosphorous compound, charring was found to be successful at lower phosphorus levels. • Addition of the fire retardant leads to the charring of flax fibers and improves the fire behavior of PBS/ flax biocomposites.	[107]
Vinyl ester/flax fabric/ chitosan/ammonium polyphosphate	Resin transfer molding	None	• The biocomposite's characteristics were improved by altering the reinforcement and introducing fire retardants into the matrix. • The following are the highest values: heat resistance (at 700°C, 28 wt% residue): resistance to fire (V0 rank in the UL 94 test).	[108]
Rice husk ash/ phosphazene/epoxy composites	Injection molding	Silane coupling	• When compared to Epoxy material alone, the experimental results revealed that flame retardant biocomposite had improved flame retardant and dielectric properties.	[109]

(*Continued*)

Table 25.3 Summary preparation of various types of flame retardant biocomposites. (*Continued*)

Flame retardant biocomposites	Preparation	Surface modification	Obtained properties	Ref.
(Phosphorus, nitrogen, boron)/polyethylene	Hot pressing	None	• The polyethylene biocomposite containing 30% phosphorus, nitrogen, and boron had the best results, with a limited oxygen index of 29.2, vertical burning of UL-94 V-0, and tensile and notched impact strengths of 18.6 MPa and 12.5 kJ m^2 respectively.	[110]

Table 25.4 Showing intended applications of different types of flame retardant biocomposites.

Type of flame retardant biocomposites	Preparation method	Intended application	Summary of result	Ref.
PLA/NTPA	• Injection molding	• Engineering applications • Advanced optoelectronic devices.	• After only 2.5 wt.% NTPA was applied, PLA/NTPA composites received UL-94 V-0 and a LOI of 28.5%. • A homogenous and compact char layer had a barrier effect in the condensed phase. The total heat release and average effective heat combustion of PLA/NTPA composites were lowered by 12.2% and 9.8%, respectively, while the flame retardant index was increased by 38%.	[114]

(*Continued*)

Table 25.4 Showing intended applications of different types of flame retardant biocomposites. (*Continued*)

Type of flame retardant biocomposites	Preparation method	Intended application	Summary of result	Ref.
PHBV/PBAT/AlPi/ nanometric iron oxide/antimony oxide Poly(3-hydroxyl-butyrate-co-3-hydroxyvalerate)/ poly(butylene adipate-co-terephthalate)	Melt blending Compression molding	Biomedical sector	• Improved fire retardancy, which was ascribed to increases in intermediate char, favored improvements in the UL 94 grade. • The flame-retardancy mechanism involves both the phosphorus and Nano filler components, with the former serving as a flame regulator in the gas phase and the latter aiding solid-phase cross-linking.	[115]
Poly(butylene succinate)/ bamboo fiber/ microencapsulated ammonium polyphosphate	Hot pressing Melt mixing	• Packaging	PBS/bamboo fiber biocomposites have a THR value of 62.0 kW m^2 and a pHRR value of 443.2 kW m^2. The flame retardant qualities of MCAPP are also the finest.	[116]
Polypropylene/ bamboo fiber/ microencapsulated ammonium polyphosphate	Melt blending	• Architectural materials • Cables	The THR of flame retardant biocomposite decreases from 75.3 to 49.2 MJ m^2, and the pHRR decreases from 540.0 to 227.5 kW m^2.	[117]

(*Continued*)

Table 25.4 Showing intended applications of different types of flame retardant biocomposites. (*Continued*)

Type of flame retardant biocomposites	Preparation method	Intended application	Summary of result	Ref.
Wood flour/polypropylene/ethanolamine ammonium polyphosphate	Hot pressing	• Transportation • Military • Building	The LOI of the flame retardant biocomposite improved by 71.6% to 43.0% and had UL-94-V-0 grade. In comparison to wood flour/polypropylene/Ammonium Phosphate, the heat release rate and total heat release of the biocomposite with ethanolamine ammonium polyphosphate are lower.	[98]
Wood/Polylactic acid/Triglycidyl isocyanurate/novel lignin based flame retardant	Melt blending	• Building applications	The higher LOI value was 28.6% and the char residue reached 31.3% at 700°C which improve the flame retardancy of the biocomposite.	[118]

25.8 Conclusions

Due to their environmental friendliness and biodegradability, bioplastics and biocomposites are usually used in various of industrial applications. They are used as construction materials as well as in the automotive, aviation and civil engineering industries just to mention a few. Even though bioplastics and biocomposites have gained much attention for use in various industrial applications, they have poor flame-retardant properties. For these biocomposites and bioplastics to be used in public spaces, their fire resistance features need to be enhanced. Various methods can be used to enhance the fire resistance properties of bioplastics and biocomposites. However, the most commonly used method is the incorporation of FR additives. To obtain great flame retardant properties, the flame retardant agents should be homogenously mixed with the biocomposites or bioplastics. Furthermore, today's environmental concerns have shifted the focus from halogen-based FR to a new generation of Nanosized flame retardants that are ecologically friendly.

Acknowledgements

The authors would like to thank the National Research Foundation (NRF) for funding.

References

1. Sienkiewicz, A. and Czub, P., Flame retardancy of biobased composites—Research development. *Materials*, 13, 5253, 2020.
2. Pilla, S., Engineering applications of bioplastics and biocomposites-An overview, in: *Handbook of Bioplastics and Biocomposites Engineering Applications*, pp. 1–15, 2011.
3. Bishop, G., Styles, D., Lens, P.N., Environmental performance comparison of bioplastics and petrochemical plastics: A review of life cycle assessment (LCA) methodological decisions. *Resour. Conserv. Recycl.*, 168, 105451, 2021.
4. Wojnowska-Baryła, I., Kulikowska, D., Bernat, K., Effect of bio-based products on waste management. *Sustainability*, 12, 2088, 2020.
5. Vinci, G., Ruggieri, R., Billi, A., Pagnozzi, C., Di Loreto, M.V., Ruggeri, M., Sustainable management of organic waste and recycling for bioplastics: A LCA approach for the Italian case study. *Sustainability*, 13, 6385, 2021.
6. Kovačević, Z., Flincec Grgac, S., Bischof, S., Progress in biodegradable flame retardant nano-biocomposites. *Polymers*, 13, 741, 2021.
7. Acquavia, M.A., Pascale, R., Martelli, G., Bondoni, M., Bianco, G., Natural polymeric materials: A solution to plastic pollution from the agro-food sector. *Polymers*, 13, 158, 2021.
8. Brizga, J., Hubacek, K., Feng, K., The unintended side effects of bioplastics: Carbon, land, and water footprints. *One Earth*, 3, 45, 2020.
9. Shah, A.U.R., Prabhakar, M., Song, J.-I., Current advances in the fire retardancy of natural fiber and bio-based composites–A review. *Int. J. Precis. Eng. Manuf. Green Technol.*, 4, 247, 2017.
10. Morgan, A.B. and Gilman, J.W., An overview of flame retardancy of polymeric materials: Application, technology, and future directions. *Fire Mater.*, 37, 259, 2013.
11. Chen, L. and Wang, Y.Z., A review on flame retardant technology in China. part I: Development of flame retardants. *Polym. Adv. Technol.*, 21, 1, 2010.
12. Hull, T.R., Witkowski, A., Hollingbery, L., Fire retardant action of mineral fillers. *Polym. Degrad. Stab.*, 96, 1462, 2011.
13. Morgan, A.B., Flame retarded polymer layered silicate nanocomposites: A review of commercial and open literature systems. *Polym. Adv. Technol.*, 17, 206, 2006.
14. Hamdani, S., Longuet, C., Perrin, D., Lopez-Cuesta, J.M., Ganachaud, F., Flame retardancy of silicone-based materials. *Polym. Degrad. Stab.*, 94, 465, 2009.
15. Arao, Y., Nakamura, S., Tomita, Y., Takakuwa, K., Umemura, T., Tanaka, T., Improvement on fire retardancy of wood flour/polypropylene composites using various fire retardants. *Polym. Degrad. Stab.*, 100, 79, 2014.
16. Ayrilmis, N., Akbulut, T., Dundar, T., White, R.H., Mengeloglu, F., Buyuksari, U., Candan, Z., Avci, E., Effect of boron and phosphate compounds on physical, mechanical, and fire properties of wood–polypropylene composites. *Constr. Build Mater.*, 33, 63, 2012.

17. Subasinghe, A. and Bhattacharyya, D., Performance of different intumescent ammonium polyphosphate flame retardants in PP/kenaf fibre composites. *Compos. Part A Appl. Sci. Manuf.*, 65, 91, 2014.
18. Kozlowski, R.M. and Mackiewicz-Talarczyk, M., *Handbook of Natural Fibres: Volume 1: Types, Properties and Factors Affecting Breeding and Cultivation*, Woodhead Publishing, Sawston, Cambridge, United Kingdom, 2020.
19. Mohanty, A.K., Misra, M., Drzal, L.T., *Natural Fibers, Biopolymers, and Biocomposites*, CRC Press, Boca Raton, United States of America, 2005.
20. Paul, S., Sen, B., Das, S., Abbas, S.J., Pradhan, S.N., Sen, K., Ali, S.I., Incarnation of bioplastics: Recuperation of plastic pollution. *Int. J. Environ. Anal. Chem.*, 101, 1, 2021.
21. Europe, P., *Plastics–The Facts 2020 An Analysis of European Plastics Production, Demand and Waste Data*, 2020.
22. Rochman, C.M., Browne, M.A., Underwood, A.J., Van Franeker, J.A., Thompson, R.C., Amaral-Zettler, L.A., The ecological impacts of marine debris: Unraveling the demonstrated evidence from what is perceived. *Ecology*, 97, 302, 2016.
23. Wilcox, C., Mallos, N.J., Leonard, G.H., Rodriguez., A., Hardesty, B.D., Using expert elicitation to estimate the consequences of marine litter on seabirds, turtles and marine mammals. *Mar. Policy*, 65, 107, 2016.
24. Tammemagi, H.Y., *The Waste Crisis: Landfills, Incinerators, and the Search for A Sustainable Future*, Oxford University Press, New York, United States of America, 1999.
25. Gironi, F. and Piemonte, V., Bioplastics and petroleum-based plastics: Strengths and weaknesses. *Energ. Source Part A*, 33, 1949–1959, 2011.
26. Marathe, S., Sayyed, R., Somani, V., Impact of bt cotton cultivation on bacterial soil micro flora in North Maharshtra. *Int. J. Biotechnol. Biosci.*, 2, 266, 2012.
27. Pilla, S., *Handbook of Bioplastics and Biocomposites Engineering Applications Vol. 81*, John Wiley & Sons, United States of America, 2011.
28. Averous, L., Fringant, C., Moro, L., Starch-based biodegradable materials suitable for thermoforming packaging. *Starch/Staerke*, 53, 368, 2001.
29. Chinthapalli, R., Skoczinski, P., Carus, M., Baltus, W., de Guzman, D., Käb, H., Raschka, A., Ravenstijn, J., Biobased building blocks and polymers—Global capacities, production and trends, 2018–2023. *Ind. Biotechnol.*, 15, 237, 2019.
30. Brigante, D., Examples of applications, in: *New Composite Materials*, pp. 157–172, Springer, Switzerland, 2014.
31. Reinhart, T.J., Overview of composite materials, in: *Handbook of Composites*, pp. 21–33, Springer, Boston, Massachusetts, United States of America, 1998.
32. Wool, R. and Sun, X.S., *Bio-Based Polymers and Composites*, Elsevier, United States of America, 2011.
33. Kaith, B.S., Mittal, H., Jindal, R., Maiti, M., Kalia, S., Environment benevolent biodegradable polymers: Synthesis, biodegradability, and applications, in: *Cellulose Fibers: Bio-and Nano-Polymer Composites*, pp. 425–451, Springer, Berlin, Heidelberg, 2011.
34. Groot, W., Van Krieken, J., Sliekersl, O., De Vos, S., Production and purification of lactic acid and lactide, in: *Poly (Lactic Acid): Synthesis, Structures, Properties, Processing, and Applications*, pp. 1–18, 2010.
35. Nakajima, H., Dijkstra, P., Loos, K., The recent developments in biobased polymers toward general and engineering applications: Polymers that are upgraded from biodegradable polymers, analogous to petroleum-derived polymers, and newly developed. *Polymers*, 9, 523, 2017.
36. Musioł, M., Sikorska, W., Janeczek, H., Wałach, W., Hercog, A., Johnston, B., Rydz, J., (Bio) degradable polymeric materials for a sustainable future–part 1. Organic recycling of PLA/

PBAT blends in the form of prototype packages with long shelf-life. *Waste Manage.*, 77, 447, 2018.
37. Suppakarn, N. and Jarukumjorn, K., Mechanical properties and flammability of sisal/PP composites: Effect of flame retardant type and content. *Compos. B Eng.*, 40, 613, 2009.
38. Cárdenas, M.A., García-López, D., Gobernado-Mitre, I., Merino, J.C., Pastor, J.M., Martínez, J.D.D., Barbeta, J., Calveras, D., Mechanical and fire retardant properties of EVA/clay/ATH nanocomposites–Effect of particle size and surface treatment of ATH filler. *Polym. Degrad. Stab.*, 93, 2032, 2008.
39. Sikoutris, D.E. and Kostopoulos, V., Fire response of polymers and polymer composites. Part a: Multistage degradation kinetics. *J. Compos. Mater.*, 49, 251–257, 2015.
40. Bourbigot, S. and Duquesne, S., Fire retardant polymers: Recent developments and opportunities. *J. Mater. Chem.*, 17, 2283, 2007.
41. Huang, G., Gao, J., Wang, X., Liang, H., Ge, C., How can graphene reduce the flammability of polymer nanocomposites? *Mater. Lett.*, 66, 187, 2012.
42. Shah, A.U.R., Prabhakar, M.N., Wang, H., Song, J.I., The influence of particle size and surface treatment of filler on the properties of oyster shell powder filled polypropylene composites. *Polym. Compos.*, 39, 2420, 2018.
43. Prabhakar, M.N., Shah, A.U.R., Rao, K.C., Song, J.I., Mechanical and thermal properties of epoxy composites reinforced with waste peanut shell powder as a bio-filler. *Fiber. Polym.*, 16, 1119, 2015.
44. Yun, I.S., Hwang, S.W., Shim, J.K., Seo, K.H., A study on the thermal and mechanical properties of poly (butylene succinate)/thermoplastic starch binary blends. *Int. J. Precis. Eng. Manuf. Green Technol.*, 3, 289, 2016.
45. Stapulionienė, R., Vaitkus, S., Vėjelis, S., Sankauskaitė, A., Investigation of thermal conductivity of natural fibres processed by different mechanical methods. *Int. J. Precis. Eng. Manuf.*, 17, 1371, 2016.
46. Hirschler, M., Heat release from plastic materials, in: *Heat Release in Fires*, p. 375, 1992.
47. Ke, C.H., Li, J., Fang, K.Y., Zhu, Q.L., Zhu, J., Yan, Q., Wang, Y.Z., Synergistic effect between a novel hyperbranched charring agent and ammonium polyphosphate on the flame retardant and anti-dripping properties of polylactide. *Polym. Degrad. Stab.*, 95, 763, 2010.
48. Maqsood, M. and Seide, G., Development of biobased socks from sustainable polymer and statistical modeling of their thermo-physiological properties. *J. Clean. Prod.*, 197, 170, 2018.
49. Zaikov, G. and Lomakin, S., Ecological issue of polymer flame retardancy. *J. Appl. Polym. Sci.*, 86, 2449, 2002.
50. Abou-Donia, M.B., Salama, M., Elgamal, M., Elkholi, M., Organophosphorus flame retardants (OPFR): Neurotoxicity. *J. Environ. Health Sci.*, 2, 1, 2016.
51. Maqsood, M. and Seide, G., Biodegradable flame retardants for biodegradable polymer. *Biomolecules*, 10, 1038, 2020.
52. Hobbs, C.E., Recent advances in bio-based flame retardant additives for synthetic polymeric materials. *Polymers*, 11, 224, 2019.
53. Tawfik, S., Flame retardants: Additives in plastic technology, in: *Polymers and Polymeric Composites: A Reference Series*, Palsule, S. (Ed.), p. 1, 2017.
54. Kim, Y., Lee, S., Yoon, H., Fire-safe polymer composites: Flame-retardant effect of nanofillers. *Polymers*, 13, 540, 2021.
55. Zhou, X., Qiu, S., Cai, W., Liu, L., Hou, Y., Wang, W., Construction of hierarchical MoS2@ TiO2 structure for the high performance bimaleimide system with excellent fire safety and mechanical properties. *Chem. Eng. J.*, 369, 451, 2019.

56. Wang, X., Hu, Y., Song, L., Xuan, S., Xing, W., Bai, Z., Lu, H., Flame retardancy and thermal degradation of intumescent flame retardant poly (lactic acid)/starch biocomposites. *Ind. Eng. Chem. Res.*, 50, 713, 2011.
57. Shi, X., Peng, X., Zhu, J., Lin, G., Kuang, T., Synthesis of DOPO-HQ-functionalized graphene oxide as a novel and efficient flame retardant and its application on polylactic acid: Thermal property, flame retardancy, and mechanical performance. *J. Colloid Interface Sci.*, 524, 267, 2018.
58. Lei, Z., Cao, Y., Xie, F., Ren, H., Study on surface modification and flame retardants properties of ammonium polyphosphate for polypropylene. *J. Appl. Polym. Sci.*, 124, 781, 2012.
59. Jeencham, R., Suppakarn, N., Jarukumjorn, K., Effect of flame retardants on flame retardant, mechanical, and thermal properties of sisal fiber/polypropylene composites. *Compos. B Eng.*, 56, 249, 2014.
60. Chen, J. and Han, J., Comparative performance of carbon nanotubes and nanoclays as flame retardants for epoxy composites. *Results Phys.*, 14, 102481, 2019.
61. Kaynak, C., Gunduz, H.O., Isitman, N.A., Use of nanoclay as an environmentally friendly flame retardant synergist in polyamide-6. *J. Nanosci. Nanotechnol.*, 10, 7374, 2010.
62. Gao, Y.Y., Deng, C., Du, Y.Y., Huang, S.C., Wang, Y.Z., A novel bio-based flame retardant for polypropylene from phytic acid. *Polym. Degrad. Stab.*, 161, 298, 2019.
63. Yu, Y., Chen, Z., Zhang, Q., Jiang, M., Zhong, Z., Chen, T., Jiang, J., Modified montmorillonite combined with intumescent flame retardants on the flame retardancy and thermal stability properties of unsaturated polyester resins. *Polym. Adv. Technol.*, 30, 998, 2019.
64. Perez-Salinas, C., Castro, C., Patin, G., Peña-Jordán, F.A., Paredes, D., Application of flame retardants aluminum hydroxide and magnesium hydroxide in a material composed of glass fiber with polymeric matrix and its influence on the flammability index, in: *Key Engineering Materials*, vol. 818, pp. 128–133, Trans Tech Publications Ltd, Stafa-Zurich, Switzerland, 2019.
65. Laoutid, F., Karaseva, V., Costes, L., Brohez, S., Mincheva, R., Dubois, P., Novel bio-based flame retardant systems derived from tannic acid. *J. Renew. Mater.*, 6, 559, 2018.
66. Hapuarachchi, T.D. and Peijs, T., Multiwalled carbon nanotubes and sepiolite nanoclays as flame retardants for polylactide and its natural fibre reinforced composites. *Compos. Part A Appl. Sci. Manuf.*, 41, 954, 2010.
67. Xiong, Z., Zhang, Y., Du, X., Song, P., Fang, Z., Green and scalable fabrication of core–shell biobased flame retardants for reducing flammability of polylactic acid. *ACS Sustain. Chem. Eng.*, 7, 8954, 2019.
68. Wang, Y., Zhang, S., Wu, X., Lu, C., Cai, Y., Ma, L., Yang, L., Effect of montmorillonite on the flame-resistant and mechanical properties of intumescent flame-retardant poly (butylene succinate) composites. *J. Therm. Anal. Calorim.*, 128, 1417, 2017.
69. Hu, S., Song, L., Pan, H., Hu, Y., Effect of a novel chitosan-based flame retardant on thermal and flammability properties of polyvinyl alcohol. *J. Therm. Anal. Calorim.*, 112, 859, 2013.
70. Zhang, R., Xiao, X., Tai, Q., Huang, H., Yang, J., Hu, Y., Preparation of lignin–silica hybrids and its application in intumescent flame-retardant poly (lactic acid) system. *High Perform. Polym.*, 24, 738, 2012.
71. Jiang, P., Zhang, S., Bourbigot, S., Chen, Z., Duquesne, S., Casetta, M., Surface grafting of sepiolite with a phosphaphenanthrene derivative and its flame-retardant mechanism on PLA nanocomposites. *Polym. Degrad. Stab.*, 165, 68, 2019.
72. Chen, S., Lin, S., Hu, Y., Ma, M., Shi, Y., Liu, J., Zhu, F., Wang, X., A lignin-based flame retardant for improving fire behavior and biodegradation performance of polybutylene succinate. *Polym. Adv. Technol.*, 29, 3142, 2018.

73. Wu, K., Hu, Y., Song, L., Lu, H., Wang, Z., Flame retardancy and thermal degradation of intumescent flame retardant starch-based biodegradable composites. *Ind. Eng. Chem. Res.*, 48, 3150, 2009.
74. Tawiah, B., Yu, B., Ullah, S., Wei, R., Yuen, R.K.K., Xin, J.H., Fei, B., Flame retardant poly (lactic acid) biocomposites reinforced by recycled wool fibers-Thermal and mechanical properties. *Express Polym. Lett.*, 13, 697–712, 2019.
75. Stec, A.A. and Hull, T.R., Assessment of the fire toxicity of building insulation materials. *Energy Build.*, 43, 498, 2011.
76. Liu, W., Chen, D.Q., Wang, Y.Z., Wang, D.Y., Qu, M.H., Char-forming mechanism of a novel polymeric flame retardant with char agent. *Polym. Degrad. Stab.*, 92, 1046, 2007.
77. Tawiah, B., Yu, B., Wei, R., Yuen, R.K., Chen, W., Xin, J.H., Fei, B., Simultaneous fire safety enhancement and mechanical reinforcement of poly (lactic acid) biocomposites with hexaphenyl (nitrilotris (ethane-2, 1-diyl)) tris (phosphoramidate). *J. Hazard. Mater.*, 380, 120856, 2019.
78. Xue, Y., Shen, M., Zheng, Y., Tao, W., Han, Y., Li, W., Song, P., Wang, H., One-pot scalable fabrication of an oligomeric phosphoramide towards high-performance flame retardant polylactic acid with a submicron-grained structure. *Compos. B. Eng.*, 183, 107695, 2020.
79. Wu, N., Fu, G., Yang, Y., Xia, M., Yun, H., Wang, Q., Fire safety enhancement of a highly efficient flame retardant poly (phenylphosphoryl phenylenediamine) in biodegradable poly (lactic acid). *J. Hazard. Mater.*, 363, 1, 2019.
80. Saba, N., Jawaid, M., Alrashed, M.M., Alothman, O.Y., Oil palm waste based hybrid nanocomposites: Fire performance and structural analysis. *J. Build. Eng.*, 25, 100829, 2019.
81. Szustakiewicz, K., Kiersnowski, A., Gazinska, M., Bujnowicz, K., Piglowski, J., Flammability structure and mechanical properties of PP/OMMT nanocomposites. *Polym. Degrad. Stab.*, 96, 291, 2011.
82. Lenża, J., Merkel, K., Rydarowski, H., Comparison of the effect of montmorillonite, magnesium hydroxide and a mixture of both on the flammability properties and mechanism of char formation of HDPE composites. *Polym. Degrad. Stab.*, 97, 2581, 2012.
83. Wang, J., Su, X., Mao, Z., The flame retardancy and thermal property of poly (ethylene terephthalate)/cyclotriphosphazene modified by montmorillonite system. *Polym. Degrad. Stab.*, 109, 154, 2014.
84. Im, J.S., Lee, S.K., In, S.J., Lee, Y., Improved flame retardant properties of epoxy resin by fluorinated MMT/MWCNT additives. *J. Anal. Appl. Pyrolysis*, 89, 225, 2010.
85. Isitman, N.A. and Kaynak, C., Nanoclay and carbon nanotubes as potential synergists of an organophosphorus flame-retardant in poly (methyl methacrylate). *Polym. Degrad. Stab.*, 95, 1523, 2010.
86. Enescu, D., Alongi, J., Frache, A., Evaluation of nonconventional additives as fire retardants on polyamide 6, 6: Phosphorous-based master batch, α-zirconium dihydrogen phosphate, and β-cyclodextrin based nanosponges. *J. Appl. Polym. Sci.*, 123, 3545, 2012.
87. Carosio, F., Laufer, G., Alongi, J., Camino, G., Grunlan, J.C., Layer-by-layer assembly of silica-based flame retardant thin film on PET fabric. *Polym. Degrad. Stab.*, 96, 745, 2011.
88. Utracki, L.A. and Wilkie, C.A., *Polymer Blends Handbook*, Kluwer Academic Publishers, Dordrecht, 2002.
89. Feldman, D., Polyblend compatibilization. *J. Macromol. Sci. A Pure Appl. Chem.*, 42, 587, 2005.
90. Groeninckx, G., Vanneste, M., Everaert, V., Crystallization, morphological structure, and melting of polymer blends, in: *Polymer Blends Handbook*, p. 203, 2003.
91. Tran, L.Q.N., Fuentes, C.A., Dupont-Gillian, C., Van Vuure, A.W., Verpoest, I., Understanding the interfacial compatibility and adhesion of natural coir fibre thermoplastic composites. *Compos. Sci. Technol.*, 80, 23, 2013.

92. Wang, Y., Qi, R., Xiong, C., Huang, M., Effects of coupling agent and interfacial modifiers on mechanical properties of poly (lactic acid) and wood flour biocomposites. *Iranian Polym. J.*, 20, 281, 2011.
93. Matko, S., Toldy, A., Keszei, S., Anna, P., Bertalan, G., Marosi, G., Flame retardancy of biodegradable polymers and biocomposites. *Polym. Degrad. Stab.*, 88, 138, 2005.
94. Ding, L., Jia, Z., Sun, H., Pan, Y., Zhao, J., Estimation of mechanical performance, thermal stability and flame retardancy of high-impact polystyrene/surface-modified APP/carboxylic-functionalized MWCNTs nanocomposites. *Polymers*, 11, 615, 2019.
95. Shumao, L., Jie, R., Hua, Y., Tao, Y., Weizhong, Y., Influence of ammonium polyphosphate on the flame retardancy and mechanical properties of ramie fiber-reinforced poly (lactic acid) biocomposites. *Polym. Int.*, 59, 242, 2010.
96. Idumah, C.I. and Hassan, A., Emerging trends in flame retardancy of biofibers, biopolymers, biocomposites, and bionanocomposites. *Rev. Chem. Eng.*, 32, 115, 2016.
97. Shukor, F. *et al.*, Effect of ammonium polyphosphate on flame retardancy, thermal stability and mechanical properties of alkali treated kenaf fiber filled PLA biocomposites. *Mater. Des. (1980-2015)*, 54, 425, 2014.
98. Guan, Y.H., Huang, J.Q., Yang, J.C., Shao, Z.B., Wang, Y.Z., An effective way to flame-retard biocomposite with ethanolamine modified ammonium polyphosphate and its flame retardant mechanisms. *Ind. Eng. Chem. Res.*, 54, 3524, 2015.
99. Sun, Y., Sun, S., Chen, L., Liu, L., Song, P., Li, W., Yu, Y., Fengzhu, L., Qian, J., Wang, H., Flame retardant and mechanically tough poly (lactic acid) biocomposites via combining ammonia polyphosphate and polyethylene glycol. *Compos. Commun.*, 6, 1, 2017.
100. Rane, A.V., Kanny, K., Abitha, V.K., Thomas, S., Methods for synthesis of nanoparticles and fabrication of nanocomposites, in: *Synthesis of Inorganic Nanomaterials*, pp. 121–139, Elsevier, Sawston, Cambridge, 2018.
101. Zhang, Y., Jing, J., Liu, T., Xi, L., Sai, T., Ran, S., Fang, Z., Huo, S., Song, P., A molecularly engineered bioderived polyphosphate for enhanced flame retardant, UV-blocking and mechanical properties of poly (lactic acid). *Chem. Eng. J.*, 411, 128493, 2021.
102. Jing, J., Zhang, Y., Fang, Z.P., Wang, D.Y., Core-shell flame retardant/graphene oxide hybrid: A self-assembly strategy towards reducing fire hazard and improving toughness of polylactic acid. *Compos. Sci. Technol.*, 165, 161, 2018.
103. Bajpai, P.K., Ahmad, F., Chaudhary, V., Processing and characterization of bio-composites, in: *Handbook of Ecomaterials*, pp. 1–18, 2017.
104. Young, W.-B. and Tsao, Y.-C., The mechanical and fire safety properties of bamboo fiber reinforced polylactide biocomposites fabricated by injection molding. *J. Compos. Mater.*, 49, 2803, 2015.
105. Tawiah, B., Yu, B., Cheung, W.Y., Chan, S.Y., Yang, W., Fei, B., Synthesis and application of synergistic azo-boron-BPA/polydopamine as efficient flame retardant for poly (lactic acid). *Polym. Degrad. Stab.*, 152, 64, 2018.
106. Qian, Y., Wei, P., Jiang, P., Zhao, X., Yu, H., Synthesis of a novel hybrid synergistic flame retardant and its application in PP/IFR. *Polym. Degrad. Stab.*, 96, 1134, 2011.
107. Dorez, G., Taguet, A., Ferry, L., Cuesta, J.M.L., Phosphorous compounds as flame retardants for polybutylene succinate/flax biocomposite: Additive versus reactive route. *Polym. Degrad. Stab.*, 102, 152, 2014.
108. Prabhakar, M.N., Kumar, C.N., Woo, L.D., Jung-IL, S., Hybrid approach to improve the flame-retardant and thermal properties of sustainable biocomposites used in outdoor engineering applications. *Compos. Part A Appl. Sci. Manuf.*, 152, 106674, 2022.

109. Krishnadevi, K. and Selvaraj, V., Development of halogen-free flame retardant phosphazene and rice husk ash incorporated benzoxazine blended epoxy composites for microelectronic applications. *New J. Chem.*, 39, 6555, 2015.
110. Liang, B., Hong, X., Zhu, M., Gao, C., Wang, C., Tsubaki, N., Synthesis of novel intumescent flame retardant containing phosphorus, nitrogen and boron and its application in polyethylene. *Polym. Bull.*, 72, 2967, 2015.
111. Ritchie, H. and Roser, M., Plastic pollution. *Our World Data*, 2018. Plastic Pollution. Published online at OurWorldInData.org. Retrieved from: https://ourworldindata.org/plastic-pollution
112. Thompson, R.C., Moore, C.J., Vom Saal, F.S., Swan, S.H., Plastics, the environment and human health: Current consensus and future trends. *Philos. Trans. R. Soc. Lond. B Biol. Sci.*, 364, 2153, 2009.
113. Chen, Y., Geever, L.M., Killion, J.A., Lyons, J.G., Higginbotham, C.L., Devine, D.M., Review of multifarious applications of poly (lactic acid). *Polym. Plast. Technol. Eng.*, 55, 1057, 2016.
114. Liu, L., Xu, Y., Pan, Y., Xu, M., Di, Y., Li, B., Facile synthesis of an efficient phosphonamide flame retardant for simultaneous enhancement of fire safety and crystallization rate of poly (lactic acid). *Chem. Eng. J.*, 421, 127761, 2021.
115. Gallo, E., Schartel, B., Acierno, D., Russo, P.J.E.P.J., Flame retardant biocomposites: Synergism between phosphinate and nanometric metal oxides. *Eur. Polym. J.*, 47, 1390, 2011.
116. Nie, S., Liu, X., Dai, G., Yuan, S., Cai, F., Li, B., Hu, Y., Investigation on flame retardancy and thermal degradation of flame retardant poly (butylene succinate)/bamboo fiber biocomposites. *J. Appl. Polym. Sci.*, 125, 485, 2012.
117. Nie, S., Liu, X., Wu, K., Dai, G., Hu, Y., Intumescent flame retardation of polypropylene/bamboo fiber semi-biocomposites. *J. Therm. Anal. Calorim.*, 111, 425, 2013.
118. Hu, W., Zhang, Y., Qi, Y., Wang, H., Liu, B., Zhao, Q., Zhang, J., Duan, J., Zhang, L., Sun, Z., Liu, B., Improved mechanical properties and flame retardancy of wood/PLA all-degradable biocomposites with novel lignin-based flame retardant and TGIC. *Macromol. Mater. Eng.*, 305, 1900840, 2020.

26

Biobased Thermosets for Engineering Applications

Bhargavi Koneru, Jhilmil Swapnalin, Hanumanthrayappa Manjunatha and Prasun Banerjee*

Multiferroic and Magnetic Materials Research Laboratory (MMMRL), Gandhi Institute of Technology and Management (GITAM) University, Bengaluru, India

Abstract

Increasing demands on the development of polymer made from renewable bioenergy or organic compounds can help to reduce the increased worries about the global energy crisis and carbon emissions. The thermosetting polymers are more challenging to remould and recycle than thermoplastics because they are generated by chemicals' permanent curing (hardening) reaction. The thermosetting polymers are now prepared based on natural materials termed biobased thermoset polymers, which offer lightweight, low cost, and better durability. They can produce sustainable bridge polyester blends by polycondensing a naturally existing functionalized carboxyl group (citric acid) with 1,8-octane diol and triglyceride. Hence, various engineering applications can be achieved with the production of biothermoset from renewable sources with added biocompatibility.

***Keywords*:** Thermosetting polymers, biobased thermosets, biobased monomers, renewable resources, biomass, cross-linked polymers, recycling, controlled degradation

26.1 Introduction

A thermoplastic polymer is an impermeable, chemically bound crosslink that forms a three-dimensional network chain polymer. Thermosetting polymers are more difficult to remould and recycle than thermoplastics because they are generated by a permanent curing (hardening) reaction of chemicals in an aqueous solution, mushy solids, and viscoelastic condition [1, 2]. Thermoset polymers show high stability with outstanding thermal and mechanical properties, favorable for making adhesives, coatings, and composites. These polymers are also extensively used in engineering due to their versatile nature, in particular to pliable strength, long-term durability, heat, and chemical resistance as a product of cross-linked bonds, thus allowing controlled final outcomes [3]. Thermosets constitute 14% of the total production of polymers, of which nearly 82% are thermoplastics and elastomers [4].

In the view of environmental responsibility and sustainability, the growing demand for green industries is evident. Moreover, the demand for developing new biodegradable polymer matrices increased due to its ability of sustained and control the release of drugs with no hindrance of surgical removal. New materials are coming up using natural feedstock.

**Corresponding author*: pbanerje@gitam.edu

Inamuddin and Tariq Altalhi (eds.) Handbook of Bioplastics and Biocomposites Engineering Applications: 2nd Edition, (575–588) © 2023 Scrivener Publishing LLC

Thermosetting polymers are now prepared based on natural materials termed biobased thermoset polymers, which offer lightweight, low cost, and better durability. Widely used thermosets are Epoxy resins.

Here, we explained different biodegradable, nontoxic, and inexpensive polymers achieved through polycondensation reaction. Various applications with renewable sources to produce the biothermoset and biocompatibility of the newly synthesized elastomer are described.

26.2 Sustainable Covalently Bonded Polyamides are Produced by Polycondensing a Naturally Present Functionalized Carboxyl Group (Citric Acid) with 1, 8-Octane Diol

A 3-D scaffold network is a base for cells to grow and differentiate, hence a prior need in the scope of modified tissues. A scaffold is required to be biodegradable, biologically acceptable, with adequate potential for cell loading [5, 6]. As many tissues in our body are elastic in nature, the elastomeric biodegradable scaffold is the point of keen research [7, 8].

In the realm of bioengineering, poly(1,8-octanediol-co-citric-acid) (POC) has been found to be an efficient disposable elastomer. It is made by controlled condensation polymerization of a naturally present n-heterocyclic carboxylic acid with a constitutional monomer. As a remedy to cardiovascular illness affecting blood vessels, this material is commonly employed in comparatively tiny vasculature biochemical engineering. Literature provides the advantages of the POC as a nontoxic monomer that avoids adverse body reactions. It can be synthesized under normal conditions and without using any toxic catalysts and reagents. It is also attributed to having good physical and mechanical stability with cell surface affinity making it useful in drug delivery mechanisms.

The synthesis of the resulting polymer, as per Jian Yang *et al.*, was designed to polycondense difunctional monomer (1,8-octane diol) with polyfunctional monomer citric acid to produce a well-linked polyamide network with a controlled amount of crosslinks. The resultant new polymer was also under controlled elastic nature and biodegradability. The reagents include citric acid, a reactive monomer that offers the advantages of being cheap, readily available, and nontoxic in nature. It is helpful in forming hydrogen bonds while the other reagent 1,8-octane diol facilitates the formation of ester bonds with citric acid, offering hydrophilic behavior. The biodegradable elastomer (POC) shows an inflammatory response that is similar to standard PLGA [9].

The synthesis process includes condensation in the ratio of 1:1 of reagents in an ambient environment, then purification and freeze-drying. Further, the C=O stretch in FTIR confirms the formation of ester bonds. The POC scaffolds well known in the scope of vascular tissue engineering. Furthermore, using salt leaching techniques and simple moulds, tubular and sponge-like POC can also be formed [10].

The tensile strength of the POC polymer was evaluated through many tensile mapping tests. Other various aliphatic diols were also infused in place of 1,8-octanediol and observed. The biodegradable elastomeric polymer prepared at different experimental conditions is shown through the stress-strain curve. POC degradation rate is observed by changing the time, temperature, and molar ratio of reagents.

Researchers have reported that POC scaffolds are capable of different tissue regeneration like bone [11], blood vessels [12, 13], nerve tissue [14], and cartilage multiplication of chondrocytes and production of extracellular environment *in vitro* inside the porous POC scaffold may also be achieved by manufacturing the polymer by a controlled polyesterification process using two nontoxic chemicals [15]. Fusing with hydroxylapatite (HA) or PLLA, elasticity of POC can also be controlled, and the thermal processability on top makes it a potential candidate for tissue regeneration [16].

It is now well proved that the POC is a reliable easy to prepare biodegradable elastomer compared to other counterparts. Due to its advantageous characteristic, as discussed above, it is a suitable substitute for elastic body tissue. Hence, an active participant in the scope of tissue engineering. Where precisely, the demand for biodegradable elastomer implants is high.

26.3 Biodegradable Crosslinked Polyesters by Polycondensation of a Naturally Occurring Citric Acid and Glycerol

Experiments were carried out as soon as the necessity for novel biodegradable polyesters for drug administration and tissue regeneration became apparent. In one of such experiments, researchers could prepare inexpensive, simple, and biocompatible polymer. Glycerol and citric acid as the reagents were polycondensed to form a crosslinked polyester with the required characteristics. The synthesis process of the polymer was found to be easy, non-complex, and notably could be performed at normal temperature and pressure conditions. On top of it, there is no need for any catalyst to carry out the reaction.

Literature shows that synthesizing such thermoset polymer to be beneficial for biomedical and pharmaceutical applications, and polyesterification is a good media for crosslinking. As ester bonds are known to be providing the required hydrolytic cleavage, thus two ester-bond forming reagents are needed. According to reports, the US Department of Energy considers citrate and glycerine to be among the top renewable compounds [17], and the US Food and Drug Administration considers them to be physiologically safe [18–20]. As citric acid has one hydroxyl and three carboxyls functional groups while glycerol has three hydroxyl functions, it is proved to be not harmful to the body.

Crosslinking of both citric acid and glycerol can be moderated as the amount of crosslinking is found to be inversely proportional to the degradation rate. Both of the reagents are also found to be the building block of the proper drug delivery mechanism allowing better integration of chemical functionalities [21–24]. It was observed in a synthesis process that when a catalyst is not present if pressure is decreased or temperature is increased, the equilibrium gets shifted toward the products by water removal. The bond formation between citric acid and glycerol at 90°C can be better understood schematically [25].

Synthesis of the biodegradable copolyester of poly(citric acid)-block-poly(glycerol) hybrid elastomer is adequately depicted in an earlier paper by Pramanick and T.T. Ray as, after 4 hours of synthesis, the polymer-like floating solid started separating. Crosslinking was optimum with better thermal stability as the polymer was insoluble in water and organic solvent. But was observed over the course of time, the polymer showed hydrophilic nature, possibly due to partial hydrolysis of the crosslinks The stretching frequency owing to C=O shifted from 1700/cm to 1730/cm for band at 1190/cm, indicating the creation of ester bonds, according to IR spectral analysis.

In contrast to citric acid separately, According to FTIR studies, bridging between triglyceride and lemon juice results in a two-stage breakdown profile. The ester linkages created during the condensation step raised the degradation temperature, according to the TGA analysis.

Thus, the study reveals that the interaction between citric acid and glycerol is biocompatible, resulting in a hybrid elastomer that can mimic body tissue well. The reaction can occur without any catalyst, and varying reaction conditions can moderate the crosslinking. The final product is the biodegradable, inexpensive polymer that is useful for biomedical applications with better integration of functionality.

26.4 Sugar-Based Lactones to Produce Degradable Dimethacrylates

Marking the wide range usage of polymers like plastics, the modern era can be best described as the plastic era. With the increasing population, the demand for plastic is also surging. The primary concern in using plastic is its nonrenewable source and low degradability rate. On this point, a biobased polymer is suitable to serve the demand by substituting petroleum, known to be the source of plastic manufacture. The three main classes of biobased sources are carbohydrates, lignin, and triglycerides, where carbohydrates include natural polymers like sugar, cellulose, and starch. Sucrose or simply sugar is a biomass raw material with a high production rate [26] with better functional stability and a rigid ring structure. Furthermore, unlike petroleum sources, sugar offers better functional incorporation due to its high heteroatom content and stereochemistry.

Sugar derivatives are known to be excellent candidates serving as the building blocks of natural polymers. In the advent of the era of green chemistry, new monomers are synthesized from the bicyclic sugar derivatives via polymerization. Literature shows as sugar-based raw materials like glucose and mannose can give rise to di-methacrylates monomers, namely glucuronolactone methacrylate (GDMA) and mannarodilactone methacrylate (MDMA) (Figure 26.1). These new polymers, condensed by thermally initiated free radicals, proved highly crosslinked and thermally stable. It shows high tensile strength is a rigid thermoset material [27]. The bulk polymerization of the resultant dimethacrylate monomers results in proper crosslinking helpful in preparing a stable and rigid thermoset polymer network; di-methacrylate is a widely used polymer in paints, sealants, chromatographic packing, and biomaterials, such as dental materials [28, 29].

Lactone derivatives of the di-anhydrohexitols isosorbide and isomannide are sugar-based di-lactones produced from carbohydrates and oligosaccharides. The dilactones formed with the conjunction of various diamines results in a polymer called polyhydroxy polyamides. Due to the existence of diol and dilactone reactivity, GDMA and MDMA can be polymerized in two ways, Or else the dilactones can also be polymerized using secondary alcohols. Dilactones possess an interesting heteroatom ring system offering susceptibility for further functionalization or degradation, as shown in Figure 26.2.

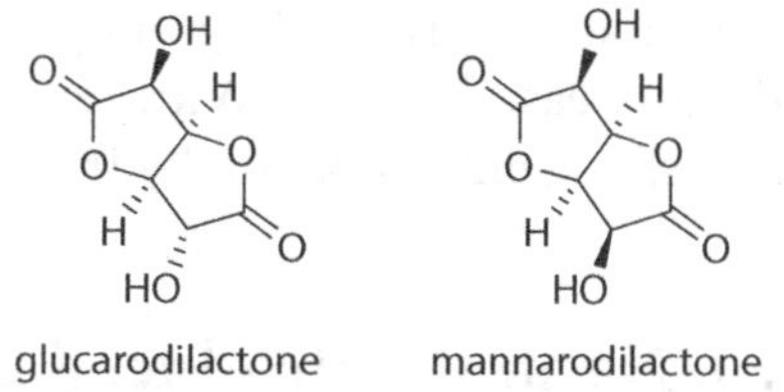

Figure 26.1 Structures of GDMA and MDMA [27].

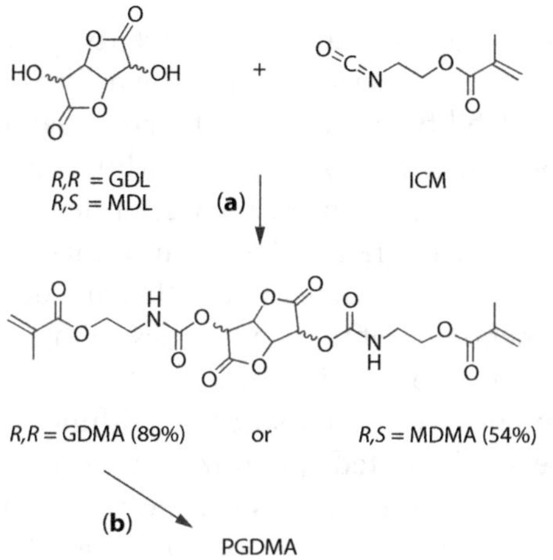

Figure 26.2 Structure of (a) dibutyltin dilaurate and (b) dicumyl peroxide PGDMA [30].

It is seen that copolymer can achieve microsphere morphology by using low concentrations of monomers and increasing the reaction times deposition polymers in Tetrahydrofuran produced P(GDMA-co-MMA) microspheres (Figure 26.3).

Dimethacrylates produced showed high crosslinking, which led to other characteristic features of the product like gel point at the low conversion and autoacceleration. Methacrylates groups were converted partially but showed exceptionally long macromolecular radical lifetimes [31–33]. The final polymer depends on many parameters like the chemical structure of the monomer reagents and different reaction conditions. It has got

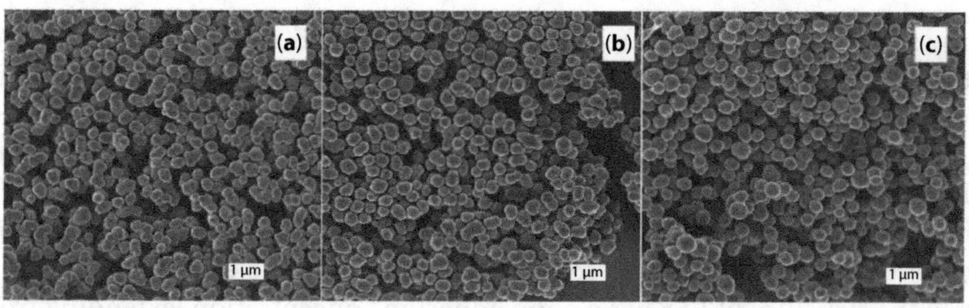

Figure 26.3 SEM photos of microspheres made of P(GDMA-MMA) co-polymer at various monomer concentrations [30].

a wide range of applicability like nanoimprint lithography, reworkable electronic components other than controlled drug release [34, 35]. With renewable sugar, the source is the future scope of interest in synthesizing environmentally friendly polymers.

26.5 Water Facilitated, Naturally Produced Difunctional or Trifunctional Carboxyl Groups and Epoxidized Sucrose Soyate Are Made (With Sugars and Soybean Oil Lipids)

Researchers observed that adding aromatic structures to epoxy resins with secondary or tertiary esters linkage improve their performance. The α-terp-epoxide monomer was used to investigate heat biodegradability. Because the thermal liability of the amide bond will change by breaking the copolymer and producing carboxyl group and anhydrides, which completely breaks the reaction process, the tertiary ester degradation, thermal degradation mechanism, and -terp-epoxide network are all similar to each other. This breakdown allows us to understand the zero crosslinking density. The thermal degradation of α-terp-epoxide method is the most recommended as it takes a very less temperature of 193°C to break the epoxide reaction at rework conditions. With the help of water, entirely biobased degradable thermosets of high-performance were also made from difunctional or trifunctional carboxylic acid groups with soybean oil assisted epoxidized sucrose soyate [36].

The main drawback of these thermosets is during the hardening process, the organic materials (VOCs) are released. The life span of a thermoset depends upon the concentration of VOCs present in it, making VOC free epoxies that can give better thermoset properties. When an epoxidized sucrose soyate (ESS) is further epoxidized with soybean oil, increased thermoset properties are observed and also resemble superior transition temperature in glass, fine curing properties and solvent resistance properties compared to other thermosets. The VOCs free thermoset epoxidized sucrose soyate, when reacted with naturally occurring acids, show better reactivity and curing process. Songri *et al.*, synthesized ESS with soybean oil epoxidation as shown in Figure 26.4 and the functionalities were studied with different naturally occurring acids, mainly fruit juices [37]. The influence of water on the curing reaction was investigated using naturally occurring Critical acid, mannose, and stearic acid.

26.5.1 Learning More About the Significance of Water in the Curing Process

Water shows two different effects respectively when treated with natural acids and in curing reactions. All the natural available fruit juices, such as malic, tartaric, citric, oxalic, and malonic acids, will dissolve completely into the ESS with the addition of water. Similarly, H_2O for the production of H^+ ions, which catalyzes the reaction between the epoxides and carboxylic acid groups to open the rings for reaction. The H^+ ions produced by the protonation of these oxygen atoms in epoxy groups trigger the adhesives; these crosslinked ESS will stay together as long as the organic compounds react with the ester bonds. Organic acids crosslinked the structure of ESS has a good adhesion concentration, extending from 94.4% to 98.7%, indicating that ESS are highly excellent in curing natural acids with water.

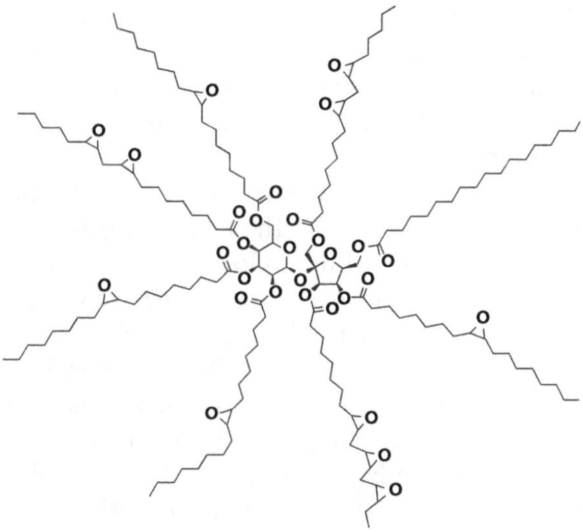

Epoxidized sucrose soyate (ESS)

Figure 26.4 Epoxidized Sucrose Soyate (ESS) [37].

These natural acids, which are crosslinked with ESS network, have superior coating features. These thermosets with better coating properties show greater hardness, solvent resistance, better strength and are more malleable than other thermosets. These ESS thermosets with natural acid crosslinks have more potential to be employed as thermosetting resins [37].

To make these ESS thermosets industrial friendly for large scale production, the curing process time should decrease. To make this practically possible Curtiss S. Kovash. Jr *et al.*, used two methods: adding the catalyst at higher or lower boiling points and adding the catalyst to deblock the ring-opened reaction at a very low temperature. The thermosets show better and greater performance under physical-chemical and mechanical stress at these conditions. In the coating process, the deblocking reaction plays a greater role by increasing the hardeners of the system [38].

26.6 Isosorbide Was Employed as a Bridge in an Adhesive System After Being Introduced Into a Carbonyl Group

Isosorbide integrated to dicarboxylic acid as crosslinker for the construction of an epoxy network to generate biobased degradable thermosets with an excellent balance of hardness and flexibility. Forever crosslinked materials offer excellent mechanical qualities and are resistant to solvents, but they cannot be treated or moulded once they have been made. Non–crosslinked and reversible crosslinked polymers can be processed, but they are soluble to overcome all these diplomatic conditions epoxy networks are designed. Which modify their structure without depolymerization through exchange reactions, and that they are insoluble and transmutable.

Damien Montarnal *et al.* synthesized epoxy networks with the mixture of di and tri-carboxylic groups so as to obtain the ester groups in the end product material. At ambient temperature, these crosslinked networks behave like epoxy resins, and when the reaction is completed at high temperatures, the number of ester linkages in the sample does not vary, according to infrared spectroscopy [39].

The collision strength of these thermoplastic materials can be improved by adding cyclohexylene moieties into the main chain. The greater impact strength is attributable to increased molecular movements caused by cyclohexylene conformational changes in the main chain. These changes in the thermosets that occurred by cyclohexylene moieties are responsible for observing energy, which can be seen as a relaxation peak.

Songai Ma *et al.*, synthesized a crosslinker for ESS, a carboxylic oligomer derived from the interaction between maleic anhydride and isosorbide was used in the reaction. The high reactive cyclic anhydrides and carboxylic groups show greater acidic nature due to their coexistence between carbon-carbon bonds. To make the crosslinker longer, maleic anhydride and isosorbides were treated at 105°C for 5 hours resulting from attaining the required properties; maleic anhydride was also used to alter isosorbide and produce a dicarboxylic acid trimer. This modified synthesis technique enrooted giving better hardness, flexibility

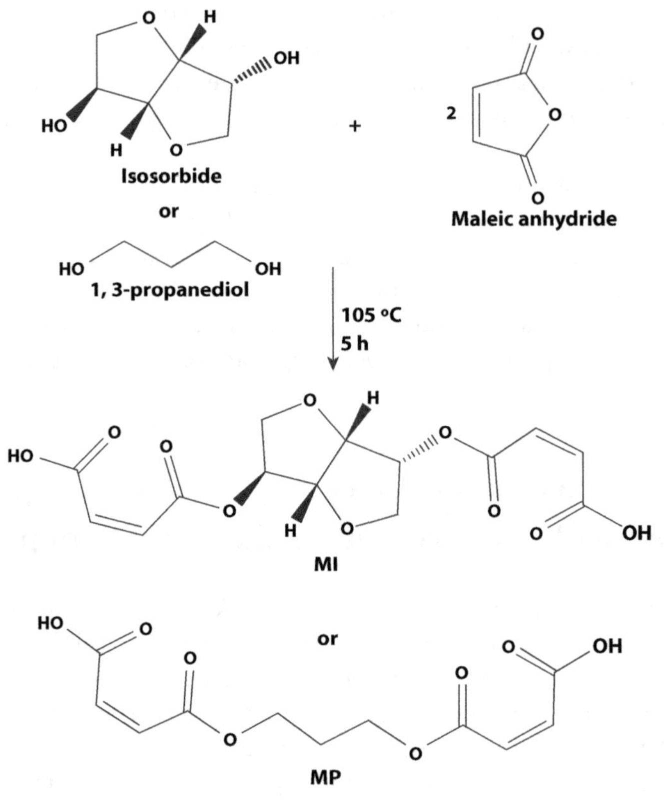

Figure 26.5 Maleic anhydride isosorbide polymer [40].

and potential transitions, also increased the crosslinker length of the ESS network for the strong isosorbide structure shown in Figure 26.5 [40].

The processes that occurred during each cure step were demonstrated in the curing research. Because of the rigidity of these prednisolone rings, thermoplastic materials have a high modulus, which is linked to high hardness. When submerged in NaOH solvents for a short time under normal circumstances, these thermosets degrade thermally [40].

26.7 Thermoplastic Polymers Based on a Spiro Diacetyl Trigger Generated From Lignin

Manufacturing biodegradable thermoplastics using responsible bonding or links might be viable for recycling thermosets under moderate conditions while lowering environmental impact. The most divisible acetals are used in thermoplastic materials based on a diacetyl trigger. These acetals are quickly absorbed into the reaction mixture of these resins, are not disseminated in nonacidic environments, and do not interfere with the resins' curing process. Because these diacetyl groups include methacrylate, thermosets are superior UV curing coating materials [41].

Using Spiro diacetyl structure degradation mechanism was the most determined method to develop the high strength thermosets by adding the acetal into the curing reaction to make it occur in a very short time period. Phenolic hydroxyl groups of CN and BA were put to ethanoxyethers for thermosets of high strength and ductility. Heating was used to perform the curing and moulding operations on the homogenous mixes of TEPA, BA-CHDMVG or BA-VBGE at 100°C for 4 hours with a pressure of 10 MPa to observe the curing of the acetal in the epoxy resin with this acetal addition thermosets got better coting properties with greater hardness [42].

26.8 Properties of Epoxy Resin Thermosets With Acetal Addition

26.8.1 Mechanical Properties

The cured epoxy resins exhibit brittle phase, crystallization, and rubbery pinnacle in the range of temperature of 120 to 160 C [43]. The polymer skeleton structure, as well as the glycidyl ether moieties, influence the transition temperature [44–50]. These structural variations in the epoxy resins by acetal addition give the path to greater critical temperatures making the material harder and strong [51–55].

26.8.2 Thermal Properties

The thermal properties can be studied based on the critical temperature that these materials can withstand greater the critical temperature [56–60], greater thermal properties are observed, and the curing reaction can be controlled at the required temperature to make it a better thermoset material [61–65].

26.9 Conclusions

In response to increased worries about the global energy crisis and emission of carbon, development of polymer from renewable bioenergy as well as organic compounds is increasing. Thermosetting polymers are more difficult to remould and recycle than thermoplastics because they are generated by a permanent curing (hardening) reaction of chemicals in an aqueous solution, mushy solids, and viscoelastic condition. In current history, biodegradable thermoplastic polymers constructed from renewable natural materials have indeed been reported. New materials are coming up using natural feedstock. Thermosetting polymers are now prepared based on natural materials termed biobased thermoset polymers, which offer lightweight, low cost, and better durability. Widely used thermosets are Epoxy resins. They can be produced sustainable bridge polyester blends by polycondensing a naturally existing functionalized carboxyl group (citric acid) with 1,8-octane diol and triglyceride. Reports also suggest the development of biodegradable dimethacrylates using carbonyl compounds derived from carbohydrates. Water was formerly used to make perfectly organic, increased, biodegradable thermoplastic materials using natural compound ringed carboxyl group. Isosorbide was then combined with a carbonyl group and employed as a bridge to create an adhesive matrix, resulting in microbial, biodegradable thermosets with an outstanding hardness-to-flexibility ratio. Hence, various engineering applications can be achieved with the production of biothermoset from renewable sources with added biocompatibility of the newly synthesized elastomer.

Acknowledgements

P. Banerjee would like to thank SERB, India, for the TARE fellowship with grant No. TAR/2021/000032.

References

1. Ma, S. and Webster, D.C., Degradable thermosets based on labile bonds or linkages: A review. *Prog. Polym. Sci.*, 76, 65, 2018.
2. Ma, S., Li, T., Liu, X., Zhu, J., Research progress on bio-based thermosetting resins. *Polym. Int.*, 65, 164, 2016.
3. Ramon, E., Sguazzo, C., Moreira, P.M., A review of recent research on bio-based epoxy systems for engineering applications and potentialities in the aviation sector. *Aerospace*, 5, 110, 2018.
4. Shen, L., Product overview and market projection of emerging biobased plastics-PRO-BIP 2009, in: *Final report*, 2009, http://www.european-bioplastics.org/media/files/docs/en-pub/PROBIP2009_Final_June_2009. pdf.
5. Lutolf, M.P., Raeber, G.P., Zisch, A.H., Tirelli, N., Hubbell, J.A., Cell-responsive synthetic hydrogels. *Adv. Mater.*, 15, 888, 2003.
6. Mooney, D.J., Mazzoni, C.L., Organ, G.M., Puelacher, W.C., Vacanti, J.P., Langer, R., Stabilizing fiber-based cell delivery devices by physically bonding adjacent fibers. *MRS Online Proceedings Library (OPL)*, vol. 331, 1993.
7. Stegemann, J.P. and Nerem, R.M., Phenotype modulation in vascular tissue engineering using biochemical and mechanical stimulation. *Ann. Biomed. Eng.*, 31, 391, 2003.

8. Nerem, R.M., Role of mechanics in vascular tissue engineering. *Biorheology*, 40, 281, 2003.
9. Yang, J., Webb, A.R., Pickerill, S.J., Hageman, G., Ameer, G.A., Synthesis and evaluation of poly (diol citrate) biodegradable elastomers. *Biomaterials*, 27, 1889, 2006.
10. Yang, J., Webb, A.R., Ameer, G.A., Novel citric acid-based biodegradable elastomers for tissue engineering. *Adv. Mater.*, 16, 511, 2004.
11. Qiu, H., Yang, J., Kodali, P., Koh, J., Ameer, G.A., A citric acid-based hydroxyapatite composite for orthopedic implants. *Biomaterials*, 27, 5845, 2006.
12. Gao, J., Crapo, P.M., Wang, Y., Macroporous elastomeric scaffolds with extensive micropores for soft tissue engineering. *Tissue Eng.*, 12, 917, 2006.
13. Yang, J., Motlagh, D., Webb, A.R., Ameer, G.A., Novel biphasic elastomeric scaffold for small-diameter blood vessel tissue engineering. *Tissue Eng.*, 11, 1876, 2005.
14. Prabhakaran, M.P., Venugopal, J., Chan, C.K., Ramakrishna, S., Surface modified electrospun nanofibrous scaffolds for nerve tissue engineering. *Nanotechnology*, 19, 455102, 2008.
15. Shi, R., Chen, D., Liu, Q., Wu, Y., Xu, X., Zhang, L., Tian, W., Recent advances in synthetic bio-elastomers. *Int. J. Mol. Sci.*, 10, 4223, 2009.
16. Webb, A.R. and Kumar, V.A., G.A., Biodegradable poly, diol citrate nanocomposite elastomers for soft tissue engineering. *J. Mater. Chem.*, 17, 900, 2007.
17. Manzer, L.E., *Biomass derivatives: A sustainable source of chemicals*, pp. 40–51, ACS Publications, 2006.
18. Witowski, J. and Knapowski, J., Glycerol toxicity for human peritoneal mesothelial cells in culture: Comparison with glucose. *Int. J. Artif. Organs*, 17, 252, 1994.
19. Sousa, S.M.G.d., Bramante, C.M., Taga, E.M., Biocompatibility of EDTA, EGTA and citric acid. *Braz. Dent. J.*, 16, 3, 2005.
20. Porter, W.L., *Autoxidation in food and biological systems*, pp. 295–365, Springer, 1980. https://doi.org/10.1007/978-1-4757-9351-2
21. Zhang, X.Q., Tang, H., Hoshi, R., De Laporte, L., Qiu, H., Xu, X., Shea, L.D., Ameer, G.A., Sustained transgene expression via citric acid-based polyester elastomers. *Biomaterials*, 30, 2632, 2009.
22. Pramanick, D. and Ray, T., Synthesis and biodegradation of copolyesters from citric acid and glycerol. *Polym. Bull.*, 19, 365, 1988.
23. Naeini, A.T., Adeli, M., Vossoughi, M., Poly (citric acid)-block-poly (ethylene glycol) copolymers—New biocompatible hybrid materials for nanomedicine. *Nanomed.: Nanotechnol. Biol. Med.*, 6, 556, 2010.
24. Namazi, H. and Adeli, M., Dendrimers of citric acid and poly (ethylene glycol) as the new drug-delivery agents. *Biomaterials*, 26, 1175, 2005.
25. Halpern, J.M., Urbanski, R., Weinstock, A.K., Iwig, D.F., Mathers, R.T., Von Recum, H.A., A biodegradable thermoset polymer made by esterification of citric acid and glycerol. *J. Biomed. Mater. Res. A*, 102, 1467, 2014.
26. Salman, S.M., Straight and branched sucrose monoamide derivatives as new surfactants. *Chem. Sci.*, 5, 295, 2016.
27. Gallagher, J.J., *Renewable polymer materials from bicyclic sugar derivatives*, University of Minnesota, 2016. https://hdl.handle.net/11299/191358
28. Li, Y., *Development of biocompatible polymer monoliths for the analysis of proteins and peptides*, Brigham Young University, 2009. http://hdl.lib.byu.edu/1877/etd3161
29. Sideridou, I.D., Achilias, D.S., Karava, O., Reactivity of benzoyl peroxide/amine system as an initiator for the free radical polymerization of dental and orthopaedic dimethacrylate monomers: Effect of the amine and monomer chemical structure. *Macromolecules*, 39, 2072, 2006.
30. Gallagher, J.J., Hillmyer, M.A., Reineke, T.M., Degradable thermosets from sugar-derived dilactones. *Macromolecules*, 47, 498, 2014.

31. Andrzejewska, E., Photopolymerization kinetics of multifunctional monomers. *Prog. Polym. Sci.*, 26, 605, 2001.
32. Sideridou, I., Tserki, V., Papanastasiou, G., Effect of chemical structure on degree of conversion in light-cured dimethacrylate-based dental resins. *Biomaterials*, 23, 1819, 2002.
33. Sideridou, I., Tserki, V., Papanastasiou, G., Study of water sorption, solubility and modulus of elasticity of light-cured dimethacrylate-based dental resins. *Biomaterials*, 24, 655, 2003.
34. Anseth, K.S. and Quick, D.J., Polymerizations of multifunctional anhydride monomers to form highly crosslinked degradable networks. *Macromol. Rapid Commun.*, 22, 564, 2001.
35. Palmieri, F., Adams, J., Long, B., Heath, W., Tsiartas, P., Willson, C.G., Design of reversible cross-linkers for step and flash imprint lithography imprint resists. *ACS Nano*, 1, 307, 2007.
36. Wang, S., Ma, S., Li, Q., Xu, X., Wang, B., Yuan, W., Zhou, S., You, S., Zhu, J., Facile *in situ* preparation of high-performance epoxy vitrimer from renewable resources and its application in nondestructive recyclable carbon fiber composite. *Green Chem.*, 21, 1484, 2019.
37. Ma, S. and Webster, D.C., Naturally occurring acids as cross-linkers to yield VOC-free, high-performance, fully bio-based, degradable thermosets. *Macromolecules*, 48, 7127, 2015.
38. Kovash Jr., C.S., Pavlacky, E., Selvakumar, S., Sibi, M.P., Webster, D.C., Thermoset coatings from epoxidized sucrose soyate and blocked, bio-based dicarboxylic acids. *Chem. Sus. Chem.*, 7, 2289, 2014.
39. Montarnal, D., Capelot, M., Tournilhac, F., Leibler, L., Silica-like malleable materials from permanent organic networks. *Science*, 334, 965, 2011.
40. Ma, S., Webster, D.C., Jabeen, F., Hard and flexible, degradable thermosets from renewable bioresources with the assistance of water and ethanol. *Macromolecules*, 49, 3780, 2016.
41. Ma, S., Wei, J., Jia, Z., Yu, T., Yuan, W., Li, Q., Wang, S., You, S., Liu, R., Zhu, J., Readily recyclable, high-performance thermosetting materials based on a lignin-derived spiro diacetal trigger. *J. Mater. Chem. A*, 7, 1233, 2019.
42. Saithai, P., Lecomte, J., Dubreucq, E., Tanrattanakul, V., Effects of different epoxidation methods of soybean oil on the characteristics of acrylated epoxidized soybean oil-co-poly, methyl methacrylate copolymer. *EXPRESS Polym. Lett.*, 7, 910, 2013.
43. Hashimoto, T. *et al.*, Degradable and chemically recyclable epoxy resins containing acetal linkages: Synthesis, properties, and application for carbon fiber-reinforced plastics. *J. Polym. Sci. Part A: Polym. Chem.*, 50, 3674–3681, 2012.
44. Naidu, K.C.B. *et al.*, 12 Surface morphology induced inorganic materials for supercapacitors, in: *Inorganic Nanomaterials for Supercapacitor Design*, vol. 213, 2019.
45. Banerjee, P. *et al.*, Advancement in electrolytes for rechargeable batteries, in: *Rechargeable Batteries: History, Progress, and Applications*, pp. 87–98, 2020.
46. Banerjee, P., Franco Jr., A., Boddula, R., Naidu, K.C.B., Pothu, R., Carbon nanomaterials for zn-ion batteries, in: *Zinc Batteries: Basics, Developments, and Applications*, 2020.
47. Banerjee, P. *et al.*, Active materials for flexible K-Ion batteries, in: *Potassium-Ion Batteries: Materials and Applications*, pp. 137–145, 2020.
48. Prakash, M. *et al.*, Electrode materials for K-Ion batteries and applications, in: *Potassium-Ion Batteries: Materials and applications*, pp. 123–136, 2020.
49. Reddy, B.V.S. *et al.*, 2D Perovskite materials and their device applications, in: *Layered 2D Advanced Materials and Their Allied Applications*, pp. 131–140, 2020.
50. Naidu, K.C.B. *et al.*, Recent advances in nanomaterials for Li-ion batteries, in: *Lithium-Ion Batteries: Materials and Applications*, vol. 80, pp. 148–160, 2020.
51. Banerjee, P., Kumar, N.S., Franco Jr., A., Swain, A.K., Chandra Babu Naidu, K., Insights into the dielectric loss mechanism of bianisotropic FeSi/SiC composite materials. *ACS Omega*, 5, 25968, 2020.

52. Kumar, N.S. *et al.*, for Microwave absorber advanced ceramics applications, in: *Applications of Advanced Ceramics in Science, Technology, and Medicine*, vol. 3, p. 51, 2020.
53. Banerjee, P., Franco, A., Naidu, K.C.B., Advanced Ceramics for Ferroelectric Devices, in: *Applications of Advanced Ceramics in Science, Technology, and Medicine*, vol. 3, p. 95, 2020.
54. Banerjee, P., Franco Jr., A., Xiao, R., Effects of Y and Ni co-doping in Bi2Fe4O9-BiFeO3 based multiferroic ceramics. *Mater. Today: Proc.*, 46, 4716, 2021.
55. Banerjee, P., Franco Jr., A., Naidu, K.C.B., Khan, A., Asiri, A.M., Natarajan, S., Metal-organic framework–based materials and renewable energy, in: *Metal-Organic Frameworks for Chemical Reactions*, pp. 153–166, Elsevier, 2021. https://doi.org/10.1016/B978-0-12-822099-3.00008-3
56. Banerjee, P., Franco Jr., A., Naidu, K.C.B., Kumar, N.S., Water-borne polyurethane-metal oxide nanocomposite applications, in *Sustainable Production and Applications of Waterborne Polyurethanes*, pp. 155–169, Springer, Cham, 2021. DOI: 10.1007/978-3-030-72869-4_10
57. Jedla, M.R., Koneru, B., Franco Jr., A., Rangappa, D., Banerjee, P., Recent developments in nanomaterials based adsorbents for water purification techniques. *Biointerface Res. Appl. Chem.*, 12, 5821, 2021.
58. Rao, K.R.M., Haripriya, K., Banerjee, P., Franco, A., Microbiologically influenced corrosion, in: *Corrosion science: Modern trends and applications*, vol. 121, 2021.
59. Naidu, K., Kumar, N.S., Banerjee, P., Reddy, B., A review on the origin of nanofibers/nanorods structures and applications. *J. Mater. Sci.: Mater. Med.*, 32, 1, 2021.
60. Melo, R.S., Franco Jr., A., Banerjee, P., Nanoscale-driven single-domain structure in nickel substituted superparamagnetic cobalt ferrites. *Solid State Commun.*, 341, 114560, 2022.
61. Lima, R., Banerjee, P., Araujo, E., Franco, A., Effect of ferrite phase addition on the functional properties of $K_{0.5}Na_{0.5}$) NbO_3 ceramics. *Eur. Phys. J. Plus*, 134, 404, 2019.
62. Banerjee, P., Mondal, S., Biswas, S.K., Broadening of tunable range of resonant frequency of cylindrical cavity resonator of a particular mode avoiding mode overlapping. *Am. Res. J. Phys.*, 1, 1, 1–8, 2015, DOI:10.21694/2380-5714.15007
63. Banerjee, P., Ghosh, G., Biswas, S.K., Amendment of cavity perturbation method for measuring dielectric properties of medium loss sample at microwave frequencies. *Optoelectron. Adv. Mater. Rapid Commun.*, 6, 623, 2012.
64. Oliveira, F., dos Santos, C., Machado, A., Banerjee, P., Franco, A., Structure and dielectric properties of $Ba_2Cu_xY_{1-x}TaO_{6-y}Ba_2Cu_xY_{1-x}TaO_{6-y}$ double perovskite. *SN Appl. Sci.*, 1, 1, 2019.
65. Swapnalin, J., Suresh Kumar, N., Chandra Babu Naidu, K., Banerjee, P., Recent developments in electron paramagnetic resonance for spectroscopic applications. *Biointerface Res. Appl. Chem.*, 13, 1, 45, 2023. https://doi.org/10.33263/BRIAC131.045

27

Public Attitude Toward Recycling Routes of Bioplastics—Knowledge on Sustainable Purchases

Farhan Shaikh and Sunny Kumar*

Department of Chemical Engineering, Institute of Chemical Technology Mumbai, Marathwada Campus, Jalna, India

Abstract

Bioplastic is a vital material seeing technological advancements at a rapid pace. This natural monomer sourced polymer has even achieved physical properties at par with traditional petroleum-based plastics. Although they have a minor market share and production, their costs and awareness are low among the public. In this chapter, we will discuss how bioplastics have replaced traditional plastics. The application and recycling are discussed briefly. Public behavior and awareness is explained. The frequent use of bioplastics in our lives necessitates strategizing their recycling and sustainable purchase among the public. There are various routes of recycling, such as collection, separation, mechanical, chemical recycling, and organics recycling. This chapter will increase public awareness of recycling waste. The age group that has more inclination to recycle and also the countries, which have been surveyed, are discussed herein. This chapter encourages the reader to develop innovative schemes for increasing awareness of sustainable purchases and recycling. Some innovative strategies have been mentioned in this chapter.

***Keywords*:** Attitude, bioplastics, purchase, recycling

27.1 Introduction

Plastics play an important role in our lives, and this is because of the rapid increase in the use of plastic due to its high strength, attractive colors, and cost-effectiveness [1]. Plastic production has also improved exponentially [2]. The long-term stability of these plastics and their majorly nonbiodegradable nature threaten the environment. The current standard of living makes single-use plastics an almost indispensable part of our lifestyle [3]. Plastics litter on the macroscopic scale and disperse on the microscopic scale. The large quantities of plastics used brought up the challenge of their disposal and the microplastics.

Microscale plastic particles have invaded all spheres of our world [4–6] and have now gotten even smaller on the nanoscale. These particles are an upcoming concern for air pollution [7, 8] and can even act as vectors for the spread of organisms, which is a cause of

**Corresponding author*: chemicalsunnyraj@gmail.com

Inamuddin and Tariq Altalhi (eds.) Handbook of Bioplastics and Biocomposites Engineering Applications: 2nd Edition, (589–604) © 2023 Scrivener Publishing LLC

concern for the post-COVID pandemic setting [9]. Microplastics acting as vectors open new scenarios for researchers to tackle [10]. Traditionally, such particles have polluted our freshwater sources worldwide [11]. Even infiltrating soil [12], these are indirect routes to enter our bodies. The increasing use of plastics in agriculture is one of the sources [13]. This effect is not limited to local geography but has a long-range impact [14]. These particles cause serious health concerns [15, 16]. The general motto of "Reduce, Reuse, Recycle, and Recover" is no longer viable. With the sheer size of plastics around us, we must shift our focus to a more robust term like "Replacement." We should look into avenues to replace traditional plastics with more environmentally friendly and degradable plastics. Therefore, there is a dire necessity to shift toward biobased plastics and products thereof to minimize the already compounding effects of plastics.

Biobased plastics are relatively new to the scene and are made with renewable sources, unlike traditional plastics, sourced from nonrenewable petroleum products. Bioplastics also completely mineralize on biodegradation, whereas plastics degrade into microplastics. One way to reduce bioplastic waste is to recycle it. The recycling process must be accessible or cost-effective. Low-cost bioplastic production technology should be developed. Public awareness is also essential toward the bioplastics purchases and uses. Government policies should be introduced to encourage awareness of bioplastics. On a local scale, the recycling route can reduce the bioplastics waste. Here, this book chapter will explain the bioplastics demands, the recycling route of the bioplastics, public awareness, government policies, and sustainable purchase.

27.2 Production of Plastics

Bioplastics are sourced from waste and natural, renewable polymers [17]. The polymers used for the manufacture of bioplastics are biopolymers. The global demand for the polymer is around 400 million tonnes and 302 million tonnes of waste generation [2], as shown in Figure 27.1. These biopolymers are defined as involving living organisms in their synthesis processes [18]. Hence, bioplastics contain biopolymers in varying percentages. The

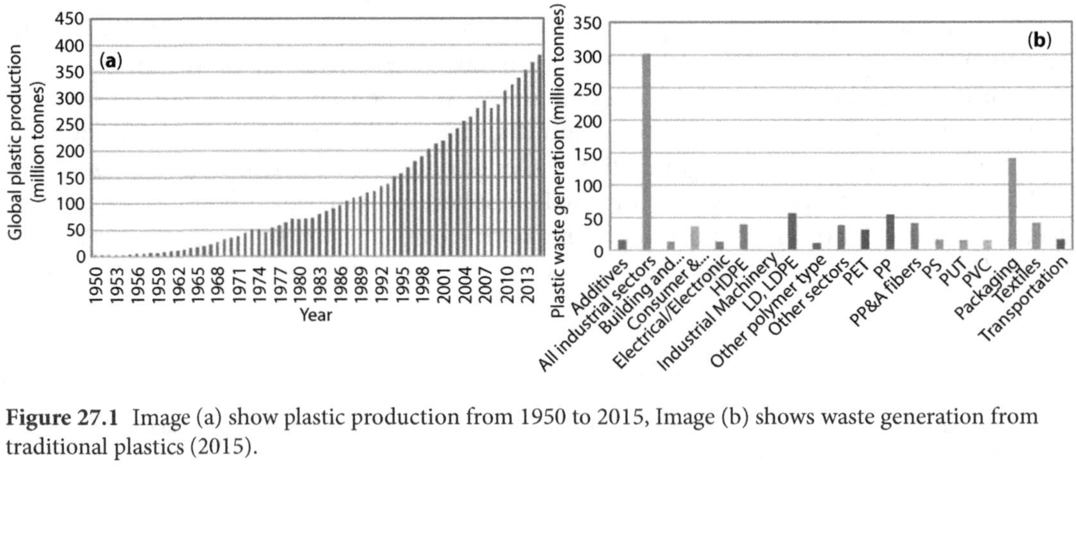

Figure 27.1 Image (a) show plastic production from 1950 to 2015, Image (b) shows waste generation from traditional plastics (2015).

generally accepted that the definition of bioplastics is any plastic that tends to mineralize after disposal completely. Few petrochemical-based bioplastics are polycaprolactone, polybutylene adipate/terephthalate. Some bioplastics are similar to their traditional plastics. Biopolyethylene is one such bioplastic mimicking polyethylene [19]. The raw materials of bioplastics range from lignocellulose [20, 21] to bacteria [22, 23]. Lignocellulose is responsible for a wide range of bioplastics. Lignocellulosic sources are renewable and composed of agricultural waste, and are nonedible. These sources do not require demarcated land for their specific farming and can be sourced from agricultural farmland.

Rice straw has no economic value and is burned after harvesting the crop, a severe cause of pollution. Recently, this is used to produce bioplastic with an appreciable shape memory function [24]. Food waste can also be used to produce bioplastics. Tsang *et al.* has proposed the setup of a biorefinery-like platform specifically for this purpose [25]. The fermentation of food waste to produce polyhydroxyalkanoates (PHA) by chemical pretreatment is followed by biological conversion. This will reduce the emissions and energy required to dispose of food waste. Conventional production of bioplastics requires a pure and sterile bacteria substrate. This increases the cost of production. Jiang and coworkers used untreated wastewater from paper mills to produce PHA [26]. The wastewater was fermented in two open bioreactors in series. The process showed a simple integration into the already established paper mill. The use of wastewater of various industries for producing bioplastics is a well-known area of research [27–29].

27.3 Application of Bioplastics

Bioplastics have over the years been developed to have similar properties to traditional plastics [30, 31] some widespread applications are shown in Figure 27.2. Additives for bioplastics are also catching a craze [32–35]. The use of bioplastics as films is also reaching commercial manufacturing feasibility [36–39]. Bioplastic films have shown appreciable barrier properties. Disposable plates and cutlery in the food industry is a major consumer of bioplastics [40–44]. Electrical properties of bioplastics, specifically Polylactic acid, show

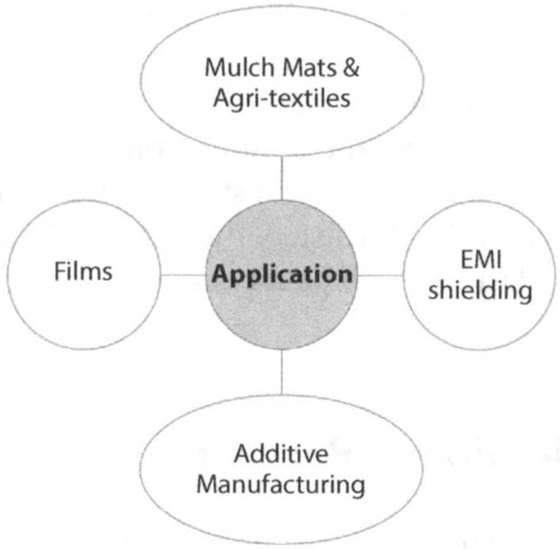

Figure 27.2 Application of plastic and bioplastics.

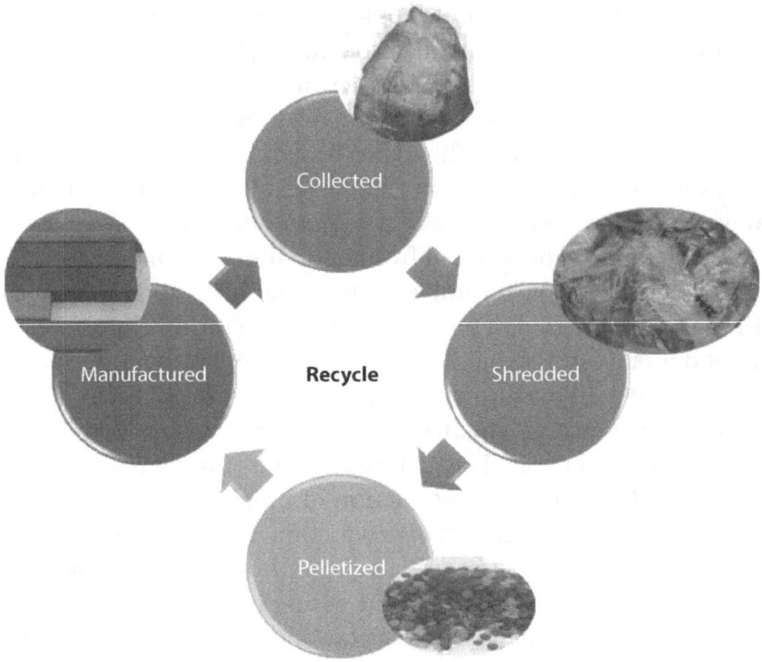

Figure 27.3 The recycling process for bioplastic.

hope of replacing plastics in the future [45, 46]. With the incorporation of carbon nanotubes, these properties could be of further use in additive manufacturing [47] and even electromagnetic interference (EMI) shielding [48]. Soy protein-based plastics are a recent phenomenon with potential use as agroplastics and food packaging [49–51].

27.4 Recycle Route of Bioplastics

Biodegradation of bioplastics is aimed to mineralize the plastics completely. This is in contrast with recycling, which aims to reintroduce bioplastics into our life. Mechanical and chemical routes conventionally address recycling. These routes simplify waste management and preservation of resources [52]. The similarity of bioplastics to traditional plastics makes strategizing their end-of-life collection easier.

The local community should collect the plastics from different sources for the recycling process, as shown in Figure 27.3. In the second step, plastic waste was processed for shredded into almost similar sizes. In the third step, the shredded plastic was sent to pelletizing or briquetting machine. The last step will be for the manufacturing of different application plastics.

27.5 Public Contribution of Recycling

Plastic recycling can be summarized as a triangle, with each corner representing stakeholders responsible directly or indirectly for the environment, as shown in Figure 27.4.

Figure 27.4 The stakeholder triangle indicates each corner without which the recycling of bioplastics will not be possible.

The industries are the source of plastics, and this supply is to quench the demand from the citizens who use the end-products. The role of the government is brought into the picture, as the citizens' awareness will ensure the implementation and discussion of recycling laws in the government. Currently, recycling plastics is mainly based on legislation. Meaning recycling plastics is the duty of the local government. This shifts the responsibility to just one corner of the triangle, which is a significant problem.

A circular bioplastic economy is one way to close the loop in the life cycle. This strategy eliminates the use of disposal infrastructure and focuses on collecting and sorting bioplastics. This concept adds to developing countries' economies, where sorting centers for bioplastics can generate employment in the region. These centers will be set up at places generating appreciable quantities of plastic waste. The responsibility of sorting bioplastics can be shared between the residents and the recycling center, thus making the citizens play a productive part toward their country and indirectly toward global climate change and plastic pollution [53]. Bioplastic sorting is not an easy commodity, with polymers manufactured from more than two polymers, creating complexity in separating single plastic fractions. The incoming waste plastics are usually coated with multiple films. The technology for their separation is still in the nascent stages. Contamination of waste plastics with food, washing products, and printing ink is also a concern. Decontamination of such waste adds to the complexity of the recycling center [54]. Here, the role of industries is brought into the picture, and they can work on sustainable packaging options with researchers to create technoeconomic solutions [55]. Recyclers have to be subsidized by the government, and potential public-private tie-ups can be viable. The private operation of recycling centers is not profitable as no large-scale manufacturers are using recycled plastic as raw material [56]. There are various ways of recycling bioplastics. These routes have various impacts. The primary route of recycling is only applicable to bioplastics having a specific end life, which makes knowing the bioplastic's history necessary. The chemical breakdown of bioplastics can be possible, but it is energy-intensive. This route is for nonbiodegradable bioplastics, and biodegradable plastics can be completely mineralized by a natural process; this aspect requires further refining. Mineralization of bioplastics is ultimately achieved only in laboratory conditions [57]. A well-known study from the Netherlands found that efficient sorting routes can increase the circulation of recycled plastics. Recycled bioplastics have an open-loop life cycle, where disposal is necessary after recycling [58]. This open loop can only be closed by the biodegradable nature of bioplastics. Bioplastics are both nonbiodegradable as well as biodegradable [59]. Polyhydroxyalkanoates are promising due to their biodegradable nature, but their production and physical properties should be developed more [60, 61].

The role of citizens in closing the loop on recycling is of paramount importance. After all, these citizens will be potential industrialists and members of the government. The habit of recycling is like any other habit and can be introduced into the lifestyle of citizens with some form of training in schools, industry, and households, as shown in Figure 27.5. This training has to be imparted at each stage of the citizen's life.

The idea of recycling can be conveyed to primary school students. The influence of such communication at a young and developing age will be effective. At any stage of learning, students (preschool, school, college) are very much like absorbents. They are the future consumers and must therefore be exposed to awareness campaigns. A questionnaire was circulated among college students to examine their knowledge of recycling practices. The questionnaire also analyzed the relationship between behavior, social rules, and attitude. The expectation-maximization algorithm then modeled the answers, then tested by structural equation modeling. This behavioral study showed a positive relationship between attitudes toward recycling and knowledge about recycling. Students who have more recycling knowledge tend to practice it in their daily lives. Societal norms play an essential role in encouraging recycling. The collective behavior of societies can enforce good recycling behavior. A major driving force is government initiatives, which could disseminate recycling information in schools and colleges. This will repeatedly enforce a recycling mindset into the children's lifestyle [62]. A survey of residents of Sharjah, United Arab Emirates, showed that students had the most knowledge of recycling. The city of Sharjah showed less awareness of recycling as compared to Malaysia [63], Thailand [64], and India [65]. Indians showed high awareness of recycling but little to no knowledge of color segregation and disposal.

Industries are a necessary growth engine for many countries. These industries propel economic prosperity to new heights and create waste. This waste produced may be side

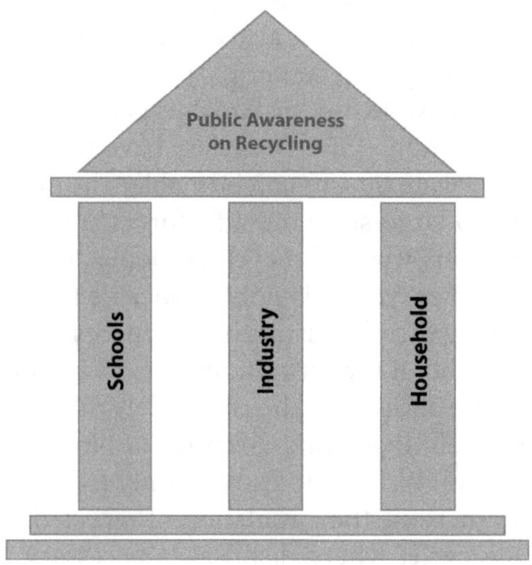

Figure 27.5 The pillars supporting recycling awareness in the public.

products or the actual end products. Side products are due to industry process parameters, while the wastage of actual end products, especially plastics, is crucial. Since they have no real-world application, they are not valuable. Efforts to generate ideas and incubate new ventures to innovate and add value to these wastages. Shoes made from waste plastics are one such innovation that has achieved commercial success in India. An integrated waste management model is one such strategy that can recycle plastic wastes from industries. Recycling industries can be set up to help segregation point sources of plastics. These could provide saplings as a motivational gift to its customers. Governmental policies should encourage such industries, with incentives, such as lower tax brackets or tax exemptions and special loan facilities from banks for setting up such industries [66]. The textile industry can be seen as a possible route for the reuse of recycled plastics. Marketing strategies should generate consumer behavior toward recycled clothes and textiles. This will have a long-lasting impact on the reduction of plastics in landfills. Textiles themselves are incinerated or landfilled; this leaves little scope for the inclusion of recycled plastics in the textile industry. As these may still have end of life at the same place as textiles [67]. Recycling on an industrial scale in an unregulated market has its risks. Developing countries have yet to pass legislation pertaining to plastic recycling and have a steep task in front of them that can directly skip setting up the initial baseline policies. Critical barriers to this process, as seen from segregation centers, are:

- Inclusion of stabilizers into plastic commodities: this makes it immensely difficult to mineralize plastic commodities.
- Lack of labeling: Many plastic articles do not have the proper symbols printed or embossed on them, which leads to confusion for both the citizens and the recycling center.

Effective communication plays a significant role in the motivation of groups. Here, we are devising strategies not for single individuals but for a group of individuals. Families can be motivated by posting a recycling calendar, wherein the schedule for collecting recycle bins can be mentioned. This activity, coupled with positive feedback letters, increased household participation. Households should also be informed of the benefits of recycling and the types of materials that can be recycled. The most preferred method of conveying this information is leaflets [68]. With the increase in internet usage, physical leaflets may be minimized. The leaflets can also be sent digitally, such as social media posts and short stories. This can increase interaction with households all year long. Linking the latest trends on social media with recycling information will be a great way to influence households. Women's participation is higher within household recycling schemes. This can be attributed to the lifestyle of the particular dataset, indicating higher women participation in homemaker-women compared with working-class women. Thus, the use of women-centric pamphlets and posters will be more effective. The crux of any household recycling scheme will always be the location of the recycling bin and recycle bin schedule. This can be a major demotivator if not appropriately publicized, and the trend of backyard recycling is also shaky [69]. This can only be practiced in well-off households having large backyards. This idea rules out traditional urban centers and residential complexes.

Recent surveys show a bleak outlook on backyard composting and recycling. Due to lack of space, information and time. Innovation in backyard composting can be brought about by

Table 27.1 The attractive design of recycle bin.

Recycle bin modification	Reasons
The user easily handles the recycling bin	It should be a reasonable size and shape for users and reduce users' effort. It is also convenient for disabled users.
The recycling bin is attractive	It should be encouraging the recyclers with a phrase and good bright color
The recycling bin does not pose a risk to human health and safety	It does not invite the microbes and prevent the odor.
The recycling bin protests the recyclables	The bin should be protected from rain, snow and high humidity.
The recycling bin is handled by staff of the waste management services.	The waste can be quickly transferred

rooftop composting in residential buildings and community composts. This can be implemented in future city plans. Major demotivators to address are malodor, inaccessibility to infrastructure, and age. The old-age population has little knowledge about recycling. The working-class population, specifically white-collared workers, has little motivation toward recycling. The manual workers with higher income showed an eagerness to recycle [70]. An attractively modified recycle bin can also attract the public to throw the waste in the recycle bin, as described in Table 27.1 [71].

27.6 Awareness of Sustainable Purchase

Sustainable bioplastic purchases can happen when the bioplastic has low cost, good properties, and a large amount of production compared to the other materials. Researchers and industries can play a good role in purchasing sustainable biopolymers by developing low-cost biopolymers. This approach focuses on the purchase intention of bioplastic, green consumer value, product experience, and knowledge about bioplastics [72]. The bioeconomy of plastics should be promoted, in which bioplastics are used and even produced from traditional industries. Encouragement of setting up Industrial-scale fermenters within the industrial premise will lower the price of bioplastics. The production can be made sustainable when locally available raw material is used to produce bioplastics. Thus, researchers need to concentrate on the feedstock used for bioplastic production. Second-generation biorefineries, specifically lignocellulosic biorefineries, are potential for integration with bioplastic production facilities (polyhydroxyalkanoates). Although bioplastics save 2.9 kg CO_2 (compared to HDPE) and 15 kg CO_2 (compared to polystyrene), their production cost is 15 to 17 times higher than conventional plastics [73].

Figure 27.6 explains the environmental responsibility purchase intention. Major factors responsible for steering the intention of purchase are depicted. Willingness to pay is an essential factor, as bioplastics are comparatively more expensive than traditional plastics. This, however, has not lowered the intention of Indians; in contrast, Indians showed a positive approach toward a greener and sustainable purchase [74]. The recommended attitude and policies toward bioplastic and traditional plastics as shown in Table 27.2 [75].

Globally, researchers have surveyed local populations for their awareness toward environmental friendly options. The results are encouraging. Businesses can launch products based on bioplastics and expect returns similar to traditional plastics. The marketing done by such business will further bring bioplastics into trend. Thus catalyzing a shift toward their consumer acceptability, which will further bring in price parity from the increased market demand. A compilation of such surveys is shown in Table 27.3 below.

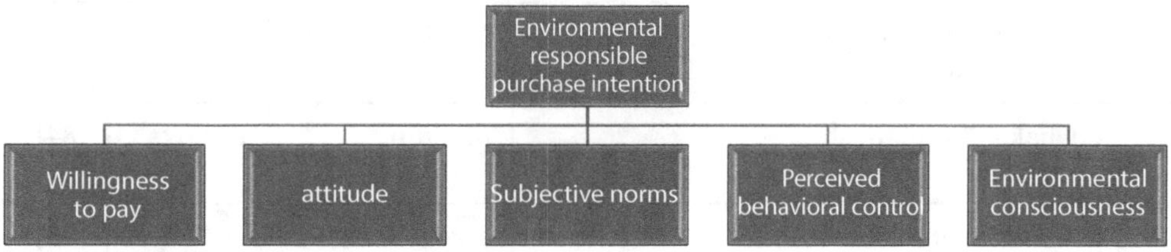

Figure 27.6 Environmental responsible purchase intention.

Table 27.2 Attitude and policies towards bioplastic and regular plastics.

Bioplastic use	**Regular plastic use**
Easy availability of information and products	Raise awareness about environmental issues
Raise awareness about the environmental issues	Regulatory legislations
Full support from the government, producers, and brands	Economic policies
Reasonable price	Refuse, reuse and recycle

Table 27.3 Comparison of awareness of recycling and awareness towards green products (bioplastics).

Sr. no.	**Geography**	**Awareness**	**Attitude**	**Reference**
1	New Delhi, **India**	High (83%)	Positive	[65]
2	Massey University, **New Zealand**	High (96%)	Positive	[76]
3	Muharraq Governate, **Kingdom of Bahrain**	High (90.3%)	Positive	[77]

(Continued)

Table 27.3 Comparison of awareness of recycling and awareness towards green products (bioplastics). (*Continued*)

Sr. no.	Geography	Awareness	Attitude	Reference
4	Kuwait City, **State of Kuwait**	High (89%)	Positive	[78]
5	Eskişehir, **Turkey**	High (89%)	Positive	[79]
6	University of Hong Kong, **Hong Kong**	High (77%)	Positive	[80]
7	University of California, Berkeley, **United States of America**	High (76%)	Positive	[81]
8	Royal Borough of Kensington and Chelsea, Greater London, **United Kingdom**	High (72.8%)	Positive	[82]
9	Dammam Metropolitan Area, **Kingdom of Saudi Arabia**	Medium (65%)	Positive	[83]
10	Dhaka, **Bangladesh**	Medium (61.94%)	Positive	[84]
11	Nablus, **Palestine**	Low (45.4%)	Positive	[85]
12	Nibong Tebal town, Penang, **Malaysia**	Low (32.5%)	Positive	[86]
13	Minsk, **Belarus**	Low (30%)	Positive	[87]
14	Hatyai City, **Thailand**	Low (14%)	Negative	[88]
15	Ras Al Khaima, **United Arab Emirates**	Low (10%)	Negative	[89]

27.7 Conclusion

This chapter sheds light on the rising appearance of microplastics across the world, in addition to plastic generation and wastage. The evolution of bioplastics from being lab scale to their commercial synthesis for real-world application is elucidated. This rise in bioplastics will also result in the waste generation of nonbiodegradable bioplastics. These nonbiodegradable bioplastics have little-known analysis methods and recycling technologies. This seems a distant but achievable task with the recent interest in recycling and segregation of traditional plastics. The latter part of the chapter cites surveys on behavior analysis out the how much the production of bioplastic and produced waste of it. It was found that penetration of recycling awareness has increased over time, but accessibility and motivation have yet to reach appreciable levels. If provided with adequate psychological reinforcement bolstered by infrastructure, citizens tend to take pride in recycling and segregating their waste. The three most important aspects of awareness have also been addressed, with each playing an essential role in increasing public awareness. The technical aspects of segregating plastics and the difficulties faced in the process are significant blockades in the path of progress. Public contribution is significant as, the triangle figure mentions, each stakeholder can exchange places, and no one is permanent. Therefore, it is a basic necessity to train the public from a young age. Surveys reported herein have also modeled behavioral algorithms to derive the value of motivation toward recycling from each class of citizenry. Finally, sustainability, a new term that is garnering interest in all aspects of engineering, is explained concerning the purchasing power of the public. This helps link the technical engineering tactics with economics and is a make or break for the commercialization of the technology.

References

1. Andrady, A.L. and Neal, M.A., Applications and societal benefits of plastics. *Philos. Trans. R. Soc. Lond., B, Biol. Sci.*, 364, 1977, 2009.
2. Ritchie, H. and Roser, M., Plastic Pollution. *Our World Data*, 2018.
3. Chen, Y., Awasthi, A.K., Wei, F., Tan, Q., Li, J., Single-use plastics: Production, usage, disposal, and adverse impacts. *Sci. Total Environ.*, 752, 141772, 2021.
4. Shim, W.J. and Thomposon, R.C., Microplastics in the ocean. *Arch. Environ. Contam. Toxicol.*, 69, 265, 2015.
5. Allen, S., Allen, D., Moss, K., Le Roux, G., Phoenix, V.R., Sonke, J.E., Examination of the ocean as a source for atmospheric microplastics. *PloS One*, 15, e0232746, 2020.
6. Zhang, Q., Xu, E.G., Li, J., Chen, Q., Ma, L., Zeng, E.Y., Shi, H., A review of microplastics in table salt, drinking water, and air: Direct human exposure. *Environ. Sci. Technol.*, 54, 3740, 2020.
7. Amato-Lourenço, L.F., dos Santos Galvão, L., de Weger, L.A., Hiemstra, P.S., Vijver, M.G., Mauad, T., An emerging class of air pollutants: Potential effects of microplastics to respiratory human health? *Sci. Total Environ.*, 749, 141676, 2020.
8. Sridharan, S., Kumar, M., Singh, L., Bolan, N.S., Saha, M., Microplastics as an emerging source of particulate air pollution: A critical review. *J. Hazard. Mater.*, 418, 126245, 2021.
9. Shen, M., Zhu, Y., Zhang, Y., Zeng, G., Wen, X., Yi, H., Ye, S., Ren, X., Song, B., Micro nanoplastics: Unignorable vectors for organisms. *Mar. Pollut. Bull.*, 139, 328, 2019.
10. Allen, S., Allen, D., Baladima, F., Phoenix, V.R., Thomas, J.L., Le Roux, G., Sonke, J.E., Evidence of free tropospheric and long-range transport of microplastic at Pic du Midi Observatory. *Nat. Commun.*, 12, 7242, 2021.
11. Strungaru, S.A., Jijie, R., Nicoara, M., Plavan, G., Faggio, C., Micro-(nano) plastics in freshwater ecosystems: Abundance, toxicological impact and quantification methodology. *Trac. Trend Anal. Chem.*, 110, 116, 2019.
12. Hurley, R.R. and Nizzetto, L., Fate and occurrence of micro(nano)plastics in soils: Knowledge gaps and possible risks. *Curr. Opin. Environ. Sci. Health*, 1, 6, 2018.
13. Rodríguez-Seijo, A. and Pereira, R., Microplastics in Agricultural Soils: Are They a Real Environmental Hazard?, in: *Bioremediation of agricultural soils*, CRC Press, Boca Raton, 2019.
14. Zhang, Y., Gao, T., Kang, S., Allen, S., Luo, X., Allen, D., Microplastics in glaciers of the Tibetan Plateau: Evidence for the long-range transport of microplastics. *Sci. Total Environ.*, 758, 143634, 2021.
15. Prata, J.C., da Costa, J.P., Lopes, I., Duarte, A.C., Rocha-Santos, T., Environmental exposure to microplastics: An overview on possible human health effects. *Sci. Total Environ.*, 702, 134455, 2020.
16. Rodrigues, M.O., Abrantes, N., Gonçalves, F.J.M., Nogueira, H., Marques, J.C., Gonçalves, A.M.M., Impacts of plastic products used in daily life on the environment and human health: What is known? *Environ. Toxicol. Pharmacol.*, 72, 103239, 2019.
17. Kumar, S., Madan, S., Bariha, N., Suresh, S., Swelling and shrinking behavior of modified starch biopolymer with iron oxide. *Starch/Stärke*, 73, 2000108, 2021.

18. Standard test methods for determining the biobased content of natural range materials using radiocarbon and isotope ratio mass spectrometry analysis, ASTM, D6866-05, 2017.
19. Agrawal, P., Araújo, A.P., Brito, G.F., Cavalcanti, S.N., Alves, A.M., Freitas, D.M., Mélo, T.J., Rheological and mechanical properties of poly(lactic acid)/bio-based polyethylene/clay biocomposites containing montmorillonite and vermiculite clays. *J. Polym. Environ.*, 29, 1777, 2021.
20. Yang, J., Ching, Y.C., Chuah, C.H., Applications of lignocellulosic fibers and lignin in bioplastics: A review. *Polymers*, 11, 751, 2019.
21. Reshmy, R., Thomas, D., Philip, E., Paul, S.A., Madhavan, A., Sindhu, R., Sirohi, R., Varjani, S., Pugazhendhi, A., Pandey, A., Binod, P., Bioplastic production from renewable lignocellulosic feedstocks: A review. *Rev. Environ. Sci. Biotechnol.*, 20, 167, 021.
22. Witholt, B. and Kessler, B., Perspectives of medium chain length poly(hydroxyalkanoates), a versatile set of bacterial bioplastics. *Curr. Opin. Biotechnol.*, 10, 279, 1999.
23. Krüger, A., Schaefers, C., Schroeder, C., Antranikian, G., Towards a sustainable biobased industry – Highlighting the impact of extremophiles. *New Biotechnol.*, 40, 144, 2018.
24. Bilo, F., Pandini, S., Sartore, L., Depero, L.E., Gargiulo, G., Bonassi, A., Federici, S., Bontempi, E., A sustainable bioplastic obtained from rice straw. *J. Clean. Prod.*, 200, 357, 2018.
25. Tsang, Y.F., Kumar, V., Samadar, P., Yang, Y., Lee, J., Ok, Y.S., Song, H., Kim, K.H., Kwon, E.E., Jeon, Y.J., Production of bioplastic through food waste valorization. *Environ. Int.*, 127, 625, 2019.
26. Jiang, Y., Marang, L., Tamis, J., van Loosdrecht, M.C., Dijkman, H., Kleerebezem, R., Waste to resource: Converting paper mill wastewater to bioplastic. *Water Res.*, 46, 5517, 2012.
27. Bosco, F. and Chiampo, F., Production of polyhydroxyalcanoates (PHAs) using milk whey and dairy wastewater activated sludge: Production of bioplastics using dairy residues. *J. Biosci. Bioeng.*, 109, 418, 2010.
28. Kumar, M., Ghosh, P., Khosla, K., Thakur, I.S., Recovery of polyhydroxyalkanoates from municipal secondary wastewater sludge. *Bioresour. Technol.*, 255, 111, 2018.
29. López Rocha, C.J., Álvarez-Castillo, E., Estrada Yáñez, M.R., Bengoechea, C., Guerrero, A., Orta Ledesma, M.T., Development of bioplastics from a microalgae consortium from wastewater. *J. Environ. Manage.*, 263, 110353, 2020.
30. Sun, S., Song, Y., Zheng, Q., Morphology and mechanical properties of thermo-molded bioplastics based on glycerol-plasticized wheat gliadins. *J. Cereal Sci.*, 48, 613, 2008.
31. Song, Y., Zheng, Q., Zhang, Q., Rheological and mechanical properties of bioplastics based on gluten- and glutenin-rich fractions. *J. Cereal Sci.*, 50, 376, 2009.
32. Zárate-Ramírez, L.S., Romero, A., Martínez, I., Bengoechea, C., Partal, P., Guerrero, A., Effect of aldehydes on thermomechanical properties of gluten-based bioplastics. *Food Bioprod. Process.*, 92, 20, 2014.
33. Jansens, K.J., Bruyninckx, K., Redant, L., Lagrain, B., Brijs, K., Goderis, B., Smet, M., Delcour, J.A., Importance of crosslinking and disulfide bridge reduction for the mechanical properties of rigid wheat gluten bioplastics compression molded with thiol and/or disulfide functionalized additives. *J. Appl. Polym. Sci.*, 131, 1–9, 2014.
34. Tummala, P., Liu, W., Drzal, L.T., Mohanty, A.K., Misra, M., Influence of plasticizers on thermal and mechanical properties and morphology of soy-based bioplastics. *Ind. Eng. Chem. Res.*, 45, 7491, 2006.
35. López-Castejón, M.L., Bengoechea, C., García-Morales, M., Martínez, I., Effect of plasticizer and storage conditions on thermomechanical properties of albumen/tragacanth based bioplastics. *Food Bioprod. Process.*, 95, 264, 2015.

36. Abotbina, W., Sapuan, S.M., Sultan, M.T.H., Alkbir, M.F.M., Ilyas, R.A., Development and characterization of cornstarch-based bioplastics packaging film using a combination of different plasticizers. *Polymers*, 13, 3487, 2021.
37. Ramakrishnan, N., Sharma, S., Gupta, A., Alashwal, B.Y., Keratin based bioplastic film from chicken feathers and its characterization. *Int. J. Biol. Macromol.*, 111, 352, 2018.
38. Sudhakar, M.P., Magesh Peter, D., Dharani, G., Studies on the development and characterization of bioplastic film from the red seaweed (Kappaphycus alvarezii). *Environ. Sci. Pollut. Res.*, 28, 33899, 2021.
39. Zhao, X., Anwar, I., Zhang, X., Pellicciotti, A., Storts, S., Nagib, D.A., Vodovotz, Y., , Thermal and barrier characterizations of cellulose esters with variable side-chain lengths and their effect on PHBV and PLA bioplastic film properties. *ACS Omega*, 6, 24700, 2021.
40. Peelman, N., Ragaert, P., De Meulenaer, B., Adons, D., Peeters, R., Cardon, L., Van Impe, F., Devlieghere, F., Application of bioplastics for food packaging. *Trends Food Sci. Technol.*, 32, 128, 2013.
41. Jabeen, N., Majid, I., Nayik, G.A., Bioplastics and food packaging: A review. *Cogent. Food Agric.*, 1, 1117749, 2015.
42. Zhao, X., Cornish, K., Vodovotz, Y., Narrowing the gap for bioplastic use in food packaging: An update. *Environ. Sci. Technol.*, 54, 4712, 2020.
43. Jariyasakoolroj, P., Leelaphiwat, P., Harnkarnsujarit, N., Advances in research and development of bioplastic for food packaging. *J. Sci. Food Agric.*, 100, 5032, 2020.
44. Keränen, O., Komulainen, H., Lehtimäki, T., Ulkuniemi, P., Restructuring existing value networks to diffuse sustainable innovations in food packaging. *Ind. Mark. Manage.*, 93, 509, 2021.
45. Oi, T., Shinyama, K., Fujita, S., Electrical properties of heat-treated poly-lactic acid. *IEEE J. Trans. Fundam. Mater.*, 131, 395, 2011.
46. Ravenstijn, J., Bioplastics in consumer electronics. *Ind. Biotechnol.*, 6, 252, 2010.
47. Yang, L., Li, S., Zhou, X., Liu, J., Li, Y., Yang, M., Yuan, Q., and Zhang, W., Effects of carbon nanotube on the thermal, mechanical, and electrical properties of PLA/CNT printed parts in the FDM process. *Synth. Met.*, 253, 122, 2019.
48. Wang, G., Zhao, G., Wang, S., Zhang., L., Park, C.B., Injection-molded microcellular PLA/graphite nanocomposites with dramatically enhanced mechanical and electrical properties for ultra-efficient EMI shielding applications. *J. Mater. Chem. C*, 6, 6847, 2018.
49. Han, J., Shin, S.-H., Park, K.-M., Kim, K.M., Characterization of physical, mechanical, and antioxidant properties of soy protein-based bioplastic films containing carboxymethylcellulose and catechin. *Food Sci. Biotechnol.*, 24, 939, 2015.
50. Jiménez-Rosado, M., Perez-Puyana, V., Cordobés, F., Romero, A., Guerrero, A., Development of superabsorbent soy protein-based bioplastic matrices with incorporated zinc for horticulture. *J. Sci. Food Agric.*, 99, 4825, 2019.
51. Jiménez-Rosado, M., Perez-Puyana, V., Rubio-Valle, J.F., Guerrero, A., Romero, A., Evaluation of superabsorbent capacity of soy protein-based bioplastic matrices with incorporated fertilizer for crops. *J. Polym. Environ.*, 28, 2661, 2020.
52. Fredi, G. and Dorigato, A., Recycling of bioplastic waste: A review. *DOAJ*, 4, 159, 2021.
53. d'Ambrières, W., Plastics recycling worldwide: Current overview and desirable changes, in: *Field Actions Science Reports. The journal of field actions*, vol. 19, p. 12, 2019.
54. Horodytska, O., Valdés, F.J., Fullana, A., Plastic flexible films waste management – A state of art review. *Waste Manage.*, 77, 413, 2018.
55. Roosen, M., Mys, N., Kleinhans, K., Lase, I.S., Huysveld, S., Brouwer, M., van Velzen, E.U.T., Van Geem, K.M., Dewulf, J., Ragaert, K., Dumoulin, A., Expanding the collection portfolio of plastic packaging: Impact on quantity and quality of sorted plastic waste fractions. *Resour. Conserv. Recy.*, 178, 106025, 2022.

56. Bassi, S.A., Boldrin, A., Faraca, G., Astrup, T.F., Extended producer responsibility: How to unlock the environmental and economic potential of plastic packaging waste? *Resour. Conserv. Recy.*, 162, 105030, 2020.
57. Lamberti, F.M., Román-Ramírez, L.A., Wood, J., Recycling of bioplastics: Routes and benefits. *J. Polym. Environ.*, 28, 2551, 2020.
58. Brouwer, M., Picuno, C., Thoden, V., Velzen, E.U., Kuchta, K., Meester, S., Ragaert, K., The impact of collection portfolio expansion on key performance indicators of the Dutch recycling system for post-consumer plastic packaging waste, a comparison between 2014 and 2017. *Waste Manage.*, 100, 112, 2019.
59. Rahman, M.H. and Bhoi, P.R., An overview of non-biodegradable bioplastics. *J. Clean. Prod.*, 294, 126218, 2021.
60. Alcântara, J.M.G., Distante, F., Storti, G., Moscatelli, D., Morbidelli, M., Sponchioni, M., Current trends in the production of biodegradable bioplastics: The case of polyhydroxyalkanoates. *Biotechnol. Adv.*, 42, 107582, 2020.
61. Meereboer, K.W., Misra, M., Mohanty, A.K., Review of recent advances in the biodegradability of polyhydroxyalkanoate (PHA) bioplastics and their composites. *Green Chem.*, 22, 5519, 2020.
62. Ramayah, T., Lee, J.W.C., Lim, S., Sustaining the environment through recycling: An empirical study. *J. Environ. Manage*, 102, 141, 2012.
63. Desa, A., Kadir, N.B.A., Yusooff, F., A study on the knowledge, attitudes, awareness status and behaviour concerning solid waste management. *Proc. Soc Behav. Sci.*, 18, 643, 2011.
64. Singhirunnusorn, W., Donlakorn, K., Kaewhanin, W., Household recycling behaviours and attitudes toward waste bank project: Mahasarakham municipality. *AjBeS*, 2, 17, 2017.
65. Govind, M. and Mahongnao, M., Recycling of municipal solid waste in India: Empirical findings from some select colonies in Delhi. *Afr. J. Sci. Technol. Innov. Dev.*, 13, 899, 2021.
66. Gunarathne, A.D.N., Tennakoon, T.P.Y.C., Weragoda, J.R., Challenges and opportunities for the recycling industry in developing countries: The case of Sri Lanka. *J. Mater. Cycles Waste Manage.*, 21, 181, 2019.
67. Hole, G. and Hole, A.S., Recycling as the way to greener production: A mini review. *J. Clean. Prod.*, 212, 910, 2019.
68. Mee, N. and Clewes, D., The influence of corporate communications on recycling behaviour. *Corp. Commun. An. Int. Jnl.*, 9, 265, 2004.
69. Babaei, A.A., Alavi, N., Goudarzi, G., Teymouri, P., Ahmadi, K., Rafiee, M., Household recycling knowledge, attitudes and practices towards solid waste management. *Resour. Conserv. Recy.*, 102, 94, 2015.
70. Wang, H., Liu, X., Wang, N., Zhang, K., Wang, F., Zhang, S., Wang, R., Zheng, P., Matsushita, M., Key factors influencing public awareness of household solid waste recycling in urban areas of China: A case study. *Resour. Conserv. Recy.*, 158, 104813, 2020.
71. Keramitsoglou, K.M. and Tsagarakis, K.P., Public Participation in designing the recycling bins to encourage recycling. *Sustainability*, 10, 1240, 2018.
72. Klein, F., Emberger-Klein, A., Menrad, K., Möhring, W., Blesin, J.-M., Influencing factors for the purchase intention of consumers choosing bioplastic products in Germany. *Sustain. Prod. Consum.*, 19, 33, 2019.
73. Dietrich, K., Dumont, M.-J., Del, Rio, L.F., Orsat, V., Producing PHAs in the bioeconomy — Towards a sustainable bioplastic. *Sustain. Prod. Consum.*, 9, 58, 2017.
74. Kumar, A., Prakash, G., Kumar, G., Does environmentally responsible purchase intention matter for consumers? A predictive sustainable model developed through an empirical study. *J. Retail. Consum. Serv.*, 58, 102270, 2021.

75. Leal Filho, W., Salvia, A.L., Bonoli, A., Saari, U.A., Voronova, V., Klõga, M., Kumbhar, S.S., Olszewski, K., De Quevedo, D.M., Barbir, J., An assessment of attitudes towards plastics and bioplastics in Europe. *Sci. Total Environ.*, 755, 142732, 2021.
76. Kelly, T.C., Mason, I.G., Leiss, M.W., Ganesh, S., University community responses to on-campus resource recycling. *Resour. Conserv. Recy.*, 47, 42, 2006.
77. Abbas, S.Y., Kirwan, K., Lu, D., Measuring the public awareness toward household waste management in Muharraq Governorate-Kingdom of Bahrain. *IJEP*, 11, 196, 2020.
78. Koushki, PARVIZ A., Al-Humoud, J., Al-Duaij, U., Municipal solid waste in Kuwait: Trends and attitudes on collection, separation and willingness to pay. *Kuwait J. Sci. Eng.*, 31, 173, 2004.
79. Arı, E. and Yılmaz, V., A proposed structural model for housewives' recycling behavior: A case study from Turkey. *Ecol. Econ.*, 129, 132, 2016.
80. Wan, C., Cheung, R., Qiping Shen, G., Recycling attitude and behaviour in university campus: A case study in Hong Kong. *Facilities*, 30, 630, 2012.
81. Abramson, R., Plastic recycling awareness among students at UC Berkeley, in: *Plastic recycling awareness*, vol. 20, 2008.
82. Robinson, G.M. and Read, A.D., Recycling behaviour in a London Borough: Results from large-scale household surveys. *Resour. Conserv. Recy.*, 45, 70, 2005.
83. Almulhim, A.I. and Abubakar, I.R., Understanding public environmental awareness and attitudes toward circular economy transition in Saudi Arabia. *Sustainability*, 13, 10157, 2021.
84. Afroz, R., Hanaki, K., Tuddin, R., Ayup, K., A survey of recycling behaviour in households in Dhaka, Bangladesh. *Waste Manage. Res.*, 28, 552, 2010.
85. Al-Khatib, I.A., Monou, M., Abu Zahra, A.S.F., Shaheen, H.Q., Kassinos, D., Solid waste characterization, quantification and management practices in developing countries. A case study: Nablus district – Palestine. *J. Environ. Manage.*, 91, 1131, 2010.
86. Isa, M.H., Asaari, F.A.H., Ramli, N.A., Ahmad, S., Siew, T.S., Solid waste collection and recycling in Nibong Tebal, Penang, Malaysia: A case study. *Waste Manage. Res.*, 23, 565, 2005.
87. Miafodzyeva, S., Brandt, N., Olsson, M., Motivation recycling: Pre-recycling case study in Minsk, Belarus. *Waste Manage. Res.*, 28, 340, 2010.
88. Charuvichaipong, C. and Sajor, E., Promoting waste separation for recycling and local governance in Thailand. *Habitat Int.*, 30, 579, 2006.
89. Assaf, H., Idwan, S., Farhat, M., Assessing recycling attitude and behaviour in Ras Al Khaimah, UAE. *Environ. Eng. Sci.*, *14*, 218, 2018.

28

Applications of Bioplastic in Composting Bags and Planting Pots

Sonica Sondhi

Department of Biotechnology, Chandigarh College of Technology, CGC Landran, Mohali, Punjab, India

Abstract

Plastics made from nonrenewable petroleum sources are increasingly being used in agriculture, posing a significant environmental danger. Plastic garbage from transplanting pots is generated in huge quantities by horticultural operations, as well as landscaping plant cultivation, and is rarely recycled. Biodegradable pots and bags are a good alternative to nondegradable pots as they are safe to the environment. Despite this, only a few studies have been conducted on the preparation and characterization of biodegradable planting pots. Planting pots made of biodegradable materials eliminate the need for transplanting and, as a result, the need to discard a container. When placed in the ground, pots made of biodegradable polymers will decompose. This chapter summarizes the most recent research on biopots (biodegradable planting pots) made of bioplastics. Bioplastics, which are commonly reinforced with plant or wood fibers, could be a viable substitute to traditional petro-based pots for horticulture and floriculture containers in terms of plant development and functionality.

Keywords: Bioplastics, compostable bags, planting pots

28.1 Introduction

The tremendous manufacture of plastics over the last six decades has resulted in massive amounts of garbage all over the world. In 2019, global plastics output reached 368 million metric tons [1]. The COVID-19's effects on the industry are expected to reduce production by about 0.3% in 2020 [2]. Petroleum-based polymers, which are nonrenewable, are employed in a broad series of applications [3]. Plasticulture involving use of plastic in agriculture, e.g., in direct coverings including greenhouse covering films, plastic film mulching in soil to increase soil productivity, solar heat and fumigation films, silage wraps, shading nets, materials used in harvesting and postharvesting operations, drainage and irrigation pipes, are just some of the plastics used in modern agriculture [4].

While there are numerous benefits of numerous uses of plastics in agriculture, there are also numerous dangers. It is estimated that the container crops horticulture sector in the

Email: sonicasondhi@gmail.com

Inamuddin and Tariq Altalhi (eds.) Handbook of Bioplastics and Biocomposites Engineering Applications: 2nd Edition, (605–618)

United States produces about 4 billion single-use plastic containers each year [5]. Plastic debris can limit porosity of soil and, as a result, air movement. It can also have an impact on microbial populations, potentially lowering farmland fertility [6]. Plastic fragments have also proven to liberate phthalate acid esters (possibly carcinogenic substances) into land, where they are able to concentrate in crops and constitute a risk to individual health [7]. Other harmful agrochemicals that are commonly applied to crops may also accumulate in plastics [8]. Because cattle have the potential to swallow plastic or other toxins that seep from it, this is a very dangerous situation for them [9]. Plastic pollution can make its way into rivers and oceans, posing a threat to aquatic life [10]. As a result, reducing the usage of plastic materials in agriculture is a key problem.

In agriculture, transplanting is particularly based on popular petroleum-based polymers [11]. Transplanting is the process of shifting a plant from one location to another. It is a technique that is carried out all over the world [12]. Farmers and gardeners utilize a variety of pots, containers, bags and cell trays made of different shapes, colors and sizes, to meet diverse harvest variety, rising system, and advertising strategy [13]. The great bulk of transplanting pots and bags are made of nonrenewable oil-based plastics such as polyethylene, polypropylene, and polystyrene. These resources have demonstrated acceptable involuntary qualities, chemical and microbiological degradation resistance, durability, and cheap cost [13].

According to some studies, use of biodegradable plastics made up of polypropylene and polyethylene even worsens the situation as when they are released into the environment, they degrade into hundreds of small fragments of plastic, posing a serious threat to our environment [14]. Smaller fragments of plastic are easier for animals to digest than complete bags. Removing hundreds of small fragments of plastic from the environment is also far more difficult than removing a single bag [15]. Petro-pots and bags lead to unhygienic soil,

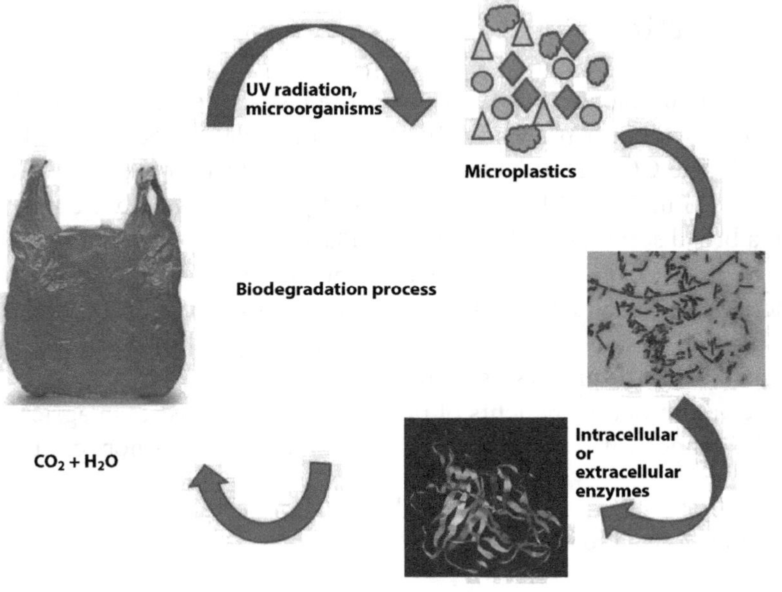

Figure 28.1 Mechanism of biodegradation of planting pots.

organic debris, and agrochemicals after use (which is usually single use) [16]. Nonetheless, the expenses of properly collecting, disposing, and recycling discarded plastic pots and bags are substantial, leading to their abandonment in landfills or uncontrolled combustion, resulting in harmful chemicals being released into the atmosphere and land. The great mass of plant containers and seed trays are either buried or burned in an unregulated manner. Plastic pots and bags are made from a considerable amount of fossil fuel, which takes roughly 500 years to degrade [2]. The chemical structure, molecular weight, water solubility, and xenobiotic properties of plastics are all factors that limit their biodegradation.

Biodegradable plastic pots and bags, also known as biocontainers or biopots, could be a viable alternative to oil-based plastic pots [17]. The breakdown of biopots and compositing bags by naturally existing microorganisms to inert products and biomass, primarily water and carbon dioxide, makes them appropriate for agricultural uses [17] (Figure 28.1). These biopots for nursery plants are being developed in an effort to encourage further study and application in green business.

28.2 Biodegradable Pots (Biopots)

Alternative containers for nursery and greenhouse production have been developed to replace standard petroleum-based plastic containers. Bioplastics and biocomposites are emerging as environmentally friendly alternatives to petroleum-based polymers that may even outperform them in some applications [18]. Containers made of plastic take up landfill space and can last an infinite amount of time in the environment. Vessels that are meant to decay rather than add to landfill garbage are called sustainable containers [19]. Biocontainers are distinguished from their standard plastic counterparts by their capacity to breakdown when planted or composted. These biobased materials can be blended and compounded to generate a wide range of qualities [18]. Recycled plastic alternative containers are characterized depending on their requirements as well as decomposition after harvesting. Plantable and compostable biopots are the two basic types of biopots (Figure 28.2).

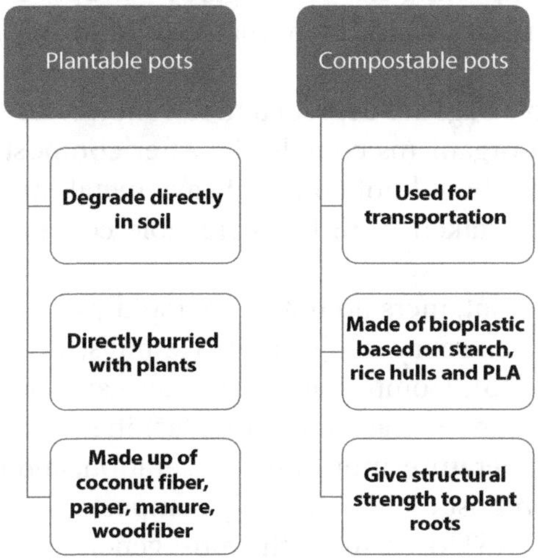

Figure 28.2 Types of biopots: plantable and compostable pots.

28.2.1 Plantable Pots

Biocontainers that can be planted directly into the ground are known as plantable biocontainers. These pots are designed to endure handling and watering during manufacturing and transportation for a limited duration. Because they are designed to decompose fast once planted, containers allow plant roots to reach the soil. With plantable pots, root damage and transplant shock can be prevented [20]. To work as intended, plantable containers must break down fast after installation to allow root formation in the surrounding soil [21]. Many factors influence the rate of container biodegradation in landscapes. Degradation can be influenced by a variety of factors, including the container material, the bacteria present, nitrogen, temperature, moisture, and pH. Furthermore, regional variances may develop as a result of varying soil types and climates.

As there is no way to dispose of plant pots, plantable biopots can be buried in the ground with juvenile plants or saplings intact, resulting in a significantly faster transplanting operation and easier field cleanup [22]. This resulted in a reduction in farm labor, costs, and pollution. Whether in an open field or in larger containers indoors, plantable biopots allow roots to develop more naturally in the growing media. Biomass and inorganic chemicals are produced as plantable biopots breakdown in the ground (i.e., water and carbon dioxide).

28.2.2 Composting Bags

In a compost pile, compostable bags decompose biologically, with the plastic breaking down into carbon dioxide (CO_2), water (H_2O), biomass, and inorganic compounds at a rate comparable to other compostable materials (e.g., cellulose) [13]. Composting may transform biodegradable waste, such as ecological plastics, into helpful soil improvement elements [13]. Composting is defined as the controlled decomposition of heterogeneous organic materials in a wet, warm, aerobic atmosphere by a mixed microbial population [23]. Useful compost, as well as water and CO_2, will be created as a result of the biodegradation of these natural resources. There is no increase in greenhouse gas emissions as a result of CO_2 production because it is already in the biological carbon cycle. In addition to biodegradable plastics, composting can accept garden and food waste, as well as nonrecyclable paper goods, which account for more than half of MSW. This is why composting is so important [24].

Initially, the compostable bags are degraded to low molecular weight compounds, which are then utilized by microorganisms completely. After composting for 12 weeks at 50°C, the product of composting should not contain heavy metals and other toxic materials. At installation, plants must be taken from biodegradable containers, which are composted separately [23].

In the landscape, these containers do not decay rapidly or entirely. The vast majority of bioplastics fall within this category, as do hard rice hulls, thick-walled paper, peat, and/or wood fiber containers [25]. Compostable materials can also be classified according to whether or not they require industrial composting facilities to thoroughly decompose [23]. Because of insufficient temperature, moisture, pH, aeration, and microbial populations in a conventional backyard compost pile, industrially compostable containers may not decompose. In the United States, ASTM D6400 is the most generally used standard for certifying industrially biodegradable polymers, and it is the most widely used standard in the world [26]. Bioplastics must breakdown at least 60% in 90 days at or above 140 degrees Fahrenheit to be considered compostable, according to this criterion.

Biodegradable and compostable bioplastic polymers, unlike typical petrochemical-based polymers, can be composted. There are methods that can aid in this regard: aerobic waste management systems that produce carbon and nutrient-rich compost for soil usage Each year, more than 2 million tonnes of garbage are decomposed at more than 300 composting sites in the United Kingdom (about 75% home waste, 5% is community nonhousehold waste, and 20% is commercial waste). For BDPs, aerobic biodegradation mechanisms are essential. Some BDPs can also be used in anaerobic digestors to convert biowaste to methane, which can then be used to power energy generators [13].

28.3 Biodegradable Planting Pots

Natural materials can be used to create alternative containers. Typically, these containers are comprised of renewable materials that are often left over from industrial processes. By repurposing waste from another process, their usage in container manufacturing can drastically minimize landfill trash. Bioplastics have been tested as a coating for commercially available fiber containers in hopes of increasing their performance. Using bioderived polyamides, polylactic acids, or polyurethanes to coat fiber containers during crop production, increases the strength, durability, and water-use efficiency of these containers while also improving the health and quality of plants cultivated in coated containers [27].

28.3.1 Biodegradable Planting Pots Based on Pressed Fibers

On the market, there are various distinct types of hot-pressed fiber containers. The materials used to manufacture these include rice hulls (*Oryza sativa*), wheat (*Triticum aestivum*), spruce fibers (*Picea* sp.), coconut palm (*Cocos nucifera*) and composted cow manure [28]. Semiporous fiber containers allow water and air to readily pass between the rooting substrate and the environment. Depending on the material and manufacturing process, the containers may be biodegradable or compostable. For example, cow manure can be used as a binding agent in some containers that use natural or synthetic binding materials. Other containers do not use a binding agent [29].

For long-term storage containers, the inherent structural stability and extended life of the material are critical considerations. Pressed fiber containers have varying degrees of rigidity, material strength, and decay resistance, depending on the source material and production technique [30].

The strength of pressed fiber containers is controlled by the container itself, unlike plastic, which performs fairly consistently in a mechanical production system (source material, material moisture content, binder, etc.). Irrigation, usage of shade/supplemental illumination, ambient temperature, and other production practices have an impact on the environment in which the containers are placed. Container performance and lifetime are further influenced by plant rooting pattern, pot spacing, and production length [31]. When filled with water, fiber containers can weigh up to three times as much as thin-walled plastic containers, which has an effect on both container weight and shipping costs.

28.3.2 Biodegradable Planting Pots Based on Bioplastics

Because they are made from non–petroleum-based biopolymers or a combination of both, bioplastics are equivalent to conventional plastics in many ways [19]. Bioplastics can be used in the same applications as regular plastics. In order to define a biopolymer or bioplastic, it is necessary to consider one or more of the following criteria: a) the source of the raw material, which must be "biobased," that is, derived from natural resources; b) the degree to which the polymer is biodegradable. A polymer must be biobased, biodegradable, or both to be classified as a bioplastic. As illustrated in Figure 28.3, biodegradable polymers can be derived from fossil fuels or from renewable sources. Bioplastics are currently employed in the production of nursery containers in two primary types: 1) starch-based plastics; 2) polylactic acid-based plastics (PLA). To increase physical and chemical qualities, starch blends are created by mixing 20% to 80% starch with biobased polymers [19]. Because of its slow biodegradability in soil, poly lactic acid is mostly used in starch blends [20]. It is manufactured anaerobically from feedstock and is primarily utilized in starch blends. Microorganisms, animals, and plants can all be used to make bioplastics. It is important to remember that biobased polymers are not necessarily sustainable; the material's long-term

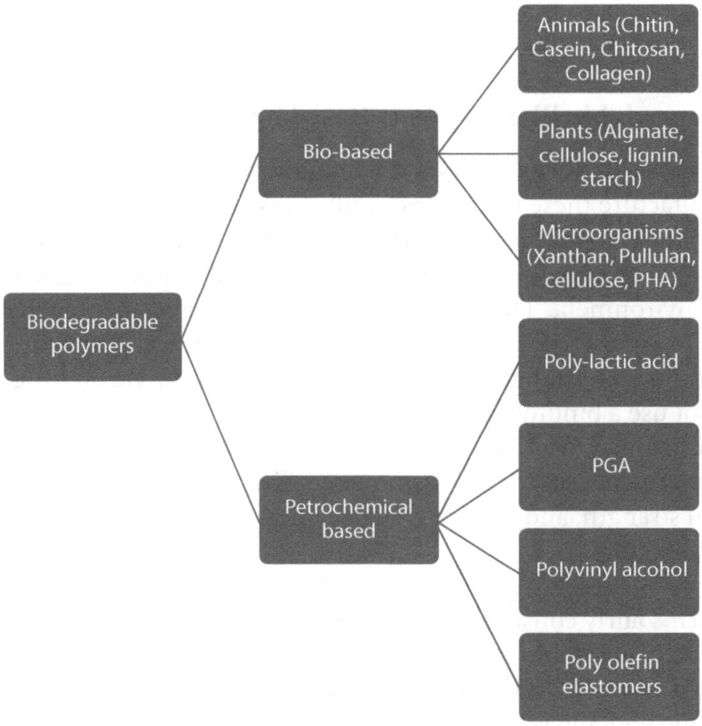

Figure 28.3 Classification of biodegradable polymers.

viability is determined by a range of factors, including the material's source, manufacturing process, and how the product is managed after it has been used. Renewable raw materials are used to make biopolymer-based polymers. Organic feed stocks (such as *Beta vulgaris* [32], *Zea mays* [33], *Solanum tuberosum* [34], *Manihot esculenta* [35], *Saccharum officinarum* [36], palm fiber [37]) provide starch or cellulose. Lipids (plant oils and animal fats), polysaccharides (cellulose, starch, lignin), and proteins (gelatine and wheat gluten) are all biodegradable in nature. Soybeans (Glycine max) and keratin from discarded poultry feathers are the two most common sources of plant-derived protein [38]. Plant oils and animal fats are both sources of lipids [39]. To lower the cost, improve performance, or both, these essential ingredients are routinely combined with fossil fuel-based polymers generated through petrochemical refining. Materials that can be composted include natural rubber and polyesters obtained from microbes or plants (such as polyhydroxyalkanoates and poly-3-hydroxybutyrate) or synthesized from bioderivative monomers (such as PLA) [13]. If the container was not removed before transplantation, the delayed degradability of bioplastics could stifle root growth. Some containers, like the Ball Horticultural Co.'s SoilWrap, a bioplastic-based sleeve design, decompose in the soil and can be used as plantable pots [17].

28.3.3 Biopots Based on Industry and Agriculture

An estimated 1.3 billion metric tonnes of solid waste are generated annually by the world's population and industries combined [40]. Solid waste can help achieve the primary goal of the circular economy, which is the sustainability of systems and industrial processes as well as environmental friendliness. It is feasible to increase crop yields by integrating trash into agricultural systems by recycling industrial or agricultural waste to create new materials. An increasing number of companies across the world are now producing and selling biopots made from various types of organic materials like cow manure, coconut husks, and wood pulp. Biodegradable pots and trays made from a variety of industrial and agricultural wastes are included in Table 28.1.

In a long-term greenhouse crop, Lopez & Camberato [41] looked at quality parameters like appearance and durability of a variety of biopots. Wood pulp with Canadian sphagnum peat moss, composted cow dung, rice straw with coco fiber, coconut husk, wheat starch-derived bioresin, rice hull, and recycled paper pots were all compared to a normal plastic pot. The first four were more susceptible to algae growth and pot breakage, whereas the rest were less so. Another crucial factor to examine is how effectively water is used in biopots. Biobased containers, according to Wang *et al.* [42], have a lower water consumption efficiency than conventional containers. Researchers studied the amount of water plants used daily in recycled paper or coir fiber containers in different regions of the country. Their findings emphasize the need of pots capable of producing high-performing plants. For the "Cooler Grape" vinca (*Catharanthus roseus*) crop, researchers in Arkansas and Mississippi examined irrigation cycles and water requirements both with and without plastic shuttle trays [43]. (USA). When compared to plastic containers, biopots made of water-permeable materials (such as straw, manure, coconut fiber, slotted rice hulls, wood fiber, and peat pots) lose water more quickly, necessitating more frequent irrigation in order to grow a vinca plant. Water is lost at a similar rate in solid rice hulls and bioplastic pots (both of which are normally water-resistant). This suggests that the water requirements and irrigation cycles of

Table 28.1 Some plantable and compostable materials available in the market.

Name of product	Material
Plantable	
Bio pot	Bamboo fiber
Coir pot	Coconut coil fiber
Cow pot	Composted cow manure
DOT pots	Spruce fiber, peat moss
Elle pot	Paper
Fertil pot	Spruce wood fiber & peat moss
Jiffy pot	Peat
KordFiber pot	Wood and paper
Net pot	Rice hull
Soil Wrap	Mirel
Straw pot	Rice straw
Western pulp plot	Moulded wood pulp
Compostable	
Carbon lite	Starch
Ecotainer	Plant starch (PSM)
Kordfiber grow	Recycled paper or cardboard
Large pulp pots	Wax permeated wood pulp
Terra shell	Poly lactic acid
Speedy pot	Peat and (PLA)
Wax tough pot	Wood and paper coated in wax

these two materials are similar. It is possible to limit water loss through the container walls by using permeable biopots, although these biodegradable containers require more water than their plastic counterparts.

It has been observed that plants grown in containers made of PLA and soy polymers are healthier and of a higher grade than those cultivated in ordinary petroleum-plastic containers at commercial scale. They also found that containers constructed from various bioplastics can perform extra activities above and beyond those provided by petroleum-plastic containers [44]. Additionally, soy-based polymers can improve root shape (less circling, more fibrous roots) during manufacture of bioplastics including soy-based polymers [45].

Nambuthiri and Ingram [46] investigated the influence of biopots on the tallness and exposure of various plant species when compared to conventional plastic and paper pots. The experiment used biopots made of rice husk and polyhydroxyalkanoate polymers created through fermentation of renewable carbon-based feedstocks, predominantly corn. Biodegradable containers did not need to be transported; instead, plants were cultivated in a variety of pots before being transplanted. Using planting containers, instead of transplanting, saved a significant amount of time. Ten weeks into production, paper and biopolymer pots began to crack and dissolve almost totally in terms of stability. It is possible that root circling is caused by deteriorating pots, which might impact water, nutrient flow and uptake. Container biodegradation in the field may be hindered by soil N and C concentrations [47]. One way to help modern agriculture progress is to incorporate a variety of leftovers into the production chain of unique, sustainable items. Biopots must be rapidly degraded in soil to reduce biopot construction and root looping, while increasing the efficiency of biopot water consumption when growing plants.

28.4 Growth and Quality of Plants in Biopots

In addition to environmental and sustainability concerns, root circling is a practical consequence of utilizing petroleum-based plastic horticultural containers, especially when the plastic is black. A tree's lifespan can be shortened by stem girdling, which is caused by water and nutrient flow being restricted by circling roots. When compared to nonbiodegradable plastic containers, biodegradable pots are an excellent alternative for growing plants because they disintegrate completely in the soil and, in many cases, promote plant growth and development. Every day, the same amount of fertilizer was applied to azalea plants that were irrigated with the same amount of water. In terms of plant growth, leaf mass and area, and overall biomass, paper biocontainers outperformed plastic pots. Plants grown in biopots had much larger root length and surface area. Plants developed in petroleum-derived polymer pots against PLA-based containers were evaluated for growth metrics of "Arizona Sun" flowers (Gaillardia grandiflora) [47]. The plant's size is unaffected by the container. Plants cultivated in bioplastic containers generated shoots and roots in a comparable manner to plants grown in control pots. When it comes to biocontainers, the type and materials utilized can either strengthen or weaken them depending on the plant's life cycle.

For future biocontainer studies, using plants with production cycles that are longer than the 8–9 week time period employed in this experiment, is recommended. *Cyclamen persicum* was grown in a range of biopots, including rice husk pots, dairy manure and coconut fiber as well as paper and peat pots as a control. With the exception of wood fiber pots, plants grown in biopots had higher dry shoot and root weights than those grown in plastic pots. Container production timelines, on the other hand, were remarkably consistent.

Using alginate-based pots, Schettini *et al.* [48] examined the effects on pepper roots and height. The pots were able to maintain the substance alive for 35 days before dissolving into the soil after 16 days in all cases. Instead of having long roots, it showed a complex branching arrangement with minor ramifications. The plant heights of tomato and hemp fiber biopots were much higher than those of the control biopots (0.67 m).

Seven biopots were tested in 2011 and 2012 for lantana, cleome and the New Guinea Impatiens, which included manure, coir, peat and straw in addition to the rice husk pots [49].

In this case, the research was carried out in the USA (in five diverse locations). It was then determined how various biopots affected plant growth and quality. There was just a small effect on plant growth in the first year regardless of container choice. In contrast, the second year's plant development varied based on the position and variety, as did the container type. This type of pot had a smaller impact on plant growth than the location effect, which was surprising. As for pot degradation, plant species have a smaller impact than actual pot material does.

In a recent study, Ewa [50] concluded that seed germination was not affected by the use of bioplastics. They could, however, cause a delay in the germination process. There were both inhibiting and promoting effects on root and stem growth. The biological activity of nitrifiers and the transformation of carbon compounds was unaffected by bioplastic microparticles.

Different biopots can be considered innovative alternatives to petroleum-based plastics typically used in horticulture and floral containers, since they have a similar effect on root systems and plants, along with their relevant sustainable features for soil and the environment.

28.5 Future Trends and Challenges

Nonrenewable petroleum-based plastics, which are rarely recycled, are largely used in modern agriculture. It is critical to replace these resources with eco-friendly alternatives that can mitigate the harmful effects of the former. Seed trays, pots and containers in particular, utilize a lot of plastic in agriculture, and biodegradable alternatives are becoming more popular. Because they are planted concurrently with the sapling and biodegrade directly in soil, planting biopots offers a range of benefits to the producer, including low cost, reduced agricultural labor, and no waste. On a laboratory and industrial scale, petroleum-based polymers have been widely lauded for their favorable properties in agriculture. Planting pots produced from bioplastics and industrial waste look to be just as successful as petroleum-based containers. Biodegradable vessels can be a good option to petroleum-based plastics used in horticulture and floriculture containers in terms of plant development and function, because nonrenewable biopots have an impact on root architecture before and after transplanting.

28.6 Conclusion

This trend is expected to continue in the packaging business in the years to come. Postuse biodegradable plastics and other biowastes, such as paper, food, and garden waste, can release methane under anaerobic circumstances, making landfilling undesirable and conflicting with regulations such as the EU Landfill Directive for their disposal. To date, the most promising bioplastics for composting and anaerobic digestion systems have been biodegradable bioplastics, which can be used without additional research. As with any other organic garbage, they should be collected with other nonbiodegradable items in the home. Composts made from these biological processes can be utilized to improve soil quality, reducing the amount of waste sent to landfills. Clear certification and labeling techniques are necessary to apply appropriate biological treatments to the growing spectrum of biodegradable bioplastics. In industrial composting systems, biodegradable polymers that meet the compostability standards degrade quickly. However, as previously stated, only a subset of those plastics would biodegrade adequately in the ambient, mesophilic conditions found in residential composters in the United Kingdom, and this distinction must be made public. Bioplastic polymers have a lot of potential for assisting with material recovery, landfill reduction, and the use of renewable resources. To fully realize these benefits, extensive public awareness of these materials will be necessary, as well as an adequate infrastructure for strict documentation, compilation, separation, and composting regulations.

References

1. https://www.statista.com/statistics/282732/global-production-of-plastics-since-1950
2. Shams, M., Alam, I., Mahbub, S., Plastic pollution during COVID-19: Plastic waste directives and its long-term impact on the environment. *Environ. Adv.*, 5, 100119, 2021.
3. Ashter, S.A. and S. A. B. T.-I. @ to B. E. Ashter, Commercial applications of bioplastics, in: *Plastics design library*, vol. 9, pp. 227–249, William Andrew Publishing, Oxford, 2016.
4. Scarascia-mugnozza, G., Sica, C., Russo, G., Plastic materials in European agriculture, in: *Actual use and perspectives*, 2011.
5. Schrader, J.A., McCabe, K.G., Srinivasan, G., Haubrich, K., Grewell, D., Madbouly, S., and Graves, W.R., Development and evaluation of bioplastic containers for sustainable greenhouse and nursery production. *Acta Hortic.*, 1104, 79, 2015.
6. Zhou, Y., Wang, J., Zou, M., Jia, Z., Zhou, S., Li, Y., Microplastics in soils: A review of methods, occurrence, fate, transport, ecological and environmental risks. *Sci. Total Environ.*, 748, 141368, 2020.
7. Kelly, B.C.O., Microplastics in soils: An environmental geotechnics perspective. *Environ. Geotech.*, 8, 586, 2021.
8. Vox, G., Loisi, R.V., Blanco, I., Mugnozza, G.S., Schettini, E., Mapping of agriculture plastic waste. *Agric. Agric. Sci. Proc.*, 8, 583, 2016.
9. Priyanka, M. and Dey, S., Ruminal impaction due to plastic materials - An increasing threat to ruminants and its impact on human health in developing countries. *Vet. World*, 11, 1307, 2018.
10. https://www.pewtrusts.org/en/research-and-analysis/articles/2018/09/24/plastic-pollution-affects-sea-life-throughout-the-ocean
11. Kasirajan, S. and Ngouajio, M., Polyethylene and biodegradable mulches for agricultural applications: A review. *Agron. Sustain. Dev.*, 32, 2, 501, 2012.
12. Candido, V., Miccolis, V., Gatta, G., Margiotta, S., Picuno, P., Manera, C., The effect of soil solarization and protection techniques on yield traits of melon in unheated greenhouse. *Acta Hortic.*, 559, 705, 2000.
13. Song, J.H., Murphy, R.J., Narayan, R., Davies, G.B.H., Biodegradable and compostable alternatives to conventional plastics. *Philos. Trans. Royale Soc B*, 1, 2127, 2009.
14. Madbouly, S.A., Schrader, J.A., Srinivasan, G., Liu, K., McCabe, K.G., Grewell, D., Graves, W.R., Kessler, M.R., Biodegradation behavior of bacterial-based polyhydroxyalkanoate (PHA) and DDGS composites. *Green Chem.*, 16, 1911–1920, 2014.
15. Ferronato, N. and Torretta, V., Waste mismanagement in developing countries: A review of global issues. *Int. J. Environ. Res. Public Health*, 16, 1060, 2019.
16. Alabi, O.A., Ologbonjaye, K.I., Awosolu, O., Alalade, O.E., Public and environmental health effects of plastic wastes disposal: A review. *J. Toxicol. Risk Assess.*, 5, 6, 2019.

17. Geneve, R. and Niu, G., Moving toward sustainability with alternative containers for greenhouse and nursery crop production: A review and research update. *Hortic. Technol. Hortte*, 25, 8, 2015.
18. Schrader, J. *et al.*, Evaluation and classification of bioplastic plant containers. *Acta Horticulturae*, 2014. DOI: 10.17660/ActaHortic.2015.1104.13
19. Shah, A.U., Nur, S., Safri, A., Hasfanizam, M., Yazik, M., Overview of bioplastic introduction and its applications in product packaging. *Coatings*, 2, 1423, 2021.
20. Khan, A.A., Mahmood, T., Bano, B., Development of bio-decomposable (Jiffy) pots for raising and transplanting nursery plants. *Int. J. Agric. Biol.*, 2, 380, 2000.
21. Evans, M.R., Taylor, M., Kuehny, J., Physical properties of biocontainers for greenhouse vrops production. *Hortic. Technol. Hortte*, 1, 549, 2010.
22. Tomadoni, B., Merino, D., Alvarez, V.A., Chapter 4 biodegradable materials for planting pots. In Book: *Advanced Applications of Bio-degradable Green Composites*, pp. 85–103, Materials Research Forum LLC., 2020. 10.21741/9781644900659-4.
23. Harrison, R.B., Composting and formation of humic substances current approaches and future trends in compost quality criteria for agronomic, environmental, and human. In Book: *Encyclopedia of Ecology*, pp. 713–719, 2008.
24. Kulikowska, D. and Bernat, K., Effect of bio-based products on waste management. 12, 2088, 2020.
25. Evans, M. and Karcher, D., Properties of plastic, peat, and processed poultry feather fiber growing containers. *Hortic. Sci.*, 39, 1008, 2004.
26. ASTM D6400-04, in: *Standard specification for compostable plastics*, pp. 1–2, 2018.
27. McCabe, K., Schrader, J., Madbouly, S., Grewell, D., Graves, W., Evaluation of biopolymer-coated fiber containers for container-grown plants. *Horttechnology*, 24, 439, 2014.
28. Rencoret, J., Ralph, J., Marques, G., Gutie, A., Mart, A.T., Jose, C., Structural characterization of lignin isolated from coconut *(Cocos nucifera)* coir fibers. *J. Agric. Food Chem.*, 61, 2434, 2013.
29. Lungu, M.G., Munteanu, N., Stoleru, V., Teliban, G., Caba, C., Preliminary studies regarding the selection of species and types of pot and container for vegetable growing. *Horticulture*, 76, 1, 58, 2019.
30. Appleton, B.L., Nursery production alternatives for reduction or elimination of circling tree roots. *Arboriculture & Urban Forestry*, 1993. https://doi.org/10.48044/jauf.1993.060
31. Ho, M.P., Wang, H., Lee, J.H., Ho, C.K., Lau, K.T., Leng, J., Hui, D., Critical factors on manufacturing processes of natural fibre composites. *Compos. Part B Eng.*, 43, 3549, 2012.
32. Coppola, G., Teresa, M., Catia, G., Lopresto, G., Calabro, V., Curcio, S., Bioplastic from renewable biomass: A facile solution for a greener environment. *Earth Syst. Environ.*, 5, 231, 2021.
33. Marichelvam, M.K. and Mohammad Jawaid, M.A., Corn and rice starch-based bio-plastics as alternative packaging materials. *Fibers*, 7, 32, 2019.
34. Bezirhan, E., Production of bioplastic from potato peel waste and investigation of its biodegradability production of bioplastic from potato peel waste and investigation of its biodegradability. *Int. Adv. Res. Eng. J.*, 4, 2019.
35. Olivares, C.A., Castañeda, C.A., Jhonny, W., Flores, V., Benites-alfaro, E., Valverde, Y., Bioplastic made from manihot esculenta (Cassava) and ficus benjamina as an ecological alternative for food products. *Chem. Engg. Trans.*, 87, 67, 2021.
36. Cheavegatti-Gianotto, A., de Abreu, H.M.C., Arruda, P., Bespalhok Filho, J.C., Burnquist, W.L., Creste, S., di Ciero, L., Ferro, J.A., de Oliveira Figueira, A.V., de Sousa Filgueiras, T., Grossi-de-Sá, M.D.F., Sugarcane (Saccharum X officinarum): A reference study for the regulation of genetically modified cultivars in brazil. *Trop. Plant Biol.*, 62, 2011.

37. Yang, J., Ching, Y.C., Chuah, C.H., Hai, N.D., Singh, R., Nor, A.R.M., Preparation and characterization of starch-based bioplastic composites with treated oil palm empty fruit bunch fibers and citric acid. *Cellulose*, 28, 4191, 2021.
38. Álvarez-Castillo, E., Felix, M., Bengoechea, C., Guerrero, A., Proteins from agri-food industrial biowastes or co-products and their applications as green materials. *Foods (Basel, Switzerland)*, 10, 981, 2021.
39. Chiellini, E., Chiellini, F., Cinelli, P., Polymers from renewable resources, in: *Degradable polymers*, G. Scott (Ed.), Springer, Dordrecht, 2002, https://doi.org/10.1007/978-94-017-1217-0_7.
40. https://www.worldbank.org/en/news/press-release/2018/09/20
41. Lopez, R.G. and Camberato, D.M., Growth and development of 'Eckespoint Classic Red' poinsettia in biodegradable and compostable containers. *HortTechnology*, 21, 4, 1000–1005, 2011.
42. Wang, X. *et al.*, Multistate evaluation of plant growth and water use in plastic and alternative nursery containers. *Hortech*, 25, 1023, 2015.
43. Evans, M.R., Taylor, M., Kuehny, J., Physical properties of biocontainers for greenhouse crops production. *Hortic. Technol. Hortte*, 20, 549, 2021.
44. Schrader, J., Report on the annual consumption of plastics for specialty-crop containers in the United States, 2013. http://dx.doi.org/10.13140/RG.2.2.29992.39689/1
45. Schrader, J., Srinivasan, G., Grewell, D., McCabe, K., Graves, W., Fertilizer effects of soy-plastic containers during crop production and transplant establishment. *Hortic. Sci. Publ. Am. Soc. Hortic. Sci.*, 48, 724, 2013.
46. Nambuthiri, S.S. and Ingram, D.L., Evaluation of plantable containers for groundcover plant production and their establishment in a landscape. *Hortic. Technol. Hortte*, 1, 48, 2014.
47. Flax, N., Currey, C., Mccabe, K., Schrader, J., Greenhouse cultural practices for emerging bioplastic and biocomposite containers, in: *Bioplastic Container Cropping Systems: Green Technology for the Green Industry*, J.A. Schrader, H.A. Kratsch, and W.R. Graves (Eds.), pp. 141–168, Sustainable Hort. Res. Consortium, Ames, IA, USA, 2016.
48. Schettini, E., Santagata, G., Malinconico, M., Immirzi, B., Scarascia, G., Vox, G., Recycled wastes of tomato and hemp fibres for biodegradable pots: Physico-chemical characterization and field performance. *Resour. Conserv. Recycl.*, 70, 9, 2013.
49. Niu, G., Sun, Y., Bi, G., Koeser, A., Jacobsen, K., Anderson, V., Verlinden, S., Conneway, R., Impact of biocontainers on plant performance and container decomposition in the landscape. *Hortech*, 25, 2014.
50. Liwarska-Bizukojc, E., Effect of (bio)plastics on soil environment: A review. *Sci. Total Environ.*, 795, 148889, 2021.

29

Bioplastics, Biocomposites and Biobased Polymers—Applications and Innovative Approaches for Sustainability

V. P. Sharma[1]*, Anurag Singh[2], Neha Srivastava[3], Prachi Srivastava[2] and Inamuddin[4]

[1]*CSIR – Indian Institute of Toxicology Research, Lucknow, India*
[2]*Amity Institute of Biotechnology, Amity University Uttar Pradesh – Lucknow Campus, Lucknow, U.P., India*
[3]*Excelra Knowledge Solution Pvt Ltd, NSL Arena, Uppal, Hyderabad, Telangana, India*
[4]*Advanced Functional Materials Laboratory, Department of Applied Chemistry, Zakir Husain College of Engineering and Technology, Faculty of Engineering and Technology, Aligarh Muslim University, Aligarh, India*

Abstract

Increased environmental hazard caused by plastics and other nonbiodegradable polymers are a major issue of international concern and there are continued efforts for development of items which are eco-friendly and cost-effective. Bioplastic, biocomposites, and biobased polymeric products are the most innovative eco-friendly materials developed to overcome the limitations of presently available products. In spite of the post–COVID-19 implications on the economy the bioplastics market is anticipated to reach USD 5643 million by 2028 from the present approximately 4000 million indicating a CAGR of 5.9%. Similarly, biocomposites market too is expected to reach up to USD 36.76 billion by next 2 years. In view of compliances and effects toward sustainable development goals (SDGs), we need to increase awareness among the masses with reference to sustainable and cost-effective products. The novel materials and translational science may also contribute in development of packaging solutions, biomaterials, medical implants with desired properties and minimal toxicological effects upon administration.

The ISO 17088:2021, European Union, ASTM D6400-04 etc. specification need to be enforced in view of regulatory requirements and infrastructure to be standardized, validated as well as established for monitoring and effective management. The innovative and sustainable products have an array of application ranging from agriculture and food packaging to automotive industry. These materials are not only a boon to solve the current environmental crisis but also they are highly versatile in their application, which makes them the choice material. The futuristic smart and intelligent technologies are a ray of hope for better future, sustainable holistic developments in terms of environmental safety.

Keywords: Awareness, bioplastics, compliance, environmental, sustainable, toxicity

---

**Corresponding author*: vpsitrc1@rediffmail.com

Inamuddin and Tariq Altalhi (eds.) Handbook of Bioplastics and Biocomposites Engineering Applications: 2nd Edition, (619–628) © 2023 Scrivener Publishing LLC

29.1 Introduction

Polymeric products are integral and indispensable products of the modern generation, and thus, we are weaved around with them. On an average, in India, a person consumes 11 kg of plastics every year, compared to any other western countries, this figure is fairly low. It is astonishing that an average American consumes approximately 109 kg plastics per year [1]. It is worthless to discuss on data or figures, yet there is a need to overcome this problem by all means as it is well evident that plastics are nonbiodegradable material and are major cause of environmental pollution. Overuse of such material is causing serious societal issues; hence, the development of "environment friendly" plastics and polymers has attracted a lot of interest like "bioplastics," "biocomposites," and other "biobased polymers" [2].

Bioplastics are the polymers of the current era and are an important contribution of the material world [3, 4]. Bioplastics, in general, are made up of sugars, corn, potatoes, etc. [5] and are made by utilization of a range of microorganisms [6]. When we discuss about bioplastics, there are a few types of bioplastics:

- **Photodegradable plastics**—these types of bioplastics have a light sensitive group attached directly onto their backbone as an additive [7].
- **Compostable bioplastics**—these are biologically decomposed during a standard composting process without leaving any visible toxic remains [8].
- **Biodegradable bioplastics**—these are fully degraded by the actions of degrading microorganisms without leaving any toxic left overs. These types of bioplastics are very promising due to their actual utilization by microbes [7].

Biocomposites, on the other hand, are broadly defined as materials developed by conglomeration of natural/biofibers and petroleum-derived nonbiodegradable polymers. The consciousness towards environment and sustainability has promoted the need for biofibers rich in cellulose from grasses, leaves, fruits, and other plant regions (coconut, hemp, banana, sisal, bamboo, rice, etc.) to enable fabrication of fast moving consumable goods

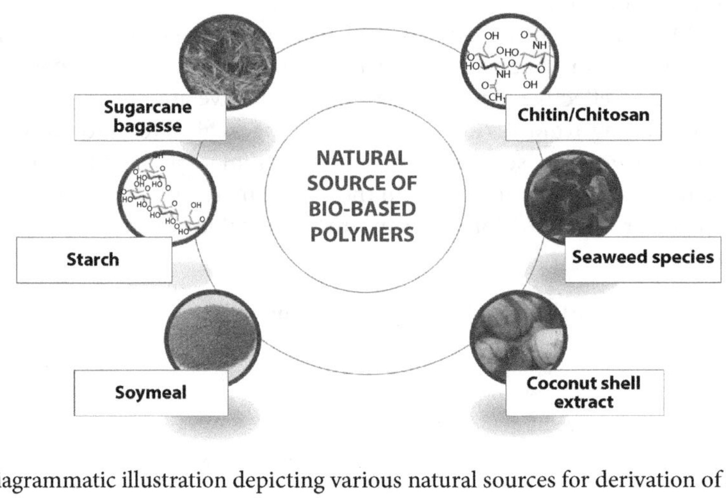

Figure 29.1 A diagrammatic illustration depicting various natural sources for derivation of bio-based polymers.

(FMCG) product for day to day applications including innovative packaging (Figure 29.1) [9–11]. Few bioproducts have limitations of moisture resistance, fiber/matrix incompatibility, thermal stability, processing, and anisotropic properties.

Although the biobased polymers and biodegradable polymers are usually commonly used terms, there are distinct differences between the two polymers. In biodegradable polymers, the physical and compound properties of molecules disintegrate or decomposed in exposure with microorganisms, oxygen-consuming carbon dioxide measures, methane (anaerobic cycles), and water (oxygen consuming and anaerobic measures), whereas, biobased polymers may be either biodegradable or nondegradable [12].

The beginning material for the creation of the biobased polymers and bioplastics (PLA, PHA) is derived from plant materials that are sustainable and renewable every year. This implies that, on a basic level, both these aliphatic polyesters can be delivered in a reasonable way. Since they are biodegradable, these biobased plastics can be changed over back to CO_2, which is then fixed by plants through photosynthesis. In this way, the creation of PLA and PHA can be viewed as, carbon neutral or, zero-discharge measures, whereby the net measure of carbon in the climate stays consistent over the long haul and on a worldwide scale [13].

29.2 Characteristics of Biobased Polymers

In general, any biobased polymer, like bioplastics and biocomposites, is used as an alternate of plastic and other environmentally hazardous materials, and these materials have a lot of functionality thus, biobased polymers should be efficient, effective, clean and recyclable. That is the first and foremost requirement that anyone looks for. The packaging industry too is looking for such materials which generate minimal waste throughout the life cycle of the product. A material ready to use in any food and beverage packaging industry must have a few properties and only then it will be considered safe for use. These properties include UV resistance, antifogging property, water resistance, printability, thermoforming properties, etc. The cost of the material is also an important factor. Thus, a biopolymer must have mechanical strength, temperature and gaseous barrier, permeability, and easy to dispose and biodegradability. The biopolymers have varied degree of plasticity based on chemical structure, molecular weight and functional groups. The thermoplastic starch (TPS) is combined with PLA and improved for mechanical properties through processing aids and are made compatible for applications in biomedicine and food packaging. These products must comply with the regulatory specifications of International Standardization Organization, as well as ASTM D6400; D6868 [14].

29.3 Biobased Polymers and Bioplastics Sustainability

The commonly used additive viz. plasticizers, stabilizers, etc., which are used to upgrade the properties, may have adverse effect to the environment and human health, if they are above the permissible limit. The microparticles present in plastics are easily spread in environment either through wind, water, birds, or by organic entities. Even sometimes these micro particles also block the drain and channels causing water and sewage choke problem. Broad littering in the climate leads to serious health hazard that result in serious health diseases, injuries, and even in some cases, death. These adverse consequences can be overcome by using recyclable sacks produced from regular filaments or utilizing bad quality and single use articles [15, 16].

29.4 Biodegradation and Standardization of Bioplastics and Biobased Polymers

Based on evaluation of biodegradation of various bioplastics, biopolymers get the more remarkable attention in the market as they match with the standards and fundamental requirements for ecological effect or evaluations of the lifecycle for its legitimate removal [17]. In addition to this, some innovative and smart plastics do not require long time span for degradation but unlikely, these are known to be an expensive bioplastics. To overcome cost limit, the bioplastics are mixed with normal biomaterials. [18]. Standard protocols are available for assessing the compliance to complete aerobic/anaerobic degradation. Standardization leads to two different systems for evaluation. First, the testing method that described the methodology criteria and the protocol for performing that should be followed. Second, the specification that has defined a set of parameters and functions for pass and fail that is required for product or material in order to meet desired standards.

29.4.1 Standard EN 13432

This standard is required for packaging recoverable via compositing and biodegradable. The testing for certification as per EN 13434 and EN 14995 requires chemical test, ecotoxicity test, disintegration, and biodegradability in control composting conditions. The biodegradability studies must be scientifically evidenced to reveal at least 90% of the organic material conversion into carbon dioxide within duration of 6 months (180 days approximately). The certification of EN ensures that the products maybe industrially composted, including associated components, i.e., colors, labels, glues in case of packaging products. This standard is basically designed for industrial composting and anaerobic digestion.

29.4.2 Standards for Oxodegradation

Oxodegradation is described by CEN in TR15351 as a degradation caused by the oxidative cleavage of macromolecules. This category includes ordinary plastics that can be degraded abiotically and cannot be biodegradable unless for a long period of time. At present, there are no internationally set parameters, certification, and standards available for oxodegradation. It is just merely appealing marketing. The oxodegradation may help to reduce the problems of microplastics, littering, and adverse implications due to climatic changes. It does not fulfill the requirement for the international standards for plastics packaging and recovery due to its long and time-taking process for biodegradation, which adversely affects the quality and market value of the compost and is unable to release plastics.

29.4.3 Australasian Bioplastics Association

The Australasian Bioplastics Association (ABA) association provides two verification programs for industry and individuals who want to verify accordance to bioplastics appropriate for home and commercial use. In order to recognize compostable packaging and disposal with authenticity and verification of claims of compliance ABA has launched the seedling logo certification system. The two salient specifications in this aspect are AS4736-2006 and European Norm(s) EN13432. The compostable bags made from cornstarch may be used as garbage bags to eliminate regular plastics dumping and leaching of new chemical entities, which are responsible for adversely affecting the biosystem. The multiple requirements of standards include disintegration testing, biodegradable testing, heavy metals limits, and quality of compost and toxicity prediction as well. At present, over 120 companies globally are certified by ABA.

29.4.4 Australian Packaging Covenant Organization

The Australian Packaging Covenant Organization (APCO) works in collaboration with government and businesses to decrease the effect of the environment on packaging in Australia. It is a coregulatory initiative for businesses by APCO to manage packaging waste in a sustainable manner. The APCO provides detailed guidelines for managing environmental wastes. Additionally, it also shares responsibilities and commitment toward reducing the effect of packaging over the environment. In 2018, Australia established National Packaging Targets 2025 for creating targets for managing packaging. The major targets by 2015 are 100% of reusable, recyclable or compostable packaging, 70% of recycled or composted plastics packaging, average 50% of packaging will include recycled content, and discontinue of single-use plastic packaging. It is strictly applied to all packaging manufactured, used and sold in Australia.

29.5 Application of Bioplastics, Biocomposites, and Biobased Polymers

Since plastics and other nonbiodegradable products pose serious health hazards to terrestrial and aquatic environments, an immediate replacement is needed for plastics and other nonbiodegradable products. With the emergence of biobased and biodegradable products has given the renewable resources to society and also reduces the effect of harmful gases during the production process [19]. In the last few years, the implementation and worthiness of biodegradable products have made an extraordinary impact on society. The biobased polymeric materials have shown the leading edge in research and development in various areas including medicine, packaging, agriculture, and the automotive industry.

29.5.1 Application in Medicine

In the modern scenario, the medical industry is growing with an entirely new perspective for net zero targets in place and environmental sustainability. The medical implants, devices, packaging solutions, and delivery pathways are exploring the new directions for environmental compliances and minimization of waste.The recent application of biodegradable polymers is in surgical implants especially in orthopedics, cardiac and other surgical approaches, noninvasive techniques. The scaffolds used in tissue engineering for vascular and neural tissues, bone, skin, ligament, ocular fields, etc. are designed for artificial implants, etc. They may be used for regeneration of new tissues and creating temporary barriers as per requirements. These are generally required to be biocompatible, bioabsorbable, and meet biological as well as mechanical properties for the specific situations. The material of choice may vary from poly-D, L-lactic acid (PDLLA), poly (lactic-co-glycolic acid) (PLGA) to poly-hydroxyalkanoates (PHA) for sutures, dental supports, vascular tissues, etc. [20–23]. Due to their falsified property to change into various shapes, the synthetic polymers are commonly used in medical implantation. PGA fabrics are widely used in tissue regeneration as scaffolding matrices [24]. In addition, chitin poly-L-leucine and chitin are generally used in preparation of skin substitutes and in dressing of wounds [25]. The alginate gel is frequently used in drug delivery systems as a controller [26]. Besides this, the polyanhydrides are extensively used in treatment of eye diseases [27]. The PHA-based nanomedicine is extensively implemented in cancer cell adhesion as an extracellular matrix ligand in a drug delivery system [28]. While the utilization of nanoparticle-based poly hydroxy octanoate/polyethylene glycol (PHO/PEG) loaded with anticancer drug paclitaxel using nanoprecipitation reflects depletion in colon carcinoma in mice cells [29]. Thus, PHA-based nanoparticles have proven activity, as well as bioavailability in drug delivery systems and therapeutic intervention.

29.5.2 Application in Packaging

Packaging is known to be the most focused area of the biodegradable polymers. It has been widely used in packaging products [30]. Internationally, regulatory agencies are taking special attention in view of sustainable development goals target of 2030 that undue recycling, dumping and littering of plastics is avoided. This is to encourage better strategies for environmental control and management. Most widely and extensively used biodegradable plastics packaging material is starch and its blends [31]. Starch is used in food products as edible coating material protect from external risk. The use of sunflower antioxidant extract maybe used as an additive in starch-based film utilized in making of renewable food packaging materials [32]. The polylactic acid-based products from renewable sources viz. sugarcane/corn starch are used as packaging materials alternatives to petroleum-based plastics [33]. Besides this PCL is used as soft compostable packaging. The chitosan is used in paper-based packaging and is frequently employed in the food industry as an oil barrier or fat barrier packaging [34]. In addition, the chitosan-based film showed its effectiveness as a food preservative and in microbial packaging, whereas PBAT, a biodegradable plastic of petrochemical origin, has proven mechanical and rheological properties widely employed in films to improve its antioxidant and anti-microbial activities [35–37].

29.5.3 Application in Agriculture

In agriculture, starch, cellulose, chitin, alginic acid, and lignin have been widely used as the natural polymers that controlled the release systems [38]. In marine agriculture, the biobased polymers are widely used in making fishing nets and ropes. In continuation, the biodegradable polymers have proven to be effective in low tunnel cultivation and mulching that leads to enhanced sustainability and eco-friendly agricultural activities. The starch-based films are generally used as mulch film in agriculture because of its property to degrade into nontoxic products [39].

29.5.4 Other Applications

In continuation, the biodegradable polymers have been also widely used in many other applications as well. PLA-based material is popularly used in 3D printing. The PLA is a very appealing material having low melting point and with less energy requirement for printing [40–42, 47, 51]. PLA is a perfect option for 3D printing because of some of its very useful features, such as low cost, low co_2 emission, and reduced wastage of raw materials. *Mcl*PHAs, a genesis of chiral (*R*)-3-hydroxyalkanoic acids (R3HAs), used as antecedent in the synthesis of pharmaceutical drugs, vitamins, and antibiotics. The state-of-the-art polymeric products used in fabrication of medical implants or devices must ensure biocompatibility, toxicological studies as per ISO 10993. This is recommended as per OECD, FDA, and ISO for obtaining compliance adherences, and certification. Carboxylated *mcl*PHAs may be widely used in making stable nanoparticle suspensions. It is generally employed in xerographic tonners, coating, hot adhesive, and also in cosmetics and toothpaste as microbeads [43–51].

29.6 Conclusion

Moving ahead in the future, we cannot underestimate the risks and challenges posed by plastics and artificial and synthetic nonbiodegradable polymeric materials. It is a major challenge in front of our community and we can only win over this challenge by acknowledging it and working in the right direction with scientific and community support. It is high time that we all start using biobased polymers and replace plastics and other hazardous materials. Bioplastics, biocomposites, and biobased polymers are not only safe and biodegradable but also have the potential to be multifunctional, from agricultural sector to medicine to food packaging, they can be employed almost in every sphere of life.

The changing climatic variations have impacted the socioeconomic system across the globe. The usage pattern with changing lifestyle and modern amenities are interlinked for behavioral changes of the mankind. It is anticipated that in view of fast changing regulations at internationals platforms, the tradition plastic items applications have a paradigm shift. The market for flexible packaging have been positively impacted due to COVID-19 and situations thereafter, due to increased usage of personal and healthcare products, packed food and beverages, distribution and supply through online systems, nutraceuticals, etc. The demand for bioplastics is also associated to changing demands in information technology tools and electronic products. There is mounting demand for usage of nonrenewable resources and development of eco-friendly materials by scientific and industrial communities.

Organizations, like ASTM, ISO, AORA, APCO, etc., are working tirelessly in maintaining the correct standards and SOPs for their efficient production and infrastructure development and close monitoring. Biobased polymers will also be helpful in achieving the sustainable development goals (SDGs) and reduce the emission of harmful toxicants in the environment. The toxicants that are currently being released due to the manufacturing and usage of plastics and other significantly harmful polymers have been known to cause serious disorders in people especially children thus, replacing them with biopolymers may overcome these health risks. In view of holistic approaches and development, the bio-based products have a bright future to serve as alternatives and future needs.

References

1. Kalia, V.C., Raizada, N., Sonakya, V., Bioplastics. *J. Sci. Ind. Res.*, 59, 433, 2000.
2. Arikan, E.B. and Ozsoy, H.D., A review: Investigation of bioplastics. *J. Civ. Eng. Arch.*, 9, 188, 2015.
3. Mohanty, A.K., Misra, M., Drzal, L.T., Sustainable bio-composites from renewable resources: Opportunities and challenges in the green materials world. *J. Polym. Environ.*, 10, 19, 2002.
4. Bioplastics Market Size, Share & Trends Analysis Report by Product (Biodegradable, Non-biodegradable), By Application (Packaging, Automotive & Transportation, Textile), By Region, And Segment Forecasts, 2021–2028, 2021. https://www.grandviewresearch.com/industry-analysis/bioplastics-industry.
5. Karana, E., Characterization of natural and high-quality materials to improve perception of bio-plastics. *J. Clean. Prod.*, 37, 316, 2012.
6. Luengo, J.M., Garcia, B., Sandoval, A., Naharro, G., Olivera, E.R., Bioplastics from microorganisms. *Curr. Opin. Microbiol.*, 6, 251, 2003.
7. El-Kadi, S., Bioplastic production from inexpensive sources: Bacterial biosynthesis, cultivation system, in: *Production and biodegradability*, VDM Publishing, Germany, 2010.
8. Sarasa, J., Gracia, J.M., Javierre, C., Study of the bio-disintegration of a bioplastic material waste. *Bioresour. Technol.*, 100, 3764, 2009.
9. Wilkinson, S.L., Plants to bugs: Buzz off. *Chem. Eng. News*, 79, 42, 2001.
10. John, M.J. and Thomas, S., Biofibres and biocomposites. *Carbohydr. Polym.*, 71, 343, 2008.
11. Biocomposites Market by Fiber (Wood Fiber and Non-Wood Fiber), Polymer (Synthetic and Natural), Product (Hybrid and Green), End-Use Industry (Transportation, Building & Construction and Consumer Goods), and Region - Global Forecast to 2022, 2021. https://www.marketsandmarkets.com/Market-Reports/biocomposite-market-258097936.html.
12. Babu, R.P., O'connor, K., Seeram, R., Current progress on bio-based polymers and their future trends. *Prog. Biomater.*, 2, 1, 1–16, 2013.
13. Sudesh, K. and Iwata, T., Sustainability of biobased and biodegradable plastics. *Clean–Soil, Air, Water*, 36, 433, 2008.
14. Pandit, P., Nadathur, G.T., Maiti, S., Regubalan, B., Functionality and properties of bio-based materials. *Bio-based Mater. Food Packag.*, 81–103, 2018.
15. Shamsuddin, I.M., Jafar, J.A., Shawai, A.S.A., Yusuf, S., Lateefah, M., Aminu, I., Bioplastics as better alternative to petroplastics and their role in national sustainability: A review. *J. Biosci. Bioeng.*, 5, 63, 2017.
16. Edwards, S., A conceptual framework for sustainable product design, in: *Beyond child's play: Sustainable product design in the global doll-making industry*, pp. 29–44, Baywood Publishing Company, Incorporated, New York, 2010.
17. Hottle, T.A., Bilec, M.M., Landis, A.E., Sustainability assessments of bio-based polymers. *Polym. Degrad. Stab.*, 98, 1898, 2013.
18. Wu, C.S., Preparation, characterization, and biodegradability of renewable resource-based composites from recycled polylactide bioplastic and sisal fibers. *J. Appl. Polym. Sci.*, 123, 347, 2012.

19. Andreopoulos, A.G. and Theophanides, T., Degradable plastics: A smart approach to various applications. *J. Elastomers Plast.*, 26, 308, 1994.
20. Malafaya, P.B., Silva, G.A., Reis, R.L., Natural–origin polymers as carriers and scaffolds for biomolecules and cell delivery in tissue engineering applications. *Adv. Drug Deliv. Rev.*, 59, 207, 2007.
21. Yamaoka, T., Tabata, Y., Ikada, Y., Body distribution of intravenously administered gelatin with different molecular weights. *J. Control Release*, 31, 1, 1994.
22. Hokugo, A., Ozeki, M., Kawakami, O., Sugimoto, K., Mushimoto, K., Morita, S., Tabata, Y., Poster 27-Potentiality of gelatin hydrogel in promoting the bone repairing activity of platelet-rich plasma (PRP); An experimental study in rabbits. *J. Oral. Maxillofac. Surg.*, 8, 95, 2003.
23. Nair, L.S. and Laurencin, C.T., Biodegradable polymers as biomaterials. *Prog. Polym. Sci.*, 32, 762, 2007.
24. Muzzarelli, R.A., Chitin and its derivatives: New trends of applied research. *Carbohydr. Polym.*, 3, 53, 1983.
25. Khor, E. and Lim, L.Y., Implantable applications of chitin and chitosan. *Biomaterials*, 24, 2339, 2003.
26. Augst, A.D., Kong, H.J., Mooney, D.J., Alginate hydrogels as biomaterials. *Macromol. Biosci.*, 6, 623, 2006.
27. Tiainen, J., Veiranto, M., Suokas, E., Törmälä, P., Waris, T., Ninkovic, M., Ashammakhi, N., Bioabsorbable ciprofloxacin-containing and plain self-reinforced polylactide-polyglycolide 80/20 screws: Pullout strength properties in human cadaver parietal bones. *J. Craniofacial Surg.*, 13, 427, 2002.
28. Kim, H.N., Lee, J., Kim, H.Y., Kim, Y.R., Enzymatic synthesis of a drug delivery system based on polyhydroxyalkanoate-protein block copolymers. *Chem. Commun.*, 7104–7106, 2009.
29. Kim, H.Y., Ryu, J.H., Chu, C.W., Son, G.M., Jeong, Y.I., Kwak, T.W., Kim, D.H., Chung, C.W., Rhee, Y.H., Kang, D.H., Kim, H.W., Paclitaxel-incorporated nanoparticles using block copolymers composed of poly(ethylene glycol)/poly(3-hydroxyoctanoate). *Nanoscale Res. Lett.*, 9, 1, 1–10, 2009.
30. Petersen, K., Nielsen, P.V., Bertelsen, G., Lawther, M., Olsen, M.B., Nilsson, N.H., Mortensen, G., Potential of biobased materials for food packaging. *Trends Food Sci. Technol.*, 10, 52, 9(1), 1–10, 2009.
31. European Bioplastics, in: *Bioplastics market development update 2019*, European Bioplastics, Berlin, Germany, 2019.
32. Menzel, C., Gonzalez-Martinez, C., Chiralt, A., Vilaplana, F., Antioxidant starch films containing sunflower hull extracts. *Carbohyd. Polym.*, 214, 142, 2019.
33. Auras, R., Harte, B., Selke, S., An overview of polylactides as packaging materials. *Macromol. Biosci.*, 4, 835, 2004.
34. Sinclair, R.G., The case for polylactic acid as a commodity packaging plastic. *J. Macromol. Sci. - Pure Appl. Chem.*, 33, 585, 1996.
35. Da Silva, J.B.A., Santana, J.S., Lucas, A.D., Passador, F.R., Costa, L.A.D., Pereira, F.V., Druzian, J.I., PBAT/TPS-nanowhiskers blend preparation and application as food packaging. *J. Appl. Polym. Sci.*, 136, 47699, 2019.
36. Ferreira, R.R., Souza, A.G., Rosa, D.S., Essential oil-loaded nano-capsules and their application on PBAT biodegradable films. *J. Mol. Liq.*, 337, 116488, 2021.
37. Ellingford, C., Samantaray, P.K., Farris, S., McNally, T., Tan, B., Sun, Z., Huang, W., Ji, Y., Wan, C., Reactive extrusion of biodegradable PGA/PBAT blends to enhance flexibility and gas barrier properties. *J. Appl. Polym. Sci.*, 139, 51617, 2022.
38. Rustgi, C. and Rustgi, R., Biodegradable polymers. *Prog. Polym. Sci.*, 23, 1273, 1998.

39. Briassoulis, D., An overview on the mechanical behaviour of biodegradable agricultural films. *J. Polym. Environ.*, 12, 65, 2004.
40. Benwood, C., Anstey, A., Andrzejewski, J., Misra, M., Mohanty, A.K., Improving the impact strength and heat resistance of 3D printed models: Structure, property, and processing co relationships during fused Deposition modeling (FDM) of poly (lactic acid). *ACS Omega*, 3, 4400, 2018.
41. Harris, A.M. and Lee, E.C., Improving mechanical performance of injection molded PLA by controlling crystallinity. *J. Appl. Polym. Sci.*, 107, 2246, 2008.
42. Saeidlou, S., Huneault, M.A., Li, H., Park, C.B., Poly(lactic Acid), crystallization. *Prog. Polym. Sci.*, 37, 1657, 2012.
43. Możejko-Ciesielska, J., Szacherska, K., Marciniak, P., Pseudomonas species as producers of eco-friendly polyhydroxyalkanoates. *J. Polym. Environ.*, 27, 1151, 2012.
44. Radivojevic, J., Skaro, S., Senerovic, L., Vasiljevic, B., Guzik, M., Kenny, S.T., Maslak, V., Nikodinovic-Runic, J., O'Connor, K.E., Polyhydroxyalkanoate-based 3-hydroxyoctanoic acid and its derivatives as a platform of bioactive compounds. *Appl. Microbiol. Biot.*, 100, 161, 2016.
45. Peprah, B.A., Ramsay, J.A., Ramsay, B.A., Stabilization and characterization of carboxylated medium-chain-length poly(3-hydroxyalkanoate) nanosuspensions. *Int. J. Biol. Macromol.*, 108, 902, 2018.
46. Andrew, J. and Dhakal., H.N., Sustainable biobased composites for advanced applications: Recent trends and future opportunities–A critical review. *Composites Part C*, 7, 100220, 2022.
47. Biswal, T., Biopolymers for tissue engineering applications: A review. *Mater. Today: Proc.*, 41, 397, 2021.
48. Mazitova, A.K., Aminova, G.K., Buylova, E.A., Zaripov, I.I., Vikhareva, I.N., Biodegradable polymer materials and modifying additives: State of the art. Part III. *Development*, 73, 78, 2021.
49. Vu, D.H., Åkesson, D., Taherzadeh, M.J., Ferreira, J.A., Recycling strategies for polyhydroxyalkanoate-based waste materials: An overview. *Bioresour. Technol.*, 298, 122393, 2020.
50. Cameron, R.E. and Kamvari-Moghaddam, A., Synthetic bioresorbable polymers, in: *Degradation rate of bioresorbable materials*, pp. 43–66, Woodhead Publishing, Sawston, Cambridge, 2008.
51. Boddu, S.H., Bhargav, P., Karla, P.K., Jacob, S., Adatiya, M.D., Dhameliya, T.M., Ranch, K.M., Tiwari, A.K., Polyamide/poly (amino acid) polymers for drug delivery. *J. Funct. Biomater.*, 12, 58, 2021.

30

Recycling of Bioplastics: Mechanism and Economic Benefits

Nadia Akram*, **Muhammad Saeed, Muhammad Usman, Tanveer Hussain Bokhari, Akbar Ali and Zunaira Shafiq**

Department of Chemistry, Government College University Faisalabad, Faisalabad, Pakistan

Abstract

The bioplastic industry is emerging continuously, the public awareness is a key factor in the flourishing of bioplastic products. The large consumption of bioplastics is generating a huge pile of waste material. It is very difficult to dispose of this waste material of bioplastic, moreover, the single-use consumption is not suitable for biomaterials. Hence, there is a great need to recycle the bioplastics in order to attain the maximum utilization of these bioplastics. The chapter discusses the overview of available bioplastics and emphasis on the need of the recycling of the bioplastics and associated economic and environmental advantages with recycling. It concisely narrates all the associated features of bioplastics as well.

Keywords: Market of bioplastics, mechanical recycling, chemical recycling, economic aspects

30.1 Overview of Popular Bioplastics

The most popular commercial material of the current age is the fossil-derived or petroleum-based polymers, which are commonly known as plastics, whereas, the bioplastics are the emerging materials derived from renewable resources, including plants or vegetable or biominerals. Bioplastic are the material obtained from biomass making an ecologically suitable impact. Numerous bioplastics of starch, cellulose, alginate, chitin are in development that can compete with synthetic plastic such as polyethylene furanoate (PEF). As the bioplastics are made from naturally occurring polymers these are biodegradable which can be disposed of biologically by the bacterial decomposition. The biodegradable plastics can be entirely of biomaterials or these may be the synthetic polymers containing a small fraction of biomaterial as well. Whereas, the synthetic originated bioplastics such as polyamides, polyhydroxy alkanoates or polysaccharides are decomposed through composting activity of bacteria and microorganisms in a controlled and systematic manner. Whereas, the oxo-degradable are the synthetic plastics which are degradable naturally. The oxo-degradable plastics are decomposed by heat, sunlight and oxygen [1]. There are several applications of

**Corresponding author*: nadiaakram@gcuf.edu.pk

Inamuddin and Tariq Altalhi (eds.) Handbook of Bioplastics and Biocomposites Engineering Applications: 2nd Edition, (629–648)

the bioplastics. Bioplastic packaging is gaining interest now a days. At present, 300 different types of bioplastics have been produced. Bioplastics are different in nature from the materials they are synthesized, their processing, mode of decomposition and degradation makes it difficult to estimate that which factor effect the handling and recycling of the wastes of these products conveniently. Bioplastics are potential alternative to petrochemicals and they are ecofriendly plastics. As plastics has a number of applications and are an important part of lives. Its usage has been enhanced twenty times and is expected to increase twice within the next 20 years. Globally, annual production of plastics is 330 million tons. Production, usage, wastage and disposal of plastics have been evolved as a nuisance in the environment. If plastics are disposed of improperly then it would result in the loss of lives of millions of animal lives and decreases their fertility rates as well. In such situation, the bioplastics which are delivering the similar products and functions as that of conventional plastics are convenient to use. The bioplastics decrease the release of greenhouse gases and enhance the sustainability of environment. Bioplastics are also a great solution to the pollution and a long-term public awareness will be helpful to utilize the material in true spirit [2]. Biodegradable plastics decompose naturally in the environment. This is achieved due to the ability of environmental microbes to rupture the structures which make the bioplastics. The resultant plastics which are created are ecofriendly as compared to conventional plastics that are thrown every day. Bioplastics make biodegradable plastics that are the products manufactured from renewable raw materials as compared to the virgin items that are harvested at a potential cost for the environment. Two forms of bioplastics are used today, which are injection and solid molded bioplastics. Solid form of bioplastics is used for manufacturing bags, food containers and water bottles, which are consumed. Injection molded bioplastics manufactures other plastics that are consumed [3].

Bioplastics arc thc groups of products each with different applications, properties, manufacturing procedures and raw materials. Some of the types of bioplastics are listed below;

30.1.1 Starch-Based Bioplastics

Initially these bioplastics were invented from the starch of maize, which were commercially marketed as well. Such plastics were developed by the mixing of biodegradable starch polymer compounds with petrochemical polymer plastic. At present, polymers that are based on starch are manufactured from corn, potato, tapioca and wheat. Bioplastics of such type can also be produced from starch by microorganism fermentation technique. Examples of such bioplastics that are produces from starch through fermentation of microorganisms are polylactic acid (PLA) and polyhydroxyalkanoates (PHA). However, raw materials like starch and carbohydrates are used for making new products of bioplastic and products. Examples of such new generation of bioplastics are polyvinyl chloride (PVC), polyolefins, polyethylene (PE), and polyethylene terephthalate (PET) that is partially biobased [4]. When these bioplastics are disposed, polymer starch molecules are degraded by microorganisms and the plastic polymer then disintegrates. Chemical and physical properties of substituted starch bioplastic is used for the purpose of various applications. Additionally, the collection of plastic residue in water and soil that is non degradable causes pollution in environment. Such types of bioplastics that are manufactured directly from starch offers a better replacement of conventional plastics.

30.1.2 Cellulose-Based Bioplastic

Glucose monomers combine in the form of beta 1-4 glycosidic linkage to form cellulose as polymers. Cellulose is primarily present in plant's cell wall. Cellulose is produced through chemical modification or extraction of natural occurring cellulose. At present, twenty percent of chemical paste is used in the preparation of cellulosic esters. Cellulosic esters are manufactured through esterification process of organic acid and cellulose. At industrial level, cellulose esters are used in the form of packing films, textile fibers, cigarette filters, pharmaceutical and other packaging applications. Regeneration of cellulose is carried out through dissolution of cellulose and is structured in the form of films or fibers.

30.1.3 Polylactic Acid (PLA)-Based Bioplastics

Polylactic acid (PLA) is the most important bioplastics in today's market. The basis of PLA is lactic acid and these are manufactured through starch microbial fermentation that is obtained from potato, cassava, sugar beet, maize and sugar cane. Limitations which are encountered from starch substituted bioplastics are overcome by PLA bioplastics. Initially, the starch obtained from plant is changed by microorganism into lactic acid monomer and then these monomers are treated chemically in order to link these monomers to form long chain of polymer. PLA plastics appears like petrochemical plastics but these are truly biodegradable plastics. Advantages of PLA plastics are higher stability, rigidity, thermoplasticity, transparency and high performance as compared to the conventional plastics. PLA has a number of applications in packaging of food because of the properties that are similar to polypropylene and polyethylene. Furthermore, manufacturing of PLA plastics secures two third of energy as compared to the energy used for making fossil fuel-based bioplastics. During landfill degradation PLA emit 70% less greenhouse gases than conventional plastics. Recently PLA blends and PLA occur in granular form for making molded parts, containers of cups, drinks, bottles and other items of everyday usage. Applications of PLA plastics is spreading in a variety of fields like textile, cosmetics, medical and household appliances. Furthermore, PLA plastics are used in making dashboards, door tread plates in automobile industries.

30.1.4 Polyhydroxy Alkanoate-Based Bioplastics (PHA)

PHA is produced through microbial fermentation of starch that is based on plant origin. Physical and chemical properties of PHA plastic are similar to polyethylene, polypropylene and polyesters. Different groups of polyhydroxylalkanoates are used for making plastics such as . polyhydroxy valerate (PHV). Generally specific microorganisms collect PHA polymer granules as 80% of their body weight in the cell's cytoplasm under specific conditions. PHA granules are formed through carbohydrates transgenic fermentation in their body. Purification of PHA polymer is done through breakage of the cell wall and then it is harvested by the extraction method that is aqueous based with organic solvents. Then, PHA biopolymer is converted in the lattice form after purification for manufacturing of plastics. Several plants are modified through genetic engineering to produce PHA. Scientists have produced genetically modified PHA from genetically modified maize and the plants are cultivated for plastic crops. At present, bioscience engineering company that is European based, named as Metabolix has launched PHA production plant which is biobased. At present, biopolymers like alginates and polyhydroxyalkanoates are produced through anaerobic digester of digestate and aerobic wastewater treatment produce sludge as well. Biopolymer production from sludge produce biopolymer and it has the advantage of lower costs from inexpensive waste sludge. Plastics that are based on PHA are biodegradable and they are degraded within a one year period through varied species and genera of fungi and bacteria under anaerobic and aerobic conditions. PHA's biodegradation produce water and carbon dioxide and is natural and non-toxic polymer. PHA has a number of applications in medical implants, agriculture and food packaging. Globally, PHA are present as injection molding grades and films along with the blow molding and extrusion grades [5].

30.1.5 Organic Polyethylene

Organic polyethylene is produced through fermentation process. In this fermentation process, raw agricultural materials like sugar cane and corn ferment to form a polyethylene which in turn form bioplastics materials.

30.1.6 Protein-Based Bioplastics

Bioplastics containing proteins are made from the sources of protein like albumin, soy, casein, milk and wheat gluten. It is used for making uncontaminated materials and edible films. Thermal, antimicrobial and viscoelastic properties of proteins are studied with plasticizers like glycerol, rubber and water. Albumin has similar viscoelastic and thermal properties with bioplastics while soy has different thermal and viscoelastic properties. For antimicrobial property, albumin-glycerol is best bioplastics as there are no bacterial growth observed in plastics when inoculated for 24 hours. So, protein-based bioplastics are used for biomedical and food packaging materials.

30.1.7 Drop-In Bioplastics

These bioplastic are composed of partially or completely biomass or fossil materials however they produce the replica of petrochemical-based plastics. Commonly, these plastics are manufactured from sugar beet, sugarcane, corn and are hybrid of petrochemical plastics. Recently the utilization and production of biobased drop-in materials are increasing greatly. This class of plastic is similar to regular plastics and have exactly similar applications. Bio-PET is used in making bottles and bio-LDPE is used in making films. Succinic acid that is biobased is used for many practical applications in footwear and sports, packaging, automotive, non-woven, agriculture and fibers. Main benefit of biobased material is the reduction in carbon footprint because it is manufactured from renewable biomass. These drop-ins can also be manufactured, processed and recycled in the available facilities of traditional plastics that in turn reduces the production cost of plastics [6].

30.1.8 Fossil Fuel-Based Bioplastics

A new class known with the name of polybutyrate. Poly butyrate is used in conjunction with bioplastic materials like starch for enhancement of its applications and performance because of its rigidity and biodegradability. In recent years, development of partially biobased and new biobased version has been increased, and PBAT plastic is used for renewable resources. Examples of this are poly (butylene adipate-co-terepthalate) PBAT is biodegradable and has efficient properties like fracture resilience, higher elasticity and flexibility. It can thus be used as a substitute for the production of wraps, bags, rubbish bags and disposable packaging materials. PBAT has faster decomposition rate [7].

30.2 Recycling of Bioplastics

Recycling is the procedure of assembling of trash and thrown away materials and processing of these waste materials into brand new products. The recycling of bioplastics has become essential due to mass production of this material. Bioplastics when thrown in the trash bin along with the conventional plastics, it becomes difficult to degrade with ease. For example, most commonly used plastic is PET, which when recycled produces PET flakes. These flakes are purer and can be reused but a little amount of bioplastic in recycling of PET can contaminate it [8].

30.2.1 Background of Bioplastics Recycling

From past when recycling system was established, a number of different and innovative packing materials through bioplastics and plastics have been introduced. Different types of lightweight and new packaging plastics such as wraps and pouches are made. It decreases the amount and weight of packaging materials which are needed to supply and deliver drinks and foods. It in turn also has great impact on the environment providing benefit to it by providing less energy, decreasing the wastage of food and reduces the emission of greenhouse gas emissions. Although all these benefits complicate the recycling rate.

At present, some plastics are recycled at a rate as fragment of their counterparts. The recycling rate of plastic bottles is 29% which is a little bit greater than glass bottles whose recycling rate is 27%. This recycling rate is a little bit less than cans of aluminum whose recycling rate is 49% and the bottles used for packing detergent and milk bottles are recycled at a rate of 31%. Overall, the recycling rate of plastic bottles is very low which is approximately 13%. Now most of the plastic making companies are more committed to create more economy through reuse of plastics rather than discarding it [9]. In waste hierarchy, suitable proposal are given for reuse, reduction, and recycle for waste disposal and suitable treatment of bioplastics. Although the bioplastics that enter into municipal waste stream complicate the recycling system of plastics although there is variation of base materials. An example of it is present in America in which America's plastic making economy is committed to get circular economy by setting two goals for plastics. The first goal is that 100% of packaging through plastic will be recoverable and recyclable by 2030 and will be recycled, reused and recovered by 2040. To meet this goal, expansion of both advanced and mechanical recycling method is essential for making plastics.

Alternatives to petrochemical plastics are biochemical plastics. The space of landfill is decreased for the reduction of landfill space which in turn increases the level of CO_2 and pollution of environment. At global level, plastics are dumped incorrectly because of decrease in the waste management which causes a number of environmental as well as economic loss. Bioplastics and plastics are collected and accumulated in marine ecosystem, which in turn bioaccumulate in the bodies of humans when marine animals are eaten by them.

30.2.2 Options of Recycling

A number of opportunities are present to replace plastics that are based on petroleum with bioplastics whether these plastics are biodegradable or biobased. These opportunities are in the form of chemical or mechanical recycling of bioplastics. Routes that are optimal for recycling after their reuse and recycling are alcoholysis, pyrolysis, biodegradation and glycolysis. Mechanical recycling (MR) is given preference than chemical recycling (CR). The reason behind this is that MR consumes low cost for processing, least acidification and eutrophication, has less potential of global warming and consume low energy that is non-renewable. On the other hand, CR is diverse and depends on the polymer type but it produce additional value materials [10].

Recycling is a preferred option for the proper bioplastic management. Furthermore, incineration process is suitable than landfilling. Recycling methods involve materials, organic and chemical recycling is feasible for bioplastics. Material recycling is also called mechanical recycling (MR), and in it, waste material is reprocessed into a new material. MR is suitable for all bioplastics and its types but large supply of bioplastic waste that is unreliable make such recycling economically less efficient than that for conventional plastic recycling. MR companies transform the sorted products from waste into recyclates after passing through the processes of density separation, compounding and washing. At present, bioplastics that are produced are recycled easily with traditional plastics and it depends upon the base materials. PLA polymers can also be recycled through MR for a few times without changing its properties and recyclates are transformed into products of piping, plastics, pallets and plastic lumber. On the other hand, chemical recycling is the process in which the biopolymers wastes are regranulated and remelted for producing new product. In some cases, wastes of biopolymers are transformed into monomers which can be reused for biopolymer production [11].

Composting recycle biodegradable plastics into soil amendments that are rich in nutrients and is implemented in most of the parts of the globe. Naturally existing microorganisms transform biodegradable polymers into simple compounds under aerobic conditions and then these complexes are disintegrated into water and CO_2 under anaerobic conditions decomposed into carbon dioxide and methane. Composting is also refereed as organic recycling because it is one form of recycling of material. Composting is ecologically suitable option for bioplastic management of wastes.

30.2.3 Generation of Energy From Recycling Process

Wastes of bioplastics produce energy through many processes like pyrolysis, anaerobic digestion and incineration. All this occur after retrieval of components that are recycled from bioplastic's wastage. Bioplastics can produce energy from traditional petroleum-based bioplastics. Some bioplastics like natural cellulose starch and fiber is similar to wood containing the property of less gross calorific value (GCV). Energy is recovered from this process through incineration. Carbon dioxide released from wastes can be used to produce new products. Anaerobic decomposition of bioplastics produces the waste components which are treated with municipal waste present in solid form. In anaerobic decomposition, bioplastics are broken down into carbon dioxide and methane in anaerobic conditions. In this decomposition, biogas and digestate is also produced, biogas is utilized as renewable energy source while digestate is used for making organic manure. It is reported that biodegradable plastic causes anaerobic decomposition while biopolymers changes into methane. Furthermore, landfill gases that contain methane are further utilized for producing energy [12].

The different recycling routes are primary, secondary, tertiary and quaternary. Primary recycling is a method of recycling on scraps of plastic of high quality. It reuses the material or recycle the material mechanically and produce the original structures plastic scrap. Secondary recycling on the other hand transform the plastic waste into less demanded product mechanically through injection molding, screw extrusion and blow molding. Tertiary recycling is a chemical recycling process which degrade and depolymerize the plastic into its monomers or less useful materials. In quaternary recycling energy is recovered through incineration of waste plastics containing low grades. The process also release harmful chemical like dioxins in the atmosphere [13, 14].

Although there are a number of alternatives that are ecofriendly but for recycling of bioplastics, the sustainability and proper disposal is very important. Otherwise, all the bioplastics will end up in landfill like other waste trash or it will disturb the recycling process of conventional petroleum-based plastics. When a material for recycling is considered, footprint of plastic is reduced but it is difficult to choose a product that is degradable or compostable and can break in a shorter lifespan. Business owners of plastics and bioplastics considers the materials which they use in their business operations and their way of disposal. As bioplastics are important alternative of conventional plastics so their recycling and manufacturing must be carefully treated for all types of wastes. Although bioplastics are not the alternative to save the planet from pollution but they are suitable and sustainable alternatives of plastics. The steps involved in the recycling of bioplastics is given in Figure 30.1 [15].

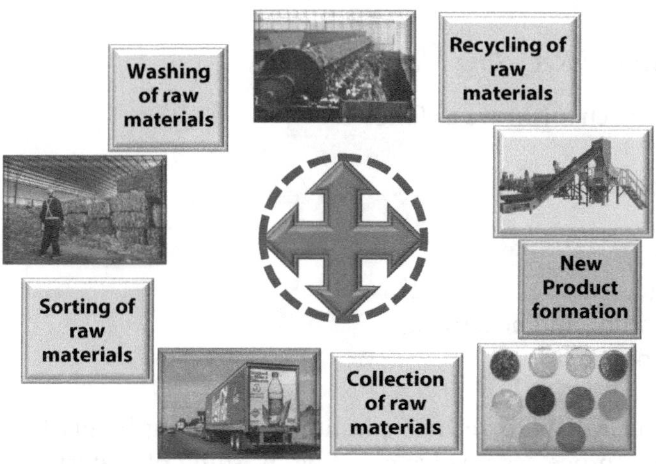

Figure 30.1 Recycling of bioplastic materials.

30.3 Types of Recycling

Following types are important for the recycling;

1. Mechanical recycling (MR)
2. Chemical recycling (CR)

30.3.1 Mechanical Recycling

30.3.1.1 Method of Mechanical Recycling

The method that is used in MR is extrusion method. Regranulated materials are produced from the plastic waste. MR method is practical, solvent free, large scale and applicable to most of the polymers. Extruder uses rotating screws and heat to induce plasticization or thermal softening. After it, material is fed through barrel section which is controlled by temperature to produce cross section extrudate. Viscous sheering and thermal conduction is applied to the polymers within the extruder which in turn lead to thermo-induced or shear induced chain scission, branching of chain or cross linking of polymers. This form of degradation of chain decreases the length of polymeric chain and also reduces its processibility and mechanical properties. Effect of extrusion process depends upon the chosen extrusion condition and characteristics of polymers. Major process of degradation involve the radical formation along with the polymeric chain due to oxygen induced per oxy radical and thermally induced hydrogen atoms abstraction. Radicals formed cause β-scissions of chains, apply shear forces and decrease the length of chain. Degradation is controlled through the choice of extrusion conditions. Screw speed and temperature has direct effect on the stability process as well as quality of product. Extrusion at excessive screw speed and temperature accelerates chain scission and make unprocessable polymers [16]. Though recycling energy is recovered and it is end of life option for bioplastics. The volume of bioplastic materials which is producing nowadays can be recycled easily along with all the counterparts such biobased poly ethylene or poly ethylene terephthalate. Thus, bioplastics contribute to higher implementation of circular economy through recycling.

30.3.1.2 Mechanical Recycling Mechanism

Main steps of MR mechanism are as follows:

- Separating and sorting
- Balling
- Washing
- Grinding
- Compounding and pelletizing
- Target product

The main aim of MR is to recover plastic by operating bioplastic through mechanical processes. These processes involve grinding, washing of material, then separating the components, drying, then regranulating and finally compounding of counterparts. Hence, recyclates that are produced are converted into plastic products after undergoing through these processes and also through the substitution of plastics. This phenomenon is identified as plastic recovery or plastic recycling.

30.3.1.3 Mechanical Recycling in Landscape

MR type of plastic are sorted on the basis of type of plastic is milled, washed, passed through chamber for separation followed by drying. The fragmented plastics are used in the original condition or processed further into granulates. MR is used currently for the recovery of post-industrial or pre-consumer material recovery. Thermoplastic materials are subjected to MR in which polymeric plastics are re-processed and re-melted into various products through extrusion or injection molding techniques. This is an efficient technology for the recycling of various thermoset materials of PP, PE and PET. The various steps of mechanical recycling has been shown in Figure 30.2.

30.3.1.4 Sorting

Wastes of plastics that are post-consumer are incinerated and in homogenous in nature. It consists of a wide range of materials like multilayer films, composites and blends with different shape, sizes and color. Initially, wastes pass through huge automated or manual mechanical process of sorting in special facilities which is made to isolate different materials. Material identification in a proper way is essential for getting highest purity of materials of recyclates using advanced techniques. For this purposes, many techniques laser and X-rays have been used. Mostly in many European countries for post-customer packaging sorting near infrared technique is used. Although accuracy is enhanced in the technology of sorting to a greater level but efficiency of sorting never approached to ideal condition because of various draw backs in separation, blending and lamination. In most of the cases, this contaminates the recycled materials with other additives and plastics. Quality of sorting process depends upon that how efficient collection schemes are and collection schemes changes widely with EU member states [17].

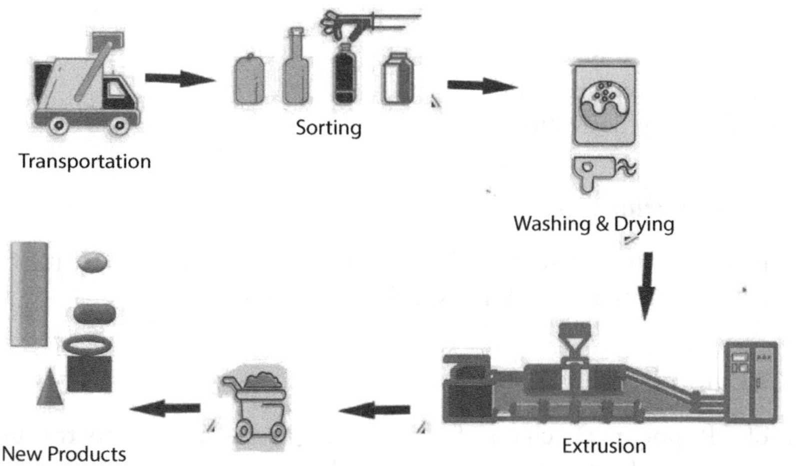

Figure 30.2 Mechanical recycling process.

When grinding and cleaning is done, recovery of materials occur through regranulating and remelting. Resultant recyclates are handled with the basic transformation into plastic products. In the reprocessing and melting, shear forces and high temperatures causes mechanical and thermal degradation of polymers which effect the length of polymeric chain and distribution of polymers. Reprocessing also effect the properties of materials like mechanical strength or crystallinity. If sufficient quality is obtained after sorting or reprocessing, materials of recycling can be reused in similar products. After the recovery of plastic materials for its reusage, replacement of virgin plastic material occur which contribute to the economy. However, quality of resulting material is often used in lower value applications which is also known as down cycling.

During 2018, the proportion of post-consumer plastic waste that was collected was 32.5 % which was to be recycled in Europe. Norway and Switzerland seem to be the only two nations that recycle through mechanical means. In Europe, it is essentially the only way to reuse the plastics. The percentage of post-consumer plastic trash that is recycled varies greatly. Finland and Bulgaria has the highest percentage of EU member nations, at roughly 20%. In several Member States, the amount of garbage generated has been increased. Pretreatment, such as sorting, is already determined when the data is entered. Other nations count garbage that is not recycled as well as being recycled and shipped. EU has started legislative efforts to initiate uniform reporting of recycling rates. The amount of data of wastes that is recycled as well as collected is rare. In Netherland in 2018, 47 percent of used plastic was collected, from which recycled waste was 33%. So, total recycling rate in Netherland is 16%. Huge variety of plastic remains unrecycled due to complexity and lower volume like composite items, blends, multilayer films, items contaminated with food and other residues. With the improvement in products design and collection of wastes, the recycling of products will grow continuously [18].

30.3.2 Chemical Recycling

The main aim of chemical recycling (CR) is to recycle the products by changing the structure of materials. Three ranges of categories of CR are suitable for different forms of plastics. CR is done through one of these mechanisms

- Solvent purification
- Chemical depolymerization
- Thermal depolymerization

30.3.2.1 Solvent Purification

In this form, plastics is dissolved in a solvent. Thus, it is done to separate it from other polluted and contaminated materials. It is used for polystyrene.

30.3.2.2 Chemical Depolymerization

In this form of CR, polymeric chain of plastic breaks into small units through chemicals which break down into their original monomers. Monomers are purified and then they are used to make new plastics. Such type of CR is done with polyester, PET and other polymers.

30.3.2.3 *Thermal Depolymerization*

In this recycling, polymer of plastic are heated and then breaks down into smaller units. Heat treatment is given with the limited supply of oxygen and this form of treatment is known as pyrolysis. Resultant oil after pyrolysis is a chemical mixture which is purified and further used as a feedstock to produce polymer. Oil is being used to replace oil derived naphtha which moves into petrochemical cracker of steam. Polyolefins plastics like PE and PP are used for this treatment. In few plants, pyrolytic oil is burnt rather than forming new polymers but burning the oil is not under the EU law of recycling [19].

30.3.2.4 *Benefits of Chemical Recycling*

Following are the benefits of chemical recycling method;

- Form complementary products
- Solution oriented to increase the consumption of materials
- Virgin quality output of products
- Ease of use
- Certified transparency
- Saving of emission and resources

30.3.3 Textile Fibers Recycling Through MR or CR

In MR waste is taken and converted into secondary material without any change in its structure. Some techniques of textile recycling are melting, extruding and shredding of fibers such as polyesters.

Textile wastes can be recycled mechanically into cotton through following ways. Initially, the fiber is sorted into many blends of fibers. Fibers or garments are then isolated by its colors, then the fabric is cut, shredded and pulled apart into fibers. Recycled fibers are disintegrated and aligned through carding process before spinning of fibers. Recycled fibers of cotton are then spun with polyester or virgin cotton for quality improvement because fiber length of recycled cotton is very less and weaker than its counterparts [20].

30.3.4 Recycled Polyester From Plastic Bottles

Mostly recycled polyester originates from plastic bottles and not from polyester wastes of textiles. It is because that both plastic water bottles and polyester are made from plastic named as polyethylene terephthalate (PET). Recycled polyester is made from plastic bottles by separating the bottles on the basis of plastic color and type. Only PET are utilized to make polyester hence sorting can be considered a crucial process. The washing of fragmented plastics is done to remove contaminants. Shredded plastic is then dried completely to remove surplus water. The fragments of plastics are converted into tiny particle, which are heated till its melting stage converts it into long filaments as a result of recycling. CR through solvents convert fibers of cellulose into pulp. Once the pulp is dissolved, it can be used in viscous process to form cellulose fibers that are manmade. Nowadays, pulp of wood is preferred for viscous process due to quality and technical reasons [21]. The recycling mode of some selected samples is provided in Table 30.1.

Table 30.1 Selected biopolymers and their mode of recycling.

Polymer	Mode of recycling	Source
Cellulose acetate ester waste	Hydrolysis	[22]
Polyhydroxybutyrate	Enzymatic recycling	[23]
Poly(trimethylene carbonate)	Enzymatic recycling	[24]
Poly(butylene adipate co-succinate)	Enzymatic recycling	[24]
Poly(e-caprolactone)	Lipase-catalyzed chemical recycling	[25]
Polyhydroxyalkanoates	Thermal degradation	[26]
(3-hydroxybutyrate-co-3-hydroxyvalerate)s	Catalytic recycling	[27]
Poly lactide	Dry-heat depolymerization	[28, 29]
Lactic acid-based polymer	Alcoholysis	[30]
Hydroxycarboxylic acid-derived polyester	Hydrothermal Depolymerization	[31]
Poly(trimethylene carbonate)	Enzymatic recycling	[32]

30.3.5 Significance of Recycling

30.3.5.1 Significance of MR

- MR preserve the molecular structure of molecules and mechanically crushes the polymers. It remelts the plastics into granules.
- Plastics are required in MR which are recycled. This need much information for example which additives and dyes are used in plastics that are recyclable.
- Designing of product plays an important role in MR. For example, if a cheese packet consist of seven layers of film, practically it is not recyclable. It includes great effort than producing everything from scratch.
- MR granulate cannot be used in packaging of food. Chemical substances are present in granulate which is dangerous to health and it can enter the food. PET bottles that are brought back by consumers are reused for new packaging of food.
- Other challenge of recycled plastic is MR color. When waste plastic is remelted and shredded into granular form, the product form is greyish black in color. Products of plastic that contain brilliant colors are not produced from colors. Additionally, granular recycled plastic contain unpleasant order and this smell is transferred to other products. This problem can be resolved through improved recycling process machine. The machine separate the plastic flakes into three compartments according to the brightness such as on the basis of white, colorless and colorful. Machine can also sort up the flakes up to one tons flakes per hour.

30.3.5.2 Significance of CR

- CR breaks the polymeric chains and provide products like naphtha, crude oil or fuels
- Recently most of the companies are supporting CR although it needs energy and doesn't save most of the CO_2
- Many facilities of CR face difficulty with continuous operation. The product quality decreases if the initial mixture contains different materials or if it is too dirty and the whole process become economically unviable.

30.4 Economic Aspects of Bioplastic Recycling Industry

According to the report from US Department of Agriculture, USDA product industry that are biobased create 4.2 million jobs and contribute $393 billion economy in 2014. USDA defines it economically favorable as this includes biobased chemicals, biorefinery, enzymes, bioplastics packaging and bottles, textiles, and forest products. Packaging sector was responsible for $11.4 million in adding direct value. In the report, bioplastic bottles and packaging account for 138000 oil barrels in savings of petroleum and 68% to 80% reduction in greenhouse gases emission as compared with plastics that are based on petroleum. The bioplastic packaging and bottles sector is very small as compared to the overall industry, which shows growth since USDA started tracking in 2013 and aimed to reach to the new levels by 2020. America has a great demand for everyday products like textiles, plastic bottles, supplies of cleaning, and many more that are manufactured from renewable resources. The demand of renewable resources is supporting millions of jobs and thus bringing the manufacturing products back to the rural communities and thus decreasing the nation's carbon footprints. This demand has given many packaging ideas that are biobased, and USDA is conducting research on products like packaging film that is edible and is made from milk protein. Surprisingly, France has taken a stand to ban the use of even a single plastic cup, cutlery, and dishes by 2020 [33].

Majority of bioplastics are producing today than they were ever produced in human history. These bioplastics are also finding their way into the aquatic environment, including ocean and thus contaminating the drinking water. There is an estimation by the scientists that by 2050 the plastic wastes could be more in the ocean than the fishes and tap water contain up to 80% of microplastics at that time. In the University of Bath, researches showed that the plastics, which contains only carbon dioxide and sugar results in the production of polycarbonates. These bioplastics are broken down naturally, thus depositing the gas back into the environment, and this gas is further used to form bioplastics.

Few greenhouse gasses are released into the environment when biodegradable plastics are used rather than refined plastics. The annual consumption of plastics is > 0.1 billion tons, which is responsible for liberating of 0.5 billion tons of CO_2 by industry each year based on normal production ratio of 5:1. It is equivalent to as 19 million tons of carbon dioxide released by 19 million automobiles. Recycling of plastics could save 30% net emission of carbon dioxide, estimating the saving percentage to be as high as 80%. Rather than

Table 30.2 Application of recycled polymers.

Polymer	Conventional applications	Recycled products	Reference
PET	Packaging	Fiber applications/packaging	[35]
PVC	Construction	Construction	[36]
Polyglycolide	Suture manufacturing	Biomedical application	[37]
PHA	Packaging, Biomedical	biohydrogen	[38]
PE	Packaging	Packaging	[39]
Starch-based polymers	Packaging	Biogas yield	[40]
PLA	Biomedical	Industrial composting/ biogas	[41]
PP	Packaging	Insulation packaging	[42]
PLA-PMMA blend	Durable & high performance applications	Recovered feed stock	[43]

initial financial costs, plastic making industries are contributing to increase the economy. Switching to biodegradable plastics assist to reduce the amount of greenhouse gases emissions as compared to that produced on commercial basis [34].

As already known that biodegradable plastics consumes greater cost in its production cycle but it utilize less energy in the manufacturing process. Through this technology, there is no need to go through various processes of acquiring, locating, and transferring hydrocarbons for making plastic items. In other words, few fossil fuels are used throughout the process of manufacturing and when the polymers reach to the end of life, few toxins would be released. Due to the energy saving in making bioplastics, using biodegradable plastics are less costly than standard plastics in the long run. These are useful specifically when greater expenses are included during the cleanup procedure of plastic pollution. Currently, waste stream contain roughly 13% plastics. It is equivalent to 32 million tons of wastes per year. Of these wastes, only 9% goes to the recycling system. The remaining waste amount is disposed of in the landfills and other disposal facilities of trash where wastes cover space from a century. Biodegradable plastics can be completely degraded in 18 to 36 months, with proper composting technology, depending on the method used. The recycled products of polymers are given in Table 30.2.

30.4.1 New Market and Economic Benefits

In 2016, production of biodegradable plastics in China was over 290000 tons. During that time, around 130,000 tons of bioplastics was used locally and exported. Year over year, 13% of local sales were increased which brought value to the category to more than $350 million. Most of the mature markets of industrialized world of plastic items were searching for the solution to overcome the trash and carbon footprints. Switching to bioproducts can

Figure 30.3 Economic growth in future involving recycling of bioplastics.

be beneficial because it can reduce pollution afterwards. If manufacturing biodegradable plastics become a major priority then biodegradable plastics can be moneymakers for the various countries [44].

The claim that biodegradable plastics are completely safe, C-level executives and consumers generally regard the bioplastics as products to promote sustainable business operations. Those organizations which uses bioplastics frequently are regarded as desirable because they are considered as environmental conscious. As a result, there is possibility to obtain benefit from the earnings of executives, shareholders and employees. Previously, Coca Cola Company developed bioplastic bottle to be used with its products. The company named as Poland Spring is reducing the quantity of components of plastic which they use in their packaging materials. Through this modification, there is a great impact on the interaction of people as well as on other businesses [45]. The future glimpse with economic growth has been shown in Figure 30.3.

30.4.2 Disadvantages of Biodegradable Plastics for Economy

30.4.2.1 Usage of Specific Disposal Procedure

The main advantage of bioplastic is that they decompose easily without causing pollution in the environment. This benefit is, however, only achieved if the disposal of bioplastic occur properly and items are treated similar to the compostable materials. If bioplastics are dumped along with the traditional plastics then they will be disposed of in a way similar to nonbiodegradable plastic. It is necessary to be aware of waste reduction initiatives and recycling program in order to ensure the benefits taken from the natural products which they offer.

30.4.2.2 Metallic Contamination

Although the items in the bioplastics are biodegradable with the usage of technology, some items of the bioplastic during decomposition process releases metals. The manufacturing cycle create polyethylene through natural things which may contain manganese at high level. When composting is done for these items then it can stop breaking down of polyethylene. In 2009, it was reported that a brand of biodegradable plastic contained higher level of lead and cobalt, which was left behind after the decomposition process [46].

30.4.2.3 *Environmental Cooperation for Disposal*

In 2000s, a study of the Biodegradable Product Institute indicated that breakdown of materials present in biodegradable items depends upon the weather conditions of the area. Humidity and temperature plays an important role for this. During cold weather, the process of composting becomes slower. At high humidity level, composting process completely stops. It indicates that many of benefits of bioplastic product disappear in far northern and equatorial climate [47].

30.4.2.4 *High Capital Cost*

Currently at technology level, costs of manufacturing biodegradable plastic is 50% more than following the manufacturing cycle of traditional plastic. There will be reduction in price as the resources will become cheaper and with the improvement in technology. It is assumed that if the production volume of bioplastics is at low cost of manufacturing, then its demand continue to increase, and the consumer price will be similar to the traditional plastic that is used nowadays. This indicates that this is an affordable solution in the future but not at present.

30.4.2.5 *Usage of Cropland to Produce Items*

Currently, technologies are using cropland to produce natural materials in order to manufacture biodegradable plastics rather than using these materials to create food. Due to hunger and scarcity problem in about one in five families in the developing countries and in the developed world, there is a need to expand this bioplastic industry in the land, which is not reserved for food. Resources of water could be used to grow crops in order to obtain raw materials. As products of corn indicate 40% of U.S.–based bioplastics industry, so, 3000 gallons of water are required for one acre for each bush to produce yield. One metric tons of PLA are produced from 100 corn bushes through the usage of current technologies whereas the farmers produce 168 bushes from an acre [48].

30.4.2.6 *Marine Pollution Problems*

The environment of oceanic water is too cold for bioplastic degradation, so bioplastic do not decompose in oceanic water. As bioplastics do not break down, hence it either floats like conventional plastics on the surface of water or form microparticles which are dangerous for aquatic life. Pollution problem will not resolve after switching to the bioproduct.

30.4.2.7 *Guarantee of Net Savings*

Most of the benefits of bioplastics is based on the net savings created by decreasing the amount of energy, water, and emission of greenhouse gases. As there is no general overview from the crop growth to the final distribution manufacturing cycle of bioplastics, so, there is no reliable data which suggest that bioplastics are beneficial due to their ability to create compost fodder or to offer profits to some industries. Although most of the reports that are cited include data that are produced by the industry [49, 50].

The fear, which the recyclates face, is that most of the nonpetroleum-based plastics or bioplastics corrupt the recycling streams. Most of the bioplastics have low heat of resistance and biodegrading is not desirable for recycled plastic manufacturers due to the fear that plastic that is recycled will be permanently degraded. Most of the manufacturers of the bioplastics are of the view that the concern of the recyclers are not found. In fact, in many conventional recycling stream, biopolyethylene is accepted, and bioplastics are recyclable. For PHA and PLA, municipal recycling collection does not accept it, but some of the manufacturers of bioplastics take their products back for recycling, although there is no infrastructure for the collection and transportation of plastics back to the manufacturers of bioplastics for those who are not willing to return the bioplastic back to the manufacturers dump the bioplastic into the trash. At present, recycling facilities are not found for bioplastics. If the incentives are given to the composters and recyclers in order to accept the bioplastic in their facilities or streams, then there will not be greater volume of bioplastics for composting or recycling. The predicted and current rate of growth for the bioplastic industry is up to 40% within the next 4 years, which is according to the prediction of some experts [41].

30.5 Conclusion

The wide consumption of bioplastics is beneficial and getting popular continuously. The environmental pollution has urged the use of bioplastics instead of petroleum-based plastics. As the bioplastics are formed by the contribution of bioingredients, it is very important to recycle the bioplastics not only to control the waste production but also to gain the optimum benefits from the synthesized products. Two methods are involved in the recycling process known as mechanical and chemical recycling, which involve a series of complicated processes to recycle the bioplastics. The recycling can be performed on various paper and fabric products. There are numerous advantages of environmental and economic concerns. Like the bioplastic production, the recycling industry is also flourishing. There is no restriction of bioplastic product's recycling, both the fabric and the paper products can be recycled with equivalent ease. It is launching a new industry in the market hence making the bioplastic and it is recycling a publicly acceptable process both for the manufacturer and the consumer. The only major drawback associated with this flourishing industry is the consumption of bioingredients rapidly than expected ratio. Even recycling is far better than generating the waste material, it is even better to make the public aware about the reuse of the bioplastic to its maximum possible shelf life to get the best results.

References

1. Wojnowska, B, I., Kulikowska, D., Bernat, K., Effect of bio-based products on waste management. *Sustainability,* 12, 2088, 2020.

2. Shen, M., Song, B., Zeng, G., Zhang, Y., Huang, W., Wen, X., Tang, W., Are biodegradable plastics a promising solution to solve the global plastic pollution. *Environ. Pollut.*, *263*, 114469, 2020.
3. Lambert, S. and Wagner, M., Microplastics are contaminants of emerging concern in freshwater environments: An overview. *Freshwater Microplastics.*, 1, 23, 2018.
4. Abdullah, A.H., Sri, P., Myrtha, K., Oceu, D.P., Rani, H.F., Fabrication and characterization of sweet potato starch-based bioplastics plasticized with glycerol. *Online J. Biol. Sci.*, 19, 57, 2018.
5. Sohail, R. and Jamil, N., Aliphatic biopolymers as a sustainable green alternative to traditional petrochemical-based plastics, in: *Bioplastics for sustainable development*, K.M. Roohi (Ed.), pp. 295–306, Springer, Singapore, 2021.
6. Befort, N., The promises of drop-in vs. functional innovations: The case of bioplastics. *Ecol. Environ. Sci.*, *181*, 106886, 2021.
7. Karan, H., Funk, C., Grabert, M., Oey, M., Hankamer, B., Green bioplastics as part of a circular bioeconomy. *Trends. Plant Sci.*, 24, 237, 2019.
8. Yee, F.Y., Kee, D.M., Jye, L.M., Jingyi, L., Wen, L.J., Almutairi, S., Save environment by replace plastic, the case of pepsi cola. *Int. J. Appl. Bus. Int. Manage.*, 5, 51, 2019.
9. Wielgosiński, G., Czerwińska, J., Szufa, S., Municipal solid waste mass balance as a tool for calculation of the possibility of implementing the circular economy concept. *Energies.*, 14, 1811, 2021.
10. Milios, L., Christensen, L.H., McKinnon, D., Christensen, C., Rasch, M.K., Eriksen, M.H., Plastic recycling in the Nordics: A value chain market analysis. *Waste. Manage.*, 76, 180, 2018.
11. Sherwood, J., Closed-loop recycling of polymers using solvents. *Johnson Matthey Technol. Rev.*, 64, 1, 2020.
12. Purmessur, B. and Surroop, D., Power generation using landfill gas generated from new cell at the existing landfill site. *J. Environ. Chem. Eng.*, 7, 103060, 2019.
13. Lamberti, F.M., Román-Ramírez, L.A., Wood, J., Recycling of bioplastics: Routes and benefits. *J. Polym. Environ.*, 28, 2551, 2020.
14. Niaounakis, M., Recycling of biopolymers–the patent perspective. *Eur. Polym. J.*, 114, 464, 2019.
15. Atiwesh, G., Mikhael, A., Parish, C.C., Banoub, J., Le, T.A., Environmental impact of bioplastic use: A review. *Heliyon.*, 07918, 1, 2021.
16. Schyns, Z.O. and Shaver, M.P., Mechanical recycling of packaging plastics: A review. *Macromol. Rapid Commun.*, *42*, 2000415, 2021.
17. Jönsson, C., Wei, R., Biundo, A., Landberg, J., Schwarz Bour, L., Pezzotti, F., Syrén, P.O., Biocatalysis in the recycling landscape for synthetic polymers and plastics towards circular textiles. *Chem. Sus. Chem.*, 14, 1, 2021.
18. Faraca, G. and Astrup, T., Plastic waste from recycling centres: Characterisation and evaluation of plastic recyclability. *Waste. Manage. Res.*, 95, 388, 2019.
19. Lee, R.P., Tschoepe, M., Voss, R., Perception of chemical recycling and its role in the transition towards a circular carbon economy: A case study in Germany. *Waste. Manage.*, 125, 280, 2021.
20. Subramanian, K., Chopra, S.S., Cakin, E., Li, X., Lin, C.S., Environmental life cycle assessment of textile bio-recycling–valorizing cotton-polyester textile waste to pet fiber and glucose syrup. *Resour. Conserv. Recycl.*, 161, 104989, 2020.
21. Pillai, C.K.S. and Sharma, C.P., Review paper: absorbable polymeric surgical sutures: Chemistry, production, properties, biodegradability, and performance. *J. Biomater. Appl.*, 25, 291, 2010.
22. Al-Salem, S.M., Lettieri, P., Baeyens, J., Recycling and recovery routes of plastic solid waste (PSW): A review. *Waste. Manage.*, 29, 2625, 2009.

23. Kobayashi, S., Uyama, H., Takamoto, T., Lipasecatalyzed degradation of polyesters in organic solvents. A new methodology of polymer recycling using enzyme as catalyst. *Biomacromolecules.*, 1, 3, 2000.
24. Kaihara, S., Nishikawa, K., Toshima, K., Doi, Y., Matsumura, S., Enzymatic transformation of bacterial polyhydroxyalkanoates into repolymerizable oligomers directed towards chemical recycling. *Macromol. Biosci.*, 5, 644, 2005.
25. Tsuji, H., Daimon, H., Fujie, K., A new strategy for recycling and preparation of poly(L-lactic acid): Hydrolysis in the melt. *Biomacromolecules.*, 4, 835, 2003.
26. Ariffin, H., Nishida, H., Hassan, M.A., Shirai, Y., Chemical recycling of polyhydroxyalkanoates as a method towards sustainable development. *Biotechnol. J.*, 5, 484, 2010.
27. Mueller, R.J., Biological degradation of synthetic polyesters enzymes as potential catalysts for polyester recycling. *Process. Biochem.*, 41, 2124, 2006.
28. Nishida, H., Mori, T., Hoshihara, S., Fan, Y., Shirai, Y., Endo, T., Effect of tin on poly(L-lactic acid) pyrolysis. *Polym. Degrad. Stab.*, 81, 15, 2003.
29. Fan, Y., Nishida, H., Mori, T., Shirai, Y., Endo, T., Thermal degradation of poly(L-lactide): Effect of alkali earth metal oxides for selective L, lactide formation. *Polymer.*, 45, 205, 2004.
30. Saeki, T., Tsukegi, T., Tsuji, H., Daimon, H., Fujie, K., Depolymerization of poly(L-lactic acid) under hydrothermal conditions. *Hiroyuki Fujie, Koichi.*, 61, 561, 2004.
31. Matsumura, S., Enzyme-catalyzed synthesis and chemical recycling of polyesters. *Macromol. Biosci.*, 2, 105, 2002.
32. Zhu, Y., Romain, C., Williams, C.K., Sustainable polymers from renewable resources. *Nature.*, 540, 354, 2016.
33. Benavides, P.T., Lee, U., Zarè-Mehrjerdi, O., Life cycle greenhouse gas emissions and energy use of polylactic acid, bio-derived polyethylene, and fossil-derived polyethylene. *J. Clean. Prod.*, 277, 124010, 2020.
34. Mazhandu, Z.S., Muzenda, E., Mamvura, T.A., Belaid, M., Integrated and consolidated review of plastic waste management and bio-based biodegradable plastics: Challenges and opportunities. *Sustainability.*, 12, 8360, 2020.
35. Barman, N.K., Bhattacharya, S.S., Mandot, A., Research article mechanical properties of melt-spun monoflaments produced feom virgin and recycyled poly (ethylene terephthalate) blends. *Int. J. Recent. Sci. Res.*, 6, 4517, 2015.
36. Burat, F., Güney, A., Olgaç, K.M., Selective separation of virgin and post-consumer polymers (PET and PV.C) by fotation method. *Waste. Manage.*, 29, 1807, 2009.
37. Malik, N., Kumar, P., Shrivastava, S., Ghosh, S.B., An overview on PET waste recycling for application in packaging. *Int. J. Plast. Technol.*, 21, 1, 2017.
38. Amulya, K., Jukuri, S., Mohan, S.V., Sustainable multistage process for enhanced productivity of bioplastics from waste remediation through aerobic dynamic feeding strategy: Process integration for up-scaling. *Bioresour. Technol.*, 188, 231, 2015.
39. Morschbacker, A., Eduardo, C., Campos, S., Cassiano, L.C., Bio-polyethylene, in: *Handbook of green materials*, pp. 89–104, World Scientifc Publishing, Singapore, 2010.
40. Hamad, K., Kaseem, M., Deri, F., Recycling of waste from polymer materials: An overview of the recent works, polymer degradation and stability. *Polym. Degrad. Stab.*, 98, 2801, 2013.
41. Piemonte, V., Sabatini, S., Gironi, F., Chemical recycling of PLA: A great opportunity towards the sustainable development? *J. Polym. Environ.*, 21, 640, 2013.
42. Madhavan, N.K., Nair, N.R., John, R.P., An over view of the recent developments in polylactide (PLA) research. *Bioresour. Technol.*, 101, 8493, 2010.
43. Auras, R., Harte, B., Selke, S., An overview of polylactides as packaging materials. *Macromol. Biosci.*, 4, 835, 2004.

44. Lzuhairi, M.A.H., Khalil, B., II, Hadi, R.S., Nano ZnO catalyst for chemical recycling of polyethylene terephthalate (PET). *Eng. Technol. J.*, 35, 831, 2017.
45. Samper, M.D., Bertomeu, D., Arrieta, M.P., Ferri, J.M., López, M.J., Interference of biodegradable plastics in the polypropylene recycling process. *Materials.*, 11, 1886, 2018.
46. Beriot, N., Zomer, P., Zornoza, R., Geissen, V., A laboratory comparison of the interactions between three plastic mulch types and 38 active substances found in pesticides. *Peer. J.*, 8, 9876, 2020.
47. Mohan, S.K. and Srivastava, T., Microbial deterioration and degradation of polymeric materials. *J. Biochem. Technol.*, 2, 210, 2010.
48. Markus, F. and Ramani, N., Biodegradable plastic as integral part of the solution to plastic waste pollution of the environment. *Curr. Opin. Green Sustain. Chem.*, 100490, 30, 2021.
49. De, C.R., Dusselier, M., Sels, B.F., Heterogeneous catalysis for bio-based polyester monomers from cellulosic biomass: Advances, challenges and prospects. *Green. Chem.*, 19, 5012, 2017.
50. Zembouai, I., Bruzaud, S., Kaci, M., Benhamida, A., Yves-Marie, C., Grohens, Y., Mechanical recycling of poly(3-hydroxybutyrate-co-3-hydroxyvalerate)/polylactide based blends. *J. Polym. Environ.*, 22, 449, 2014.

Index

Printed in the USA/Agawam, MA
April 26, 2023

809111.073